Alexander von Humboldt: Writings in English

Part I: 1789–1824

Alexander von Humboldt, *Self-portrait*, Paris, 1814.

Alexander von Humboldt: Writings in English

Part I: 1789–1824

Edited by

Oliver Lubrich and Thomas Nehrlich

American Philosophical Society Press
Philadelphia

Published by
The American Philosophical Society Press
Philadelphia, Pennsylvania 19106-3387
www.amphilsoc.org

Printed in the United States of America on acid-free paper
10 9 8 7 6 5 4 3 2 1

Library of Congress Cataloging-in-Publication Data
Names: Humboldt, Alexander von, 1769–1859 | Lubrich, Oliver, 1970– editor. | Nehrlich, Thomas, editor.
Title: Alexander von Humboldt : writings in English / edited by Oliver Lubrich, & Thomas Nehrlich.
Other titles: Writings in English
Description: Philadelphia : American Philosophical Society Press, 2024– | Includes bibliographical references and index. | Contents: Part 1. 1789–1824 — part 2. 1825–1859. | Summary: "Alexander von Humboldt's articles and essays give us a more complete picture of the scientist as a writer. During his first ten years of publishing, from 1789 to 1799, Humboldt mainly wrote scientific papers and academic reviews. His endeavors as a writer changed when he embarked on his American journey, from 1799 to 1804, a time during which his reputation grew considerably. The text presents all the 250 texts known to be translated into English during Humboldt's lifetime – a historical sample from the corpus to be newly discovered by English–speaking readers"— Provided by publisher.
Identifiers: LCCN 2023042552 (print) | LCCN 2023042553 (ebook)
Subjects: LCSH: Humboldt, Alexander von, 1769–1859—Correspondence. | Humboldt, Alexander von, 1769–1859—Travel. | America—Description and travel. | Plant ecology.
Classification: LCC Q143.H9 A25 2024 (print) | LCC Q143.H9 (ebook) | DDC 508.8—dc23/eng/20230929
LC record available at https://lccn.loc.gov/2023042552
LC ebook record available at https://lccn.loc.gov/2023042553

Hardback two-volume set ISBN: 978-1-60618-080-8

Hardback ISBN: 978-1-60618-016-7

eBook ISBN: 978-1-60618-017-4

Cover illustration reprinted from Alexander von Humboldt, *Recueil d'observations de zoologie et d'anatomie comparée, faites dans l'Océan Atlantique, dans l'intérieur du Nouveau Continent et dans la Mer du Sud pendant les années 1799, 1800, 1801, 1802 et 1803*, 2 vols. Paris: F. Schœll/G.EL Dufour, 1811 [1812]; J. Smith/Gide, [1813–]1833, volume 1, plate VIII, "Vultur gryphus Lin."

Cover design by Ellen B. Zanolle.

Frontispiece: Alexander von Humboldt, *Self-portrait*, 1814, "von mir selbst im Spiegel."

About This Edition

This edition contains the writings of Alexander von Humboldt (1769–1859) as they were published in English translations during his lifetime—complemented by the first translation of his earliest article. These writings consist of his nonbook publications, that is, articles, essays, letters, excerpts of his books as well as other smaller pieces included in journals, newspapers, and the works of other authors or editors. The texts are unabridged and unaltered reproductions of the historical versions as they appeared—most in the United Kingdom and the United States—but also, for instance, in Australia, the Virgin Islands, New Zealand, India, South Africa, and Panama. These texts thus show his influence in the English-speaking world during the times of Jefferson, Darwin, and Thoreau, contributing to the history of scientific thought. The titles used are those of the English publications. Texts that appeared without a title have been given an editorial title in brackets. The order of the texts follows the chronology of the respective original publications in German or French (see Sources).

The present volume includes Part I of the writings (1789–1824); a second volume, published simultaneously, contains Part II of the writings (1825–1859). The contents of Part II are included herewith for the reader's convenience.

Contents

CONTENTS OF ALEXANDER VON HUMBOLDT: WRITINGS IN ENGLISH, PART II: 1825–1859

The Whole World in One Thousand Essays: Introduction to Alexander von Humboldt's "Other *Cosmos*"

Oliver Lubrich and Thomas Nehrlich

I read and re-read Humboldt.
—Charles Darwin[1]

Undeniably, Alexander von Humboldt is famous. No other scientist's name is inscribed on the Earth's geography more often. Cities, rivers, and mountains are named after Humboldt; as is an ocean current flowing near Chile, a district in California, and even a crater on the moon.[2] Humboldt inspired writers and scientists such as Goethe and Lord Byron, Charles Darwin, Ralph Waldo Emerson, Henry David Thoreau, Edgar Allan Poe, Walt Whitman, Aldous Huxley, Ezra Pound, and Saul Bellow.[3] Yet, no other canonical author's fame appears to be so disproportionate to the availability of his works. Alexander von Humboldt was celebrated much more than he was read. The reasons are linguistic, political, and historical. Humboldt was both a French and a German author. He was a European cosmopolitan in the age of nationalism, who did not belong to any single culture. Humboldt resolutely opposed colonialism, racism, and slavery. His writings defied the appropriation of his name by the German Empire. And because Humboldt combined diverse forms of knowledge, no individual discipline felt responsible for his legacy.

---

[1] *The Life and Letters of Charles Darwin, Including an Autobiographical Chapter*, ed. Francis Darwin, 3 vols. (London: John Murray 1887), 1:190.

[2] A systematic compilation lists about 1,000 designations on 150 pages: Ulrich-Dieter Oppitz, "Der Name der Brüder Humboldt in aller Welt," in *Alexander von Humboldt. Werk und Weltgeltung*, ed. Heinrich Pfeiffer (Munich: Piper, 1969), 277–429. An installation on the name "Humboldt" marked the beginning of the exhibition *Networks of Knowledge* in Berlin and Bonn: cf. the catalogue, *Alexander von Humboldt—Netzwerke des Wissens*, conceived by Frank Holl (Berlin: Haus der Kulturen der Welt 1999), 21–22.

[3] Cf. *Transatlantic Echoes. Alexander von Humboldt in World Literature* (100 literary texts), ed. Rex Clark and Oliver Lubrich (New York: Berghahn Books, 2012); *Cosmos and Colonialism. Alexander von Humboldt in Cultural Criticism* (50 essays), ed. Rex Clark and Oliver Lubrich (New York: Berghahn Books, 2012).

For many decades, particularly in the late nineteenth and twentieth centuries, these qualities obstructed his reception. But these same qualities have favored his rediscovery in recent years. Alexander von Humboldt's multilingualism and internationality, his postcolonial engagement and interdisciplinary research practices have made him a figure of present-day interest. He is now being rediscovered in the English-speaking world within the context of contemporary debates.[4] Humboldt lived to see his most important books made available in English: *Relation historique du Voyage aux régions équinoxiales du Nouveau Continent* (1814–31),[5] *Essai politique sur le royaume de la Nouvelle-Espagne* (1808–11),[6] *Essai politique sur l'île de Cuba* (1826),[7] *Ansichten der Natur* (1808, 1826, 1849),[8] *Vues des Cordillères et monumens des peuples indigènes de l'Amérique* (1810–13),[9] *Essai géognostique sur le gisement des roches*

[4] Cf. Stephen T. Jackson, "Humboldt for the Anthropocene," in *Science* 365, no. 6458 (13 September 2019): 1074–76.

[5] The travel account was translated into English twice during Humboldt's lifetime, Williams's translation being the only complete, unabridged, and authorized one: *Personal Narrative of Travels to the Equinoctial Regions of the New Continent*, trans. Helen Maria Williams, 8 vols. (London: Longman, Hurst, Rees, Orme, and Brown, 1814–31); *Personal Narrative of Travels to the Equinoctial Regions of America*, trans. Thomasina Ross, 3 vols. (London: Henry G. Bohn, 1852–1853 [incomplete]). Cf. the more recent selections in two Penguin series: *Personal Narrative of a Journey to the Equinoctial Regions of the New Continent*, trans. Jason Wilson (London: Penguin Classics, 1995); *Jaguars & Electric Eels*, trans. Jason Wilson (London: Penguin Great Journeys, 2007). Cf. Alison E. Martin, *Nature Translated. Alexander von Humboldt's Works in Nineteenth-Century Britain* (Edinburgh: Edinburgh University Press, 2018).

[6] *Political Essay on the Kingdom of New-Spain*, trans. John Black, 4 vols. (London: Longman, Hurst, Rees, Orme, and Brown, 1811).

[7] *The Island of Cuba*, trans. J. S. Thrasher (New York: Derby & Jackson, 1856). Humboldt protested against the omission of the antislavery chapter in this translation. Cf. *Political Essay on the Island of Cuba*, ed. and trans. Vera M. Kutzinski and Ottmar Ette (Chicago: University of Chicago Press, 2011); cf. Oliver Lubrich, "In the Realm of Ambivalence: Alexander von Humboldt's Discourse on Cuba," in *German Studies Review* 26, no. 1 (2003): 63–80.

[8] The third edition of 1849 was translated twice: *Aspects of Nature*, trans. E. J. Sabine, 2 vols. (London: Longman, Brown, Green, and Longmans, 1849); *Views of Nature*, trans. E. C. Otté and Henry G. Bohn (London: Henry G. Bohn, 1850); cf. *Views of Nature*, trans. Mark W. Person, eds. Stephen T. Jackson and Laura Dassow Walls (Chicago: University of Chicago Press, 2014). Numerous excerpts from the first and second editions appeared earlier, making the book virtually available for English-speaking readers of newspapers and journals well before publication of the complete book translations.

[9] *Researches Concerning the Institutions & Monuments of the Ancient Inhabitants of America, with Descriptions and Views of some of the Most Striking Scenes in the Cordilleras*, trans. Helen Maria Williams, 2 vols. (London: Longman, Hurst, Rees, Orme & Brown, J. Murray & H. Colburn, 1814); cf. *Views of the Cordilleras and Monuments of the Indigenous Peoples of the Americas. A Critical Edition*, eds. Vera M. Kutzinski and Ottmar Ette, trans. J. Ryan Poynter (Chicago: University of Chicago Press, 2012).

dans les deux hémisphères (1823),[10] and finally, *Kosmos* (1845–62).[11] A large part of his work, however, has yet to be translated.[12]

The Berlin-Brandenburg Academy of Sciences and Humanities[13] published editions of Humboldt's correspondence (1973–2013)[14] as well as his American travel journals (1982–2000). His graphic œuvre became available in a complete edition in 2014.[15] Drawings from his travel journals followed in 2018,[16] and sketches from his personal papers were published in 2019.[17]

Until recently, scholars had not collected and edited Alexander von Humboldt's *writings*, that is, those publications that did not appear independently as books, but were included in journals, magazines, and newspapers, and in the works of other authors and editors.[18] This extensive and varied corpus was available only in fragments and documented rudimentarily. Although a detailed inventory of Humboldt's books has been available since 2000,[19] attempts to bibliographically catalog the nonbook publications have been limited.[20]

---

[10] *A Geognostical Essay on the Superposition of Rocks in Both Hemispheres* [no translator named] (London: Longman, Hurst, Rees, Orme, Brown, and Green, 1823).

[11] During Humboldt's lifetime, *Cosmos* was translated twice, almost completely: *Cosmos. A Sketch of a Physical Description of the Universe*, trans. E. J. Sabine, 4 vols. (London: Longman, Brown, Green, Longmans & Roberts, and John Murray, 1846–58); *Cosmos. Sketch of a Physical Description of the Universe*, trans. E. C. Otté, 5 vols. (London: Henry G. Bohn, 1849–58).

[12] Since 2009, the University of Chicago Press has been publishing new translations of Humboldt's major books: *Essay on the Geography of Plants* (2009), *Political Essay on the Island of Cuba* (2011), *Views of the Cordilleras* (2012), *Views of Nature* (2014), *Political Essay on the Kingdom of New Spain* (2019).

[13] Alexander von Humboldt, *Reise durch Venezuela* (first part of the journey), transcribed by Gisela Lülfing and Margot Faak, ed. and trans. Margot Faak (Berlin: Akademie, 2000); *Reise auf dem Río Magdalena, durch die Anden und Mexico* (second part of the journey), transcribed by Gisela Lülfing and Margot Faak, ed. and trans. Margot Faak, 2 vols. (Berlin [GDR]: Akademie 1986/90); *Lateinamerika am Vorabend der Unabhängigkeitsrevolution* (selection from the journals), transcribed by Gisela Lülfing and Margot Faak, ed. and trans. Margot Faak (Berlin [GDR]: Akademie, 1982); cf. *Ueber einen Versuch den Gipfel des Chimborazo zu ersteigen*, eds. Oliver Lubrich and Ottmar Ette (Berlin: Eichborn Berlin, 2006).

[14] A list of about seventy editions of Alexander von Humboldt's letters can be found at: www.humboldt.unibe.ch/briefausgaben.html.

[15] Alexander von Humboldt, *Das graphische Gesamtwerk*, ed. Oliver Lubrich (Darmstadt: Lambert Schneider, 2014).

[16] Alexander von Humboldt, *Bilder-Welten. Die Zeichnungen aus den Amerikanischen Reisetagebüchern*, eds. Ottmar Ette and Julia Maier (Munich: Prestel, 2018).

[17] Alexander von Humboldt, *Das zeichnerische Werk*, eds. Dominik Erdmann and Oliver Lubrich (Darmstadt: Wissenschaftliche Buchgesellschaft, 2019).

[18] Cf. Oliver Lubrich, "Alexander von Humboldt (1769–1859). Zum 150. Todestag des Naturforschers und Reiseschriftstellers. Dossier," in *Zeitschrift für Germanistik* 19, no. 2 (2009): 396–402.

[19] Horst Fiedler and Ulrike Leitner, *Alexander von Humboldts Schriften. Bibliographie der selbständig erschienenen Werke* (Berlin: Akademie, 2000).

[20] The bibliography published by Julius Löwenberg in 1872 covered just under 10 percent of the writings catalogued today, the Berlin–Brandenburg Academy of Sciences and Humanities listed about 20 percent: Julius Löwenberg, "Alexander von Humboldt. Bibliographische Übersicht seiner Schriften und zerstreuten

Humboldt's nonbook publications are difficult to survey. His writings appeared in over 1,200 periodicals worldwide, in more than 400 places, and in various languages. Because editors customarily printed texts without requesting the writer's consent or without clear identification, authorship is not always easy to determine.[21] At most recent count, some 750 different texts appeared during Humboldt's lifetime, that is, between 1789 and 1859, in a total of about 3,600 versions, including republications, adaptations, and translations. These articles and essays can be said to constitute Humboldt's "Other *Cosmos*." While in his monumental *Kosmos* (1845–62) he describes "the whole world in one book," in his minor contributions he describes the world in one thousand writings.[22] The following selection of Humboldt's nonbook œuvre presents all the 250 texts known to be translated into English during his lifetime—a historical sample from the corpus to be newly discovered by English-speaking readers. (In order to cover the span of Humboldt's publications completely, a new translation of his very first article from 1789 has also been included.) What do we learn from Humboldt's writings? In what way are they significant? How do they change our assessment of their author?

THE MASTER OF SMALL FORMS

Alexander von Humboldt's articles and essays give us a more complete picture of the scientist as a writer. During his first ten years of publishing, from 1789 to 1799, Humboldt mainly wrote scientific papers and academic reviews. His endeavors as a writer changed when he embarked on his American journey, from 1799 to 1804, a time during which his reputation grew considerably. Humboldt's repertoire expanded. He found new forms for new topics. During his American expedition he communicated with the European public by means of reports on his travels and fieldwork.

Abhandlungen," in *Alexander von Humboldt. Eine wissenschaftliche Biographie*, ed. Karl Bruhns (Leipzig: F. A. Brockhaus, 1872), 2:485–552; avh.bbaw.de/uns. There are specific bibliographies of the publications in Russian and Polish: Natal'ja Georgievna Suchova, "Alexander von Humboldt in der russischen Literatur. Eine annotierte Bibliografie," in *Alexander von Humboldt und Russland. Eine Spurensuche*, eds. Kerstin Aranda, Andreas Förster, and Christian Suckow (Berlin: Akademie, 2014), 411–503; Krzysztof Zielnica, *Alexander von Humboldt in der polnischen Literatur: eine kommentierte Bibliographie* (Berlin: Dietrich Reimer, 1989).

[21] Cf. Oliver Lubrich, "Von der ersten bis zur letzten Veröffentlichung. Alexander von Humboldts 'Sämtliche Schriften' in der 'Berner Ausgabe'," in *Zeitschrift für Germanistik* 28, no. 1 (2018): 119–30; Thomas Nehrlich, "Sensationsfund oder falsche Fährte? Über einen 'Brief an Kleist' in der 'Berner Ausgabe' von Alexander von Humboldts *Schriften*," in *Zeitschrift für Germanistik* 28, no. 3 (2018): 604–15.

[22] Cf. Alexander von Humboldt, *Der Andere Kosmos. 70 Texte – 70 Orte – 70 Jahre*, eds. Oliver Lubrich and Thomas Nehrlich (Munich: dtv, 2019).

When he returned home, Humboldt published academic presentations, forewords to the works of other authors, excerpts in anthologies and textbooks, political articles in the daily press, and letters of thanks for honors bestowed upon him in different countries. Humboldt chose new formats to relate his scientific observations and his social concerns. As his fame grew, so did the demand for his texts. Many were reprinted, translated, and extracted from his books without his knowledge. After his sensational expedition, many wanted to "listen to Humboldt tell his stories" ("einmal Humboldten erzählen hören"), as Ottilie confesses in Goethe's novel *Elective Affinities* (1809).[23]

The result is an œuvre of an extreme diversity and of various genres, as the titles make apparent: "Addendum," "Additions," "Advertisement," "Analysis," "Announcement," "Chronology," "Collection," "Comments," "Comparison," "Contributions," "Correspondence," "Corrigendum," "Counterstatement," "Description," "Discovery," "Draft," "Essay," "Expert Opinion," "Explanation," "Extract," "Fragments," "History," "Ideas," "Inaugural Address," "Introduction," "Investigations," "Lecture," "Letter," "Map," "Measurements," "Message," "Narrative," "Notes," "Observations," "Opening Speech," "Outline," "Overview," "Painting," "Preface," "Presentation," "Report," "Reply," "Request," "Results," "Sketch," "Speech," "Travel," "Trial," "Views," "Words of Thanks," "Work," "Writing," and "Writings." Humboldt experimented with numerous modes of writing, thereby refining his means of expression. The author of the twenty-nine-volume *Voyage aux régions équinoxiales du Nouveau Continent* (1805–38) was also a master of small forms.[24]

THE MOST INTERNATIONAL WRITER OF HIS TIME

Humboldt's earliest writings appeared mainly in German-speaking countries. When he traveled to the Americas, he began to publish his work in the international media. In the five decades from the 1810s until the late 1850s, Humboldt's articles appeared on all five continents, for example, in Berlin, Vienna, Zurich, Paris, London, Edinburgh, Manchester, Milan, Madrid, Prague, Warsaw, Caracas, Bogotá, Havana, Mexico, Rio de Janeiro, Curação, New York, Chicago, Boston, Washington, San Francisco, St. Petersburg, Moscow, Sydney, Bombay, Shanghai, and

[23] Johann Wolfgang von Goethe, *Die Wahlverwandtschaften. Ein Roman*, 2 vols. (Tübingen: J. G. Cotta, 1809), 2:150.

[24] Cf. Oliver Lubrich, "Alexander von Humboldt als Essayist und Publizist," in Alexander von Humboldt, *Das große Lesebuch* (Frankfurt: Fischer Klassik, 2009), 315–40, esp. 319–24.

Pietermaritzburg in South Africa. Quite likely, Alexander von Humboldt was the most international writer of his time.[25]

Humboldt published so widely that in every single one of his 70 years of writing, at least one text appeared in a new place each year,[26] a claim that few, if any, other authors can credibly make. His publishing success is owed in equal measure to his long career and high productivity, as well as to a worldwide market and a demand unparalleled by that of any of his contemporaries, including Goethe or Darwin, nor probably by any other writer or scientist in the English-speaking world.

As the geographical distribution of Humboldt's work grew, so did the linguistic composition of the corpus. Humboldt was always a polyglot. His Latin publications only play a marginal role; after the ancient language lost its status as the scientific lingua franca in the late eighteenth century, outside of descriptive botany and zoology, Humboldt changed course.[27] He wrote articles in French, the native language of his mother, a Huguenot, and in German, the native language of his father, a Prussian. He published his first article in French in 1789 (for the *Gazette littéraire de Berlin*) and his last article in German in 1859 (for various newspapers, initially in Berlin). He wrote about his two transcontinental expeditions in French (*Voyage aux régions équinoxiales du Nouveau Continent* as well as *Asie centrale*) but drafted most of his articles in German. During the nineteenth century, French gave way to German as an international language of science, a shift reflected in the publications and translations of Humboldt's texts. At the same time, his choice of language also reflects his locations, his scientific networks, and his political engagements.

Humboldt's writings have been translated into twelve other languages: English, Spanish, Portuguese, Italian, Dutch, Danish or Norwegian, Swedish, Hungarian, Russian, Polish, and Hebrew, making Humboldt one of the most widely translated authors of his time. The number of translations and the number of languages his works were translated into significantly increased the reach of his writings.

---

[25] Cf. Oliver Lubrich and Thomas Nehrlich, "Alexander von Humboldt als internationaler Publizist. Zur Edition seiner sämtlichen Schriften," in *Iberoamerikanisches Jahrbuch für Germanistik* 9 (2015): 71–88.

[26] Cf. Humboldt, *Der Andere Kosmos.*

[27] Alexander von Humboldt, "Observatio critica de Elymi hystricis charactere," in *Zeitschrift für die Botanik* 3, no. 7 (1790): 3–6; 3, no. 9 (1790): 32; "Plantas subterraneas descripsit Fr. A. ab Humboldt," in *Annalen der Botanick* 1, no. 3 (1792): 53–58. Humboldt also published some of his books in Latin, including *De distributione geographica plantarum secundum coeli temperiem et altitudinem montium, prolegomena* (Paris: Librairie grecque-latine-allemande, 1817).

In addition to his many Hispanophone translations in Spain and Latin America, an overview of the corpus reveals an unexpected number of publications in English, more than 1,540, surpassing even the number of German and French originals. The distribution in English is likely a result of Humboldt's reputation in the United States,[28] his anticolonial views, and the relatively free press in the Anglophone world, which was advantageous to his reception. Humboldt voiced his opinion on American politics, generating a dialog between Europe and the Americas. He sent open letters to his correspondents on the other side of the Atlantic.[29] Although he did not write his texts in English, they were widely translated and reprinted in the United States in the 1840s and 1850s—so much so in fact, that based on the number of published texts, by the end of his life, by all intents and purposes, he was an English-language author.

THE POSTDISCIPLINARY RESEARCHER

Humboldt's writings chronicle his intellectual development. The 750 articles, essays, letters, and so on, constitute, on average, almost one text per month over 70 years. These frequent, shorter publications represent a more continuous publication biography than Humboldt's twenty-five books, which averaged one written every three years. By an early age, Humboldt was creatively combining different forms of knowledge, an approach constitutive of his later work. However, many of Humboldt's early articles focus on individual subjects, particularly geology, physiology, and botany. The monodisciplinary nature of these contributions is evident in the agendas of the journals in which they were published: *Magazin für die Botanik, Bergmännisches Journal, Jahrbücher der Berg- und Hüttenkunde, Medicinisch-chirurgische Zeitung, Astronomisches Jahrbuch* oder *Allgemeine Geographische Ephemeriden* [*Magazine for Botany, Mining Journal, Yearbooks of Mining and Metallurgy, Journal for Physicians and Surgeons, Yearbook of Astronomy, Journal of General Geography*].

The American journey opened up new fields of research, most important of which were ethnology and anthropology (the indigenous

[28] Cf. Laura Dassow Walls, "'The Napoleon of Science': Alexander von Humboldt in Antebellum America," in *Nineteenth Century Contexts* 14, no. 1 (1990): 71–98; *Seeing New Worlds. Henry David Thoreau and Nineteenth-Century Natural Science* (Madison: University of Wisconsin Press, 1995); *The Passage to Cosmos: Alexander Von Humboldt and the Shaping of America* (Chicago: University of Chicago Press, 2009).

[29] Cf. *Alexander von Humboldt und die Vereinigten Staaten von Amerika. Briefwechsel*, ed. Ingo Schwarz (Berlin: Akademie, 2004).

peoples of the "New World," the "unity of the human race" in its diversity[30]), archaeology (the monuments of pre-Hispanic civilizations), and demography (the composition of colonial societies). Added to this were a smaller number of occasional studies on zoology used to investigate exotic species (mosquito, manatee, condor, New World monkeys, etc.).

Humboldt's expedition spurred a new kind of research that surpassed disciplinary boundaries. The nature of the tropics and the reality of life in the colonies impelled him to think beyond the realms of the individual disciplines, which by then had become institutionalized. In his study of plant geography, for example, he combines botanical, geographical, historical, and climatological knowledge so as to trace the distribution and migration of species as a function of natural as well as anthropogenic factors.[31] In his report on his field research with electric eels, he integrates zoological, ethnographic, geographical, and physiological interests.[32] In his examination of plants and animals, we witness Humboldt developing an ecological thinking *avant la lettre*. Instead of isolating individual life forms, he observed them in their environment and in their interaction.[33]

Following Humboldt's writings chronologically, we watch his research transform. The number of disciplines that inform each text increases notably during his American journey, and as a result, the texts' spectrum broadens.[34] Humboldt moved from mono- to inter- and transdisciplinary practices. This finding is not an *ex–post* projection but a historical peculiarity. Humboldt's "Combination von Wissenschaften"

[30] Alexander von Humboldt, "De l'unité native de l'espèce humaine," in *Nouvelles annales des voyages, de la géographie et de l'histoire* 3 (1846): 112–20; cf. *Kosmos. Entwurf einer physischen Weltbeschreibung*, 5 vols. (Stuttgart: Cotta, 1845–62), vol. 1 (1845), 378–86 (notes: pp. 490–493, here: p. 385).

[31] The series of articles on plant geography includes the studies "Geografía física. Ideas sobre el límite inferior de la nieve perpétua, y sobre la geografía de las plantas" (1804), "Ideen zu einer Physiognomik der Gewächse" (1806), "Geografía de las plantas" (1809). and many other writings in which Humboldt continuously elaborated his program. His most prominent book publication on this subject, containing his famous "Tableau physique des Andes," was *Essai sur la géographie des plantes, accompagné d'un tableau physique des régions équinoxiales* (Paris: F. Schœll/Tübingen: J. G. Cotta, 1807); cf. *Essay on the Geography of Plants*, trans. Sylvie Romanowski, ed. Stephen T. Jackson (Chicago: University of Chicago Press, 2009).

[32] Alexander von Humboldt, "Jagd und Kampf der electrischen Aale mit Pferden," in *Annalen der Physik* 25, no. 1 (1807): 34–43.

[33] Cf. Aaron Sachs, *The Humboldt Current: Nineteenth-Century Exploration and the Sources of American Environmentalism* (New York: Viking, 2006); Stephen T. Jackson, "Introduction: Humboldt, Ecology, and the Cosmos," in *Essay on the Geography of Plants*, trans. Sylvie Romanowski, ed. Stephen T. Jackson (Chicago: University of Chicago Press, 2009), 1–46; Stephen T. Jackson, "Alexander von Humboldt and the General Physics of the Earth," in *Science* 324, no. 5927 (May 1, 2009: 596–97); Oliver Lubrich, "Humboldtian Landscapes," in *Natura. Environmental Aesthetics After Landscape*, eds. Jens Andermann, Lisa Blackmore, and Dayron Carrillo Morell (Zurich: Diaphanes, 2018), 73–109.

[34] Sarah Bärtschi, "Layered Reading: Wie kann man das Gesamtwerk eines Autors lesen? Quantitative und qualitative Methoden am Beispiel der unselbständigen Schriften Alexander von Humboldts" (PhD thesis, University of Bern, 2017).

[combination of sciences], "welchen man ehedem keine engere Verwandtschaft zutraute" [which had not been thought to be closely related before], impressed his contemporaries, as it led to "unerwarteten Ergebnissen" [unexpected results]. This statement comes from the entry on Alexander von Humboldt in Brockhaus's *Conversations-Lexikon* of 1853.[35] The publication history of his writings contradicts the widely held belief that Alexander von Humboldt was the "last universalist," a Renaissance-style scientist who combined everything with everything else, *before* the differentiation of disciplines. By contrast, his writings reveal that he can be seen not only as a *trans*disciplinary but indeed as an early *post*disciplinary researcher.

THE PUBLIC INTELLECTUAL

Humboldt's research and publications shaped his reputation and discursive presence. His public role changed. He was already communicating strategically and managing his image in the media during his American expedition of 1799 to 1804.[36] During his stay in the United States of America, he received an honorary membership certificate from the American Philosophical Society. It bore the signature of Thomas Jefferson, president of the Society, and at the time, the country's president as well. Shortly before leaving for Europe, Humboldt wrote the first comprehensive account of his voyage, which was published in Philadelphia in July 1804. The American Philosophical Society Archives hold the manuscript of the text to this day.[37] Upon his return to Europe, he gained more attention and greater access to international forums and debates. The researcher and travel writer became a celebrity and a public intellectual.[38]

Though his fame culminated in the 1840s and 1850s, Humboldt's experience in the Americas around 1800 remained at the core of his

[35] "Humboldt (Friedr. Heinr. Alexander, Freiherr von)," in *Allgemeine deutsche Real-Encyklopädie für die gebildeten Stände. Conversations-Lexikon*, 10th ed., 15 vols. (Leipzig: Brockhaus, 1853), VIII:126–33, esp. 130.

[36] Cf. Michael Strobl, "Alexander von Humboldts Pico del Teide-Aufstieg als mediale Selbstinszenierung um 1800," in *Orbis Litterarum* 73, no. 1 (2018): 52–79.

[37] "Baron Humboldt," in *The Literary Magazine, and American Register* 2, no. 10 (July 1804): 321–27; cf. Rex Clark, "Alexander von Humboldts erster Reisebericht. Ein Spiel mit der Anonymität," in *Zeitschrift für Germanistik* 29, no. 1 (2019): 118–29; Gerhard Casper, "A Young Man from 'ultima Thule' Visits Jefferson: Alexander von Humboldt in Philadelphia and Washington," in *Proceedings of the American Philosophical Society* 155, no. 3 (September 2011): 247–62; Sandra Rebok, *Humboldt and Jefferson: A Transatlantic Friendship of the Enlightenment,* (Charlottesville: University of Virginia Press, 2014).

[38] Cf. Michael Strobl, "Alexander von Humboldt als *Public Intellectual*: Seine Beiträge in der *Neuen Zürcher Zeitung* (1825–1859)," in *Zeitschrift für Germanistik* 28, no. 2 (2018): 368–75.

research and of his public perception. His highly publicized expedition afforded him symbolic capital. He used it for matters that went well beyond science—for social goals and political concerns and to promote talented peers. Besides publishing scientific articles in specialist journals, Humboldt increasingly released texts written for high-circulation periodicals. They include the *Neue Berlinische Monatschrift, Morgenblatt für gebildete Stände, Allgemeine Zeitung, Vossische Zeitung,* and *Spenersche Zeitung* in the Prussian capital; *Wiener Zeitung* in Austria; *El Siglo diez y nueve* in México; the *Boston Post* and the *National Intelligencer* in Washington, DC. Some of the widely read journals and newspapers that were the leading media of their time remain influential to this day, for example, the (then liberal) *Neue Zürcher Zeitung* of Switzerland, *The Times* of London, and the *New York Times*. The *Economist* printed an excerpt from Humboldt's *Cosmos*.

Humboldt's political opinions, hinted at behind closed doors and only posthumously published in his letters[39] and diaries,[40] are apparent in the writings he issued during his lifetime. They display a considerable public engagement. Humboldt fought slavery.[41] He expressed his solidarity with the Jews as a "persecuted and oppressed people," and he opposed efforts to rescind the rights granted them in Prussia.[42] (Symbolically, one such appeal appeared in Hebrew.[43]) Humboldt took part in transatlantic debates. In the 1856 US presidential election campaign, he endorsed the liberal, Republican candidate, John C. Frémont (albeit in vain).[44]

[39] For example, *Briefe von Alexander von Humboldt an Varnhagen von Ense aus den Jahren 1827 bis 1858* (Leipzig: F. A. Brockhaus, 1860).

[40] Especially in *Lateinamerika am Vorabend der Unabhängigkeitsrevolution.*

[41] For example, "African Slavery," in *The Liberator* (26 March 1831): [3]; "Alexander Humboldt on Negro Slavery," in *The Friend* (February 19, 1831): 145–46, (February 26, 1831): [153]; "Baron von Humboldt versus the American Proslavery Censorship," in *Anti-Slavery Advocate* (September 1, 1856): 394–95; "Baron Humboldt on American Slavery. A private letter to Mr. Julius Froebel. Communicated to the Tribune with Humboldt's consent," *New York Daily Tribune*, May 27, 1858, 4.

[42] Humboldt on the Law on Jewish Citizens, *Allgemeine Zeitung des Judenthums,* (7 May 1842), 276–77; (14 May 1842), 302; (21 August 1843), 505–9; a letter from Humboldt to Hirsch, *Allgemeine Zeitung des Judenthums* (9 February 1857), 86; a letter from Humboldt to Slonimski, *Allgemeine Zeitung des Judenthums* (13 September 1858), 516–17; a letter from Humboldt to Lewysohn, *Allgemeine Zeitung des Judenthums* (21 June 1858), 358–59.

[43] Humboldt's letter to the author, in Israel, Joseph Benjamin, *Eight Years in Asia and Africa. From 1846 to 1855* (Hannover: self-published, 1858), III; in Hebrew: in Yiśra'el ben Yosef Binyamin, *Sefer Masa'ei Yisra'el, Bo Yesupar Me'aheinu Benei Yisra'el Ha-Nefutsim Be'artsot Asiyah Ve-Afrikah* (Lyck/Elk: 1859): [n.p.].

[44] First published in John Bigelow, *Memoir of the Life and Public Services of John Charles Fremont* (New York: Derby & Jackson, 1856), 327–29, reprinted in *Albany Evening Journal*, July 5, 1856, 2; *New York Daily Tribune*, July 12, 1856, 4; *Chicago Daily Tribune*, July 15, 1856, 3; *National Era*, August 7, 1856, 127; *Daily Atlas*, August 7, 1856, [1]. Cf. Strobl, "Alexander von Humboldt als *Public Intellectual*," 370–71.

Programmatically, he reflected on the independence of the colonies, on free trade, on a channel connecting the Atlantic and Pacific Oceans (which Humboldt had thought of long before its realization in Panama in 1914), and on the "future relations between Europe and America" ["die künftigen Verhältnisse zwischen Europa und Amerika"].[45] His commitment was unremitting. Humboldt was the most active German writer (at least of his era) in questions of foreign and intercultural issues, as evidenced by his anthropological and ethnographic studies[46] as well as his political and historiographical contributions,[47] among them: "Ueber die Urvölker von Amerika" [On the Native Peoples of America] (1806), "Sur les peuples qui mangent de la terre" [On Earth-Eating Peoples] (1809) or "War Poison of the Indians" (1821). From his first to his final text (1789–1859), Humboldt was critical of colonialism.[48]

THE AUTOBIOGRAPHER

Alexander von Humboldt did not publish an autobiography.[49] And he refrained from using an autobiographical style in his travel accounts.[50] But he noted that his texts *include* autobiographical elements: "Seek my life in my writings."[51] His collected essays allow us to follow this advice.

Humboldt's travel reports and political contributions, and to a lesser extent, his scientific articles, indeed divulge thousands of biographical details. Though he rarely relates comprehensive life stories, his texts are peppered with dates, facts, and episodes: with numerous (auto-) *biographèmes* that shed new light on his life. Of the more than seventy Humboldt biographies published to date, however, most ignore this corpus. Even Andrea Wulf's much-read biography lists Humboldt's books,

---

[45] Alexander von Humboldt, "Ueber die künftigen Verhältnisse von Europa und Amerika," in *Morgenblatt für gebildete Stände* 20, no. 33 (February 8, 1826): 129–30; 20, no. 34 (February 9, 1826): 134–35.

[46] Alexander von Humboldt, *Ueber die Urvölker von Amerika und die Denkmähler welche von ihnen übrig geblieben sind. Anthropologische und ethnographische Schriften*, ed. Oliver Lubrich (Hannover: Wehrhahn, 2009).

[47] Alexander von Humboldt, *Ueber die künftigen Verhältnisse von Europa und Amerika. Politische und historiographische Schriften zur Neuen Welt*, ed. Oliver Lubrich (Hannover: Wehrhahn, 2010).

[48] See Lubrich, "Von der ersten bis zur letzten Veröffentlichung."

[49] As a posthumous compilation cf. *Aus meinem Leben. Autobiographische Bekenntnisse*, ed. Kurt-R. Biermann (Munich: C. H. Beck, 1987).

[50] See Oliver Lubrich, "Alexander von Humboldt: Revolutionizing Travel Literature," in *Monatshefte* 96, no. 3 (2004): 360–87.

[51] Cited by Hermann Klencke, *Alexander von Humboldt. Ein biographisches Denkmal* (Leipzig: Otto Spamer, 1851), vii.

diaries, and correspondence in the bibliography, but not a single article or essay.[52]

In his writings, we learn *en passant* about those with whom Humboldt collaborated and exchanged letters, about a close friend or colleague; we find out where and when he took measurements and collected data, whose letters he received and whose works he studied, which events in contemporary history he took interest in and how his political views developed, which regions he crossed, as well as which invitations he declined due to other obligations or illness; which honors he received, how his service as personal advisor to the Prussian king kept him from his research, and even to what extent he was burdened by correspondence and administration. These incidental bits of information, put together, form a dense microchronology that complements our knowledge of Humboldt's life.

Moreover, Humboldt's writings include his curriculum vitae, which he wrote in 1799 for the Spanish king;[53] his first account of his American journey, which he wrote in 1804 in the United States;[54] and an encyclopedia article about himself, which consists largely of his own account[55]—in a narrower sense all important documents for the narrative of his life. Still, it remains to be seen how Alexander von Humboldt's biography will be rewritten based on his writings.

IMPACT

If it were possible to assign to Humboldt a modern *impact factor*, that is, the importance of scientists' and scholars' research contributions measured by the quantity and the ranking of their publications, his hundreds of writings would have been more influential than his books, however numerous, extensive, and important. With their continuity and their low-threshold accessibility in periodicals large and small, Humboldt's writings ensured his discursive presence and wide reception. They were the texts his contemporaries most readily followed.

---

[52] Andrea Wulf, *The Invention of Nature. Alexander von Humboldt's New World* (New York: Alfred A. Knopf, 2015), 435–37. Some better known writings have nevertheless been included in this account.

[53] In Dutch translation: "Korte Levens-Schets van Frederik Alexander van Humboldt," in *Algemene Konst- en Letter-Bode voor het Jaar 1800* 13, no. 316 (1800): 17–20.

[54] "Baron Humboldt," among others in *The Literary Magazine and American Register* 2, no. 10 (July 1804): 321–27.

[55] "Alexander von Humboldt," in *Die Gegenwart. Eine encyklopädische Darstellung der neuesten Zeitgeschichte für alle Stände*, 12 vols. (Leipzig: Brockhaus, 1848–56), 8:749–62 (1853).

Of Humboldt's works, only a small number of his books are known to the nonspecialized public. The majority of his books played only a minor role in his contemporary reception. Apart from the German-language bestsellers *Ansichten der Natur* (published in three editions: 1808, 1826, and 1849) and *Kosmos* (published in five volumes from 1845 to 1862), most of his works were too demanding, too extensive, and too expensive for the general public and only available in small print runs. Merely 600 copies of *Vues des Cordillères* (1810–13), Humboldt's most splendid and costly volume, were printed in the large-format folio edition.

The limited accessibility of many of his books accounts for the large number of excerpts, short passages, and entire chapters that circulated in newspapers and magazines. Some excerpts were the result of strategic efforts by Humboldt's publishers to promote his books. Johann Friedrich Cotta, for example, included numerous sections from *Ansichten der Natur* (*Views of Nature*) and from the German translation of Humboldt's work on Mexico, the *Essai politique sur le royaume de la Nouvelle-Espagne* (1808–11) in his widely read *Morgenblatt.* Some excerpts appeared in daily newspapers in remote places, probably without Humboldt's knowledge. They made his works known in regions that Humboldt never visited and where his books were hardly available, for example, in South Africa, India, Australia, as well as in the West and Midwest of the United States. In his time, Humboldt's books were most often read in international periodicals.

Humboldt's popularity in the United States in the nineteenth century is a strong case study for the relevance of his writings within contemporary publication practices—a phenomenon that has been largely underestimated. Almost against all logic—the foreigner who had never spent significant time in the United States, the scientist espousing progressive theories contrary to widely held American opinions concerning race and religion, and the author who never wrote in English and was only known in translation—Humboldt became a celebrity perhaps without equal in the mid-nineteenth century. In contrast to the historical memory in German and South American culture, however, the United States's embrace of Humboldt has been forgotten. Only occasionally have specialists in American history recalled the enthusiasm that named hundreds of towns, parks, streets, rivers, mountains, and other features for a distant Prussian. And there are few accounts of why Americans gathered by the thousands to dedicate statues, hear poems and speeches, and march in parades for Humboldt's centenary birthday across the country in 1869. Seldom have literary scholars traced the recurring appearance of Humboldt in American literature.[56] As a result of the decline of Humboldt's historical reputation in the United States, there has been little interest in the question of how such an unlikely figure was able to achieve such fame.

This is how an article written in 1990 by Laura Dassow Walls described the state of research around that time: Before the publication of *Cosmos* in 1845, Walls claims, Humboldt was not "a prominent cultural figure in America" and "there seems to be relatively little about him in the popular press." Finding a review article from 1823, she concludes that it "seems to be the first (and for some time the only) public, indigenous American response."[57]

The current collection of Humboldt's writings greatly expands the documentation of his influence in the United States and shows the emerging publication practices of the time. Humboldt's articles were often copied repeatedly from city to city, which magnified the number of readers. Perhaps there is no better example of this than the manuscript in French given by Humboldt in 1804 to the secretary of the American Philosophical Society to be translated and published (see "Baron Humboldt" in this collection). The American reception of this first comprehensive travel narrative was underappreciated for a long time—not until 1958 were the original and the resulting first American printing in an obscure monthly magazine discussed in the secondary literature.[58] Not until research was done for the 2019 *Sämtliche Schriften* edition was it discovered that the article was also printed in newspapers up and down the American East Coast. And due to its length, the travel account was often serialized over several days. Instead of one impression in a literary magazine, from July to November readers of thirteen newspapers could find news from Humboldt in at least twenty-seven different issues. This pattern of extensive newspaper distribution of Humboldt's writings continued in the

[56] For a concise account of Humboldt in US popular culture, see Laura Dassow Walls, "'Hero of Knowledge, Be Our Tribute Thine:' Alexander von Humboldt in Victorian America," in *Northeastern Naturalist* 8 [Special Issue 1: *Alexander von Humboldt's Natural History Legacy and Its Relevance for Today*] (2001): 121–34; cf. Sandra Nichols, "Why Was Humboldt Forgotten in the United States?," in *Geographical Review* 96, no. 3 (2006): 399–415. For 100 examples of literary receptions of Humboldt, many from America, see *Transatlantic Echoes. Alexander von Humboldt in World Literature*, eds. Rex Clark and Oliver Lubrich (New York: Berghahn Books, 2012). For Humboldt's influence on American environmental writers, see Aaron Sachs, *The Humboldt Current*, and Andrea Wulf, *The Invention of Nature*.

[57] Walls, "The Napoleon of Science," 79.

[58] Helmut de Terra, "Studies of the Documentation of Alexander von Humboldt: The Philadelphia Abstract of Humboldt's American Travels, Humboldt Portraits and Sculpture in the United States," in *Proceedings of the American Philosophical Society* 102, no. 6 (1958): 560–89.

following decades. An example from the end of Humboldt's career shows how periodicals amplified his support of the abolitionist cause and how this awareness of his moral position increased his standing as a public figure. Abolitionists were eager to print his works. For example, in this volume there are texts from *The North Star*, edited by Frederick Douglass; antislavery articles in the Quaker publication, *The Friend*; and the letter to Julius Froebel in the *New-York Daily Tribune* from 1858. A now-well-known incident occurred when Humboldt signaled his displeasure about the censorship of his antislavery views in an English translation of his work on Cuba—the news went around the world in more than thirty newspapers.[59] A manuscript letter in the Boston Public Library[60] shows the political efforts of some abolitionists to receive support from Humboldt. Its author, Anne Warren Weston, was one of the six Weston sisters who were prolific publicists, powerful organizers, and founders of the Boston Female Anti-Slavery Society.[61] They also published an annual abolitionist literary book for fund-raising, *The Liberty Bell*, and Weston included a copy of this with her letter to Humboldt. Initiated by a newspaper article and no doubt with the hope of future support or participation in their book project, Weston's efforts illustrate that there were unique topics in the American political sphere that expanded Humboldt's audience (in this case to politically active women's organizations) well beyond perhaps typical assumptions of readers mainly within the European scientific community.

SIGNIFICANCE

To be sure, the corpus of Humboldt's writings consists of more than just fragments of his books. Two thirds have no equivalent in his major works. They represent independent texts with new information. These writings contain groundbreaking insights. Among them are the first proven connection between the two largest river systems in the world, the Amazon and the Orinoco; the concept of zones of equal temperature extending across the globe, which Humboldt depicted as isothermal lines, an innovative

[59] "Humboldt and the Americans," July 29, 1856, in this volume.

[60] [Letter To] Monsieur De Humboldt, Boston Public Library. [Anne Warren Weston sends a copy of the *Liberty Bell*. Glued to page 3 of her letter is a clipping from the *Journal des Débats*, Aug. 1, 1856.] MS A.9.2 v.1, p.11, 11 March 1858.

[61] Clare Taylor, *Women of the Anti-Slavery Movement: The Weston Sisters* (London: Palgrave Macmillan, 1994); Lee V. Chambers, *The Weston Sisters: An American Abolitionist Family* (Chapel Hill: The University of North Carolina Press, 2014). The editors are grateful to Rex Clark for his advice and contributions.

form of presentation in climatology; the nocturnal increase of sound propagation in the tropics, which is now known as the *Humboldt effect*; or the defense of electric eels against horses by means of electric shocks, confirmed by biological experiments in 2016.[62]

Humboldt made significant contributions to a wide range of disciplines, from anatomy and anthropology to botany, cartography, chemistry, geology, physiology, and zoology. Even Humboldt specialists were largely unaware of these texts until his 250th birthday. They now complement our knowledge of Humboldt's research and literature.

In addition to writings of biographical and scientific relevance, the corpus contains other noteworthy contributions, such as Humboldt's very first text, in which, at the age of nineteen, he reported in French on a Javanese poison tree; or his very last publication, a sarcastic "Call for Help," in which the almost-90-year-old asks his readers to abstain from sending him unnecessary correspondence and allow him to finish his work in peace. Among the stylistic and literary highlights is Humboldt's only fictional text, the story "Die Lebenskraft oder der Rhodische Genius" ["The Force of Life, or the Rhodian Genius"], which was first published in Friedrich Schiller's journal *Die Horen* in 1795. On the most famous episode of his travels, the ascent of the Chimborazo, Humboldt never gave a detailed account in book form—but in an essay: "Ueber zwei Versuche den Chimborazo zu besteigen" ["On Two Attempts to Ascend Chimborazo"]. In 1802, Humboldt and his travel companions had set a world altitude record by climbing this volcano in present-day Ecuador, which was then considered the highest mountain on the planet. He waited to publish his narrative until 1837, when his mountain-climbing feat was surpassed by another expedition.[63]

Of singular importance are texts originally conceived as books but abandoned in the intended form. For example, the text "Géographie des Plantes" (1826) is the only published testimony to a planned second edition of Humboldt's *Essai sur la géographie des plantes* (1807), which he prepared in the 1820s but never executed. Similarly, the oceanological essay of 1837 on the current off the west coast of South America,

---

[62] Kenneth C. Catania, "Leaping eels electrify threats, supporting Humboldt's account of a battle with horses," in *Proceedings of the National Academy of Sciences* 113, no. 5 (2016): 6979–84.

[63] Humboldt depicted the Chimborazo in the "Tableau physique" of his *Essai sur la géographie des plantes* (1807) and in other illustrations. And he briefly mentions the episode in two short chapters of his *Vues des Cordillères* (1810, 1811). But the story of the ascent remained untold until 1837. Cf. Oliver Lubrich, "Fascinating Voids: Alexander von Humboldt and the Myth of Chimborazo," in *Heights of Reflection: Mountains in the German Imagination from the Middle Ages to the Twenty-First Century*, eds. Sean Ireton and Caroline Schaumann (Rochester: Camden House, 2012), 153–75.

later named the Humboldt Current, is only a fragment of an extensive manuscript intended for the second volume of his *Kleinere Schriften*, which never materialized (the first volume appeared in 1853).

Another important type of text within his published writings is Humboldt's correspondence with friends, scientists, and officials on both sides of the Atlantic. Originally, his written accounts, which he sent from America to acquaintances and editorial offices in Europe, were a means of reporting live, as it were, from his current location to his home audience, and of making research results public even before his return. Later, toward the end of his life, his fame had grown to such an extent that even the briefest of his notes was published in daily newspapers, often framed with solemn introductions, to please the readers' curiosity. Acknowledgments taken from private correspondence, thank-you notes and short communiqués, opinion pieces to international addressees intended for the public and published accordingly all contributed to Humboldt's transnational fame.

ECHOES

Finally, Humboldt's articles and essays were the source of his scientific influence and literary resonance. Goethe was inspired by Humboldt's seminal essay "Über den Bau und die Wirkungsart der Vulkane in verschiedenen Erdstrichen" ["On the Structure and the Effects of Volcanoes in Different Parts of the World"], which resonates in scenes of *Faust II*, which juxtapose different geological theories. Humboldt's writings pose many questions for future research: Which of his articles were important to writers such as Achim von Arnim and Adelbert von Chamisso, inspiring generations of scientists such as Carl Friedrich Gauß and Charles Darwin? Which of his essays prompted literary adaptations and fictional adventures such as those of Jules Verne, who quotes Humboldt several times in his science fiction novels? Which public statements and political interventions of the *auteur engagé* influenced American opponents of slavery, South American independence fighters, Prussian Jews, and German revolutionaries in 1848? Where, by contrast, was Humboldt appropriated and misrepresented in manipulated quotations and tendentious adaptations? With his writings, scholars from Miami and Mexico to Manchester and Moscow can trace the echoes of Humboldt's articles and essays all over the world.

Published more than 250 years after his birth, this edition of Humboldt's writings makes the most public part of his works newly accessible. His complete œuvre, manuscripts included, might not be edited until his 300th birthday. May the day come soon when we read Humboldt more than we celebrate him.

Editorial Note

The present selection of texts is based on the ten-volume edition of Alexander von Humboldt's *Sämtliche Schriften: Aufsätze, Artikel, Essays* [*Complete Writings: Papers, Articles, Essays*], edited by Oliver Lubrich and Thomas Nehrlich in Berne and published by dtv in Munich in 2019.[1] Two hundred fifty of Humboldt's texts were translated into English during his lifetime. These historical translations are assembled here for the first time. If various versions of the same text exist, the selection was made on the basis of the following criteria: date (first print or earliest possible print), integrity (as complete a version as possible), relevance (prominence of the publication medium). The order of the texts follows the chronology of the respective original publications as established in the German edition of 2019. A new translation of Humboldt's first article of 1789 complements the historical English versions.

The orthography and the punctuation of the edited texts correspond to the historical originals. Only obvious errors that distort the meaning of the text have been emended. All emendations are documented in the emendation list, separated by volume. Textual and typographic characteristics are preserved as far as possible. Footnotes are numbered for each article. The layout of the headings has been standardized. The page numbers and page breaks of the original are marked. The historical vocabulary (*negro, mulatto*) and the historical units of measurement (*toise, lieue*) have not been altered.

[1]Alexander von Humboldt, *Sämtliche Schriften: Aufsätze, Artikel, Essays* (*Berner Ausgabe*), eds. Oliver Lubrich and Thomas Nehrlich, 10 vols. (Munich: dtv, 2019). www.humboldt.unibe.ch/text.

Acknowledgments

The editors thank the persons involved in the making of the 2019 edition of Humboldt's *Sämtliche Schriften*. Above all, we thank the co-editors of the text volumes, Sarah Bärtschi, Rex Clark, Joachim Eibach, Bernhard Metz, Jutta Müller-Tamm, Michael Strobl, Jobst Welge, Norbert Wernicke, and Yvonne Wübben; the copy editors of the supplement volumes, Corinna Fiedler, Johannes Görbert, and Norbert Wernicke; and the board of advisers, Michael Hagner, Eberhard Knobloch, Alexander Košenina, and Hinrich C. Seeba.

We are especially grateful to the persons who helped us with the present edition: We thank Michael Strobl for his contribution to the conception of this compilation and for facilitating the necessary communication among those involved; Rex Clark for his research in the archives of the American Philosophical Society, his suggestions and contributions to the Introduction, and his translation of Humboldt's earliest article; Stephen T. Jackson, Alison E. Martin, and Sylvie Romanowski for their helpful comments; Naomi Lubrich for copyediting the Introduction; Paula Steck and Matthias Stöckli for their technical assistance with the manuscript and the text files; and Pamela Lankas for her careful copyediting and composition of the index.

Alexander von Humboldt: Writings in English

Part I: 1789–1824

Letter to the Author of this Journal; on the Bohon-Upas, by a Young Gentleman of This City

Sir.

Some time ago you published in your Gazette Littéraire the description of a toxic tree that grows in India and is known by the name of *bohon upas*. This description included so many marvelous things that one was almost tempted to call them fables. You yourself, Sir, did not appear to regard them as established truths. However, a very recent pamphlet, which a friend of mine has just received from Sweden and which, given its author, can be considered trustworthy, shows us clearly that the effects of this terrible poison surpass with more cruelty anything that the imagination can produce. This recent work is entitled:

Arbor Toxicaria Macassariensis: Upsaliae 1788

The author of this treatise is Professor *Thunberg*, Knight of the Order of Vasa, a member of most of the academies of Europe, renowned for his talents and his universal knowledge of nature. He is a pupil of *Linnaeus*, in whose footsteps he walks with achievement that earns him no less honor than his great teacher. After completing his studies at the Academy in Uppsala, he went to Holland, where, equipped with letters of recommendation from Swedish scholars, he was able, thanks to his intelligence and his knowledge, to win the favor of some of the most distinguished officials in the republic. His ardent desire to prove himself useful to the human race through significant discoveries drew his attention to the parts of our globe that are still too little known. His illustrious *patrons*, who noticed his noble passion, took advantage of it for the benefit of the republic of letters. It was at their expense that Mr. *Thunberg* undertook trips to Cape Town, to Java and Japan over the course of several years. He visited all the most remarkable places of East India, and few botanists among our contemporaries can boast of such brilliant successes. After returning to Europe, he was honored in his homeland with all the distinctions that a citizen of his kind deserves. The King of Sweden, who promotes the sciences and fine arts, honored him with his order. The most celebrated academies in Europe hastened to accept him as one of their members, and in Uppsala he presently holds the same position which was previously distinguished by the immortal *Linnaeus*.

Mr. *Thunberg* was the first to dare to make beneficial changes to his teacher's system—which no one before him had successfully revised. He found that the last four classes were not useful and put them into twenty classes: a simpler method that Herr *Willdenow*, a young, but insightful botanist, has followed in his excellent *Flora Berolinensis*.

Since Sir *Thunberg* has spent several years on those immense islands that lie to the east of Asia, he knows the magnificent natural products of these favorable climates better than any other scholar in Europe. His description of the plants of Japan, his treatises on the carnation and nutmeg and so many other learned works testify that hardly a botanist before him has traveled with such attention to everything that concerns nature. We can therefore rely on his judgment if we are misled by any unfamiliar or gullible traveler about something. The stories of the poison tree of India will show us once again how things are exaggerated, which, by their own nature, offer us enough wonders without the need to add imaginary ones to make them more interesting.

Mr. *Thunberg* begins with a botanical description of the *bohon-upas*, or, more correctly spelled, the *boa-upas*, which, in the language of the Malays, means *poison tree*. These peoples distinguish two varieties with the names *Macan-Cavul.* and *Djato matti*; of which the latter is more dangerous than the former. The famous *Rumphius* describes it in his *Herbarium Amboinense* as *female* and *male* in the ridiculous manner of earlier botanists. Both have a strong, thick trunk and protruding branches, the bark is furrowed and brownish gray; yellowish, black-spotted wood. The leaves are oval, two inches wide and a hand span long. Since the flowers and fruits of this dangerous tree have never been able to be examined by any botanist, the genus cannot be determined with certainty. Mr. *Thunberg*, however, has reason to presume that it belongs to the genus *Cestrum Lin.*[1] as the ancients called a type of jasmine. This opinion seemed all the more correct to him when he saw at the Cape of Good Hope how the Hottentots mixed the sap of a *Cestrum* with the terrible venoms that they take from their snakes.

The *boa-upas* grows mainly on the islands of Java, Sumatra, Borneo, Bali, and Makassar. It thrives on bare mountains and deserts. An infertile,

[1] Other botanists believed that the *boa-upas* was a species of the iron tree *Sideroxilon Lin.* I do not know why. *Rumphius* says, "The Indians hide this tree with such care that even after the conquest of Celebes in 1670, no description of it could be given." However in 1694 he managed to obtain a branch and a drawing was done. *Herbarium Amboinense, Volume II, Plate LXXXVII.*

arid or ash-like soil indicates its presence. No tree, not even grass, can grow in its shade. Within a stone's throw of this tree, our author writes, the earth seems to be burnt. "However," he adds parenthetically, "*pro certo venditatur*, or as they say," thereby demonstrating rather clearly that this is just a popular rumor.

Here a reminder of the trust one should have in superstitious or malicious travelers: some have written that within a surrounding 10 to 12 miles neither tree nor shrub, not even grass grows. Others report, "that within the area of 15–18 miles, no human or beast can be found, not even a fish in the water." What a huge difference is there between the space of a stone's throw and that of the surrounding 18 miles? Mr. *Thunberg* noted that the priests in particular were interested in spreading errors that the common people in India were all too happy to accept. *Voltaire* would say this proves that the priests do not change below the equator. We suspend judgment of this and are content to deplore the sad fate of mortals who are duped by their so-called brethren. The priests of Mohammad could refrain from making the commoners believe that God gave in to the Prophet's insistence and created this tree to punish the peoples for their sins. The evils of humanity are numerous enough without the need to augment them by such harmful ideas.

One can doubt that the infertility that all authors mention is rightly attributed to the toxic vapors of the *boa-upas*. It may well be that this tree sustains itself in soil from which no other plant could be nourished. A lone juniper tree that grows in a crevice certainly doesn't prove that it suppresses all vegetation around it. Even though Mr. *Thunberg* reports that travelers had not seen a single animal or plant within a radius of several miles, this is nothing unusual for the hot zone of our globe. The excessive heat in summer causes all plants to die; animals do not find any more food in these deserts and flee to the densest woods. The frequent rains in winter revive nature and bring it, so to speak, out of its lethargic slumber; the earth turns its infertility into carpets of grass, and you can see herds grazing in the same places that were previously believed uninhabitable.

The sap of this awful tree is a blackish resin that dissolves in the heat of the fire. It goes for a steep price among Indians. The peoples who own it have a clear advantage over their enemies. *Rumphius*, formerly Consul in Ambon, reports that his compatriots, the Dutch, before effective antidotes were discovered, feared the arrows poisoned with this sap more than all other dangers and hardships of war.

This sap is very difficult to collect; it is not without risk. Since the tree's vapors are extremely pernicious, one can only approach it with utmost caution; and these difficulties and dangers drive up the price of the poison. Those who collect it must wrap their heads, hands and feet in cloths. No one dares to touch the deadly tree trunk. One stays at a certain distance, "because," the author writes, "it seems that death has put its foot down and found its place near the tree."

It is with long bamboo tubes that the Indians collect this deadly sap. They sharpen these tubes at one end and push them into the tree. The bark split by this process immediately releases its blackish sap, which flows into the tubes in thick drops. Fifteen or twenty of the bamboo canes are drilled into the tree in this way; three or four days later they are pulled out filled with the deadly poison. As long as the sap is fresh, it is soft and malleable like a paste; small rolls are then made from it, which are put back into the bamboo tubes, and since this poison is very volatile, the tubes are wrapped eight to ten times with double canvas. From the type of precautions I have described, which must be taken to approach this tree, where death seems to await those who dare to brave it, you can easily conclude that nothing is more dangerous than this harvest; and you, just as I, could not imagine that there are men who are so bold to take this dangerous adventure. On the one hand, what can thirst for revenge do, on the other hand, the greed for gold?

The people, as superstitious as they are in all of India, believe that if the trunk of the tree is cut off, the poison becomes even more effective, even more terrible. The people don't think this through; they believe it blindly; otherwise the question would be what connection exists between the sap of a tree and the trunk from which it is extracted. And if touching this tree is so fatal, how can you cut off its trunk? And if it were possible, such a procedure, often repeated, would soon destroy the *boa-upas*.

The poison trees seem to be a public good of the state. *Rumphius* writes that the inhabitants of the mountains give all the sap they collect to a dignitary of the country called *Creyn Sumana*. He keeps this national treasure in his castle *Boerenbourg*, in rooms that should not be too warm nor too cold; both extremes are equally harmful to the poison. Every week the sap and bamboo tubes are rubbed and cleaned. Only women are requested and allowed to do this work because men are not believed to be honest enough to entrust it to them; others allege foolish reasons[2] that don't merit our attention.

[2] We quote them in Latin because they cannot be expressed in French or English in a decent way. *Menstruum nempe muliebre huic misceri veneno dicitur, atque in eum finem Macassariensium feminas bracteis indutas esse, in quibus istud colligebant.* [It is said that the female monthly blood is mixed with this poison, and for this purpose the women of the Macassarians wear metal plates in which they collected it.] – What absurdities!

The poison of the *boa-upas* surpasses by far the most terrible of the perfidious arts that were once used for revenge by some ultramontane princes. The tree's vapors alone stiffen the limbs and cause the most violent convulsions. *Rumphius*, the only botanist who up until now has had the advantage to receive a branch of this sinister tree, reports that its destructive effects were felt through the bamboo in which it was enclosed. If you put your hand on the tube, you felt a kind of cramp, similar to that caused by a sudden change of cold and heat. Those who risk standing bareheaded under this malignant tree lose their hair. A drop of the toxic sap, even if it only touches the skin, causes thick swellings. The air around the tree is so poisoned that all animals avoid approaching it. A bird that strays on its branches falls dead in an instant. Mr. *Thunberg* did not consider it appropriate to quote a fact, which, it seems, should not be passed over in silence. Here it is according to the report from *Rumphius*.

The only animal that likes to stay in the shadow of the *boa-upas* is a snake that is no less dangerous than the place it inhabits. The Indians say it displays a horn or, more likely, a large comb. At night its eyes shine with a bright glow; its voice mimics the crowing of a rooster. They can sometimes be heard near people's dwellings. The snake's exhalations are very toxic and prevent it from being approached; they are only killed from a distance. It is too much that in one place there are two similar beings, as dangerous and as fatal for that which they exhale, yet one can admire the wisdom of nature, which, by giving each being a special constitution, makes the same place pernicious for one and beneficial for the other.

The poison of the *boa-upas* deserves the full attention of the physician and the naturalist; the cause and effects are equally surprising. If you want to believe the natives in the places where this extraordinary tree is found, its pure, unmixed sap is almost harmless; it even serves as an antidote to the toxic exhalations of some fish. *Rumphius* confirms this by reporting that he has used it internally as a remedy, which seems to be quite astonishing. How can the vapors of this terrible tree be so deadly and its sap produce such a healing result? After all, the sap of the *boa-upas*, mixed with the juice of the *zerumbet*,[3] is the most effective poison that art and nature have ever produced, and at the same time the same *zerumbet* is a healing remedy that is used in India as an antidote. Such

---

[3] *Zerumbet* is the root of a plant called *Amomum zerumbet Lin.* Ginger, *Amomum Zingiber*; cardamom, *Amomum Cardamomum*, and other aromatic species belong to the same genus.

are the difficulties which need to be overcome for those who want to know the true causes of these extraordinary phenomena!

The conclusion follows.

The Conclusion of the Letter to the Author of this Journal; on the Bohon-Upas, by a Young Gentleman of This City

The sap of the *boa-upas*, once it has entered the human body, does not seem to come out anytime soon. The infected patient must be particularly careful not to eat the root of the *zerumbet*, because three years after consuming *boa-upas*, the *zerumbet* would cost him his life. What is this poison that stays in the body for so long without changing and without losing its deadly property, slumbering within and concentrating in the body despite its volatility, to develop its malignancy when mixed with another sap, which itself is an antidote to other poisons of various kinds? In particular, it has been reported that those who have been cured of *boa-upas* by antidotes feel the venom in their veins reappearing every year.

When the residents of Celebes are attacked by an enemy, they have all of the bamboo tubes containing the deadly poison brought from their arsenal or depot and these are distributed into different grades. Since the eye cannot judge the quality of the poison with certainty, it has to be tested with chemical means. A small grain of the hardened sap is used for this sample. It is tossed into the sap of the *zerumbet*, and if the poison is of great strength, it brings the *zerumbet* to a boil and causes an intense effervescence. After this test, the *boa-upas* is prepared with the *zerumbet*; the arrows are dipped in it and just a scratch from them is fatal if aid is not promptly given; but there is not an instant to lose.

The arrows dipped in this mixture retain their deadly power for two years if the poison is good; others lose it within two to three months. Indians who want to know about their weapons often test them with the sap of the *zerumbet*. Those who are unfortunate to be wounded by these poisoned arrows succumb to a death both quick and cruel. First they experience convulsions throughout the body, the face swells, the mouth fills with foam, the eyes come out of their head, and they expire, wailing, some within a quarter of an hour, others in half an hour; sometimes even faster, depending on the potency of the poison.

The Macassarians' arrows are no longer as threatening to Europeans as they used to be. *Rumphius* reports that these people have unlimited confidence in their weapons and that the Europeans have tried to emulate them. The Macassarians once devastated the island of Ambon, but the Dutch, by means of special clothing made of Spanish leather, could protect themselves from the terrible effects of their poisoned arrows.

In the past, Europeans knew no other antidote to these fatal wounds than human excrement, taken internally. This disgusting remedy, which naturally caused nausea, served as a vomitive to the unfortunate victim and thus deprived the poison of its deadly energy. However, it had to be administered immediately. In the battles against these peoples, European soldiers were often seen asking some of their comrades for mercy and to administer to them this necessary antidote; and anyone who could help the miserable one on the spot was regarded as a guardian angel. *Rumphius* quotes the example of a Dutch soldier who escaped death five times thanks to this impure means. Later on it was learned that the Indians themselves had more effective and less repugnant antidotes, such as the roots of *Crinum Asiaticum*, the bark of *Ficus ramosa* etc., etc. Amputation, this universal cure in surgery, is of no utility for those who have been injured by poisoned arrows; internal remedies are necessary. The kings of Celebes have made many attempts by injuring slaves who have been condemned to death with these deadly arrows. The injured limb was immediately separated from the body, but the slave perished nonetheless.

All of this shows you, Sir, that the *boa-upas* is one of the strangest phenomena in nature; it would be desirable for a naturalist to have the means to thoroughly examine this dangerous tree. But how can you research an object that harbors so many dangers from different sides for those who wanted to attempt this enterprise? We are still very much obliged to Sir *Thunberg* for sharing his observations on this. Educated travelers will want to add to these observations, either on their own, or by collecting all that will be found on this object so worthy of the attention of naturalists.

I am honored to be, etc. etc.
Berlin, 1st of January, 1789.

Extract of a Letter from Mr. Humboldt to Mr. Blumenbach, containing new Experiments on the Irritation caused by the Metals with Respect to their different Impressions on the Organs of Animals[1]

Mr. Humboldt is one of the philosophers who has made the most numerous observations on the phenomenon discovered by Galvani concerning the irritability produced by the contact of different metals with the parts of animals in which the principle of life is apparently extinguished. As long ago as the year 1795 he observed that the animal irritability was augmented by the oxygenated muriatic acid. Not having discontinued his attention to this object, the perusal of the physiological writings of Reil, and his correspondence with Scarpa and Volta, afforded him indications for new enquiries, of which he has occasionally had the courage to make himself the subject. | 257 |

"In a conversation," says he, "with M. Scarpa, at Pavia, on the effects which Galvanism produced upon myself, nothing surprised him more than the appearance of a lymphatic and serous humour on my back. 'What can be the nature,' said he, 'of this stimulant, which in a few instants changes the nature of the vessels to such a degree as to cause them to prepare humours, which, the instant they touch the epidermis, excite inflammation, and mark their course by a redness which lasts for whole hours?'" M. Humboldt promised to repeat the experiment; and the account he gives of the facts constitutes one of the most interesting articles in his letter.

For this purpose he caused two blistering plasters to be applied on the deltoid muscle of both shoulders. When the left blister was opened, a liquor flowed out which left no other appearance on the skin than a slight varnish, which disappeared by washing. The wound was afterwards left to dry up: this precaution was necessary, in order that the acrid humour which the Galvanic irritation would produce might not be attributed to the idiosyncrisis of the vessels. This painful operation was scarcely commenced on the wound, by the application of zinc and silver, before

[1] This Letter forms part of Gren's (German) Journal of Natural Philosophy for the month of October last. The extract was read to the Institute of France, by Citizen Guyton.

the serous humour was discharged in abundance: its colour became visibly dark in a few seconds, and left on the parts of the skin where it passed traces of a brown inflamed red. This humour having descended towards the pit of the stomach, and stopped there, caused a redness of more than an inch in surface. The humour, when traced along the epidermis, left stains, which after having been washed appeared of a blueish red. The inflamed places having been imprudently washed with cold water, increased so much in colour and extent, that Mr. Humboldt, as well as his physician Dr. Schallern, who assisted at these experiments, entertained some apprehension for the consequences.

Mr. Humboldt has not undertaken to determine the nature of the fluid which produces such astonishing effects; but he applies himself to circumscribe the phenomena in the real circumstances which produce them. He judiciously varies the preparations, and carefully notes all the results; being persuaded that the cause of Galvanism cannot be explored with success, but by observing the proportions in which the chain of metals either irritates or has no effect: and to extend still more this vast field of observation, he employs various means to raise or diminish the irritable capacity of the animal organs.

What is the sensation which the Galvanic irritation produces? Mr. Humboldt has discussed this question. "No one (says he) can speak more decidedly on this subject than myself, having made several experiments on my own person, the seat of which, in some instances, was the socket of a tooth which I had caused to be extracted; in others, certain wounds which I made in my hand; and in others, the excoriations produced by four blistering plasters." The following is his answer:

The Galvanic irritation is always painful, and the more so in proportion as the irritated part is more injured, and the time of irritation more prolonged. The first strokes are felt but slightly; the five or six following are much more sensible, and even scarcely to be endured, until the irritated nerve becomes insensible from continued stimulus. The sensation does not at all resemble that which is caused by the electric commotion and the electric bath; it is a peculiar kind of pain, which is neither sharp, pungent, penetrating, nor by intermissions, like that which is caused by the electric fluid. We may distinguish a violent stroke, a regular pressure, accompanied by an unintermitting glow, which is incomparably more active when | 258 | the wound is covered with a plate of silver, and irritated by a rod of zinc, than when the plate of zinc is placed on the wound, and the silver pincers are used to establish the communication.

When the communication is made by the contact of the epidermis, it produces no effect; it appears to insulate like glass, when interposed between the wound and the metal: but if the skin be removed, by making two wounds at eight inches distance, and a plate of zinc be placed on one of them, and on the other a leg of a frog prepared, this last is seen to contract itself when it communicates with the zinc by the silver wire; which proves that the Galvanic fluid then passes beneath the epidermis.

This fluid produced in some circumstances a very sensible acid taste. The two wounds of Mr. Humboldt having been covered, one with silver, the other with zinc, an iron wire of several feet in length, attached to the zinc, was conveyed between his upper lip and the spongy substance of the teeth, and thence to the tongue of another person. When the iron wire was made to touch the silver, a strong contraction of the scapular muscle took place, and at the same instant the person whose tongue formed part of the chain of communication perceived the sensation of acidity. There are also cases in which the fluid acts on the organs of taste without producing any sensible effect on the organs of motion: such is that where the epidermis serves as the conductor from zinc to the frog; for there is not then any contraction, but merely an acid taste on the tongue.

The author, having learned from Mr. Volta that he employed the solution of pot-ash (*oleum tartari per deliquium*) in order to augment the conducting power, availed himself successfully of this means to raise the capacity of the animal organs. He moistened one of his wounds with this liquor, which produced little pain; but the Galvanic irritation was more violent, and accompanied with more heat; sparks appeared and disappeared before his eyes; the tongue moistened with the same distinctly perceived the acid sensation, although the communication was established only between zinc and zinc. The thigh of the frog, moistened with the alkaline solution and laid upon a plate of glass, without touching either metal or carbonic matter, fell of itself into violent convulsions, the antagonist muscles of the legs and toes being incessantly agitated. Irritability has been re-established by this application in the animal parts, where it had been extinguished by warm solutions of the oxide of arsenic. Lastly, the irritation (which does not commonly take place when the nerve and the muscle are armed with the same metal, the different metals being between the coatings) becomes manifest after this preparation; which seems to indicate that the alkali not only irritates the nerve, but likewise adds to its irritability.

The author applied this method to amphibious animals, which he roused from their winter's sleep, and in which he perceived a peculiar symptom of irritability.

These observations led him to distinguish two states of the animal organ. The first, of irritability naturally or artificially raised or excited; the second, irritability in a less degree. These two states, which he calls *positive* and *negative*, are merely, as he remarks, different degrees, and not phenomena absolutely distinct from each other.

In individuals naturally sensible, the effects produced by alkaline solutions, by the oxygenated muriatic acid, by the solution of oxide of arsenic, are very rarely of the same intensity.

In the case of increased irritability, muscular motions are observed without metal or | 259 | carbonic matter. They may be obtained with metals, though without communication between the nerve and the muscle; that is to say, without the regular connection or chain. They may be also obtained by forming the chain of similar metals.

Let the crural nerve of an animal naturally tenacious of life be placed upon glass. Let a small piece of fresh muscular flesh be fixed on a stick of sealing-wax, and then brought into contact with the crural muscle. The result will be a violent convulsion at the instant when the chain of communication is completed. The same thing happens if, instead of the small piece of muscular flesh, a detached piece of the crural nerve be fixed on the stick of sealing-wax. The connection is therefore formed of two things, nerve and muscular fibre. How in this simple case can the fluid which passes from the nerve into the muscle cause it to be contracted? Mr. Humboldt thinks that it becomes stimulant, merely because it returns from the nerve into the nerve by a foreign animal matter; that is to say, not organically connected with the nerve.

The disparity of the metals forming the chain has hitherto appeared as a necessary condition to produce Galvanic irritation. This hypothesis, however, is overturned by the experiments of Mr. Humboldt. If it be true that, in the state of less irritability, there is very rarely contraction with similar metals (as Volta affirms, contrary to Aldini), this circumstance becomes indifferent in the case of increased irritability. Mr. Humboldt put into a china cup some mercury exactly purified; he placed the whole near a warm stove, in order that the entire mass might assume an equal temperature: the surface was clear, without the appearance of oxidation, humidity, or dust. A thigh of a frog, prepared in such a manner that a crural nerve and a bundle of muscular fibres of the same length hung down separately, was suspended by two silken threads above the mercury. When the nerve alone touched the surface of the metal, no irritation was manifested; but as soon as the muscular bundle and the nerve touched the mercury together, they fell into convulsions so brisk that the skin was extended as in an attack of tetanus.

We ought not to be surprised at the precaution here taken by Mr. Humboldt to heat the mercury. This is required in consequence of the opinion which he announces, that the parity of the metals does not depend on the homogeneity of their chemical constituent parts, but of their heat, polish, hardness, and form.

Gold, placed between two armatures of zinc, produces irritation only when the gold is moistened by some volatile fluid, or by the moisture of respiration. Lastly, Mr. Humboldt has attempted to include all the cases in the following formulæ:

1. In the State of increased Irritability.

Positive cases.

Frog—muscular flesh.
Frog—zinc—zinc.
Frog—zinc—muscular flesh—silver.
Frog—zinc—silver—zinc.
Frog—muscular flesh—silver—zinc.
Frog—zinc—muscular flesh—silver—muscular flesh—zinc.

2. In the State of diminished Irritability.

Positive cases.

Frog—zinc—silver.
Frog—zinc—muscular flesh—silver—zinc.
Frog—zinc—muscular flesh—silver—muscular flesh—silver—zinc. | 260 |

Negative cases.

Frog—zinc—zinc.
Frog—zinc—muscular flesh—silver.
Frog—zinc—muscular flesh—silver—zinc.

Mr. Humboldt finishes this letter by some observations which he has collected in the course of his experiments on the *sthenic* or *asthenic* virtue of chemical agents; that is to say, their energy or their inefficacy to produce irritation. Alkalis appear to be to the sensible fibres what acids are to muscular groups. The muriatic acid augments the iritability of the muscle while it extinguishes that of the nerves, which does not re-appear even after the acid has been saturated with alkali.

By continuing to bathe the nerve with an alkaline solution, an entire atony is at length produced by excess of irritation; but if a few drops of muriatic acid be let fall on the part, the irritability is re-established.

A thigh of a frog, irritated even to total relaxation by a warm solution of oxide of arsenic, has exhibited new convulsions, after having been immersed for two minutes in a solution of pot-ash.

The sthenic virtue of the oxygenated muriatic acid is not less remarkable. Thighs of frogs naturally flaccid, and weakened still more by the Galvanic process for seven hours, which afforded no sign of motion when silver served as a conductor between zinc and the nerve, exhibited violent contractions when the nerve was moistened with oxygenated muriatic acid. The author refers to this subject the experiment which he published in 1793, in his *Flora Fribergensis*, by which it is ascertained that ordinary muriatic acid retards the germination of plants, but that oxygenated muriatic acid had caused a plant to germinate in seven hours, which required thirty-eight in pure water in order to arrive at the same development. This fact appears to him to indicate some relation between the vegetable and animal organization.

A judgment may be formed from this extract of the number of important facts contained in this letter, and of the interest they will excite when they shall be collected, arranged, and amplified, in the large work which the author is preparing.

A Letter from M. de Humboldt to M. Pictet, on the Magnetic Polarity of a Mountain of Serpentine[1]

At the beginning of the eighteenth century the attention of natural philosophers was entirely fixed on the phenomena of magnetism. The progress which has since been made in the theory of electricity, and the preponderance which chemistry has acquired over all the other branches of natural history, have diminished the interest with which enquiries into the nature of the magnetic fluid ought to have been pursued. It is true that your celebrated countrymen Messrs. de Saussure and Prevôt have given vigour to this pursuit by discoveries worthy of their sagacity: the first, by inventing an instrument capable of measuring the comparative intensity of the magnetic forces in different regions of the globe; and the other, by reducing the laws of polarity to the simple laws of attraction. But these discoveries have not afforded inducement sufficient to lead philosophers into a path so honourably explored. The most valuable work on the origin of magnetic forces has been neglected, together with the calculations of the ingenious Coulomb, and his experiments with the balance of Torsion.

Having traversed with the compass in my hand great part of the mountains of Europe, I became convinced that declinations caused by masses of iron in beds or in veins are infinitely less frequent than naturalists affirm. The observations which Messrs. de Saussure and Trembley have made on the summit of Cramont,[2] appear to me the more curious, as | 98 | it stands alone, and presents to our knowledge a very extended image of the dimensions of magnetic spheres. It is among the Alps of Sweden and Norway, those northern regions which nature has enriched with an enormous deposit of iron less oxided than in our country, that we were entitled to expect similar phenomena.

I hasten to communicate to you a discovery I made in the month of November, and which appears to me of considerable importance in the progress of geology. You are acquainted, Sir, with the laws and the

---

[1] I received this communication in manuscript from the Right Hon. Sir Joseph Banks, Bart. P. R. S. & c. It is written in French, under the title of "Lettre quatrième de M. de Humboldt à M. Pictet, sur la Polarité Magnetique d'une Montagne de Serpentine." The same liberal promoter of science has favoured me with a specimen of this rock, with permission to make experiments upon it. A few observations on this specimen are added at the end of this memoir. N.

[2] Voyage dans les Alpes, T. I. p. 375.—T. II. p. 343.

harmony which I have observed in the direction and inclination of the primitive strata, from the banks of the Mediterranean to those of the Baltic Sea. You have even condescended, jointly with our friend Dolomieu, to express an interest with regard to this laborious undertaking; which, in more skilful hands than mine, would, I am well assured, throw great light on the construction of the globe. I traversed the chain of mountains of the High Palatinate and the margraviate of Bayreuth; and I found, in the bottom of the Fichtelgebirge, between Munichberg and Goldcronach, an isolated hill, which rises to the elevation of fifty toises above the surrounding plain. Its height above the level of the sea may be estimated at two hundred and eighty, or three hundred, toises. This hill extends in length from west to east, and forms a pyramid extremely obtuse. The rocks which crown the summit or ridge are composed of serpentine of considerable purity, which, by its colour and foliated fracture, approaches in various parts to the chlorithschiefer of Werner (*schistous chlorite*). This serpentine is divided into strata rather distinct, of which the inclination to the north-west presents an angle between 60 and 65 degrees. It reposes on a foliated granite, mixed with hornblende; a mixture which we distinguish by the name of syenite. I approached this serpentine with the compass, in order to determine more accurately the angle it formed with the meridian. The magnetic needle was in a state of continual agitation. I advanced two steps farther, and beheld that the north pole was entirely turned to the south. I called two friends, Messrs. Godeking and Killinger, who assisted me in my geological pursuits; and we were alike penetrated with that joy which the contemplation of interesting phenomena produces in the minds of thinking men. I shall not detain your attention by a full recital of our observations; but shall merely present the results, to which I may hereafter make additions, if my occupations should not lead me from this part of Germany.

The action of this mountain of serpentine upon the magnet shews itself in a very curious manner. The uncovered rocks which are seen on the northern slope, and those on the declivity towards the south, have poles directly opposite. The former exhibit only south poles, and the latter north poles. The whole mass of foliated serpentine does not therefore possess a single magnetical axis, but presents an infinity of different axes perfectly parallel to each other. This parallelism also agrees with the magnetic axis of the globe, though the poles of the serpentine are inverted; so that the northern pole of the hill is opposed to the south pole of the earth. The east and western slopes present what in the theory of magnetism would be called points of indifference. The magnet does not at this part appear to be in any respect affected, though the substance of the rock differs in no external character from the other parts. It is the same on the south side of the summit.[3] I have observed not only that the magnetic axes are not disposed in the same horizontal plane; but I | 99 | have likewise remarked, that two points, of which the action is very strong, are joined by rocks which do not exert the least attraction. The chemical analysis of these compounds affords the same results; and it would be no less difficult to discover any difference of aggregation between them, than between iron which has received the touch, and other iron which had never acquired the magnetic power.

On this occasion a question presents itself which cannot be resolved in less than half a century. The tables founded on the observations of Picard, La Hire, Maraldi, Cassini, and Le Monnier, shew that the needle has declined since 1660 towards the west; and that this declination continues to increase, though the oscillations caused by the heats of the south, and the temperature of the seasons, often produce a retrograde course. If the magnetic axis of our mountain were astronomically determined by the culmination of the stars, whether its direction would remain the same until the year 1850, or whether its south-pole would turn towards the west, in connection with the variation of the magnetic needle? From our profound ignorance of the causes of the declination, as well as of most geological phenomena, it is not in our power to resolve so complicated a problem.

Other observations equally interesting may be made on the identity of magnetic forces. I have discovered a mass of rocks which affect the needle at the distance of twenty-two feet. With an apparatus similar to the magnetometer of M. de Saussure, we might observe whether the intensity of the forces of magnetic action remains the same in winter and in summer; whether it be stronger in the morning, at noon, at the solstices, during the aurora borealis, or in an atmosphere loaded with electric fluid? I suppose that these same rocks might act on the needle sometimes at 16, and sometimes at 28 feet distance.

It has been observed, that metals exposed to the air gradually imbibe the magnetic fluid. A slight oxidation of the iron seems to favour this effect. I have myself observed, that in a magnetic bed of iron those parts only which were in contact with the air affected the needle. This phenomenon is considered as the effect of atmospheric electricity. I am

[3] I suspect an error of the copyist, as the words "Il en est de même du côté meridional de la sommité" contradict the general description immediately preceding. N.

aware that lightning converts a bar of iron into a magnet; that the discharge of the Leyden vial sometimes increases the intensity of magnetic forces; but I do not see why the atmospheric electricity should act simply on the external surface of a bed of magnetic iron, which is a good conductor of the electric fluid. Does not the oxygene of the atmosphere rather act a part in this operation? Without wandering in the sphere of probabilities, I have chosen to adhere to enquiries respecting facts. I have observed the rocks which were covered with turf, from which I detached pieces that had not been in contact with the air. I found that the magnetic force was constantly the same.

The mountains of the Harz present a granite rock called the schnarcher, which is elevated in the form of a tower, or broken pyramid. This granite likewise affects the needle; but it acts only in the mass, and in a single band or perpendicular vein. Detached pieces shew no action upon the needle. It is to Mr. de Trebra, celebrated for his researches concerning the internal parts of mountains, that we are indebted for this important discovery.

Some philosophers pretend that the schnarcher contain in their bowels a mass of magnetic iron; others presume that a stroke of lightning has caused the magnetic vein in these mountains.

The nature of the rocks which I have the honour to present to your notice in this paper does not admit of similar explanations. The serpentine not only acts in a mass, in its na- | 100 | tural situation, but all the pieces, when broken, to infinity, still exhibit two very distinct poles. Pieces of five inches diameter act on the needle at the distance of half a foot. The examination of the magnetic axes affords an object of curious enquiry. They are mostly found in a direction parallel to that of the foliated grain; nevertheless, I have found some which cross it perpendicularly. Fragments extremely small, of the magnitude of 0.01 of a cubic line, shew a very strong polarity in proportion to their masses. You see them turn very suddenly when the poles of the weakest magnet are successively presented to them. It is a very striking phenomenon, that a stone possessed of so high a degree of polarity should exhibit no attraction for iron which is not magnetised. I have never observed the smallest particle of filings of iron adhere to the serpentine; but the serpentine, reduced to powder, attaches itself very readily to the magnet.

You will enquire with impatience, if it be well proved that my serpentine is not mixed with magnetic iron; whether this mixture may not be sufficiently intimate to enter into the composition of each particle of the rock? I can assure you, that I have made the most assiduous enquiries in this respect. All my experiments were made in conjunction with Mr. Godeking, whose knowledge and abilities are a sufficient assurance against error; but we were decidedly convinced, that if the magnetic force cannot adhere to the earthy substances which form the base of the serpentine, it can be attributed only to the oxide of iron with which it is coloured. These are our reasons: The rock has no mixture of metallic substances. It presents only here and there a few fragments of talc or amianthus; but neither pyrites, nor schoerl, nor octahedrons of magnetic iron. When reduced to very fine powder, it resembles pounded chalk. The microscope discovers only earthy parts, of a clear whitish green. The specific gravity of this serpentine is very small. I find it only from 1.901 to 2.04 assuming water to be 1.0. There are not consequently any minerals but pumice-stone, mountain-leather, and some varieties of the opal, which do not equal our serpentine in density. The chemical experiments we have hitherto made, prove that it contains, like the jade or lapis ollaris, oxided iron; but not iron capable of attraction by the magnet. The solutions in muriatic acid, mixed with the nitric acid, are yellow, and not green like those which the micaceous iron, and all the ores which contain pure or metallic iron, afford.

Here then is a very striking phenomenon, namely, the polarity of super-oxigenated iron. We learn by the valuable experiments of my celebrated countrymen Klaproth and Wenzel, that pure nickel and cobalt are attracted by the magnet; we know that iron slightly oxided (the black oxide) is also affected; but how great the difference between this state of oxidation, and that of the iron which colours the serpentine, various calcareous stones, and perhaps even certain vegetable matters! What difference between a substance which acts alike on the two extremities of the needle, and a stone of which the smallest portions exercise a spontaneous polarity! Let us pursue the path of observation; let us collect indubitable facts. By this method the theories of natural philosophy will be established on solid and durable foundations.

A Letter from Mr. Von Humboldt to M. H. Van Mons on the Chemical Process of Vitality; together with the Extract of a Letter from Citizen Fourcroy to Citizen Van Mons on the same Subject[1]

I have lately addressed several letters to Messrs. Dolomieu and Fourcroy at Paris, and perceive by those I have received from the former that mine have miscarried. Permit me, Sir, to address myself to you. By your means I may perhaps succeed in forwarding to Paris some explanations respecting facts which, as I understand, employ part of the time of the National Institute. Be pleased to accept my assurances of the great respect which your zeal and your chemical discoveries have inspired me with. The natural philosophers of Europe ought to form a single family. They are in pursuit of the same interesting objects; and this is a sufficient motive to produce that useful degree of intimacy which is calculated to promote their researches.

You are probably acquainted with my Essays on the Vegetable Philosophy, such as my Aphorismi ex doctrinâ physiologiæ chimicæ plantarum, annexed to my *Flora subterranea Fribergensis*, and several memoirs which I have presented to the National Institute.—The memoir on the action of oxygenated muriatic acid upon the vegetable and animal fibre, which is printed in the *Magazin Encyclopédique* of Millin, Noël, and Warens, seems to have | 360 | had more success. I am happy to hear that Messrs. Vauquelin and my friend Dolomieu have begun to repeat my experiments. As the memoir which was read to the National Institute related principally to the germination of vegetables, I have thought it my duty to announce to you certain facts more striking respecting the animal fibre. The strongest stimulus of the nervous fibre is that of the alkalis. It appears that these salts affect the irritable and sensible system by means of their azote. Let the thigh of a frog be thrown into the oxygenated muriatic acid, or the nitric acid, and it will remain motionless. Let it be put into a solution of pot-ash, or of soda, and it will undergo contractions no less strong than when irritated by the metals. These motions always

[1]Annales de Chimie, XXII. 64.

commence at the lower extremities. The toes move first, afterwards the musculus gastrocnemius, and then the thigh. If the nerve be very sensible (for nothing more is required than simply to immerse the extremity of the crural nerve in the oleum tartari per deliquium), the contractions will end in an universal tension or rigidity. The leg rises up perpendicularly, the membrane of the feet extends itself, and the tetanus appears. In this situation all the irritability of the fibre appears to be extinguished; and if an electric stroke be passed through the limb, the exhaustion becomes real. It is a striking phenomenon to see the last remaining signs of tetanus disappear in an instant. But there is another method by which the tension disappears, and by which I am able to restore the irritability to the organs. It seems that the acidifiable bases of the alkali, principally the azote, have consumed all the oxygene contained in the fibre. The chemical process of vitality ceases. If I pour an acid, for example the nitric acid, upon the nerve, an effervescence will take place; part of the alkali becomes latent, and the rest will have a proper proportion with respect to its oxygene. From this moment the contraction with zinc and silver is again produced. Increase the quantity of acid, and the movements are again weakened. In this manner it is, that by forming an equilibrium between the azote of the alkali and the oxygene of the acid applied to the animal fibre, the irritability of the organs may be taken away or restored three or four times in succession. You may easily perceive, Sir, that these experiments require steady attention. The degree of insensibility to which the nerve is reduced by repeating them may be very different. It is possible to determine exactly the quality of the chemical agents, their weight and temperature; notwithstanding which, many experiments do not succeed. The reason is, that there are conditions which depend on the individuality of the organization, and concerning which we must still confess our total ignorance. The influences of the oxygenated muriatic acid upon the animal fibre are less marked than those of the alkalis; but they are nevertheless of much importance. I steeped the feet of a frog (I mention this animal by preference, though I have made the same experiments on other species) in a solution of opium in alcohol. The metals, or galvanism, excited no motion. I threw one leg into pure water, and the other into the oxygenated muriatic acid; the first remained motionless, the second gave very strong contractions, and shewed that its irritability was restored. The common acids depress the irritability of the nervous fibre. A crural nerve, rendered insensible by the ordinary muriatic acid, remains so though it has been steeped in the solution of pot-ash: but the mineral acids exhaust the forces of the muscles, by condensing the elements of the muscular fibre. These acids act in the same manner as cold, which depresses the nerves, and is beneficial to the muscles. The muscles and the nerves have specific stimuli, agreeable to the diversity of the elements. The terrible action which the alkalis exercise on the | 361 | nerves appears to explain the effect of the secretion of the seminal liquor on the blood. It is this alkali which, distributed throughout the system, answers the purpose of a stimulus beneficial to the animal fibre. By this action I account for the ferocity of the ichthyophagi.

My eldest brother, who is very skilful in the study of anatomy, applied zinc and silver to the mouth and the brain of a dead fish; it afforded no motion. I poured oxygenated muriatic acid on the nerves, and at that instant the contractions became very strong. Mr. Herz and several learned men of Berlin were present at these and many other experiments. The heart of the same fish, which had entirely ceased to palpitate, began to perform this movement with regularity when I threw it into the oxygenated muriatic acid. The same experiment succeeded very often with the hearts of frogs: when a heart is immersed in a solution of pot-ash, it loses its irritability for ever; so that azote is not the specific stimulus of the heart.

Mr. Pfaff, while employed in my experiments respecting germination in the oxygenated muriatic acid, has discovered that frogs suffocated in the oxygenated muriatic acid gas exhibit a very high degree of irritability after their death. I beg you will fix the attention of Mr. Vauquelin on the action of sulphate of pot-ash upon the nerves. I have been astonished at every thing I beheld. Two legs of frogs in a very lively state were steeped in the solution of the sulphate of pot-ash. I tried them three or four minutes afterwards with the metals. The contractions had increased in force, and were even convulsive. It appeared that the three acidifiable bases contained in the solution hydrogene, azote, and sulphur, acted strongly on the oxygene conveyed by the arterial blood. This action revives the process of vitality. After fourteen or sixteen minutes the whole thigh became of a blackish brown. All the oxygene of the blood was absorbed, and the carburet of hydrogene appeared in a disengaged state. The zinc and the silver are not then capable of exciting the smallest motion.

Yet it would be a great mistake to conclude that all irritability is exhausted in this case. I have seen the contractions re-appear several times on restoring oxygene to the fibre by means of a solution of the oxide of arsenic. The flame is thus renewed which seemed ready to expire. The oxide of arsenic produces a tetanus and perfect insensibility if the nerve remains long immersed. It seems then that the too great quantity of oxygene absorbs as it were the acidifiable bases which support the chemical process of vitality. I have thrown the whole thigh into the solution of pot-ash, and I observed that galvanism afterwards had the power of exciting motion.

You see, Sir, what an immense number of experiments remain to be made on these objects of vital chemistry. It is enough that a method has been pointed out of measuring the degree of irritability of the organic parts by means of galvanism. I shall have the honour to send you my work on the nervous and muscular fibre, and on the chemical process of vitality. I collect facts, and mistrust my own hypothetical ideas. You will perceive with me how mistaken the notion is that oxygene performs the principal part in this process. My experiments prove that the irritability or tone of the fibre depends only on the mutual equilibrium between all the elements of the fibre, azote, hydrogene, carbone, oxygene, sulphur, phosphorus, &c. The chemical combinations of phosphorus and of azote, for example, appear to be in no respect less important than those of oxygene with the acidifiable bases. How much light may we not expect from the advances of yourself, Fourcroy, and Vauquelin on these objects!

Von Humboldt. | 362 |

Addition to the foregoing Letter.

Having preserved some frogs for the winter, I have this morning repeated some experiments, of which I venture to send you an account. In the preceding letter I have remarked that, as we are only superficially acquainted with the principles of vital chemistry, we ought not to be surprised if we do not always obtain the same results. A negative experiment proves nothing against another of an affirmative nature.

I am very sure that a nerve rendered insensible by alcohol will not recover its irritability by sulphate of pot-ash. But it may very well happen that a thigh, of which the tetanus has been caused by the oxide of arsenic, should remain in a state of tension notwithstanding the action of the solution of pot-ash.

I have seen the following facts within this quarter of an hour. I took the four extremities of a very lively frog. The right arm and the right leg leaped on zinc and silver. I steeped them for four minutes in alcohol. The hydrogene acted strongly on the fibre. The toes of the foot trembled during the first minute. Soon afterwards a total rigidity came on; the muscle became white, the blood having apparently lost its oxygene. I replaced the arm and the leg on the zinc and silver, but there was not the slightest contraction. I then quickly threw them into the oxygenated muriatic acid, which I had shaken strongly before it was poured out; the limbs remained in it for three minutes. A slight tremulous motion shewed, even in the cup, that the vital forces were restored. I replaced the arm and the leg on the metals; the contractions were again produced, not only with zinc and silver, but with zinc and iron. Here I think is a very simple and decisive experiment. I then changed the method in order to observe the effect. I took the left thigh, and immersed it for nine minutes in alcohol. It lost all irritability, and the oxygenated muriatic acid was no longer capable of restoring the vital force. The left arm had remained untouched for fifteen or eighteen minutes. I prepared its nerve, but it shewed only very weak and slow contractions with zinc and silver. I threw it into alcohol. After the first minute its irritability was increased, the galvanism acted more strongly; but after three minutes all the irritability was exhausted, and I applied in vain the remedy of oxygenated muriatic acid. I steeped the arm in the solution of the oxide of arsenic, and it then afforded contractions, though very weak. Here are four experiments, two of which succeeded, and in the two others the vital forces were not restored. I think, nevertheless, that in good logic we ought to admit the affirmative experiments. Examine the conditions, and you will see they are very different. The left leg remained too long, nearly nine minutes, in the alcohol. The right arm was already very weak when the experiment began. Who can boast of reviving the dead?—If of two chemists the one should obtain oxygene gas by heating the red oxide of mercury, while the other did not obtain it, we should always believe that the apparatus of the latter was not hermetically closed. I never saw an organ rendered insensible by alcohol which recovered its irritability by being left to itself. It necessarily follows, therefore, that in the experiments I have ventured to relate, and of which my work contains a very great number, the oxygene of the muriatic acid must have been a principal agent. The art of medicine will be infinitely benefited if we should succeed in observing the phenomena which the several elements produce in contact with the irritable fibre. It is proper to begin | 363 | with simple combinations, and afterwards proceed to combinations of two, three, and four principles.

I have sent to the National Institute a memoir on the nature of light, and its chemical combinations. Mr. Wedgwood pretends that the phosphorescence of calcined bodies is not altered in hydrogene and azotic gas. I think he did not purify these gases by means of phosphorus, as I did. I have seen luminous wood extinguished in the azotic and hydrogene gases. A small quantity of oxygene, being admitted into the vessel, causes the whole of the phosphorescence[2] to revive. I have also converted morilles (Phallus Esculentus) into a substance which resembles tallow, by means of the sulphureous acid. I have made soap of it.

Bayreuth, December 29, 1796.
Von Humboldt.

[2] Spallanzani observed the same phenomena; and, what is more remarkable, he observed the phosphoric or shining animals ceased to emit light in the azote, hydrogene and carbonic gases, and that they emitted a light infinitely more vivid in oxygene gas than in the atmospheric air. Chimico Esame degli Esperimenti de Goettling. Modena, 1796.

On the Effects of Oxygen in accelerating Germination

From the *Journal de Physique*, 1798.

Mr. Humboldt discovered, in 1793, that simple metallic substances are unfavourable to the germination of plants, and that metallic oxydes favour it in proportion to their degree of oxydation. This discovery induced him to search for a substance with which oxygen might be so weakly combined as to be easily separated, and he made choice of oxygenated muriatic acid gas mixed with water. Cresses (*lepidium sativum*) in the oxygenated muriatic acid showed germs at the end of six hours, and in common water at the end | 310 | of 32 hours. The action of the first fluid on the vegetable fibres is announced by an enormous quantity of air bubbles which cover the seeds, a phenomenon not exhibited by water till at the end of from 30 to 45 minutes. These experiments announced in Humboldt's *Flora Subterranea Fribergensis*, and in his Aphorisms on the chemical physiology of Plants, have been repeated by others.[1] They were made at a temperature of from 12 to 15 Reaumur. In the summer of 1796, Humboldt began a new series of experiments, and found that by joining the stimulus of caloric to that of oxygen he was enabled still more to accelerate the progress of vegetation. He took the seeds of garden cresses (*lepidium sativum*), peas (*pisum sativum*), French beans (*phaseolus vulgaris*), garden lettuce (*lactuca sativa*), mignonette (*reseda odorata*); equal quantities of which were thrown into pure water and the oxygenated muriatic acid at a temperature of 88° F. Cresses exhibited germs in three hours in the oxygenated muriatic acid, while none were seen in water till the end of 26 hours. In the muriatic, nitric[2] or sulphuric acid, pure or mixed with water, there was no germ at all: the oxygen seemed there to be too intimately united with bases of azot or sulphur,

---

[1] See Uslar's Fragments of Phythology, Plenck's Physiology, Villdenow's Dendrology, and *Dictionnaire de Physique* par Gehler.

[2] The nitric acid, however, diluted with a great deal of water, accelerates germination also, according to the experiments of Candolle, a young naturalist, who has applied with great success to vegetable physiology. This phenomenon is the more interesting, as chemistry affords other analogies of the oxygenated muriatic acid and the nitric acid. Professor Pfaff, at Kiel, by pursuing Humboldt's experiments, has found that frogs suffocated in oxygenated muriatic acid gas increase in irritability, while those which perish in carbonic acid gas are less sensible of Galvanism.

to be disengaged by the affinities presented by the fibres of the vegetable. The author announces that his discoveries may one day be of great benefit in the cultivation of plants. His experiments have been repeated with great industry and zeal by several distinguished | 311 | philosophers. Professor Pohl at Dresden caused to germinate in oxygenated muriatic acid the seed of a new kind of *euphorbia* taken from Bocconi's collection of dried plants, 110 or 120 years old. Jacquin and Van der Schott at Vienna threw into oxygenated muriatic acid all the old seeds which had been kept 20 or 30 years at the botanical garden, every attempt to produce vegetation in which had been fruitless, and the greater part of them were stimulated with success. Even the hardest seeds yielded to this agent. Among those which germinated were the yellow bonduc or nickar tree (*guilandina bonduc*), the pigeon cytisus or pigeon pea (*cytisus cajan*), the *dodonæa angustifolia*, the climbing mimosa (*mimosa scandens*), and new kinds of the *homæa*.— There are now shown at Vienna very valuable plants which are entirely owing to the oxygenated muriatic acid, and which are at present from five to eight inches in height. Humboldt caused to germinate the *clusia rosea*, the seeds of which had been brought from the Bahama islands by Boose, and which before had resisted every effort to make them vegetate. For this purpose he employed a new process, which seems likely to be much easier for gardeners who have not an opportunity of procuring the oxygenated muriatic acid: He formed a paste by mixing the seeds with the black oxyde of manganese, and then poured over it the muriatic acid diluted with water. Three cubic inches of water were mixed with half a cubic inch of the muriatic acid. The vessel which contains this mixture must be covered, but not closely shut; else it might readily burst. At the temperature of 95° the muriatic acid becomes strongly oxygenated; the oxygenated muriatic gas which is disengaged passes through the seeds; and it is during this passage that irritation of the vegetable fibres takes place.

Remarks on the above Observations

The memoir of Count Morozzo treats of a very important subject, with which I have been engaged for some months. The Italian philosopher has observed the difference between natural atmospheric air and an artificial compound of azot and oxigen; but, in my opinion, he goes too far, in asserting that atmospheric oxygen is not vital gas; though I agree with him that we are not able to form an aeriform fluid, such as the illustrious Lavoisier announces in his Elements of Chemistry, perfectly similar to that of the atmosphere. I however find that this difficulty consists neither in our ignorance of the *quantity*, nor of the *quality* of the two gaseous bases. The difference found between the effect of natural and artificial atmospheric air, is to be ascribed to the *state of the combination* in which the oxygen is joined to the azot. The atmosphere is not a *mixture*: its constituent parts ought to be considered as in a state of *chemical combi-*
| 420 | *nation.*

It is for this reason that the two bases of a specific gravity so different, that of azot and that of oxygen, do not entirely separate from each other, though the high regions are most charged with azot. Hence azot retains so strongly the last portion of the oxygen, which phosphorus, the sulphure of potash, and other acidifiable bases ought to take from it; and I have for this reason seen that the nitrous gas acts in proportions very different in decomposing the natural atmospheric air, or a compound of 27 of oxygen and 73 of azot. But how could Morozzo think to imitate the atmosphere by mixing the carbonic acid and oxygen? He has confounded azot with the carbonic acid gas. But he assures us that his atmospheric air with a carbonic acid base, gave the same specific weight as atmospheric air. Here is a very striking experiment: one cubic inch of azot weighing 0,46624 grains, while the same cubic inch of carbonic acid weighed 0,67500 grains! Morozzo tells us that a candle burnt in his artificial mixture with a very brilliant flame (*con fiamma lucidissima*). By combining 0,25 of the carbonic acid with 0,75 of oxygen, I have seen a wax taper extinguished. The same thing took place when in company with Tasseart in Vauquelin's laboratory, I mixed two parts of the carbonic acid gas with ten parts of atmospheric air. There is then formed a new chemical combination: the carbonic acid lays hold so strongly of the oxygen of the atmosphere, that the affinity presented by the lighted wax taper is not powerful enough to take it away. In my work on the analysis of the mephitic vapour of mines, which is about to be translated into French, it will be seen that there exists airs not unfit for respiration, which are composed of 0,27 of oxygen, 0,70 of azot, and 0,03 of carbonic acid gas. It is the state of the combination, and not always the quantity of oxygen, which renders air more or less capable to destroy animal life, or to extinguish flame.

Von Humboldt's Expedition to Spanish America

The two following letters from this indefatigable philosopher to one of his friends in Germany, have been published in one of the foreign journals:—Corunna, June 3, 1799.

I wrote you from Marseilles, that I had been disappointed in my hopes of undertaking a voyage round the world with Captain Baudin (to which I had been invited by the French Government) just at the moment when I was going down to the port to embark. After this I proposed to go out to Bonaparte with the second expedition from Toulon, and my friends were anxiously expecting me; but this expedition was rendered impossible by the battle of Aboukir. Firm to my purpose, I then wished to go by a Swedish frigate, expected at Marseilles, to Algiers, in order that I might undertake, with the caravan of Mecca, that dangerous journey through the desart of Selima to Cairo. The frigate, however, did not arrive; and, after waiting two months to no purpose in Provence, as war had broken out between France and Algiers, I proceeded to Spain. I had procured recommendations to the king, and received from him what no foreigner ever obtained—letters of recommendation to all the viceroys, and permission to traverse all the Spanish settlements with my instruments. You must acknowledge that I have been exceedingly fortunate. I have been provided with every thing necessary, and in a few hours shall sail, in the Spanish frigate Pizarro, for the Havannah; from which I intend to proceed to Peru, Mexiko, and Chili. I shall be absent for several years, but I flatter myself with the hopes of accomplishing something of importance. Bonpland, a young French botanist; accompanies me. I shall write you from the Havannah.

Alex. Von Humboldt. | 95 |

Orotava in the Island of Teneriffe,
June 24, 1799.

We left Corunna on the 5th, and arrived, without any accident, at Lancerotta on the 16th, and St. Croix, in Teneriffe, on the 17th. We were in sight of four English frigates, which we escaped, but we cannot tell how.

I have examined the peak with great attention; I was almost in the crater at the height of 11,500 feet. This excursion was attended with more fatigue than danger.

We found the heat of the crater on the ground 70° Reaum. and the air at 2°. The pumice-stone, which has occasioned so much dispute, is obsidian-stone fused and decomposed. It is here as clear as the day. I am so tired that I must conclude my letter. We are just going to set out for Caraccas and the Havannah.

Alex. Von Humboldt.

Extract of a Letter from M. Von Humboldt to Lalande

Caraccas, Dec. 14. 1799.

A few weeks after my arrival in South America, I transmitted to Delambre an extract from my astronomical observations, because I hoped that some of them might be interesting to the Board of Longitude: as I have, however, learned that the brig by which I sent my letters was wrecked near Guadaloupe during the storm which lately prevailed in these regions, I think it necessary to transmit to you a copy of them.

After a passage of about six weeks, in the frigate Pizarro, I arrived at the coast of Paria. My plan was to proceed first to the Havannah, and thence to Mexiko; but I could not withstand the strong desire I had of seeing the wonders of the Oronoko, and the lofty Cordilleras, which extend from the high land of Quito to the rivers of Guarapeche and Arco. As I have now returned from a very interesting journey to the interior of Paria, through the Cordilleras of Cocolar, Tumeri, and Guiri, and to the settlement there of the Capuchins, never before visited by any naturalist, during which I carried with me, on three mules, my astronomical instruments, viz. a quadrant by Bird, and sextants, telescopes, and micrometers, by Ramsden and Troughton, you will perhaps expect that I have done a great deal for astronomy; but you know that this science is too remote from my principal objects— philosophy in general, geology, eudiometry, and the physi- | 356 | ology of plants and animals; and under the latitude of 10° it is impossible to labour so incessantly as under 49°. I chose rather to make a few observations with accuracy, than a great many imperfect ones; and to insert them with great minuteness in my journals, that in case I should die in the course of my travels, no doubt might remain in regard to their exactness.

In the two papers which I transmitted to Delambre, you will find my observations made in Europe with Borda's new *inclinatorium*. I found by it, that local circumstances have a greater influence on the inclination of the magnetic needle than on its declination; and that no connection seems to exist between the situation of the place of observation and the inclination. This I find confirmed in the new world in the interior of New Andalusia, and therefore the observations made by Nouet in Egypt seem to be established. The declination, however, is affected by local causes, but much less: at sea they are much more regular, and the variations are much more uniform.

I have here given you my observations of the inclination only, which are true to 15′. During calms I obtained them at sea with more accuracy, and the periods of the vibrations of the needle could be determined very exactly. If the same number is counted in the same time during five or six repeated trials, and if this is still the case when the instrument is moved from its place, the results, in my opinion, may be considered as correct. Though calms are not uncommon between the tropics, during the course of forty days I was able to make only ten very accurate observations. | 357 |

Places of Observation, 1799.	Latitude.	Longitude from the first Meridian.	Inclination.		Magnetic Power. Number of Vibrations in Ten Minutes.
			Centigrade Degrees	Old Degree	
Paris – – – –	48° 50′15″	20° 0′ E	77·15°	69° 28′	245
Nîmes – – – –	43 50 12	21 59	72·65	65 23	240
Montpellier– –	43 36 29	21 32·5	73·20	65 53	245
Marseilles – – –	42 17 49	23 3·5	72·40	63 10	240
Perpignan – – –	42 41 53	20 33·5	72·55	65 18	248
Barcelona – – –	41 23 8	19 52	71·80	64 37	245
Madrid – – –	40 25 18	13 58	75·20	67 41	240
Valencia – – –	39 28 55	17 29	70·70	63 38	235
Medina del Campo			73·50	66 9	240
Guadarama – –			73·50	66 9	240
Ferrol – – – –	43 29 0	9 24·5	76·15	68 32	237
In the Atlantic	38 52 15	3 40	75·18	67 40	242
Ocean between	37 14 10	3 30	74·90	67 30	242
Africa and	32 15 54	2 52·5	71·50	64 21	—
America.	25 15 0	0 36 W	67	60 18	239
	21 36 0	5 39	64·20	57 49	237
	20 8 0	8 34	63	56 42	236
	14 20 0	28 3	58·80	52 55	239
	12 34 0	33 14	50·15	45 8	234
	10 46 0	41 24	46·40	41 46	229
	10 59 30	44 31·5	46·50	41 57	237

In the year 1776, according to Cavallo, the inclination of the needle in the undermentioned latitudes and longitudes was as follows:

Latitude.	Longitude.	Inclination.
24° 24′	18° 11′	59°
10	22 52	44 12
0	37 38	30 3

Since Coulomb and Cassini no longer employ themselves with observing the declination, I know no place in the earth where the declination has been determined with certainty to 10″, and not ten where it has been determined to 1′. What uncertainty still prevails respecting the real declination of the needle at Paris! The ten observations which I made at sea, will serve hereafter to determine whether the inclination speedily changes. The longitude and latitude of the places of observation were always accurately determined at the time by means of a Ramsden's sextant, divided from 15 to 15 seconds; and a Berthoud's chronometer. You will see by them that the inclination decreases very rapidly from the latitude of 37°, and that it increases less towards the east than to the west from the latitude of 37° to 48°.

It appears to me, that in the higher chain of the calcareous mountains of this province, not far from the equator, small elevations above the level of the sea derange the inclination much more strongly than the higher mountains in the Pyrenees and Old Castile. As a proof, I shall here give observations made at four places which lie pretty nearly in an arch from north to south (comprehending 24′.)

Places	Height above the Sea, Paris Feet.	Inclination.		Vibrations in Ten Minutes.
		Centigrade Degrees.	Old Degrees.	
Cumana	24	44·20°	39° 47′	229
Zueteppe	1111·2	43·30	38 58	229
Impossibile	1470·	43·15	38 50	233
Cumanacoa	636	43·20	38 53	228
Cocollar	2352	42·60	38 20	229

Borda is of opinion, as appears by the memorandum drawn up for the men of science who accompanied La Perouse, that the intensity of the magnetic power is the same all over the earth; as he ascribes the small variation which he found in it at Cadiz, Teneriff, and Brest, to the imperfection of the compass. He requested me to verify this circumstance.

You | 358 | here see that the magnetic power is so different, that between Paris and Cumana it decreased from 245 to 229 vibrations in ten minutes, though it does not decrease with the inclination. This decrease cannot certainly be ascribed to any change in the goodness of the needle, or to other accidental causes: for the same needle made in the same time at Paris 245 vibrations; at Girona, 232; at Barcelona, 245; at Valentia, 235; and, after a journey of several months, gave at the same places exactly the same number of vibrations as before my departure. These are always the same, in the open fields, in the house, or in a cavern, so that the magnetic power in any place is always the same, and remains a long time without change; and seems to be a general power, like that of gravity.

I had the mortification of not being able at sea to make any good observations of the declination. Notwithstanding all the trouble I took, I could not find an azimuth compass on which I could depend within 40′. This is the true reason why I have made no mention to you of the declination at sea. The point, however, where the inclination vanishes, certainly lies further to the west than marked in Lambert's chart, in the Berlin astronomical almanac for 1779. A good observation, made in the year 1775 aboard an English ship from Liverpool, places this vanishing point in latitude 29° north, long. 66° 40′ west. I observed, with great care, the declination at two places on the coast of America with a compass by Lenoir, in which the needle is suspended by a thread, according to the method of Prony and Von Zach. At noon, October 1799, it was at Cumana, and twenty leagues further east, 4° 13′ 45″ east: at Caripe, the chief place of the Capuchin mission among the Chaimas and Caribs, 3° 15′ east.

During the earthquake at Cumana, on the 4th of November 1799, the inclination of the magnetic needle was altered, but not the declination. The former, before the shock, was 44° 20 of the new division; after the shock it remained at 43° 35. The number of the vibrations of the dipping needle, however, was the same as before, viz. 229 in ten minutes. This, united to other observations, seems in my opinion to prove, that during the earthquake this small part of the ball of the earth was changed, and not the needle; for, in districts where no signs of an earthquake ever appeared in the primitive chain of accumulated masses of granite, the inclination was as great after as before.

On account of the interest which you take in every thing | 359 | that relates to navigation, the following observations will not be unacceptable.

I have carefully proved what Dr. Franklin and Captain Jonathan Williams assert in the Transactions of the American Society[1] respecting

[1] Vol. iii. p. 32.

the use of the thermometer for discovering shallows at sea; and am able to confirm, in the fullest manner, what they have said. I was astonished to see how the water became evidently colder as its depth decreased, and how the neighbourhood of shallows, and of the coasts, could thereby be announced. The worst spirit of wine thermometer, if only sensible, will therefore, in the hands of the most ignorant mariner, be a very useful instrument in the night-time, and during storms, or when it is difficult or impossible to heave the lead. This observation I cannot too strongly recommend to the attention of the Board of Longitude. Our whole crew were astonished to see how speedily the thermometer fell, when we approached the large bank which extends from Tobago to Grenada, and east from Margarita. These observations may be made with more ease, as the water of the sea, in an extent of 12000 square miles, has always the same temperature day and night; so that the most sensible thermometer, during from four to six days sail, does not rise or fall above 0 3°. In the neighbourhood of shallows it was from 2 to 3 degrees, and more, colder. This observation of Franklin, hitherto forgotten, may at some period be of great use to navigation; not that seamen should throw aside the lead entirely, and trust to the thermometer, for this would be folly; but because the observations may be so easily repeated, and because the thermometer will announce the danger much sooner than the lead, as the colder water above the shallows lessens the temperature in the water in the neighbourhood. I can assert that this new mean is not more uncertain than the log, and the helps already employed in navigation. If the thermometer does not fall, navigators ought not to depend on that they are entirely secure from shallows; but if it falls, they must be on their guard. This warning is certainly more valuable than our sea charts, where the shallows are in general laid down in a very incorrect manner; and to immerse a thermometer in a bucket filled with sea water is certainly much easier. I have also measured several times the specific gravity and temperature of the sea water, at the surface and at certain depths, by means of Dollond's balance and thermometers, which are placed in cases furnished with a valve. As my instruments were compared with the best Parisian ones, and as I could be more | 360 | certain in regard to my longitudes than is usual, the small chart in which I intend to make known the result of these experiments will be interesting. In the latitude of 17° and 18° north, there is a zone in the sea between Africa and the West Indies, where, without an extraordinary current prevailing, the water is denser than under a less latitude. The following are some of my results respecting the temperature of the sea water:

		Temperature	
North Lat.	Long. From the First Meridian.	of the Sea Water at the Surface.	of the Air.
43° 29′	9° 29′ E.	12°	18°
39 10	3 41·5	12	13
36 3	2 57	12	14
35 8	2 15	13	16·5
32 15	2 52·5	14·2	13·5
30 35	3 6	15	16
28 55	2 37·5	15	17
26 51	— 47	16	15
20 8	8 33 W.	17	16
18 53	10 5	17·4	17
18 8	13 2	17·9	19
17 26	15 26	18	16
15 22	22 49	18·5	20
14 57	24 40	19	17
13 51	30 2·5	19·8	18·9
10 46	41 24	20·7	20·3
10 28	46 31	21	17 to 27
10 29	46 35	17·8	23

This letter was written at Cumana, but for want of a proper opportunity I was obliged to bring it with me to this large capital of the province of Caraccas, which is situated in a fertile valley abounding with cacao, cotton, and coffee, 2400 feet above the level of the sea, and in a climate perfectly similar to that of Europe. The thermometer in the night-time falls to 11°, and in the day never rises above 17° or 18°. The cloudy weather, which here daily follows the sun's passage over the meridian, makes the observation of corresponding altitudes very difficult and uncertain, as the afternoon observation is often lost. The cloudy weather after the earthquake of Cumana deprived me of an opportunity of observing the immersions of the second satellite of Jupiter on the 2d and 9th of November. I shall here give you the prin- | 361 | cipal determinations of the longitude, which I made by means of Berthoud's chronometer, from the observed horary angles.

	Long. west.
Cumana, castle of St. Antonio, lat. 10° 27′ 37″ and (taking the long. of Madrid at 13° 58′)	46° 31′
Puerto Espana, in the island of Trinidad	43 49 30″
Tobago, the eastern extremity –	42·47 30
Macanao, the west part of the island St. Margarita	46 35 30
Punto Araya, in New Andalusia –	46 35 30
Island of Coche, eastern extremity –	46 12
Bocca de Drago (not certain) –	44 23
Cabo de tres Puntas – –	44 54 30
Caraccas à la Trinité 10° 31′ 4″ latitude, very good.	

I shall here also mention, that neither wind, storms, nor the earthquake, have had the least influence on the regular daily variation in the state of the barometer; and that, according to C. Richard, this daily variation amounts at Surinam to two lines.

These observations are the more interesting, as all the charts of this part of the world are very incorrect; and the agreement of my longitude of Teneriff and Tobago, with the observations of Borda and Chabert, within from 2″ to 5″ of time, is a proof of the excellence of my chronometers.

The serenity of the nights between the tropics gave me an opportunity of comparing with each other the strength of the light of the austral stars; which in some of them, such as the Crane, Altar, Toucan, and the Foot of the Centaur, seems to have changed since the time of La Caille. I employed the method proposed by Dr. Herschel, and diaphragms of the same kind as those used for the satellites. If the light of Sirius be supposed equal to 100 parts, and that of Procyon 88, the light of the following stars, according to my observations, are as below expressed.

	Parts.		Parts.
Canopus –	98	α in the Phœnix –	65
α in the Centaur –	96	α in the Peacock –	78
Acharnar – –	94	α in the Crane –	81
α in the Indian –	50	β – – –	75
β – – –	47	γ – – –	58
α in the Toucan –	70		

Humboldt's travels through Spanish America

The following letter from this enterprising philosopher to Fourcroy has lately been published in one of the French journals: "Guyara, January 25, 1800.

"As the yellow fever, which desolates this part of South America, obliges us to make our stay here as short as possible, I snatch this opportunity of sending you a few lines, and repeating, from the middle of the torrid zone, how much my thoughts are occupied with you and your illustrious colleagues, among whom I met with such a flattering reception during my last stay at Paris. Since our departure from Saint Croix in Teneriffe, where I descended into the crater of the volcano, the atmospheric air being there at 0·8 of R. and at 0·19 of oxygen, I have written to you twice. I have sent to Delambre and Lalande an extract of my astronomical labours; the longitude of some important places; an observation of the eclipse on the 28th of October last; immersions | 276 | of the satellites, and researches respecting the intensity of light of the austral stars, measured by means of diaphragms. I have addressed to the Institute a chemical memoir on the phosphorescence of the sea; observations on a particular gas furnished by the fruit of the *coffea arabica* when exposed to the sun; on a snow white feld-spar, which, when moistened, absorbs all the moisture of the atmosphere; experiments on the milk of the *cecropia peltata*and the *euphorbia curassavica*, which will form a supplement to your excellent memoir on the *cahout-chouc*, and to that of our friend Chaptal; and on the air which circulates in vegetables.

"The cruisers which cover the seas here make me apprehend that a part of my letters may not have reached you; though I sent them sometimes by way of Guadaloupe, and sometimes by that of Spain. These few lines I have sent by an American vessel, which will sail in a few days for Boston; and, though they cannot reach you but through Hamburgh, they will, perhaps, be less liable to miscarry. People here are accustomed to make four or five duplicates of their letters. But how can I find time, when I have so many things to observe, arrange, and calculate?

"I shall confine myself, therefore, to letting you once more know that I still enjoy the best health possible, and that I am treated with the utmost kindness by the inhabitants of these countries: that the passports and letters of recommendation from the Spanish government procure me every facility for making researches useful to the sciences: that none of my instruments, even the most delicate, such as barometers, thermometers, hygrometers, Bordas' dipping needle, &c. are deranged: and that, at the extremity of the missionary establishments among the Chayma Indians, in the mountains of Toumiriquiri, I have my laboratory mounted as if I were in the Hotel Boston, in the Rue du Colombier. My fellow-traveller, Bonpland, educated in the *Jardin des Plantes*, becomes every day more valuable to me. To an extensive knowledge of botany and comparative anatomy he unites indefatigable zeal. I hope one day to restore him to his country, worthy of attracting public attention. Never did any foreigner enjoy such permission as that | 277 | granted to me by the king of Spain. This idea alone is capable of exciting us to redouble our activity. During the seven months we have been in this beautiful continent, we have dried (with duplicates) nearly 4000 plants, written more than 800 descriptions of new species or species little known, particularly new genera of palms, cryptogamia, bifaria, melastoma; collected insects and shells, and made many drawings respecting the anatomy of marine worms; with observations on magnetism, electricity, humidity, the temperature of the atmosphere, and the quantity of oxygen it contains. We have measured that immense and high chain of mountains extending to the coast of Paria, and examined their volcanoes, which vomit forth kindled inflammable air, sulphur, and hydro-sulphurous water. We have also collected many seeds, which we shall send off in three or four weeks for Europe, addressed to the *Jardin des Plantes*. We have spent five months in the interior of New Andalusia and on the coasts of Paria, where we experienced very violent earthquakes in the end of last year. One part of these countries is still inhabited by the savage Indians, and others have began to be cultivated only within five or six years. In what words shall I describe to you the majesty of the vegetation here; the forests of Ceiba, Hura, Hymenea, which the rays of the sun never penetrate; the variety of the animals; the superb plumage of the birds; the apes, the tigers; the hideous aspect of the crocodiles (*caimans*) which swarm in the rivers, and of which some are thirty feet in length? From Cumana we proceeded to Caraccas, where we remained during the month of November and part of December: A charming capital situated in a valley which has 426 toises of elevation; and, though in latitude 10° 31′, enjoys the coolness, and I might say the cold, of Paris! From this place we ascended to the summit of the famous Silla de Caraccas, or Sierra de Avila, where, at the height of 1316 toises, we discovered beautiful crystals of titanium. I found also dendrites similar to those of manganese, which are oxyd of titanium.

We shall proceed hence for Varina and the snow-covered mountains of Merida, the cataracts of Rio Nigro, and the | 278 | unknown world of Oronoco, in order to return by Guyana to Cumana, from which we shall set out for the Havannah and Mexico.

"We shall take care to transmit the seeds we have collected for the *Jardin des Plantes* at Paris, the Museum, and Sir Joseph Banks, as we agreed with Jussieu – – – –.

"How much I lament the fate of Dolomieu, detained a prisoner in Sicily! If he should return, communicate to him the following fact: — It is more than three years since I announced to Lametherie, that, in the primitive mountains of Italy, France, Swisserland, Poland, and I can now add Spain, there exists a *parallelism of direction* between the strata of foliaceous granites, slate, micaceous schists, corneous schists: that these strata incline (sink down) towards the north-west, and that their direction makes with the axis of the globe an angle of 45° 57′: that this inclination and direction in no manner depend on the form and direction of the mountains: that it is not any way affected by the valleys, but that it announces a cause much greater, and more general: that it depends on a phenomenon of attraction which has acted at the time of the consolidation of the globe. Having travelled over the greater part of Europe on foot, and with sextants and compasses, I have a very extensive collection of observations on that subject. My manuscript on the identity of strata in the construction of the globe, is in the hands of my brother. I was employed on it since 1791, but it will not appear till I have seen more of the globe. To my great astonishment, I have observed in the Cordillera of Paria, New Andalusia, New Barcelona, and Venzuela, that in the new world, near the equator, the strata follow the same laws and the same parallelism.

"You remember the last ingenious observations of Coulomb on the air which issues with explosion from the trunks of trees when they are pierced. I have made experiments on the *clusea rosea*, in which (in the interior of the *pneumato-chimiser*, vessels of Hedwig, the *vasa cochleate* of Malpighi,) there circulates an immense quantity of air. This air contains as far as 35/100 of oxygen. The leaves of the same tree, when exposed to the sun under water, do not give a cubic | 279 | mellimetre of air. This air, which circulates, certainly serves, as in the animal body, to coagulate the fibrous part by absorption of oxygen. The *clusea* is a milky plant, and elastic gluten is formed in it.

"Though the purity of the atmospheric air amounts here, particularly in the night, to 0·305 of oxygen, I have found that the air contained in the pods and capsules of equinoctial plants, for instance the *paullinia*, is more azotous than our atmospherie air. It does not exceed 0·24 or 0·25 of oxygen. The air in the *culmi geniculate* here, has only 0·15 of oxygen. All this proves that the air which circulates is purer, and that the air which is in a state of rest, deposited in the capsules or *utriculi*, is less pure than atmospheric air. The former is recently produced by the organs which decompose the water: it proceeds to those parts where it ought to serve, by its abundance of oxygen, to precipitate the fibrous principle to form the fibrous tissue. The other is the residuum of a gas which has already discharged these functions.

ALEX. HUMBOLDT."

Sketch of a Geological Delineation of South America[1]

Since I sent to Madrid the two first sketches of a geological delineation of South America, from the Caraccas and Nueva Valencia, I have travelled 1200 miles, and described a square between Caribe, Portocabello, Pimichin, and Esmeralda, a space comprehending above 59000 square miles, for I am not acquainted with the land between the mountain Parea and Portocabello, and between the northern coast and the valley of the Black river. In consequence of the great circumference of this district, I must content myself with delineating it in a general manner, and to avoid details, with describing the construction of the earth, the declivity of the land, the direction and inclination of the mountains, their relative ages, their similarity with the formation of those in Europe. These are the circumstances most necessary to be known in this science. We must proceed in mineralogy as in geography; we are acquainted with stones, but not with mountains; we know the materials, but we are ignorant of the whole of which they form component parts. I wish I may be able, amidst the variety | 348 | of the objects which occupy my attention during my travels, to throw any light on the structure of the earth. The laborious journeys which, for eight years, I have made through Europe, had no other object; and if I have the good fortune to return to Europe, and to recover my geological manuscripts which I left behind me in France and Germany, I shall venture to give a sketch of the structure of the earth. What I have long said, that the direction and inclination, the rising and falling of the primitive strata, the angles which they form, with the meridian of the place, and with the axis of the earth, are independent of the direction and depression of the mountains; that they depend on laws, and that they observe a general parallelism which can be founded only in the motion and rotation of the earth. What Freiesleben, Von Buch and Gruner have proved better than I will be found confirmed, namely, that the succession of the alluvial strata, which was considered as a peculiarity of certain provinces, such as Thuringia and Derbyshire, takes place generally, and that there appears an identity in the order of the strata (see

[1] This sketch is an extract from a paper transmitted by M. Von Humboldt from South America, together with a geological collection, to the directors of the cabinet of natural history at Madrid. It was sent also by M. Von Humboldt to Delametherie, and inserted by him in the *Journal de Physique*, vol. 53, p. 30.

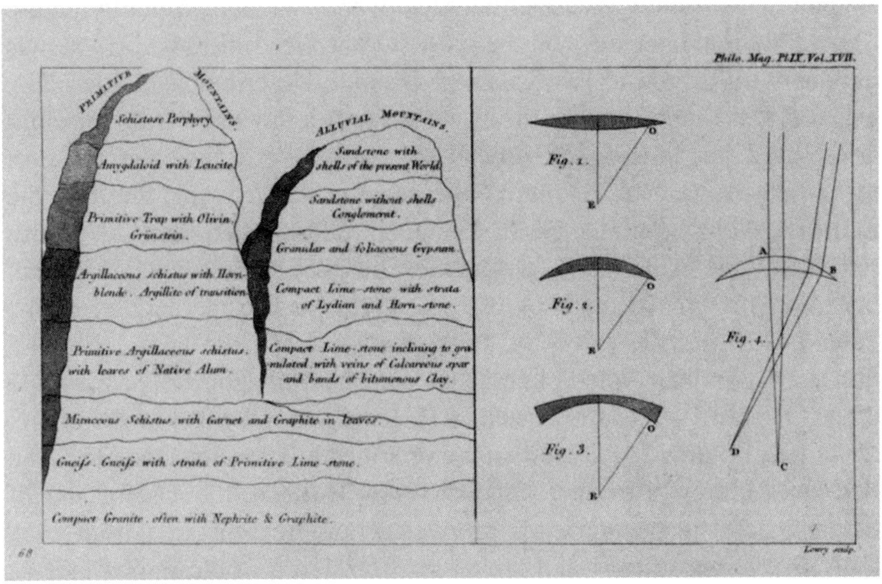

Plate IX.

Plate IX.); from which there is reason to conclude that the same deposition has been effected at the same time over the whole surface of the earth. All these ideas are of the greatest importance, not only to the philosopher, who endeavours to elevate himself to general principles, but also to the miner, who must conceive in his mind what he has not before his eyes, and guide himself by analogy deduced from actual experience.

Before I describe the situation of the mountains which I have observed from the coast to the province of Venezuela, I shall give a general view of the form of this continent. Unfortunately there are no early observations to serve as a ground for this description. For half a century past many accidental observations respecting this land have been collected, but not a single idea relating to its geology has been made known. The great genius of Condamine, the zeal of Don George Juan de Ulloa, would certainly not have left us in the dark on this subject, had mineralogy been more cultivated at the time when they wrote. All that could then be done was to measure and to take levels. As they were employed on the high cordillera of the Andes, which extends north and south from Zitara, as far as Cape Pilar, and beheld with wonder the immense height of the mountains, they forgot that South America exhibits other cordilleras, which extend east and west parallel to the equator, and which, on account of their height, deserve as | 349 | much the attention of naturalists as the Carpathians, Caucasus, the Alps of the Valais, and the Pyrenees. The whole immense tract on the west side of the Andes, which extends obliquely to the coast of Guiana and Brasil, is described as a low plain, exposed to the inundation of the rivers. As only a few Franciscan missionaries and a few soldiers have been able to penetrate over the cataracts to Rio Negro, the inhabitants of the coast of Caracas imagine that the immense plains (the Llanos de Calabozo, del Guarico, and de Apure) which they see to the south, beyond the valleys of Aragua, extend without interruption to the Pampas of Buenos-Ayres, and to the country of the Patagonians; but the extent of these plains is far from being so great; they are not uninterrupted plains, they are rather phenomena of the same kind as those presented by Canada and Yucatan, the Island of St. Domingo, the north of Sierra de St. Martha, the province of Barcelona, and the land between Monte-Video and Mendoza, New Holland, the eastern part of Hungary, and the country of Hanover. They are separated from each other by the cordilleras, and are as far from lying in the same plane as the deserts of Africa, and the steppes of Tartary, which rise by gradations, according to the distance from the sea coast.

When one considers the irruptions which the North Sea, the Mediterranean, &c. have made into the old world, the direction of its cordilleras appears not to be very different from that of those in the new world, as most naturalists have asserted. We are acquainted also with the traces of several high chains of mountains which extend from north to south, and run out from those which extend east and west. The garnet and micaceous schistus of Norway, Scotland, Wales, Brittany, the province of Gallicia, Alemtego, Cape Bogador, (I have found the same with granite on Teneriff,) the upper part of Guinea, Congo, and the Table Mountain, as also the original mountains of Orenburg, Caucasus, Lebanon, of Abyssinia and Madagascar, seem at first to have formed nothing else than two large cordilleras parallel to the meridian.

In the new world these cordilleras run parallel to the meridian from Cape Pilar to the north of California beyond Nootka and Prince William's sound towards the Aleganhey mountains, which were discovered in 1792 by Mr. Stewart on his journey to the sources of the Missoury, the northern part of the Andes, which is inhabited by Indians nearly as much civilized as the Peruvians were fifteen hundred years ago. From this cordillera proceed ramifications of the | 350 | original mountains which extend from west to east. With those of North America I am not acquainted, but it appears that some exist in Canada under the latitude of 50° and 42° north latitude, as in the destroyed continent of the Gulph of Mexico under 19°

and 22°, as is proved by the mountains of Cuba and Saint Domingo. In South America there are three chains of original mountains which run parallel to the equator: the chain of the coast under 9° and 10°; that chain which is in the great cataracts of Autures (in latitude 5° 39′) is between latitude 3° and 7°; and that in Maipure in 5° 12′ 50″, which I therefore call the chain of the cataracts or that of Parime, and the chain of Chequitos under 15° and 20° south latitude.

These chains in the old continent on this side of the western ocean can be traced, and it is seen how the original mountains of Fernambouc, Minas, La Bahia, and Janeiro, correspond, under the same latitude, to those of Congo, as the immense plains near the river Amazon lie opposite to the plains of Lower Guinea, the cordillera of the cataracts opposite to those of Upper Guinea, and the Llanos of the Mississippi, since the irruption of the Gulph of Mexico, a property of the sea, opposite to the Desart of Serah. This view will appear to be less hazarded when one reflects in what manner the old continent has been separated from the new one by the force of the water. The form of the coasts, and the salient and re-entering angles of America, Africa, and Europe, are a sufficient proof of this catastrophe. What we call the Atlantic ocean is nothing else than a valley scooped out by the sea. The pyramidal form of all the continents with their summits turned southwards, the great flattening of the earth at the south pole, and other phænomena, observed by Dr. Forster, seem to show that the influx of the water was from the south. On the coast of Brasil, from Rio Janiero to Fernambouc, it found resistance, and taking a direction from the latitude of 50° north towards the north-east, where it scooped out the Gulf of Guinea near Loango, Benin and Minc, it was obliged by the mountains of Upper Guinea to direct itself north-west, and separated, to the latitude of 23° north, the coast of Guinea from Mexico and Florida. The force of the waters was still broken by the cordillera of the United States of America, and once more turned towards the north-east, and seems to have spared less the western coast of Europe than the northern of America. The least breadth of this channel is at the Brasils and Greenland; but, agreeably to the geographical history of plants and animals, it seems to have | 351 | formed at a time when the organic creation had not been properly expanded. It would be of great importance to geology if a sea voyage were undertaken, at the expense of some government, to examine the rising and depression and the relative situation of the mountains to the salient and re-entering angles of America and Africa. The same analogy would be found here as is observed in the English Channel, in the Sound, the Straits of Gibraltar, and the Hellespont;

small creeks which are as new as the secondary formation of the chalk rocks of Jura, of Pappenheim, La Mancha, Marseilles, Derbyshire, and Suez, which have all been produced at the same time by precipitation.

Of the three cordilleras of primitive mountains which traverse South America from west to east, the most northern, that of Venezuela, is the highest, but the narrowest. The real chain of the Andes extends from the large plain of Quito, through Popayan and Choco, to the western side of the river Atrato (or Rio San Juan), between the valley of Tatabé, in the provinces of Zitara and Biruguete, towards the isthmus, where it forms a mountainous district of not more than two or three hundred toises in height on the bank of the Chagre. From these Andes arises the cordillera on the coast of Venezuela. Rows of mountains higher, but forming groups less regular, extend on the east side of the Rio Atrato under the name of the Sierra de Abibé and the Montes de Cauca, through the high savannahs of Jolu towards Magdalen river and the province of St. Martha. The cordillera of the coast contracts itself like that of the Gulph of Mexico, approaches nearer to Cape Vela, and then proceeds first from south-south-west to north-north-east, and then from west to east to the ridge of Paria, or rather to the Punta de la Galera in the Island of Trinidad. Its greatest height is found at that place where it has the name of Sierra de Nevada de St. Martha in latitude 10° 2′, and of Sierra Nevada de Merida in latitude 8° 30′; the former is about 5000 the latter 5400 Spanish ells (*varas*), or 2350 toises in height. The Paramo de la Rosa and de Macuchi, and also the mountain of Merida, are continually covered with snow: boiling water (with hydrogenated sulphur) issues from their sides, and they exceed in height the Peak of Teneriff, and are, perhaps, equal to Mont Blanc, which has been more accurately measured. These colossal masses and St. Martha stand almost insulated, being surrounded by few high ridges. To the west of Santa Fé, or as far as the Sierra of Zuindiu, no snow-clad peaks are seen, and the Sierra Nevada de Merida | 352 | stands at the edge of the plain of Caracas, which is scarcely forty toises above the level of the sea. Mont Blanc, which terminates the high ridge of the Alps, exhibits the same phænomenon. The altitude of the highest mountains, however, is so very small in proportion to the magnitude of the earth that it would appear that very small local causes ought to have accumulated more matter in these points. That part of the cordillera of the coast which lies to the west of Maracaybo-Sees, and joins the Andes, has large valleys extending from north to south, such as that of Magdalena, of Cauca, of Saint George, of Sinu, and Atrato. They are very long and narrow, but covered with wood.

On the other hand, that part of the cordillera which extends from Merida to Trinidad incloses three valleys lying east and west, which show by certain signs, like Bohemia, or the Haslithal of Swisserland, that they have formerly been lakes the water of which has evaporated or run off by opening for itself a passage. These three valleys are inclosed by the two parallel rows of mountains into which the cordillera of the coast divides itself, from Cape Vela to Cape Codera; the northern row is a continuation of Saint Martha, the southern a prolongation of Sierra Nevada de Merida. The first extends through Burburuta, Rincon del Diablo; through the Sierras de Mariara, the mountain Aguasnegras, Monte de Arila, and the Silla de Caracas, to Cape Codera. The second from three to four miles more to the south extends through Guigui, La Palma, the high summits of Guairaima, Tiara, Guiripa, and the Savana de Ocumare, as far as the mouths of the Tuy. These two chains unite with two arms which run from north to south, like, as it were, dykes, by which these old lakes were confined within their boundaries. These dykes are, on the west the mountains of Carora, Torito, Saint Maria, Saint Philips, and Aroa; they separate the Llanos de Monai from the valleys of Aragua: on the east they are the naked summits of Los Teques, Coquiza, Buena Vista, and the Altos de S. Pedro, by which the valley of Aragua or the sources of the Tuy (for there is only one valley between the bottom of Coquiza or the Hacienda de Brisenno to Valencia) from the valley of Caracas. On the east from Cape Codera the greater part of the cordillera of the coast of Venezuela was destroyed and laid under water by the great catastrophe which formed the Gulph of Mexico. The rest of it is distinguished in the high mountain peaks of the Island of Margaretha (Macanao and the Valle S. Juan) and in the cordillera of the Isthmus of Araya, which contains the | 353 | micaceous schistous mountains of Maniguares, Chuparipari, Distilador, Cerro-Grande, the mountain of St. Joseph and of Paria: the remainder I have accurately examined, and found in them the same structure, the same direction, and the same inclination of the strata. The three hollows, or valleys of Caracas, Aragua, and Monai, are remarkable on this account, that the level of them is above the surface of the sea; they become lower by gradations, and the highest step is the eastern, which may serve as a proof that they were formed at an earlier period than the Llanos, whose declivity proceeds from east to west, like the whole continent of South America. By repeated barometric measurement I found the height of the valleys of Caracas to be 416 toises, of Aragua 212 toises, above the surface of the sea; the Llanos of Monai, the western bason, appears to have an elevation of no more than 80 or 100 toises.

The valley of Caracas has once been a lake, which formed for itself an efflux through the Quebrada de Tipe, Catia, and Rio Mamon; the bason of Aragua appears, on the other hand, to have become dry by gradual evaporation; for the remains of the old water (loaded with muriate of lime) are still seen in the lake of Valencia, which becomes less every year, and discovers islands which are known under the name of Aparecidas. The height of the cordillera of the coast is commonly from 600 to 800 toises; the highest peaks, Sierra de Nevada de Merida and the Silla de Caracas, (to which we undertook a laborious journey with our instruments) are 2350 and 1316 toises in height. To the west they always become lower, and the height of Cape Codera is only 176 toises. The Macanao, on the island Margaretha, which I measured trigonometrically, is not more in height than 342 toises; but this speedy depression takes place only in the primitive mountains of the cordillera. On the eastern coast secondary accumulations of lime rise from Cape Unare to a more considerable height than the gneis and micaceous schistus; these calcareous rocks, which are covered with sandstone of a calcareous base, and which accompany the cordillera of the coast in its southern declivity, are very low on the side towards Cura, but rise in a mass towards the eastern extremity of the continent.

In Bergantin they are 702 toises high, in Coccollard 392, in Cucurucho du Tuminiquiri (the highest summits of the province of Cumana) 976 toises, and the pyramid of the Guacharo rises above 820 toises: from Cape Unare they form a separate ridge of mountains, in which the original ridge totally disappears; they are connected also with the | 354 | micaceous schistous cordillera of Maniquare and Paria only by the Cerro de Meapire, which, analogous to the branches of Torito and los Teques, which separate the basons of Monai, Aragua, and Caracas, extends north and south from Guacharo and Catouaro, to the mountain Paria, and separates the valley of Cariaco (the dried up bank of the Gulph of Cariaco) from the valley of St. Boniface, which formerly belonged to the Golfo Triste. It will be seen hereafter, that the accumulation of calcareous formation on the eastern part of the coast of this country seems to have been more exposed to earthquakes; and that the Cerro de Meapire, at the time of the irruption of the Gulph of Cariaco, and the Golfo Triste, prevented the water from converting the land of Araya and the ridge of Paria into an island.

The declivity of the cordillera of the coast of Venezuela is gentler towards the south than towards the north, which is particularly striking when one descends from the heights of Guigue, through St. Juan, Parapara, and Ortiz towards the Mera de Paja, which belongs to the great Llano de Calabozo. The northern declivity is every where very steep, and there is scarcely found, Mont Blanc excepted, above Courmayeur, a more frighful precipice than the perpendicular wall of Silla de Caracas, beyond Caravalledo, which rises to the height of 1300 toises. An accurate measurement of this wall of rock was of great importance to navigators, as they could find its distance from the coast only by taking the angle of its elevation: its longitude, therefore, of 60° 37′ 32″ west from Paris, will enable them to discover it.

The phenomenon of a more gentle declivity towards the south seems to contradict the observations made in other cordilleras of the earth, as it is asserted that they all decline more abruptly towards the south and west. This contradiction, however, is only apparent as the northern part of the cordillera, during the great catastrophe which produced the Gulph of Mexico, was torn away by the force of the water; and therefore the northern declivity might at that time be gentler than the southern.

If the form of the coast be considered, it appears to be pretty regularly indented. The headlands of Tres Puntas, Codera, S. Roman, and Chichibacoa, on the west, from Cabo de la Vela, form a row of promontories, the western of which runs more to the north than the eastern. To the windward of each of these capes a creek has been formed; and one cannot help seeing, in this singular formation, the | 355 | action of the tropical currents, which may be called the currents of the earth's rotation; an action which shows itself also in the direction of the coast from Cuba, St. Domingo, Porto Rico, Yucatan, and Honduras, as in the series of the windward islands Grenada, Orchila, Rocca, Aves, Buenos-Ayres, Curaçoa, and Aruba, the ruins of the cordillera from Cape Chichibacoa, which are all parallel to the equator. It was this headland of Chichibacoa, notwithstanding its inconsiderable height, which by its resistance to the influx, preserved the kingdom of New Grenada from losing so much land as the general government of Caracas.

The second original cordillera of South America, which I have called the cordillera of the Cataracts of Orinoco, is yet very little known. During the journey which we made on the Black River, to the borders of the Great Bara, we travelled more than 200 leagues, first from north to south, from Cerro de Uruana to Atabapo and Tuamini; then from west to east, from the mouths of the Ventuari to Vulcan de Duida, which I have found to be in latitude 3° 13′ 26″, and longitude 60° 34′ 7″ west from Paris. Since the journey of Messrs. Ituriaga and Solano, a passage over these cordilleras, which may be called also Parima or Dorado (golden), a name which has occasioned so much misfortune in America, and so much ridicule in Europe, has been possible; but as all the European settlements on the Alto Orinoco, and the Rio Negro (Black River), contain at this time no more than 400 Indian families, and as the way from Esmeralde to Erevato and Caura has been totally lost, our researches in a land so little civilized, presented more difficulties than Condamine experienced during his tedious navigation on the river Amazon, the banks of which for many years have been inhabited.

The cordillera of the Cataracts or of Parima separates itself from the Andes of Quito and Popayan, in the longitude of from 3° to 6°. It extends from west to east from Paramo de Tuquillo and St. Martin, or the sources of the Guaviare, the theatre of the gallant deeds of Philip de Urre, and the old residence of the Orneguas, through Morocote, Piramena, and Macuco, stretching through the country of the Indians of Guajibos, Sagi, Dagueres, and Poigraves, according to the direction of the great rivers Meta, Vichada, Zama, Guaviare, and Ymerida, in the longitude of 70° west from Paris, between the high summits of Uniama and Cunavami. They form the Raudals of Atures and Maypuré, tremendous waterfalls, which afford the only passage by | 356 | which one can penetrate into the interior of the land in the valley of the River Amazons.

These cordilleras of the Cataracts rise from the longitude of 70°, and spread out in such a manner that they comprehend the whole immense tract of country between the rivers Caura, Erevato, Cavony, Paraguamusi, Ventuari, Jao, Padamo, and Manariche, and then ascend south towards the sources of the Pasimona, Cachevayneris, and Cababury, towards the forests, where the Portugueze, penetrating into the Spanish district, collect the best sarsaparilla known (*Smilax Sarsaparilla*, Linn.). In this district the cordilleras of the Cataracts are above 120 miles in breadth. Their continuation more towards the east, between the longitude of 68° and 60° west from Paris is little known. I proceeded with astronomical instruments only, as far as Rio Guapo, which discharges itself into the Orinoco, opposite the Cerro de la Cauclilla, in longitude 68° 33′ west from Paris. The Indians of Catarapeni and Maquiritares, who reside in the small mission of Esmeralde came fifteen miles further east over the mountains Guanaja and Yamariquin to the Canno Chiguire; but neither the Europeans, nor Indians with whom Europeans have had any intercourse, are acquainted with this source of the Orinoco, which is here called Canno Paragua, and is scarcely 150 or 200 toises in breadth, whereas at Boca de Apuré, in latitude 7° 32′ 20″, it is 4632 toises, as I myself found. The wildness of the Indians of Guaicas, who are only four feet in height, but who are a very white and warlike people, and particularly the savage state of the Guajaribos, greater men-eaters than any of the other nations which we visited, prevent any one from penetrating over the small cataracts (Raudal de Guajaribos) east from Chiguire, unless a military expedition were undertaken on purpose. But by the wonderful journey undertaken by D. Antonio Santos, who married Onotho, and who dressed sometimes as a Carib, and sometimes as a Macacy, whose languages he spoke, from Orinoco (the mouth of the Rio Caronis) to the small lake Parima and the river Amazon, we have obtained information respecting the continuation of the cordillera of the Cataracts. Under the latitude of from 4° to 5° and longitude 63°, it becomes so narrow that it is scarcely 60 miles in breadth. It assumes here the name of Cerrania de Quimiropaca and Pacaraimo, and forms a chain of not very high ridges, by which the waters were divided. The water of the northern declivity, the Nocapray, Paraguamuci, Benamo, and Mazurini, flow towards the Orinoco and Rio Esquibo; the | 357 | waters of the southern, the Rio Cururicana, Parime, Madari, and Mao, pour themselves into the River Amazon. Some degrees further towards the east, the cordillera again extends in breadth as it ascends southwards towards the Canno Parara along the Mao. It is here that the Dutch give to the Cerro d'Ucuamo, the magnificent name of the GoldMountain, or Dorado, because it consists of a very shining micaceous schistus, a fossil which has brought into celebrity the small island of Ypamucena in the lake of Parima.

On the east from Rio Esquibo, or on the other side of the land of the Aturajo Indians the cordillera turns south-east as it unites with the garnet mountains of the Dutch and French Guiana, which are inhabited by a mixture of Negroes and Caribs, and give an origin to the rivers Berbice, Surinam, Marony, Aprouague, and Oyapock. The last mentioned ridge of mountains extends very much: its gneis appears at Baxo Orinoco, in latitude 8° 20′, between the mouths of the Upata and Acquire, and in latitude 2° 14′ on the north side of the river Amazon, in the mountains of Fripoupon and Maya.

Such is the form of the great cordilleras of the Cataracts, which are inhabited by a great number of uncivilized savages, little known to the Europeans. I must here observe, that in this description I have followed my own observations only, and the notices we obtained from the Indians, as also the observations of D. Antonio Santos, and the companions of his journey, who dictated to their friends. The maps of this part of the continent are entirely false, and the map added to the history of the Evircoco by P. Caulin, a work in other respects meritorious, is by our last observations some degrees more wrong in longitude and latitude than the map published thirty years before by d'Anville. All the Indian names in it also are mutilated, and mountains and rivers are delineated where none exist; a defect the more pardonable as the author was never beyond the waterfalls of Orinoco, nor at the Rio Negro.

[To be continued.] | 26 |

Sketch of A Geological Delineation of South America. By F. A. Von Humboldt

[Continued from our last volume, p. 357.]

The cordillera of Parima never reaches to the same height as the Sierra Nevada in the province of Caraccas, which is 2350 toises. Their highest summit seems to be the Cerro de la Esmeralda, or the mountain Duida, which, by trigonometrical measurement, I found to be 1323 toises above the surface of the sea, which is the height also of the Canigou. This mountain is situated in a delightful plain covered with ananas and palms: the monstrous mass which it exhibits towards the Mission and the rivers Canu-canuma and Tamatama, and the flames it vomits up towards the end of the rainy season, give it a romantic and majestic appearance. No Indian is able to clamber up to the top of this mountain and the rocks of its summit without a week's labour, because the luxuriance of vegetation in this climate impedes the progress of travelling. Next to the Duida, the Maraguaca, more towards the east of the river Simirimóni, and the high cordillera of Cunarami and Calitamini, which at Maypuré and St. Barbara is known under the false name of Sipapo, are the highest summits of the chain; they are from 1000 to 1100 toises in height. The common height of the cordillera, however, does not exceed 600 toises, and sometimes it is less, as the part situated between the left bank of the Cassiguiaré, an arm of the Orinoco which connects together the Rio Negro and the river Amazon, and the sources of the cataracts and Piramena between Carichana and Morocote, is destroyed, and still exhibits insulated rocks rising from the ground. The cause of this destruction seems to have been an eruption of water from the bason of the Amazon river towards the bason of Calabozo and Baxo-orinoco, which differ in height about 160 toises.

The geological chart of this district which I have constructed represents an immense valley which unites the Llanos of the Rio Negro,

Cassiguiaré, and Amazon, with those of the province of Caraccas, Barcelona, and Cumana; a valley which sinks down towards the north, and is intersected by a large series of single rocks which show the direction of the old cordillera on the banks of the Guaviare and Nuta in the province of Cassemora. The eastern extremity of this valley is the lowest part of it, and therefore the remains of the water of the Orinoco cut out for itself a | 27 | bed in this place. This cordillera has two remarkable properties. In the first place, as has been remarked in other ridges, the southern declivity is much steeper than the northern: the high summits of Caravami, Jao, of the volcano of Duida, Maraguaca, &c., all lie towards the south, and are there cut into perpendicular precipices. In the second place, this cordillera does not seem to contain a single rock of alluvial mountains, and consequently has borrowed nothing from the organized kingdom. On our passage over this ridge we observed nothing but granite, gnieis, micaceous schist, and hornblend schist; nowhere a covering of sand-stone or alluvial chalk, which on the cordillera of Venezuela on the coast rises to the height of 976 toises above the level of the sea. Had the proximity of the equator and the rotation of the earth any influence on this phænomenon?

The third chain of original mountains, the cordillera of Chiquitos, is known only from the accounts of some persons who have resided at Buenos-Ayres and travelled through the Pampas. It unites the Andes of Peru and Chili with the ridges of Brasil and Paraguay as it stretches from La Paz, Potosi, and Tucuman, through the provinces of Maxos, Chiquitos, and Chaco, towards the government of the Mines and of St. Paul in Brasil. Their highest summits seem to be situated between the latitude of 15° and 20° south, as the streams between the rivers Amazon and La Plata divide themselves at that height.

Between the three cordilleras, the direction of which we have hitherto followed, lie three broad and deep valleys. 1st, The valley between the south side of the cordillera of Venezuela, on the coast, and the cordillera of the Cataracts, or the valley of Orinoco and Apuré, between latitude 8° and 10°. 2d, The valley of the rivers Negro and Amazon, bordered by the Parima ridge and the cordillera of Chiquitos, between latitude 3° north and 10° south. 3d, The valley of Pampas of Buenos-Ayres, which extends from Saint Cruz of Sierra to Cape Virgin, between 19° and 52° south latitude. The first and second valley are in some measure united by the destruction of a part of the Parima cordillera. I do not know whether this be the case also with the Pampas and valley of the Amazon; it, however, appears that it is not, though the Llanos of Monso form a sort of canal which descends from north-west to south-east. All these immense valleys or plains are entirely open towards the east, as they run out into a low sandy coast: towards the west they are shut by the chain of the high Andes. There are some creeks | 28 | (*anses*) which proceed from east to west in the direction of the tropical current, and on that account extend further into the land the broader the continent is. The valleys of Apuré and Orinoco are closed by the ridge which extends from Pampelona to Merida in longitude 73°, and the valley of Pampas in longitude 70°: they both fall together a little towards the east, and seem to be covered by one and the same formation of alluvial strata.

Tralles says, that in Swisserland there is more reason to wonder at the depth of the lakes than at the height of the mountains: I will venture to make a similar observation in regard to the *Llanos* or plains of South America. How astonishing it is to see a continent which in its interior parts several hundred miles from the coast, and in the neighbourhood of mountains 3000 toises in height, is elevated scarcely fifty toises above the surface of the sea! If the flux in these places should rise to as great a height as at St. Malo and Bristol, and if more motion should be communicated to the ocean by earthquakes, the greater part of these valleys would be laid under water. The highest Llano which I have measured is that between the rivers Ymirida, Temi, Pimichia, Cassiguiaré, and Guiainia (Rio Negro); it is 180 toises in height; but it sinks down towards Atures in the north, as towards the river Amazon in the south. The valley of Orinoco and Apuré is still much lower than that of Cassiguiare and Calabozo in the middle of the Llano where I made observations, in latitude 8° 56′ 56″ and longitude 70° 9′ west from Paris. At Angostura, the capital of Guyana, latitude 8° 8′ 24″, longitude 66°, it is only 33 toises, and eighty miles from the coast scarcely eight toises above the level of the sea. The plains of Lombardy, in Europe, have the greatest resemblance to the Llanos on account of their small elevation. Pavia is only 34, and Cremona 24 toises in height; the other plains of Europe have a much greater elevation. In Saxony and Lower Silesia the plains are only from 87 to 120 toises in height; those of Bavaria and Swabia are from 230 to 250. The declivity of the Llanos of America is so gentle, their inequalities are so imperceptible, that no large river flows to either side. The Orinoco appears in the longitude of about 70°, as if about to discharge itself in the sea towards Portobello; but at Cabrouta it turns to the east without the least obstacle being discovered either there or at St. Fernando de Atabapo, in latitude 7° 55′ 8″, to oppose its course. In the large valley of Rio Negro, and of the Amazon river, is a tract of land, in 2° or 3° north latitude, of not less than | 29 | 1600 square miles, which is bordered by the large rivers Atabasso, Cassiguiaré, Guiainia, and Orinoco, and represents a parallelogram, in which the water flows on the four opposite sides in opposite directions. In regard to the Orinoco, I found a fall of 151 toises in the distance of 70 miles from the mouth of Guaviare to the Apuré; but from the capital to the sea not more than eight toises. La Condamine observed the same thing in regard to the river Amazon, from the narrow pass of Paucis to Para, where it runs through a district of 240 miles, but falls not more than 14 toises. It is not improbable that there might have been on the north side of the cordillera of the coast of Venezuela a plain as much lower than the plain of Orinoco as the plain of Rio Negro is higher than that of Orinoco, and on this account the former plain was covered by the water of the bay.

The two Llanos or plains which lie at the opposite extremities of America exhibit a striking difference from that which lies between them, namely, the vale of the river Amazon. The latter is covered by so impenetrable forests that rivers alone can force a passage through them, and that scarcely any other animals but such as frequent trees can live in that district; so much is vegetation favoured by the continual rains under the equator. The case is quite different with the plains of Orinoco and Pampas; they are level valleys covered with herbs, and savannahs which contain only a few scattered palm-trees. The same heat, the same want of water, and the same phænomena of refraction, that is to say, the inverted image of objects seen floating in the atmosphere, are observed here as in the deserts of Africa and Arabia. But plains so perfect are nowhere else to be found; for the Mesa de Pavone and the Mesa de Guanipa in 800 square miles contain no eminence of eight or ten inches in height. The plains of Lower Hungary, on the west of Presburgh, have the greatest resemblance to them; for the flat land of La Mancha, Champagne, Westphalia, Brandenburgh, and Poland, is hilly when compared with the Llanos of South America. Nothing but a long stagnation of water could have produced so horizontal a bottom. Traces of old cities are found here, but seldom are any seen which rise like castles (La Piedra Guanan, longitude 69° 3′, latitude 1° 59′ 48″) in the Llano of Cassiguiaré and of Rio Negro. But from St. Borja to the mouth of the Black river Condamine observed no eminence; and the Llano of Orinoco is also without islands. As the Morros of San Juan belong to the southern declivity of the cordillera of Vene- | 30 | zuola, an impetuous current of water must have swept every thing along with it; and the present sea presents large spaces without islands: instead of islands there are in the Llanos whole uninterrupted portions of from 200 to 300 square miles of surface which rise from two to five feet above the plain, and which are called *mesas* or *bancos*; which is as much as to say, that they were shoals or sand-banks in the antient sea. I must here observe, that the middle of the plain of Orinoco is the most beautiful and levellest part of it. The bottom of this immense bason rises up and becomes unequal at the edge; the plains therefore which one traverses between Guyana and Barcelona are less perfect and level than those of Calabozo and Uritucu.

This remarkable difference which we found between the cordillera of Venezuola and that of the Cataracts, which is that the latter consist of alluvial mountains entirely bare, is observed between the northern Llano of the Orinoco and that of the Rio Negro and river Amazon. In the former, the original mountains are every where covered with compact limestone, gypsum, and sandstone: in the latter the granite every where appears. The more one approaches the equator the thinner is the stratum of sand which covers the crust of earth on the original mountains: in a land where vegetation is so luxuriant, there is seen in the middle of forests spaces of 40,000 square toises scarcely covered with a few lichens, and which do not rise two inches above the rest of the surface. Will the same be discovered in Africa? for it is only in America and Africa that there is land under the equator.

Having taken a view of the direction of the mountains and valleys, or the form of the inequalities of the earth, let us now turn our attention to objects of more importance which have been less examined, namely, the rising and falling of the strata of the original mountains which form this part of the earth I have traversed. I have been convinced since 1792 that the rising of the original mountains follows a general law, and that, making allowance for those inequalities which may have been produced by trifling local causes, and particularly veins and strata in mines, or by very old valleys, the stratified coarse-grained granite, the foliated granite, and particularly the micaceous schist and argillaceous schist, rise in the league 3 4/8 by the miner's compass, as they form with the meridian of the place an angle of 52 1/2°. The falling of the strata is towards the north-west; that is to say, they fall parallel with a body that might be thrown in the same direction, or the aperture of | 31 | the angle of inclination (less than 90°) which it makes with the earth's axis stands towards the north-east. The rising is more constant than the falling, especially in the simple mountains (argillaceous schist, hornblend schist), or in the compound mountains with fewer crystallized grains, such as the micaceous schist. In granite (it is, however, found very regularly stratified rising in the league 3 — 4, and falling towards the north on the Schneekopfe, the Ochsenkopfe, the Siebengebirge, and the Pyrenees,) and in the gnieis the attraction of the crystallized mixed parts to each other seems to have prevented the regular stratification; therefore more coincidence is found among the micaceous and argillaceous schist, and these first led me to the idea of the law of rising during my tour to the Fichtelberg and the Thuringian forest. Since that time I have examined with great care the angle of the strata of other original mountains in other parts of Germany, in Swisserland, Italy, the southern parts of France, and the Pyrenees, and lately in Gallicia. Mr. Freiesleben, whose labours have been of so much service to geology, assisted me in this examination; and we were astonished at the uniformity in the rising and falling of the mountains which we found at each step on one of the highest cordilleras of the earth, the Alps of Savoy, the Valais, and the Milanese.

An examination of this phænomenon, and of the identity of the strata, was one of the principal objects when I undertook a voyage to America. A measurement of the angles which I have hitherto made on the cordillera of Venezuola and Parima gave again the result of my observations in Europe in the chain of the micaceous schist mountains of Cavaralleda as far as Rio Mamon; on the Silla de Caracas at the height of 1000 toises; of the Rincon del Diablo, on mount Guigue; in the islands in the beautiful lake of Valencia, which has almost the same elevation as the lake of Geneva, at the boundaries of the isthmus of Maniguaré and Chupariparu; on the hornblend schist which appears uncovered in the streets of the capital of Guyana, and also in the Cataracts, and on the stratified granite at the foot of the Duida. Every where the strata form an angle of 50° with the meridian (in the league 3 — 4 by the Saxon compass) as they rise from the north-east to south-west, and fall about from 60 to 80 towards the north-west.

This great coincidence in the old and new world must excite serious considerations. It exhibits a very important geological fact. After so many observations which I have made in places so far distant from each other, it can no | 32 | longer be believed that the rising of the strata follows the direction of the cordillera, and that the falling follows the declivity of the mountains. The profile of many of the mountains, particularly a section of the mountains, such as that of Genoa through the Bochetta, and of St. Gothard as far as Franconia in Germany, which I intend to publish at a proper time, proves exactly the contrary. The rising and declivity of the cordillera, the form of the small inequalities of the earth, seem to be newer phænomena. A stream might scoop out a valley in this or in that direction; might tear asunder a part of the cordillera, and give it apparently one direction or another. The strata of the original mountains appear, amidst all these angles of rising and falling observed at present, to have existed before these changes at the surface of the earth. They are the same at the summit of the Alps, and in the mines into which we descend. When one travels for 15 miles over strata of argillaceous schist, which are inclined parallel to each other, at an angle of 70° towards the north-west, one can no longer believe that they are deranged strata, which once stood horizontal. We must suppose mountains that were once 15 miles in height, and that the whole mass had an uniform fall, and then reflect on the space which such a mass would occupy: and one must remember the strata on the heights of Genoa, or on the heights of Bochetta, or on St. Maurice, which are exactly parallel; and on the strata of the Fichtelberg of Gallicia, the Silla de Caracas of Robolo on the isthmus of Araya of Cassiguiare, in the neighbourhood of the equator. One must allow that this coincidence gives evidence of a cause which has acted at a very early period, and in a general manner; a cause which must have arisen from the first attraction by which matter was forced together to form a spherical planet.

This grand cause does not exclude local causes, by which individual smaller parts of matter were determined to arrange themselves in this or in that manner, according to the laws of crystallization. Delametherie has made an ingenious remark on this subject: he shows the influence of a large mountain (as a small nucleus) on the neighbouring small mountains. One must not forget that, besides the general attraction towards the centre, all matters exercise a mutual attraction on each other.

The crust of the earth, for I will venture to speak only of this part, must be the result of an immense action of powers of attraction of affinities, which determined, put in equilibrium, and modified each other. M. Klugel | 33 | thought he found, by calculation, that the great flattening of the earth must be on the west side of the north pole. Has the axis of rotation been changed? What will be the inclination of the strata in the southern hemisphere? We are not acquainted with the cause; let us rather continue to examine the phænomena.

This falling of the strata of the original mountains in the cordillera of Venezuola has a great and melancholy influence on the fertility of the provinces of Caracas, Cumana, and Barcelona; the water which filtres through at the summit of the mountain flows down according to the direction of the strata, and for this reason there is great want of water in the whole large district which lies on the south side of the cordillera, and therefore so many springs and small streams burst forth on the northern declivity, which, by this great quantity of moisture, and the superabundance of wood, which shelters it almost the whole day from the sun's rays, is rendered as unhealthful as it is fruitful.

The alluvial mountains which I have hitherto observed are almost under the same circumstances as in Europe. The oldest seem to have experienced the action of the same causes which determined the strata of the original mountains, as they rise in the league 3—4, or as the seamen express it, N. 50 E. They often fall towards the south-east, as in the Alps of Bern, the Valais, Tyrol, and Steyermark; but the greater part of them, and particularly the newest, which where I have been are the most numerous, follow no certain law; their strata often lie horizontally, or rise towards the edge of the large dried-up basons, which in America are called Llanos, and in Africa Deserts.

La Condamine says that in Peru and Quito he observed no petrifactions. The cordillera of Quito, however, is not like that of Parima, naked granite, for at Cuença, and on the south side, there is gypsum and alluvial chalk. Buffon dwells much, in his *Epoques de la Nature*, on the question whether South America contains petrifactions? I have found an immense quantity of them in calcareous alluvial sandstone which covers the northern and southern declivity of the coast of Venezuola, from the summit of St. Bernandin, and the Altos de Conoma, to the Cerro de Meapiré, or the headland of Puria and Trinidad. The same stratum is found also in Tobago, Guadaloupe, and St. Domingo. An immense quantity of sea and land shells, which in Europe are seldom found mixed together, cellulariæ, madrepores, cerallines, and astroites, are found in- | 34 | terspersed in this sandstone. The shells themselves are half broken: whole rocks consist merely of such remains reduced to powder. My fellow-traveller, Bonpland, discovered in them shells of the genus Pinna, Venus, and Ostrea, of which living specimens are still met with on that coast; an observation of great importance to geology. Every thing shows that this stratum, which I have seen only at the distance of nine or ten miles from the present coast, is of very modern origin, and that the fluid in which it was produced had been in a state of violent motion. The petrified shells in a much older stratum of compact limestone are scarcer and much differently stratified: they are anomia, terebratulites, &c. placed together in families, and in such a manner that it is seen that they have lived (as those of Mount Salive, the Heinberg near Göttingen, of Jena, and Geneva) on the spot where they are now found petrified. They are not interspersed throughout the whole mass of the limestone; they are only peculiar to certain strata. Many rocks may be examined without finding any of these petrifactions; but where found they are in great quantity, and present themselves chiefly on great heights; peculiarities which they have in common with the shells found in the limestone of the high Alps of Swisserland and Salzburg, which is identic with the hardened marl of Thuringia, a limestone which lies above the very old sandstone.

I must observe also, that, besides the new sandstone stratum with a calcareous base, of which I have already spoken, the petrifactions do not often occur; and I was particularly astonished to find no single belemnites or ammonites which are so common in all the mountains of Europe. The Llano of Orinoco, and that even of Rio Negro, are covered with a coarse grained breccia (*nagelfluhe*) which contains no petrified shells, and perhaps covers the other alluvial strata with petrifactions. But this breccia contains on the other hand petrified trunks of trees, which are sometimes found of the length of a toise, and of the diameter of two feet. They seem to belong to a kind of Malphigia.

The sandstone which contains all kinds of marine animals (the quarry of Punta del Barrigon near Araya is of this sort) never exceeds the height of from 30 to 40 toises. In several places it forms the bottom of the Gulph of Mexico (Cabo Blanco, Punta Araya). In the compact limestone I never saw petrified shells above the height of 800 toises; but other very new testimonies prove the residence of the water at much greater heights. Slate found on the Silla de | 35 | Caracas, at the height of 1130 toises, proves that the water once, as on the Bonhomme in Savoy, formed this aperture between the two peaks or pyramids of the Avila, an aperture which is much older than the five counted in the cordillera of the coast, namely, those of Rio Neveri, Unare, Tuy, Mamon, and Guyaca. Among the mountains of the province of Cumana, there are very singular valleys of a perfect circular form, which seem to be dried up lakes. Of this kind are the valleys of Cumanacoa and St. Augustine, 507 toises in depth, which are celebrated for the refreshing coolness which travellers experience in them.

When the modern action of water is considered, two opposite effects are observed: one recollects a very distant epoch, when the irruption of the sea formed the Gulph of Cariaco and the Golfo Triste; separated Trinidad and Margaretha from the main land, and convulsed the coast of Mochima and Santa Fé, where the islands of la Boracha, Picua, and Caracas, form a heap of ruins. The sea then attacked the land; but the contest did not long continue: the ocean again begins to draw back. The islands Coche and Cuagua are shoals which emerged from the water; the large plain of Salado, lying in Cumana, belongs to the Bay of Cariaco, and is only 5 1/2 toises above the level of the sea. The hill on which the castle of St. Antonio is situated was an island in this gulph, as an arm of the sea passed to the north of Tatoraqual through the Charas towards Punta Delegada, as is proved by a multitude of unaltered shells. It is observed here and at Barcelona that the sea is daily retiring: in the harbour of Barcelona it has lost in 20 years above 900 toises. Is this decrease of the sea in the Gulph of Mexico general, or is it the case here, as in the Mediterranean Sea, that it gains in one point and loses in another? This retreat of the sea must not be confounded with another real phænomenon easy to be explained, namely, the decrease of fresh water, of rain, and of the rivers in this continent. The Orinoco, as we see it at present, is no longer the shadow of what it was 1000 years ago, according to the evidence of the traces which the water has left on both banks at the height of 70 or 80 toises. These traces have long attracted the notice of learned Europeans who have seen the Barraguan, the Cueva de Ataruipe (the burying place of the Atures Indians, who formed a kind of mummies), the Cerro Cuma, the Daminari, the Keri, Oco, and Ouivitari, the bottom of which at present is scarcely covered by the foam of the Cataracts of Maypuré. These traces remind the Indians of a great inundation, during | 36 | which many persons saved themselves on rafts of Agave, and afterwards cut out inscriptions and hieroglyphics, with which the granite of Urnana, of Incaramada, and the banks of Cassiquiaré, are seen covered, but of which no one at present has the key. This tradition, common among the Indians of Erovato and of Parima, shows great analogy with the mythology of the antients. People think they read the history of Deucalion, and Pauw would find the remembrance of this flood not uninteresting.

[To be concluded in our next.]

Sketch of a Geological Delineation of South America. By F. A. Von Humboldt.

[Concluded from p. 36.] | 172 |

Having already given a cursory view of the general appearance which the mountains of South America exhibit to the eye of the geologist,

I shall now enumerate the different kinds of mountains which I have hitherto discovered in that country, beginning with the oldest.

I. PRIMITIVE MOUNTAINS.

Granite.—The whole cordillera of Parima, and particularly the neighbourhood of the volcanoes of Duida and Marcielago, consist of granite, which does not form a transition into gneiss. In the cordillera of the coast it is almost every where covered and mixed with gneiss and micaceous schist. I saw it disposed in strata of from two to three feet in thickness, exceedingly regular, declining from three to four per league, towards the north-west between Valencia and Portocabello. I found it on the Rincon del Diablo south-east from Portocabello, with large and beautiful crystals of feldspar an inch and a half in diameter, like the large grained granite on the high summits of the Schneegebirg and the Fichtelberg, those of Scotland and Chamouni. It is here split into regular prisms; and I saw it on Calavera du Cerro de Mariana beyond Cura, and on the Silla de Caracas, in this prismatic form, which the learned mineralogist M. Karsten observed on the Schneekoppe in Silesia. The northern part of Germany, and the lands on the Baltic in Europe, but not the plain to the south of the Fichtelberg in Swabia and Bavaria, are full of monstrous blocks of granite which have rolled down from the heights. In neither of the llanos of South America, that of Orinoco, and that of the Amazon river, did we find any such masses, and no fragments of primitive mountains. The granite mountains of Los Mariches near Caracas, and those of Torrito be- | 173 |
tween Valencia and St. Carlos, and that of Sierra Neveda de Merida, contain, like that of St. Gothard, fissures which are covered with very beautiful and large rock crystals.

The granite is covered with gneiss and micaceous schist, particularly on the cordillera of the coast of Venezuela. Gneiss is abundant in particular from Cape Chichibocoa to Cape Codera in the Tequez, Cocuiza, and the mountain Guigue, as well as in the islands of the Lake of Valencia, where I found (on Cape Blanc, opposite to Guacara,) blackish quartz in the gneiss which passes into Lydian stone, or rather into the schistous state of Werner. The Macanao on Margaret's island, and the whole cordillera on the isthmus of Cariaco, is nothing else than micaceous schist full of red garnets; and at Maniquarez it is combined with a little cyanite. Green garnets are intermixed with the gneiss of the mountain Avila. In the gneiss of the rock Calamicari in Cassiquiare, and in the granite of Las Trincheras near Valencia, I saw round masses, from three to four inches in diameter, interspersed, which consisted of finer grained granite, yellow feldspar, a great deal of quartz, and scarcely any mica. Is this old granite contained in some of later formation, or are these masses, which have the appearance of accumulations, merely the effect of attraction, which here and there made the particles to approach nearer to each other, but at the same time that the whole mountain was formed? This phænomenon of one kind of granite interspersed in the other is observed also in Silesia, at Wunsiedel, on the Fichtelberg, in Chamouni, on St. Bernard, on the Escurial, and in Galicia. Nature is uniform in her natural productions, even to the small variations in proportions.

The micaceous schist passes into talc schist in the cordillera of the coast, on the mountain Capaya, and on the Quebrada Secca, in the valley del Tuy. In the cordillera of Parima talc is found in very large shining masses, and this has contributed so much to the celebrity of the Dorado, or Cerro Ucucuamo, between the river Esquivo and Mao, in the island Pumacena. The bright fiery appearance exhibited sometimes by the truncated pyramids of the large Cerro Calitamini, near Cunavami, at sunsetting, seems also to proceed from a stratum of talcy schist cut perpendicularly towards the west.

Small idols of nephrite, which I saw brought from Erovato, show that to the south of Raudal de Mura there are nephrite rocks in gneiss like those I found at the bottom of St. Gothard, near Ursern. This formation was repeated by | 174 | nature in the land of the Tupinambaros Indians. La Condamine discovered this variation of the hard nephrite, which is known under the name of the Amazon stone.

The granite, gneiss, and micaceous schist, contain here, as in Europe, strata of chlorite schist arranged under each other in the sea at Cape Blanc west from Guayra. Very pure and beautiful hornblend schist is found in the streets of Guyana; and, still more south, in the cordillera of Parima, feldspar effloresces into porcelain earth in the Silla de Caracas; strata of quartz, with magnetic iron-stone, is found at the sources of the Cutuche, near Caracas; grained foliaceous, primitive limestone, without tremolite, but with a great deal of sulphureous pyrites and sparry iron-stone, on the Quebrada de Topo on the road from Caracas to Guayra. This limestone is entirely wanting in the cordillera of Parima, where it has been sought for many years. Zeichen schist, a kind of carbonaceous iron, and pretty pure graphite, are found in the Quebrada de Tocume near Chacao, in the Quebrada Secca near Tuy, and north from the Laguna Chica; on the difficult road which leads across the isthmus of Cariaco to Chiparipara, there are found veins of quartz, which contain auriferous sulphureous pyrites and antimony, native gold, gray silver ore, mountain blue, malachite, &c.

The copper ore of Aroa is the only kind here taken from the earth: sixty or seventy slaves obtain yearly 1500 quintals at most of refined copper. The quintal is sold for twelve piastres. The valley in which this ore is dug up is less unhealthful than the valleys near the sea where the Indians wash gold; namely, Urama, Maron, and Alpagoton, where the air appears to be poisonous, as is the case in the fertile valley of Cararinas between Nirgua and Rio Jaracuy. The gold is dispersed throughout the whole province, particularly in the strata of quartz at Baruta, Catia, Guigue, Quebrada del Oro near Tuy, and on the Cerro de Chacao, and Real de Santa Barbara near St. John, where I found barytic spar, the only instance I ever met with in this country. All the rivers of the province of Characas, wash down gold. It however does not thence follow that this province is rich, and contains veins of gold not yet discovered: the gold may be interspersed in whole masses of granite; and I am acquainted with no high granite cordillera, either here or in Europe, the rivers of which do not wash down gold. The Cerro Duida of Esmeralda in Dorado, the Quebrada du Tigre near Encaramada, and the Cerros de Amoco, the Real de S. Barbaro near St. John, | 175 | the Quebrada de Catia, the alum ore of Chuparuparu, some traces of iron ore in the llano of St. Sebastian, and particularly the Aroa abundant in copper, seem to call for the industry of the miners.

Argillaceous schist is very scarce: it covers the micaceous schist on the southern declivity of Venezuela, in the neighbourhood of the Llanos, in the Quebradas de Malparo, and Piedra Azul: there is blue argillaceous schist, with veins of quartz, on the isthmus of Cariaco, near Chuparuparu, in the Distillador Arroyo du Robola, and also on Macanao. In the four last-mentioned places there are found in the argillaceous schist alum and vitriolic schist, in strata of two or three feet in thickness, which effloresce sulphate of alumine, or natural alum, with which the Indians of Guayqueries carry on a little trade.

Serpentine is found on the cordillera of Venezuela above micaceous schist, on the surface of Villa de Cura, at the height of 245 toises; between the Cerro de Piedras Negras and the Rio Tucutunemo, here and there green olivin mixed with glimmer, without garnets, schillerspath, or hornblend, but with veins of bluish lardstone.

Grunstein (green rock), original trapp, an intimate union of hornblend and feldspar, sometimes intermixed with sulphureous pyrites and quartz, often confounded with basaltes, and very little known in Europe, is found in strata of two fathoms in thickness, or balls of from three to four feet in diameter, composed of concentric strata united with micaceous schist or original argillaceous schist, in several places of the northern and southern declivity of the cordillera of the mountain Avila, in the sea near Cape Blanc, in a real vein which traverses the strata of gneiss, but intermixed with newer granite, which fills up the vein between Antimano and Carapa near Caracas. The gray stone contains here red garnets which I have never seen in Europe. I have sent specimens of them to Madrid in the first box which I transmitted to the captain-general of Caracas.

II. KIND OF MOUNTAINS WHICH FORM THE TRANSITION FROM PRIMITIVE TO ALLUVIAL MOUNTAINS. FORMATION OF THE TRANSITION OF WERNER.

This formation is found in particular to the north of the Parima cordillera, opposite to Caccara, and in large masses on the southern declivity of the Venezuela cordillera. Between the llanos and Morros of S. Juan, between the Villa de Gura and Parapara, between longitude 9° 33′ and 9° 55′, one seems to enter a land of basaltes, on descending from | 176 | the height of 300 to 63 toises above the level of the sea. Every thing reminds one here of the mountains of Bilin in Bohemia, or of Vienza in Italy. The primitive serpentin on the banks of the Tucutunemo, which like that of Silesia contains copper veins, becomes gradually mixed with feldspar and hornblend, and makes the transition into trapp or grunstein. This trapp is found in stratified masses declining 70° towards the north, or in balls with concentric strata, which, interspersed in calcareous clay, form pyramidal hills; sometimes the transition argillaceous schist of Werner is interspersed in green and very heavy argillaceous schist, which consists of hornblend and argillaceous schist intimately mixed together. The same argillaceous schist makes a transition near the Quebrada de Piedras Azules into the primitive argillaceous schist above which it lies. The trapp or grunstein contains also foliaceous olivin, crystallized in pyramids of four faces, a fossil which M. Friesleben discovered on our tour into Bohemia, and described in the Mineralogical Journal of Freyberg, augite with a shelly fracture, leucite in dodecaedra, the sides of the holes and cavities of which are covered with green earth like that of Verona, and a substance which has the splendour of mother-of-pearl, and which I consider as zeolite. All these interspersed fossils increase towards Parapara, and the trapp there forms real amygdalite. Above this amygdalite, near the hill Florez, at the entrance into the large valley of Orinoco, lies that remarkable stone which is scarce in Europe, and which Werner describes under the name of porphyry schist. The hornschist of Charpentier, a kind of rock which accompanies basaltes, forms groups of irregular columns, and by the impression of the ferns which it contains in the middle of the mountains, as discovered by M. Reuss, proves that it is not of volcanic origin. The porphyry schist of Parapara is a green mass of sonorous stone, which is very hard, acute angled, and has transparent fragments on the edges: it strikes fire with steel, and contains vitreous feldspar. I did not expect to find this stone again in South America; it however does not form here such groups of grotesque appearance as in Bohemia, and on mount Eugoneide in the Venetian territories, where I have seen it.

III. ALLUVIAL MOUNTAINS.

These secondary formations, which are of later origin than the organic bodies of the earth, follow each other in the order of their relative age, as in the plains of Europe, | 177 | and as has been mentioned by that excellent geologist M. Von Buch, in his Mineralogical Description of the County of Glatz in Silesia, a small work, which contains valuable ideas and interesting observations.

I found here two formations of compact limestone. The one makes a transition into the small grained and imperceptibly foliaceous limestone, and is identic with the limestone of the high Alps; the other is compact, exceedingly homogeneous, with several petrifactions of shells, and analogous to the limestone of Jura, Pappenheim, Gibraltar, Verona, Dalmatia, and Suez; a formation of foliaceous gypsum, and another mixed with clay, containing common salt and rock oil. The saline clay which I always found accompanied with rock salt in the Tyrol, Steyermark, and Salzbourg in Swisserland; marl schist stratified in limestone of the Alps, and two formations of sandstone, one of which is older and almost without petrifactions, sometimes small and large-grained sandstone of the llanos, and the other full of the remains of marine animals, which forms the transition into the compact limestone.

The blue limestone of the Alps, with white veins of calcareous spar, is found on the micaceous schist lying upon the Quebrada Secca near Tuy to the east from the Punta Delgada, on the road from Cumana, on the Impossible towards Bordones, on the island of Trinidad, and on the mountain Paria. This limestone contains here, as in Swisserland, three formations arranged under each other:— 1st, Repeated strata of black marl schist; marl schist, or cupreous schist of Thuringia, mixed with pyrites, and earth pitch on the Cuchivana near Cumanacoa. This clay contains carbon, and absorbs the oxygen of the atmospheric air. 2d, Saline clay mixed with rock salt and crystallized gypsum, in which the salt pits of Araga, Pozuelas, and Margaret's Island are placed. 3d, Small-grained sandstone, with a calcareous base, almost without petrifactions of shells, always penetrated by water, and sometimes with brown strata of ferruginous earth on the Cocollard, Tamirquiri. I am not certain whether the last-mentioned stone lies on the limestone, or is not sometimes covered by it.

This limestone serves as the base for a newer one. It is exceedingly white and compact, full of holes (Cueva del Guacharo, in which thousands of birds reside, and among which is a new genus of Caprimulgus, from which a kind of fat much used in the country is obtained, Cueva del S. Juan, Cueva del Cuchivano); sometimes porous like the Franconian, and forms grotesque rocks (Morros de | 178 | S. Juan, de S. Sebastian). It contains strata of black hornstone, which passes into siliceous schist or Lydian stone (Morro de Barcelona) and Egyptian jasper to the south of Curataquiche. Over this compact limestone is placed, as on Jura, very beautiful alabaster in large masses at Soro, in Golfo Triste. All this gypsum contains sulphur as well as the gypsum of Bex and Kretzetzow, and in the Carpathians. This formation of limestone, with black hornstone and gypsum, seems also to occur in the valley of the Amazon and Rio Negro, where they were found by la Condamine near Cuença, between Racam and Guyausi, on the east side of the Andes.

This limestone and gypsum (the latter in the llano of Barcelona near Cachipé) are often covered in the valleys of Orinoco, and the Amazon river, by a conglomeration or sandstone, with large strata, in which the remains of limestone, quartz, Lydian stone, all of greater antiquity than the sandstone itself, occur. This conglomeration, breccia, which has a similarity to that of Aranjuez, Salzburg, &c., is extended over more than 18000 square miles in the llanos. It contains strata with small grains and traces of brown and red iron ore. I have never seen petrifactions in it.

The sandstone full of shells and coral, without any traces of crocodiles in a country which unfortunately contains so many, and which passes into limestone, but on closer examination is intermixed with grains of quartz, is of newer formation, and always nearer the coasts: P. Araya, Cabo Blanco, Castillo, S. Antonio de Cumana.

It may perhaps be expected that I should close this description with an enumeration of the volcanie productions of this country, which has been convulsed by the most terrible earthquakes, the high summits of which (Duida), and lately some of its caverns (Cueva du Cuchivano), vomit forth flames, where boiling springs are thrown up from Golfo Triste to the Sierra Nevada de Merida (the springs of Triachevar I found to be 72° 3 of Reaumur), where, on the coast of Paria, near Cumacator, there is an air volcano, the noise of which is heard at a great distance, and sulphureous pits in several places as at Guadaloupe—a country where, in the extent of several square miles, the whole surface is undermined and hollow (Tierra Hueca de Cariaco), where, in the year 1766, the earth, after being agitated eleven months by violent shocks, opened on all sides, and poured forth sulphureous water and bitumen; and where, in the midst of the driest plains in the Mera de | 179 | Guanipa and du Cary, flames burst from the earth. But nature discharges me from this task. The effects of the volcanoes in this part of the world are very different from those seen in Europe. Great and melancholy in their consequences, they change the rocks which are exposed to their action. The immense revolution of Pelileo and Tonguragua de Zuito has not only covered the earth with lava, but with clayey mud, deposited by the sulphureous water which spouted from the earth. The sulphureous gypsum, the mixture of sulphureous pyrites in all the rocks, even in granite, the bituminous saline clay, the rock oil, or asphaltum, which every where floats on the water or lies on the ground, the immeasurable quantity of rainwater, and the lakes which penetrate into the earth heated by the sun, the aqueous vapours and immense quantities of hydrogen gas every where disengaged, seem to be the principal causes which contribute to produce these volcanic effects.

The sulphureous pits of Guadaloupe, of Montmisene, St. Christopher de l'Oualiban, St. Lucia, and Montserrat, are in all probability connected with those on the coast of Paria. These volcanoes, however, belong rather to the province of natural philosophy than of mineralogy; and I must visit other countries before I can venture to form any opinion on so difficult a subject. May Heaven avert from the eastern side of New Andalusia such a catastrophe as that which has convulsed the plains of Pelileo!

Letter from C. Humboldt to C. Fourcroy, Member of the French National Institute

Cumana, October 16, 1800.

The capture of the island of Curacao, by the English and the Americans, having obliged the agent of the republic, C. Bressot, and general Jeannet to re-embark their troops, in order to return to Guadaloupe, they have put into this port for want of provisions; and though they intend to remain only twenty-four hours, I have endeavoured to collect for you some objects worthy of your attention, and which I hope will reach you in safety. You are well enough acquainted with the nature of my travels, and the difficulty and expense attending conveyance in the centre of a vast continent, to know that my object is rather to collect ideas than things. A society of naturalists sent out by government, accompanied with painters, collectors, packers, &c., might be able to embrace all the detail of the descriptive part of natural history, and, no doubt, would do so; but a private person, of a very moderate fortune, who undertakes a voyage round the world, ought to confine himself to objects more interesting. To study the formation of the globe, and the strata which compose it; to analyse the atmosphere; to measure, with the most accurate instruments, its elasticity, its temperature, its humidity, its electric and magnetic charge; to observe the influence of climate on the animal and vegetable œconomy; to compare, on a grand scale, the chemistry and physiology of organized beings; — such is the labour which I have proposed. But, without losing sight of this principal object of my voyage, you may readily conceive that, with much zeal and a little activity, two men, who traverse an unknown continent, may at the same time collect a great many things, and make a great many observations. | 4 |

During the sixteen months we have been traversing the vast territory situated between the coast, the Orenoquo, Rio-Nigro, and the river of the Amazons, C. Bonpland has dried, with duplicates, more than six thousand plants. I have described with him on the spot twelve hundred species, great part of which appeared to us to belong to genera not described by Aublet, Jacquin, Mutis, or Dombey. We have collected insects, shells, and different kinds of wood proper for dyeing; we have dissected crocodiles, lamantins, apes, and the gymnotus electricus, the fluid of which is absolutely galvanic and not electric; and have described a great many serpents, lizards, and fish.

I have made drawings of a great number of objects; in a word, I flatter myself that if I have erred it is rather through ignorance than want of activity. What enjoyment to live in the midst of these riches of nature, so majestic and grand! Behold, then, the dearest and most ardent of my wishes gratified! Amidst the thick forests of the Rio-Nigro; surrounded by ferocious tygers and crocodiles; my body tormented with the stings of the formidable moskitos and ants; having had for three months no other aliment than water, bananas, and manioc, among the Otomaqua Indians, who eat earth; or on the banks of the Casquiara, under the equator, where, in the course of a hundred and thirty leagues, no human being is seen; — in all these embarrassing situations I never repented of my undertakings: my sufferings have been great, but they were only momentary.

When I left Spain I intended to proceed directly to Mexico, thence to Peru and the Philippines; but a malignant fever, which broke out in our frigate, induced me to remain on this coast of South America; and, thinking it possible to penetrate thence into the interior, I undertook two journeys, one to the missions of the Chayma Indians of Paria, and the other to that vast country situated to the north of the river of the Amazons, between Popayan and the mountains of the French part of Guyana. We twice passed the grand cataracts of the Orenoquo, and those of Atures and Maypura, in lat. 50° 12′ and long. 5° 39′, W. dep. from Paris 4° 43′ and 4° 41′ 40″. From the mouth of the Guaviara and the rivers Atabapo, Temi, and Tuamini, I caused my pirogua to be carried by land as far as the Rio-Nigro, while we followed on foot through forests of Hevea, Cinchona, and Canella Wintertona. I descended the Rio-Nigro as far as Saint Carlos[1] that I might determine its longitude by Berthoud's | 5 | time-keeper, with which I am still well satisfied. I ascended the Casquiara inhabited by the Ydapaminares, who eat nothing but ants dried in the smoke. I penetrated to the sources of the Orenoquo, even beyond the volcano of Duida, or as far as the ferocity of the Guaica and Guaharribo Indians would permit me to venture, and I descended the whole of the Orenoquo, by the force of its current, as far as the capital of Guyana; performing a journey of 500 leagues in twenty-fix days, without counting those on which we stopped.

[1] The error in the latitude (d'Anville's chart) is more than two degrees, as it had never been determined by astronomical instruments.

My health has withstood all the fatigues of a journey of more than 1300 leagues; but my poor companion, C. Bonpland, had nearly fallen a victim to his zeal and devotion for the sciences. After our return, he was attacked by a violent fever, accompanied with a dangerous vomiting; which, however, was speedily cured.

The river of the Amazons has been inhabited for 200 years by Europeans; but on the Orenoquo and the Rio-Nigro, it was only about thirty years ago that the Europeans ventured to form a few settlements beyond the cataracts. Those which exist do not comprehend above 1800 Indians, from the eighth degree to the equator; and there are no other whites than six or seven missionary monks, who did every thing they could to facilitate our journey.

From St. Thomas, the capital of Guyana, lat 8° 8′ 24″, long. 4° 25′ 2″, we crossed once more the great desert called Elanos, inhabited by wild cattle and horses. I am now employed in constructing a map of the country through which I have travelled. I have been so fortunate as to make astronomical observations in fifty-four places. I observed at Caraccas, Cumana, and Tuy, twelve eclipses of the satellites of Jupiter; an eclipse of the sun on the 28th of October 1799. By these means, and the chronometer, I flatter myself I shall be able to give a very exact map. We shall embark here at length for the Havannah, from which we shall proceed to Mexico. — Such is the summary of my travels. I know that you, Chaptal, Vauquelin, Guyton, are all interested in my fate; and for that reason I am not afraid of tiring you.

We have scarcely any communication here with Europe. I have often attempted to write to you, as well as to our friends Vauquelin and Chaptal. I have sent you some experiments on air, and the cause of miasmata. I have sent to Delambre and Lalande, extracts from my small astronomical observations. Have any of these reached the place of their destination? By the consul of the republic at Saint Thomas I transmitted to you the milk of a tree which the Indians call the *cow*, because they drink this milky juice, | 6 | which is not at all prejudicial, but exceedingly nourishing. By the help of the nitric acid I have made caoutchouc, and I mixed soda with that destined for you, according to the principles which you yourself fixed.

In the month of January last we sent, by the corvette Philippina, a collection of seeds for the *Jardins des Plantes* at Paris. We know they have arrived, and must have been delivered to citizens Jussieu and Thoum by the ambassador of the republic at Madrid. By the flag of truce, which we expect here from Guadaloupe, the musæum will receive other articles;

for at present we must be satisfied with presenting you a few objects for your chemical analysis.

I have procured for you the *curare*, or celebrated poison of the Indians on the Rio-Nigro. I undertook a journey to Enneralda on purpose to see the liane, which produces this juice, but unfortunately we found it without flowers; and to see the method practised by the Catarapeni and Maquiritures Indians for making this poison. I shall give you, some other time, a more ample description of it. I shall only add, that I send you the *curare* in a box of tin plate,[2] and the branches of the plant *maracury*, which produces the poison. This liane grows, but not in great abundance, among the granitic mountains of Guandia and Yumariquin, under the shade of the theobromacacao and the caryocar. The Indians take off the epidermis and make an infusion of it cold, having first expressed the juice; they then leave the water over the epidermis half expressed, and afterwards filter the infusion. The filtered liquor is yellowish: it is then baked, and concentrated by evaporation and inspissation to the consistence of molasses. This matter contains already the poison, but not being sufficiently thick to daub over the points of their arrows, they mix it with the glutinous juice of another tree, which they call *kiracaguero*. This mixture is again baked till the whole is reduced to a brownish mass. You know that the *curare* is taken internally as a stomachic: it is not noxious but when it comes into contact with the blood, which it deoxydates. It is only a few days ago that I began to make experiments upon it, and I have found that it decomposes atmospheric air. I beg you will try to de-oxydate with it the metallic oxyds, and that you will examine whether the experiments of Fontana were properly made.

I add to the *curare* and the *marecury*, the *dapitche*, the *leche de pindare*, and the earth of the Otomaquas. The *dapitche* is a state of the elastic gum, which, is, no doubt, un- | 7 | known to you. We discovered it in a place where there is no hevea, in the marshes of the mountain of Javita, lat. 2° 5′, which are samous on account of the terrible serpents, of the boa kind, found in them.

Among the Pormisano and Paragini Indians we saw musical instruments made of the caoutehouc, and the inhabitants told us they found it in the earth. The *dapitche* or *zapir* is really a spongy white mass found under the roots of two trees, which appeared to us of a new genus, the *jacio* and the *curvana*, and of which we shall one day give a description.

[2] This box, and the other articles announced here, have not yet reached C. Fourcroy.

The juice of these trees is a very aqueous milk, but it appears that it is a malady in these trees to lose the juice by the roots. This discharge causes the tree to perish, and the milk coagulates in the moist earth, where it is preserved from the contact of the air. I send you the *dapitche* itself, and a mass of caoutchouc made from it, merely by exposing it to heat or dissolving it over the fire. This production, and the milk of the *cow*, in your hands, will serve to throw new light on this substance, so curious in a physiological point of view.

The *leche de pindare*, which is the dried milk of a pindar-tree, is a natural white varnish. The Indians cover their vessels and *tacuma* with this milk when it is fresh. It dries speedily, and forms a very beautiful varnish; but, unfortunately, it becomes yellow when dried in a large mass; and it is in this state that I send it to you.

In regard to the earth of the Otomaquas, I must observe that this nation, so hideous by the paintings which disfigure their bodies, when the Orenoquo is very high, and they can find no tortoises, for three months eat scarcely any thing but a kind of fat earth. There are some of them who eat a pound and a half of it per day. Some of the monks assert that they mix with it the fat of the tails of crocodiles: but this is false. We found among the Otomaquas stores of the pure earth which they eat: they give it no other preparation than that of burning it slightly, and rendering it moist. It appears to me astonishing that people can be robust and eat a pound and a half of earth daily, while we find that earth produces a very pernicious effect among children. My own experiments on earths and their properties, however, give me reason to suspect that they may be nourishing; that is to say, that they may act by affinities.

I add for the musæum, because it has fallen into my hands, the smoking instrument of the Otomaquas, and a shirt of the Piroas, a neighbouring nation. This smoking instrument is none of the smallest, as you will see. It is a kind of plate, | 8 | on which they place the rasped and rotten fruit of a mimosa, mixed with salt and a little quick-lime. The Otomaqua holds the plate in one hand and in the other the tube, the two ends of which enter his nostrils, that he may inhale this stimulating tobacco. This instrument has a historical interest: it is common only to the Otomaquas and the Omeguas, two nations who at present are 300 leagues distant from each other, among whom Condamine saw it. It proves that the Omeguas, who, according to an old tradition, came from Guaviara, may be descended from the Otomaquas, and that the city of Manoa was seen by Phillip de Vure between Meta and Guaviara. These facts are interesting in regard to the origin of the fable of the *Dorado*.

The shirt, which one of my people wore for a long time, is the bark of the tree called *morima*, without any preparation. You see that shirts grow upon trees in this country, and near the Dorado, where I found no mineral curiosities but talc and a little titanium.

It has been impossible for us to arrange the seeds and plants of the Rio-Nigro destined for Thouin, Jussieu, and Desfontaines, who will not altogether have forgotten me. We have very uncommon things; for example, new kinds of *befaria*, new genera of palms; all which we shall soon dispatch, and be assured that we shall not lose sight of the interests of the musæum. But alas! captain Baudin has set out, and we are here! This is very hard and distressing. We shall perhaps find him in the South Sea.

I beg you will remember me to the respectable members of the National Institute: my respects to Berthollet, Chaptal, Vauquelin, Guyton, Jussieu, Desfontaines, Halley, Delambre, Laplace, Cuvier. In the letter which I send to Delambre I forgot an eclipse, which I beg you will add to it.

Immersion of the 3d satellite on the 4th of October 1800, at Cumana, at 16 h. 59′ 36″ mean time.

P. S. Repeat, I beg of you, my request to the Board of Longitude for the *Connoissances des Tems*. I regret the death of general Desaix. What a loss to the republic and all mankind!

Letter from M. A. Humboldt to C. Delambre, Member of the French National Institute

New Barcelona, Nov. 24, 1800.

During my stay in South America I dispatched several letters to you and Lalande; I know you are interested in my fate, and I never let slip an opportunity of writing to you, though I have scarcely any hopes of my letters reaching the place of their destination. I am now on the point of setting out for the Havannah and Mexico, after having performed a tour of thirteen hundred nautical leagues in this part of the New World, situated between Popayan, Quito, and Cayenne. I have slept for three months in the open air, in the woods, surrounded by tigers and hideous serpents, or on plains covered with crocodiles. Bananas, rice, and manioc, have been our sole nourishment; for all provisions soon become putrid in this damp and scorching country.

How grand and majestic is nature among these mountains! From Baraquan and Uruana, which unknown nations have covered with hieroglyphics, as far as the volcano of Duida, at the distance of sixty leagues from the small lake of Dorado, the elevation of which I have found to be 2176 metres, there is only one cordillera of granite, that descends from Quito, | 366 | and proceeds from west to east to join the mountains of the French part of Guyana. What variety among the Indian races! All free, all governing themselves and eating each other, from the Guaicas of Gehetta, a pigmy nation, the largest of whom are about four feet two inches in height, to the white Guajaribos, who have really the whiteness of Europeans; from the Otomacos, who eat a pound and a half of earth per day, to the Marivitanos and the Magueritares, who feed on ants and resin. Having already spoken of all these in a letter[1], which I dispatched from the mouths of the Orenoquo to our good friend Pommard, I shall confine myself at present to a few astronomical observations, which, I think, I have made with a considerable degree of care.

My time-keeper, by Berthoud, continues to go with great correctness. I regulate it every four, five, or six days, by corresponding altitudes, taken with my instruments, which do not err a second; viz. sextants by

[1] This letter, when this was published, had not reached France.

Ramsden and Troughton, a quadrant by Bird, and a horizon by Carroché. You know that I am not very learned in the mathematics, and that astronomy is not the object of my travels; yet with zeal and application, and by daily handling the same instruments, I have been able to do something, and to do it better. As I traversed a country never visited by Europeans till about thirty years ago, in which all the Christian missions do not amount to 1800 souls, and consequently where no one has ever yet been able to make observations, I conceived that I ought not to neglect so favourable an opportunity of enlarging our geographical knowledge. You would have laughed had you seen me amidst the Ydapamianeres Indians in the forest of Casquiara, with my instruments mounted on boxes or trunks, while the shells of tortoises served us as stools. Eight or nine apes, which we carried with us, had a strong desire to handle my hygrometers, barometers, and electrometers also: around all these ten or twelve Indians stretched out in their hammocks, together with fires to secure us from the tigers, which are no less ferocious here than in Africa. The want of nourishment, the mosquitoes, the ants; the chigers, which enter the skin and plough up the flesh; the desire of cooling ourselves | 367 | in the water, and the impossibility of doing it on account of the ferocity of the caymans, the danger of being pricked by the rajas and the teeth of the small caribfish—youth and a great deal of resignation are required to endure all these. The evil is passed, and I have reaped more than I durst venture to hope.

It is believed (see the map of father Caulin, the best extant, though all the names are wrong,) that the Spanish possessions of Guyana extend to the equator. But I have found, by very good observations of the stars called the Cross and Canopus, which I made among the rocks of Culimacari, that San Carlos del Rio Negro, the most southern establishment, is in 1° 53′ of north latitude; and that the line passes through the government of Great Para, near St. Gabriel-de-las-Cachuellas, where there is a cataract, but not so considerable as the two famous ones of Atures and Maypura.

At Cumana, before the earthquake, which we experienced on the 4th of November 1799, the magnetic inclination, measured with Borda's compass, was found to be 44° 20′ of the new division: after the earthquake it was 43° 35′; the needle made 229 oscillations in the course of ten minutes. Experiments have proved that the magnetic charge has changed in this part of the world, and not in the needle.

At Calabozo, in the centre of Uana, lat. 8° 56′ 56″, long. from Paris 44° 40′ 18″, the inclination was 39° 30′: number of oscillations 222.

At Atures, one of the cataracts of the Orenoquo, in lat. 5° 39′, long. 44° 42′ 19″, the inclination was 32° 85′: number of oscillations 221.

At St. Fernando d'Atabapo, a mission at the mouth of the Guaviara, lat. 4° 9′ 50″, the inclination was 30° 30′: number of oscillations 219.

At St. Carlos de Rio Negro, lat. 1° 53′, the inclination was 23° 20: number of oscillations 216.

According to the rules given by Messrs. Cavendish and Dalrymple, care was always taken, while observing, to turn the compass to the east and west to find the mean inclinations, and to correct the error which takes place when the axis of the needle does not pass exactly through its two points. | 368 |

During this journey, which lasted a year, I determined 54 points of South America, in which I observed the latitudes and longitudes: the former deduced, for the most part, from the meridian altitude of two stars at least; and the latter, either from the distances of the moon from the sun and stars, or from the time-keeper and horary angles. I am now employed in constructing a map of the country through which I have travelled; and as my observations fill up the vacuum found in the maps between Quito and Cayenne, to the north of the river of the Amazons, I flatter myself that they will be interesting to geographers.

My time-keepers have not given me with exactness, but the differences of meridian between the places of my departure and the Caraccas, Cumana, and St. Thomas de Nueva-Guayanna, lat. 8° 8′ 24″, long. 21 of time, east from Cumana. I am very anxious, therefore, on account of my map, to fix the position of these three places in regard to Paris, and by observations purely astronomical. Besides, it is very necessary that navigators should be able, at the time of their arrival on this coast, to find the longitude of the ports well determined, that they may know the state of their chronometers; for, except Martinico, Guadaloupe, Portorico, where M. De Churucca observed; Cayenne, and Quito, there are very few places the longitude of which can be depended on; especially in Spanish America. Carthagena, according to the *Connoissance des Temps*, is at 5 h. 12′ 12″. But the three emersions of the satellites, observed by Herrera, all give 69° 24′ 10″ west of Cadiz, or 5 h. 13′ 11″ to the west of Paris.

I observed, with a telescope of Dollond, which magnifies 95 times, at Cumana, in lat. 10° 27′ 37″:

The immersion of the second satellite Nov. 7, 1799, at 11 h. 41′ 18′ true time.

Of the second satellite, Sept. 11, at 16 h. 31′ 0″ true time.

Of the first satellite, Sept. 25, 1800, at 17 h. 10′ 21″ mean time.

The *emersion* of the 4th satellite, Sept. 26, at 17 h. 28′ 0″ mean time.

Of the third satellite, Sept. 27, at 16 h. 25′ 55″ mean time.

Of the fourth satellite, Sept. 26, at 17 h. 28′ 0″ mean time. | 369 |

I am therefore mistrustful of the longitude of Cumana, as given me by my time-keeper. When I arrived from the Canaries at the Continent, I found the longitude to be 4 h. 26′ 4″; and the observations of M. Fidalgo, who observed emersions at Trinidad, but not at Cumana, give still more; viz. 4 h. 26′ 16″. Fidalgo found Trinidad 55° 16′ 32″ to the west of Cadiz, and Cumana 2° 41′ 25″ to the west of Puerta España. But the map of Trinidad, published at London, from the excellent observations of M. De Charucca, makes Puerta España 61° 22′ west from London. I am of opinion, therefore, that, in constructing the map, the authors had before them the calculations by Lalande of the occultation of Aldebaran, observed at Porto Rico on the 21st of October 1793; for the capital of Porto Rico is by the time-keepers 4° 34 to the west of Puerta España, calculating the longitude by that of Porto Rico 63° 48 15″; and for Cumana 66° 29′ 40″ to the west of Paris. The five eclipses of the satellites which I send you, must throw light on this subject; and, in my opinion, the longitude of Cumana will not be much beyond 4 h. 25′ 20″. Unfortunately, the eclipse of the sun, which I completely observed on the 28th of September at Cumana, making the horns pass along the horizontal and vertical wires, was not visible in Europe. I observed the end at 8 h. 14′ 22″ mean time; the time certain to 1″ nearly, having taken corresponding heights the same day.

At Carras (Plaza della S. Trinidad) lat. 10° 31′ 4″, I observed:

The *immersion* of the first satellite, Dec. 7, 1799, at 16 h. 11′ 57″ true time.

Of the third satellite, Dec. 7, at 17 h. 11′ 36″ true time.

The *emersion* of the first satellite, Jan. 17, 1800, at 11 h. 14′ 8″ mean time.

Of the second satellite, Jan. 28, at 7 h. 58′ 8″ mean time.

Of the fourth satellite, Jan. 18, at 8 h. 13′ 3′ mean time.

At the Valle del Tuy al Pic della Cocuiza, lat. 10° 17′ 23″.

The *emersion* of the first satellite, Feb. 9, 1800, at 11 h. 26′ 57″ mean time.

Of the third satellite on the 10th of February, at 7 h. 58′ 50″ mean time. | 370 |

But these last eclipses were observed with a telescope of Caroché, which, though a very good one, magnifies only 58 times, not being able to carry along with me, to Rio-Negro, the large telescope by Dollond.

Declination of the magnetic needle at Cumana on the 27th of October 4° 13′ 45″; at Caraccas, 4° 38′ 45″; at Calabozo, 4° 54′ of the old division.

The port of La Guayra is exactly 29″ in time west from Caraccas; and I hope that, by giving immersions and emersions, the meridian of Caraccas will be properly fixed. I have described, with Bonpland, more than 1200 plants[2].

[2] A letter from Haspel-la-Chenaye, chemist at Guadaloupe, dated Jan. 5, states, that M. Humboldt had set out for the Havannah, after having left with the agent of the government at Guadaloupe a box for the Institute and two packets, one for Fourcroy and the other for Delambre. As the box has not yet arrived, nor the packets addressed to Fourcroy, it is to be presumed that the above letter is not that mentioned by Haspel-la-Chenaye.

Letter from M. Humboldt to C. Delambre, one of the perpetual Secretaries of the National Institute[1]

Lima, Nov. 25, 1802.

My respected friend,

I have just arrived from the interior of a country where in a large plain I made experiments on the small horary | 166 | variations of the magnetic needle; and I learn, with regret, that the frigate Astigarraga, which was not to set out before the end of a fortnight, has hastened her departure for Cadiz, and will sail this very night. This is the first opportunity we have had during five months of writing to Europe from the solitude of the South Seas; and the want of time renders it impossible for me to write as I ought to the National Institute, which has given me the most affecting marks of the interest and kindness with which it honours me. A few days before my departure from Quito for Jaën and the river Amazon, I received the letter dated Pluviose 9th, addressed to me, through you, by that illustrious body. This letter employed two years to reach me in the Cordillera of the Andes: I received it the day after a second excursion which I made to the crater of the volcano of Pinchincha to carry thither a Volta's electrometer, and to measure the diameter of it, which I found to be 752 toises, while that of Vesuvius is only 312. This reminded me that on the summit of Guaguapichincha, where I have often been, and which I love as classical ground, La Condamine and Bouguer received their first letter from the *ci-devant* academy; and I figure to myself that Pinchincha, *si magna licet componere parvis*, is fortunate to philosophers. It is impossible for me to express the joy with which I read that letter of the Institute, and the repeated assurances of your remembrance. How agreeable it is to know that one lives in the memory of those whose labours continually tend to favour the progress of the human mind! In the deserts of the plains of the Apure, in the thick forests of Casiguiare and of the Oronoko, every where, in short, your names

[1] From *Annales du Museum d'Histoire Naturelle*, No. 8.

have been present to me; and, while reviewing the different periods of my erratic life, I stopped with pleasure on that of the years 6 and 7, when I lived among you, and when Laplace, Fourcroy, Vauquelin, Guyton, Chaptal, Jussieu, Desfontaines, Hallé, Lalande, Prony, and you in particular, loaded me with kindness in the plains of Lieusaint. Receive together the homage of my tender attachment and of my constant gratitude.

Long before I received the letter which you wrote to me in your quality of secretary to the Institute, I successively addressed to the Class of the Physical and Mathematical Sciences three letters, two from Santa-Fé di Bogota, accompanied with a work on a genus of cinchona; that is to say, specimens of seven kinds of bark, with coloured drawings of these vegetables; the anatomy of the flower so different by the length of its stamina, and the skeletons carefully | 167 | dried. Dr. Mutis, who showed me a thousand marks of kindness, and for whose sake I went up the river in forty days, gave me a manuscript of nearly a hundred magnificent drawings representing new genera, and new species of his Flora of Bogota. I have thought that this collection, as interesting to botany as remarkable for the beauty of the colouring, could not be in better hands than in those of Jussieu, Lamarck, and Desfontaines; and I have presented it to the National Institute as a small mark of my attachment. This collection and cinchona were sent off for Carthagena of the Indies about the month of June this year; and Dr. Mutis took upon him to transmit them to Paris. A third letter for the Institute was dispatched from Quito with a geological collection of the productions of Pinchincha, Catopaxi, and Chimborazo. It is distressing to remain under a melancholy uncertainty in regard to the arrival of those objects, as well as to that of the collection of rare seeds, which three years ago I addressed to the *Jardin des Plantes* at Paris.

Want of time at present will not allow me to give you an account of my travels and occupations since our return from Rio Negro. You know that at the Havannah we received the false intelligence of the departure of captain Baudin for Buenos Ayres. Faithful to the promise which I made of joining him wherever I could, and persuaded that I should be more useful to the sciences by uniting my labours to those of the naturalists who accompanied captain Baudin, I did not hesitate a moment to sacrifice the little glory of finishing my own expedition; and I immediately freighted a small vessel to Bataban, that I might proceed to Carthagena of the Indies. This short passage was lengthened more than a month by stormy weather; the winds had ceased in the South Seas, where I expected to find captain Baudin; and I entered on the difficult route to Quito by Honda, Ibagué, the passage of the mountain of Quindin, Popayan, and Pastos. My health continued in a wonderful manner to withstand the change of temperature to which one is exposed in this route, descending every day from snowy regions of 2460 toises in height to scorching valleys where the thermometer does not fall below 26° or 24°. My companion Bonpland, whose knowledge, courage, and immense activity were of great assistance to me in my botanical researches and comparative anatomy, was afflicted for two months with a tertian fever. The season of the great rains came upon us in the most critical passage, the high plain of Pastos, and after a journey of eight months we arrived | 168 | at Quito, where we learned that captain Baudin had pursued his voyage from west to east by the Cape of Good Hope. Accustomed to misfortunes, we consoled ourselves with the idea of having made so great sacrifices for an intention of doing good: casting our eyes on our herbals, our barometric and geodesic measurements, our drawings, and our experiments on the air of the Cordillera, we did not regret our having traversed that country, a great part of which has never been visited by any naturalists. We were sensible that man ought never to depend on any thing but what is produced by his own energy.

The province of Quito, the highest land in the world, and torn by the grand catastrophe of the 4th of February 1797, furnished us with a vast field for physical observations. Volcanoes so enormous, the flames of which often rise to the height of 500 toises, have never been able to produce a drop of liquid lava; they vomit up fire, sulphurous hydrogen gas, mud, and carbonated argil. Since 1797 this whole part of the world has been in agitation; we every moment experience frightful shocks, and the subterranean noise in the plains of Rio Bamba resembles that of a mountain crumbling to pieces under our feet. The atmospheric air and moistened earth (all these volcanoes are in decomposed porphyry) appear to be the grand agents of these combustions and these subterranean fermentations. It has hitherto been believed at Quito that 2470 toises is the greatest height at which men could resist the rarity of the air. In the month of March 1802 we spent some days in the large plains which surround the volcano of Antisana at 2107 fathoms, where the oxen, when hunted, often vomit up blood. On the 16th of March we found out a passage over the snow, a gentle acclivity, on which we ascended to the height of 2773 toises. The air there contained 0 008 of carbonic acid, 0 218 of oxygen, and 0 774 of azote. Reaumur's thermometer was only at 15°; it was not at all cold, but the blood issued from our lips and eyes. The situation did not permit me to make a trial of Borda's compass but in a grotto lower down at the height of 2467 toises: the intensity of the magnetic forces was greater at that height than at Quito in the ratio of 230 to 218: but it must not be forgotten that the number of oscillations often increases when the inclination decreases, and that this intensity is increased by the mass of the mountain, the porphyry of which affects the magnet.

In the expedition I undertook on the 23d of June 1802 to Chimborazo, we proved that with patience it is possible to sustain a | 169 | greater rarity of the air. We ascended 500 toises higher than Condamine (on Carazon), and on Chimborazo we carried our instruments to the height of 3031 toises, where we saw the barometer fall to 13 inches 11 2 lines: the thermometer was at 1 3° below zero. We still bled at the lips; our Indians deserted us as usual; C. Bonpland, and M. Montufar, son of the marquis de Salvalegre of Quito, were the only persons who remained. We all experienced an uneasiness, debility, and desire to vomit, which certainly arose as much from the want of oxygen in these regions as from the rarity of the air. At that immense height I found only 0 20 of oxygen. A frightful chasm prevented us from reaching the summit of Chimborazo, of which we were within 236 toises. You know that a great uncertainty still prevails in regard to the height of this colossus, which La Condamine measured only at a very great distance, assigning to it the height of nearly 3220 toises, whereas Don Juan makes it 3380 toises; nor does this difference arise from the different heights which these astronomers adopted for the signal of Carabura. I measured in the plain of Tapia a base of 1702 metres. Pardon me if I speak sometimes of toises and sometimes of metres, according to the nature of my instruments. You know that in publication every thing may be reduced to the metre and centigrade thermometer. Two geodosic operations gave me for Chimborazo 3267 toises above the level of the sea; but the calculations must be rectified by the distances of the sextant from the artificial horizon and by other circumstances.

The volcano of Tunguragua has decreased a great deal since the time of La Condamine: instead of 2620 toises I found no more than 2531; and, in my opinion, this does not arise from an error in the operations, because in my measures of Cayambe, Antisana, Cotopaxi, and Iliniza, I seldom differ ten or fifteen toises from the results of La Condamine and Bouguer. The inhabitants of these unfortunate countries all say that Tunguragua has visibly decreased in height: on the other hand, I find that Cotopaxi, which has been subject to such immense explosions, is of the same height as in 1744, or rather somewhat higher. But the stony summit of Cotopaxi indicates that it is a chimney, which resists and retains its figure. The operations we made from January to July in the Andes of Quito gave to their inhabitants the dismal intelligence that the crater of Pinchincha, which La Condamine saw full of snow, burns again; and that Chimborazo, which was thought to be so peaceable and innocent, has been a volcano, and perhaps will one day be so again. We | 170 | found burnt rocks and pumice-stone at the height of 3031 toises. It will be unfortunate for the human race if the volcanic fire, for it may be said that the whole high land of Quito is one volcano with several summits, should force a passage through Chimborazo. It has often been said that this mountain is granite, but a single atom of it is not to be found: it is porphyry, here and there disposed in columns inclosing vitreous feldspar, corncerre and olivin. This stratum of porphyry is 1900 toises in thickness. On this subject I could mention a polarizing porphyry which we discovered at Voisaco near Pasto; a porphyry which, analogous to the serpentine I described in the *Journal de Physique*, has poles without attraction. I might mention other facts relative to the grand law of the parallelism of the strata, and of their enormous thickness near the equator: but this is too much for a letter, which perhaps will be lost; and besides, I shall recur to this subject another time: I shall only add, that besides the elephants teeth which we sent to C. Cuvier from the land of Santa-Fé, 1350 toises in height, we have preserved for him others more beautiful; some of the carnivorous elephant, and others of a species a little different from those of Africa, brought from the valley of Timana, the town of Ibarra, and from Chili. Here then we have confirmed the existence of that carnivorous monster from the river Ohio from 50° northern latitude to 35° south latitude.

I spent a very agreeable time at Quito. The president of audience baron de Carondelet loaded us with kindness, and for three years I have not once had reason to complain of the agents of the Spanish government, which have every where treated me with a delicacy and distinction of which I must ever retain a grateful remembrance. How much the times and manners have changed! I have paid particular attention to the pyramids and their foundation, which I do not think at all deranged in regard to the mill-stones (*pierres molaires*). A generous individual, a friend to the sciences, and to those men who have done honour to them, such as La Condamine, Godin, and Bouguer, the marquis de Salvalegre, at Quito, thinks of reconstructing them; but this leads me too far.

After passing Assonay and Cuenca, where they gave us bull-fights, we pursued our way by the Oxa to complete our labours on cinchona.

We then spent a month in the province of Jaën de Bracamorros and among the Pongos of the river Amazon, the banks of which are ornamented with the andiva and buganvillæa of Jussieu. It appeared to me of importance to fix the longitude of Tomependa and Chu- | 171 | chungat, where Condamine's chart begins, and to connect these points with the coast. La Condamine was able to determine the longitude only of the mouth of the Napa: time-keepers were not then in existence, so that the longitude of these countries requires a great many changes. My chronometer by Louis Berthoud does wonders; as I see by making observations from time to time on the first satellite, and comparing point for point my differences of meridian with those found during the expedition of M. Fidalgo, who by order of the king performed trigonometrical operations from Cumana to Carthagena.

From the river Amazon we crossed the Andes at the mines of Hualgayoc, which produce a million of piastres per annum, and where the mine of gray argentiferous copper is found at the height of 2065 toises. We descended by Casamasca (where in the palace of Atahualpa I delineated the arches of the Peruvian vaults) to Truxilla, proceeding thence by the deserts of the coast of the South Sea to Lima, where for one-half of the year the heavens are obscured by thick vapours. I hastened to Lima, that I might observe there the transit of Mercury on the 9th of November 1802.

Our collections of plants, and the drawings which I made in regard to the anatomy of *genera*, agreeably to the ideas communicated to me by Jussieu in conversations in the Society of Natural History, have greatly increased by the riches which we found in the province of Quito, at Loxa, at the river Amazon, and in the Cordilleras of Peru. We found a great many of the plants seen by Joseph de Jussieu; such as the *lloqua affinis, the quillajae*, and others. We have a new species of jussiæa which is charming, colletia, several passifiores, and the loranthus in a tree of sixty feet of height. We are particularly rich in palms and gramineous plants, on which C. Bonpland has made a very extensive work. We have at present 3784 very complete descriptions in Latin, and nearly a third more of plants in herbals which for want of time we have not been able to describe. Of every vegetable we can indicate the rock where it resides, and the height in toises at which it grows; so that in our manuscripts will be found very correct materials for the geography of plants. To do still better, I and Bonpland have often described the same plant separately. But two-thirds and more of the descriptions belong to the assiduity of Bonpland alone, whose zeal and devotion for the sciences cannot be too much admired. Jussieu, Desfontaines, and Lamarck, have in him formed a pupil | 172 | who will do great things.

We compared our herbals with those of Dr. Mutis, and we consulted a great many books in the immense library of that great man. We are convinced that we have a great many new genera and species, but much time and labour will be required to determine what is really new. We shall bring with us also a siliceous substance analogous to the tabascher of the East Indies, which M. Macé has analysed. It exists in the knots of a gigantic gramineous plant which is confounded with the bambou, but which in its flower differs from the *bambusa* of Schreiber. I do not know whether Fourcroy has received the milk of the vegetable cow, a tree so called by the Indians. It is a milk which when treated with nitric acid gave me a caout-chouc of a balsamic odour, but which, instead of being caustic and hurtful like all vegetable milks, is nourishing, and agreeable to drink. We discovered it in the road to Oronoko, in a plantation, where the negroes drink a great deal of it. I have sent also to Fourcroy by the way of Guadaloupe, as well as to sir Joseph Banks by Trinidad, our *dapiché*, or white oxygenated caout-chouc, which exudes from the roots of a tree in the forests of Pimichin in the most remote corner of the world towards the sources of the Rio Negro.

I shall not go to the Philippines. I shall proceed to Europe by Acapulco, Mexico, and the Havannah; and I hope to have the pleasure of seeing you at Paris in September or October 1803.

Health and respect, Humboldt.

P. S. I shall be at Mexico in February, and at the Havannah in June.

Letter from Mr. Humboldt to Prof. Cavanilles

Whilst we are anxiously expecting that Mr. Humboldt, who has arrived at Paris on his way home, will soon publish some account of his travels, we doubt not but that, in the mean time, the following extract of a letter from this celebrated naturalist to the late Professor Cavanilles, will be read with considerable interest. It has appeared in the 18th number of the *Anales de Ciencias Naturales*, and is dated Mexico, April the 22d, 1803.

After remarking that both M. Bompland and himself had preserved the most vigorous health, notwithstanding the want of shelter and hunger they endured, particularly in passing over the deserts, in a journey across the immense continent of South America, through the countries bordering on the Amazon, to Lima, Mr. Humboldt, so far from confirming what has been said of the enervating effects of these latitudes, supposed to disqualify for every mental exertion, asserts on the contrary that they never found their heads clearer than when contemplating the magnificent scenery which nature displays in those regions. They could | 574 | bear, without inconvenience, an exposure for three hours together to the burning sun of Acapulco and Guayaquil, tread the glaciers of the Andes, cross the arid deserts, penetrate the thick woods, and wade through the miry morasses, without feeling their courage fail, or even their spirits in the least depressed.

The travellers left Lima on the 25th of December 1802, stopped two months at Guayaquil, where they had the satisfaction of the company of two distinguished botanists, Tafalla and Manzanilla, in their botanical excursions, and arrived at Acapulco on the 22d of March, after experiencing a dreadful storm off the Gulf of Nicoya.

"The volcano of Cotopaxi, over which," says Mr. Humboldt, "I had passed calmly the preceding year, on the 5th of January made so terrible an explosion, that the noise of it was heard by us, sailing at the distance of six leagues. It vomited forth torrents of flame and showers of ashes, and the snow was detached from its summit; but it does not appear as yet to have done the smallest mischief, though, not being extinguished, the province of Quito is kept in a continual alarm.

"You know the ardent enthusiasm of my friend and companion Bompland, and can judge of the botanical riches we must have acquired in passing through a country where no botanist had ever trod before, and where nature delights in producing vegetables so different in form and organization from all that are hitherto known. Our collection exceeds 4,200 species, among which are a great number of new genera, and very many grasses and palms. It contains a hundred species of Melastoma, though all those of Linnæus are not found in it. We have described 4,200 plants, and made a great number of drawings from living specimens. We cannot ascertain the actual number of nondescript plants till we return to Europe, and have an opportunity of comparing them with those that have been already | 575 | published; but we hope to have amassed sufficient materials to form a work worthy the attention of naturalists. Comparative anatomy, as well as botany, has formed a pursuit in addition to the principal object of our journey, and we shall bring home many preparations made by my companion Bompland. I have designed a number of profiles, or geographical charts, with hygrometrical and eudiometrical tables, &c. to determine the physical qualities which have so much influence on vegetable physiology; in such a manner that I can set down in toises the elevation above the sea at which each species of the tropical trees is found.

"I am sorry to see what has been written on the Cinchonas, because science never profits by the admixture of personalities with these discussions, and am truly afflicted at the manner in which the venerable Mutis has been treated. The most unfounded notions have been propagated in Europe of this celebrated naturalist, who treated us at Santafé with that frankness which seems the peculiar character of a Banks, communicating without reserve all his possessions in botany, zoology, and physics, and permitting us to take notes of every thing we wished respecting the new genera of the Flora of Santafé. He is now old; but it is astonishing what he has done and is doing for posterity, and we cannot but admire that one man should have conceived and executed such vast designs.

"M. Lopez showed me his memoir on Cinchona before it was printed, and I pointed out to him that in it was contained evidence that Mutis had discovered the Cinchona in the mountains of Tena in 1772, whereas M. Lopez himself had only seen it near Honda in 1774.

"With regard to the tree that affords the fine bark of Loxa, having examined it in its native soil, we are of opinion that it has not yet been even described. We have compared it with the Cinchonas which we have seen in the provinces of Santafé, Payau, Peru, and Jaen; in its foliage | 576 | it resembles Cinchona glandulifera of the Flora Peruviana, but its flowers are different. We have sent to France a collection relative to the

Cinchonæ, consisting of fine specimens of the barks, with branches in flower and in fruit, and splendid coloured drawings presented to us by the generous Mutis. To these we have added fossil bones of elephants, found in the Cordilleras, at an elevation of 1400 toises."[1]

In speaking of that colossal grass called in Santafé *Guaduas*, which produces a siliceous earth similar to that of the bamboo, the writer remarks that this plant forms a new genus very different from Arundo of Linnæus, and from Bambusa of Schreber. It was not without great difficulty that they discovered the flowers of this plant which the Indians deny ever to bear any, and which even botanists who have observed it for thirty years in countries where it is very abundant, have never seen in blossom. Our travellers, however, were more fortunate, meeting with it in flower in one of the most remote corners of the world, on the Casiquiare, which unites the Oronoko with the Marignon, and afterwards in the valley of Cauca, in the province of Popayan.

[1] We understand that these subjects have been deposited in the national museum at Paris.

Account of the Travels of M. A. de Humboldt in South America, extracted from some of his Letters[1]

M. Humboldt's brother, who is now at Rome, received from him lately three letters: one dated Quito, June 3, 1802; another, Cuença, July 13, 1802; and the third, Lima, the capital of Peru, November 25, 1802. They announce that M. Humboldt will soon return, and that he expected to land in the month of August or September at Cadiz, or Corunna; but the last of his letters in particular is the most interesting. In the following extract from it care has been taken to insert every thing worthy of attention in the other two:

My Dear Brother,

Lima, Nov. 25, 1802.

You must have learned by my preceding letters that I had reached Quito, at which we arrived by traversing the snow of Quiridian and Tolima; for as the cordillera of the Andes forms three separate branches, and as we were at Santa Fé de Bogota, on that which is the most eastern, it was necessary to cross the highest to approach the coasts of the South Sea. Oxen are the only animals which can be employed in this passage for transporting baggage. Travellers in general are carried by men called *largeros*. They have a chair, in which the traveller is seated, tied to their back; they travel about four hours journey every day, and in five or six weeks earn only fourteen piastres. We preferred | 238 | travelling on foot; and the weather being exceedingly fine, we spent only seventeen days in these solitudes, which exhibit no trace of their having ever been inhabited: we slept in huts constructed of the leaves of the heliconia, which travellers carry with them on purpose. On the western side of the Andes there are marshes in which we sunk up to the knees. The weather had changed, and during the last days of our journey there fell such torrents of rain that our boots rotted on our legs; and we arrived at Carthago with our legs naked and covered with bruises, but enriched with a beautiful collection of new plants, of which I have a great number of drawings.

[1] From *Magazin Encyclopédique*, Thermidor, an. 11.

From Carthago we went to Popayan by Buga, crossing the beautiful valley of the river Cauca, and having always at our sides the mountain of Choca, and the platina mines which it contains.

During the month of November 1801 we remained at Popayan, and went to visit the basaltic mountains of Julusuito; the mouths of the volcano of Puracé, which with a horrid noise throw out vapours of hydrosulphurous water; and the porphyritic granites of Pisché, which form columns of from five to seven planes, similar to those which I remember to have seen in the Fuganean mountains of Italy, and which are described by Strange.

The greatest difficulty still remained; which was, to go from Popayan to Quito. It was necessary to cross the Paramos from Pasto, and even in the rainy season, which had already commenced. The name of Paramo is given in the Andes to every place at the height of 1700 or 2000 toises, where vegetation ceases, and where a cold which penetrates to the bones is experienced. To avoid the heats of the valley of Patia, where people in the course of one night are seized with fevers which continue three or four months, and which are known under the name of *calcuturas de Patia* (fevers of Patia), we passed the summit of the cordillera by horrid precipices, in order to proceed from Popayan to Almager, and thence to Pasto, situated at the bottom of a terrible volcano.

Nothing can be more frightful than the entrance and outlet of this valley, in which we spent the Christmas holidays, and where the inhabitants received us with the utmost hospitality. They were covered with thick forests, situated among marshes where the mules sunk half up to the backs, and we passed ravines so deep and so narrow that we thought we were entering the galleries of a mine. The roads therefore are paved with the bones of mules which have perished | 239 | here of cold and fatigue. The whole province of Pasto, comprehending the environs of Guachucal and Tuqueres, is a cold plain, almost above that point at which vegetation can take place, and surrounded by volcanoes and soufrieres which continually throw up clouds of smoke. The unfortunate inhabitants of these deserts have no other food but potatoes; and when these fail, as they did last year, they go into the mountains to eat the trunk of a small tree called *achupalla* (*pourretia pitcarnia*). As this tree however is the food also of the bears of the Andes, the latter often dispute with them the only nourishment which these elevated regions afford. To the north of the volcano of Pasto I discovered in the small Indian village of Voisaco, at the height of 1370 toises above the level of the sea, a red porphyry with an argillaceous base inclosing vitreous feldspar and hornstone, which has all the properties of the serpentine of the Fichtel-Gebirge. This porphyry has very evident poles, and shows no attractive force. After having been wet day and night for two months, and exposed to the danger of drowning by a very sudden rise of the waters, accompanied with earthquakes, we arrived on the 6th of January 1802 at Quito, where the marquis of Salvaalegre was so kind as to prepare for us a house, which after so many fatigues afforded us all the conveniences that we could have wished for at Paris, or at London.

Quito is a beautiful town, but the sky is dismal and cloudy. The neighbouring mountains exhibit little verdure, and the cold is very considerable. The great earthquake of February 4th 1797, which agitated the whole province, and destroyed in a moment from thity-five to forty thousand people, has also been fatal to the survivors. It has so changed the temperature of the air, that the thermometer generally stands at from 4° to 10° of Reaumur, and rarely ascends to 16° or 17°; while Bouguer always observed it at 15° or 16°. Since that catastrophe there have been continual earthquakes: and what shocks! It is probable that the whole elevated part is only one volcano. What are called the mountains of Cotopaxi and Pinchincha are only small summits, the craters of which form different apertures all terminating in the same hollow. The earthquake of 1797 unfortunately proved this hypothesis, for the earth everywhere opened and vomited up sulphur, water, &c. Notwithstanding these horrors and dangers with which nature has surrounded the inhabitants of Quito, they are cheerful, lively, and agreeable. Their town breathes nothing but pleasure; and no where does there appear so decided a taste for amuse- | 240 | ments. It is in this manner that man is accustomed to sleep soundly on the brink of a precipice.

We resided nearly eight months in the province of Quito; that is to say, from the beginning of January to the month of August, and employed that time in visiting the different volcanoes. We examined in succession the summits of Pinchincha, Cotopaxi, Antisana, and Iliniça, spending from a fortnight to three weeks on each of them; and always returning in the intervals to Quito, which we left on the 9th of June 1802 to proceed to the environs of Chimboraço, which is situated in the southern part of the province.

I twice ascended, viz. on the 26th and 28th of May 1802, to the edge of the crater of Pinchincha, a mountain which overlooks the town of Quito. Before, no person, as far as I know, except Condamine, ever saw it; and Condamine himself arrived there only after five or six days of fruitless researches, and without instruments; and, on account of the excessive cold, could remain on it only twelve or fifteen minutes. I succeeded in carrying thither my instruments; I made important measurements, and collected some of the air to analyse it. The first time I ascended I was accompanied only by an Indian. As La Condamine approached the crater at the lower part of its edge covered with snow, I made my first attempt by following his traces; but we were in danger of perishing. The Indian fell into a fissure up to the breast; and we observed with horror that we had walked on a bridge of frozen snow, for at the distance of some paces from us there were holes through which daylight appeared. We then found ourselves on arches which adhere to the very crater. Alarmed, but not discouraged, I changed my project. From the circumference of the crater there arise, projecting themselves as I may say over the abyss, three peaks or rocks not covered with snow; because the vapours exhaled from the mouth of the volcano continually dissolve it. I climbed up one of these rocks, and found at its summit a stone, which being supported at one end only, and hollow below, projected over the precipice in the form of a balcony. Here I stationed myself to make experiments. But this stone was only about twelve feet in length and six in breadth, and was strongly agitated by frequent shocks of an earthquake, of which I counted eighteen in less than thirty minutes. That we might examine the bottom of the crater better, we lay down on our bellies; and I do not think that the imagination can conceive any thing more gloomy and frightful than what we then saw. The mouth of the volcano forms a circular hole of nearly a | 241 | league in circumference, the edges of which, cut perpendicularly, are covered with snow at the top. The inside is very black, but the gulph is so immense, that the summits of several mountains placed there can be distinguished. These summits seemed to be 300 toises below us: you may judge then where their bases must be. I have no doubt that the bottom of the crater is on a level with the town of Quito. La Condamine found this crater extinct, and even covered with snow; but we had melancholy news to carry to the inhabitants of Quito, that the volcano in their neighbourhood was now burning. We were convinced of this beyond all doubt by the most evident signs. When we approached the mouth of it we were almost suffocated by sulphureous vapours. We even saw blue flames moving about here and there, and every two or three minutes we experienced strong shocks of an earthquake, with which the edges of the crater were agitated, and of which nothing was perceived at the distance of 100 toises. I suppose that the great catastrophe of February 7th 1797 kindled up the flames also of Pinchincha. After visiting this mountain alone I returned two days after, accompanied by my friend Bonpland, and Charles de Montufar the son of the marquis de Selvaalegre. We were furnished with more instruments than the preceding time, and measured the diameter of the crater and the height of the mountain. We found the former to be 754 toises[2], and the latter 2477. In the interval of two days which took place between our excursions to Pinchincha we had a very violent earthquake at Quito. The Indians ascribed it to some powder which I must have thrown into the volcano.

During our journey to the volcano of Antisana, the weather was so favourable that we ascended to the height of 2773 toises. The barometer fell in that elevated region to 14 inches 7 lines; and in consequence of the rarity of the air the blood flowed from our lips, gums, and even eyes. We experienced extreme weakness, and one of the persons who accompanied us fainted. It was before thought impossible to ascend higher than the summit called Corazon, which is 2470 toises in height, and which La Condamine reached. On analysing the air brought from the highest point to which we ascended, it gave 0 008 of carbonic acid for 0 218 of oxygen gas.

We paid a visit also to the volcano of Cotopaxi, but it was impossible for us to reach the mouth of the crater. It is not | 242 | true that this mountain has become lower since the earthquake of 1797.

On the 9th of June 1802 we left Quito to proceed to the southern part of the province, where we wished to examine and measure Chimboraço and Tunguragua; and to take a plan of the whole country convulsed by the grand catastrophe of 1797. We succeeded in approaching to within about 250 toises of the summit of the immense colossus of Chimboraço. A ridge of volcanic rocks, destitute of snow, facilitated our ascent. We ascended to the height of 3031 toises, and felt ourselves incommoded in the same manner as we had been on the summit of Antisana. Two or three days even after our return to the plain we were still subject to an indisposition, which we could ascribe only to the effect of the air in these elevated regions, which by analysis gave us 20 hundreds of oxygen. The Indians by whom we were accompanied deserted us before we reached that height, saying that we intended to kill them. Bonpland, Charles Montufar, and one of my domestics, who carried a part of my instruments, were the only persons who remained with me: nevertheless, we could have continued our journey to the top had we not been prevented by a fissure too deep to be crossed. We therefore thought proper to descend. Being ill secured from the cold of these elevated regions, we suffered

[2] The crater of Vesuvius is only 312 toises in diameter.

very much; and I in particular had the misfortune to lacerate my foot by a fall a few days before, which subjected me to great pain in a road where we every moment struck our toes against sharp stones, and were obliged to examine the ground at every step. La Condamine found the height of Chimboraço to be nearly 3217 toises. Trigonometrical measurement, which I made at two different times, gave 3267 toises, and I have reason to place some confidence in my operations. All this enormous colossus, as well as all the high mountains of the Andes, consists not of granite, but of porphyry, from the base to the summit; and the porphyry is 1900 toises in thickness. The short stay which we made at that enormous height was dismal and melancholy: we were enveloped by a thick fog, which only suffered us from time to time to have a glimpse of the horrid abysses by which we were surrounded. No living creature, not even the condour, which on Antisana continually hovered over our heads, was to be seen. Small kinds of moss were the only organized beings which reminded us that we were still in the neighbourhood of the earth.

It is almost probable that Chimboraço, like Pinchincha | 243 | and Antisana, is of a volcanic nature. The ridge on which we ascended consists of burnt and scorified rock, mixed with pumice stone. It resembles all the currents of lava in this country; and continues beyond that point where I was obliged to set bounds to my researches, towards the summit of the mountain. It is possible that this summit may be the crater of an extinguished volcano; and this is even probable. The idea of this possibility, however, makes one shudder—and with reason; for, if the volcano should be rekindled, this colossus would destroy the whole province.

The mountain of Tunguragua sunk down at the period of the earthquake of 1797. Bouguer makes its height to be 2620 toises; I found it to be only 2531: it has lost therefore nearly 100 toises of its height. The inhabitants of the neighbouring country assert that they have seen its summit crumble down before their eyes.

During our stay at Riobamba, where we spent some weeks with the brother of Charles Montufar, who is corregidor there, we by chance made a very curious discovery. The state of the province of Quito before the conquest of the inca Tupayupagi[3] is absolutely unknown. But the king of the Indians, Leandro Zapla, who resides at Lican, and whose mind is highly cultivated, has in his possession manuscripts written by one of his ançestors in the sixteenth century, which contain the history of that period. These manuscripts are written in the language of Paraguay, which

[3] Quito was conquered by the Peruvians in 1470.

formerly was the general language of Quito; but in the course of time it gave place to that of the Incas, or the Anichna, and is now lost. Fortunately, another of Zapla's ancestors amused himself in translating these manuscripts into Spanish. We made extracts from these valuable documents, and particularly in regard to the memorable period of the eruption of the mountain called *Nevado del Attas*, which must have been the highest in the universe, superior even to Chimboraço, and which the Indians called *Capa-Urcu*, "the chief of mountains." Ouainia Abomatha, the last independent *cochocando* (king of the country), reigned at that time at Lican. The priests informed him that this catastrophe was a sinister presage of his destruction. "The face of the universe," said they to him, "is changing: other gods will expel ours. Let us not oppose what has been ordained by fate." The Peruvians indeed introduced into the country the worship of the sun. The eruption of the mountain continued seven years, and Zapla's manuscript asserts that the shower of ashes at Lican was so abundant that continual night pre- | 244 | vailed during that period. When the quantity of volcanic matters which are found in the plain of Tapia around the enormous mountain which then crumbled to pieces, is considered, and when we reflect that Cotopaxi has often involved Quito in darkness for fifteen or eighteen hours, we may believe that the exaggeration of this account is not very great. This manuscript, the traditions which I collected at Parime, and the hieroglyphics I saw in the desert of Casiquiare, where at present no vestiges of mankind remain, added to the ideas offered by Clavigero respecting the emigration of the Mexicans towards the southern part of America, have given rise to some conjectures on the origin of these people, which I purpose to explain as soon as I can find leisure.

I have applied also with great assiduity to the study of the American languages, and I have seen how much what La Condamine says of their poverty is false. The Carib language is rich, beautiful, energetic, and polished: it is in no want of expressions for abstract ideas. It speaks of posterity, eternity, existence, &c.; and the numerical signs are sufficient to express all the possible combinations of figures. I applied in particular to the Inca language: it is generally spoken in company; and is so rich in delicate and varied phrases, that the young men, in order to say soft things to the ladies, when they have exhausted all the resources of the Castillan, begin to speak Inca. These two languages, and others equally rich, are sufficient to prove that America formerly possessed a greater degree of culture than the Spaniards found there in 1492. But I have collected still further proofs, not only at Mexico and in Peru, but even at the court of the king of Bogota, a country the history of which is absolutely unknown in Europe, and whose mythology even and fabulous traditions are highly interesting. The priests were acquainted with the art of drawing a meridian line, and observing the moment of the solstice. They reduced the lunar year to a solar by intercalations; and I have in my possession a heptagon stone, found near Santa Fé, which they employed for calculating these intercalary days. But what is still more, even at Erevato, in the interior of Parime, the savages believe that the moon is inhabited by men; and know by tradition from their ancestors, that it derives its light from the sun.

From Riobamba I proceeded by the famous paramo of Assuay towards Cuença, after having visited the large sulphur mines of Tirrau. It was this mountain of sulphur which the negroes who revolted in 1797, after the earthquake, attempted to set on fire. This no doubt was the | 245 | most desperate project ever attempted, for they hoped by these means to form a volcano which would swallow up the whole province of Alaussy. At the height of the paramo of Assuay, an elevation of 2300 toises, are the magnificent ruins of the Inca's highway. It conducted almost to Cuzco, was entirely constructed of cut stone, and very straight, and resembled the most beautiful of the Roman roads. In the same neighbourhood are found also the ruins of the palace of the inca Tupayupangi, of which La Condamine gave a description in the Memoirs of the Academy of Berlin. In the quarry which furnished the stones there are still seen several half cut. I do not know whether Condamine spoke also of the so-called Inca's billiard-table. The Indians name this place in the Quichua language, *Inca-chungana*, the Inca's game. But I much doubt whether it was ever destined for that purpose. It is a seat cut out in the rock, with ornaments in the arabesc form, in which it is believed that the ball ran. There is nothing more elegant in our gardens in the English style; and every thing proves the good taste of the inca, for the seat is so situated as to command a delightful view. In a wood not far from this place is found a round spot of yellow iron in freestone: the Peruvians have ornamented it with figures, supposing it to be the image of the sun. I made a drawing of it.

We remained only ten days at Cuença, and proceeded thence to Lima through the province of Jaen, where we spent a month in the neighbourhood of the river of the Amazons. We arrived at Lima on the 23d of October 1802.

In the month of December I purpose proceeding from this place to Acapulco, and thence to Mexico, that in the month of May 1803 I may reach the Havanna, where I shall embark without delay for Spain. I have given up, as you may see, the idea of returning by the Philippines. I should have crossed an immense tract of the ocean without seeing any thing but Manilla and the Cape; or, if I had attempted to proceed to the East Indies, I should have wanted the necessaries for that voyage, and which it was impossible for me to procure here.

We have had forty or fifty young crocodiles, on the respiration of which I have made very curious experiments. Other animals diminish the volume of the air in which they live, but the crocodile increases it. A crocodile immersed in 1000 parts of atmospheric air, which contain 274 of oxygen gas, 15 of carbonic acid gas, and 711 of azot, increases this mass in one hour and forty-three minutes 124 parts; | 246 | and these 1124 parts contain then, as I found by exact analysis, 106 8 of oxygen, 79 of carbonic acid, and 938 2 of azotic gas, mixed with other unknown gaseous substances. The crocodile then in one hour and three quarters produces 64 parts of carbonic acid, and absorbs 167 2 of oxygen; but as 46 parts are found in the 64 parts of carbonic acid, it appropriates to itself only 121 parts of oxygen; which is very little, considering the colour of its blood. It produces 227 parts of azote, or other gaseous substances, on which the acidifiable bases exercise no action.

I made these experiments in the town of Munpox with lime water and nitrous gas prepared with great care. The crocodile is so sensible to carbonic acid gas and to its own exhalations, that it dies when put into air corrupted by one of its own species. It however can live two or three hours without breathing at all. I made these experiments on crocodiles seven or eight inches in length: notwithstanding this smallness of size, they are capable of cutting off a finger with their teeth, and they have the courage to attack a dog. These experiments are troublesome, and require great circumspection. We have made very minute descriptions of the caïman or crocodile of South America; but as the descriptions of that of Egypt which I saw before my departure from Europe were not equally circumstantial, I cannot venture to determine whether they are of the same species. The Institute of Egypt must undoubtedly have obtained details which will remove all doubt in regard to this point. This much however is certain, that there are three different species of crocodile in the tropical regions of the new continent, to which the inhabitants give the names of *bava, caïman* , and *crocodile*. No naturalist has yet sufficiently distinguished these species. These monsters, as at New Barcelona, are sometimes of so peaceable a nature that people bathe before them; and sometimes, as at New Guiana, they are so mischievous and ferocious, that during the time we were there they devoured an Indian on the quay in the middle of the street. At Uritucu we saw an Indian girl, eighteen years of age, whom a crocodile seized by the arm. She had the courage to put her other hand into her pocket to pull out her knife, with which she gave the monster so many wounds in the eyes that he let her go, but cut off the arm near the shoulder. This girl's presence of mind was as astonishing as the skill displayed by the Indians in speedily curing so dangerous a wound: one might have said that the arm was amputated and dressed at Paris.

Near Santa-Fe there are found in the Campo de Gigante, | 247 | at the height of 1370 toises, an immense number of fossil elephants' bones, both of the African species and of the carnivorous kind discovered near the Ohio. We caused several to be dug up, and have sent some specimens of them to the National Institute. I much doubt whether any of these bones were ever before found at such a great height: since that time I have received two from a place of the Andes situated about two degrees of latitude from Quito and Chili, so that I can prove the existence and destruction of these gigantic elephants from the Ohio to the country of the Patagonians. I shall bring with me a fine collection of these bones for M. Cuvier. About fifteen years ago the entire petrified skeleton of a crocodile was discovered in a calcareous rock in the valley of the Magdalen: it was broken through ignorance, and it was impossible for me to procure the head, which existed not long ago.

Baron Humboldt

The following abstract of the American Travels of the celebrated Baron Humboldt and his companion Bonpland, has been drawn up from notes which the former has kindly furnished, and will supersede the many very incorrect accounts hitherto published relative to this interesting object.

Baron Humboldt, having travelled from the year 1790, as a naturalist, through Germany, Poland, France, Switzerland, and through parts of England, Italy, Hungary, and Spain, came to Paris in 1798, when he received an invitation, from the directors of the national museum, to accompany captain Baudin in his voyage round the world. Citizen Alexander Aime Gourjon Bonpland, a native of Rochelle, and brought up in the Paris museum, was also to have accompanied them; when on the point of departing, the whole plan was suspended until a more favourable opportunity, owing to the re-commencement of the war with Austria, and to the consequent want of funds.

Mr. Humboldt, who, from 1792, had conceived the plan of travelling through India at his own expence, with a view of adding to the knowledge of the sciences connected with natural history, then resolved to follow the learned men, who had gone on the expedition to Egypt.— His plan was to go to Algiers in the Swedish frigate which carried the consul Skoldebrandt, to follow the caravan which goes from Algiers to Mecca, going through Egypt to Arabia, and thence by the Persian gulph to the English East-India establishments. The war which unexpectedly broke out in October, 1798, between France and the Barbary powers, and the troubles in the East, prevented Mr. Humboldt from embarking at Marseilles, where he had been fruitlessly two months waiting to proceed. Impatient at this delay, and continuing firm in his determination to go to Egypt, he went to Spain, hoping to pass more readily under the Spanish flag from Carthagena to Algiers and Tunis. He took with him the large collection of philosophical, chemical, and astronomical instruments, which he had purchased in England and France.

From a happy concurrence of circumstances, he obtained, in February, 1789, from the court of Madrid, a permission to visit the Spanish colonies of the two Americas, a permission which was granted with a liberality and frankness, which was honourable to the government and to a philosophic age. After a residence of some months at the Spanish court, during which time the king showed a strong personal interest in the plan, Mr. Humboldt, in June, 1799, left Europe, accompanied by Mr. Bonpland, who, to a profound knowledge in botany and zoology, added an indefatigable zeal. It is with this friend that Mr. Humboldt has accomplished, at his own expence, his travels in the two hemispheres, by land and sea, probably the most extensive which any *individual* has ever undertaken.

These two travellers left Corunna in the Spanish ship Pizarro, for the Canary islands, where they ascended to the crater of the Peak of Teyde, and made experiments on the analysis of the air. In July they arrived at the port of Camana, in South America. In 1799, 1800, they visited the coast of Paria, the missions of the Chaymas Indians, the province of New Andalusia (a country which had been rent by the most dreadful earthquakes, the hottest, and yet the most healthy, in the world) of New Barcelona, of Vene- | 322 | zuela, and of Spanish Guayana—In January, 1800, they left Caraccas to visit the beautiful vallies of Aragua, where the great lake of Valencia recals to the mind the views of the lake of Geneva, embellished by the majesty of the vegetation of the tropics. From Porto Cabello they crossed, to the south, the immense plains of Caloboza, of Apure, and of the Oronoco, also los Llanos, a desert similar to those of Africa, where in the shade (by the reverberation of heat) the thermometer of Reaumur rose to 35 and 37 (111 to 115 F.) degrees. The level of the country for 2000 square leagues does not differ 5 inches. The sand every where represents the horizon of the sea, without vegetation; and its dry bosom hides the crocodiles, and the torpid boa (a species of serpent). The travelling here, as in all Spanish America, except Mexico, is performed on horseback— They passed whole days without seeing a palm-tree or the vestige of a human dwelling. At St. Fernando de Apure, in the provinces of Varinas, Messrs. Humboldt and Bonpland began that fatiguing navigation of nearly 1000 marine leagues, executed in canoes, making a chart of the country by the assistance of chronometers, the satellites of Jupiter, and the lunar distances. They descended the river Apure, which empties itself into the Oronoco, in 7 degrees of latitude. They ascended the last river (passing the celebrated cataracts of Mapure and Atures) to the mouth of the Guaviare. From thence they ascended the small rivers of Tabapa, Juamini, and Tenie. From the mission of Sarita they crossed by land to the sources of the famous Rio Negro, which Condamine saw, where it joins the Amazon, and which he calls a sea of fresh water. About 30 Indians carried the canoes through woods of Mami Lecythis and Laurus Cinamomoides to the cano (or creek) of Pemichin. It was by this small stream that the travellers entered the Rio

Negro, or Black River, which they descended to St. Carlos, which has been erroneously supposed to be placed under the equator, or just at the frontiers of Great Para, in the government of Bresil. A canal from Tenie to Pemichin, which from the level nature of the ground is very practicable, would present a fine internal communication between the Para and the province of Carracas, a communication infinitely shorter than that of Cassiquiare—From the fortress of St. Carlos on the Rio Negro, Mr. H. went north up that river and the Cassiquiare to the Oronoco, and on this river to the volcano Daida or the mission of the Esmeralda, near the sources of the Oronoco: the Indians Guaicas (a race of men almost pigmies, very white and very warlike) render fruitless any attempts to reach the sources themselves.

From the Esmeralda Messrs. H. & B. went down the Oronoco, when the waters rose, towards its mouths at St. Thomas de la Guayana, or the Angostura. It was during this long navigation that they were in a continued state of suffering, from want of nourishment, and shelter from the night rains, from living in the woods, from the mosquetoes, and an infinite variety of stinging insects, and from the impossibility of bathing, owing to the fierceness of the crocodile and the little carib fish, and finally the miasmata of a burning climate. They returned to Cumana by the plains of Cari and the mission of the Carib Indians, a race of men very different from any other, and probably, after the Patagonians, the tallest and most robust in the world.

After remaining some months at New Barcelona and Cumana, the travellers arrived at the Havanna, after a tedious and dangerous navigation, the vessel being in the night on the point of striking upon the Vibora rocks. Mr. H. remained three months in the island of Cuba, where he occupied himself in ascertaining the longitude of the Havanna, and in constructing stoves on the sugar plantations, which have since been pretty generally adopted. They were on the point of setting off for Vera | 323 | Cruz, meaning, by the way of Mexico and Acapulco, to go to the Philipine Islands, and from thence, if it was possible, by Bombay and Aleppo, to Constantinople, when some false reports relative to Baudin's voyage alarmed them, and made them change their plan. The gazettes held out the idea that this navigator would proceed from France to Buenos Ayres, and from thence, by Cape Horn, for Chili and the coast of Peru. Mr. Humboldt had promised to Mr. Baudin and to the Museum of Paris, that wherever he might be, he would endeavour to join the expedition, as soon as he should know of its having been commenced. He flattered himself that his researches, and those of his friend Bonpland, might be more useful to science, if united to the labours of the learned men who would accompany captain Baudin.

These considerations induced Mr. Humboldt to send his manuscripts, for 1799 and 1800, direct to Europe, and to freight a small schooner at Batabano, intending to go to Carthagena, and from thence, as quickly as possible, by the Isthmus of Panama, to the South Sea. He hoped to find captain Baudin at Guayaquil, or at Lima, and with him to visit New Holland, and the islands of the Pacific Ocean, equally interesting in a moral point of view, as by the luxuriance of their vegetation.

It appeared imprudent to expose the manuscripts and collections already made to the risks of this proposed navigation. These manuscripts, of the fate of which Mr. H. remained ignorant during three years, and until his arrival in Philadelphia, arrived safe, but one third part of the collection was lost by shipwreck. Fortunately (except the insects of the Oronoco and of the Rio Negro) they were only duplicates; but unhappily friar John Gonzales, monk of the order of St. Francis, the friend to whom they were entrusted, perished with them. He was a young man full of ardour, who had penetrated into this unknown world of Spanish Guayana further than any other European.

Mr. Humboldt left Batabano in March, 1801, and passed to the south of the island of Cuba, on which he determined many geographical positions. The passage was rendered very long by calms, and the currents carried the little schooner too much to the west, to the mouths of the Attracto. The vessel put into the river Sinu, where no botanist had ever before visited, and they had a very difficult passage up to Carthagena. The season being too far advanced for the South Sea navigation, the project of crossing the isthmus was abandoned; and animated by the desire of being acquainted with the celebrated Mutis, and admiring his immensely rich collections of objects of natural history, Mr. H. determined to pass some weeks in the woods of Turbaco, and to ascend (which took forty days) the beautiful river of Madalaine, of the course of which he sketched a chart.

From Honda, our travellers ascended through forests of oaks, of *melastomo*, and of *cinchona* (the tree which affords the Peruvian bark), to St. Fe de Bogota, capital of the kingdom of New Grenada, situated in a fine plain, elevated 1360 toises (of six French feet) above the level of the sea. —The superb collections of Mutis, the majestic cataract of the Tequendama (falls of 98 toises height) the mines of Mariquita, St. Ana, and of Tipaquira, the natural bridge of Scononza (three stones thrown together in the manner of an arch, by an earthquake), these curious objects stopped the progress of Messrs. Humboldt and Bonpland until the month of September, 1801.

At this time, notwithstanding the rainy season had commenced, they undertook the journey to Quito, and passed the Andes of Quindiu, which are snowy mountains covered with wax palm-trees (palmiers a cire), with passe flores (passion flower) of the growth of trees, storax, and | 324 | bambusa (bamboo). They were, during 13 days, obliged to pass on foot through places dreadfully swampy, and without any traces of population.

From the village of Carthago, in the valley of Cauca, they followed the course of the choco, the country of Palatina, which was there found in round pieces of basalte and green rock (grein stein of Werner), and fossil wood. They pass by Buga to Popayan, a bishop's see, and situated near the volcanoes of *Sotara and Purace*, a most picturesque situation, and enjoying the most delicious climate in the world, the thermometer of Reamur keeping constantly at 16 to 18 (68 to 72 Fahr.) They ascended to the crater of the volcano of Purace, whose mouth, in the middle of snow, throws out vapours of sulphureous hydrogene, with continued and frightful rumbling.

From Popayan they passed by the dangerous defiles of Almager, avoiding the infected and contagious valley of Patia, to Posto, and from this town, even now situated at the foot of a burning volcano, by Tuqueras and the province of Pastos, a flat portion of country, fertile in European grain, but elevated more than 1500 to 1600 toises above the towns of Ibarra and Quito.

They arrived, in January, 1802, at this beautiful capital, celebrated by the labours of the illustrious Condamine, of Bouger, Godin, Dr. George Juan, and Ulloa, and still more celebrated by the great amiability of its inhabitants, and their happy turn for the arts.

They remained nearly a year in the kingdom of Quito: the height of its snow-capped mountains, its terrible earthquakes (that of February 7, 1797, swallowed up 42,000 inhabitants, in a few seconds), its fertility, and the manners of its inhabitants, combined to render it the most interesting spot in the universe. After three vain attempts, they twice succeeded in ascending to the crater of the volcano of Pichincha, taking with them electrometers, barometers, and hygrometers. Condamine could only stop here a few minutes, *and that without instruments*. In his time, this immense crater was cold and filled with snow. Our travellers found it inflamed; distressing information for the town of Quito, which is distant from it only 5000 to 6000 toises.

They made separate visits to the snowy and porphyritic mountains of Antisana, Cotopaxi, Tungarague, and Chimborazo, the last the highest point of our globe. They studied the geological part of the Cordillera of the Andes, on which subject nothing has been published in Europe, mineralogy (if the expression may be used) having been created, as it were, since the time of Condamine. The geodesical measurements proved that some mountains, particularly the volcano of Tungarague, has considerably lowered since 1750, which result agrees with the observations made to them by the inhabitants.

During the whole of this part of the journey, they were accompanied by Mr. Charles Montufar, son of the marquis of Selva-alegre, of Quito, a person zealous for the progress of science, and who is, at his own expence, rebuilding the pyramids of Saraqui, the extremity of the celebrated bases of the *triangles* of the Spanish and French academicians. This interesting young man having followed Mr. Humboldt in the remainder of his journey through Peru and the kingdom of New Spain, is now on his passage with him to Europe. Circumstances were so favourable to the efforts of the three travellers, that at Antisana they ascended 2200 French feet, and at Chimborazo, on June 22, 1802, nearly 3200 feet higher than Condamine was able to carry his instruments. They ascended to 3036 toises elevation above the level of the sea, the blood starting from their eyes, lips, and gums. An opening, of 80 toises deep, and very wide, prevented them from reaching the top, from which they were only distant 134 toises.

It was at Quito that Mr. Humboldt received a letter from the Na-
| 325 | tional Institute of France, informing him, that captain Baudin had proceeded by the Cape of Good Hope, and that there was no longer any hope of joining him.

After having examined the country overturned by the earthquake of Riobamba, in 1797, they passed by the Andes of Assuay to Cuenza. The desire of comparing the barks (cinchona) discovered by Mr. Mutis, at Santa Fe de Bagota, and with those of Popayan, and the cuspa and cuspare of New Andalusia, and of the river Caroni (named falsely Cortex Augustura), with the cinchona (barks) of Loxa and Peru, they preferred deviating from the beaten track from Cuenza to Lima; but they passed with immense difficulties in the carriage of their instruments and collections, by the forest (paramo) of Saragura to Loxa, and from thence to the province of Jaen de Bracamoros. They had to cross thirty-five times, in two days, the river Guancabamba, so dangerous for its sudden freshes. They saw the ruins of the superb Ynga road, comparable to the finest roads in France, and which went upon the ridge of the Andes from Cusco to the Assuay, accommodated with fountains and taverns.

They descended the river Chamaya, which led them into that of the Amazones, and they navigated this last river down to the cataracts of Tomeperda, one of the most fertile, but one of the hottest, climates of the habitable globe. From the Amazone river they returned to the south-east by the Cordilleras of the Andes to Montar, where they found they had passed the magnetic equator, the inclination being 0, although at seven degrees of south latitude. They visited the mines of Hualguayoc, where native silver is found at the height of 2000 toises. Some of the veins of these mines contain petrified shells, and which, with those of Pasco and Huantajayo, are actually the richest of Peru. From Caxamarca they descended to Truxillo, in the neighbourhood of which are found the ruins of the immense Peruvian city, Mansiche.

It was on this western descent of the Andes that the three voyagers, for the first time, had the pleasure of seeing the Pacific Ocean. They followed its barren sides, formerly watered by the canals of the Yngas at Santa, Guerma, and Lima. They remained some months in this interesting capital of Peru, of which the inhabitants are distinguished by the vivacity of their genius, and the liberality of their ideas.

Mr. Humboldt had the good fortune to observe the end of the passage of Mercury over the sun's disk, in the port of Callao. He was astonished to find, at such a distance from Europe, the most recent productions in chemistry, mathematics, and medicine; and he found great activity of mind in the inhabitants, who, in a climate where it never either rains or thunders, have been falsely accused of indolence.

From Lima our travellers passed by sea to Guayaquil, situated on the brink of a river, where the growth of the palm tree is beautiful beyond description. They every moment heard the rumbling of the volcano of Cotopaxi, which made an alarming explosion on the 6th January, 1803. They immediately set off to visit it a second time, when the unexpected intelligence of the speedy departure of the frigate Atalanta determined them to return, after being seven days exposed to the dreadful attacks of the mosquitoes of Babaoya and Ujibar.

They had a fortunate passage, by the Pacific Ocean, to Acapulco, the western port of the kingdom of New Spain, famous for the beauty of its harbour, which appears to have been formed by earthquakes, for the misery of its inhabitants, and for its climate, which is equally hot and unhealthy.

Mr. Humboldt had originally the intention to remain only a few months in Mexico, and to hasten his return to Europe; his voyage had already been too much protracted, | 326 | his instruments, particularly the chronometers, began to be out of order, and every effort that he made to have new ones sent to him proved of no avail; add to this consideration, that the progress of science is so rapid in Europe, that, in a journey that lasts four or five years, great risk is run of contemplating the different phenomena under aspects, which are no longer interesting at the moment of publishing the result of your labours. Mr. Humboldt hoped to be in France in August or September, 1803, but the attractions of a country, so beautiful and so varied, as is that of the kingdom of New Spain, the great hospitality of its inhabitants, and the fear of the yellow fever, so fatal, from June to November, for those who come from the mountainous parts of the country, led him to stay a year in this kingdom.

Our travellers ascended from Acapulco to Tasco, celebrated for its mines, as interesting as they are ancient. They rise, by small degrees, from the ardent valley of Mescala and Papagayo, where the thermometer of Reaumur stands, in the shade, constantly from 28 to 31 (95 to 101 Fah.), in a region 6 or 700 toises above the level of the sea, where you find the oaks, the pines, and the fougere (fern) as large as trees, and where the European grains are cultivated. They passed by Tasco, by Cuerna Vacca, to the capital of Mexico. This city, of 150,000 inhabitants, is placed upon the ancient site of Texochtitlan, between the lakes of Tezcuco and Xochimilco, lakes which have lessened somewhat since the Spaniards have opened the canal of Hacheutoca, in sight of two snow-topped mountains, of which one, Hopocatepec, is even now an active volcano, surrounded by a great number of walks, shaded with trees, and by Indian villages.

This capital of Mexico, situated 1160 toises *above the sea*, in a mild and temperate climate, may doubtless be compared to some of the finest towns in Europe. Great scientific establishments, such the Academy of Painting, Sculpture, and Engraving, the College of Mines, (owing to the liberality of the Company of Miners of Mexico), and the Botanic Garden, are institutions which do honour to the government which has created them.

After remaining some months in the valley of Mexico, and after fixing the longitude of the capital, which had been laid down with an error of nearly two degrees, our travellers visited the mines of Moran and Real del Monte, and the Cerro of Oyamel, where the ancient Mexicans had the manufactory of knives made of the obsidian stone. They soon after passed by Queretaro and Salamanca to Guanaxoato, a town of fifty thousand inhabitants, and celebrated for its mines, more rich than those of Potosi have ever been. The mine of the count of Valenciana, which is 1840 French feet perpendicular depth, is the deepest and richest mine of the universe. This mine alone gives to its proprietor nearly six hundred thousand dollars annual and constant profit.

From Guanaxoato they returned by the valley of St. Jago to Valladolid, in the ancient kingdom of Michuacan, one of the most fertile and charming provinces of the kingdom. They descended from Pascuaro towards the coast of the Pacific Ocean to the plains of Serullo, where, in 1759, in one night, a volcano arose from the level, surrounded by two thousand small mouths, from whence smoke still continues to issue. They arrived almost to the bottom of the crater of the great volcano of Serullo, of which they analized the air, and found it strongly impregnated with carbonic acid. They returned to Mexico by the valley of Teluca, and visited the volcano, to the highest point of which they ascended, 14,400 French feet above the level of the sea.

In the months of January and February, 1804, they pursued their researches on the eastern descent | 327 | of the Cordilleras, they measured the mountains Merados, de la Puebla, Popocatyce, Izazihuatli, the great peak of Orizaba, and the Cofre de Perote; upon the top of this last Mr. Humboldt observed the meridian height of the sun. In fine, after some residence at Xalappa, they embarked at Vera Cruz, for the Havannah. They resumed the collections they had left there in 1801, and by the way of Philadelphia, embarked for France, in July, 1804, after six years of absence and labours. A collection of 6000 different species of plants (of which a great part are new) and numerous mineralogical, astronomical, chemical, and moral observations, have been the result of this expedition. Mr. Humboldt gives the highest praises to the liberal protection granted to his researches by the Spanish government.

Baron Humboldt was born in Prussia, on the 14th of September, 1769.

Baron Humboldt's Travels in America

The active mind of Mr. Humboldt had been engaged in experimental pursuits from a very early day. He had distinguished himself by his observations and inquiries in physics before he left Europe. In possession of juvenile vigour, a good constitution, an investigating and intrepid spirit, an understanding duly prepared by education, a facility in speaking modern tongues, amiable manners, and a communicative and social disposition, he obtained the patronage of the king of Prussia to enable him to enlarge the bounds of natural knowledge by travelling in America. For several years he has been engaged in exploring the southern part of the new world. Thence he came to Philadelphia to take a survey of some of its more northern regions. But after a few short tours through | 96 | some of the Middle States, he had been induced, by the impatience of the men of science, to return to Europe, and, with all convenient speed, publish the account of what he has done between the tropics. From this work, all the lovers of researches into nature have high expectations. By the following extract of a note to Dr. Mitchill, written just before the Baron sailed, there may be some hope entertained that he will revisit our continent: an expectation, however, which we fear almost too flattering to indulge.

„ *à Philadelphie, ce* 27 *Juin,* 1804.

"Sur le point de partir pour votre belle ville, j'ai été forcé de changer de projet et de m'embarquer pour la France, où je publierai vraisemblablement mes travaux des tropiques. Il est bien douloureux pour moi de perdre l'occasion de vous voir de près et de vous dire avec quel interêt j'ai lu les ideés ingenieuses que vous avez énoncé dans plusieurs de vos memoires physiologiques. Je me flatte qu'accoutumé de courir d'hemisphere en hemisphere ma belle etoile me ramenera dans ce pays aussi interessant sous les rapports moraux et physiques. Je souhaite que vous conservez une santé précieuse aux sciences! Agréez les sentimens de la plus haute consideration avec laquelle j'ai l'honneur d'être,

"Monsieur,

"Votre tres humble et tres obeissant serviteur,

"Humboldt."

On the Variations of the Terrestrial Magnetism in different Latitudes[1]

An inquiry into the laws of terrestrial magnetism is no doubt one of the most important questions that philosophers can propose. The observations already made on this subject have discovered phænomena so curious, that one cannot help endeavouring to solve the difficulties they present; but notwithstanding the efforts hitherto employed, it must be confessed that we are absolutely unacquainted with the causes of them.

It was difficult to obtain on this point any precise knowledge at a time when the construction of the compass was still imperfect; and so little time has elapsed since the discoveries of M. Coulomb have taught us to render them completely exact, it needs excite no astonishment that so few facts in the observations of travellers have been found worthy of confidence.

The expedition which M. Humboldt has terminated has procured for this part of philosophy a collection no less valuable than those with which he has enriched the other branches of human knowledge. Furnished with an excellent dipping-needle, constructed by Le Noir on the principles of Borda, M. Humboldt has made more than three hundred observations on the inclination of the magnet, and on the intensity of the magnetic force in that part of America which he traversed. By adding to these results those which he had already obtained in Europe before his departure, we shall have for the first time a series of correct facts on the variation of the magnetic forces in the northern part of the globe, and in some points of its southern part.

The friendship which M. Humboldt has testified for me since his return having given me an opportunity of communicating to him some experiments on this subject, which I made this year in the Alps, he immediately offered to unite his to mine in a memoir. But if friendship and a desire of making known new facts induced me to accept this offer of M. Humboldt, justice forbids me to take advantage of it to his prejudice; and I must here declare, that a very small part of it belongs to me. | 249 |

To place in order the facts and consequences which may be deduced from them, it is necessary to consider the action of terrestrial magnetism

[1] From the *Journal de Physique*.

under different points of view, corresponding to the different classes of the phænomena which it produces.

If we consider it first in general, we find that it acts on the whole surface of the globe, and that it extends beyond it. This last fact, which was doubted, has been lately proved by one of us, and particularly by our friend M. Gay-Lussac, during his two aërostatic voyages. And if these observations, made with all the care possible, have not shown the least sensible diminution in the intensity of the magnetic force, at the greatest height to which man can attain, we have a right to conclude that this force extends to an indefinite distance from the earth, where it decreases, perhaps, in a very rapid manner, but which at present is unknown to us.

If we now consider magnetism at the surface even of the earth, we shall find three grand classes of phænomena, which it is necessary to study separately, in order to have a complete knowledge of its mode of action. These phænomena are; the declination of the magnetic needle, its inclination, and the intensity of the magnetic force, considered either comparatively in different places or in themselves, paying attention to the variations which they experience. It is thus that, after having discovered the action of gravity as a central force, its variation, resulting from the figure of the earth, was afterwards ascertained in different latitudes.

The declination of the magnetic needle appears to be that phænomenon which hitherto has more particularly fixed the attention of philosophers, on account, no doubt, of the assistance which they hoped to derive from it in determining the longitude; but when it was known that the declination changes in the same place, in the course of time, when its diurnal variations were remarked, and its irregular traversing, occasioned by different meteors; in a word, the difficulty of observing it at sea, within one degree nearly, it was necessary to abandon that hope, and to consider the cause of these phænomena as much more complex and abstruse than had been at first imagined.

In regard to the intensity of the magnetic force in different parts of the earth, it has never yet been measured in a comparative manner. The observations of M. Humboldt on this subject have discovered a very remarkable phænomenon; it is the variation of the intensity in different lati- | 250 | tudes, and its increase proceeding from the equator to the poles.

The compass, indeed, which at the departure of M. Humboldt gave at Paris 245 oscillations in 10 minutes, gave no more in Peru than 211, and it constantly varied in the same direction; that is to say, the number of the oscillations always decreased in approaching the equator, and always increased in advancing towards the north.

These differences cannot be ascribed to a diminution of force in the magnetism of the compass, nor can we suppose that it is weakened by the effect of time and of heat; for, after three years' residence in the warmest countries of the earth, the same compass gave again in Mexico oscillations as rapid as at Paris.

There is no reason, either, to doubt the justness of M. Humboldt's observations, for he often observed the oscillations in the vertical plane perpendicular to that meridian; but by decomposing the magnetic force in the latter plane, and comparing it with its total action, which is exercised in the former, we may from these data calculate its direction, and consequently the direction of the needle.[2] This inclination, thus calculated, is found always conformable to that which M. Humboldt observed directly. When he made his experiments, however, he could not foresee that they would be subjected to this proof by which M. La Place verified them.

As the justness of these observations cannot be contested, we must allow also the truth of the result which they indi- | 251 | cate, and which is the increase of the magnetic force proceeding from the equator to the poles.

To follow these results with more facility it will be proper to set out from a fixed term, and it appears natural to make choice for that purpose of the points where the inclination of the magnetic needle is null, because they seem to indicate the places where the opposite action

---

[2] Let HOC (plate V. fig. 1.) be the plane of the magnetic meridian passing through the vertical OC; let OL be the direction of the needle situated in that plane, and OH a horizontal. The angle LOH will be the inclination of the needle, which we shall denote by I. If F represent the total magnetic force which acts in the direction OL, the part of this force, which acts according to OC, will be F sine of I: but the magnetic forces which determine the oscillations of the needle in any plane, are to each other as the squares of the oscillations made in the same time. If we denote then by M, the number of the oscillations made in 10′ of time in the magnetic meridian, and by P, the number of oscillations made also in 10′, in the perpendicular plane, we shall have the following proportion.

$$\frac{\text{F sin. I}}{\text{F}} = \frac{\text{P}^2}{\text{M}^2}$$

from which we deduce

$$\text{Sin. I} = \frac{\text{P}^2}{\text{M}^2}$$

The inclination then may be calculated by this formula, when we have oscillations made in the two planes.

In like manner, by making a needle oscillate successively in several vertical planes, we might determine the direction of the magnetic meridian.

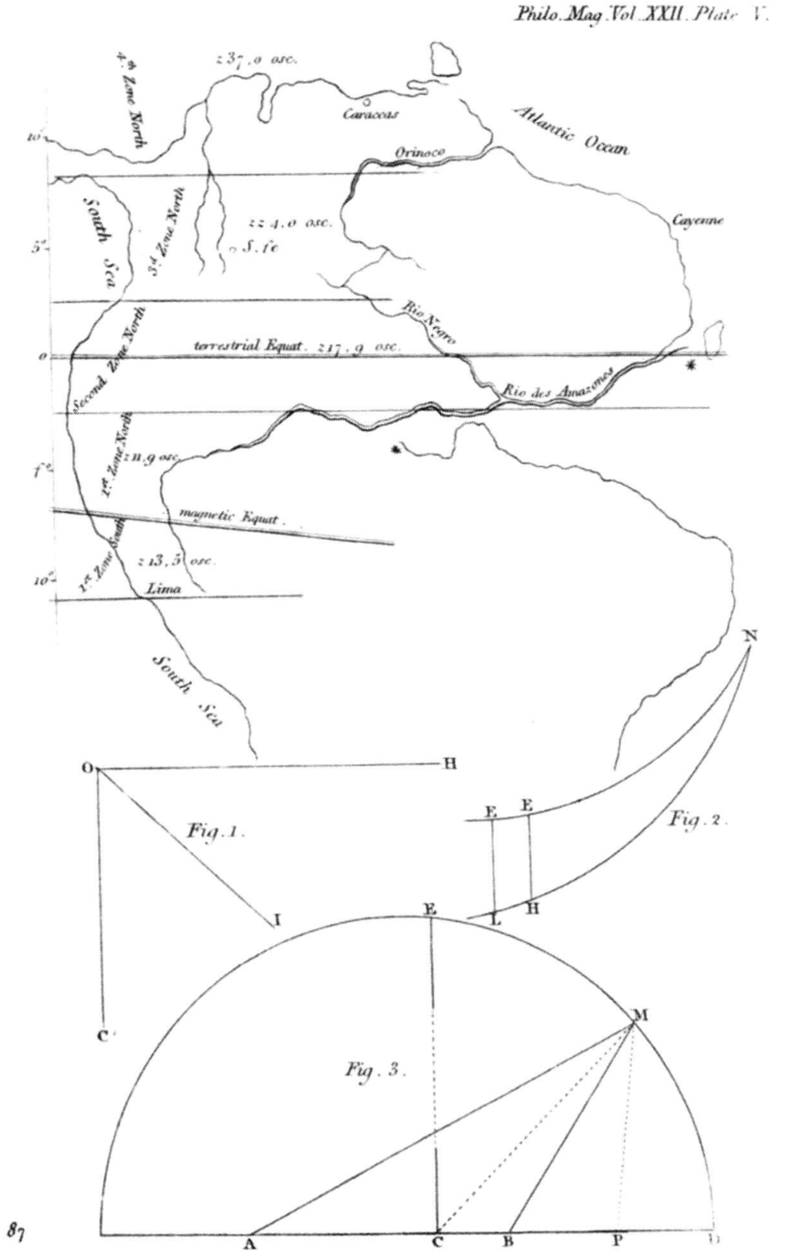

Plate V.

of the two terrestrial hemispheres is equal. The series of these points forms on the surface of the earth a curved line which differs very sensibly from the terrestrial equator, from which it deviates to the south in the

Atlantic Ocean and to the north in the South Sea. This curve has been called the *magnetic equator*, from its analogy to the terrestrial equator, though it is not yet known whether it forms exactly a great circle of the earth. We shall examine this question hereafter; at present it will be sufficient to say, that M. Humboldt found this equator in Peru about 7·7963° (7° 1′) of south latitude, which places it, for that part of the earth, nearly in the spot where Wilke and Lemonnier had fixed it.

The places situated to the north of that point may be divided into four zones; the three first of which, being nearer the equator, are about 4·5° (4°) of breadth in latitude; while the latter, more extensive and more variable, is 16° (14°). So that the system of these zones extends in America from the magnetic equator to 25·5556° (23°) of north latitude, and comprehends in longitude an interval of about 56° (50°).

The first zone extends from 7·7963° (7° 1″) of south latitude to 3·22° (2° 54′). The mean number of the oscillations of the needle in the magnetic meridian in 10′ of time is there, 211·9: no observation gives less than 211, or more than 214. From M. Humboldt's observations one might form a similar zone on the south side of the magnetic equator, which would give the same results.

The second zone extends from 2·4630° (2° 13′) of south latitude to 3·61° (3° 15′) of north latitude. The mean term of the oscillations is there, 217·9: they are never below 220, or above 226.

The fourth zone, broader than the other two, extends from 10·2778° (9° 15′) to 25·7037° (23° 8′) of north latitude. Its mean term is 237: it never presents any observation below 229, or above 240.

We are unacquainted, in regard to this part of the earth, with the intensity of the magnetic force beyond the latitude of 26° (23°) north; and on the other hand, in Europe, where we have observations made in high latitudes, we | 252 | have none in the neighbourhood of the equator: but we will not venture to compare these two classes of observations, which may belong to different systems of forces, as will be mentioned hereafter.

However, the only comparison of results, collected in America by M. Humboldt, appears to us to establish with certainty the increase of the magnetic force from the equator to the poles; and, without wishing to connect them too closely with the experiments made in Europe, we must remark, that the latter accord so far also with the preceding as to indicate the phænomenon.

If we have thus divided the observations into zones parallel to the equator, it is in order that we may more easily show the truth of the fact which results from them, and in particular to render the demonstration independent of those small anomalies which are inevitably mixed with these results.

Though these anomalies are very trifling, they are, however, so sensible, and so frequently occur, that they cannot be ascribed entirely to errors in the observations. It appears more natural to ascribe them to the influence of local circumstances, and the particular attractions exercised by collections of ferrugineous matters, chains of mountains, or by the large masses of the continents.

One of us, indeed, having this summer carried to the Alps the magnetic needle employed in one of his late aërial excursions, he found that its tendency to return to the magnetic meridian was constantly stronger in these mountains than it was at Paris before his departure, and than it has been found since his return. This needle, which made at Paris 83·9° in 10 minutes of time, has varied in the following manner in the different places to which it was carried:

Places of observation.	**Number of oscillations in ten minutes of time.**
Paris before his departure	83·9
Turin	87·2
On Mount Genêvre	88·2
Grenoble	87·4
Lyons	87·3
Geneva	86·5
Dijon	84·5
Paris, on his return	83·9

These experiments were made with the greatest care, conjointly with excellent observers, and always employing the same watch verified by small pendulums, and taking the | 253 | mean terms between several serieses of observations, which always differed very little from each other. It appears thence to result that the action of the Alps has a sensible influence on the intensity of the magnetic force. M. Humboldt observed analogous effects at the bottom of the Pyrenees; for example, at Perpignan. It is not improbable that they arose from the mass of these mountains, or the ferrugineous matters contained in them; but whatever may be the cause, it is seen by these examples that the general action of terrestrial magnetism is sensibly modified by local circumstances, the differences of which may be perceived in places very little distant from each other. This truth will be further confirmed by the rest of this memoir.

It is to causes of this kind, no doubt, that we must ascribe the diminution of the magnetic forces observed in some mountains; a diminution which, on the first view, might appear contrary to the results obtained during the last aërial voyages. This conjecture is supported by several observations of M. Humboldt. By making his needle to oscillate on the mountain of Guadaloupe, which rises 676 metres (338 toises) above Santa-Fé, he found it in 10 minutes of time give two oscillations less than in the plain. At Silla, near Caracas, at the height of 2632 metres (1316 toises) above the coast, the diminution went so far as five oscillations; and, on the other hand, on the volcano of Antisana, at the height of 4934 metres (2467 toises), the number of oscillations in 10 minutes was 230; though at Quito it was only 218: which indicates an increase of intensity. I observed, indeed, a similar effect on the summit of Mount Genêvre, at the height of 1600 or 1800 metres (8 or 900 toises), as may be seen by the numbers which I have already given; and it was on this mountain that I found the greatest intensity of the magnetic force. I saw on the hill of La Superga, in the neighbourhood of Turin, an example of these variations equally striking. Observing, with Vassali, on this hill, at the elevation of about 600 metres (300 toises), we found 87 oscillations in 10 minutes of time. On the side of the hill we had 88·8 oscillations; and at the bottom, on the bank of the Po, we obtained 87·3. Though these results approach very near to each other, their difference is, however, sensible, and fully shows that their small variations must be considered as slight anomalies produced by local circumstances.

This examination leads us to consider the intensity of magnetism on the different points of the surface of the globe, as subject to two sorts of differences. One kind are | 254 | general: they depend merely on the situation of the places in regard to the magnetic equator, and belong to a general phænomenon, which is the increase of the intensity of the magnetic forces in proportion as we remove from the equator: the other kind of variations, which are much smaller and altogether irregular, seem to depend entirely on local circumstances, and modify either more or less the general results.

If we consider terrestrial magnetism as the effect of an attractive force inherent in all the material particles of the globe, or only in some of these particles, which we are far from determining, the general law will be, the total result of the system of attraction of all the particles, and the small anomalies will be produced by the particular attractions of the partial systems of the magnetic moleculæ diffused irregularly around each point; attractions rendered more sensible by the diminution of distance.

It now remains to consider the inclination of the magnetic needle in regard to the horizontal plane. It has long been known that this inclination is not every where the same: in the northern hemisphere the needle inclines towards the north; in the southern towards the south; the places where it becomes horizontal form the magnetic equator; and those where the inclination is equal, but not null, form on each side of that equator curved lines, to which the name of magnetic parallels has been given from their analogy to the terrestrial parallels. One may see in several works, and particularly in that of Lemonnier, entitled *Lois du Magnetism*, the figure of these parallels and their disposition on the face of the earth.

It evidently results from this disposition that the inclination increases in proportion as we recede from the magnetic equator; but the law which it follows in its increase has not yet, as far as appears to us, been given. To ascertain this law, however, would be of great utility; for the inclination seems to be the most constant of all the magnetic phænomena, and it exhibits much fewer anomalies than the intensity. Besides, if any rule, well confirmed, could be discovered on this subject, it might be employed with advantage at sea to determine the latitude when the weather does not admit an observation of the sun; which is the case in various places during the greater part of the year. We have some reason to expect this application when we see the delicacy of that indication in the observations of M. Humboldt, where we find 0·65° (35′ 6″) of difference between two towns so near each other as Nismes and Montpellier. These motives have induced us to study with great | 255 | interest the series of observations made by M. Humboldt in regard to the inclination; and it appears to us that they may be represented very exactly by a mathematical hypothesis; to which we are far from attaching any reality in itself, but which we offer merely as a commodious and sure mode of connecting the results.

To discover this law, we must first exactly determine the position of the magnetic equator, which is as an intermediate line between the northern and the southern inclinations. For this purpose we have the advantage of being able to compare two direct observations; one of Lapeyrouse, and the other of M. Humboldt. The former found the magnetic equator on the coasts of Brasil at 12·1666° (10° 57′) of south latitude, and 28·2407° (25° 25′) of west longitude, counted from the meridian of Paris. The latter found the same equator in Peru at 7·7963° (7° 1′) of south latitude, and 89·6481° (80° 41′) of west longitude, also reckoned from the same meridian. These data are sufficient to calculate the position of the magnetic equator, supposing it to be a great circle of the terrestrial sphere; an hypothesis which appears to be conformable to observations.

The inclination of this plane to the terrestrial equator is thus found to be equal to 11·0247° (10° 58′ 56″), and its occidental node on that equator is at 133·3719° (120° 2′ 5″) west from Paris, which places it a little beyond the continent of America, near the Gallipagos, in the South Sea; the other node is at 66·6281° (59° 57′ 55″) to the east of Paris, which places it in the Indian Seas.[3] | 256 |

We do not give this determination as rigorously exact: some corrections might no doubt be made to it, had we a greater number of observations equally precise; but we are of opinion that these corrections would be very small; and it will be seen hereafter that, independently of the confidence which the two observations we have employed deserve, we have other reasons for entertaining this opinion.[4]

It is very remarkable that this determination of the magnetic equator agrees almost perfectly with that given long ago by Wilke and Lemonnier.

---

[3] To calculate this position let NEE′ (Plate V. fig. 2.) be the terrestrial equator; NHL the magnetic equator, supposed also to be a great circle; and HL the two points of that equator, observed by Messrs. Humboldt and Lapeyrouse. The latitudes HE, LE′, and the arc EE′, which is the difference of longitude of these two points, is known: consequently, if we suppose HE = b, LE′ = b', EE′ = v, EN = x, and the angle ENH = y, we shall have two spherical triangles NEH, NE′L, which will give the two following equations:

$$\text{Sin. } x = \frac{\text{tang. } b \text{ cot. } y}{R} \quad \text{sin.}\,(x+v) = \frac{\text{tang. } b' \text{ cot. } y}{R}$$

from which we deduce

$$\text{Sin.} \frac{(x+v)}{\text{sin. } x} = \frac{\text{tang. } b'}{\text{tang. } b}$$

and developing

$$\text{cot. } x = \frac{\text{tang. } b'}{\text{tang. } b \text{ sin. } v} - \frac{\text{cos. } v}{\text{sin. } v}$$

Let us now take an auxiliary angle φ, so that we man have

$$\text{tang. } \varphi = \frac{\text{tang. } b \text{ sin. } v}{\text{tang. } b'}$$

and we shall have

$$\text{tang. } x = \frac{\text{sin. } v \text{ sin. } \varphi}{\text{sin. } (v - \varphi)}$$

By these equations we may find x, and then y, by any of the first two.

[4] Since this memoir was read, we have collected new information which confirms these first results. Lapeyrouse, after having doubled Cape Horn, fell in a second time with the magnetic equator in 18′ north lat. and 119° 7′ of longitude west from Paris. He was therefore very near the node of the magnetic equator, such as we have deduced it from observations. This fact establishes in a positive manner two important consequences: first, that the preceding determinations require only very slight corrections; and

The latter in particular, who for want of direct observations had discussed a great number of corresponding observations, indicates the magnetic equator in Peru towards 7°1/3 of south latitude; and M. Humboldt found it in the same place at 7·7963° (7° 1′); besides, Lemonnier's chart, as well as that of M. Wilke, indicates for the inclination of the magnetic meridian 12·22° (about 11°), and they place the node about 155° 56′ (140°) of west longitude, reckoned from the meridian of Paris. Can it be by chance, then, that these elements, found more than 40 years ago, should accord so well with ours founded on recent observations? or does the inclination of the magnetic equator experience only very small variations, while all the other symptoms of terrestrial magnetism change so rapidly? We should not be far from admitting the latter opinion, when we consider that the inclination of the magnetic needle has changed at Paris 3° during 60 years since it has been observed; and that at London, according to the observations of Mr. Graham, it has not changed 2° in 200 years; while the declination has varied more than 20° in the same interval, and has passed from east to west: but, on the other hand, the observation of the inclination is so difficult to be made with exactness, and it is so short a time since the art of measuring it with precision was known, | 257 | that it is perhaps more prudent to abstain from any premature opinion on phænomena the cause of which is totally unknown to us.

To employ the other observations of M. Humboldt in regard to the inclination, I first reduced the terrestrial latitudes and longitudes reckoned from the magnetic equator. The latter, being reckoned from the node of that equator in the South Sea, I could first perceive by these calculations that the position of that plane determined by our preceding researches was pretty exact; for some of the places, such as Santa-Fé and Javita, where M. Humboldt observed inclinations almost equal, were found nearly on the magnetic parallel, though distant from each other more than 6·6666° (6°) in longitude.[5]

When these reductions were made, I endeavoured to represent the signs of the inclinations observed, and to leave as little to chance as possible. I first tried a mathematical hypothesis conformable enough to the idea which has hitherto been entertained in regard to terrestrial magnetism.

I have supposed in the axis of the magnetic equator, and at an equal distance from the centre of the earth, two centres of attractive forces, the

---

the second, that the magnetic equator is really a great circle of the earth, if not exactly at least very nearly.—*Note of the Authors.*

[5] This confirms what we have already said, that the magnetic equator is sensibly a great circle of the earth.—*Note of the Authors.*

one austral and the other boreal, in such a manner as to represent the two opposite magnetic poles of the earth: I then calculated the effect which ought to result from the action of these centres in any point of the surface of the earth, making their attractive force reciprocally vary as the square of the distance; and in this manner I obtained the direction of the result of their forces, which ought to be that also of the magnetic needle in that latitude.

[To be continued.] | 299 |

On the Variations of the Terrestrial Magnetism in different Latitudes

[Concluded from p. 257.]

The calculation is as follows:—I suppose that the point B (fig. 3.) is the north magnetic pole of the earth, and that the point A is the south magnetic pole: I suppose also that there is in the point M, at the surface of the earth, a molecula of the austral fluid which is attracted by B and repelled by A in the inverse ratio of the square of the distance; and I require what will be the direction of the power resulting from these two forces acting on that molecula. It is evident that this direction will be that also which would be assumed in the point M by the needle of a compass freely suspended: for, in consequence of the smallness of the needle in comparison of the radius of the earth, the lines drawn from its points to one centre, B or A, may be considered as parallel, especially if the points A and B are near the centre of the earth; which is the case with nature, as may be seen.

I shall first suppose that the earth has a spherical figure, and that the two poles A and B are equal in force; I shall then examine how far the latter supposition agrees with the results observed.

Let AM then = D′, BM = D, CP = x; PM = y, the angle MCP = u, CA = CB = a, and I shall make a = Kr; r being equal to the radius of the earth, and K a constant but indeterminate quantity.

Let X, Y, also be the forces which attract M in a direction parallel to the axes of the co-ordinates, and β the angle which the resulting force makes with the axis ABC.

We shall first have the following equations, in which F is the magnetic force, at a distance equal to unity.

These equations determine the direction of the magnetic needle in regard to each point M, the distance of which from the magnetic equator is known; but it is seen that this direction depends on the quantity K, which represents the distance of the magnetic centres from the centre of the earth: this distance being expressed in parts of the terrestrial radius, we must therefore first determine this quantity from observations.

To do it in the manner of approximation, and thus acquire a first idea of the value of K, I have chosen an observation made by M. Humboldt at Carichana in 7·2978° (6° 34′ 5″) of north latitude counted from the terrestrial equator, and 78·111° (70° 18′) west longitude reckoned from | 301 | the meridian of Paris; which gives 16·526° (14° 52′ 25″) of latitude counted from the magnetic equator, and 53·7390° (48° 21′ 53″) of west longitude proceeding from the node formed by that equator with the equator of the earth. The inclination of the magnetic needle was observed in that place by M. Humboldt in the month of Messidor, year 8, and found to be equal to 33·78° of the centigrade division.[6] A comparison of this result, with the other observations of M. Humboldt, shows that it may be indeed considered as agreeing to that latitude.

To make use of it I have successively given to K different values in the formula: I have calculated the inclinations resulting from that latitude; and, comparing these results with that which M. Humboldt really observed, the progress of the errors naturally led me to the most proper supposition. The following is a table of these trials:

Values of K.	**Inclinations of the Needle.**	**Errors.**
K = 1	7·73°	26·04°
K = 0·6	18·80	14·97
K = 0·5	22·04	11·73
K = 0·2	29·38	4·39
K = 0·1	30·64	3·13
K = 0·01	31·04	2·73
K = 0·001	31·07	2·7

The first value of K would place the centre of the magnetic forces at the surface of the earth and the poles of the magnetic equator. It is

[6] All the measures of inclination which I have given in this memoir will be expressed, like those of M. Humboldt, in decimal parts of a quadrant.

$$X = \frac{F}{D^2}\cos.\mathrm{MBD} - \frac{F}{D'^2}\cos.\mathrm{MAD};$$

$$D'^2 = y^2 + (x+a)^2 = r^2 + 2\,\mathrm{axis} + a^2$$

$$Y = \frac{F}{D^2}\sin.\mathrm{MBD} - \frac{F}{D'^2}\sin.MAD;$$

$$D^2 = y^2 + (x-a)^2 = r^2 - 2\,\mathrm{axis} + a^2,$$

| 300 | or by putting for the cosines their values:

$$X = \frac{F(x-a)}{D^3} - \frac{F(x+a)}{D'^3}$$

$$Y = \frac{Fy}{D^3} - \frac{Fy}{D'^3}$$

and as we have

$$\mathrm{tang}.\beta = \frac{Y}{X},$$

we shall have also

$$tang.\beta = \frac{\frac{y}{D^3} - \frac{y}{D'^3}}{\frac{x-a}{D^3} - \frac{x+a}{D'^3}} = \frac{y(D'^3 - D^3)}{x(D'^3 - D^3) - a(D'^3 + D^3)};$$

and putting for x, y, and a, their values, r cos. u; r sin. u; Kr;

$$\mathrm{tang}.\beta = \frac{\sin.\ u}{\cos.\ u - K\left(\frac{D'^3 + D^3}{D'^3 - D^3}\right)};$$

$$D'^2 = r^2(1 + 2\,K\cos.u + K^2);$$

$$D^2 = r^2(1 - 2\,K\cos.u + K^2);$$

which gives the system of the two equations,

$$\mathrm{tang}.\beta = \frac{\sin.\ u}{\cos.\ u - K\left(\frac{D'^3 + D^3}{D'^3 - D^3}\right)}$$

$$K\left(\frac{D'^3 + D^3}{D'^3 - D^3}\right) = \frac{(1 + 2\,K\cos.u + K^2)^{\frac{3}{2}} + (1 - 2\,K\cos.u + K^2)^{\frac{3}{2}}}{(1 + 2\,K\cos.u + K^2)^{\frac{3}{2}} - (1 - 2\,K\cos.u + K)^{\frac{3}{2}}}K.$$

seen that this supposition cannot be admitted, because it would give an increase of inclination much less rapid than that indicated by observations. The case is the same with the following results, which place the centres of action on the terrestrial radius at different distances from the centre of the earth: but it is seen also, in general, that they approach more and more to the truth in proportion as this distance becomes less; which evidently shows that the two centres of action of the magnetic forces are situated near the centre of the earth. All the other observations of M. Humboldt would also lead to the same consequence. The most proper supposition would be to make K null, or so small that it would be needless to pay attention to it; which amounts to the same thing as to consider the two centres of action placed, as we may say, in the same molecula. The result, indeed, obtained in this manner is the most exact of all; it is equal to 31·0843°: this value is still a little less | 302 | than that which M. Humboldt observed, and the difference is equal to 2·69; but it must be considered also that the formula from which we derive these values supposes the position of the magnetic equator to be perfectly determined; but it may not be so with the utmost exactness, according to the only two observations of Lapeyrouse and Humboldt, which we have employed. It is therefore by studying the progress of the formula, and comparing it with the observations, that we are able to appreciate it justly; after which we may think of remedying the small errors with which it may be accompanied.

To obtain the result I have here mentioned, and which is, as it were, the limit of all those which may be obtained by giving to K different values, it is to be remarked that the quantity

$$K\left(\frac{D'^3 + D^3}{D'^3 - D^3}\right)$$

or

$$K\frac{(1 + 2\,K\cos.u + K^2)^{\frac{3}{2}} + (1 - 2\,K\cos.u + K^2)^{\frac{3}{2}}}{(1 + 2\,K\cos.u + K^2)^{\frac{3}{2}} - (1 - 2\,K\cos.u + K^2)^{\frac{3}{2}}}$$

becomes 0/0 when K is null, but by applying to it the methods of known quantities it will be found that its value in this supposition is really determinate and equal to

$$\frac{1}{3\cos.\ u}$$

By substituting this in the formula we shall have

$$\text{tang.}\,\beta = \frac{\sin.\ u}{\cos.\ u - \dfrac{1}{3\cos.\ u}}$$

an equation which may be reduced to this form:

$$\text{tang.}\,\beta = \frac{\sin.\ 2u}{\cos.\ 2u + \dfrac{1}{3}}\ ;$$

which will easily give the value of β: and when this value is known we shall have the inclination I, by the following formula:

$$\text{I} = 100 + u - \beta,$$

which will serve throughout the whole extent of the two hemispheres.

From the progress I have traced out it is seen that the preceding formula is not merely an empyric construction of observations; on the contrary, it is totally independent, and only supposes the inclination of the magnetic needle to be | 303 | produced by a magnet, infinitely small, placed in the centre of the terrestrial surface; but by calculating from this formula the inclination for the different latitudes, I have found precisely the same numbers as M. Humboldt observed either in Europe or in America: and it is not his observations only that are represented in this manner; but those which have been made in Russia, and at Kola in Lapland, during the last transit of Venus, are also comprehended under the same law. This is proved by the table annexed to this memoir, in which I have calculated the observations of Mallet and Pictet, with a part of those of M. Humboldt, which I took at random, but, however, in such a manner as to include all the rest in the intervals.

It is seen that the results of the formula deviate very little from the observations; but these differences may be rendered still smaller. By examining, indeed, the progress of the errors, it is seen that the numbers given by calculation are a little too small in America for the low latitudes, and a little too great for the high latitudes; which shows that the whole may be allowed, with some slight modifications, either by changing, however little, the node and inclination of the magnetic equator, which two observations cannot determine with the utmost exactness, or by displacing ever so little our small magnet, leaving, however, its centre in the plane of the magnetic equator, and placing it in such a manner that it shall be a little nearer America than Europe. It is by the observations themselves, when we shall have a greater number, that we must be guided in these small corrections.

In a word, it must not be expected that we can represent in a rigorous manner, by a mathematical law, all the inclinations observed; for the phænomenon of the inclination, though more regular than the other magnetic effects, is not free from some anomalies: this may be easily seen on constructing the curve given by the observations themselves. Thus, for example, the inclination observed at Popayan is 0° 10′ greater than at St. Carlos del Rio Negro, though the magnetic latitude of the latter is 0·6852° (3° 7′) greater. The case is the same with observations made at Javita and Santa-Fé. Other anomalies are discovered in the comparative progress of the observations and formula. This is the case in regard to Carichana, St. Thomas de la Guyane, and Carthagena. The increase of the inclination from the first to the second of these points is by no means in harmony with the increase from the second to the third; and if we | 304 | compare together the intensities observed in these different places, the anomalies they exhibit announce in some measure those which the inclination ought to experience.

The cause of these anomalies becomes evident from what has been already remarked; they are merely the effect of local circumstances, and arise from the small systems of attraction by which the general phænomena are modified. This must be sensible in particular for that part of America which M. Humboldt travelled over, and which is traversed throughout its whole length by the grand chain of the cordillera of the Andes. It is also in these places that the most considerable differences exist. Popayan, for example, is situated near the volcanoes of Sotara and Pourace; it is joined to basaltic mountains abounding with magnetic iron. Near Sulumito, to the east of Popayan, these basaltic columns have very striking poles: in like manner Mexico is situated at the height of 1160 toises on the ridge of the grand cordillera of Lenschtitlan: the ground there is covered with porous basaltes and amygdaloids, which are almost all charged with magnetic iron. Must not all these causes have a sensible influence on the inclination of the magnetic needle; and must not the different dispositions of the ferruginous masses, or their change of state, in consequence of the action of nature, produce also variations? M. Humboldt made on this point a decisive observation: the earthquake of the 4th of November 1799 changed at Cumana the inclination of the needle. On the 1st of November it was 43° 65′; on the 7th it was only 42° 75′; and ten months after it returned to 42° 85′: but it did not regain its former value; the intensity of the magnetic force was not changed by the effect of this earthquake.

It is proved, then, by these observations, that local circumstances may have on the inclination a sensible influence; and this influence is remarked in the countries traversed by M. Humboldt.[7]

It appears, therefore, that the mathematical hypothesis which we have employed really expresses the law of nature at least to the north of the magnetic equator; for, though the first results observed towards the south seem to bend to it also, the uncertainty under which we are in regard to the true cause of these phænomena must stop our conjectures, and | 305 | prevent us from extending too far the consequences of the laws which we observe.[8]

From the preceding results, we may calculate the points where the axis of the magnetic equator pierces the terrestrial surface; for their latitudes are equal to the complements of the obliquity of that equator, and their meridian is at 100° of longitude from its nodes. The north magnetic pole is found also at 97·7975° (79° 1′ 4″) of north latitude, and at 33·3719° (30° 2′ 5″) of longitude west from Paris, which places it to the north of America. The other magnetic pole, symmetric to the preceding, is situated in the same latitude south, and at 66·6281° (149° 67′ 55″) of longitude east from Paris, which places it amidst the eternal ice: indications entirely analogous to those of Wilke and Lemonnier.

If we could reach these poles, the compass would be seen vertical; but if any confidence can be placed in the law which we have discovered,

---

[7] We can observe that the anomalies are sensible in particular in the islands.—*Note by the Authors of the Memoir.*

[8] Since this memoir was read, we can advance something more positive. Observations made at the Cape of Good Hope, Cape Horn, and New Holland, by different navigators, are very exactly represented by our formula; and it follows, that it extends also to the austral hemisphere. We hope soon to have numerous and very exact observations on the inclination of the needle in that part of the earth. But we have thought it our duty to add to our table such results as relate to it, and which we have been able to procure. We have inserted also two observations on the intensity, made with great care by M. Rossel, during the expedition of d'Entrecasteaux, which are very important, as they prove that the terrestrial magnetic force increases also in the austral hemisphere in proportion as one removes from the equator.—*Note by the Authors of the Memoir.*

this would be the only difference which would be observed in regard to the inclination, and we should be still as far distant as in Europe from the real centres which produce it. This result might appear to be of such a nature as to diminish the interest one might have in visiting these horrid regions, had we not also the hope of discovering there new phænomena in regard to the intensity of the magnetic force, and the influence of meteors.

These consequences do not entirely accord with the opinion pretty generally received, and which ascribes the increase of the magnetic effects towards the north to the great quantity of iron dispersed throughout these regions; but it appears to us that this opinion is not agreeable to the truth. The cordillera of the Andes contains an enormous quantity of magnetic iron: the native iron of Chaco, that problematic mass analogous to that of Pallas, and those of Xacateras in Mexico, is found even under the tropics.[9] | 306 |

On seeing the inclinations of the compass so exactly represented in our hypothesis, we endeavoured to discover whether it could he applied also to the intensities observed by M. Humboldt; but we found that it did not apply. It gives, indeed, an increase of the magnetic forces from the equator to the pole; but this increase, which at first is too slow, becomes afterwards too rapid: I have not yet been able to try whether the small displacement of the terrestrial magnet will contribute towards representing them better: but it must be remarked, that the series of the intensities is extremely whimsical, and contains an infinite number of anomalies; so that local phænomena may have on this phænomenon a much more sensible influence than on the inclination.

On reviewing the results which we have given in this memoir, it is seen that we have first determined the position of the magnetic equator by direct observations, which had never been done before; we have then proved that the magnetic force increases in proceeding from that equator to the poles: in the last place, we have given a mathematical hypothesis, which when reduced to a formula satisfies all the inclinations hitherto observed.

Supposing, as we have done in this formula, the small corrections of which it is susceptible, its utility becomes evident, either for making known, in the course of time, the variations which may take place in the action of the terrestrial magnetism, or to ascertain or even foresee the value of the inclination, which in a great many cases is of great importance.

---

[9] We may now add to the preceding considerations this decisive fact, that the intensity also increases when one approaches the south pole.— *Note by the Authors of the Memoir.*

For example, near the magnetic equator, the increase or diminution of the inclination will indicate to a vessel on a voyage whether she has gained or lost in latitude by currents. This knowledge of the latitude is sometimes as important as that of longitude. On the coasts of Peru, for example, the currents tend from Chiloé to the north and north-east with such force, that one may go from Lima to Guayaquil in three or four days, and two, three, and sometimes five months are necessary to return. It is consequently of the greatest importance for vessels coming from Chili which stretch along the coast of Peru, to know their latitude. If they go beyond the port to which they are bound they must work to the southward, and every day's progress requires often a month of return. Unfortunately, the fogs which prevail during four or five months on the coasts of Peru prevent navigators from distinguishing the form of the coast: nothing is seen but the summits of the | 307 | Andes, and that of the peaks which rise above that stratum of vapours; but the figure of it is so uniform that pilots fall into mistakes. They often remain twelve or fifteen days without seeing the sun or the stars, and during that interval they come to anchor, being afraid of overshooting their port: but if we suppose that the inclination of the magnetic needle in the ports to the south of Lima is known, for example at Chancay, Huaura, and Santa, the dipping needle will show whether it be, in regard to Lima, to the south or the north. It will show at the same time opposite what point of the coast a vessel is; and this indication will be attended with more exactness than one could hope for, because in these seas the inclination varies with extraordinary rapidity. M. Humboldt, to whom we are indebted for these remarks, observed in these seas the following values:

Places.		**South Latitudes.**		**Inclinations.**
Huancey	–	10° 4′	–	6,80°
Huaura	–	11 3	–	9,00
Chancay	–	11 33	–	10,35

These observations prove that the error of three or four degrees in the inclination in these seas would produce but a degree of error in latitude; and, on account of the tranquillity of the Pacific Ocean, the inclination may be observed to within a degree nearly. Frequent instances of such results may be seen in books of voyages. In like manner, if one knew exactly the inclination at the mouth of the Rio de la Plata, it would be very useful to navigators, who, when the Pamperos blow, remain fifteen or eighteen days without seeing the heavenly bodies, and go on different tacks for fear of losing the parallel of the mouth of that river.

In a word, the inclination may indicate also the longitude in these seas: and this method may be employed when others fail. A vessel which sails there in the direction of a parallel could not find its longitude either by a chronometer or the declination of Halley, unless a star could be seen in order to take an horary angle or the magnetic azimuth. The dipping needle, then, throws light on the longitude amidst the thickest fogs. We point out this method as one of those which have only a local application; but hitherto little attention has been paid to it. These ideas may be extended and rectified by able navigators.

In general, if the inclination of the needle, and the law we have tried to establish, could be depended on, to observe the inclination and the terrestrial latitude would be sufficient to determine also the longitude: but we have not yet examined the extent of the errors of which this method may | 308 | be susceptible, and consequently we confine ourselves to a mere indication of it.

The phænomenon of the inclination has in maritime observations a particular and very remarkable advantage, namely, that of not being subject to those great progressive variations which affect the declination. Without repeating what we have already said above on the supposed constancy of this phænomenon, it may be remarked that our formula even affords a new proof that it may comprehend in the same law the observations made thirty-six years ago in Lapland, those which Lacaille brought back in 1751 from the Cape of Good Hope, and those which M. Humboldt has lately made in America.

In short, when we tried to represent the inclinations in different latitudes by the supposition of a magnet infinitely small, very near the centre of the earth and perpendicular to the magnetic equator, we did not pretend to consider that hypothesis as any thing real, but only as a mathematical abstraction useful to connect the results, and proper to ascertain in future whether any changes exist. In regard to the declination and intensity, we freely confess that we are entirely unacquainted with their laws or their causes; and if any philosopher is so fortunate as to bring them to one principle, which explains at the same time the variations of the inclination, it will no doubt be one of the greatest discoveries ever made. But this research, exceedingly difficult, requires perhaps before it be attempted more observations, and in particular more precise observations than have hitherto been collected. For this reason we thought we might present to the class the preceding researches, imperfect as they are, begging it to receive them with indulgence. Should we be so happy as to find that our results appear of any utility, we propose to unite all the exact observations which have been made on this subject, in order to give the utmost degree of precision to the law we have discovered.

Northern Magnetic Hemisphere.									
Names of the Observers.	**Places of Observation.**	**Latitudes reckoned from the Terrestrial Equator.**	**Longitudes reckoned from the Meridian of Paris.**	**Latitudes reckoned from the Magnetic Equator.**	**East Longitudes reckoned from the Node of the Magnetic Equator in the South Sea.**	**Number of Oscillations in 10′ of Time.**	**Inclinations given by Theory**	**Inclinations given by direct Observation.**	**Differences in Degrees of the Centesimal Division.**
		Old Division.	Old Division.	Old Division.	Old Division.		Centig. Division.	Centig. Division.	
Humboldt –	Magnetic equator in Peru – – –	7° 1′ 0″ S	80° 41′ 0″ W	0° 0′ 0″	40° 17′ 56″	211	0,000	0,00	0,00°
Lapeyrouse	Magnetic equator at sea between Brazil and the Island of Ascension – – – –	10 57 0	25 25 0	0 0 0	95 33 56	– – –	0,000	0,00	0,00
Humboldt –	Tompenda – –	5 31 4	80 27 0	1 30 54	39 52 51	213	3,3642	3,55	0,1858
The same –	Loxa – – – –	4 0 0	81 12 0	2 54 27	38 55 0	212	6,440	6,00	+ 0,44
The same –	Cuença – – –	2 54 9	80 43 0	4 3 44	39 13 52	214	8,97	9,35	— 0,38
The same –	Quito – – – –	3 13 17	80 15 0	6 46 59	39 17 52	218	14,87	14,85	+ 0,02
The same –	St. Antonio – –	0 0 0	80 12 0	7 0 53	39 18 52	220	15,29	16,02	— 0,73
The same –	Popayan – – –	2 24 33 N	78 45 0	9 36 16	40 24 27	223	– – – –	23,20	– – – –
The same –	St. Carlos del Rio Negro – – –	1 52 4	70 10 0	10 13 14	49 6 35	216	22,0278	23,10	— 1,0722
The same –	Javita – – – –	2 49 0	70 30 0	11 7 40	48 39 6	218	23,87	27,00	— 3,13
The same –	Esmeralda – –	3 13 26	68 38 0	11 45 45	50 29 15	217	– – – –	28,85	– – – –
The same –	Santa Fe di Bagota	4 36 5	76 37 0	12 5 13	42 17 13	226	25,76	26,97	— 1,21
The same –	Carichana – –	6 34 5	70 18 0	14 52 25	48 21 53	227	31,08	33,77	— 2,69

(continued)

Northern Magnetic Hemisphere. *(continued)*

Names of the Observers.	**Places of Observation.**	**Latitudes reckoned from the Terrestrial Equator.**	**Longitudes reckoned from the Meridian of Paris.**	**Latitudes reckoned from the Magnetic Equator.**	**East Longitudes reckoned from the Node of the Magnetic Equator in the South Sea.**	**Number of Oscillations in 10′ of Time.**	**Inclinations given by Theory**	**Inclinations given by direct Observation.**	**Differences in Degrees of the Centesimal Division.**
The same –	St. Thomas of Guyana – –	8 8 24	66 26 0	16 54 18	52 7 26	222	34,77	39,00	— 4,23
The same –	Carthagena – –	10 25 57	78 2 0	17 38 43	39 55 13	240	36,07	39,17	— 3,10
The same –	Mexico – – –	19 26 2	101 22 0	22 35 14	14 36 41	242	44,87	46,85	— 1,98
De Rossel in 1791 Humboldt in 1799	St. Croix in Teneriffe – – – –	28 28 30	18 37 0	39 12 40	72 0 26	238	64,9975	69,85	— 4,35
The same –	Atlantic Ocean –	38 52 0	16 20 0	49 28 22	106 30 10	242	74,29	75,76	— 1,47
The same –	Paris – – – –	48 50 15	0 0 0	57 57 0	128 22 47	245	80,69	77,62	+ 3,07
Euler junior	Petersburgh in 1755 – – –	59 56 23	27 58 0 E	64 41 0	173 30 25	– – –	85,21	81,67	+ 3,54
Mallet – –	Kola, in Lapland, in 1769 – –	68 52 30	30 40 30 E	71 44 36	179 9 29	– – –	89,59	86,39	+ 3,20
Lord Mulgrave	In an island near Spitzbergen in 1773 – – –	79 50 00	7 38 00 E	83 9 50	127 40 5	– – –	96,1882	91,1111	+ 5,0771

Southern Magnetic Hemisphere.

Names of the Observers.	Places of Observation.	Latitudes reckoned from the Terrestrial Equator	Longitudes reckoned from the Meridian of Paris.	South Latitudes reckoned from the Magnetic Equator.	East Longitudes reckoned from the Node of the Magnetic Equator in the South Sea.	Inclinations given by Theory	Inclinations given by direct Observation.	Differences in Degrees of the Centesimal Division.	Number of Oscillations in 10′ of Time.
		Old Division.	Old Division.	Old Division.	Old Division.		Centig. Division.	Centig. Division.	
Humboldt –	Lima – – –	12° 2′ 31″ S	79° 33′ 0″ W	4° 48′ 36″ S	41° 42′ 51″	10,6145	11,10	0,4855°	219
Derossel, expedition of Entre-casteaux	Sourabaya in the Isle of Java –	7 14 23	110 21 28 E	15 37 22	228 56 50	32,4660	28,5185	+ 3,9475	204
Bayli in 1775	Cape of Good Hope	33 55 30	16 10 0 E	26 15 34	131 38 53	49,58	47,78	+ 1,8	– – –
Lapeyrouse	Bay of Talcaguara	36 42 0	75 53 0 W	28 42 14	49 0 5	52,8889	55,555	— 2,6667	– – –
The same –	In sight of the Isle of Patagonians	52 21 26	69 38 0 W	44 30 3	57 13 52	70,04	68,89	+ 1,15	– – –
Derossel, expedition of Entre-casteaux	New Holland –	43 34 30	144 36 33 W	54 12 43	263 21 18	78,7037	77,9667	— 0,737	265

The results comprehended in this table extend from 38° 55′ to 263° 21′ 18″ east longitude, reckoned from the node of the magnetic equator in the South Sea: consequently, they comprehend more than 224°, and their agreement shows that in this extent the magnetic equator is sensibly a great circle of the earth. We have not yet calculated the observations for the 36 degrees of longitude which would complete the contour of that equator.—For the southern hemisphere we have given the observations, made with great care, by M. Derossel, during the expedition of d'Entrecasteaux. It results from them that the intensity of the terrestrial magnetism increases also in that hemisphere, when one removes from the magnetic equator. The inclination observed by M. Derossel at Teneriffe being exactly the same as that observed by M. Humboldt eight years after, this agreement has allowed us to render comparable the results obtained by these philosophers in regard to the intensity: for this purpose we have multiplied the results of M. Derossel, by giving the numbers which he and M. Humboldt observed at Teneriffe. The result of this calculation will be found in the column of oscillations. It is there again seen that this phænomenon is very much modified by local circumstances, and incomparably more than the inclination. The increase of the intensity deduced from the observations of M. Humboldt is less than that which would result from our hypothesis; and that given by the observations of M. Derossel is too great: which proves that nothing can be determined in regard to the law of this increase. The influence of local circumstances on the inclination is particularly sensible in the isles. The declination and intensity of the magnetic forces experience there also sensible anomalies. This fact is indicated by several observations, and in particular by those which M. Derossel made at Sourabaya in the Island of Java. In a word, by comparing the results of our formula with the observations of different navigators, the latter must be examined with critical accuracy, and must not be admitted but when they agree with each other and with those of other navigators: without this precaution we should every moment fall into great errors, occasioned by the incoherency of the results: besides, we present the preceding only as a first approximation.

Experiments on the Torpedo, by Messrs. Humboldt and Gay-Lussac. Extracted from a Letter from M. Humboldt to M. Berthollet, dated Rome, 15 Fructid. Year 13[1]

The phænomena of electric fish ought to be the subject of our renewed researches, with a view to the opinion of many philosophers, who conceive that they are capable of being explained upon the principles of that beautiful theory with which Volta has enriched science. You well know, my respected friend, what must have been our anxiety to procure the torpedo; and you will perhaps be astonished that I should be so long in writing to you respecting it. At Genoa we found some of them, but we were then without instruments. At Civita Vecchia we searched for them in vain. At last, during our stay at Naples, we procured them very frequently, of great size, and very vigorous. I shall relate to you, in this letter, the series of experiments which M. Gay-Lussac and I have instituted upon the action of the torpedo (Raja Torpedo of Linnæus). M. de Buch, a German mineralogist, was present at our experiments. I present to you the results which they afforded, and relate the facts without introducing any theoretical notions.

Our experiments were principally intended to discover those conditions in which the torpedo is unable to commun- | 357 | icate those shocks termed electric, although the feeling is very different from that which is occasioned by the discharge of a Leyden phial. Having at hand no other work besides that of Aldini, in which he has combined into one view the beautiful researches of Geoffroy with those of Spallanzani and Galvani, we shall not have it in our power to compare our own labours with those of preceding philosophers.

1. Although the power of the torpedo cannot be compared to that of the gymnotus, it is not less capable of occasioning disagreeable sensations. A person much in the custom of receiving electric shocks, supports with some difficulty the shock of a torpedo 14 inches long, and in a state of perfect vigour. The gymnotus communicates its influence under water, unless when much weakened.

[1] From *Annales de Chimie*, no. 166.

M. Gay-Lussac has observed, that the action of the torpedo in this condition is not perceptible until it is raised above the surface of the water. It is with this fish, as with frogs on which Galvanic experiments are made: the circumstances under which the contraction takes place, vary according to the degree of excitability in the organs.

2. I have remarked, while in South America, that the gymnotus gives the most frightful shocks without making any external motion of the eyes, head, or fins: it moves no more than one person communicating an idea or sensation to another. But the torpedo, on the contrary, moves its pectoral fins in a convulsive manner before each shock; and the violence of the stroke is always proportioned to the extent of the surface of contact.

3. The organs of the torpedo, or gymnotus, cannot be discharged by us at will, like a Leyden phial or a Galvanic pile; nor does the electric fish uniformly communicate a shock when touched. It must be irritated that it may give its stroke; for this action depends upon the will of the animal, which in all probability does not always keep its electric organs charged: it charges them, however, with astonishing celerity, and is thus able to give a long series of shocks.

4. The shock is felt, provided the animal is disposed to give it, when a single finger is applied to a single surface of the electric organs; or when the two hands are placed one on the upper and the other on its under surface at the same time. And in either of these cases the shock is equally communicated, whether the person be insulated or not.

5. If a person while insulated touches the torpedo with his finger, it is indispensably necessary that the finger be in | 358 | immediate contact; for no shock is communicated when a conducting body, such as a piece of metal, is interposed between the finger and the organs of the fish. Thus we may touch the animal with a key, or other metallic instrument, and experience no shock in consequence.

6. M. Gay-Lussac having observed this important fact, we placed the torpedo on a plate of metal, so that the inferior surface of its electric organs was in contact with the metal. The hand which supported the plate felt no shock, although another person in a state of insulation irritated the animal, and when the convulsive motions of its pectoral fins plainly indicated very powerful discharges of its electric fluid.

7. If on the contrary a person support the torpedo placed on a metallic plate, with his left hand as in the preceding experiment, and with his right touches the upper surface of the electric organ, then a violent shock is felt in both arms at the same moment.

8. The same feeling is experienced when the fish is placed between two plates of metal, the edges of which do not touch, and the person applies a hand to each plate at the same instant.

9. But if in the preceding experiment there exists an immediate communication between the edges of the two plates, no shock is felt in the arms; for in this case the chain of connection between the two surfaces of the organ is formed by the plates, and the new communication established by applying the two hands to the plates becomes altogether inefficient.

10. The most delicate electrometer does not indicate the state of electricity of the organs of the torpedo: it is no way affected by any method which we can have recourse to, either by bringing it near to the organs, or by insulating the fish, covering it with a plate of metal, and then forming a communication by means of a wire between the plate and the condenser of Volta. Nothing shows here, as in the gymnotus, that the animal can modify the state of electricity of surrounding bodies.

11. As electric fish act while in a state of health with the same power under water as in the air, we examined the conducting properties of this fluid. A number of persons having formed a circle of connection between the upper and under surfaces of the organs of the torpedo, no shock was experienced till they had moistened their hands with water. The shock is equally felt when two persons who have their right hands applied to the torpedo, instead of taking hold of | 359 | each other's left hands, plunge a pointed piece of metal into a drop of water placed upon an insulating body.

12. By substituting flame instead of a drop of water, the communication is interrupted, and no sensation is experienced until the two pieces of metal touch each other within the flame.

13. We must also observe that no shock will take place either in air or under water, unless we immediately touch the body of the electric fish. They are unable to give their stroke through a layer of water, however thin; a fact which is the more remarkable, as we know that in Galvanic experiments, where the frog is placed under water, it is sufficient to bring the silver forceps near to the muscles, and that the contraction takes place when the layer of water interposed is one or two millimetres in thickness.

Such, my respected friend, are the principal observations which we have made upon the torpedo. Experiments 4th and 10th prove that the electric organs of these animals manifest no tension or excess of charge. We should rather be inclined to compare their action to a chain of small Leyden phials than to the pile of Volta. As some communication is always necessary for the occurrence of a shock, and having received strokes from the gymnotus through very dry cords, I conclude that in the case where this powerful animal appeared to give these violent shocks without the existence of any communication, it must have arisen from my imperfect insulation. If the torpedo acts by poles, by an electrical equilibrium which tends to reestablish itself, experiments 5th and 6th seem to prove that these poles exist near one another upon the same surface of the organ: for we feel the shock by touching a single surface with the finger. A plate interposed between the hand and the organ (6.) re-establishes of itself the equilibrium, and the hand which supports this plate feels nothing, because it is out of the current of the electric influence. But if we suppose a number of poles of different descriptions upon each surface of the organ, why is it that, by covering these with two metallic plates the edges of which do not touch, and placing the hands upon these plates, the equilibrium is re-established through the medium of the arms? Wherefore, it may be asked, does not the positive electricity of the inferior surface seek, at the moment of explosion, the negative electricity of the neighbouring pole? and wherefore does it find it only in the superior surface of the electric organ? These difficulties are perhaps not in- | 360 | surmountable, but the theory of these *vital actions* requires still further research. Geoffroy[2] has proved that rays, which do not exhibit any marks of electricity, possess organs very analogous to those of the torpedo. The least injury of the brain of this animal prevents its electrical action. The nerves, therefore, without doubt, act the chief part in the production of these phænomena; and the physiologist, who takes a general and enlarged view of the vital actions, would with reason oppose the ideas of the philosopher, who conceives he can explain the whole by the contact of the albumino-gelatinous pulp with the tendinous septa which nature has combined in the formation of the organs of the torpedo.

[2] For M. Geoffroy's paper, see Phil. Mag. vol. xv. p. 126.

Experiments on the Eudiometrical Means, and on the Proportion of the Constituent Principles of the Atmosphere[1]

If philosophers and chemists are at present agreed respecting the nature of the constituent principles of the atmosphere, they are not yet so with regard to their absolute quantity. Since Scheele and Lavoisier, who had found twenty-seven hundredths of oxygene in the air, numerous experiments, for which we are indebted to | 232 | Messrs. Messrs. Cavendish, Marti, Berthollet, Fourcroy, and Dery, have greatly modified this proportion, by fixing it at between twenty and twenty-three hundredths. Nevertheless it must be admitted that these proportions are still far remote from that degree of accuracy which the present state of Science allows of; or, if these limits are well established, it must be concluded that the atmosphere experiences considerable oscillations in its composition. Although with respect to most chemical phenomena, it is not necessary that we should have an accurate knowledge of the absolute quantity of its principles, this knowledge is not less interesting in itself, than important with regard to the history of our globe. If all the facts of geology tend to prove that the earth is no more what it has formerly been; that the highest mountains have been covered with water, and that the Arctic regions have nourished animals which at present are to be found only between the tropics, these very changes prove of what great utility it would be to future ages actually to ascertain the physical state of the globe at the present day; and even though the grand catastrophes which it has already experienced should never more be reiterated, it is possible that it may undergo gradual modifications which man would not be able of himself to appreciate, unless he found incontestible documents in the annals of science. It would therefore be of the highest importance to fix in an authentic manner the grand phenomena of nature, which are to be supposed subject to variation; such as the intensity of the magnetic forces, the elevation of the barometer at the level of the sea, that of the sea itself, the medium temperature of every climate, and the proportion of the constituent principles of the atmosphere. We have directed our attention to the latter question, and although we have not yet solved it in a manner entirely | 233 | satisfactory

[1] From the *Journal de Physique*.

to ourselves, we venture to give an account of the commencement of the investigation which we have undertaken upon this subject, and the researches to which it has led us.

But the eudiometrical means which must serve to determine the proportions of the constituent principles of the air are not all susceptible of equal precision, and some distinguished chemists give the preference to a means which is rejected by others. It therefore was indispensably necessary that we should put to the trial the known eudiometrical methods, in order that we might be enabled properly to appreciate their value; for we are well convinced that accuracy in experiments depends less upon the faithful observation of the divisions of an instrument, than upon the accuracy of the method itself. Although the nitrous gas appears indeed at first to be the most uncertain eudiometric means that can be chosen, we have convinced ourselves that by combining its action with that of the sulphate of iron, or of the oxygenated muriatic acid and of potash, it is capable of indicating with great precision the quantity of oxygene contained in the air. All the eudiometrical means would give the same results, if we were equally acquainted with them all, and it is only because it is very difficult to make all the corrections they allow of, that we naturally give the preference to such as present less occasion for them, although they are not always the most simple in their use. We shall therefore first give an account of the eudiometrical researches with which we have occupied ourselves, and then apply them to the analysis of the atmospheric air and of that obtained from water under different circumstances, or placed in contact with it. We find it necessary again to repeat that we shall not treat the question which we have proposed to ourselves in so extensive a | 234 | manner as it merits. Compelled to break off our researches before we were able to bring them to their termination, our purpose is only to give an account of their principal results. During nearly two months since the time that we commenced them in one of the laboratories of the Polytechnic-school at Paris, we pursued them, in spite of the coldness of the weather, which is very disagreeable in researches of this kind, with so much the greater assiduity, as M. Humboldt found a peculiar interest in them. In the year 6, he had presented to the Institute two memoirs on the analysis of the air, which contain a great number of experiments, which he now considers (it is he himself that avows it) not only as very inaccurate, but also as very justly controverted by Mr. Davy, and by a chemist with whose particular friendship we are both honoured, namely, M. Berthollet. Zealous for the advancement of science, M. Humboldt, when he commenced his researches, wished to substitute in the place of those investigations of his early youth, others founded upon more solid foundations, and as he desired to have me for his associate in them; I ought to consider myself the more highly honoured by his proposition from our having been, since his return from the tropics, linked together in the closest bonds of friendship.

OBSERVATIONS ON SEVERAL OF THE EUDIOMETRIC MEANS.

It is not our intention in this memoir to record all the researches which we have undertaken upon different eudiometric means: most of them are still in too imperfect a state; but having occupied ourselves more particularly with the alkaline sulphurets, and especially with hydrogene gas, we shall here exhibit the result of our observations on these two eudiometrical means. | 235 |

Although the alkaline sulphurets have in general a sufficiently-constant action for the analysis of the air, which has caused them to be justly preferred to the other eudiometrical means, they nevertheless present some sources of error, which it is indispensibly necessary we should be well acquainted with, in order to justify us in placing an entire confidence in their results. It has long been believed that they have no action upon the azote; and although M. Marti announced, as early as in the year 1790, that they absorb this gas, no farther attention has since been paid to this property. It is true M. Marti at the same time announced, that by saturating them with azote we might employ them to advantage for the analysis of the air, and constantly obtain for the oxygene a proportion comprehended between 0.21 and 0.23. On the other hand, this chemist not having indicated the details of his experiment with sufficient precision, M. Berthollet, who repeated it after him under different circumstances, announced in his Chemical Statics, that he had not found the alkaline sulphurets to possess the property of absorbing azote, whereby he gave new sanction to their employment for the analysis of the air. When we began to employ this means we placed great confidence in it, and had nothing to object to it but the great length of time which it requires, and which had long rendered it a desirable object that, notwithstanding its accuracy, some other might be substituted in its place that were not attended with the same inconveniences; but we soon discovered that it did not always act in an uniform manner, and herein we were favoured by accident.

Having placed 100 parts of atmospheric air in contact with a solution of sulphuret of potash, made with hot water, in three vessels of unequal capacities, we observed at the end of eight days that the air had lost twenty-three parts of its volume in one of the vessels, and 23,6 26,0 | 236 | in the two others. This great inequality gave us at first much surprise; but having remarked that the absorption had been most considerable in the largest flask, we suspected that azote had been absorbed, and, in order the better to confirm our suspicion, we repeated the same experiment, employing two vessels more unequal in capacity, but otherwise placed in the same circumstances. At the end of ten days we found that in the small flask the absorption had been only 22,5 parts, whilst in the large one it was 30,6. But the most decisive experiment which we made on this subject, was the placing of a solution of sulphuret of potash, which had been heated to ebullition, in contact with azote in unequal vessels, when it was discovered that the absorption was proportionate to their capacities. It would therefore be possible to cause a determinate quantity of atmospheric air to be absorbed by a solution of alkaline sulphuret, and to make it be considered as pure oxygene, if it were supposed that the whole diminution of volume were owing to the oxygene gas. But if, instead of employing a solution of sulphuret made with hot water, we employ one made with cold, as M. Berthollet has always done, the solution of the azote takes place no longer, at least not in a perceptible manner, and the results of the analysis of the air made by this means become then much more susceptible of comparison. This variable action of the alkaline sulphurets, dissolved at different temperatures, requires to be farther elucidated, which we are about to do by citing phenomena of an analogous but more comprehensible nature.

As water always holds in solution a certain quantity of air in which the proportion of oxygene is more considerable than it is in the atmospheric air, it happens that when we heat it, or dissolve a salt in it, a part of its air is disengaged from it whilst another is retained, which may | 237 | be separated from it by a more intense heat. If, therefore, we place this water, which has been deprived of its air by the last-mentioned means in contact with atmospheric air, it will absorb, in returning to its original temperature, a quantity equal to that which it has lost; and if we were not aware of this absorption, but judged upon appearances, we should suppose that the water alone, or charged with salt, had made the analysis of the air. Thus M. Hetter has very recently announced that a solution of muriate of soda absorbed all the oxygene of the air, though on repeating his experiment with a highly-charged solution of the same salt, but made with cold water, we did not find the slightest difference between the ordinary atmospheric air and that which had been in contact with the solation of muriate of soda for a month and a half.

The very same thing happens with a sulphuret as with a salt. At the moment of its solution in water, a portion of air is expelled, and an equilibrium of saturation is established between the water, the sulphuret, and the air which it holds in solution, so that if the circumstances are not changed, there is now no reason why it should absorb a fresh quantity of air; but if we heat the solution, there is disengaged from it a part of the gas which it contained, and, in returning to its original temperature, it is necessary that it absorb what it had lost, in order that its equilibrium may be re-established.[2] We | 238 | think, therefore, that the difference between the results of Messrs. Marti and Berthollet may be explained by the difference of the circumstances under which they operated; but it appears to us, that M. Marti believed that it was the nature of the sulphuret to absorb azote, whereas it does not absorb it all, but rather prevents the water with which it has been boiled from absorbing as much as it would do without it.

Thus if we take care to dissolve the sulphurets in cold water, and leave them for some time in contact with azote or with air, we may employ them with advantage for the analysis of the atmosphere. We shall however observe, that as they are attended with the inconvenience of requiring a great length of time before their action is completed, we are obliged to have recourse to the corrections of the thermometer and barometer, which often are very uncertain. The best way of remedying this inconvenience is undoubtedly to follow the method of Messrs. Berthollet and Marti, which consists in placing, for comparison, upon water a determinate quantity of air, in order to judge, from its variations of volume, of that of the air which we analyze; but this method has not appeared to us to be attended in practice with all the advantage which it seems to promise.

We must still remark with regard to all the eudiometrical means, where the absorbing substance is solid or liquid, that if we commit an error, either in observing the divisions of the instrument, or in the appreciation of the uncertainties of the method, this error necessarily falls altogether upon the quantity of oxygene; and since with all possible exactness we cannot answer for its amounting to much less than a hundredth part, it would follow that the proportion of oxygene contained in

[2] The absorption of which we mean to speak in this place is independent of that of oxygene by the sulphuret, which is thereby converted into sulphate. But as the sulphuret absorbs the oxygene which the water holds in solution, it will happen that the water will be able to absorb a larger quantity of azote; so that if we employed a solution made with cold water, and very recently prepared, there would be a still greater diminution of volume than that which is owing to the absorption of the oxygene. We say *very probably*; for we have not yet made the experiment.

the air cannot be determined within this quantity. In fact, we find that chemists who have employed such means | 239 | have found very considerable variations in the quantity of oxygene of the air; and M. Marti himself, who appears to have made a great number of experiments with the alcaline sulphurets, and who was acquainted with the precautions which they require, fixes it between 0,21 and 0,23. We shall see hereafter that the eudiometrical means, in which the substance which combines with the oxygene is gaseous, admit of a much greater precision.

As we had proposed to ourselves from the commencement of our investigation, to ascertain whether the eudiometer of Volta was fit to be employed for the analysis of the air, we principally directed our attention to it. This instrument has been accused of being fallacious, of indicating too small quantities of oxygene in the air; but it appeared to us that supposing it required corrections, we might by appreciating them, as well as the law of their variations, render it very exact and convenient. With this view we proposed to ourselves the following questions:

1. When a mixture of hydrogene gas and oxygene gas is inflamed in Volta's eudiometer, can the absorption of one of the gases be complete?
2. Is the product of their combination of a constant nature?
3. What is the exact proportion of the two gases for forming water?
4. What are the limits of error that Volta's eudiometer admits?

We must examine these four questions in succession, but first of all we think it incumbent upon us to give an account of the manner in which we prepared the gases which we employed in our experiments.

To be continued in our next. | 303 |

Experiments on the Eudiometrical Means, and on the Proportion of the Constituent Principles of the Atmosphere. By Messrs. Humboldt and Gay-Lussac.

(Continued from Page 239.)

The oxygene gas we obtained from the super-oxygenated muriate of mercury. To obtain it we employed a glass retort, to which had been soldered, by the flame of a lamp, the curved tube through which the gas was to pass; and, in order that we might have it as exempt from azote as possible, we filled the retort about one-fourth with water. This water being entirely reduced to vapour before the decomposition of the salt, soon expelled all the air out of the retort; but, in order to prevent the absorption which would have taken place before the disengagement of the oxygene gas, we plunged the ex- | 304 | tremity of the tube in a saucer filled with mercury, which we removed as soon as the gas began to be disengaged. In order that the oxygene may not, passing through the water, expel azote from it, we convey it directly into the upper part of the recipient which is to receive it, by means of a tube bent at a right angle, which at the one end ascends to the upper part of the recipient, and at the other is adapted to the first tube by means of a cork stopper, common to both. This process, which is very simple in its application, is particularly advantageous for the gases which are soluble in water, such as the carbonic acid gas, the azote oxyd gas, &c. We obtained our hydrogene gas by decomposing water by means of zinc and muriatic or sulphuric acid diluted with about six parts of water: we took care to fill exactly with acid the vessel from which the gas was to be disengaged, and not to make it pass through the water; but notwithstanding all these precautions, our oxygene left with the sulphuret four thousandth parts of azote, and the hydrogene, analysed by other means, shewed six thousandths. After these elucidations let us proceed to the questions which we have proposed to ourselves to resolve, beginning with this:

When we inflame a mixture of oxygene and hydrogene gas in the eudiometer of Volta, can the absorption of one of the gases be complete?

In order to ascertain whether all the oxygene or all the hydrogene could be entirely destroyed, we thought that if two gases were perfectly pure, or if we knew their degree of purity, and that their absorption must be complete, we should be able to find the same proportion for the principles of water, whether the hydrogene or the oxygene were predominant. In fact, by detonating mixtures of 300 parts hydrogene and 100 oxygene, and | 305 | of 200 of the first and 200 of the second, in which the hydrogene and oxygene alternately predominate, and making the corrections due to the impurity of the gases, we obtained very nearly the same proportion. Although the absorption of the two gases might be complete, it were however possible that the proportions obtained, in making them alternately predominate, might not be identical, and this would be the case if, according to the predominance of either of the gases, there were formed an oxygenated or hydrogenated water; but since the proportions have become identical, it must necessarily be concluded, that the hydrogene and the oxygene were entirely absorbed. But though the absorption of the two gases may be complete under certain circumstances, it must not be supposed that it is so with any quantities; there are not only such proportions of hydrogene and oxygene, or of their mixture with azote, or even with any other gas, that it is impossible to inflame them by means of the electric spark; but there are also others with which the inflammation having been commenced, stops before the combustion is completed. We proceed to cite experiments to this effect, which appear to us to be conclusive.

We mixed 100 parts of hydrogene with 200, 300—900 of oxygene, and inflamed them by the electric spark: with these different proportions the absorption constantly amounted to 146 parts; but with 1000 of oxygene it was at once reduced to 55; with 1,200 and 1,400 it was reduced to 24 and 14, and with 1,600 it was reduced to 0; that is to say, no inflammation took place. These different results are exhibited in the following table. | 306 |

Vol. VIII.—Second Series. R r 100

Hydrogene.	**Oxygene.**	**Absorption.**
100 – – –	200 – – –	146
100 – – –	300 – – –	146
100 – – –	600 – – –	146
100 – – –	900 – – –	146
100 – – –	950 – – –	68
100 – – –	1000 – – –	55
100 – – –	1200 – – –	24
100 – – –	1400 – – –	14
100 – – –	1600 – – –	0*

*The absorptions 68, 55, 24, and 14, are possibly not exact within two or three hundredths, for our instruments being too small for the corresponding proportions, we were obliged to measure them several times; but this is of no moment with respect to the phenomenon in general.

What is striking in these different experiments is to see, 1, a constant absorption, with very different proportions, change suddenly into a decreasing absorption; 2, the combustion of hydrogene gas which had commenced, stop before it was completed; 3, that there are such proportions of hydrogene and oxygene as cannot be inflamed. These different phenomena will be in some measure explained in the sequel; but in the mean time we shall remark, that there are even very extended proportions with which the combustion of the hydrogene gas may be complete.

The above-mentioned phenomena are not peculiar to the hydrogene and oxygene gases mixed together, under the circumstances of which we have been speaking: they also take place when we inflame 100 parts of oxygene with 200, 300—1000, &c. of hydrogene; only it then happens, that the term when the absorption ceases to be constant is more remote; and to comprehend the reason | 307 | of this it is sufficient to observe, that in this case there disappear about 300 parts by the inflammation, whereas there disappeared only half the quantity in the preceding experiments.

Azote gas and carbonic acid gas present also analogous results. If, for example, we inflame a mixture of 900 parts of azote, 100 of hydrogene, and 100 of oxygene, the absorption, which ought to be 146 parts, if the combustion were complete, was in one experiment, which is that which we take as example, only 50 parts, though in others we have seen it a little above or below this quantity. With inferior proportions of azote, we have constantly had the same absorption of 146. Although the azote appears here to comport itself like the oxygene, since with 100 of hydrogene and 1000 of oxygene, we had nearly the same result as with 100 of hydrogene, 100 of oxygene, and 900 of azote, we shall not draw from hence any inference, because we have not sufficiently multiplied and varied our experiments. Nevertheless, those which we have made tend to prove, that when determinate proportions of oxygene gas and hydrogene gas are mixed with different gases, the absorption may be constant as far as to a certain point, beyond which it diminishes very rapidly.

The absorption of the oxygene and of the hydrogene being complete in determinate proportions, and not so in others, it will always be possible, when a gaseous mixture is given, which alone would not be able to inflame, to reduce it to another with which the absorption of one of the gases would be complete, by adding to it oxygene or hydrogene, or even both together.

The combustion of the 100 parts of hydrogene in the preceding experiment not having been complete, we analyzed the residue. 100 parts, placed in contact with | 308 | phosphorus, diminished by 7 in the space of four hours, an evident proof that the residue contained oxygene. In order to ascertain whether it had retained hydrogene, we inflamed, in Volta's eudiometer, a mixture of 200 parts of the preceding residue, 200 of oxygene gas, and 200 of hydrogene gas; in all 600 parts. After the inflammation 312 parts had disappeared; and as, according to experiments of which we shall give an account hereafter, 100 of pure oxygene require for their saturation 200 of hydrogene gas, the absorption which, with the hydrogene gas which we have employed, ought only to have been 292 parts, amounted to 312, the residuum must necessarily have furnished a sufficient quantity of it to carry the absorption from 292 to 312; that is to say, it must have contained 13,3 parts. Now calculation shews that it ought to have contained 12; it is therefore clearly proved, that though inflammation took place, the combustion was not complete; and that all the hydrogene did not enter into combination, since we have found that which had not been absorbed in the residuum. We must observe, that in all cases in which the absorption was not complete the inflammation was languid.

From comparing, in the inflammation of hydrogene and oxygene gas, the effects of electricity with those of a high temperature, we have been led to believe that the inflammation produced by the electric shock might very likely be owing to the heat produced by the instantaneous compression which the electric spark occasions in its passage. In fact we know, from our own experience, that the inflammation of a mixture of hydrogene and oxygene gas depends solely upon the temperature when this inflammation is produced by heat. For if we cause this mixture to pass very slowly through a tube heated very gradually, from its extremity to its central part, without | 309 | opposing the free dilation of the gases, the inflammation will take place as soon as the temperature shall be raised to a sufficient degree. This being admitted as fact, that the inflammation of the oxygene and the hydrogene gas takes place only at a certain temperature, let us see what passes in their inflammation by the electric spark. When this passes through a mixture of oxygene and hydrogene, it displaces it by its rapid passage, which does not permit the gaseous particles to communicate to each other the motion as quickly as they have received it; hence results a very strong instantaneous compression, which produces an elevation of temperature superior to that which is requisite for the combination of the gases, and the inflammation being thus commenced, must be propagated very rapidly.

According to this mode of accounting for the effects of electricity, we thought that when a weak spark produces only an imperfect combustion in a mixture of hydrogene and oxygene gas, a stronger one would produce a more complete combustion; but whether it was that we did not employ a sufficient brisk electricity, or that we did not multiply our experiments sufficiently, we obtained no sensible differences in employing the spark of an electrophorus, three decimeters in diameter, or the shock of a highly-charged Leyden flask; but the construction of our eudiometer did not permit us to draw very brisk sparks, on which account we shall reserve our opinion respecting the influence of the force of the electricity in the inflammation of the hydrogene and oxygene, till we shall have made farther researches upon the subject.

In the above-described experiment on the inflammation of a mixture of 900 parts of azote, 100 of oxygene, and 100 of hydrogene, the absorption was not so considerable as it ought to have been, and we have proved | 310 | that the residuum ought to contain what had escaped combustion; that is to say, it ought to be composed of six parts of hydrogene, eight of oxygene, and eighty-six of azote in the hundred. Therefore, since the combustion was interrupted when these proportions took place, it may be concluded that another electric spark would no more be able to inflame this mixture. Consequently, in the atmosphere, which contains much less than six hundredths of hydrogene, the electric spark will not be able to inflame it, or if it does it at the place of its passage, by reason of its great force, the inflammation will not be able to propagate itself, but will be in a manner confined to the places which it traverses. Hence, finally, the inflammation of hydrogene gas, by lightning, and *à fortiori* by weaker charges of electricity, will not serve to explain the igneous meteoric phenomena; or if these phenomena are actually results of the inflammation of hydrogene gas, we must conclude that there are more than six hundredths of it in the air at the moment when they are produced.; which is contrary to all probability, especially when we recollect that air, collected at a very great elevation, presented no appreciable quantity of hydrogene above that contained in atmospheric air collected at the surface of the earth.

But if every time that we cause an electric spark to pass into a mixture of hydrogene and oxygene, or of azote, hydrogene, and oxygene, which is not capable of inflaming, there is actually produced a local and instantaneous heat by the compression which the spark occasions in its passage, it is possible that by directing a succession of sparks into one of the mixtures of which we have been speaking, a slight local inflammation might be produced each time upon the passage of the spark, and that thus it might be practicable to destroy a determin- | 311 | able quantity of hydrogene inveloped in a large proportion of azote and oxygene, or of oxygene only. What seems to confirm this supposition is, that it is well known that ether and ammoniac, which are decomposed by heat when they are made to pass in vapours through a red-hot tube, are likewise decomposed by repeated electric shocks. It would also be interesting to know, whether it be possible to inflame by the electric spark a proper mixture of oxygene and hydrogene, after having dilated it by means of the pneumatic machine. If its inflammation by the electric spark really depends upon the heat which this produces by compression, it would be natural to suppose, that when these gases are dilated, the compression by the spark being less considerable, the heat which is produced by it must also be much slighter, and that there may be a degree of dilatation of the gases at which the inflammation cannot take place. We have not yet had time to try these different experiments; but we do not abandon our design to attempt them, which we hope we shall be able to do shortly.

To recapitulate: there exist certain proportions of hydrogene and of oxygene, or of these two gases with azote, in which combustion can be complete. There exist also others, in which it stops spontaneously before being completed; and, finally there are proportions in which it cannot take place at all. The hydrogene gas which escapes the combustion is found again entire in the residue. When we cannot produce by the electric spark a complete inflammation of the hydrogene gas, or even when we cannot commence it, nothing more is necessary than to augment the proportions of the oxygene or of the hydrogene. The igneous meteoric phenomena cannot be the result of the inflammation of the hydrogene gas, because in the regions where the principal of them |312| are supposed to take place, such as the sudden and abundant torrents of rain which sometimes succeed a clap of thunder, it would be necessary that there should then be more than six hundredths of hydrogene in the atmosphere, without which the inflammation could not take place; besides which, only the quantity exceeding this proportion could pass into inflammation.

We may account for the cases in which the combustion was not complete according to the laws of the affinities, by saying, that when one of the gases becomes very predominant it may defend the other by its affinity, and guard it in part from combustion. Although this affinity may be very weak, we conceive, with M. Berthollet, how the quantity of gas may compensate for it; and if there be in the different gases peculiar properties of stopping the combustion sooner or later, this may be explained by their different nature. But when we consider the case in which the hydrogene is mixed with oxygene only, and suppose the phenomena of its combustion with different proportions of oxygene to depend upon affinity, how explain the sudden transition from a constant absorption to a decreasing absorption, when it is agreed that if the hydrogene can be prevented from the combination by the oxygene, the effect of the latter must follow a regular law? How conceive that these two gases, after having been placed in circumstances favourable to their combination, can by their affinity maintain themselves in their elastic state, when they might form a much more dense combination, namely, water? How conceive, finally, that an affinity which produces a very great condensation and saturation, can be inferior to an affinity which produces no change in the dimensions of the two gases, no saturation? Hydrogene and oxygene, in whatever state they may be, have the same degree of affinity, as | 313 | this affinity is measured by their capacity of saturation; only the state in which they are may be more or less favourable to their combination. Now, to say that hydrogene and oxygene have a greater affinity in the state of gas than in the liquid state, is to say that their molecules attract each other more when they are very remote than when they are very near to one another. These objections against an explanation founded solely upon the affinities, having appeared to us to be of some weight, we have endeavoured to present one which, in our opinion, did not involve the same difficulties.

All combustible bodies require in general a certain elevation of temperature, in order to combine with oxygene. Carbon, for example, is not converted into carbonic acid until it is red-hot; and this same substance, which at a high temperature can continue to burn when it is exposed to a current of aqueous vapour, is extinguished as soon as it is immersed in water. This principle, that bodies in general require a certain elevation of temperature in order to burn, being once admitted, let us suppose that we have a body which burns in a given volume of atmospheric air, and that the temperature necessary for the combustion is maintained solely by the heat due to the absorption of the oxygene: let us also suppose, that at the commencement of the combustion the heat due to the fixation of the oxygene contained in a cubic centimeter of air is = 1, and that the heat lost during the fixation, whether in radiating heat, or by the absorption which is made of it by the azotic gas or other bodies, is = 1/2, not taking here into consideration the law according to which it decreases. According to these premises, it is very evident that in the first moments of the combustion the temperature of the body must rise; but, in pro- | 314 | portion as the quantity of oxygene shall diminish, and that of the azote proportionally increase, the heat communicated will also diminish. A period will therefore arrive, at which the heat lost will be equal to the heat communicated, and below which, the temperature being too low, the combustion must cease. What evidently proves that the combustion stops only because the temperature is too low, is that if we artificially maintain the temperature sufficiently high, the body will continue to burn.

Now this explanation will still hold good, when, instead of azote, sulphurous gas, hydrogene gas, carbonic acid, or any other gas, is mixed with the ox- | 315 | ygene; only the combustion may cease sooner or later than with azote gas. For it is very evident, that if the sulphurous gas, or carbonic acid gas, had a capacity for caloric, much greater than that of azote, supposing them to be mixed with oxygene in the same proportions as the latter, the loss of heat would be much greater, and consequently the cessation of the combustion must take place sooner. But if the gases had equal capacities for caloric, they must all of them stop the combustion at the same period, as we have seen has nearly been done by oxygene and azote with hydrogene, and this would perhaps afford a solution of the important question, whether the gases have equal or different capacities.

Thus a combustible body, sulphur for example, would cease to burn in a determinate volume of air, not because the affinity which the azote or the gases produced have for oxygene were more powerful than that of the combustible body; but because the heat absorbed by these gases, which tend to place themselves in an equilibrium of temperature with the burning body, would be greater than the heat proceeding from the fixation of the oxygene; whence it would result, that the temperature would soon be reduced below that necessary for the combustion. In fact, we know that sulphur can continue to burn in air in which it had been extinguished if, we raise the temperature to a sufficient degree.

What takes place in the instantaneous combustion of the hydrogene in Volta's eudiometer, is perfectly analogous to what passes in its successive combustion in a given volume of air, or in that of any other body. If we place a lamp, the flame of which is supplied by hydrogene gas, under a bell-glass filled with oxygene gas, the flame will become small, bright and slightly coloured. If we replace the oxygene by atmospheric air, the flame will be more voluminous, less bright, and more coloured. In proportion, especially as the relative quantity of oxygene diminishes, the flame will increase in size, because the hydrogene will be obliged to go farther to come at the oxygene, and the flame will soon be extinguished, although the air still contain some hundredth parts of oxygene. The phenomena which take place in Volta's eudiometer are of the same nature. When the proportions of oxygene and hydrogene do not deviate much from those which constitute water, the flame is still very bright, notwithstanding its dilatation; but if we mix, for example, 1000 oxygene with 100 of hydrogene, the flame is then weak, of a blueish-green colour, and the combustion of the hydrogene is far from being complete, for we still find nearly two-thirds of it in the residue. What farther proves that the combustions' not having been complete was owing to the temperature not having been sufficiently elevated, is that if the residue is made to pass, as was done by us, through a red-hot tube of porcellain, the whole of the hydrogene will be absorbed. | 316 |

We must observe, that in the combination of the hydrogene and oxygene gases, a very singular phenomenon takes place, which has long since engaged the attention of M. Monge.

"How does it come to pass," says this distinguished philosopher, "that by elevating the temperature of the two gases, or, in other words, by increasing the proportion of the solvent, we diminish the adhesion which it had with its bases?" Far from conceiving that the present state of our knowledge is competent to give a satisfactory solution of this question, we wish to recommend it anew to the attention of philosophers. In fact, according to the idea which we are able to form to ourselves of the force which produces the combinations, and of those which are opposed to it, the elastic state indicates that the force of cohesion is destroyed, and that two bodies in this state are in the condition most favourable to combination; so that now that the attractive force of their molecules has been changed into a repulsive force, every cause which shall favour the latter will be in opposition to the former. It happens, however, that by raising the temperature of the two gases, that is to say, by augmenting their repulsive force, their attractive force is increased. It cannot be believed, that the heat does nothing more than separate their molecules to greater distances from each other; for, in this case, why should a mixture of hydrogene and oxygene gas not inflame under the receiver of an air-pump, where it may be indefinitely dilated? Neither is it to be supposed that the heat acting instantaneously can produce a compression which favours the combination of the two gases by bringing their molecules into closer contact; for it is easy to convince ourselves that a mixture of oxygene gas and hydrogene gas, heated very gradually | 317 | without opposing its dilatation, will nevertheless inflame when the temperature shall be sufficiently elevated.

Having now proved, that under determined circumstances the combustion of hydrogene and oxygene may be complete, we shall proceed to examine whether its products are constant.

According to all the experiments which have been made on the composition of water, the result has been considered to be uniform. There has, however, sometimes been obtained a small quantity of nitric acid; but it has been ascertained that this acid is not a constant product of the combustion of the hydrogene, but, on the contrary, merely accidental. Cavendish was the first who discovered this formation of nitric acid, and

Messrs. Foureroy, Seguin, and Vauquelin, have taught us how we may avoid it, and obtain water without any acidity. It has not, indeed, been demonstrated, that oxygenated or hydrogenated waters have not been formed, because, in all the exact experiments that have been made, the combustion of the hydrogene gas has always been effected in the same manner, and at most it would be proved that those which have been obtained are constant under the same circumstances. If we compare the combustion of hydrogene gas with that of nitrous gas, the products of which are so variable, we shall be still more justified in concluding, that since oxygene has always predominated in the experiments which have been made, an oxygenated water may have been formed; whereas, if hydrogene had been predominant, a hydrogenated water would have been the result. Let it then be admitted, that an oxygenated water may be formed; if, for example, we obtain it in all circumstances, and it be constant, this will be of no consequence with regard to the | 318 | proportion of its principles, which must serve for the analysis of the air; but if it be so only because the oxygene predominated, it is manifest that we shall no longer obtain the same proportions when we alternately cause both gases to predominate.

To be continued in our next. | 365 |

Experiments on the Eudiometrical Means, and on the Proportion of the Constituent Principles of the Atmosphere. By Messrs. Humboldt and Gay-Lussac.

(Continued from Page 318.)

Now as we have made a great number of experiments which prove, that when we employ them alternately in excess we constantly, obtain the same proportions, it follows, that the combustion of hydrogene gas is of an uniform nature. The Galvanic phenomena of the decomposition of water seem however to prove that water is capable of oxygenating or hydrogenating itself; and it is upon this supposition that Messrs. Laplace and Berthollet have accounted for the singular experiment of the decomposition of water by two metallic wires immersed at one end in this liquid, and communicating by the other with the two poles of a Galvanic pile. But without wishing to controvert this explanation, which appears to us the most satisfactory that has hitherto been offered, we shall observe that the complete absorption of the hydrogene at one of the wires, and of the oxygene at the other, proves that the water does not become oxygenated or hydrogenated, because, in order to be- | 366 | come so, it would be necessary that it should absorb one of the gases in a proportion greater than that required for the composition of water. If therefore it absorbs oxygene and hydrogene in exact proportions for forming water, it must be concluded that the properties of one of the gases are neutralized by those of the other. Accordingly, in the circumstances of which we are treating, the water might become instantaneously oxygenated at one of the wires, and hydrogenated at the other; but the two gases, being deprived of elasticity, and in exact proportions, must soon enter again into combination.

If it is well demonstrated that, under given circumstances, hydrogene or oxygene may be completely absorbed, and, if it is equally so, that the product of their combination is constant, nothing more remains to be done, in order to solve the third question, which we have proposed to ourselves, than to determine the proportions of oxygene and hydrogene which constitute water.

To 100 parts of oxygene gas we added 300 parts of hydrogene gas; and after having inflamed them by the electric spark, we obtained in twelve experiments the following residues:

100,8	101,0	102,0
101,4	101,7	102,0
100,5	102,0	101,0
101,0	101,5	101,5
The mean term of which is – –	101,3	

Thus 100 parts of oxygene, supposed to be very pure, required 98,7 of hydrogene: but on placing our oxygene gas in contact with sulphuret, we found that it had all been absorbed within about 0,004; hence it follows, that 99,6 of oxygene absorbed 199,1 of hydrogene, or that 100 absorbed 199,89, or, to use round numbers, | 367 | that 100 of oxygene requires 200 of hydrogene for their saturation.

In the preceding experiments it was the oxygene that disappeared. Let us now reverse the experiment, by inflaming a mixture of 200 parts of each gas; the residue of the different inflammations will be the following:

101,5	102,0	101,5
101,3	102,0	102,3
102,2	101,0	102,0
102,0	101,0	102,0
Mean term – –	101,7	
Mean absorption	298,8	

200 parts of hydrogene, supposed to be pure, required therefore 98,3 of oxygene; whereas, according to the proportion which we have just established, they ought to have had 100. But if we admit this same proportion to be exact in the 298,3 of absorption, there would be only 198,8 of hydrogene, which would indicate 0,006 of azote in this gas.

Supposing even the hydrogene to be perfectly pure, the two proportions obtained by causing the oxygene or the hydrogene to predominate, agree sufficiently with each other to confirm all that has been said in the course of this memoir: to render them identical, nothing more is necessary than to suppose 0,006 of azote in the hydrogene, in which, in fact, we can demonstrate its presence.

We have just seen, from the preceding experiments, that 200 parts of hydrogene, without making any correction, absorbed 98,3 of oxygene. Let us then take the residues 101,0 and 101,5, proceeding from the combustion of 100 of oxygene and of 300 of hydrogene, and detonate them with 200 of oxygene gas. In these two | 368 | residues there must be 0,008 of azote belonging to the 200 parts of oxygene gas; and if the remainder 201,7 was pure hydrogene it ought to absorb 99,1 of oxygene, and consequently there ought to have disappeared by the inflammation 300,8 parts, but there disappeared only 295,0: it follows, therefore, that the residuum 201,7 was not pure hydrogene; and that, according to the proportion of 100 of oxygene to 200 of hydrogene, it contained 5,0 of azote, proceeding from 600 of hydrogene; that is to say, that this last-mentioned gas contained 0,008 of azote.

We therefore think it proved, that 100 parts, in volume, of oxygene gas require very nearly 200 parts of hydrogene gas for their saturation. According to the experiment of Messrs. Fourcroy, Vauquelin, and Seguin, 100 parts of the former would require 205 of the latter; but we shall observe, that, whichever of the two proportions we adopt, the error cannot amount to more than 0,0035 upon the absolute quantity of the oxygene of the air, and when the question respects relative quantities, the error will be still much smaller.

We have ascertained, that the proportion does not vary in consequence of changes of the temperature. It is evident that it must be so, since, as the heat equally dilates both gases, and causes them to dissolve equal quantities of water, the real weights of hydrogene and oxygene contained in equal volumes remain always in the same proportion. It would therefore be more accurate to say, supposing our proportion by the volumes to be well established, that 100 parts of oxygene require 200 of hydrogene, than to indicate the proportions of water by the weight. If the oxygene and hydrogene employed for the composition of the water had been perfectly dry, or if a correction had been made according to the quantity of | 369 | moisture which they might contain, it would be a matter of indifference whether the proportions of its principles were enunciated according to their volumes or their weights; but since hydrogene combines with oxygene in double the volume of the latter, and as they both dissolve the same proportion of water, it is evident that they do not carry into the combination quantities of water which are in the same reciprocal proportions as the quantities, by weight, of oxygene and of hydrogene, and that consequently the proportion of the principles of the water must thereby be altered. Thus the proportion, according to the volumes, has the property of remaining constant, notwithstanding the changes with regard to temperature and moisture, whilst that, according to the weights, varies under the same circumstances. And let it not be imagined that this consideration is of little importance: for it would be very easy to prove that it has very considerable influence upon the proportion of the principles of water. According to the experiment of Messrs. Fourcroy, Vauquelin, and Seguin, the most accurate that has hitherto been made upon this subject, water contains by weight 85,662 of oxygene and 14,338 of hydrogene. But the experiment having been made at a temperature of about 14°, and the correction due to the water held in solution by the gases not having been made, it follows, that if we adopt their specific weight of the oxygene and hydrogene gases, as well as the proportion of their volumes in their combination, and moreover admit with Saussure, that a cubic foot of air, at the temperature of 14°, contains very nearly 10 grains of water in solution, the proportion by weight between the oxygene and hydrogene, instead of being 85,662 to 14,338, would be 87,41 to 12,59; a very remarkable difference, and which especially must be of great in- | 370 | fluence in analyses where the real weight of the hydrogene is to be determined. The same consideration applies also to the specific weight of the gases, and principally to that of the hydrogene, of which about a sixth part is due to the water which it holds in solution when the temperature, as here supposed, is 14° Reaumur. We do not therefore doubt, that if we had hydrogene gas perfectly dry, and deprived of the azote gas, which appears very often to accompany it, we should find its specific lightness to be at least 15 times greater than that of the atmospheric air.

It still remains for us to answer the last question which we have proposed to ourselves, and to shew what are the limits of the error in Volta's eudiometer; and next, what are the smallest quantities of oxygene or of hydrogene that can be estimated by his method.

The effects obtained with this instrument, being instantaneous, are independent of the thermometer and barometer. In this point of view, it has the very decided advantage over phosphorus and the alkaline sulphurets, of giving results very capable of comparison; but this is not the only one; it has also that of the eudiometrical means which give multiples of the quantity to be estimated. As in this instrument each hundredth of oxygene is represented by a three times greater absorption, the error that can be committed amounts only to a third upon this gas; and now, especially as we have instruments which divide the measure into three hundred parts, it is evident, that if we err even by one division, the accuracy, as to the quantity of oxygene, may be carried to near a thousandth part of the quantity of air analyzed.

If therefore the results of the combustion of hydrogene gas admit so well of comparison, and the errors to which | 371 | they are liable are restricted within such narrow bounds, it is evident that we may not only find the slight differences which exist between two portions of atmospheric air, but also determine less than three thousandth parts of oxygene which shall be lost in azote or hydrogene gas: but in this case, in order that the inflammation take place, it would be necessary to add a given quantity of oxygene, the absorption of which with hydrogene gas had been determined by previous experiments, and then the excess of the first absorption above the second would be attributed for a third to the oxygene gas contained in the air which is analyzed.

On the other hand, in order to determine whether one portion of hydrogene be more pure than another portion, or whether there exist very small quantities of it in a gas or in the atmospheric air, it would be necessary, in the first case, to mix 100 parts of hydrogene gas with 100 of oxygene: the quantities of real hydrogene would be in the direct ratio of the absorptions. But if the proportion of hydrogene were very small, for example one two hundredth, it would be necessary, in order to effect its combustion, to add 100 parts of this gas to 200 of the air to be analyzed, and to detonate the mixture with a sufficient proportion of oxygene. By this means, and with the practice which we have now acquired, we have been able to find again three thousandth parts of hydrogene gas which we had mixed with atmospheric air.

It might be objected against Volta's eudiometer, that the hydrogene not being always the same, we might incur errors difficult to be appreciated. We shall first observe that it is indifferent whether or not it contain azote; but if it contained oxygene, its quantity confounding itself with that which we wish to appreciate, would alter the results. In order to avoid this inconvenience, we may | 372 | first of all detonate separately 500 parts of hydrogene with 100 of oxygene; by this means its oxygene will be destroyed, and we may then employ it for the analysis of the air. With this precaution we may employ a gas prepared as inaccurately as possible. It is sufficient if it have been extracted from the water by means of zinc and sulphuric acid or muriatic acid; for it is well known that if we employ another metal, such as iron, the gas is no longer of the same nature.

After all the experiments which we have recorded, we might have been justified in concluding that the eudiometer of Volta must indicate the whole of the oxygene contained in the atmospheric air; but we wished to determine this point by direct experiment. We analyzed an air composed of 20 parts of very pure oxygene and 80 of azote, obtained from the decomposition of ammoniac by oxygenated muriatic acid, all possible precautions being taken to prevent its becoming mixed with atmospheric air. 200 parts of this air, inflamed with 200 of hydrogene gas, gave five absorptions, the greatest of which differed from the smallest only by 5,1000 and the mean term of which was 124,9. These 124,9 parts indicate 41,6 of oxygene, of which the half 20,8 corresponds to 100 of our factitious air. We find therefore 0,008 of oxygene more than we had employed, which might seem to indicate that the proportion of 100 of oxygene to 200 of hydrogene is rather too great; but we must observe that our hydrogene, though carefully prepared, still emitted light with phosphorus, and that in order to explain our result, it is sufficient to suppose that the azote contained a hundredth part of oxygene, which is sufficiently probable if we consider that the oxygenated muriatic acid is very speedily decomposed by the action of light.

It appears from what has been said, that the results afforded by Volta's eudiometer admit very well of compar- | 373 | ison, and that the limits of their differences may be reduced very nearly to a thousandth part of the air analyzed. It appears also that by its means we may estimate very minute differences between two airs or very small quantities of hydrogene mixed with the atmospheric air. Independently of the property which this instrument possesses of indicating the whole quantity of oxygene contained in an air, it is the only one with which we can ascertain the proportion of hydrogene in a gaseous mixture, and in this point of view it still merited to have its mode of acting attended to and studied.

Thus the illustrious Volta, to whom Natural Philosophy is indebted for so many beautiful discoveries, has also the merit of having furnished Chemistry with the most accurate and valuable instrument for analyses.

ANALYSIS OF THE ATMOSPHERIC AIR BY VOLTA'S EUDIOMETER.

As we have now proved that Volta's eudiometer gives results very capable of comparison, that it can indicate the whole quantity of oxygene contained in the air, and that it has over the other eudiometrical means in which the absorbing substance is solid or liquid, the advantage of giving multiples of the quantity of oxygene to be estimated; we shall proceed to apply it to the analysis of the air. If the proportion of 100 of oxygene and 200 of hydrogene which we have established be strictly correct, we shall obtain the proportion of the oxygene to the azote likewise correct; but supposing even that the quantity of hydrogene were too large or too small by 5 parts, the error would not amount to more than three thousandth parts of the air analyzed, and we should have the advantage of obtaining a greater precision than by any of the other known eudiometrical means. | 374 |

The atmospheric air which we have analyzed was collected over the middle of the Seine in cold, temperate and rainy weather, and during the prevalence of different winds. In order to obtain a greater parity between the circumstances, and better to appreciate the differences in the nature of the air, if there were any, we analyzed on the same day the different portions of air which we had collected in different weathers, and which we had preserved in glass vessels well closed and inverted over the water. For brevity's sake, we have comprized in the table annexed to this memoir, the absorptions produced by the inflammation of 200 parts of air and 200 of hydrogene gas, and have at the same time indicated the correspondent quantities of oxygene.

It will there appear that our experiments prove, in the first place, that there are no variations exceeding a thousandth part in the quantity of the oxygene of the air, though that which we analyzed, having been collected during the prevalence of different winds, came from regions very remote from each otber; and in the second place, that the proportion of the volume of the oxygene to the other gases which exist in the air, is as 21 to 79. The first result, that the air does not vary in its composition, is rigorously exact, because it is independent of the proportion of the hydrogene and oxygene gas which constitutes water; but the second result, that the air contains 21 hundredths of oxygene, can also deviate but very little from the real truth; for if we suppose that the quantity of hydrogene required for saturating 100 parts of oxygene were larger or smaller by 5 parts than that which we have assigned (and which we have reason to believe is correct within a much less difference), the error with respect to the proportion of oxygene which we have found in the air, would not amount, as we have already ob- | 375 | served, to more than three thousandth parts of the air analyzed.

But as many meteoric phenomena may be attributed to inflammation of hydrogene gas, it has been endeavoured to explain them by admitting the existence of this gas in the atmosphere. We therefore thought it a very interesting enquiry, to investigate whether the air actually contains hydrogene gas; and in order the more easily to discover it, we made a gaseous mixture, in which we were sure that there was none of it, and we made a comparative analysis of the two airs. We made a mixture of 20 parts of oxygene and 80 parts of azote, obtained from ammoniac by means of the oxygenated muriatic acid, and we detonated 300 parts of each of the two airs with 100 of hydrogne; but the result of six experiments made with the atmospheric air was exactly the same as that of six others made with the factitious air. And as we have shewn that we could appreciate less than three thousandth parts of hydrogene, it must be concluded that the atmosphere does not contain this gas, or if it contains it, its quantity does not amount to three thousandth parts. It cannot however be doubted that there exists a little hydrogene in the air; some is daily evolved from marshes, but its quantity may be so small (a thousandth for example) as to elude all our means. The proportion of carbonic acid existing in it ought to be much more considerable, if we reflect upon the abundance of the sources which furnish it, and yet, if it did not form insoluble combinations with lime and barytes, it would perhaps still remain to be ascertained by the determination of its volume whether any of it existed in the air. The carbonic acid, it is true, cannot accumulate in the air, because vegetation decomposes it; but is it proved that there do not exist causes which return the hydrogene | 376 | to the earth, and thereby prevent it from accumulating in the air?

To draw a conclusion from the preceding experiments, we shall say: 1, that the atmosphere does not vary generally in its composition; 2, that the quantity of oxygene which it contains is 21 hundredths; 3, finally, that it does not contain any hydrogene which we are able to appreciate.

This identity of composition in which the principles of the atmosphere are constantly maintained, and this absence of hydrogene which our experiments prove, must give confidence to the Geometrician with respect to the theory of refractions. The refractive power of the different gases being different, and that of hydrogene being stronger than that of oxygene and of azote, the theory of refractions, which is founded only upon the variations of the barometer and thermometer, would be very imperfect if the atmosphere changed in its constituent principles; but fortunately it is easy to prove that these changes do not take place in a sensible degree, and that the hydrogene gas, the refractive power of which is very great, does not exceed 0,003, at least as far as the greatest heights to which men have ever ascended. The Geometrician will therefore have nothing else to consider in the Theory of Refractions than the barometer, the thermometer, and the hygrometer.

In fact, a little reflexion will be sufficient to convince us that the atmosphere cannot vary very considerably in the space of some years, and still loss of some days, at least if we do not speak of some very particular local variations. For if it varied thus in so short a space of time, by what miracle should it do so, and return suddenly to its original state? How conceive a cause sufficiently powerful to change from one day to another the proportion of oxygene by a thousandth part only, unless we should admit the existence of a magnetic or electric power, or any | 377 | other equally imaginary, which could change, by unknown modifications, the oxygene into azote, and *vice versâ*? It is possible that the atmosphere varies very slowly both in the proportion of its principles and in its weight; but these variations, for being so insensible, ought not the less to engage the attention of philosophers.

If it is now well proved that in general the atmosphere does not vary in its composition, we must seek for the cause of the differences which some have imagined they had discovered in it, in the local circumstances under which it was analyzed. Volcanoes upon high mountains, particular fermentations, water issuing from a marsh or a lake, might perhaps in some degree impair the purity of the air in contact with them, either by depriving it of oxygene, or by exhaling into it non-respirable elastic fluids; but how trifling must not this diminution of the proportion of the oxygene be in so large a mass of continually agitated air, when we consider that in places where a great number of individuals is collected, or in those where there seems to exist a focus of infection, the air nevertheless experiences but very slight variations. We have analyzed two portions of air, one of which was collected in the pit of the Theatre Français, immediately before the curtain was raised for performing the after-piece, three hours and a half after a great number of spectators had assembled, and the other of which was collected three minutes after the entertainment had ended in the most elevated part of the house. These two portions scarcely rendered lime water turbid; the atmospheric air indicating 0,210 of oxygene, the air of the pit indicated only 0,202 and that of the highest part of the house 0,204. | 378 |

Analysis of the atmospheric air.	**Of the air of the pit.**	**Of the air of the highest part of the building.**
200 atmosph. air.	200 atmosph. air.	200 atmosph. air.
200 hydrogene.	200 hydr.	200 hydr.
126 air absorbed.	121,5 abs.	122,5 abs.
21 oxygene.	20,2 oxyg.	20,4 oxyg.

M. Seguin has also analyzed the air of hospital wards which he had kept closely shut up for the space of twelve hours, and found it to be almost as pure as the atmospheric air, although it had an insupportably infectious smell.

If therefore, even under circumstances the most favourable to the absorption of the oxygene, the air does not lose one hundredth part of it, we cannot thereby account for the sense of anxiety which we feel in close and crowded apartments, or the maladies which are peculiar to the vicinity of lakes and marshes, or to certain countries. Under some circumstances they will be produced by emanations which elude all our eudiometrical means, and which act in a peculiar manner upon the human body. Thus a bubble of sulphurated hydrogene gas, of oxygenated muriatic acid, a putrid exhalation, even a flower, may fill an immense space with this odour, and astonish our imagination by their extreme subtility, even when we are ready to sink under their action. The pestilential miasmata may be equally subtile without being the less deleterious, and equally elude all our means of analysis. Fortunately, if we cannot seize these atomic substances and determine their nature, we may at least, after the labours of M. Guyton, which have been productive of such great benefit to mankind, destroy their action. But under other circumstances the maladies may arise from | 379 | the humidity of the air, from its temperature, from its electric state, or in general from the state of the atmosphere with respect to the peculiar state of the individual affected; and in these cases, which may be very frequent, the malady may make great ravages without its being possible to arrest its progress: it would therefore be illusory to attribute all to a single cause, when the state of human health depends upon the concurrence of all the circumstances under which men are placed.

But let us now recapitulate the principal facts contained in this first part of our Memoir, and call to mind some of the explanations which we have offered, hoping we may be permitted to consider them as expressing the real state of the facts.

The solution of an alkaline sulphuret, when made cold, does not absorb the azote, and it may be employed with advantage for the analysis of the air: when made hot, it absorbs it, and indicates a greater diminution of volume in the air than that which proceeds from the absorption of the oxygene. It is to the water only, and not to the sulphuret, that we are to attribute this property.

There are certain proportions of oxygene and hydrogene in which the combustion produced by the electric spark may be complete; there are others in which the combustion ceases before being completed; and finally there are others in which it cannot take place at all. These last phenomena seem to depend upon the circumstance that the temperature necessary for the combustion is not sufficiently elevated, and not upon the mutual affinity of the gases: for in all cases in which the combustion is not complete, we need only raise the temperature artificially in order that it may become so. When the hydrogene and oxygene are not entirely absorbed, we | 380 | find them again in the residues, and prove that they have not formed new combinations.

When we cannot inflame a gaseous mixture which contains oxygene and hydrogene, it will be sufficient to augment the proportion of these two gases. The meteoric phenomena cannot be results of the inflammation of hydrogene gas, since even in an air consisting merely of pure oxygene, it would require more than 6 hundredths of hydrogene for combustion to take place, and even then it would be only partial. Electricity seems to act in the inflammation of oxygene and hydrogene gas by the heat arising from the compression which it produces in its passage through their mixture. These two gases, by their combination, form water, which is constant in its nature. If the galvanic phenomena seem to prove that water is susceptible of oxygenation or hydrogenation, they may be equally well accounted for without the aid of such an hypothesis.

One hundred parts by volume of oxygene require for their saturation 200 of hydrogene. This proportion is independent of the changes of temperature and moisture, whereas that determined by weight varies under the same circumstances, because the two gases do not carry into the combination quantities of water which are in the same proportion as their quantities by weight; whence it results that the proportions of the water which have been established must be modified. The Eudiometer of Volta is capable of indicating the whole quantity of oxygene contained in a given volume of air within about a thousandth part of that volume, and its results admit very well of comparison. In the present state of our knowledge, it is the most exact of our eudiometrical means; it is not only capable of indicating very small quantities of oxygene or of hydrogene, and determining the purity of the last- | 381 | mentioned gas, but it has likewise the advantage of giving multiples of the quantity to be estimated. It has therefore in all these respects a very decided superiority over the other eudiometrical means. The atmosphere contains only 0,21 in volume of oxygene, and it does not vary in its composition. It contains no hydrogene, or if it contains any, its quantity cannot amount to 3,003.

To be continued in our next. | 449 |

Experiments on the Eudiometrical Means, and on the Proportion of the Constituent Principles of the Atmosphere. By Messrs. Humboldt and Gay-Lussac.

(Concluded from Page 381.)

OF THE NATURE OF THE AIR EXTRACTED FROM WATER, AND OF THE ACTION OF WATER UPON THE GASES, BOTH PURE AND MIXED.

We have hitherto examined the eudiometrical means which lead to the exact analysis of the atmospheric air. We should undoubtedly have confined ourselves to the enunciation of the principal facts to which the first part of our investigation has conducted us, had we not remarked in the course of these experiments, and particularly of those upon the sulphurets, that water and other liquids exert an action upon the air which may frequently become a cause of error, the more important as it has hitherto been little attended to. We therefore apprehended, that we should leave our labours in a still more imperfect state than we have already done, if we had not directed our inquiries to this action of water upon the gases, both pure and mixed, which are subjected to it. The experiments which we have made with a view to this object, shall form the conclusion of our memoir. | 450 |

It is generally known that water can hold air in solution. Boyle, Huggens, and Mairan, have discussed this fact; but they did not possess the means of discovering that this dissolved air differs chemically from the atmospheric air. The celebrated Priestley was the first who observed that the air extracted from water contains more oxygene than common air. M. Hassenfratz afterwards asserted that rain-water evolved an air which contained nearly forty-hundredths of oxygene; and Messrs. Ingenhoutz and Breda, in their experiments upon nitrous gas, were led to similar results.

But if it is already known that the air contained in water is more pure than atmospheric air, it has also been maintained that water absorbs the oxygene gas more abundantly and more readily than the azote. M. Fourcroy even mentions the curious fact, which however he himself does not conceive to be sufficiently authenticated, that water charged with oxygene gas absorbs hydrogene gas, upon which common water has hardly any action. We shall see hereafter, that that which it exercises upon any gas is modified by the nature of the air which it already holds in solution.

Mr. Henry, in a memoir lately published in England, has examined the absorption of different gases by water deprived of air. He effected these absorptions under a pressure equal to that of two or three atmospheres; but he has not treated of the mixture of different gases, nor of the affinity which water has for this mixture: he confines himself to the examination of the quantity absorbed, according to the difference of temperature and of barometric pressure, without directing his inquiries to the action of water already saturated with other gases.

We conceived that we ought not to neglect a subject so intimately connected with eudiometrical inquiries, and to which chemists have hitherto paid so little attention. | 451 | We have examined the degree of affinity by which the oxygene dissolved in water is retained in it, according to the temperature and the salts which it may contain. We placed in contact with water equal quantities of gases, both pure and mixed, and we observed the changes which these mixtures undergo in their chemical composition. Finally, we have begun to examine a problem of great importance to meteorology, namely, whether rain water holds hydrogene in solution.

All these inquiries, which we intend to prosecute during the course of this year, and particularly upon the mountains which we purpose to visit, are not as yet in a very advanced state; we shall therefore content ourselves for the present with offering some leading facts, which, we trust, will not be deemed altogether destitute of interest by philosophers.

After mixing the whole mass of air which water yields by boiling, without separating the portions first disengaged from those evolved at the end of the operation, we found by Volta's eudiometer that distilled water, which has re-absorbed atmospheric air, gives an air which contains in the hundred parts 32,8 of oxygene; the water of the Seine, one which contains 31,9 oxygene; and rain-water an air containing 31,0 of the same principle.

From these experiments it follows, that we may extract from these three waters air nearly equally rich in oxygene, and purer by ten-hundredths than the atmospheric air. This quantity of oxygene is more variable in the waters of wells, which in the bowels of the earth are in contact with substances that exercise an affinity upon oxygene. Water of the Seine, collected at another period, yielded an air containing only 29,1 of oxygene, an air somewhat less pure than that of the rain water.

If distilled water which has again absorbed air, rain water, and river water, yield airs all of which are much | 452 | purer than the atmospheric air, it will be still more interesting to examine the nature of the gaseous mixtures which water yields when heated gradually. It is in these experiments that the great affinity of oxygene for this liquid displays itself in its strongest light. We gradually heated water of the Seine to ebullition, and collected the air which is disengaged by successive and unequal portions. We took 200 parts of these portions, and having detonated them with 200 parts of hydrogene gas, they gave us the following results.

Portions of air, according to the order of their evolution.	**Absorption.**	**Oxygene gas contained in 100 parts of the air evolved.**
First – – –	142,0 – – –	23,7
Second – – –	164,5 – – –	27,4
Third – – –	185,0 – – –	30,2
Fourth – – –	195,0 – – –	32,5

These experiments, several times repeated, prove that water abandons at first an air, the purity of which is only a little superior to that of atmospheric air; after which the purity of this air, or the disengagement of oxygene, progressively increases, and the last gaseous portions which the heat separates contain the most oxygene. On repeating this experiment upon snow-water, the first portions of air had 24,0; the last 34,8 of oxygene. Possibly if the mass of water were heated still more slowly, and the small portion of air which is first evolved carefully separated, we should have at the beginning of the operation an air still less pure than that which we obtained.

Water therefore does not exercise an uniform action upon oxygene and upon azote, and elevation of temperature diminishes its action upon the first less than upon the last. It is even probable that the portion of air which is disengaged towards the end of the operation would be of greater purity than that of 32 or 34 *per cent.* of oxy- | 453 | gene, if the water contained in the vessel which receives the gaseous mixture did not begin to be heated, and to evolve its air, which then is only at 23 *per cent.* of oxygene. This disengagement takes place especially when the aqueous vapour begins to pass, and it is this diminution of the purity of the air expelled last and the inequality of volume of the four separate portions, which explain how the whole mass of air extracted at once contains to the amount of 31 hundredths of oxygene.

This unequal action of water upon the oxygene and the azote manifests itself also in the solution of salts. We have observed that the pure water of the Seine gave by ebullition nearly one half more of air than the same water charged with muriate of soda. The cause of this diminution consists in the very considerable quantity of air which already disengaged from the water, still cold, whilst the solution of the salt is taking place. This air, accurately analyzed, shewed only 0,225 of oxygene, whilst the air obtained by boiling the water charged with muriate of soda, contained 0,305. Hence it appears that the water, in dissolving the salt, abandons a part of the air which it holds in solution, but that this part contains oxygene in a less proportion than that which it retains.

The condensation which water experiences in passing from the liquid to the solid state presents to us a third class of phenomena analogous to those which we have just described. Melted ice yields only about half the quantity of air which we obtain from common water, and it is to be observed that it does not begin to evolve its air until its temperature has been raised as high as above the sixtieth degree of the centigrade thermometer. The air obtained, divided into two unequal portions, shewed in Volta's eudiometer 27,5 and 33,5 of oxygene. Then the purest air was again evolved the last. | 454 |

The small quantity and the great purity of the air evolved from melted ice, proves that water, as it passes into the solid state, abandons a large portion of its air, and that this part, separated during the congelation, is air of much less purity than that which it retains. Thus three phenomena, which at the first view appear different, water at a temperature from 35° to 40° centigram, water dissolving salts cold, and pure water congealing into ice, present results entirely similar to each other in their action upon oxygene and azote. A moderate temperature acts like the solution of a salt, and this like the transition from the liquid into the solid state. The water in these three cases evolves an air of less purity than that which it holds in solution.

It is a very remarkable phenomenon that the congelation of water into the state of snow expels less air from it than the formation of ice. We melted some newly fallen snow, and heating it gradually, we obtained a volume of air nearly twice as great as that which melted ice yields. The air extracted from the snow-water was almost in equal abundance with that evolved from the water of the Seine; for the latter gave by ebullition 1940 measures of air, while the same volume of snow-water yielded 1892. These 1892 parts collected into 5 portions, according to the periods at which they were expelled by the heat, shewed successively in Volta's eudiometer.

The first – – – – –	24,0 of oxygene.
Second – – – –	26,8.
Third – – – –	29,6.
Fourth – – – –	32,0.
Fifth – – – –	34,8.

This last portion of air is the purest we have yet extracted from any water. | 455 |

The volumes of each portion being known, calculation gives us for the purity of the air considered collectively 28,7 of oxygene. The water of the Seine yielded on the same day an air which was less pure by 4/1000. Otherwise the two waters, that of melted snow and that of the river, give a volume of air equal to nearly 1/23 of their own.

These experiments upon snow-water and upon melted ice, which we design hereafter to vary in many ways, present considerations of great importance to the study of meteorology. Snow is nothing else than an aggregate of small crystals of ice which are formed in the higher regions of the atmosphere, and yet these small crystals melted give a volume of air nearly twice as great as that which the ice formed upon our rivers yields. Hence it is to be concluded that when the water dissolved in the air congeals into snow, it does not expel that large portion of air which it disengages in its congelation upon the surface of the earth, unless we might be allowed to suppose that snow retains between its small crystals a certain quantity of air which it absorbs as it melts; for it appears that it is principally at the moment of its congelation that water abandons the greater part of its air.

The beautiful vegetation which surrounds the Glaciers, the rapid developement of plants when the snow melts in the spring, and several phenomena which are supposed to have been observed in agriculture and in bleaching, have led to the notion that the waters of ice, snow and rain, produced peculiar effects by a large quantity of dissolved oxygene which was evolved from them. The experiments which we have hitherto made do not seem favourable to these conjectures. There undoubtedly exist wells, the waters of which contain air inferior in purity to the atmospheric air, and we do not doubt that these waters, charged besides with salts and car- | 456 | bonic acid, must have an influence upon vegetation and bleaching very different from that of snow-water. But the differences produced by distilled water exposed to the air, rain-water, snow-water and the water of the Seine, are not easily to be accounted for by the oxygene dissolved in them, if we recollect that all these airs contain water of nearly equal purity, and that they contain it in almost equal quantity. The phenomena of vegetation, like those of meteorology, are so complicated; they depend upon the concurrence of so many causes at once, that we must be very careful not to attribute to one what is the effect of many.

The experiments which we have recorded on the force with which the last particles of oxygene dissolved are retained by water, throw an additional light upon the state in which air exists in liquids. The specific gravity of distilled water and of that charged with air, being apparently the same, Mairan has concluded with reason that this air cannot be lodged in the fluids in an elastic state. The chemical phenomena confirm this opinion. If water deprived of its air by distillation, or by the air-pump, could be considered as a sponge the pores of which are empty, how should these pores not fill themselves on the first contact with atmospheric air? But this solution of air in water can be considered only as the effect of a chemical affinity; for how but by this affinity could the absorption of the gases by the water deprived of air be so slow, and above all, how else should the water dissolve one gas in preference to another? How should water charged with one species of air abandon a part of it, as we shall see hereafter that it does, to receive another of a different nature?

After having examined the air which may be extracted from water under different circumstances, we shall conclude our Memoir with the experiments which we have | 457 | made in placing gases, both pure and mixed, in contact with water. The gases which we employed were exactly of the same volume, and the quantity of filtrated water of the Seine was nearly equal. After a space of from 6 to 8 days, we not only measured the quantity of the volumes absorbed, but we also analyzed the residues. This analysis was the more necessary as one might often be tempted to conclude from a very small change in the volume of the gas placed in contact with the water, that the latter has no sensible action upon it, whilst the nature of the residue indicates that this action has been very strong, but disguised by the quantity of air evolved from the water in lieu of that which has been absorbed.

Of all the gases, the oxygene is that which the water of the Seine absorbs in the most considerable quantity. If we place in contact with this water, already charged with air, 100 parts of oxygene gas, 100 of azote, and 100 of hydrogene, the oxygene gas will have diminished by 40 parts when the two others have lost only 5 and 3 parts. But the real absorption of the oxygene gas is still much more considerable than its apparent diminution indicates. The 60 parts of residue, instead of being pure oxygene, contained 37 parts of azote and only 24 of oxygene; so that the 100 parts of oxygene employed, had lost upon the Seine water 77 parts, which had expelled 37 parts of azote. Thus a river water which had been exposed for a long time to the atmosphere, and which might be considered as saturated with air, absorbs a large quantity of oxygene when it is presented to it. It takes it, without abandoning a portion of azote equal in volume to the oxygene absorbed.

The action of water upon the volume of hydrogene gas appears to amount almost to nothing. The inequa- | 458 | lity of the results which we obtained prevent our saying any thing concerning the slight changes which it may undergo during their contact.

The volume of pure azote gas diminishes upon the water by between 2 and 3 hundredth parts; but the residue is no longer pure azote: we discovered in it 11 parts of oxygene which had been displaced from the water by 14 parts of azote. The azote therefore dislodges ozygene from the water, as oxygene dislodges azote. The action is analogous, but the quantities absorbed and dislodged are different.

The contact of the river-water with a mixture of hydrogene and oxygene gases was examined under various circumstances. Sometimes we mixed the two gases in equal quantities, sometimes we caused one or the other to predominate. The diminution of the volume of the gases is greater when the oxygene predominates; that is to say, when we expose to the water a mixture of 200 parts of oxygene with 100 of hydrogene. In all these experiments azote is again dislodged from the water. In analyzing the residuum of a mixture of equal parts of oxygene and of hydrogene, we found in 100 parts 20 parts azote, 50 hydrogene, and 30 oxygene. The greater the absorption of oxygene was, the more azote we found to be dislodged. Having mixed 400 parts of oxygene with 200 of hydrogene, this volume was reduced upon Seine water, in the space of ten days, from 600 parts to 562. If this residue had undergone no chemical change in its proportions, if no other gas had been dislodged, it ought to have contained 375 parts of oxygene and 187 of hydrogene; but our analysis shewed it to contain 246 azote, 142 hydrogene, and 174 oxygene.

These experiments prove that hydrogene, which, when placed alone in contact with water, is not sensibly ab- | 459 | sorbed by it, is dissolved in it, and that in a very considerable proportion, when it is mixed with oxygene. On this subject a question of great importance to Natural Philosophy presents itself; namely, whether this hydrogene absorbed by the water exists in it in the state of hydrogene, or whether it combines in it with oxygene and so forms water. We have endeavoured to solve this question by leaving a mixture of hydrogene and oxygene in contact with water recently deprived of all its air by ebullition. After the space of 12 days, we distilled this water, and on analyzing the air disengaged from it, we found the hydrogene to exist in it in such abundance that we could inflame it in Volta's eudiometer without adding to it any other gas. This experiment proves beyond a doubt that the hydrogene absorbed is found again in the water. But did this water give up again the same quantity which it had absorbed? Would not this hydrogene dissolved in the water unite with the oxygene if it had been lodged in it for several months? We design to make a long series of experiments upon this subject. If the hydrogene and oxygene contained in the water could combine in it, we should be able more readily to conceive how the hydrogene gas which rises from the surface of the earth is not to be discovered either in the air which surrounds us, or in the high regions of the atmosphere into which we have ascended. We must recollect upon this subject that having carefully examined rain-water, in order to discover hydrogene in it, we have ascertained that the air disengaged from this water did not contain a quantity of it amounting to 3/1000. We shall repeat these experiments upon rain that has fallen in different seasons, and particularly during storms.

River water placed in contact with mixtures of gases has in general acted less upon the mixtures of oxygene | 460 | and azote than those of oxygene and of hydrogene. This result will appear the less surprizing if we take a general view of what takes place in these phenomena.

We find that water has a continual tendency to place itself in a state of equilibrium with the gases presented to it. If we present oxygene to it, it abandons azote. If we place it in contact with azote, it abandons oxygene. If we present to it a mixture of oxygene and hydrogene, it absorbs a part of these two gases, and replaces them with azote. It every where tends to modify the proportions of the air which it holds in solution according to the nature of the gases which are presented to it. Now the water of the Seine being charged with a mixture of azote and of oxygene, it seems natural that it should excite more action upon a mixture of hydrogene and oxygene than upon that of azote and oxygene, which is analogous to the air which it holds in solution. In order to obtain an accurate idea of these phenomena, we shall make experiments upon water recently deprived of air, by charging it with different gases both pure and mixed, and examining the action of this water after a long space of time; for frequently it is not till after a long repose that nature is enabled to overcome the obstacles which oppose themselves to the action of the affinities.

Here we shall conclude the account of the inquiries into which we have been engaged for several months past. The more extensive the field is which we purpose to explore, the more strongly we feel the imperfection of the work which we now present to the public; but this sentiment, so far from discouraging us, will only render us the more assiduous in interrogating nature, in order that we may carry these researches to a higher degree of perfection. | 461 |

TABLE Representing the Analysis of the Air.

Days when the Air was collected.		Temperature expressed in Degrees of the centigrade Thermometer.	State of the Atmosphere.	Absorption proceeding from the Inflammation of 200 of Air and 200 of Hydrogene.	Quantity of oxygene contained in 100 Parts of Air.
Brumaire	26	7°,3	Sky overcast; wind E.	126,0	21,0
				126,0	21,0
	27	4,5	Sky overcast; wind E. S. E.	126,0	21,0
				126,0	21,0
	28	4,7	Light rains; very high wind S. W.	126,0	21,0
				126,0	21,0
	29	10,0	Light rain; wind S.	126,0	21,0
				126,5	21,1
	30	12,5	Sky overcast; wind S. W.	126,8	21,2
				126,0	21,0
Frimaire	1	6,7	Sky cloudy, slight rain; wind S. W.	126,0	21,0
				126,0	21,0
	2	1,5	Cloudy sky; wind W.	126,0	21,0
				126,0	21,0
				126,3	21,0
	3	8,5	Rain; wind S.	126,5	21,1
				126,2	21,0
	4	10,6	Sky overcast; wind S. W.	126,5	21,1
				126,5	21,1
	5	3,3	Cloudy sky; wind E.	126,0	21,0
	6	—1,6	White frost; wind N.	126,0	21,1
	7	—1,3	Snow; wind N.	126,5	21,0
	10	—4,1	Fog; wind N. N. E.	126,0	21,0
	12	—2,3	Cloudy sky, mists; wind E.	136,5	20,9
	14	4,2	Rain; wind S.	126,0	21,0
	16	3,1	Thick fog	126,0	21,0
	22	9,6	Rain; wind S. S. W.	126,0	21,0
	28	—2,2	Sky overcast; wind N. E.	126,0	21,0
Nivose	2	1,0	Hoar frost, thick fog; wind S. E.	126,0	21,0

Universal Diffusion of Life

When the active curiosity of man is engaged in interrogating nature, or when his imagination dwells on the wide fields of organic creation, among the multifarious impressions which his mind receives, perhaps none is so strong and profound as that of the universal profusion with which life is everywhere distributed. Even on the polar ice the air resounds with the cries or songs of birds, and with the hum of insects. Nor is it only the lower dense and vaporous strata of the atmosphere thus filled with life, but also the higher and more ethereal regions. Whenever Mont Blanc or the summits of the Cordilleras have been ascended, living creatures have been found there. On the Chimborazo, 8,000 feet higher than Etna, we found butterflies and other winged insects, borne by ascending currents of air to those almost unapproachable solitudes, which man, led by a restless curiosity or unappeasable thirst of knowledge, treads with adventurous but cautious steps; like him, strangers in those elevated regions, their presence shows us that the more flexible organization of animal creation can subsist far beyond the limits at which vegetation ceases. The condor, the giant of the vulture tribe, often soared over our heads above all the summits of the Andes, at an altitude higher than would be the Peak of Teneriffe if piled on the snow-covered crests of the Pyrenees. The rapacity of this powerful bird attracts him to these regions, whence his far-seeing eye may discern the objects of his pursuit, the soft-woolled vicunas, which, wandering in herds, frequent, like the chamois, the mountain pastures adjacent to the regions of perpetual snow.—

Humboldt.

Memoir on a new Species of Monkey found in the eastern Declivity of the Andes[1]

In the vast plains which extend from the eastern declivity of the Andes towards the shores of the Brazils, in the thick forests on the Amazons, Rio Negro (Black River), and Oronoco, the *cavia capybara, sus tajassu*, and the monkeys, are the quadrupeds most common. The marmose and alouate prevail over all the others, whether for variety of species or number of individuals. Some of these monkeys, such as the capuchin of Oronoco, very different from the simia capucina Linn., the tiger-monkey or cusicusi, and the widow, (three new species which I have discovered,) live in pairs, melancholy, mistrustful, flying (like man in a savage state) their proper species. Others, especially the sagouin and howling monkey, are seen in troops, from 80 to 100, springing from branch to branch in search of nourishment. The *saimiri* of Buffon, that are the *titi* of Atures (*simia sciurea*), so esteemed on account of their gaiety, their mildness, and extreme littleness, assemble together when it begins to rain. A fall of temperature of three or four degrees of the centigrade thermometer disturbs them so much, that they mutually embrace each other and form balls or knots, of which each individual seeks to occupy the middle, in order to find shelter. The Indian hunters, advertised by the cry of the titi, direct their arrows towards these flocks. Notwithstanding the great number of monkeys that naturalists have described, it is probable that we are still ignorant of the tenth part existing. In Africa, and even in South America, there are vast plains of twenty thousand square leagues which have not yet been visited by any European. On the other hand, the monkeys the most common are still so imperfectly represented even in the most recent works, executed with the greatest elegance, that those who have seen the living individuals would have difficulty to recognise them in the drawings published. Of this I might cite as examples the *simia sciurea*, varieties of the *simia capucina*, and even the *simia paniscus*, which is the game commonly eaten in the Upper Oronoco.

Among the great number of new *sapajous*, or marmoses, that I had the opportunity of describing in my voyage to the tropics, I have chosen

[1] From *Recueil d'Observations de Zoologie et d' Anatomie comparé,* 1re livraison.

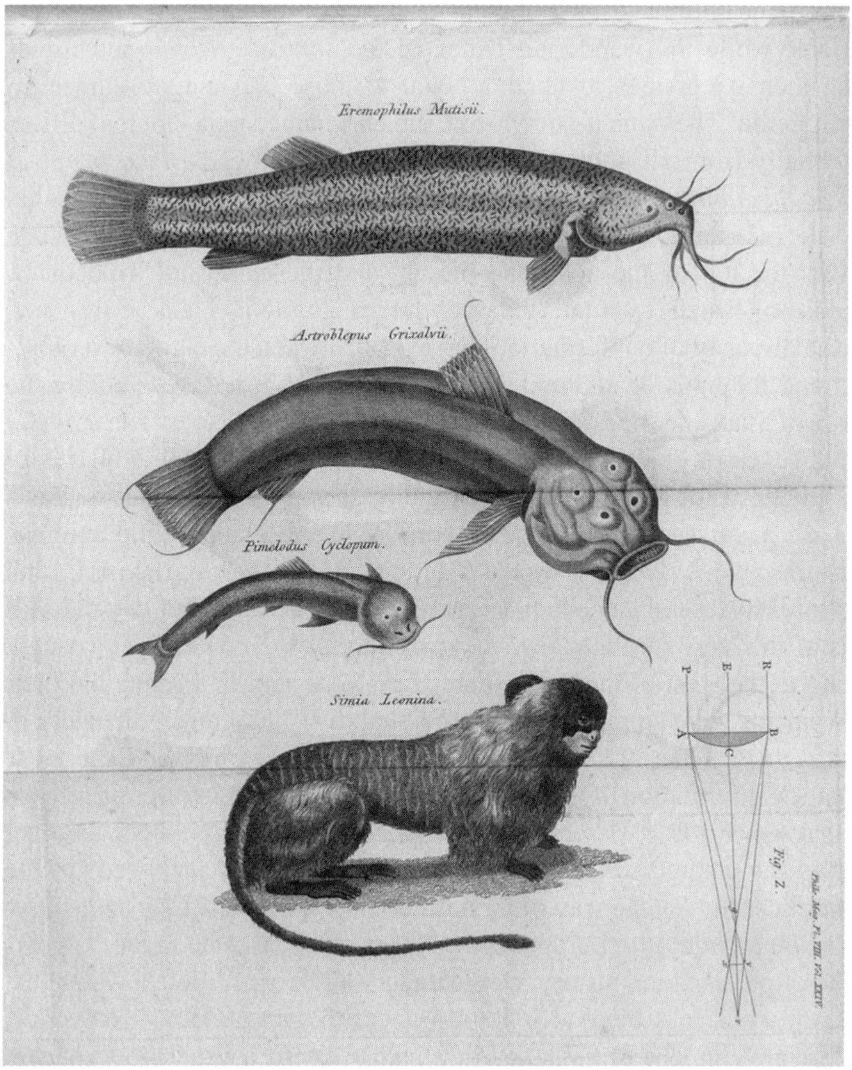

Plate VIII.

a monkey of the plains of Mocoa, remarkable for its resemblance with the lion of Africa, of which I made a drawing during my residence at Popayan. My sketch has been copied and improved by M. Turpin. (Plate VIII.)

The *leoncito*[2] is very rare, even in its native country. It inhabits the plains which border on the eastern declivity of the Cordelliers, the fertile

[2] *Leoncito*, from *leon* (lion) a diminutive of endearment more common even with the Spaniards than the Italians.—*Translator*.

banks of the Putumayo and Caqueta: it never ascends even to the temperate regions, while the wandering bands of the *simia beelzebul* sometimes push their excursions to heights equal to those of Canigou, and even Mont Perdu. The *leoncito* of Mocoa, which I name simia leonina, differs essentially from all known species. It has *not the white head of the s. leucocephala figured in the work of Audebert. It differs from the s.* rosalia, and the saki or fox-tailed monkey (*s. pithecia*), by a white spot which covers the top of the nose, the mouth, and the chin; and from the *s. iacchus* of Brazil, by a tail without white rings, by its black visage, and by the disparity of conformation that exists between the claws of its fore-feet and the nails on its hinder ones; the former almost resembling the claws of a cat, and the latter having nails like the human toes. The *leoncito* is but seven or eight inches long, without counting the tail, which is of the length of the body. It is one of the least and most elegant monkeys that we have seen. It is gay and sportive, but, like most little animals, very irascible. When it is vexed it bristles up the hair on its neck, that increases its resemblance to the African lion. I have seen but two individuals of this very rare monkey: they were the first that had been brought living to the west of the Cordellier. They were kept in a cage; and their movements were so rapid and so continual, that I had much difficulty to design them. Their hissing imitates the song of little birds, and I suppose that the conformation of their larynx (having a particular sac) is analogous to that which I have described in the *simia œdipus*. I have been assured, that in the cottages of the Indians of Mocoa the leoncito breeds in the domestic state. By the way of Grand-Para and the river of Amazons they might be brought into Europe. If a government, interested in the progress of descriptive natural history, would undertake an expedition in which that interest would not be rendered subservient to geographical discoveries; if that government sent canoes or small boats to ascend the Oronoco, and to penetrate by the Casiquiaré and Rio Negro to the river of Amazons; in short, if, after having explored the mouths of the Caqueta and Putumayo, it would make these same boats descend to Grand-Para, it would in a little time unite collections the most precious to the study of zoology and botany. Such an expedition would be of little expense, and its success certain.

Simia leonina.

Ex olivaceo fuscescens, facie atra, ore albo, dorso striis albo-flavescentibus notato.

Caput parvum, depressum, nigrescens. Facies anthropomorpha, atra; macula albo-cærulescens circa os et nares. Auriculæ subtriangulares distantes, margine superiori deflexo, magnæ, aterrimæ, pilosæ. Corpus ex badio olivaceum, pilis nigro-annulatis, in collo longioribus. Dorsum maculis et striis albo-flavescentibus variegatum. Cauda non prehensilis, longitudine corporis, superne atra, inferne badia, apice incurva et incrassata. Manus et pedes aterrimi, inferne nudi, pollice in manibus anterioribus et posteriorbus distante. Ungues acuti, incurvi, atri; pollicis ungue in manibus anterioribus oblongo, acuto, in manibus posterioribus (pedibus) obtuso, anthropomorpho.

Memoir on a new Species of Pimelodus thrown out of the Volcanoes in the Kingdom of Quito; with some Particulars respecting the Volcanoes of the Andes[1]

The chain of the Andes, from the Straits of Magellan to the northern shores bordering on Asia, extending over more than 2000 leagues, presents above fifty volcanoes still active, of which the phænomena are as various as their height and local situation. A small number of the least elevated of these volcanoes throw out running lava. I have seen, at the volcano of Zurullo, in Mexico, a basaltic cone that sprung from the earth the 15th September 1759, and at present rising 249 toises (1595 2/3 feet) above the surrounding plain. The volcanic ridges of Guatimala cast out a prodigious quantity of muriate of ammonia. Those of Popayan and the high plain of Pasto present either *solfatares*, which exhale sulphureous acid, or little craters filled with boiling water, and disengaging sulphurated hydrogen, which decomposes by contact with the oxygen of the atmosphere. The volcanoes of the kingdom of Quito throw out pumice-stone, basaltes[2], and scorified porphyries; and vomit enormous quantities of water, carburetted argil, and muddy matter, which spreads fertility from eight to ten leagues around. But, since the period to which the traditions of the natives ascend, they have never produced great masses of running melted lava. The height of these colossal mountains, that surpasses five times that of Vesuvius, and their inland situation, are, without doubt, the principal causes of these anomalies. The subterranean noise of Cotopaxi, at the time of its great explosions, extends to distances equal to that from Vesuvius to Dijon. But, notwithstanding this inten- | 334 | sity of force, it is known, that if the volcanic fire was at a great depth, the melted lava could neither raise itself to the edge of the crater, nor pierce the flank of these mountains, which to the height of 1400 toises (8971 2/3 feet) are fortified by high surrounding plains. It appears, therefore, natural, that volcanoes so elevated should discharge from their mouth but isolated stones, volcanic

[1] From *Recueil d'Observations de Zoologie et d'Anatomie comparé*, 1re livraison.

[2] It would have been of some use to geology had the author here mentioned whether the stone which he calls basaltes has been submitted to the action of fire or water; or whether, in addition to the other well known characters of this mineral, it yielded hydrogen gas on distillation, the latter being the peculiar characteristic of what is properly denominated basaltes.—*Translator*.

cinders or ashes, flames, boiling water, and, above all, this carburetted argil impregnated with sulphur, that is called *moya*[3] in the language of the country.

The mountains of the kingdom of Quito occasionally offer another spectacle, less alarming, but not less curious to the naturalist. The great explosions are periodical, and somewhat rare. Cotopaxi, Tungurahua, and Sangay, sometimes do not present one in twenty or thirty years. But during such intervals even these volcanoes will discharge enormous quantities of argillaceous mud; and, what is more extraordinary, an innumerable quantity of fish. By accident, none of these volcanic inundations took place the year that I passed the Andes of Quito; but the fish vomited from the volcanoes is a phænomenon so common, and so generally known by all the inhabitants of that country, that there cannot remain the least doubt of its authenticity. As there are in these regions several very well informed persons, who have successfully devoted themselves to the physical sciences, I have had an opportunity of procuring exact information (*renseignemens*) respecting these fishes. M. de Larrea, at Quito, well versed in the study of chemistry, who has formed a cabinet of the minerals of his country, has been, above all others, the most useful to me in these researches. Examining the archives of several little towns in the neighbourhood of Cotopaxi, in order to extract the epochs of the great earthquakes, that fortunately have been preserved with care, I there found some notes on the fish ejected from the vol- | 335 | canoes. On the estates of the marquis of Selvalegre the Cotopaxi had thrown a quantity so great, that their putrefaction spread a fetid odour around. In 1691 the almost extinguished volcano of Imbaburu threw out thousands on the fields in the environs of the city of Ibarra. The putrid fevers which commenced at that period were attributed to the miasma which exhaled from these fish, heaped on the surface of the earth and exposed to the rays of the sun. The last time that Imbaburu ejected fish was on the 19th of June 1698, when the volcano of Cargneirazo sunk, and thousands of these animals enveloped in argillaceous mud were thrown over the crumbling borders.

The Cotopaxi and Tungurahua throw out fish, sometimes by the crater which is at the top of these mountains, sometimes by lateral vents, but constantly at 2500 or 2600 toises above the level of the sea: the

[3] M. Humboldt seems not to have been aware that this name has been affixed to it in consequence of its having some resemblance to a kind of blackish coarse bread made of grits or pollard, and used in Spain by some very poor but proud people, or for purposes of penitence in cases of a pecado mortal. —*Translator*.

adjacent plains being 1300 toises high, one may conclude that these animals issue from a point which is 1300 toises more elevated than the plains on which they are thrown. Some Indians have assured me that the fish vomited by the volcanoes were sometimes still living in descending along the flank of the mountain: but this fact does not appear to me sufficiently proved: certain it is, that among the thousands of dead fish that in a few hours are seen descending from Cotopaxi with great bodies of cold fresh water, there are very few that are so much disfigured that one can believe them to have been exposed to the action of a strong heat. This fact becomes still more striking when we consider the soft flesh of these animals, and the thick smoke which the volcano exhales during the eruption. It appeared to me of very great importance to descriptive natural history to verify sufficiently the nature of these animals. All the inhabitants agree that they are identical with those which are found in the rivulets at the foot of these volcanoes, and called *prennadillas*[4]: they are even the only species of fish that is discovered at the height of above 1400 toises | 336 | in the waters of the kingdom of Quito. I have designed it, with care, on the spot, and my design has been coloured by M. Turpin. I have observed that the *prennadilla* is a new species of the genus *silurus*. M. Lacepede, who has also examined it, advised me to place it in that division of *silurus* which, in the fifth volume of his Natural History of Fishes, he has described under the name of *pimelodes*.

This new species of *pimelodus* has a depressed body of an olive colour mixed with little black spots. The mouth, which is at the extremity of the nose, is very large, and furnished with two barbillons or whiskers attached to the jaws. The nostrils are tubulous; the eyes are very small, and placed towards the middle of the head. The skin of the body and the tail is covered with an abundant mucus, and the mouth is furnished with very small teeth. The branchial membrane has four radii, like the *pimelodus chilensis*; the pectoral fin has nine; the ventral five; the first dorsal six; the fin of the anus seven; and that of the tail, which is bifid, has twelve radii. The first radius of all the fins is indented on the outside: the second dorsal fin is adipose, and placed near the tail. This little *pimelodus*, which is found in lakes even to the height of 1700 toises, is, without doubt, the fish that lives in the most elevated regions of our globe. Its common length scarcely amounts to ten centimetres (four

[4] This word is an indifferent or contemptuous diminutive, indicating abundant, pregnant, fruitful, easily taken, but not a pleasing or desirable object. The name is purely Spanish and not Indian, of course could never have been applied to any fish used as food by Spaniards.—*Translator*.

inches); but there are varieties which do not appear to reach five centimetres (two inches) in length.

In the system of ichthyology this new species of *pimelodus* should be ranged in the first sub-genus established by Lacepede, among the forked-tailed pimelodes. It must be in the first species, before the *pimelodus bagre*. As it is the only one of that division that has but two whiskers, I give it the name of PIMELODUS Cyclopum. (Plate VIII.) *Cirris duobus, corpore olivaceo nigro-punctato.*

This little fish lives in rivulets at the temperature of 10° of the centigrade thermometer, while other species of the same genus exist in rivers in the plains the water of which is at 27°. The *pimelodus* is but very rarely eaten, and then | 337 | only by the most indigent race of Indians; its aspect and the sliminess of its skin render it very disgusting.

From the enormous quantity of *pimelodes* that the volcanoes of the kingdom of Quito occasionally discharge, one cannot doubt that country contains great subterranean lakes which conceal these fishes; for the individuals that exist in the little rivers around are very few in number. A part of those rivers may communicate with the subterranean pits: it is also probable that the first *pimelodes* which have inhabited these pits have mounted there against the current. I have seen fish in the caverns of Derbyshire, in England; and near Gailenreuth, in Germany, where the fossil heads of bears and lions are found, there are living trouts in the grottoes, which at present are very distant from any rivulet, and greatly elevated above the level of the neighbouring waters. In the province of Quito, the subterraneous roarings that accompany the earthquakes; the masses of rocks that we think we hear crumbling down below the earth we walk on; the immense quantity of water that issues from the earth in the driest places during the volcanic explosions; and numerous other phænomena, indicate that all the soil of this elevated plain is undermined. But, if it is easy to conceive that vast subterranean basins may be filled with water which nourishes fishes, it is more difficult to explain how these animals are attracted by volcanoes that ascend to the height of 1300 toises, and discharged either by their craters or by their lateral vents. Should we suppose that the *pimelodes* exist in subterranean basins of the same height at which they are seen to issue? How conceive their origin in a position so extraordinary; in the flank of a cone so often heated, and perhaps partly produced by volcanic fire? Whatever may be the source from which they issue, the perfect state in which they are found induces us to believe that those volcanoes, the most elevated and the most active in the world, experience, from time to time, convulsive movements, during which the disengagement of caloric appears less considerable than we should suppose it. Earthquakes do not always accompany those phænomena. Perhaps, in the different concamerations that may be ad- | 338 | mitted in the interior of a volcano, the air is found occasionally condensed, and that it is this condensed air which contributes to raise the water and fish; perhaps they issue from a concavity distant from those which emit volcanic fire; possibly, in fine, the argillaceous mud in which those animals are enveloped defends them from the action of great heat. Notwithstanding all the researches that have been recently made on volcanoes, there is nothing but the study of volcanic productions that has made any progress. As to the nature of the combustibles which nourish those subterranean fires, and the mode of action of those fires themselves, I believe that all persons who have visited the borders of craters, and who have lived a long time in the vicinity of volcanoes, will sincerely avow, with me, that we are still very far from being able to give an explication, which, without being contrary to the principles of chemistry and of physics, could account for the great phænomena which volcanic explosions present.

The corregidor of the city of Ibarra, don José Posse Pardo, has communicated to me an interesting observation on the *pimelodes*. "It is known (says he, in a letter which I have still preserved,) that the volcano of Imbaburu, at the time of its great eruption on the side next our city, threw out an enormous quantity of *prennadillas*: it even continues still occasionally to do so, especially after great rains. It is observed that these fishes actually live in the interior of the mountain, and that the Indians of S. Pabla fish[5] for them in a rivulet at the very place whence they issue from the rock. This fishery does not succeed either in the day or in moonlight: a very dark night is therefore necessary, as the *prennadillas* will not otherwise come out of the volcano, the interior of which is hollow." It appears, then, that the light is injurious to those subterranean fishes, which are not | 339 | accustomed to so strong a stimulus: an observation so much the more curious, that the pimelodes of the same species, which inhabit the brooks in the vicinity of the city of Quito, live exposed to the brightness of the meridian sun.

[5] This is an assertion somewhat contrary to that of their being very bad food, and disagreeable in appearance. It is within the particular knowledge of the translator, that the Spaniards of South America are both very sceptical and very witty, and that to play upon the philosophical faith of Europeans would be their highest delight. He must therefore be pardoned for regarding the letter of *el Senor Corregidor* as a *jeu d'esprit en revanche* for the sarcastic observations of French travellers on the Spaniards. —*Translator.*

Memoir on the Eremophilus and Astroblepus, two new Genera of the Order of Apodes[1]

When we ascend the chain of the Andes to the height of 2600 toises (16661 2/3 English feet) and upwards, great level plains and lakes of a considerable extent are seen. It is singular to observe, that, while the soil is still covered with a beautiful vegetation, the woods filled with quadrupeds, and the air with a great variety of birds, the water alone, the lakes and the rivers, are so little inhabited. The cause of this phænomenon relates, without doubt, to geological facts; it pertains to the grand mystery of the origin and migration of species.

The considerable lakes which surround the city of Mexico, at the height of 1160 toises,[2] nourish but two species of fish, of which one, the *axalotl*, belongs rather to the genera *sirenus* and *proteus*. M. Cuvier, to whom we brought this extraordinarily organized animal, unknown in Europe, is engaged with its anatomy, which he will shortly publish. In the kingdom of New Granada, in the beautiful valley of Bogota, about 1347 toises high, there also exist but two species, which the inhabitants of that country call *capitan* and *guapucha*. The one is an *atherine*, and the other a new genus of *apodes*, that I am about to describe in this memoir. The form of its tail and its anal fin distinguish it sufficiently from the genus *trichiurus*, which is also found in the fresh waters of South America. I have designed this non-descript fish at the place; and Messrs. Lacepede and Cuvier, who have willingly examined my descriptions, like me, consider it a new genus well characterized. I have named it *eremophilus* on accout of the solitude in which it lives at so great | 330 | an elevation, and in waters which are inhabited by almost no other living being. The naturalists, who fear that new species of the same genus may be discovered in very different situations, may change the name of *eremophilus* into that of *thrichomycterus*, taken from the barbillons or whiskers attached to the nose of this fish.

[1] From *Recueil d'Observations de Zoologie et d'Anatomie comparé*, 1re livraison. Communicated by a correspondent.

[2] The French toise is about six feet four inches nine-tenths English.

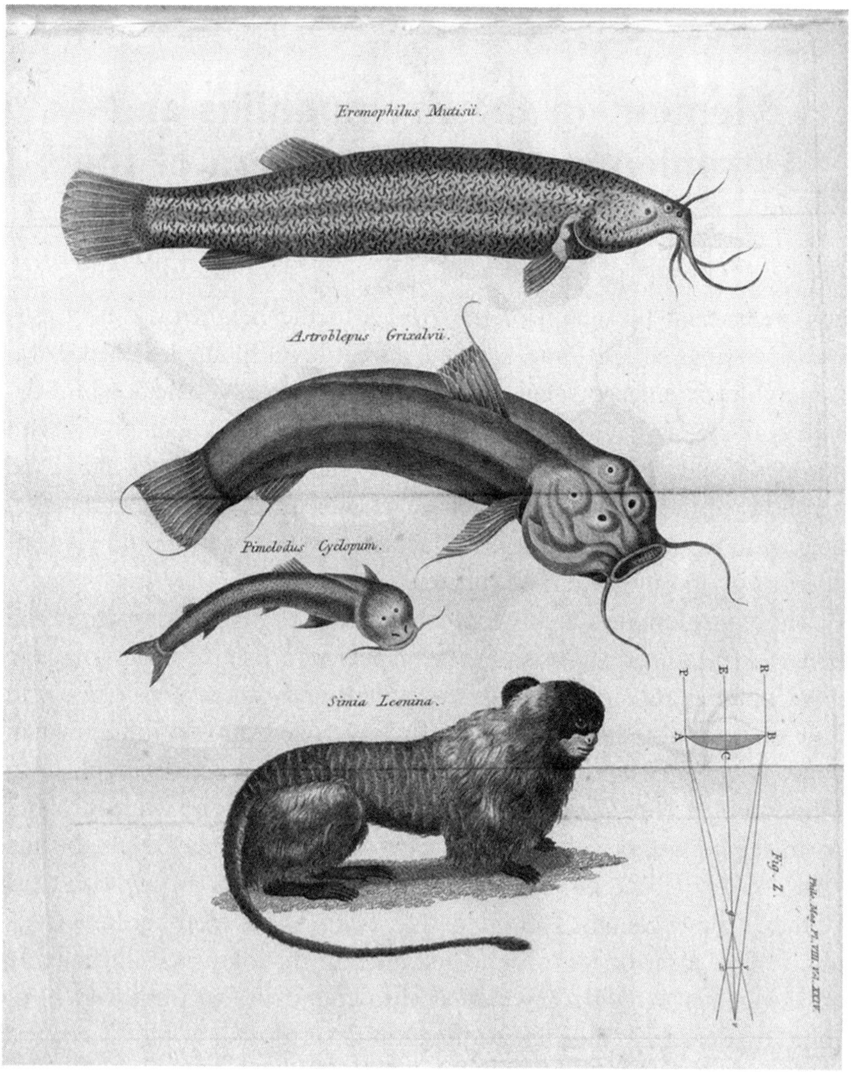

Plate VIII.

EREMOPHILUS. (See Plate VIII.)

Apod. Character Genericus Essentialis.

Corpus elongatum. Cirri maxillares 4, nasales semitubulosi 2. Pinna dorsalis et analis. Membrana branchiostega radiis 1—2.

E. Mutisii.

Corpore elongato, plumbeo, cærulescenti, maculis dædaleis olivaceis variegato; operculi branchiostegi; duplicatura spinuloso-serrata.

The body of the *captain* of Bogota is long, and has some analogy with that of the eel. It is compressed, of a blueish gray colour, and spotted with olive green. These spots, the outline of which forms very striking sinuosities, assume in some individuals a yellowish tint. The head is little and flat. The mouth, situated at the extremity of the nose, is straight. The upper jaw projects over the under one: the first, very long and double, is furnished with six fleshy barbillons or whiskers, of which the two exterior are the longest. Two other barbillons, shorter and semi-tubulous, are placed on the nostrils. It has very little eyes, which are veiled by a semi-transparent membrane, like the gymnotes and lampreys. The extremity of its lips is furnished with little teeth resembling hairs. The tongue is very fleshy, but short. The operculum or uvula forms a very narrow branchial opening, and it is very difficult to distinguish its folds (lames). In the most part of the individuals which I have examined, it appears to me that the captain, similar to the *cyclopterus dentex* and a few other fishes, has but two radii or furrows, which are as if soldered the one on the other. The edge of the operculum or uvula is indented. The dorsal fin has eight radii, the pectoral six, that of the anus six, and that of the tail, which is round, twelve radii: | 331 | it has no swim or air-bladder. The length of this fish is from 10 to 11 inches, and its body is covered with a mucus common to the greater part of the apodes. It inhabits the little river of Bogota, that forms the famous cataract of Tequendama. The captain is a very agreeable aliment, and so much the more precious, that without it the inhabitants of the capital of Santa Fé, in the time of Lent, would be reduced to the use of only salted sea-fish brought from a great distance. I have given this species the trivial name *Mutisii* in houour of the celebrated naturalist, whose rich collections are preserved in the great valley of Bogota.

The little river of Palacé, near Popayan, nourishes another fish, which, by its mucosity and the position of its fins, has some relation with the *eremophilus*, but which ought also to constitute a new genus of apodes. The breadth of its head is greater than that of the body; its eyes placed on the upper part of the head, and turned so that the pupils are directed, like as in the *uranoscopus mus*, towards the surface of the water; the indenting of the first radii of the fins, the branchial membrane of four radii; the tongue; the want of barbillons or whiskers on the nostrils; and the dorsal fin, which approaches more to the head than the tail; sufficiently distinguish the *pescado negro* (black fish) of Popayan from the *capitan* of Santa Fé. I have given the name of *astroblepus* to this genus, in allusion to the extraordinary situation of its eyes.

ASTROBLEPUS. (See Plate VIII.)

Apod. Character Genericus Essentialis.

Corpus plagioplateum. Membrana branchiostega radiis 4. Oculi verticales. Cirri 2 maxillares, nasales nulli.

A. Grixalvii.

Corpore ex olivaceo nigrescenti, capite subtruncato, radiis pinnarum exterioribus serratis.

Corpus plagioplateum, oblongum, nudum, olivaceo-nigrescens, caudam versus angustatum, subcompressum. Caput obtusum, magnum, subtruncatum. Cirri 2, apice recurvi et sublati, ricto in labio superiori adnati. Maxilla labiata, labio superiori majori plicatili. Lingua nulla. Nares 2 | 332 | magnæ, margine membranaceo. Oculi verticales, minuti. Operculum simplex, convexum, nudum. Membrana branchiostega radiis 4, ossiculo anteriori subserrato. Pinua pectoralis radiis 10. Pinna analis radiis 7. Pinna caudalis integra radiis 12; radiis duobus exterioribus (ut omnium pinnarum) extrorsum serratis. Longitudo 14 pollicaris.

I have given this fish the specific name Grixalvii[3] to perpetuate the memory of a respectable philosopher, Don Mariano Grixalva, who has disseminated at Popayan a taste for the physical sciences, which he himself cultivated with success.

The *pescado negro*, so much eaten at Popayan, is found but in that part of the river Cauca which is most contiguous to the city. The physical cause of this phænomenon is sufficiently striking. From the volcano of Purasé descends a rivulet impregnated with sulphuric acid, that the inhabitants call *Rio Vinagre* (Vinegar River); it is known by the beautiful cascade which it forms at the foot of the volcano. From the point where

[3] *Mutisii* and *Grixalvii* are doubtless very scientific names. Linnæus, to gratify his puerile vanity, introduced the custom of giving arbitrary unmeaning names of men to plants: Werner embraced the same unphilosophical system of pitiable ambition in baptizing minerals (some wits have asserted, indeed, that such is his attachment to water, that he actually performed the ceremony of sprinkling certain stones, giving them at the same time the favourite name of some of his followers): and M. Humboldt now transfers men's names to the very opposite abodes of fire and water, in his volcanic fish! All the labours of these men have done much less to disseminate a taste for the natural sciences, than the introduction of such an absurd practice has effected in obstructing the advancement of real knowledge and true philosophy. Posterity, so far from venerating such names, will execrate the being, who, to conceal his real ignorance by the assumption of universal knowledge, could thus deliberately bury true science and much accurate knowledge under the ruins of a Babylonish jargon! Peace to the manes of Lavoisier: although he himself made no real discoveries, yet the philosophical use which he made of those of the English and other philosophers will not speedily be forgotten by succeeding generations.—*Translator.*

the waters of Vinegar River unite with those of Cauca until four leagues lower down the latter is without fish, although in the upper part they are plentiful enough. The small quantities of acid that might escape our chemical analysis are often sufficiently great to injure the organization of fishes.

Baron Humboldt's Description of the Mode of Catching the Gymnotus, or Electrical Eel

While traversing the immense plains of the province of Caraccas, in order to embark at St. Fernando de Apure, and commence our voyage on the Orinoco, we stayed five days at Calabozo, a small town situated, according to my observations, in 8° 56′ 56″ north latitude. The object of this visit was to study the electric eels (Gymnoti), an innumerable quantity of which are found in the neighbourhood. Indeed I was assured, that near Urituca a road, once much frequented, had been entirely abandoned on account of these electric fishes; travellers being obliged to ford a stream in which a number of mules were annually drowned, stunned by the shocks they received from the Gymnoti.

In order to perform our experiments with more precision, we were desirous of having the electric eels in the house which we inhabited at Calabozo. Our host took all possible pains to gratify our desires. He sent Indians on horseback to fish in the pools. It was easy enough to procure plenty of dead Gymnoti; but an almost puerile fear prevented the natives from bringing them alive. We were afterwards, it is true, convinced of the difficulty of handling this fish whilst it still retained all its powers: but the dread of the common people | 77 | is the more extraordinary in these countries, as they pretend that any one with tobacco in his mouth may touch a gymnotus with impunity. If the Indian had full confidence in this preservative, which is altogether false, why did he not avail himself of it to gain the ten francs which we had promised for every living electric eel that was brought to us? The love of the marvellous is so great among the natives, that they often maintain and spread abroad as facts what they themselves are far from putting faith in. It is thus that man thinks he must needs add to the wonders of Nature, as if Nature was not of herself sufficiently mysterious, grand and imposing.

After three days of fruitless attempts in the town of Calabozo, during which we received only one eel alive, and that in a very weak state, we resolved to transport ourselves to the spot, and make our experiments in the open air, on the banks of the pools in which the gymnoti abounded. We reached first a little village called Rastro de Abaxo; thence the Indians conducted us to Caño de Bera, a muddy stagnant pool of water, but surrounded with a beautiful vegetation of the *Clusia rosea*, the *Hymenea Courbaril*, the large Indian figs, and *Mimosæ* with odoriferous flowers. We were much surprised on being told that thirty wild horses must be caught in the neighbouring savannahs, to serve for the fishery of the electric eels. The idea of this mode of fishing, which they call *embarbascar con caballos*, (to stupify by means of horses,) is truly very odd. The word *barbasco* signifies the roots of *Jacquinia*, of *Piscidia*, or of any other poisonous plant, by the contact of which a great extent of water receives in an instant the power of killing, or at least of intoxicating or benumbing, the fishes. The latter rise to the surface of the water when they are poisoned in this way. As the horses driven here and there in a pool cause the same effect with respect to the frightened fishes, the natives, confounding cause and effect, call the two modes of fishing by the same name.

Whilst our host was explaining to us this strange mode of catching fish in this country, the troop of horses and mules arrived. The Indians formed a sort of *battue*, and encircling the poor quadrupeds closely on every side, compelled them to enter the pool. I can but imperfectly depict the interesting spectacle which the attack of the eels upon the horses presented to us. The Indians, holding long canes and harpoons, place themselves around the pool; some of them climb the trees whose branches project over the surface of the water; and all of them by their shouts and their weapons prevent the | 78 | horses from gaining the bank. The eels, confounded by the hubbub and splashing of the horses, defend themselves by the reiterated discharge of their electric batteries.

For a long time they seem victorious over the horses and mules; everywhere we see the latter, stunned with the frequency and force of the electric shocks, disappearing beneath the water. Some of the horses recover themselves, and, despite the active vigilance of the Indians, gain the bank; overcome by fatigue, and their limbs paralyzed by the power of the electric commotions, they lay outstretched upon the earth.

How I wished that a clever painter could have seized the moment when the scene was at its greatest animation. The groups of Indians surrounding the pools,—the horses, with mane erect, and eyes of fright and pain, struggling to escape from the storm which had surprised them;—those yellow livid eels, like great aquatic serpents, swimming upon the surface of the water, and pursuing their enemy;—all these objects presented, in truth, a most picturesque ensemble. I remembered the superb painting which represents a horse entering a cavern and scared at the sight of a lion! The expression of terror was not stronger than we witnessed in this unequal conflict.

In less than five minutes, two horses were already drowned. The eel, being more than five feet in length, glides beneath the body of the horse or mule; it then gives a discharge from the whole length of its electric organ. It attacks at the same time the heart, the digestive viscera, and above all the *plexus* of the gastric nerves. One cannot feel astonished, then, that the effect produced by the fish upon a great quadruped much exceeds that produced on man, which it only touches by one of the extremities. I doubt, however, that the Gymnoti kill the horses immediately; I rather imagine that the latter, stunned by the electric shocks which they receive in rapid succession, fall into a profound lethargy. Deprived of all sensibility, they disappear beneath the water, the other horses and mules pass over their bodies, and a few minutes suffice to make them perish.

After this commencement, I began to fear that this conflict would terminate very tragically. I did not doubt but that I should in a little time see all the mules drowned. We pay, however, only eight francs for each, if the master of it is known. But the Indians assured us that the fishing would soon be over, and that it was only the first assault of the gymnoti that was formidable. In fact, whether it is that the galvanic electricity accumulates during a state of rest, or | 79 | that the electric organ ceases to perform its functions when fatigued by too long usage, the eels, after a certain time, resemble discharged batteries. Their muscular motions continue equally lively, but they have no longer the power of discharging very energetic shocks. When the conflict had lasted a quarter of an hour, the mules and horses seemed less alarmed; they did not erect the mane, their eye expressed less pain and terror; we saw them no longer fall over. On the other hand the eels, swimming half out of the water, and endeavouring to avoid the horses in place of attacking them, in their turn approached towards the bank. The Indians assured us, that no horse is killed when driven two days afterwards into the pool. These electric fishes require rest and abundant nourishment to produce and to accumulate a great quantity of the galvanic electricity. We know, by the experiments that have been made upon the torpedos (electric rays) of Italy, that on cutting or tying the nerves which go to the electric organs, the functions of these parts cease as the movement of a muscle is suspended, as long as the ligature of the main artery or nerve remains. The organs of the torpedo or gymnotus depend upon the nervous system and the vital functions: they are not mere electro-motive apparatuses, which attract from the surrounding bed of water the electricity they have lost. One cannot, therefore, be surprised that the power of the electric shocks of the gymnoti depends upon their health, and consequently upon rest, nourishment, age, and probably a great assemblage of physical and moral conditions.

The eels, making towards the bank, are taken with great facility. Small harpoons, attached to cords, are cast at them; the harpoon sometimes brings up two at a time. By this means they are drawn out of the water without the cord, which is very dry and of some length, communicating the shock to the person who holds it. In a few minutes five large eels were on dry land. One might have caught twenty, if as many had been wanted for our experiments. Many were only slightly wounded in the tail; others severely in the head. We were enabled to observe the natural electricity of these fishes, modified by the different degrees of the vital force which they enjoyed. I shall describe in this Memoir not only the experiments made, in conjunction with M. Bonpland , on the gymnoti taken in our presence, but also those which we had an opportunity of making upon an eel of an enormous size, which we found in our house on our return from Rastro to Calabozo. This latter had been caught in a net; it had not received any wound. Immediately on being | 80 | drawn out of the water, it had been put into the same tray in which it was brought to Calabozo. Remaining thus in the same water it was accustomed to, galvanic electricity had not been altered. We shall see, however, in the course of this Memoir, that the wounded gymnoti, that is to say, those of less power, are much more instructive in the inquiry into galvanic phænomena than the very active gymnoti, for many shades escape the eye of the observer when the electric torrent takes as impetuous a course across good conductors, as across those which are more imperfect.

When one has seen the eels knock down a horse, and deprive it of all sensibility, he is naturally afraid to touch them the moment they are brought out of the water. So strong, in fact, is this dread among the people of the country, that none of them could muster up resolution to disengage the gymnoti from the cords of the harpoon, or to transport them to the little holes filled with water which we had made along the bank of the Caño de Bera. It needed all our fortitude to receive ourselves the first shocks, which certainly were not very agreeable. The most energetic surpassed in power the most painful electric shocks I ever remembered to have received, accidentally, from a large and completely charged Leyden jar. From that time we readily conceived that, without doubt, there was no exaggeration in the story of the Indians, when they assert that persons who are swimming are drowned when one of these eels attacks them by the leg or the arm. A discharge so violent is fully capable of depriving a man for many minutes of all use of his limbs. If the gymnotus should glide along the belly or the chest, death might even follow instantaneously the electric shock; for, as we have before observed, the most noble parts, as the heart, the gastric system, the cœliac plexus, and all the nerves that depend upon it, would at once be deprived of their irritability. A feeble electricity augments the vital forces,—a strong degree extinguishes them entirely.— *Voyage de Humboldt et Bonpland*; *Zoologie*, p. 54.

On the Cinchona Forests of South America

SECTION I

The present Essay is written with a view to examine the Cinchona tree as an object of physical or botanical geography. Amongst the numerous writers mentioning the Cinchona, there are none but La Condamine, Ruiz, Pavon, and Zea, who themselves have observed this beneficial tree upon the South American continent. Only the first of these gives a physical description of this plant; the others, as well as Jacquin and Swartz, who saw the Cinchonæ in the West-India Islands, and Vahl and Lambert, who occupied themselves with dried specimens, have merely treated on the natural history and the botanical diagnosis. During my stay of four years in South America, I had occasion to reside a long time in countries where the Cinchona trees are indigenous. M. Bonpland and myself have observed them north and south of the equator, in the kingdom of New Granada, betwixt Honda and Santa Fe de Bogota, in the province of Popayan, in the corregiment of Loxa, on the Amazon river, in the province of Jaen de Bracamoros; and in the northerly part of Peru. During our abode in the house of Don Jose Celestino Mutis, in Santa Fe, the botanical treasures of that great natural philosopher were opened to us. In Spain, | 20 | also, we were enabled to collect, from the editor of the *Flora Peruviana*, in Guayaquil, (the harbour of Quito on the coast of the South Sea) from M. Tafalla, a pupil of Ruiz, in the little town of Loxa, from Don Vincente Olmedo, royal inspector of the Cinchona forests, many interesting accounts respecting objects which, but for the obliging communications from those friends, would have remained unknown to us. Respecting the very violent controversy on the question, whether the orange-coloured Cinchona bark of New Granada, or the Peruvian *Cinchona nitida*, described by Ruiz and Pavon, be identical with the genuine Cinchona of Uritusinga, famed already since 1638, he only can decide who has himself explored the regions producing these three plants. But of the contending parties, neither Mutis, Zea, nor Ruiz and Pavon, have ever set their feet in the corregiment of Loxa. Thence it is, that each party has, with equal want of foundation, asserted that the most efficacious Cinchona bark of their respective districts was the genuine one from Uristusinga. In the second fasciculus of our Æquinoctial Plants[1] we have shown, that this latter, the *Cascarilla fina de Loxa*, is entirely different from *Cinchona lancifolia* of Mutis, and from all those Peruvian Cinchona barks described in Ruiz's *Quinologia*, in the *Flora Peruviana*, and in the recent *Supplement* to the *Quinologia*. Averse as we are from entering into competition with the above-named excellent botanists, yet the accidental advantage has fallen to our lot, of having ourselves seen the Cinchona forests near Santa Fe, as also those of Loxa. In fact, for the last sixty years, since the time of Joseph de Jussieu, whose observations were moreover never published, no travelling naturalist has preceded us in visiting the beautiful mountain plains of Loxa. Favoured by these circumstances, I think myself enabled to speak with | 21 | some confidence on so difficult a subject, which, by a variety of controversies, has become more and more perplexed.

It would be superfluous to repeat the fictions concerning the history of the discovery of the medicinal powers of the Cinchona bark. Some say a patient had drank out of a lake the waters of which had acquired a bitter taste from Cinchona trunks which had lain in them; others, that a lion had cured himself of the ague by chewing Cinchona bark, and had thus directed the attention of the Indians to this tree. Lambert, in his *Monograph of Cinchona*,[2] has collected all these opinions. That animals have taught men, is a very common form of the traditions of nations. The valuable antidote Vijuco del guaco, a plant described by Mutis, which is related to the *Mikania*, and has been erroneously confounded with the *Ayapana* of Brazil, is also said to have attracted the notice of the Indians (as is affirmed of the *Falco serpentarius*) by the *Falco guaco* of New Granada fighting with serpents. However, that the great American lion without mane, *Felis concolor*, should be subject to the ague, is just as bold an hypothesis as the assertion of the inhabitants of the pestilential valley Gualla Bamba,[3] that even the vultures (*Vultur aura*) in their neighbourhood were subject to that disorder. Indeed, in the regions of the Cinchona forests there is not even a *Felis concolor* so fond of warmth to be found; but at the most the cat *Puma*, not yet properly described, (La Condamine's *Petit lion du volcan de Pinchincha*, which I should be inclined to call *Felis andicola*,) and which we have met with in heights of 2,500 toises.

[1] Plantes Æquinoctiales, par Messrs. Bonpland et Humboldt, Troisieme livraison, p. 39.

[2] A Description of the Genus Cinchona, 1792, p. 39.

[3] Near to the town of Quito.

The story, so often copied, respecting the Countess Chinchon, vice-queen of Peru, is probably still more doubtful than it is generally supposed to be. There certainly was a Count Chin- | 22 | chon, Don Geronimo Fernandez de Cabrera Bobadella y Mendoza, who was Viceroy in Lima from 1629 to 1639. It is very probable that his wife, after her return to Spain in 1640, was the first who introduced the Cinchona bark into Europe. The name of *Pulvis Comitissæ* appears even more ancient than that of *Pulvis Jesuiticus* or *Pulvis Patrum*. But I do not believe (and M. Olmedo in Loxa is of the same opinion with me) that the corregidor of Loxa, Don Juan Lopez de Cannizares,[4] who is said to have cured the Countess of the ague, received this remedy from the Indians. In Loxa there is no tradition whatever of this kind; nor is it probable that the discovery of the medicinal power of the Cinchona belongs to the primitive nations of America, if it is considered that these nations (like the Hindoos) adhere with unalterable pertinacity to their customs, to their food, and to their nostrums, and that, notwithstanding all this, the use of the Cinchona bark is entirely unknown to them in Loxa, Guancabamba, and far around. In the deep and hot valleys of the mountains of Catamago, Rio Calvas, and Macara, agues are extremely common. But the natives there, as well as in Loxa, of whatever cast, would die rather than have recourse to Cinchona bark, which, together with opiates, they place in the class of poisons exciting mortification. The Indians cure themselves by lemonades, by the oleaginous aromatic peel of the small green wild lemon, by infusions of *Scoparia dulcis*, and by strong coffee.[5] In Malacatis only, where many bark-peelers live, they begin to put confidence in the Cinchona bark. In Loxa there is no document to be found which can elucidate the history of the discovery of the Cinchona: an old tradition, how- | 23 | ever, is current there, that the Jesuits at the felling of the wood had distinguished, according to the custom of the country, the different kinds of trees by chewing their barks, and that on such occasions they had taken notice of the considerable bitterness of the Cinchona. There being always medical practitioners among the Missionaries, it is said they had tried an infusion of the Cinchona in the tertian ague, a complaint which is very common in that part of the country. This tradition is less improbable than the assertion of European authors, and among them the late writers Ruiz and Pavon, who ascribe the discovery to the Indians.

[4] Flora Peruviana, tom. ii. p. 2.

[5] Among the Indians on the Orinoco, particularly in Atures and Maypura, we have found an excellent febrifuge, the *frutta de Burro*, the fruit of a new species of *Uvaria*, which we have described by the name of *Uvaria febrifuga*.

The medicinal powers of the Cinchona was likewise entirely unknown to the inhabitants of the kingdom of New Granada.

A century elapsed before any botanical description was obtained of the tree whose pulverized bark yielded the Jesuit's Powder. The astronomer La Condamine, who ranged with indescribable vivacity through all departments of human knowledge, and by whom there are several neat botanical drawings in the collection of Jussieu in Paris, was the first man of science who examined and described the Cinchona tree. In the year 1737[6] he travelled through Loxa to Lima, and his description of the Cinchona appeared in 1738 in the *Mem. de l'Academie*. Afterwards, in the year 1739, Joseph de Jussieu explored the country in the vicinity of Loxa. There, and in the neighbourhood of Zaruma, he gathered a great number of specimens, which are still to be found in Jussieu's collection at Paris, and which we have compared with our own, collected sixty years later on the same spot. Amongst these was the *Cinchona pubescens*, which Vahl has described as new, but which, as we shall subsequently prove, is the first *Cinchona officinalis* of Linné's *Systema Naturæ* (12th edition). In the year 1743 La Con- | 24 | damine was a second time in Loxa, from whence he travelled, as we did in the year 1802, to Tomepinda and the Amazon river. At that time the first and (what is singular enough) the last attempt was made to bring young Cinchona trees alive to Europe. After the astronomer had carefully nursed them for eight months during a passage of 1200 leagues, they were swallowed up by a wave, which washed over the boat, near Cape Orange, north of Para.

Botanists for a long time were acquainted with only one species of Cinchona, which Linnæus called *officinalis*, and in the description of which, he united, without knowing it, our *C. Condaminea* and *C. cordifolia* Mutis; for the specimen sent him from Santa Fe was yellow Cinchona, and totally different from that drawn, though imperfectly, by La Condamine. At last Jacquin's voyage made us acquainted with another species, viz. the *C. caribœa*. The West India Islands, the South Seas, even the East Indies, offered from time to time more species of Cinchona to the traveller, but the most efficacious and the most remarkable ones of the continent of South America remained longest unnoticed.

From 1638 to 1776 no other Cinchona bark was met with in commerce, except that of the corregiment of Loxa and its neighbourhood. La Condamine makes mention of the bark from Riobamba and Cuenca in the province of Quito, as also of that from Ayavanca and Jaen de

[6] Voyage à l'Equateur, p. 31, 75, 186, and 203.

Bracamoros. But the bark from the interior of Peru (around Huanuca, and in the province La Paz) or even the bark from the kingdom of New Granada was entirely unknown to him.

They did not suppose it possible for Cinchona trees to exist north of the Equator, and consequently in our hemisphere, till a fortunate accident led a man, who had a long time lived in | 25 | Loxa, in some department connected with the peeling of barks, on his return to Spain, across Popayan to Santa Fe de Bogota. This observing traveller was the upper Mint Director (*Superintendente general de moneda de Santa Fe*) Don Miguel de Santistevan, who, without any botanical knowledge, discovered physiognomically, that is to say, by mere habitus, the Cinchona trees from Loxa up to 2 1/2° N. lat. In a memorial concerning the royal administration of the whole trade of Cinchona bark (*Estanco de Cascarilla*) which in 1753 he addressed to the viceroy Marquis de Villar, he expressly says that he had found Cinchona trees not only betwixt Loxa and Quito, for instance, easterly from Cuenca near the villages Paute and Gualasco, westerly from Riobamba on the declivity of the Chimborazo near Angas, and on the Cuesta de S. Antonio, but also betwixt Quito and Santa Fe in all situations, where the ground is of an equal height with Loxa, consequently 800 toises above the level of the sea. The estimate of the height according to modern measurements, and even according to the earlier ones of La Condamine,[7] is certainly too low by at least 250 toises; but the acute observation, respecting the mean height in which the Cinchona trees are always met with on the mountainous declivity, is the more striking, since even learned philosophers at that time paid attention to the geography of plants, or to the height of their situation. It is also to be observed, that although M. Santistevan, according to the manuscript accounts which I procured of him, speaks generally of Cinchona trees betwixt Quito and Santa Fe, yet we can perceive from his enumeration of particular places, that he discovered this precious produce only in the valley of Rio Tuanamba north of Pasto, in the forests of Beruccos, and in the vicinity of Popayan, near Guanacas, the dangerous pass of | 26 | the Andes, betwixt the village of this name and the Sitia de los Corrales.

Such was the state of the discovery of Cinchona north of the equator until the year 1772. All the Cinchona bark of commerce was from Loxa, Gauancabamba, and Jaen, perhaps even from Riobamba and Cuenza. The whole was shipped at the ports of the Pacific. No advantage was derived

[7] Voyage de la Rivière de l'Amazone, p. 25.

from the important discovery in the provinces of Pasto and Popayan. In the year 1772 Don Jose Celestino Mutis discovered the Cinchona about Santa Fe, and since this epoch Europe received Cinchona bark which did not double Cape Horn, and which came by way of Carthagena de Indias to Cadiz.

M. Mutis had resided already twelve years in the kingdom of New Granada. He had travelled twice through the forests between Guaduas and Santa Fe, where the Cinchona tree is surrounded by the beautiful Granada oaks. If we consider the diversity of plants which engage the attention of the botanist in these countries; if we reflect that in the tropics the height of the trunks withdraws from our eyes both leaves and blossoms; we shall be the less surprised that M. Mutis discovered the Cinchona only in 1772, when he found it in blossom. This excellent explorer of nature, who is a native of Cadiz, studied three years in Madrid, and was induced by a love of botany to accompany the viceroy Don Pedro Misia de la Cerda, as his physician, to Santa Fe. He lived a long time in the districts of Pampelona and de la Montuosa, a name which, to the greatest dissatisfaction of M. Mutis, Linnæus has construed into Mexico; so that this Swedish botanist has quoted all the New Granada specimens which he received from la Montuosa, as Mexican ones.[8] This error is the more singular, since Linnæus, who corresponded with Mutis always by way of Carthagena de Indias, must have | 27 | perceived that he never resided in Mexico. The absence of M. Mutis in the mines north of Santa Fe had kept him far from the Cinchona forests of Mave, Gascas, and the Aseradero. Mutis, in a report to the viceroy Don Manuel Antonio Florez, alleges as a ground of the later discovery of the Cinchona, that up to 1772 he had directed all his botanical excursions out of the limits of the first 5 degrees of N. lat., which he held to be the exclusive country of the Cinchona in the northern hemisphere. This great naturalist did not at that time suspect, that soon afterwards the Cinchona tree would be found to exist even at the mouth of the Rio Opon, and as far as Santa Martha, consequently in the 10th degree of N. lat.

Mutis had procured the first dried specimens of the yellow Cinchona of Loxa (*C. cordifolia*) from M. Santistevan, director of the Mint. According to these, the genus Cinchona was established in such manner as he communicated it to Linnæus. In the year 1772, when M. Mutis, in company with his friend Don Pedro Ugarte, rode through the forest of Tena, not far from the mountainous declivity of Santa Fe, he discovered

[8] For instance, *Mannettia reclinata*.

Cinchona trees. A year afterwards he also found them betwixt Honda and Guaduas; and presented to the viceroy Don Manuel de Guirior, who had just embarked on the Magdalen river, a flowering branch of Cinchona, as a newly-discovered valuable product of that country, which nature had also enriched with aromatic Nutmegs (*Myristica Otoba*), with an excellent Cinnamon tree (*Laurus cinnamomoides* Mut.) with aromatic *Puchery* or *Todaspuie* (*Laurus*[9] *Putseri* Mut.), with Almonds (*Caryocar amygdaliferum* Mut.), with four kinds of Styrax, with the Bal- | 28 | sam of Tolu (*Toluifera*[10] indica), with a Tea tree (*Alstonia theæformis* Mut.), with Ipecacuanha (*Psychotria emetica* Mut.), with Wax-palms (*Ceroxylon andicola* Humb.), with Carannia gum (*Æginetia cannifera* Mut.), with Winter's bark (*Wintera granadensis*), with *Quassia simarouba*, and with the valuable dyeing woods.

In the history of sciences, it often happens that the person who knows how to diffuse, with a certain degree of boldness, the discovery of another, passes for the discoverer himself, instead of him who made that discovery. M. Mutis, a man of a liberal and enlightened mind, asked no reward from the Government. He occupied himself without ostentation in botanical examinations of the kinds of Cinchona which he discovered, and in the application of their barks through an extensive medical practice. In the year 1783 only, he obtained a royal salary, when the botanical expedition of Santa Fe was organized by M. Gongora, who was both archbishop and viceroy.

In the year 1776, four years after M. Mutis's discovery, Don Sebastian Jose Lopez Ruiz, a cunning and petulant physician at Santa Fe, a native of Ganama, found means to persuade the Spanish Government that he had first discovered Cinchona trees in New Granada. He sent samples of the new Cinchona to Madrid, spoke a great deal of the importance of this new article of commerce, and obtained a yearly pension of 2000 piastres for his reward. From records which M. Lopez remitted to me in the year 1802, by his brother, a canon in Quito, in order to prove to me the priority of his discovery, I have found that he knew the Cinchona about Honda only in the year 1774, and that he made the first medicinal experiment with it in the year 1775. M. Lopez did not long

---

[9] The *Laurus Putseri* of Mutis is the same as the *Laurus Pucheri of Fl. Peruviana*, vol. iv. t. 352. ined. The number of species of this genus figured in the fourth vol. ined. of the *Flora Peruviana*, amounts to thirty. The fruit of the *Laurus Pucheri* is frequently sold in the London shops under the name of Sassafras nuts. *Caryocar amygdaliferum, Fl. Peruv*. vol. v. ined. t. 470. *Myrospermum balsamiferum, Fl. Peruv*. 4, ined. t. 373.

[10] *Myrospermum balsamiferum.*

enjoy his full salary. The viceroy Gongora, who besides esteemed M. Mutis greatly, and his first secretary Don Zenon de Alonzo, who was a zealous pro- | 29 | moter of the sciences, represented to the Court, that M. Lopez was not the first discoverer of the New Granada Cinchona bark. They immediately withdrew one half of the royal pension, ordered M. Lopez to travel to the Darien, where it was also pretended that Cinchona had been discovered; and as he refused to undertake a journey to such a pestilential climate, the viceroy discontinued the other half of his salary. Since this epoch a violent dispute has arisen respecting the priority of the discovery. Lopez made a voyage to Europe, and again contrived to procure for himself a salary of 1000 piastres. He ingratiated himself with M. Mutis's botanical opponents, and these have mentioned him frequently since as co-discoverer. It is still more remarkable, that Colonel Don Antonio de la Torre Miranda wishes to prove, in his Topography of the province of Carthagena, (*Noticia individual de los Poblaciones nuevas fundadas en la Provincia de Carthagena,*) by means of testimonies, that to him belongs the honour of discovering the Cinchona bark in New Granada, because in the year 1783 (consequently eleven years after M. Mutis) he had discovered it near Fusagasuga. M. Mutis had begun, in Mariquita, a plantation of Cinchona and of Cinnamon[11] of the Andaquia Missions, the remains of which we also saw. In the year 1800 the Spanish Government commissioned a French physician, M. Louis Derieux, to continue these plantations; to cultivate the indigenous Myristica,[12] and to superintend generally the packing of the Cinchona bark in New Granada. He received a salary of 2000 piastres, with the title of *Commissionado y Encargada de Investigaciones de Historia Natural en el Nuevo Reyno de Granada.* He possessed as little botanical knowledge as M. Lopez, but was a man of strong mind and intellectual capacity. He had long before lived in Santa Fe, from whence he | 30 | was dragged in chains to Carthagena, and thence to a prison in Cadiz, under the false accusation of revolutionary principles. After his innocence had been acknowledged, the Minister of State Don Mariano de Urquijo conferred on him the superintendance of the Cinchona forests. I travelled with him upon the Magdalen River, during which time his amiable son made several drawings of plants for me. The father stepped forth betwixt M. Mutis and M. Lopez. As the specific characters of the species of Cinchona excited the bitterest disputes between Zea, Ruiz, and Pavon, at Madrid,

[11] *Laurus cinnamomoides*, Mutis.—Edit.

[12] *Myristica otoba*, Humb. et Bonpl. Plantæ Æquin. 2. p. 78. tab. 103.—Edit.

so the Cinchona bark, ever since its first discovery, has been an odious object of persecution in Santa Fe. I have learned with great regret, that soon after I left South America, M. Derieux had lost his salary, and had even been compelled to leave the vice-royalty, so that the Cinchona trees again grow without any superintendance, which, indeed, has hitherto not promoted their increase or preservation.

In this simple historical narrative we have shown, that till the year 1772 all Cinchona bark was collected in the forests of Loxa, Ayavaca, and Jaen de Bracamoros, consequently between the 3rd and 5th degrees of south latitude, and that only from the year 1772 the medicinal Cinchonæ on the South American continent became used in the northern hemisphere; which species of Cinchona were discovered between the 4th and 5th degrees of south latitude. Until then, none were known in Peru Proper, especially in the mountains situated nearer to Lima, the capital. The vale of Rio Calvas, and the village Ayavaca, in whose neighbourhood the *Cinchona Condaminea* grows, famed since the year 1738, belong indeed in a political respect to Peru, but both are situated close to the confines of the corregiment of Loxa; and the bark of Ayavaca, like that of Jaen, was sold by the name of *Cascarilla fina de Uritusinga*, as well as that which was shipped in Payta. | 31 |

It was only in 1776 that the real commerce in Peruvian Cinchona bark began. Don Francisco Renquifo discovered near Huanuco, on the mountain San Christo val de Cuchero, the *C. nitida* of Ruiz, a species very nearly related to the orange-coloured one of Mutis (*Cinchona lancifolia*). An enterprising man, Don Emanuel Alcarraz, brought the first sample of it to Lima, and turned the use of it to advantage. The editors of the *Flora Peruviana* did certainly not penetrate, in 1779, as far as the Amazon River itself, but only to those rivers which flow immediately into it. They visited the beautiful valleys of Tharma, Xauxa, and Huamalies, and in 1779 determined the botanical characters of the North Peruvian species. This was consequently seven years after M. Mutis began his labours on the Cinchonæ of New Granada. Shortly afterwards, medicinal Cinchona bark was discovered at almost one and the same time in the most northern and in the most southern part of South America, in the mountains of Santa Martha, and in the kingdom of Buenosayres, near La Paz and Cochabamba, where a naval officer, Rubin de Celis, and the German botanist Taddæus Haenke, drew the attention of the inhabitants to this valuable produce.

After the year 1780, therefore, Europe was superabundantly supplied from the ports of Payta, Guayaquil, Lima, Buenosayres, Carthagena, and

Santa Martha, with Barks of various medicinal powers. Of these barks, some went direct to Spain, and some were transmitted by the smuggling trade to North America and England. West-Indian Cinchona barks were also occasionally mixed with those of the continent. They gave the name of Cinchona to barks which indeed possess great febrifuge powers, but which are derived from trees which do not even belong to the genus *Cinchona*. Thus they spoke in Cadiz of *Cascarilla* or *Quina de* | 32 | *Cumana* and of *Quina de la Angostura*. They divided all bark into genuine and into spurious, without considering, that, although true Cinchona barks possess equal medicinal power, yet that they are capable of displaying specific differences in the manner of their efficacy. They asked for bark like that of Loxa, without considering that three or four kinds of Cinchona bark had ever since 1738 come from Loxa itself to Europe, which were the produce of quite different species of Cinchona. They forgot that the quality of the bark did not depend merely on its being from the *C. lancifolia* or from *C. macrocarpa*, but that locality of growth, the age of the tree, quick or slow drying, determine its efficacy. They mistook the same species, if the bark was, instead of *canutillos, i. e.* in thin quills, in thick *cortizones*, or even powdered. They mixed, sometimes through mistake, sometimes intentionally, the bark of *Wintera granadensis* and of the tanning *Weinmannias*, with the Cinchona bark, and even stained them with an infusion of Brazil wood.

These circumstances gave rise to very singular prejudices in judging of Cinchona bark. Certain mercantile houses in Spain, which half a century since were in possession of the exclusive trade in Cinchona bark, endeavoured to throw disrepute on that from New Granada and southern Peru. They found complaisant botanists, who, by boldly exalting varieties to species, proved that all Peruvian Cinchonæ were specifically different from those which grow about Santa Fe. Physicians, like the Popes, drew lines of demarcation on the map. They insisted, that beyond a certain degree of latitude in the northern hemisphere no efficacious Cinchona could grow. But as the commerce with Cinchona bark from Huamalies and Huanuco, which Ortega, Ruiz, Pavon, and Tafalla recommended, soon fell into the hands of those who had formerly carried on the South Sea trade in the | 33 | Cinchona bark of Loxa, the new Peruvian Cinchona barks naturally gained easier access into Spain than those from Santa Fe. The latter, on the contrary, which the English and North Americans could easily procure in Carthagena, as a port more accessible to the smuggling trade, obtained a preferable fame in London, Germany, and Italy. The effect of mercantile cunning went so far, that, at the royal command, a quantity of the best orange-coloured Cinchona bark from New Granada, which M. Mutis had caused to be peeled at the expense of the king, was burned, as a decidedly inefficacious remedy, at a time when all the Spanish field-hospitals were in the greatest want of this valuable product of South America. A part of the Cinchona bark condemned to destruction was secretly bought by English merchants in Cadiz, and publicly sold in London at high prices. Since M. Zea, the present director of the botanic garden at Madrid, has maintained, in the *Annales de Ciencias Naturales*, against the editors of the *Flora Peruviana*, that their Peruvian species of Cinchona are identical with those of M. Mutis, but that they have described one and the same species under two or three names, the dispute concerning the quality of Cinchona bark from Santa Fe has again become very animated. The Supplement of the *Quinologia*, by Ruiz and Pavon, is written with a bitterness which ought always to remain foreign to the calm course of scientific inquiries.

Before we proceed from the history of the discovery of the Cinchona to its geographical diffusion, and their remaining physical relations, we must cast a glance upon the specific differences of the several kinds of Cinchona. A properly complete botanical disquisition is foreign to the purpose of this Treatise. M. Bonpland and myself will attempt it on another occasion, | 34 | viz. in the description of the two thousand new species of plants discovered during our expedition, and partly determined already by our excellent friend M. Willdenow. As almost every species of Cinchona is peculiar to its own region, to its own altitude on the mountainous declivity of the chain of the Andes, it is unavoidably necessary, for the satisfactory treatment of the subject, to adjust at least the synonymy of the most important officinal species. I shall certainly make mention of that only, which I have had the opportunity of observing with my own eyes. | 35 |

SECTION II.

The genus *Cinchona* belongs to those tribes of plants, whose species have been considerably multiplied of late. Linnæus knew but two of them, viz. *C. officinalis* and *C. Caribæa.* Vahl,[13] in his treatise on Cinchona Bark, enumerates nine; Lambert,[14] in his English Monograph, eleven; Persoon, in his little *Enchiridium Botanicum*,[15] one-and-twenty

[13] Skrivter of Naturhistorie Selskabet, B. i. H. i. p. 16.

[14] Description of the genus *Cinchona*, 1797.

[15] Synopsis Plantarum, P. i. p. 196.

species. If we yet add to these, two *Cosmibuenæ* of the *Flora Peruviana*, belonging formerly to the genus *Cinchona*, the *Cinchona excelsa* of Roxburgh, found in the East Indies, my *C. Condaminea*, Vavassour's *C. spinosa*, and Willdenow's yet undescribed small-leaved *C. brasiliensis*, for which we are indebted to Count Hoffmannsegg in the expedition instituted by him for objects of natural history, then the number of species of Cinchona appears to have increased to twenty-seven. The authors of the *Flora Peruviana* alone have entertained the notion of describing thirteen new species, while M. Mutis has reduced all the *Cinchonæ*, examined by him in South America, to seven only. Even Professor Zea, in | 36 | the *Annales de Ciencias Naturales de Madrid*,[16] has ventured to prove that almost all the efficacious species, enumerated by Ruiz and Pavon, can be reduced to four, viz. *C. lancifolia, C. oblongifolia, C. cordifolia*, and *C. ovalifolia*, described by Mutis, in the year 1793, in the literary news of Santa Fe de Bogota.[17]

Indeed I hardly know any one tree varying more in the shape of its leaves than the Cinchona. Whoever determines single specimens of dried collections, and has no opportunity to examine or observe them in their native forests, will, as is the case with the *Broussonettia papyrifera*, be led to discover different species by leaves which are of one and the same branch. The yellow bark, *C. pubescens*, Vahl, we have found at one and the same time with *fol. ovato-oblongis, ovato-lanceolatis*, and *ovato-cordatis*. Mutis calls it *C. cordifolia*, because it is the only kind on which sometimes cordate leaves are found. The same species varies like the white Cinchona, *C. ovalifolia*, Mut. (*C. macrocarpa*, Vahl) *foliis utrinque levibus* and *foliis utrinque pubescentibus*. These varieties are represented in those well executed coloured drawings which M. Mutis presented me during my residence in Santa Fe, and which have been deposited, together with a complete *hortus siccus* of my expedition to the tropics, in the *Jardin des Plantes* at Paris. Even the laurel-leaved *C. Condaminea*, the finest bark from Uritusinga, has very diversified leaves, according to the altitude at which it grows, and which equals that of Saint Gothard's or Mount Ætna. It would deceive the bark-peelers (*cascarilleros*) themselves, if they did not know the tree by the glands, left so long unobserved by botanists. In Gonzanama, not far from Loxa, we made a great number of im- | 37 | pressions, by means of printer's ink, from these heterogeneous forms of leaves, in order to prove how unsafe all those distinctions are,

[16] Anno 1801, No. 5.
[17] Papel Periodico de Santa Fe, 1793, No. 111.

which have been derived from the leaves only. The long known but yet very imperfect method of ectypa is particularly advantageous for this and similar purposes, as it offers to much-occupied travellers the means of procuring in a few minutes the most correct outlines.

The more the Cinchona trees vary in the shape and smoothness of the leaves, according to the altitude in which they grow; to the severity or mildness of the climate; to the trees standing singly, or being closely surrounded by other plants; to the luxuriance of growth, and greater or less humidity of the soil; the more necessary is it, with regard to the *diagnostic* indications, to pay attention to the form of the flowers, particularly to the length of the anthers, to the proportion between the stamens and anthers, as also between the free and the adherent part of the filaments. It is not sufficient to examine the species in such as have a smooth or hirsute corolla, or the stamens exserted, or inclosed in the tube of the corolla. An attentive observer finds in almost every species a striking difference in the structure of the corolla. Thus, the *C. parviflora*, Mut. has pubescent filaments, and dilated at the base. *C. macrocarpa*, Vahl, anthers nearly sessile, placed in the upper part of the tube of the corolla. *C. oblongifolia*, Mut. filaments very short, anthers situated below the middle of the tube of the corolla. The *Cinchona ovalifolia*, Mut., or the white Cinchona, varies frequently with from six to seven, the *C. Condaminea* with from three to four stamens only. In the first, the limb of the corolla is frequently found divided into six or seven, in that of the latter, mostly into four segments. In the *Cascarilla fina de la Provincia de Jaen*, which M. Bonpland intends shortly to describe, I found the anthers always shorter than | 38 | the free part of the filaments, and this free part again longer than the adherent one. On the other hand, I oberved that in the *Cascarilla fina de Uritusinga*, or the *C. Condaminea*, the anthers are twice the length of the free portion of the filaments, and the free parts are two-thirds shorter than the adherent one. There is scarcely any mention of these proportions in the otherwise excellent descriptions of Cinchonæ for which we are indebted to Vahl, Swartz, and the authors of the *Flora Peruviana*. In the mercantile world, several barks are called Peruvian bark which do not belong to the genus Cinchona. Thus, the excellent remedy which the Catalan Capuchin friars of the missions on the River Carony first made known, was called in Spain *Quina de la Guayana*, or *de la Angostura*. M. Mutis became acquainted with this bark in 1759 in Madrid, at the house of Don Vincente Rodriguez de Rivas:[18] he employed it in his medical practice, and even then supposed

[18] Pupel periodico de Santa Fe, No. 95. p. 337.

that it did not belong to the genus *Cinchona*. Lœfling died in the missions of Carony without knowing this valuable substance. It was afterwards ascribed, sometimes to the *Brucea ferruginea*, which however grows in Abyssinia; sometimes to the *Magnolia glauca*; sometimes (which certainly was more probable) to the *Magnolia Plumieri*. In our expedition we had an opportunity of examining botanically the Cuspare tree, which yields the *cortex Angosturæ*. We discovered it to be a new genus, on which our excellent friend Willdenow, in the Transactions of the Royal Academy of Berlin, has conferred the name of *Bonplandia*.[19] This name of my travelling companion has been retained for the Cuspare plant, since we have changed the Mexican *Bonplandia geminiflora*, described by Cavanilles, to *Caldasia heterophylla*.

The bark of Cumana, which for the last four or five years has | 39 |
been sent to Spain, through the exertions of Governor Don Vincente Emparan, under the name of *Cascarilla de Nueva Andalusia*, is likewise different from Cinchona. A chemist would hardly be able to distinguish this Cuspa bark from true Cinchona bark. It is an excellent remedy in the ague. Although we observed for almost a twelvemonth the Cuspa trees of Rio Manzanario near Cumana, yet it never fell to our lot to meet with its flowers. We do not know, therefore, by what distinctive mark it differs from the genus *Bonplandia* and *Cinchona*. The want of stipulæ, however, the situation of the leaves, and the whole habitus, make it more than probable that the *Cuspa* is not a *Cinchona*. The absence of stipulæ is particularly striking. Yet notwithstanding its alternate leaves, the Bark-tree of Cumana might still be a Cinchona, for the same reason that *Cornus alternifolia* stands isolated amongst twelve species of *Cornus* with opposite leaves. It has likewise remained doubtful to us, whether the bark of Acatamez, a village situated westward of Ville de Ibarra on the coast of the South Sea, betwixt Rio Verde and Rio Esmeraldita, be the produce of a species of Cinchona. The flower of this Acatamez bark-tree, with which we became acquainted during our stay in the town of Popayan, has not been hitherto examined by botanists. Mr. Brown, who long before us was in the South Sea, (in 1793,) has already given some account in Lambert's Monograph of Cinchona[20] of this new species of the torrid zone. Either from want of geographical information, or by corruption of the name, he calls it *Bark of Tecamez*, instead of *Cascarilla of Acatamez*.

---

[19] Samml. Deutscher Abhandl. für 1801 und 1802, S. 36.
[20] Lambert, p. 30.

A fourth tribe of plants producing Peruvian bark, although of less medicinal power, is the genus *Cosmibuena* of the *Flora Peruviana*. To this belongs *Cinchona longiflora*, Mut. or C. *gran-* | 40 | *diflora*, Ruiz. It is a tree of great beauty, which we have frequently seen in deep hot valleys exhibiting its beautiful fragrant blossoms. The stamens lie deep and hidden in the tube of the corolla; and the fructification is so similar to that of the other species of Cinchona, that the *Cosmibuena* can hardly be admitted to constitute a distinct genus.

On the other hand, it might be advisable to form the *Cinchonæ* with long stamina far projecting from the tube of the corolla, such as Jacquin's *C. Caribæa*, Swartz's *C. angustifolia, C. brachycarpa*, and *C. floribunda*, into a separate genus nearly allied to Cinchona. The seven species belonging to it possess this peculiarity, that all of them, except one, inhabit islands, viz. the Philippine, the West Indian, and the South Sea Islands, and that they prefer hot valleys, or even plains, to a high mountainous situation. I know but two species upon the South American continent which have stamina exserta, Lambert's *C. longiflora* from French Guiana, and the yet undescribed *Cinchona dissimiliflora*, Mut. (*staminibus longe exsertis, corollæ laciniis tubo longioribus, foliis cordato-oblongis*) which, in the kingdom of New Granada, descends from the declivity of the mountains towards the plains as low as 200 toises above the level of the sea. *C. Caribæa* and *C. angustifolia* are found in the West Indies in still lower spots, even in regions which are sufficiently warm for plantations of sugar canes. All these Island Cinchonæ with projecting stamens have a smooth corolla. All of them have a capitate or obtuse stigma, the *C. Philippina*[21] alone excepted, which M. Nee discovered at Santa Cruz de la Laguna, near Manilla. A divided stigma is, on the contrary, observed in all Cinchonæ with inclosed stamens. The corolla of the latter is sometimes smooth, sometimes hairy. M. | 41 | Mutis has already proposed, in the literary News of Santa Fe, to separate the Cinchona with long projecting stamina from the rest. "I know not," says he, "what my friend Linnæus thought of the Cinchona of the South Sea, for its reception in the Supplement only proves the favour of the son, whose opinion has not with me the weight of the opinions of the father." The authors of the *Flora Peruviana* wish to make the Island-Cinchonæ, *Portlandiæ*[22]; but M. Swarz, in Schrader's *Journal für die Botanik*[23], proves, that in the Island-Cinchonæ, as in those of the continent, the capsule is a dissepimentum

[21] Cavanilles Icones, t. iv. p. 15. t. 329.
[22] Flor. Peruv. t. ii. praef. and p. 49.
[23] Band. I. p. 358.

loculorum exacte parallelum, and in *Portlandia* a dissepimentum vere contrarium. Ruiz's *Portlandia corymbosa* is therefore no *Portlandia*, but belongs to the *Cinchonæ filamentis e basi tubi ortis*, to *C. Caribæa, C. floribunda*, and *C. brachycarpa*, a groupe of plants which M. Swarz also wishes to unite into a separate genus on account of the flower, but not on the score of fructification. The *C. excelsa*, with enormous leaves, frequently of twelve inches length and fifteen inches breadth, discovered in the East Indies, stands almost in the middle, betwixt the West-Indian and South-American Cinchona, and its existence seems to dissuade us, as it were, from the proposed separation of the two tribes. However, the *C. excelsa* Roxb. approaches less to the Island-Cinchona than to the New Granada and Peruvian ones, corolla pubescenti, staminibus medio tubi insertis, nec e basi tubi nascentibus, antheris nec filamentis exsertis, margine seminum lacero, haud integro. The antheræ in this East India species are eight times longer than the filaments. It is difficult to find reasons for uniting the Island-Cinchonæ into a separate genus, in the formation of the fruit. They differ[24] from the Cinchonæ of the continent of South | 42 | America, "valvulis minus extrorsum divergentibus et receptaculo ovato nec lineari seminumque margine integro nec lacero." But except the smooth unindented coat of the seed wings, which I mostly find, the remaining forms of the fruits exhibit gradations, which link together, as it were, all the Cinchonæ. For the new genus of Island-Cinchonæ, delighting in hot plains, there would consequently remain: corolla glabra, filamentis longe exsertis ex basi tubi nascentibus. Semina margine integro cincta. Stigma simplex capitatum. But 1°. Many Cinchonæ staminibus inclusis, and *C. grandiflora* Ruiz, have corollam glabram. 2°. *C. Phillippina* has far projecting filaments, stigma bilamellatum, and yet, as it appears, semina margine integro cincta. 3°. *C. excelsa* has stigma subcapitatum leviter emarginatum, the seed not indented, and the filaments not projecting. Under these exceptions, it would certainly be bold to separate tribes of plants so nearly allied.

The singular prickly *C. spinosa* of St. Domingo appears at first sight to belong least to the genus Cinchona. It is wonderfully small-leaved, and has frequently folia terna verticillata. Another prickly Cinchona differs still more in colour from the genuine Cinchona bark trees; it grows near Guayaquil, on the coast of the Pacific, and M. Tafalla showed it to us in the winter of 1803, during our stay there. This undescribed species is a creeper, and on that account in some measure related to the genus

---

[24] Schrader, a. a. O. S. 359.

Danais from Madagascar, which Persoon ranks next in succession to the *Portlandia*, since the *Pædeira fragrans*, more resembling the Cinchona, has been separated from *Pædeira fœtida*. This new *C. scandens* of Tafalla has in other respects the complete fructification of the ague-curing Cinchona, and belongs indisputably to the most remarkable phenomena of the physiognomy of plants.

The very same fruit of the genuine Cinchona is also produced | 43 | by *Pinkneya pubens* Michaux,[25] a tree which I found cultivated together with *C. Caribæa* in the excellent botanic garden of Mr. Hamilton near Philadelphia. The *Pinkneya* grows on Mary's River, in the province of Georgia, and is already described by Bartram, propter calycis laciniam unicam foliaceam bracteæformem, by the name of *Mussuenda bracteolata*. The medicinal powers for the cure of ague possessed by this plant, nearly allied to the genus Cinchona, and growing without the tropics, have not yet been investigated. On the other hand Mr. Walker has shown in two excellent treatises, that the bark of *Cornus florida* from Virginia, and of *C. sericea* from Pensylvania and South Carolina, and even the Tulip tree (*Liriodendron Tulipifera*) may be used with advantage in North America as remedies against agues.[26] In the kingdom of New Spain, where hitherto no species of Cinchona has been discovered, as the curator of the Academical Botanic Garden at Mexico has assured me, the yet undescribed *Portlandia mexicana*, discovered by M. Sesse, may supply the place of the Cinchona bark of Loxa. In the East Indies (according to D. Klein in Tranquebar) the *Swietenia febrifuga*, figured by Roxburgh, a plant of Swarz's and Jacquin's, *Portlandia hexandra* (Aublet's *Coutaria Speciosa*), nearly allied to Cinchona, produces the bark of French Guyana, known in France by the name of *Ecorce fébrifuge de Cayenne*,[27] and which is no more derived from a Cinchona, than is the bark of Cumana or the Cuspare of Angostura.

Thus much respecting the generic characters of the plants which approximate to Cinchona, and all of which belong to the | 44 | great family of *Rubiaceæ*. We see that as Caoutchouc[28] is obtained in abundance from the juices of the most diversified plants on the Orinoco and in Cayenne; from the *Hevea* on the Canno Pimichin, a branch of the Negro;

[25] Flor. Americana, I. p. 105.

[26] Walker on the virtues of the *Cornus* and the *Cinchona* compared. Philad. 1803. Rogers's Diss. on the properties of the *Liriodendron*. Phil. 1802.

[27] Ventenat Tableau du Regne Vegetal, t. ii. p. 578.

[28] The Cecropia *peltata* is frequently mentioned as a tree yielding a part of the American caoutchouc. But I doubt whether any part of the new continent makes use of a juice so difficult to inspissate.

from the tree *Jacio* in the kingdom of New Granada; from a new species of *Ficus* in the province of Popayan, near the Indian village La Cruz; from a *Lobelia* (to be described by us) in Bengal; from the *Urecola elastica*, figured in the 5th volume of Asiatic Researches; in Madagascar, from the *Commiphora madagascarensis*; so does nature also offer to us the ague-curing principle, or that mixture containing tannin and absorbing oxygene, which we obtain of a preferable quality from *Cinchona Condaminea, C. pubescens*, Vahl. and *C. lancifolia*, Mut. in plants which do not even belong to one and the same genus. A chemist would perhaps find greater differences between the West-Indian and South-American Cinchona barks, than between the Cuspa of Cumana and the Cinchona bark of Loxa; and yet the Cuspa tree, *foliis alternis, stipulis nullis*, is most probably a very remote genus from Cinchona.

After we have separated with care, partly what in a botanical point of view is nearly related to Cinchona, partly what passes in commerce amongst different nations by the name of China, Cascarilla, Quinquina, or Ecorce fébrifuge; after we have separated the Cinchonæ with inclosed filaments, not growing from the lower end of the flower tube with divided stigma and indented margins of the seeds, from the Island-Cinchona, whose long projecting filaments grow from the bottom of the flower tube, and which have, together with unindented seed wings, an | 45 | undivided stigma; after we have examined the relation and supposed similarity of mixture of Cinchona, Portlandia, Coutarea, Cosmibuena, Pinkneya, Danais, Bonplandia, Cuspa, and the Acatamez tree, we pass to the definition of those species of Cinchona which have become an object of great importance in the practice of physic and in the intercourse of nations. Without the fundamental exposition of the specific characters, and without adjusting some part of the synonymy, every thing which I am going to state respecting the geographical diffusion of the Cinchonæ, and their physical relations, would remain indistinct and dubious, since (as I have mentioned above) a peculiar region has been destined for almost every species, and some botanists have, to the great detriment of science, given one and the same name to the most heterogeneous species. Thus, for instance, *Cinchona longiflora*, Mut. is totally different from *C. longiflora*, Lambert. It is true, they both have a smooth corolla, and belong to the Cinchonæ which are fond of heat and possess fewer medicinal powers. But the first, from New Granada, has inclosed stamens, and is probably identical with *C. grandiflora, Flor. Peruv*. On the other hand, the *C. longiflora*, Lambert, from French Guiana, belongs to those species which have long projecting filaments and very short capsules. *Cinchona*

Caribœa, Jacq. is totally different from that *Cinchona Caribœa* described in the *Journal de Physique*, Oct. 1790. The diagnoses which I add are not borrowed from works already published, but arise partly from my own observations made from nature itself, partly from an instructive intercourse with M. Mutis. | 46 |

Characteristics of some Species of Cinchona.

Vahl, in his excellent Monograph, augmented by Lambert, divides all the species into two groupes of plants, floribus tomentosis, staminibus inclusis, and floribus glabris, staminibus exsertis. This division possesses this fault, that two characters are placed opposite each other, which are by no means observed at one and the same time in all the species at present known. Certainly no Cinchona with tomentose flowers has long projecting stamens, for in the East Indian species the anthers are merely visible; but there are Cinchonæ which have, like *C. parviflora*, Mut. and *C. grandiflora, Flor. Peruv*. a smooth corolla and inclosed stamens. With more, although not with perfect justice, we might separate Cinchonæ staminibus inclusis, stigmate bilamellato, seminum ala denticulatâ vel lacerâ, and Cinchonæ filamentis insertis ex imo tubi nascentibus, seminibus membranâ integrâ cinctis. However, it seems more correctly logical to divide the Cinchonæ into those with smooth and into those with hairy corollas. The first division merely subdivides itself, according to the length of the stamens, into two smaller tribes, and (what is certainly an important object) all the useful and ague-curing species associate into one groupe.

A. Cinchonæ corollis tomentosis.

1. *C. Condaminea, corollæ tubo hirto, foliis ovato-lanceolatis utrinque glaberrimis, in axillis nervorum inferne scrobiculatis. Humb. et Bonpl. Plant. Æquin. fasc.* ii. p. 29. tab. 10.

This species, the fine bark of Uritusinga, could only be taken | 47 | for the *C. glandulifera, Flor. Peruv.*; but this latter differs corollâ solummodo intus lanuginosâ, tubo externe glaberrimo, foliis inferne villosis.[29] The inhabitants also enumerate the *C. glandulifera*, which is called (at Chicoplaya) Cascarilla negrilla, among the less

[29] Flor. Peruv. t. iii. p. 1. t. 224.

efficacious species of Cinchona.

If any one species deserved exclusively the name *C. officinalis*, it would be the tree which produces the Cascarilla fina de Uritusinga, a bark which has always been held in Spain as the most efficacious in tertian agues, and which at present is gathered only for the Royal Apothecaries' Hall, and is therefore never met with in trade by lawful channels. Notwithstanding these preferences, we have, for several reasons, preferred giving it a new name, not derived from its quality or medicinal powers. 1°. Not one species, but all provided with hairy and woolly blossoms, are Cinchonæ of the shops, and no species deserves an absolute preference, since different species are to be applied according to the difference and form of the disease: for instance, in intermittent fevers of long standing, the *C. Condaminea* and *C. lancifolia*, Mut.; in diseases of the muscles or suppurating ulcers, the *C. oblongifolia*, Mut.; in the after treatment, to prevent relapses, the more mild *C. cordifolia*, Mut. 2°. In botanical writings, species of Cinchona totally distinct have been described by the name of *C. officinalis*. Had we bestowed the same name on the Cinchona of Uritusinga, it would have been confounded with the yellow *C. cordifolia*, Mut., the white *C. macrocarpa*, Vahl, or even with the *C. nitida*, Ruiz, which at different periods have been called *C. officinalis*.

This latter point, equally important to the botanical synonymy and to the materia medica, merits a more circumstantial explanation. It is asked, What plant did Linnæus, in the 12th edition | 48 | of the *Systema Naturæ*, call *C. officinalis*? Vahl maintains that it was his *C. macrocarpa*[30] from the kingdom of New Granada, which he received from Ortega. But since *C. macrocarpa*, Vahl is nothing else but our white large-flowered Cinchona of Santa Fe, *C. ovalifolia*, Mut.; and as, according to M. Mutis's own testimony, it had never been seen by Linnæus, then the *C. macrocarpa*, Vahl cannot be quoted as synonymous with *C. officinalis*, Linn. *Syst. Nat.* ed. 12. The great botanist of Copenhagen, whose early death is so justly deplored by all the friends of science, was misled to an erroneous synonymy in the following manner: 1°. He knew that Linnæus had at a later period founded his description of *C. officinalis* on specimens which he received from Santa Fe; and 2°. he erroneously presupposed that all the Cinchona forests in the neighbourhood of Santa Fe,

[30] Act. Havn. I. p. 19. Lambert, p. 22.

discovered by M. Mutis, consisted of white Cinchona, or *C. macrocarpa*.

Linnæus united, as already observed, two quite different plants under the denomination of *C. officinalis*. The dried specimen of which he made use for establishing the diagnosis, was (as M. Mutis has repeatedly and orally assured me) yellow Cinchona, *C. cordifolia*, Mut., and the same species which Vahl calls *C. pubescens*, but of which one variety has entirely smooth leaves, *folia utrinque glabra*. Linnæus quotes as synonymous the species described by Condamine in the *Mem. de l'Academie*, 1738: he consequently united one species from Santa Fe with another which grows exclusively in the neighbourhood of Loxa.

Ruiz, in his *Quinologia*,[31] calls a species *C. officinalis*, which he afterwards describes in the *Flor. Peruv.* by the name of *C. nitida*. He maintained at the time, that this tree, which | 49 | grows in the forests of Huamalies and Xauxa, consequently far from Loxa, between the 10th and 12th degree South lat., was the Cinchona described by La Condamine. In the *Supplemento a la Quinologia*, p. 68, a botanical disputation which appeared against M. Zea, Mutis, and Cavanilles, this assertion is very justly withdrawn. Indeed the *C. nitida* or *C. officinalis* Ruiz is no other than the Cascarilla naranjanda from Santa Fe, or *C. lancifolia* Mut.

Since therefore four different species, the *Cascarilla fina de Uritusinga*, which Condamine has figured, *C. pubescens* Vahl, *C. nitida* Ruiz, and *C. macrocarpa* Vahl, have already received the name of *C. officinalis*, we have called the Cinchona of Uritusinga, in commemoration of its first discoverer, *C. Condaminea*. It is true that M. Ruiz, in his *Supplemento a la Quinologia*, gives it as his opinion, that the plant called at present Cascarilla fina at Loxa, was not the plant described by the French astronomer; but not only the unanimous testimony of all inhabitants of Loxa, Caxanuma, and Uritusinga, speaks against this, but also Jussieu's *Hortus Siccus* at Paris. M. Bonpland has carefully compared our *C. Condaminea* with the specimens which were collected by Joseph de Jussieu and La Condamine. No doubt remained concerning the identity of the species.

The *C. Condaminea*, like *Myristica, Caryocar amygdaliferum*, and many precious products of the tropics, is confined to a very

[31] Cascarilla officinal. Quinolog. Arct. II. p. 56.

small space, and it has been hitherto most imperfectly described. No botanists, neither Ruiz and Pavon, nor Tafalla, nor Nee, nor Hänke, nor Mutis, have observed it before us at its place of growth. The following may be considered as imperfect figures of the C. Condaminea: Mem. de l'Acad. de 1738, p. 114.; Lamarck Encyclopédie, pl. 164. fig. 1.; Vahl Skrivt. af Naturh. Selskabet I. tab. 1., and Lambert. Monogr. | 50 | tab. 1. The true character of the leaves has been missed every where; and it would be bold to quote these synonyms, if it were not possible to verify them by examining the specimens which have served for the drawings.

Our *C. Condaminea* grows under the 4th degree south latitude, on the mountainous declivity in the mean altitude between 900 and 1200 toises. It requires a milder climate than the orange-coloured Cinchona, *C. lancifolia* Mut., from Santa Fe. It is exposed to a mean temperature from 15 to 16 degrees Reaumur, which is about the mean warmth of the Canary Islands.

I here insert an exact diagnosis of the *C. Condaminea*, which I drew up at Gonzanama, and of which (as it remained buried beneath astronomical manuscripts) M. Bonpland could not avail himself in the publication of the second fasciculus of *Plantæ Æquinoctiales*.

Calyx tubulosus basi angustatus sub-5-gonus subhirsutus ore 5-dentato, dentibus ovatis acuminatis patentibus. *Cor.* hypocrateriformis tubo cylindrico rubro lævissime hirto 5-gono (ad basin persæpe fisso) limbo 5-fido sæpissime 4-fido, laciniis ovatis acutis apice et margine ciliatis, vel tomentosis ciliis albis. *Faux* corollæ et totius tubi pars interior rubra glabra, nec ciliata. Stamina quinque, rarius tria et quatuor. In corollâ 4-fidâ sæpius stamina quinque numeravi. *Filamenta* ex rubro albescentia imo tubi adnata, cum eo cohærentia, tertiam tubi partem æquantia, eademque tantum tertiâ suæ longitudinis parte liberâ. *Antheræ* planæ lineares, parte liberâ filamenti duplo longiores. *Germen* rotundum subdepressum rubescens, sæpe punctatum et 5-sulcatum. *Stylus* fere longitudine tubi, crassus, teres. *Stigma* tubum vix superans, viridescens, compressum, bifidum sæpe bipartitum. *Capsula* calyce coronata, oblonga, flore tertiâ parte longior, bi- | 51 | partibilis, striato-costata, de medio hiscens, dissepimento parallelo. *Semina* plura compressa alâ membranaceâ crenulatâ cincta. *Rami* cicatrisati post casum foliorum, sub-4-goni; juniores glaberrimi, subpulverulenti. *Folia* petiolata decussatim opposita lanceolata acuta; integerrima, utrinque viridia, nullis venis rubris picta, fere laurina, glaberrima, in axillis nervorum infernè screbiculata. *Glandulæ* nullis pilis obsitæ, convexitate in paginâ superiori folii conspicua, venas altitudine superantes. Pagina folii inferior scrobiculum demonstrat. *Petioli* sæpè rubescentes, supernè plani, infernè teretes. *Stipulæ* deciduæ, oblongæ, carinatæ. *Panicula* axillaris et terminalis, folio longior floribus brevè pedicellatis.

Size of the parts in a tree flowering for the first time:— Calyx, 1 7/10 lines long; corolla, 5 4/10 lines; capsule; 8 lines long, 3 1/2 lines broad; according to Parisian measure. Full-grown leaves exclusive of the petioles, 4 inches 3 lines long, and 1 inch 9 lines broad. The young leaves frequently have a length of 5 inches, and the great breadth of 4 inches 7 lines.

The *C. Condaminea* varies amazingly in the leaves before the tree comes into flower. In the shoots and very young trees we frequently find folia latè ovata and ovato-lanceolata. The older the tree is, the narrower are its leaves. In great luxuriance of growth, the little grooves frequently vanish, which appear on the upper side of the leaf as convex glands. On very broad leaves, in which the parenchyma is considerably extended, they are almost entirely wanting. However, even then, we always meet with single folia scrobiculata upon the same branch.

2. *C. lancifolia foliis lanceolatis acutis utrinque glaberrimis*. Mutis, Period. de S. Fe, p. 465. (et Flor. Bogot. Mss.) In Santa Fe it is known by the names of Quina naranjanda, Quin- | 52 | quina orangé, or orange-coloured Bark. Next to *C. Condaminea*, it is the most efficacious febrifuge of all the kinds of Cinchona; the species which M. Mutis, in his *Quinologia*, calls the Quina primitiva directamente febrifuga, because he prefers it to the three following species, and because he thinks (what is erroneous, however) the fine Cinchona of Uritusinga is the same species as Quina naranjanda of New Granada. The *C. lancifolia* has smaller leaves than the others with tomentose corollas. They are also continually smooth, when on the contrary the place of growth produces, in the yellow and white Cinchonæ, varieties with hairy leaves.

The Quina naranjanda loves a rough climate. It grows between the 4th and 5th degree north lat. on mountainous declivities from 700 to 1500 toises high. The mean temperature of this place of growth is about equal with that of Rome. It amounts to 13° Reaumur; however the Cinchona trees ascending highest towards the summit of the mountains are mostly exposed to a temperature of from 8° to 9°. During the cold at nights, the thermometer falls in these alpine forests for hours as low as the freezing point; however, as far as 1500 toises high no snow falls in this latitude.

The Quina naranjanda, together with the *C. Condaminea*, belongs to the more scarce species. Nature herself has produced them in the kingdom of New Granada in a much smaller number than those of the yellow and red Cinchonæ, which latter ones form here and there almost closely-connected shrubberies. *C. lancifolia*, on the contrary, stands always single; and what is to be regretted in so valuable a produce is, that it does not increase so easily by shoots from the root, as the *C. cordifolia* and *C. oblongifolia*. In the Monographs of Vahl and Lambert, no mention is made of the species called Naranjanda of Santa Fe. An | 53 | indisputable synonym on the contrary is *Cinchona angustifolia* Ruiz, *Suppl. à la Quinologia*, p. 21, where an excellent figure is given. It is indeed surprising, that so exact a botanist as M. Ruiz should change the old Mutisian name *C. lancifolia* for *C. angustifolia*, since that name has previously been given by Swartz[32] to an Island-Cinchona with a smooth corolla and long projecting stamens.

Professor Zea thinks, and, as it appears to me, with perfect propriety, that several species of the Flora Peruviana denote merely different states of the Quina naranjanda, such as depend on the age, the climate, and the place of growth. The following appear to be varieties of the *C. lancifolia* Mut.: 1°. *C. nitida Flor. Peruv. II. Icon. t.* 191. (Ruiz, *Quinol. II.* p. 56.) Ruiz's Cascarilla officinal. 2°. *C. lanceolata Flor. Per. II.* p. 51. and *C. glabra* Ruiz *Quin. II.* p. 64. Cascarilla lampina, of which no figure is given. M. Zea thinks he may venture to add to these, the *C. rosea Flor. Peruv. II. Ic.* 199. a species which is said to be the most scarce in Peru, and (what agrees little with the nature of *C. lancifolia*) to descend from the mountains into the lowest regions.[33]

The Cinchona Bark so famous in Cadiz by the name of Calisaya, and of such particular medicinal power, belongs, according to Mutis, indisputably to *C. lancifolia*. Ruiz considers it, in his *Quinologia*, as synonymous with his *C. glabra*. But in his disputation against Zea, he withdraws this opinion, and assures us that there is no species growing in the neighbourhood of Huanuco which produces a bark similar to the Calisaya.[34] The name Calisaya is that of the

[32] Flor. Ind. occ. I. p. 380. Lambert, p. 29. Pl. 9.

[33] Ruiz Supplem. à la Quinol. p. 54.

[34] P. 73 and 95.

province producing this bark, | 54 | which is situated in the most southerly of Peru, in the Intendencia de la Paz.

The second edition of a modern French work, Alibert's *Traité de Fièvres intermittentes*, contains very exact figures of the orange-coloured Cinchona, as well as of the three following Mutisian species. They have been made from dried specimens, determined by M. Mutis, and supplied by M. Zea from his collection during his residence at Paris.[35]

3. *C. cordifolia fol. orbiculato-ovatis sæpe subcordatis subtus tomentosis supra pubescentibus*, Mut. Mss. Quina amarilla, Quinquina jaune, yellow Bark from Santa Fe, the species, as observed above, described by Linnæus in *Syst. Nat. t.* ii. *ed.* 12. p. 64. under the name of *C. officinalis*. The anthers in *C. cordifolia* and *C. lancifolia* reach as far as the upper parts of the flower-tube; when on the contrary, in the red Cinchona (*C. oblongifolia*) they are deeply hidden in the middle of the tube. *C. cordifolia* has two varieties. Var. β *foliis vix cordatis utrinque glabris. γ foliis utrinque hirsutis.* By the common people, in the kingdom of New Granada, it is called Velvet Bark. It grows under the 4th degree North latitude, in heights betwixt 900 and 1440 toises. Cordate leaves occur but seldom: however, almost every branch exhibits some of them. *C. cordifolia* Mut. is, according to Bonpland's examination, identical with *C. pubescens* Vahl, as proved by Jussieu's collection, from which Vahl received his specimen. Joseph Jussieu had collected, in 1738, this species of Cinchona and *C. Condaminea* in the forests of Loxa.

The *C. ovata Flor. Peruv. II. t.* 195. *Cascarilla pallida* Ruiz, *Quinol. Art.* 7. p. 74. called in the neighbourhood of | 55 | Pozuzo *Pala de Guallerata*, is likewise a synonym of *C. cordifolia* Mut. Ruiz and Pavon themselves have latterly acknowledged this identity.[36]

The *C. hirsuta Flor. Peruv. II. Ic.* 192. Cascarillo delgado, or *C. tenuis* Ruiz, *Quinol. II.* p. 56. is, according to Zea, a variety of *C. cordifolia* Mut. Does *C. purpurea Flor. Per. II. t.* 193. or Cascarilla morado Ruiz, *Quinol. Art.* v. p. 67. also belong here? This species varies surprisingly in its leaves, and on one and the same tree too.

4. *C. oblongifolia foliis oblongis acuminatis glabris, filamentis brevissimis, antheris infra medium tubi latentibus*. Mut. Mss.

Quina roxa, Quinquina rouge de Santa Fe, differt a *C. lancifoliâ*,

[35] Some of these figures are evidently copied from those in *Flora Peruviana*.—Edit.

[36] Supplem. à la Quinol. p. 18.

1°. foliis latioribus, majoribus oblongis nec lanceolatis; 2°. antheris haud in summo tubi latentibus.

It grows under the 5th degree North lat. in heights from 600 to 1300 toises, and is particularly common in the neighbourhood of Mariquita, a small town, which was for a long time the seat of M. Mutis's botanical expedition. It frequently bears much larger fruit than the white Cinchona, *C. ovalifolia*, for which reason it would deserve the name of *macrocarpa* with more propriety than the latter one. Its bark is less efficacious than that of *C. Condaminea* and *C. lancifolia*, yet more so than the yellow Cinchona, (*C. cordifolia.*) It is more stimulating, for weak constitutions, in inflammatory diseases frequently dangerous, but the more beneficial when applied externally in diseases of the muscles, suppurating and sphacelous ulcers.

The yellow Cinchona, *Cascarilla amarilla Quinol. Art.* vi. p. 71. or *C. magnifolia Flor. Per. II. Ic.* 196. which, on ac- | 56 | count of the fragrant and orange-flower smell of its blossoms, is called in Peru, Flor. de Azahar, and in Popayan, Palo de requeson,[37] is, according to the latter confessions of Ruiz, identical with *C. oblongifolia* Mut. or with the red Cinchona of Santa Fe.

5. *C. ovalifolia fol. ellipticis supra glaberrimis subtus pubescentibus antheris in parte tubi superiori latentibus filamentis vix ullis.* Mut. Mss.

Quina blanca, Quinquina blanc: White Cinchona of Santa Fe.

Var. β fol. utrinque pubescentibus.

γ. fol. utrinque lævibus.

Both varieties, particularly the first, have frequently a corolla with 6 or 7 divisions, and 6 or 7 stamens.

It grows under the 3d to the 6th degree North lat. in heights from betwixt 700 and 1400 toises. The variety with smooth leaves is frequent near San Martha. The *Cinchona macrocarpa* Vahl. (Lambert, p. 22. t. 3.) is a true synonym of this, acknowledged by Mutis and Ruiz themselves.[38] Amongst the Cinchonæ with hairy corollas it is the largest-flowering one of all. It must not, however, be confounded with *C. grandiflora Flor. Peruv. II.* p. 54. (*Cosmibuena obtusifolia Flor. Peruv. III. t.* 198.) having a quite smooth corolla.

6. *C. brasiliensis foliis oblongis acuminatis, venis subtus pubescentibus paniculâ terminali, tubo calycis longitudine.* Willd. Mss.

[37] Period. de Santa Fe, p. 335.

[38] Suppl. p. 18.

A very small-flowering species, for which we are indebted, as observed already, to Count Hoffmannsegg, together with Aublet's | 57 | and Lambert's *C. longiflora* from French Guiana, the only Cinchona which grows on the easterly coast of the South American continent. Nothing decisive is known about the height of its place of growth; but as it has been sent from the neighbourhood of the town of Gran Para, at the mouth of the Amazon river, and as in this region there are only low hills found, we are allowed to suppose that *C. Brasiliensis* belongs to the hot regions.

The character of this species by M. Willdenow, tube of the corolla the length of the calyx, distinguishes this Cinchona from every one hitherto described. Throat of the corollæ hairy; hairs few, short, appressed, situated on the interior surface of the corolline laciniæ.

7. *C. excelsa corollâ pubescente, filamentis e medio tubi nascentibus, antheris exsertis, foliis oblongis subtus pubescentibus*. Roxb. Plant. of the Coast of Coromandel, ii. t. 106.

The only Cinchona hitherto discovered on the continent of the ancient world, about whose medicinal use and its bitter no trials have however as yet been made. It has very small greenish-white flowers, and of all Cinchonæ the largest leaves, sometimes one foot long and five inches broad.

The *C. excelsa* (*Bundarvo* of the Felinga Indians) grows in the mountain chain of the Circars, which runs along the north-easterly coast of the great peninsula of Hindostan. Retzius[39] has at an earlier date, from accounts communicated to him by König, mentioned a Cinchona which grows in Malacca, opposite to the coast of Coromandel, and which produces the genuine *terra japonica*, called *Cotta Cambar*, a vegetable produce, which for a long time was erroneously ascribed to *Mimosa spicata* Pluk. Might not this Cinchona from Malacca be a different species from *C. excelsa*? | 58 |

B. *Cinchonæ corollis glaberrimis*.

a. *staminibus inclusis*.

8. C. grandiflora, tubo corollæ longissimo, fol. lanceolatooblongis utrinque glabris.

I have retained the former name of the *Flor. Per.* M. Ruiz calls this species at present *Cosmibuena obtusifolia*. (*Flor. Per.* vol. iii.) It is identical with *C. longiflora* Mut., a name which would cause confusion, since Lambert enumerates as *C. longiflora* the Island-Cinchona staminibus longe exsertis, described as *C. Caribæa* in

[39] Observ. Bot. fasc. iv. p. 6.

Journ. de Phys. Oct. 1790.

It is fond of warm regions, and descends from the mountains in heights from two and three hundred toises. It grows in regions whose mean temperature is from 18 to 19 degrees.

9. *C. parviflora foliis ovatis glabris, filamentis basi dilatatis et pubescentibus*. Mut. Mss.

It has the smallest fruit of all Cinchonæ.

b. *staminibus exsertis*.

10. *C. dissimiliflora foliis cordato-oblongis glaberrimis, limbo corollæ tubo longiori, capsulis sublinearibus angustissimis*. Mut. Mss.

Next to *C. longiflora* Lamb. the only species of the continent which has stamina exserta. Grows in heights betwixt 200 and 700 toises in warm regions.

11. *C. Caribæa* Swartz.

12. *C. longiflora* Lamb.

13. *C. lineata* Vahl.

14. *C. floribunda* Swartz. | 59 |

15. *C. angustifolia* Swartz.

16. *C. brachycarpa* Vahl.

These six latter species grow all in the West-India Islands, and love a temperature of from 17 to 22 degrees R.

17. *C. corymbifera* Forster.

Native of the Friendly Islands.

18. *C. Philippica*, discovered near Manilla by Nec.

I do not venture to assert that all Cinchonæ hitherto known are comprehended within the eighteen species arranged here. I have merely wished to enumerate those which are known to me, partly in their natural state partly from good figures, and which to me appear indisputably specific from each other. *C. acutifolia, C. micrantha, C. glandulifera, C. dichotoma, C. (Cosmibuena) acuminata*, and *C. spinosa*, deserve a closer investigation. The genus might perhaps increase to twenty-four species.

Account of the Ottomacs, a People who eat Clay[1]

On the coasts of Cumana, of new Barcelona, and of Caraccas, visited by the Franciscan monks of Guayana in returning from their missions, there was a tradition prevalent, that tribes inhabiting the banks of the Orinoco eat earth.

On the 6th of June 1800, when, on our return from Rio Negro, we came down the Orinoco, where we had remained thirty-six days; we spent a day in a mission inhabited by the Ottomacs, who eat earth. The village called La Concepcion of Uruana, is situated, in a very picturesque manner, upon the declivity of a rock of granite. Its latitude I determined to be 7° 8′ 3″ north, and its longitude 4° 38′ 38″ west of Paris. The earth which the Ottomacs eat is a fat and unctuous clay, a real potter's earth, of a greyish-yellow tint, coloured by a little oxide of iron. They select it with much care, and gather it on particular banks on the sides of the Orinoco, and of the Meta. They distinguish, by the taste, one kind of earth from another, for all kinds of clay are not equally agreeable to their palates. They knead this earth into balls, of from four to six inches in diameter, and bake it at a slow fire, till its outer surface becomes reddish. Before they eat these balls, they moisten them anew. These Ottomacs are for the most part a very savage race, and have an aversion to improvement. The nations of the Orinoco, that are farthest from that canton, say proverbially, when they wish to speak of any thing very filthy, "It is disgusting enough for an Ottomac to eat it." When the waters of the Orinoco and Meta are low, the Ottomacs support themselves on fish and tortoises. When the fish appear at the surface of the water, they kill them by means of darts, with a dexterity which we have often admired. When the rivers experience their periodical overflow, the fishing ceases; for it is as difficult to fish in rivers become too deep, as in the open sea. | 317 |

During this inundation, which lasts two or three months, the Ottomacs swallow prodigious quantities of earth. We have found in their huts immense stores of it, heaped up in pyramids. Each individual consumes

[1] Tableaux de la Nature, par A. de Humboldt, 2 vol. 12mo. Paris, 1808. vol. i. p. 191. For an account of the practice of eating clay among some tribes of Africans, the reader may consult Mr William Hunter's Essay on the Diseases incident to Indian seamen, or Lascars, on long voyages, Appendix, No. X. and the several authorities referred to by him.

daily three-fourths or four-fifths of a pound of earth, according to the account which Fray Ramon Bueno, a very intelligent monk, a native of Madrid, who has lived twelve years among these Indians, gave us. The Ottomacs themselves say, that, in the rainy season, this earth is their principal food. In addition, they eat little fishes, lizards, and fern-roots, when they can procure them. They are so fond of this earth, that every day they eat a little of it after their repast, to regale themselves, even in the dry seasons, and when they have abundance of fish. These people are of a very dark copper colour. Their features are as ugly as those of the Tartars. They are fat, but they have not a large belly. The missionary who resides among them, assured us, that he has remarked no difference in the health of these savages, during the time that they eat this earth.

Such is the simple narration of facts. The Indians eat great quantities of earth, without their health suffering from it. They consider it as a nourishing kind of food; that is to say, they find that this food satisfies them for some time. They attribute this satiety to the earth, and not to the other very bad articles of diet which they are enabled to add to it. If they are asked what is their provision in winter, the season, which in the warm parts of South America is called the rainy-season, they point out the heaps of earth piled up in their huts. But these partial facts do not decide the following questions. Can earth, in reality, be a nutritive substance? Can earth assimilate itself to our nature? Or is it but a load upon the stomach? Does it not serve only to keep its sides distended, and in this manner contribute to alleviate hunger? It is very singular, that Father Gumila, in other respects so credulous, and whose work is so void of sound criticism, should absolutely deny that the Indians eat earth.[2] He pretends that the balls of clay are mixed with the flour of Indian corn and the fat of the crocodile. But the missionary Fray Ramon Bueno, and brother Fray Juan Gonzales, our friend and travelling companion, whom the sea on the coast of Africa swallowed up along with a part of our collections, both assured us that the Ottomacs do not mix this earth with the fat of the crocodile. At Uruana we never heard a word about | 318 | the mixture of flour. The earth which we brought home with us, and which Vauquelin analyzed, was pure, and without any mixture. May not Gumila, confounding strange facts, have wished to allude to the bread which is prepared with the long pods of a species of *inga*? This fruit is buried in the earth, that it may ferment the sooner.

But what surprises me more is, that the use of so great a quantity of earth does not occasion any disease among the Ottomacs. Has this

[2] History of the Orinoco, vol. i. p. 283.

colony been habituated to this kind of food for many generations? In all the countries of the torrid zone, men have an astonishing, and almost irresistible desire of eating earth; not an alkaline or calcareous earth to neutralize the acid juices, but a clay which is very fat, and of which the odour is very strong. They are often obliged to tie children, to hinder them from going out and eating the earth when the rains have ceased. At the village of Banco, on the banks of the river Magdalena, the native women, who make earthen pots, put large pieces of clay into their mouth, as I have often seen with surprise.[3] The other tribes of America never fail to become diseased when they give themselves up to the singular desire of eating earth. In the mission of San Borgia, we saw a child, who, according to what its mother told us, would not eat any thing but earth, which diet had made it as lean as a skeleton. Whence is it, that, in the cold and temperate zones, the desire of eating earth is so rare, and exists only among children and pregnant women? It may be asserted, that, in all the regions of the torrid zone, this desire for earth has been observed. In Guinea, the negroes eat a yellowish earth which they call *caouac*. The slaves which are brought into America, endeavour to procure a similar gratification; but it is always to the detriment of their health.

"Another very general cause of the *mal d'estomac*," says a modern traveller, "is, that many of the slaves who come from Guinea eat earth. This does not arise from a depraved taste; it is a custom contracted in their own country, where they say that they habitually eat a certain kind of earth, the taste of which pleases them, and which does not hurt them. They search among us for earth most resembling it. That which they prefer is a reddish yellow tufa, very common in our islands. This they even sell secretly in our public markets, under the | 319 | name of *caouac*.[4] (M. Thibaut was at Martinique in the year 1751.) Those who are accustomed to the use of it are so fond of it, that no chastisement can prevent them from eating it."

In the villages of the isle of Java, between Sourabaya and Samarang, M. La Billardière saw little square reddish cakes exposed to sale. The natives call them *tanaampo*. Upon examining them more narrowly, they were found to be cakes of reddish earth, which they eat.[5] The inhabitants of New Caledonia, eat, to satisfy their hunger, pieces as large as the fist, of friable *lapis ollaris*.

---

[3] Gily has made the same remark, Saggio di Storia dell' America, t. ii. p. 311. In winter, the wolves eat earth, but, above all, clay. In general it would be interesting to analyze the feces of man, and all those animals who eat earth.

[4] Thibaut de Chanvallon, Voyage à la Martinique, p. 85.

[5] Voyage à la recherche de La Peyrouse, Vol. II. p. 322.

M. Vauquelin, upon analyzing it, found in it a considerable quantity of copper.[6] At Popayan, and in many parts of Peru, the inhabitants purchase at the market calcareous earth, with other commodities. In using it, they mix with it the cocca, or the leaves of the erythroxilon peruvianum. Thus we find, that this taste for eating earth, which it would seem that nature ought to have reserved to the inhabitants of the ungrateful regions of the north, extends through all the torrid zone, among the indolent races of men, who live in the most beautiful and the most fertile countries in the world.

[6] Ibid. p. 205.

The Cataracts of Orinoco

—A foaming surface of four miles in length presents itself at once to the eye: iron-black masses of rock resembling ruins and battlemented towers rise frowning from the waters. Rocks and islands are adorned with the luxuriant vegetation of the tropical forest; a perpetual mist hovers over the waters, and the summits of the lofty palms pierce through the cloud of spray and vapour. When the rays of the glowing evening sun are refracted in these humid exhalations a magic optical effect begins. Coloured bows shine, vanish, and reappear; and the ethereal image is swayed to and fro by the breath of the sportive breeze. During the long rainy season the streaming waters bring down islands of vegetable mould, and thus the naked rocks are studded with bright flower-beds adorned with Melastomas and Droseras, and with small silver-leaved mimosas and ferns. These spots recall to the recollection of the European those blocks of granite decked with flowers, which rise solitary amidst the glaciers of Savoy, and are called by the dwellers in the Alps "Jardins," or "Courtills." In the blue distance the eye rests on the mountain chain of Cunavami, a long extended ridge which terminates abruptly in a truncated cone. We saw the latter (Calitamini in its Indian name) glowing at sunset as if in roseate flames. This appearance returns daily: no one has ever been near the mountain to detect the precise cause of this brightness, which may perhaps proceed from a reflecting surface produced by the decomposition of talc or mica slate. During the five days which we passed in the neighbourhood of the cataracts, it was striking to hear the thunder of the rushing torrents sound three times louder by night than by day. In all European waterfalls the same phenomenon is remarked. What can be its cause in a wilderness where there is nothing to interrupt the repose of nature? Perhaps the currents of heated ascending air by causing irregular density in the elastic medium, impede the propagation of sound during the day, by the disturbance they may occasion in the waves of sound; whereas during the nocturnal cooling of the earth's surface the upward currents cease.— *Humboldt's Aspects of Nature.*

Scenes in the Pampas

—At length, after a long drought, the welcome season of the rain arrives, and then how suddenly is the scene changed. The deep blue of the hitherto perpetually cloudless sky becomes lighter; at night the dark space in the constellation of the Southern Cross is hardly distinguishable; the soft phosphorescent light of the Magellanic clouds fades away; even the stars in Aquila and Ophiucus in the zenith, shine with a trembling and less planetary light. A single cloud appears in the south, like a distant mountain rising perpendicularly from the horizon: gradually the increasing vapours spread like mist over the sky, and now the distant thunder ushers in the life-restoring rain. Hardly has the surface of the earth received the refreshing moisture, before the previously barren steppe begins to exhale sweet odours, and to clothe itself with kyllingias, the many panicules of the paspalum, and a variety of grasses. The herbaceous mimosas, with renewed sensibility to the influence of light, unfold their drooping slumbering leaves to greet the rising sun; and the early song of birds, and the opening blossoms of the water plants, join to salute the morning. The horses and cattle now graze in full enjoyment of life. The tall spring grass hides the beautiful spotted jaguar, who, lurking in safe concealment, and measuring carefully the distance of a single bound, springs, cat-like, as the Asiatic tiger, on his passing prey.— *Humboldt's Aspects of Nature.*

Observations upon the Face of the Country of Spain and its Climate

No country of Europe presents so singular a configuration as Spain. It is this extraordinary form which accounts for the aridity of the soil in the interior of the Castiles, the power of evaporation, the want of rivers, and that difference of temperature which is observable between Madrid and Naples, two towns situated in the same degree of latitude. We shall only be able to give a rough sketch of this meteorological view of Spain. Very few observations have hitherto been made on the mean temperature, or on the height of the barometer. A great deal of valuable materials perhaps remains unknown in the manuscripts of enlightened persons, who, without communicating with one another, or with the learned of other nations, have given themselves up to researches of this kind. When we do not possess exact observations, we must content ourselves with the analogy seen in neighbouring countries. It is easier to trace the natural aspect of New Spain than that of Old; and in this respect we are better acquainted with the colonies than with the mother country. | clxiv |

The interior of Spain is an elevated plain, and is the highest of any of the same kind in Europe which occupies a large extent of country. Switzerland, the Tyrol, and Scotland, contain ranges of mountains close to one another. These are masses furrowed with deep valleys, and surrounded with low plains. Switzerland is not really a raised plain. The cantons of Berne, Fribourg, Zuric, and all those countries covered with a new formation of free-stone, are plains, the height of which is only from 240 to 280 fathoms above the level of the ocean. They form part of the grand longitudinal valley which extends from the south-west to the north-east, between the chain of the upper Alps and Mount Jura, as appears by the beautiful geological maps just published by M. Ebel. In France, and particularly in Germany, there are raised plains, not of very great extent certainly, but well worth being mentioned. In France, the highest plain is that of Auvergne, in which Mont-d'Or, Cantal, and the Puy de Dome stand. It is 370 fathoms above the sea, according to the barometrical calculation of a celebrated mineralogist, M. de. Buch. Lorraine forms a raised plain that extends between the Vosges and the chain of mountains, which, passing by Epinal and Saint-Mihel, joins the Ardennes. This elevated plain, however, is only from 130 | clxv | to 140 fathoms high.

The centre of the plains of France, the department of Loir and Cher, is from eighty to ninety fathoms high.

Bavaria is the most extensive and the highest level land of Germany. A vast plain, the bed of an ancient lake, extends from the granite mountains of the upper Palatinate (*Fichtel Gebürge*) to the foot of the Alps in the Tyrol. These plains (and this fact is very curious and hitherto little known), like the small plain of Auvergne, are from 250 to 260 fathoms above the level of the ocean.

The interior of the two Castiles presents a raised plain which exceeds in height and extent all those that we have just mentioned. Its mean elevation appears to be three hundred fathoms. The height of the barometer at Madrid is twenty-six inches two lines and two-fifths, according to a note communicated by M. Bauza, a distinguished astronomer, employed in the depôt of charts for the navy at Madrid. It is therefore two inches or one-fourteenth lower than the mean height of the mercury at the level of the ocean. This is the difference of the pressure of the atmosphere which is experienced by all bodies exposed to the open air at Madrid, Cadiz or Bordeaux. At Madrid the Barometer falls as low as twenty-five inches six lines, and even lower. The *Diario de los nuevos descubrimientos de* | clxvi | *todas las Ciencias fisicas*, volume iii. pages 56, 200, 407, contains a series of very interesting meteorological observations, but which unfortunately do not include a whole year.

The following is a table of the variations of the pressure of the air in the nine first months of the year 1793—

1793. Months.	**Maximum.**	**Minimum.**	**Mean Height of the Barometer.**
	Inch. Lines.	**Inch. Lines.**	**Inch. Lines.**
January,	26 5 8	25 9 8	26 2 6
February,	25 5 3	26 6 2	26 1 6
March,	26 4 7	25 6 0	25 11 6
April,	26 2 4	25 6 9	25 11 6
May,	26 4 6	25 10 5	26 0 8
June,	26 4 0	25 11 8	26 1 6
July,	26 4 3	26 0 7	26 2 4
August,	26 3 2	25 11 5	26 1 4
September,	26 4 3	25 11 0	26 1 7

The mean height of the barometer at Madrid observed by Don Felipe Bauza shows that capital to be elevated three hundred and nine fathoms three-fifths[1] above the level of the ocean, according to M. de Laplace and the new coefficient of M. Ramond, allowing the barometer on the | clxvii | coasts, with Shuckburgh and Fleurieu Bellevue, to be at three hundred thirty-eight and twenty-four lines. Madrid consequently stands as high as the town of Inspruck, which is situated in one of the very high defiles of the Tyrol. The elevation of Madrid is fifteen times greater than that of Paris, three times greater than that of mount Valerian, and also three times greater than that of Geneva.[2]

Lalande, was the first who made known the elevation of Madrid, according to the observations which were communicated to him by the celebrated geometrician Don George Juan, (*Memoires de l'Academie des Sciences de Paris*, for the year 1776, page 148). He says, that in the street of *los Presiados*, near the *portijo de San Martin*, the town is 294 fathoms higher than Paris; which makes it three hundred and fourteen fathoms above the level of the ocean. According to M. Thalacker, the mineralogist, who has taken several heights with the barometer in the environs of Madrid, the king's palace at St. Ildefonso is five hundred and ninety-three | clxviii | fathoms, which is higher than the edge of the crater of Mount Vesuvius. No other monarch in Europe is possessed of a palace in the regions of the clouds: in our countries the heavy summer clouds are from five hundred and fifty to six hundred fathoms high.

The height of the plain of the Castiles has an effect upon its temperature. We are astonished at not finding oranges in the open air in the latitude of forty, the same as that of Tarentum, part of Calabria, Thessaly, and Asia Minor. The mean temperature[3] of Madrid appears to be 59 degrees of Fahrenheit, while that of Petersburgh is 39 degrees 52 minutes and 30 seconds; that of Berlin 46 degrees 57 minutes and 30 seconds; that of Paris 53 degrees 56 minutes and 15 seconds; that of Marseilles, 58 degrees 33 minutes; that of Toulon 61 degrees 15 minutes; that of Naples 63 degrees 30 minutes; and that of the countries situated under the equator and on the level of the ocean from 79 to 81 degrees. Genoa is 4 degrees more to the north than Madrid, and yet the | clxix | temperature

[1] Recueil d'Observations Astronomiques, by M. de Humboldt, page 18.

[2] The level of the Seine at the Pont Royal, at No. 13 of the old scale, was elevated nineteen fathoms five feet above the surface of the Ocean. The gallery of the church of Mount Valerian is elevated seventy-four fathoms above the mean height of the Seine. (Cotte Journal des Mines, April 1808, No. 313) Geneva is one hundred and eighty-eight fathoms above the level of the Sea.

[3] Naturalists find the mean temperature of the year by adding together all the heights of the Thermometer observed in the course of the year, and dividing the total by the number of observations. The mean heat of a place in the temperate zone differs sensibly from the medium taken between the maximum and the minimum of the thermometer.

of Genoa raises the glass almost 2 degrees higher than that of the capital of Spain. Such is the influence of local causes, of the elevation of site, the proximity to the sea; a chain of mountains which keeps off the cold northerly winds, and a great number of little circumstances, the combination of which moderates the temperature of places.

Rome, which is 2 degrees 32 minutes to the south of Genoa, but 1 degree 29 minutes to the north of Madrid, has almost the same mean temperature as the latter town. It is between 60 degrees 7 minutes and 30 seconds, and 61 degrees 15 minutes of Fahrenheit's, according to a great number of very exact observations made by M. Calandrelli and the elder M. de Humboldt, minister of the king of Prussia in Italy. The following table shows the mean temperature for nine months observed at Madrid and at Rome in 1793 and 1807. I have not been able to procure observations made at the same period; but in mean temperature we know the variation of one year from another is hardly perceptible. | clxx |

Mean temperature.

At Madrid.				**At Rome.**			
	deg.	min.	sec.		deg.	min.	sec.
Latitude	40	25	18	Latitude	41	53	54
Elevation 309 fathoms.			Elevation				
	Fahrenheit.	Fahrenheit.					
January	39	3	0	40	11	15	January
February	43	24	0	47	49	30	February
March	47	54	0	50	15	45	March
April	52	19	30	54	34	30	April
May	59	4	30	65	56	15	May
June	72	32	15	72	30	0	June
July	77	13	30	79	15	0	July
August	81	34	30	79	15	0	August
September	65	45	0	72	34	30	September

Even at Rome orange-trees are not able to endure the rigours of winter, and it is necessary to house them. The mean temperature certainly does not alone decide what kind of cultivation, is proper for different climates, yet it has the greatest influence upon cultivation, and the following table points out with sufficient certainty what is the mean temperature below which particular productions cannot be cultivated with success. | clxxi |

	Fahrenheit.		
	min.	deg.	sec.
Vineyards yielding wine –	48	52	30
Olive-trees – – – – – –	55	37	30
		Fahrenheit.	
	deg.	min.	sec.
Orange-trees – – – – –	62	22	30
Coffee – – – – – – – –	64	37	30
Sugar-canes – – – – –	68	0	0

If the mean temperature of the elevated plains of Spain is 59 degrees of Fahrenheit, that of the coasts from the 41st degree of latitude to the 36th, is between 63 and 68 degrees Fahrenheit. Thus we see banana-trees, heliconias, and sugar-canes growing upon these coasts in situations that are sheltered from the cold winds.

The geological profile annexed to this work is drawn upon the same principles as I thought it necessary to follow in my natural atlas of the new continent. This profile (Pl. 2.) presents to the view of the observer the extraordinary structure of this country, part of the coasts of which appear to have been swallowed up by the waves, while the central elevated plain resisted the irruption of the ocean. Recollecting the traditions of the Samothracians, and the great catastrophe which occurred in the Mediterranean, we do not pretend to decide a question which has recently engaged the attention of learned men of distinguished merit.

The analogy of form and geological structure presented by the peninsula and Mexico, has led to a comparison which may be interesting to | clxxii | naturalists. We have given an outline of Old and one of New Spain, engraved upon the same scale (Pl. 3) These are the outlines of countries, the central elevations of which enjoy different climates from that on the coasts. The capitals of both are placed in the middle of the interior plain; but the plain of the mother-country may be said to be the miniature of that of the colony. The difference in height between the highest mountains of the old and new world is only 900 fathoms. The Chimborazo is only one-fourth higher than Mount Blanc, whilst the high plains of the Castiles are four times lower than the central one of Mexico. The mean temperature of Mexico is eight degrees lower than that of Vera Cruz and Acapulco. That of Madrid is probably no more than from two to three degrees below that of the coasts of Valencia. The climate of the capital of New Spain, like that of Madrid, is not quite so cold as might be supposed from the height of the two towns, as the extent of the elevated plains imparts a degree of warmth to the air. The mean temperature of Mexico is 62 deg. 22 min. 30 sec. Fahrenheit; it is below that of Cadiz, and is almost a degree and a half within that of Madrid. The height of the barometer at Mexico is 21 inches 7 lines. The pressure of the air is indicated by a column of mercury five inches shorter than at Madrid. | clxxiii |

European Spain, situated in a latitude under which palm trees (*phœnix dactilifera, chamœrops humilis*) grow upon the plains, presents the majestic spectacle of a chain of mountains, the tops of which shoot up into the regions of everlasting snows. Don Clemente Roxas[4] has discovered by a levelling survey, executed with the greatest care, that in the Sierra Nevada of Granada, the Pico de Venleta is elevated 1781 fathoms 16, and the Mulahacen 1824 fathoms 47, above the level of the ocean. None of the mountains of the Pyrenees are of so great a height; for Mount Perdu, the highest ridge of the Spanish Pyrenees, is only 1763 fathoms, and the highest of the French Pyrenees is only 1722 fathoms. The peak of Mulahacen in the Sierra Nevada of Granada wants only 76 fathoms of being as high as the peak of Teneriffe. Yet even this summit, if situated in the same latitude as the town of Mexico, would not be perpetually covered with snow; for the never melting snows begin[5] under the equator at | clxxiv | 2460 fathoms high; under the 20th degree of latitude at 2350 fathoms; under the 45th at 1300 fathoms; and under the 62d at 900 fathoms. Such is the depression of the curve from the equator to the pole.

---

[4] Manuscript note communicated by M. Bauza. M. Thalacker, in the Annals of *Ciencias Naturales*, published by Cavanilles, has estimated the Peak of Veleta to be only 1154 fathoms high. The barometer of this traveller must have been out of order.

[5] Essay upon the Refractions in the Torrid Zone, by A. de Humboldt, page 31.

Cow Pock in Mexico

(From Humboldt's Account of New Spain.)

The small-pox, introduced since 1520, appears only to exercise its ravages every seventeen or eighteen years. In the equinoxial regions it has, like the *black vomiting* and several other diseases, its fixed periods, to which it is very regularly subjected. We might say that in these countries the disposition for certain miasmata is only renewed in the natives at long intervals; for though the vessels from Europe frequently introduce the germ of the small-pox, it never becomes epidemical but after very marked intervals; a singular circumstance, which renders the disease so much the more dangerous for adults. The small-pox committed terrible ravages in 1763, and especially in 1779, in which year it carried off in the capital of Mexico alone more than nine thousand persons. Every evening tumbrels passed through the streets to receive the corpses, as at Philadelphia during the yellow fever. A great part of the Mexican youth was cut down that year.

The epidemic of 1797 was less destructive, chiefly owing to the zeal with which inoculation was propagated in the environs of Mexico, and in the bishopric of Mechoachan. In the capital of this bishopric, the city of Valladolid, of 6800 individuals inoculated, only 170, or 2 1/2 per cent. died; and we must also observe, that several of those who perished were inoculated at a time when they were probably already infected in the natural manner. Fifteen in the hundred died of individuals of all ages, who without being inoculated were victims of the natural small pox. Several individuals, particularly among the clergy, displayed at that period a very praiseworthy patriotism, in arresting the progress of the disease by inoculation. I shall merely mention the names of two enlightened men, M. de Reano, intendant of Guanaxuato, and Don Manuel Abad, penitentiary canon of the cathedral of Valladolid, whose generous and disinterested views were constantly directed towards the public good. There were then inoculated in the kingdom between 50 and 60,000 individuals.

But in the month of January 1804, the vaccine inoculation was even introduced at Mexico through the activity of a respectable citizen, Don

Thomas Murphy, who brought several times the virus from North America. This introduction found few obstacles; the cow-pox appeared under the aspect of a very trivial malady; and the small-pox innoculation had already accustomed the Indians to the idea that it might be useful to submit to a temporary evil for the sake of evading a greater evil. If the vaccine inoculation, or even the ordinary inoculation, had been known in the new world in the sixteenth century, several millions of Indians would not have perished victims to the small-pox, and particularly to the absurd treatment by which the disease was rendered so fatal. To this disease the fearful diminution of the number of Indians in California is to be ascribed. The ships of war commissioned to carry the vaccine matter into America and Asia arrived at Vera Cruz shortly after my arrival.

Don Antonio Balmis, physician general of this expedition, visited Porto Rico, Cuba, Mexico, and the Philippine islands; and his stay at Mexico, where nevertheless the cow-pox was known before his arrival, contributed singularly to facilitate the propagation of this salutary preservative. In the principal cities of the kingdom vaccine committees were formed (*juntos centrals*,) composed of the most enlightened individuals, who, by vaccinating monthly, preserve the miasma from being lost. It is so much the less liable to be lost, as it exists in the country. M. *Balmis* discovered it in the environs of Valladolid, and in the village of Atlisco, near la Puebla, in the udders of the Mexican cows. The commission having fulfilled the beneficent views of the king of Spain, we may indulge a hope that through the influence of the clergy, and especially of the religious missionaries, vaccination will be gradually introduced into the very interior of the country. The voyage of M. Balmis will thus remain for ever memorable in the annals of history. The Indies saw for the first time those vessels, which were formerly freighted only with instruments of carnage and destruction, bearing about the germ of relief and consolation to distressed and suffering humanity.

The arrival of the armed frigate in which M. Balmis made the circuit of the Atlantic and Pacific Oceans gave rise on several coasts to one of the most simple, and therefore most affecting, ceremonies. The bishops, military governors, and persons of greatest distinction, repaired to the shore, where they took in their arms the children who were to carry the cow-pox to the indigenous Americans and the Malays of the Philippine islands, and, followed with public acclamations, they laid at the foot of the altar those precious preservative deposits, returning thanks to the supreme being for having been the witnesses of so happy an event. We must have some knowledge of the ravages occasioned by the small-pox in the torrid zone, and especially among a race of men whose physical constitution seems adverse to cutaneous eruptions, in order to feel all the importance of M. Jenner's discovery. It is a much greater blessing for the equinoxial part of the new continent than for the temperate climate of the old.

It may be useful to relate here a fact of some importance for those who take an interest in the progress of vaccination. It was unknown at Lima till the month of November 1802. At that period the small pox prevailed on the coast of the South Sea. A merchant vessel, *Santo Domingo de la Calzada*, put into Lima in the passage from Spain to Manilla. An individual had had the good sense to lend by this vessel vaccine matter to the Philippine islands. They availed themselves of this opportunity at Lima; and M. Unanue, professor of anatomy, and author of an excellent physiological treatise on the climate of Peru, vaccinated several individuals by means of the matter brought by the merchant vessel. No pustule appeared; and the virus appeared either altered or too weak. However, M. Unanue having observed that all the vaccinated individuals had a very mild small-pox, employed this variolous matter to render, if possible by the ordinary inoculation, the disease less fatal. He thus perceived in an indirect way the effects of a vaccination supposed to have sailed.

It was accidentally discovered in the course of the same epidemic in 1802, that the beneficent effect of vaccination had been long known to the country people among the Peruvian Andes. A negro slave had been inoculated for the small-pox in the house of the Marquis de Valleumbroso who shewed no symptom of the disease. They were going to repeat the inoculation, when the young man told them that he was certain of never having the small-pox, because in milking cows in the Cordillera of the Andes, he had had a sort of cutaneous eruptions, caused, as the Indian herdsmen said, by the contract of certain tubercules sometimes sound on the udders of cows. Those who have had this eruption, said the negro, never take the small-pox. The Africans, and especially the sudians, dilplays great sagacity in observing the character, habits, and diseases of the animals with which they live. We need not therefore be astonished, that, on the introduction of horned cattle into America, the lower people remarked that the pustules on the udders of cows communicated to the herdsmen a species of benignant small-pox, and that those once infected are secure from the general contagion during the epochs when the disease is epidemical.

The *matlazahuatl*, a disease peculiar to the Indian race, seldom appears more than once in a century. It raged in a particuliar manner in

1545, 1576, and 1786. It is called a plague by the Spanish authors.— As the latest epidemic took place at a time when medicine was not considered a science, even in the capital, we have no exact data as to the *matlazahuatl*. It bears certainly some analogy to the yellow fever or black vomiting; but it never attacks white people, whether Europeans or descendants from the natives. The individuals of the race of Caucasus do not appear subject to this mortal typhus, while, on the other hand, the yellow fever or black vomiting very seldom attacks the Mexican Indians. The principal site of the *vomito prieto* is the maritime region, of which the climate is excessively warm and humid; but the *matlazahuatl* carries terror and destruction into the very interior of the country, to the central table-land, and the coldest and most arid regions of the kingdom.

Father Torribio a Franciscan, better known by his Mexican name of Motolinia, asserts that the small-pox at its introduction in 1520, by a negro slave of Narvaez, carried off the half of the inhabitants of Mexico. Torquemada advances the hazardous opinion that in the two *matlazahuatl* epidemics of 1545 and 1576, 800,000 Indians died in the former and 2,000,000 in the latter. But when we reflect on the difficulty with which we can at this day estimate, in the eastern part of Europe, the number of those who fall victims to the plague, we shall very reasonably be inclined to doubt if the viceroys Mendoza and Almanza governors of a recently conquered country, were able to procure an enumeration of the Indians cut of by the *matlazahuatl*. I do not accuse the two monkish historians of want of veracity; but there is very little probability that their calculation is founded on exact data.

A very interesting problem remains to be resolved. Was the pest, which is said to have desolated from time to time the Atlantic regions of the United States before the arrival of the Europeans, and which the celebrated Rush and his followers look upon as the principle of the yellow fever, identical with the *matlazahuatl* of the Mexican Indians?— We may hope that this last disease, should it ever re-appear in New Spain, will be hereafter carefully observed by the physicians.

On the Volcanoes of Jorullo[1]

The grand catastrophe in which this volcanic mountain issued from the earth, and by which the face of a considerable extent of ground was totally altered, was perhaps one of the most extensive physical changes, that the history of our globe exhibits. Geology points out spots in the ocean, where, within the last two thousand years, volcanic islets have arisen above the surface of the sea, as near the Azores, in the Archipelago, and on the south of Iceland: but it records no instance of a mountain of scoriæ and ashes, 517 met. [563 yards] above the old level of the neighbouring plains, suddenly formed in the centre of a thousand small burning cones, thirty-six leagues from the seashore, and forty-two leagues from any other volcano. This phenomenon remained unknown to the mineralogists and natural philosophers of Europe, though it took place | 82 | but fifty years ago, and within six days journey of the capital of Mexico.

Descending from the central flat toward the coasts of the Pacific ocean, a vast plain extends from the hills of Aguasarco to the villages of Toipa, and Patatlan, equally celelebrated for their fine cotton plantations. Between the picachos del Mortero and the cerras de las Cuevas and de Cuiche, this plain is only from 750 to 800 met. [820 to 880 yards] above the level of the sea. Basaltic hills rise in the midst of a country, in which porphyry with base of greenstone predominates. Their summits are crowned with oaks always in verdure, and the foliage of laurels and olives intermingled with dwarf fan palms. This beautiful vegetation forms a singular contrast with the arid plain, which has been laid waste by volcanic fire.

To the middle of the eighteenth century fields of sugarcanes and indigo extended between two rivulets, called Cuitimba and San Pedro. They were skirted by basaltic mountains, the structure of which seems to indicate, that all the country, in remote periods, has several times experienced the violent action of volcanoes. These fields, irrigated by art, belonged to the estate of San Pedro de Jorullo (Xorullo, or Juvriso), one of the largest and most valuable in the country. In the month of June, 1759, fearful rumbling noises were accompanied with frequent shocks of an earthquake, which succeeded each other at intervals for fifty or sixty days, and threw the inhabitants of the estate into the greatest consternation.

[1] Extracted from his Essay on New Spain. Journal de Physique, vol. LXIX, p. 149.

From the beginning of the month of September, every thing seemed perfectly quiet, when in the night of the 28th of that month a terrible subterranean noise was heard anew. The frightened Indians fled to the mountains of Aguasarco. A space of three or four square miles, known by the name of Malpays, rose in the shape of a bladder. The boundaries of this rising are still distinguishable in the ruptured strata. The Malpays towards the edge is only 12 met. [13 yards] above the former level of the plain, called las playas de Jorullo; but the convexity of the ground increases progressively toward the centre, till it reaches the height of 160 met. [175 yards]. | 83 |

They who witnessed this grand catastrophe from the top of Aguasarco assert, that they saw flames issue out of the ground for the space of more than half a league square; that fragments of red hot rocks were thrown to a prodigious height; and that through a thick cloud of ashes, illumined by the volcanic fire, and resembling a stormy sea, the softened crust of the earth was seen to swell up. The rivers of Cuitimba and San Pedro then precipitated themselves into the burning crevices. The decomposition of the water contributed to reanimate the flames, which were perceptible at the city of Pascuoro, though standing on a very wide plain 1400 met. [1530 yards] above the level of the playas de Jorullo. Eruptions of mud, particularly of the strata of clay including decomposed nodules of basaltes with concentric layers, seem to prove, that subterranean waters had no small part in this extraordinary revolution. Thousands of small cones, only two or three yards high, which the Indians call ovens, issued from the raised dome of the Malpays. Though the heat of these volcanic ovens has diminished greatly within these fifteen years, according to the testimony of the Indians, I found the thermometer rise to 95° [if centig. 203° F.] in the crevices that emitted an aqueous vapour. Each little cone is a chimney, from which a thick smoke rises to the height of ten or fifteen met. [11 or 16 yards]. In several a subterranean noise is heard like that of some fluid boiling at no great depth.

Amid these ovens, in a fissure, the direction of which is from N. N. E. to S. S. E., six large hummocks rise 400 or 500 met. [440 or 550 yards] above the old level of the plain. This is the phenomenon of Monte Novo at Naples repeated several times in a row of volcanic hills. The loftiest of these huge hummocks, which reminded me of the country of Auvergne, is the large volcano of Jorullo. It is constantly burning, and has thrown out on the north side an immense quantity of scorified and basaltic lava, including fragments of primitive rocks. These grand eruptions of the central volcano continued till February, 1760. In the succeeding years they became gradually less frequent. The Indians, alarmed by the horrible noise of the new volcano, at first deserted the villages for seven or eight leagues | 84 | round the plain of Jorullo. In a few months they became familiar with the alarming sight, returned to their huts, and went down to the mountains of Aguasarco and Santa Ines, to admire the sheaves of fire thrown out by an infinite number of large and small volcanic openings. The ashes then covered the houses of Queretoro, more than 48 leagues [120 miles] in a right line from the place of the explosion. Though the subterranean fire appears to be in no great activity[2] at present, and the Malpays and the great volcano begin to be covered with vegetables, we found the air so heated by the little ovens, that in the shade, and at a considerable distance from the ground, the thermometer rose to 43° [109 4° F.]. This fact evinces, that there is no exaggeration in the report of some of the old Indians, who say, that the plains of Jorullo were uninhabitable for several years, and even to a considerable distance from the ground raised up, on account of the excessive heat.

Near the cerro of Santa Ines the traveller is still shown the rivers of Cuitimba and San Pedro, the limpid waters of which formerly refreshed the sugarcanes on the estate of Don Andrew Pimantel. These springs were lost in the night of the 29th of September, 1759: but 2000 met. [near 2200 yards] to the westward, in the soil that has been elevated, two rivulets are seen to break out of the clayey dome of the furnaces, exhibiting themselves as thermal waters, in which the thermometer rises to 52 7° [126 86° F.]. The Indians still give these the names of San Pedro and Cuitimba, because in several parts of the Malpays large bodies of water are supposed to be heard running from east to west, from the mountains of Santa Ines to the estate of the Presentation. Near this estate is a brook, that emits sulphuretted hidrogen gas: it is more than 7 met. [near 8 yards] wide, and is the most copious hidrosulphurous spring I ever saw. | 85 |

In the opinion of the natives these extraordinary changes I have described, the crust of earth raised and cracked by volcanic fire, the mountains of scoriæ and ashes heaped up, are the works of monks; the greatest, no doubt, they ever produced in either hemisphere. Our Indian host, at the hut we inhabited in the plain of Jorullo, told us, that some missionary capuchins preached at the estate of San Pedro, and, not meeting a favourable reception, uttered the most horrible and complicated imprecations against this plain, then so beautiful and fertile. They prophesied,

---

[2] In the bottom of the crater we found the heat of the air 47° [116 6° F.], and in some places 58° and 60° [136 4° and 140°]. We had to pass over cracks exhaling sulphurous vapours, in which the thermometer rose to 85° [185°]. From these cracks, and the heaps of scoriæ that cover considerable hollows, the descent into the crater is not without danger.

that the estate should first be swallowed up by flames issuing out of the bowels of the Earth; and that the air should afterward be cooled to such a degree, that the neighbouring mountains should remain for ever covered with ice and snow. The first of these maledictions having been so fatally verified, the common people foresee in the gradual cooling of the volcano the presage of a perpetual winter. I have thought it right to mention this vulgar tradition, worthy a place in the epic poem of the jesuit Landivar, because it exhibits a striking feature of the manners and prejudices of these remote countries. It shows the active industry of a class of men, who, too frequently abusing the credulity of the people, and pretending to possess the power of suspending the immutable laws of nature, know how to avail themselves of every event for establishing their empire by the fear of physical evil.

The situation of the new volcano of Jorullo leads to a very curious geological observation. It has already been observed in the 3d chapter, that there is in New Spain a line of great heights, or a narrow zone included between the latitudes of 18° 59′ and 19° 12′, in which are all the summits of Anahuac that rise above the region of perpetual snow. These summits are either volcanoes still actually burning; or mountains, the form of which, as well as the nature of their rocks, renders it extremely probable, that they formerly contained subterranean fire. Setting out from the coast of the Gulf of Mexico, and proceeding westward, we find the peak of Oribaza, the two volcanoes of la Puebla, the Nevado de Toluca, the peak of Tancitaro, and the volcano of Colima. These great heights, instead | 86 | of forming the ridge of the cordillera of Anahuac, and following its direction, which is from S. E. to N. W., are on the contrary in a line perpendicular to the axis of the great chain of mountains. It is certainly worthy remark, that in the year 1759 the new volcano of Jorullo was formed in the continuation of this line, and on the same parallel as the ancient Mexican volcanoes.

A view of my plan of the environs of Jorullo will show, that the six large hummocks have risen out of the earth on a vein, that crosses the plain from the cerro of las Cuevas to the pichaco del Montero. The new mouths of Vesuvius too are found ranged along a fissure. Do not these analogies give us reason to suppose, that there exists in this part of Mexico, at a great depth within the Earth, a fissure stretching from east to west through a space of 137 leagues [343 miles], and through which the volcanic fire has made its way at different times, bursting the outer crust of porphyritic rocks, from the coasts of the Gulf of Mexico to the South Sea? Is this fissure prolonged to that little groupe of islands, called by Collnet the Archipelago of Regigedo, and round which, in the same parallel with the Mexican volcanoes, pumice stone has been seen floating? Naturalists who distinguish the facts offered by descriptive mineralogy from theoretical reveries concerning the primitive state of our planet will pardon me for having consigned these observations to the general Map of New Spain, contained in the Mexican Atlas.

Account of the Character and present Condition of the different Classes of Inhabitants in Mexico, or New Spain

(From Humboldt's Travels.)

The Mexican population is composed of the same elements as the other Spanish colonies. They reckon seven races: 1st, the individuals born in Europe, vulgarly called Gachupines; 2d, the Spanish Creoles, or whites of European extraction born in America; 3d, the Mestizos, descendants of whites and Indians; 4th, the Mulattos, descendants of whites and negroes; 5th, the Zambos, descendants of negroes and Indians; 6th, the Indians, or copper-coloured indigenous race; and, 7th, the African Negroes. Abstracting the subdivisions, there are four casts: the whites, comprehended under the general name of Spaniards; the negroes; the Indians; and the men of mixed extraction, from Europeans, Africans, American Indians, and Malays—for, from the frequent communication between Acapulco and the Philippine Islands, many individuals of Asiatic origin, both Chinese and Malays, have settled in New Spain.

A very general prejudice exists in Europe, that an exceeding small number of the copper-coloured race, or descendants of the ancient Mexicans, remain at this day. The cruelty of the Europeans has entirely extirpated the old inhabitants of the West Indies. The continent of America, however, has witnessed no such horrible result. The number of Indians in New Spain exceeds two millions and a half, including only those who have no mixture of European or African blood. What is still more consolatory, and we repeat it, is, that the indigenous population, far from declining, has been considerably on the increase for the last fifty years, as is proved by the registers of capitation or tribute.

In general, the Indians appear to form two-fifths of the whole population of Mexico. In the four intendancies of Guanaxuato, Valladolid, Oaxaca, and La Puebla, this proportion amounts even to three-fifths. The enumeration of 1793 gave the following result.

Names of intendancies.	Total population.	No. of Indians.
Guanaxuato......	398,000......	175,000
Valladolid........	290,000......	119,000
Puebla............	633,000......	416,000
Oaxaca...........	411,000......	363,000

The Indians of New Spain bear a general resemblance to those who inhabit Canada, Florida, Peru, and Brazil. They have the same swarthy and copper-colour, flat and smooth hair, small beard, squat body, long eye, with the corner directed upwards towards the temples, prominent cheek-bones, thick lips, and an expression of gentleness in the mouth, strongly contrasted with a gloomy and severe look. The American race, after the hyperborean, is the least numerous, but it occupies the greatest space on the globe. Over a million and a half of square leagues, from the Terra del Fuego Islands to the river St Lawrence and Bering's Straits, we are struck at the first glance with the general resemblance in the features of the inhabitants. We think we perceive that they all descend from the same stock, notwithstanding the enormous diversity of language which separates them from one another. However, when we reflect more seriously on this family likeness, after living longer among the indigenous Americans, we discover, that celebrated travellers, who could only observe a few individuals on the | 917 | coasts, have singularly exaggerated the analogy of form among the Americans.

Intellectual cultivation is what contributes the most to diversify the features. In barbarous nations, there is rather a physiognomy peculiar to the tribe or horde, than to any individual. When we compare our domestic animals with those which inhabit our forests, we make the same observation. But an European, when he decides on the great resemblance among the copper coloured races, is subject to a particular illusion. He is struck with a complexion so different from our own, and the uniformity of this complexion conceals, for a long time, from him the diversity of individual features. The new colonist can hardly, at first, distinguish the indigenous, because his eyes are less fixed on the gentle, melancholic, or ferocious expression of the countenance, than on the red coppery colour, and the dark, luminous, and coarse and glossy hair, so glossy, indeed, that we should believe it to be in a constant state of humectation.

In the portrait which we draw of the different races of men composing the population of New Spain, we shall merely consider the Mexican

Indian in his actual state. We perceive in him neither that mobility of sensation, gesture, and feature, nor that activity of mind, for which several nations of the equinoxial regions of Africa are so advantageously distinguished. There cannot exist a more marked contrast than that between the impetuous vivacity of the Congo negro, and the apparent phlegm of the Indian. From a feeling of this contrast, the Indian women not only prefer the negroes to the men of their own race, but also to the Europeans. The Mexican Indian is grave, melancholic, and silent, so long as he is not under the influence of intoxicating liquors. This gravity is particularly remarkable in Indian children, who, at the age of four or five, display much more intelligence and maturity than white children. The Mexican loves to throw a mysterious air over the most indifferent actions. The most violent passions are never painted in his features; and there is something frightful in seeing him pass all at once from absolute repose to a state of violent and unrestrained agitation. The Peruvian Indian possesses more gentleness of manners; the energy of the Mexican degenerates into harshness. These differences may have their origin in the different religions and different governments of the two countries in former times. This energy is displayed particularly by the inhabitants of Tlascala. In the midst of their present degradation, the descendants of these republicans are still to be distinguished by a certain haughtiness of character, inspired by the memory of their former grandeur.

The Americans, like the Hindoos, and other nations who have long groaned under a civil and military despotism, adhere to their customs, manners, and opinions, with extraordinary obstinacy. I say opinions, for the introduction of Christianity has produced almost no other effect on the Indians of Mexico than to substitute new ceremonies, the symbols of a gentle and humane religion, to the ceremonies of a sanguinary worship. This change from old to new rites was the effect of constraint, and not of persuasion, and was produced by political events alone. In the new continent, as well as in the old, half civilized nations were accustomed to receive from the hands of the conqueror new laws and new divinities; and the vanquished Indian gods appeared to them to yield to the gods of the | 918 | strangers. In such a complicated mythology as that of the Mexicans, it was easy to find out an affinity between the divinities of Aztlan and the divinity of the east. Cortez even very artfully took advantage of a popular tradition, according to which the Spaniards were merely the descendants of king Quitzalcoatl, who left Mexico for countries situated in the east, to carry among them civilization and laws. The ritual books, composed by the Indians, in hieroglyphics, at the beginning of the conquest, of which I possess several fragments, evidently shew that, at that period, Christianity was confounded with the Mexican mythology: — the Holy Ghost is identified with the sacred eagle of the Aztecs. The missionaries not only tolerated, they even favoured to a certain extent, this amalgamation of ideas, by means of which the Christian worship was more easily introduced among the natives. They persuaded them that the gospel had, in very remote times, been already preached in America; and they investigated its traces in the Aztec ritual with the same ardour which the learned, who, in our days, engage in the study of the Sanscrit, display in discussing the analogy between the Greek mythology and that of the Ganges and the Barampooter.

These circumstances, which will be detailed in another work, explain why the Mexican Indians, notwithstanding the obstinacy with which they adhere to whatever is derived from their fathers, have so easily forgotten their ancient rites. Dogma has not succeeded to dogma, but ceremony to ceremony. The natives know nothing of religion but the exterior forms of worship. Fond of whatever is connected with a prescribed order of ceremonies, they find, in the Christian religion, particular enjoyments. The festivals of the church—the fireworks with which they are accompanied— the processions, mingled with dances and whimsical disguises, are a most fertile source of amusement for the lower Indians. In these festivals, the national character is displayed in all its individuality. Every where the Christians have assumed the shades of the country where they have been transplanted. In the Philippine and Mariana islands the natives of the Malay race have incorporated them with the ceremonies which are peculiar to themselves; and, in the province of Pasto, on the ridge of the Cordillera of the Andes, I have seen Indians, masked and adorned with small tinkling bells, perform savage dances around the altar, while a Monk of St Francis elevated the host.

Accustomed to a long slavery, as well under the domination of their own sovereigns as under that of the first conquerors, the natives of Mexico patiently suffer the vexations to which they are frequently exposed from the whites. They oppose to them only a cunning, veiled under the most deceitful appearances of apathy and stupidity. As the Indian can very rarely revenge himself on the Spaniards, he delights in making a common cause with them for the oppression of his own fellow-citizens. Harassed for ages, and compelled to a blind obedience, he wishes to tyrannize in his turn. The Indian villagers are governed by magistrates of the copper coloured race; and an Indian alcalde exercises his power with so much the greater severity, because he is sure of being supported by the priest or the Spanish subdelegado. Oppression produces, every where, the same effects; it every where corrupts the morals.

As the Indians, almost all of them, belong to the class of peasantry and low people, it is not so easy to judge | 919 | of their aptitude for the arts which embellish life. I know no race of men which appear more destitute of imagination. When an Indian attains a certain degree of civilization, he displays a great facility of apprehension—a judicious mind—a natural logic—and a particular disposition to subtilize or seize the finest differences in the comparison of objects. He reasons coolly and orderly; but he never manifests that versatility of imagination—that glow of sentiment—and that creative and animating art, which characterize the nations of the south of Europe, and several tribes of African negroes. I deliver this opinion, however, with great reserve. We ought to be infinitely circumspect in pronouncing, on the moral or intellectual dispositions of nations, from which we are separated by the multiplied obstacles which result from a difference in language, and a difference of manners and customs. A philosophical observer finds what has been printed in the centre of Europe, on the national character of the French, Italians, and Germans, inaccurate. How, then, should a traveller, after merely landing in an island, or remaining only a short time in a distant country, arrogate to himself the right of deciding on the different faculties of the soul, on the preponderance of reason, wit, or imagination, among nations.

The music and dancing of the natives partake of this want of gaiety which characterizes them. M. Bonpland and myself observed the same thing in all South America. Their songs are terrific and melancholic. The Indian women shew more vivacity than the men; but they share the usual misfortunes of the servitude to which the sex is condemned among nations, where civilization is in its infancy. The women take no share in the dancing; but they remain present to offer fermented draughts to the dancers, prepared by their own hands.

The Mexicans have preserved a particular relish for painting, and for the art of carving on wood or stone. We are astonished at what they are able to execute with a bad knife on the hardest wood. They are particularly fond of painting images, and carving statues of saints. They have been servilely imitating, for these three hundred years, the models which the Europeans imported with them at the conquest. This imitation is derived from a religious principle of a very remote origin. In Mexico, as in Hindostan, it was not allowable in the faithful to change the figure of their idols in the smallest degree. Whatever made a part of the Asiatic or Hindoo ritual, was subjected to immutable laws. For this reason, we shall form a very imperfect judgment of the state of the arts, and the natural state of these nations, if we merely consider the monstrous figures under which they represent their divinities. The Christian images have preserved in Mexico a part of that stiffness and that hardness of feature, which characterize the hieroglyphical pictures of the age of Montezuma. Many Indian children, educated in the college of the capital, or instructed at the academy of painting founded by the king, have no doubt distinguished themselves; but it is much less by their genius than their application. Without ever leaving the beaten track, they display great aptitude in the exercise of the arts of imitation, and they display a much greater still for the purely mechanical arts. This aptitude cannot fail of becoming some day very valuable, when the manufactures shall take their flight to a country where a regenerating government yet remains to be created. | 920 |

The Mexican Indians have preserved the same taste for flowers which Cortez found in his time. A nosegay was the most valuable treat which could be made to the ambassadors who visited the court of Montezuma. This monarch and his predecessors had a great number of rare plants in the gardens of Istapalapan. The famous hand-tree, the *cheirostemen* described by M. Cervantes, of which, for a long time, only a single individual was known of very high antiquity, appears to indicate, that the kings of Toluca cultivated also trees, strangers to that part of Mexico. Cortez, in his letters to the Emperor Charles the Fifth, frequently boasts of the industry which the Mexicans displayed in gardening; and he complains, that they did not send him the seeds of ornamental flowers and useful plants, which he demanded for his friends of Seville and Madrid. The taste for flowers, undoubtedly, indicates a relish for the beautiful; and we are astonished at finding it in a nation in which a sanguinary worship, and the frequency of sacrifices appeared to have extinguished whatever related to the sensibility of the soul, and kindness of affection. In the great market-place of Mexico, the native sells no peaches, nor onions, nor roots, nor pulque, (the fermented juice of the agave), without having his shop ornamented with flowers, which are every day renewed. The Indian merchant appears seated in an entrenchment of verdure. A hedge, of a metre in height, with delicate leaves, surrounds, like a semicircular wall, its fruits offered to public sale. The bottom, of a smooth green, is divided by garlands of flowers, which run parallel to one another. Small nosegays, placed symmetrically between the festoons, give this inclosure the appearance of a carpet strewn with flowers. The European, who delights in studying the customs of the lower people, cannot help being struck with the care and elegance the natives display in distributing the fruits which they sell in small cages of very light wood. The sapatolles (achras), the mammea pears, and raisins, occupy the bottom, while the top is ornamented with odoriferous flowers. This art of entwining fruits and flowers, had its origin, perhaps, in that happy period, when, long before the introduction of inhuman rites, the first inhabitants of Anachuac, like the Peruvians, offered up to the great spirit Teoil the first fruits of their harvest.

(To be continued.)

Account of the Character and present Condition of the different Classes of Inhabitants in Mexico, or New Spain. (From Humboldt's Travels.)

(Continued from p. 920.)

Among the inhabitants of pure origin, the whites would occupy the second place, considering them only in the relation of number. They are divided into whites born in Europe, and descendants of Europeans born in the Spanish colonies of America, or in the Asiatic islands. The former bear the name of Chapetones or Gachupines, and the second that of Creoles. The natives of the Canary islands, who go under the denomination of Islenos, (islanders) and who are the *gerans* of the plantations, are considered as Europeans. The Spanish laws allow the same rights to all whites; but those who have the execution of the laws, endeavour to destroy | 18 | an equality which shocks the European pride. The government, suspicious of the Creoles, bestows the great places exclusively on the natives of Old Spain. For some years back, they have disposed, at Madrid, even of the most trifling employments in the administration of the customs and the tobacco revenue. At an epoch, when every thing tended to an uniform relaxation in the springs of the state, the system of venality made an alarming progress. For the most part, it was by no means a suspicious and distrustful policy, it was pecuniary interest alone which bestowed all employments on Europeans. The result has been a jealousy and perpetual hatred between the Chapetons and the Creoles. The most miserable European, without education, and without intellectual cultivation, thinks himself superior to the whites born in the new continent. He knows that, protected by his countrymen, and favoured by chances common enough in a country where fortunes are as rapidly acquired as they are lost, he may one day reach places to which the access is almost interdicted to the natives, even to those of them distinguished for their talents, knowledge, and moral qualities. The natives prefer the denomination of Americans to that of Creoles. Since the peace of Versailles, and, in particular, since the year 1789, we frequently hear proudly declared: 'I am not a Spaniard, I am an American', words which betray the workings of a long resentment. In the eye of law, every white Creole is a Spaniard; but the abuse of the laws, the false measures of the colonial government, the example of the United States of America, and the influence of the opinions of the age, have relaxed the ties which formerly united more closely the Spanish Creoles to the European Spaniards. A wise administration may re-establish harmony, calm their passions and resentments, and yet preserve for a long time the union among the members of one and the same great family, scattered over Europe and America, from the Patagonian coast to the north of California.

The Spanish laws prohibit all entry into the American possessions, to every European not born in the peninsula. The words European and Spaniard are become synonimous in Mexico and Peru. The inhabitants of the remote provinces have, therefore, a difficulty in conceiving that there can be Europeans that do not speak their language; and they consider this ignorance as a mark of low extraction; because, every where around them, all except the very lowest class of people; speak Spanish. Better acquainted with the history of the 16th century than with that of our own times; they imagine that Spain continues to possess a decided preponderance over the rest of Europe. To them the peninsula appears the very centre of European civilization. It is otherwise with the Americans of the capital. There of them who are acquainted with the French or English literature, fall easily into a contrary extreme, and have still a more unfavouarble opinion of the mother country, than the French had at a time when communication was less frequent between Spain and the rest of Europe. They prefer strangers from other countries, to the Spaniards; and they flatter themselves with the idea that intellectual cultivation has make more rapid progress in the colonies than in the peninsula.

This progress is indeed very remarkable at the Havanah, Lima, Santa Fe, Quito, Popayan, and Caraccas. Of all these great cities, | 19 | the

Havanna bears the greatest resemblance to those of Europe in customs, refinements of luxury, and the tone of society. At Havanna, the state of politics and their influence on commerce, is best understood. However, notwithstanding the efforts ot the patriotic society of the island of Cuba, which encourages the sciences with the most generous zeal; they prosper very slowly in a country where cultivation and colonial produce engross the whole attention of the inhabitants. The study of the mathematics, chemistry, minerallogy, and botany, is more general at Mexico, Santa Fe, and Lima. We every where observe a great intellectual activity, and among the youth a wonderful facility of seizing the principles of science. It is said, that this facility is still more remarkable among the inhabitants of Quito and Lima, than at Mexico and Santa Fe. The former appear to possess more versatility of mind, and a more lively imagination, while the Mexicans, and the natives of Santa Fe, have the reputation of greater perseverance in the studies to which they have once addicted themselves.

No city of the new continent, without even excepting those of the United States, can display such great and solid scientific establishments as the capital of Mexico. I shall content myself here with the school of Mines, directed by the learned Elhuyar, to which we shall return when we come to speak of the Mines, the Botanic Garden, and the Academy of Painting and Sculpture. This Academy bears the title of Academia de los Nobles Artes de Mexico. It owes its existence to the patriotism of several Mexican individuals, and to the protection of the minister of Galvez. The government assigned it a spacious building, in which there is a much finer and more complete collection of casts than is to be found in any part of Germany. We are astonished on seeing that the Apollo of Belvedere, the group of Laocoon, and still more colossal statues, have been conveyed through mountains, at least, as narrow as those of St. Gothard; and we are surprised at finding these masterpieces of antiquity collected together, under the torrid zone, in a table-land higher than the convent of the great St. Bernard. The collection of casts brought to Mexico, cost the king 200,000 francs. The remains of the Mexican sculpture, those colossal statues of basaltes and porphyry, which are covered with antique hieroglyphics, and bear some relation to the Egyptian and Hindoo style, ought to be collected together in the edifice of the Academy, or rather in one of the courts which belong to it. It would be curious to see these monuments of the first cultivation of our species, the works of a semi-barbarous people, inhabiting the Mexican Andes, placed beside the beautiful forms produced under the sky of Greece and Italy.

The revenues of the academy of fine arts at Mexico amount to 125,000 francs, of which the government gives 60,000, the body of

Mexican miners nearly 25,000, the consulado, or association of merchants of the capital, more than 1,500. It is impossible not to perceive the influence of this establishment on the taste of the nation. This influence is particularly visible in the symmetry of the buildings, in the perfection with which the hewing of stone is conducted, and in the ornaments of the capitals and stucco relievos. What a number of beautiful edifices are to be seen at Mexico, nay, even in provincial towns like Guanaxato and Queretaro. These monuments, which frequently cost a million and a half | 20 | of francs, would appear to advantage in the finest streets of Paris, Berlin, and Petersburg. M. Tolsa, Professor of Sculpture at Mexico, was even able to cast an equestrian statue of King Charles the Fourth; a work which, with the exception of the Marcus Aurelius, at Rome, surpasses in beauty, in purity of stile, every thing which remains in this way in Europe. Instruction is communicated gratis at the Academy of Fine Arts. It is not confined alone to the drawing of landscapes and figures; they have had the good sense to employ other means for exciting the national industry. The academy labours successfully to introduce among the artisans a taste for elegance and beautiful forms. Large rooms, well lighted by Argand's lamps, contain, every evening some hundreds of young people, of whom some draw from relievos or living models, while others copy drawings of furniture, chandeliers, or other ornaments in bronze. In this assemblage, (and this is very remarkable in the midst of a country where the prejudices against the casts are so inveterate), rank, colour, and race is confounded; we see the Indian and the Mexican sitting beside the white, and the son of a poor artisan in emulation with the children of the great lords of the country. It is a consolation to observe, that, under every zone, the cultivation of science and art, establishes a certain equality among men, and obliterate, for a time at least, all these petty passions, of which the effects are so prejudicial to social happiness.

If, in the present state of things, the cast of whites is the only one in which we find almost exclusively any thing like intellectual cultivation, it is also the only one which possesses great wealth. This wealth is, unfortunately, still more unequally distributed in Mexico than in the capitania general of Caraccas. There the heads of the richest families possess a revenue of 200,000 livres. In the island of Cuba, we find revenues of more than 6 or 700,000 francs. In these two industrious colonies, agriculture has founded more considerable fortunes than has been accumulated by the working of the mines in Peru. At Lima, an annual revenue of 80,000 francs is very uncommon. I know in reality, of no Peruvian family in the possession of a fixed and sure revenue of

100,000 francs. But, in New Spain, there are individuals who possess no mines, whose revenue amounts to a million of francs. The family of the Count de la Valenciana, for example, possesses alone, on the ridge of the Cordillera, a property worth more than 25 millions of francs, without including the mine of Valenciana, near Guanaxuato, which, communibus annis, yields a net revenue of a million and a half of livres. This family, of which the present head, the young Count de Valenciana, is distinguished for a generous character, and a noble desire of instruction, is only divided into three branches; and they possess altogether, even in years when the mine is not very lucrative, more than 20,000,000 francs of revenue. The Count de Oregla, whose youngest son, the Marquis de San Christobal, distinguished himself at Paris for his physical and physiological knowledge, constructed at the Havanah, at his own expence, in acajou and cedar (cedrella) wood, two vessels of the line of the largest size, which he made a present of to his sovereign. It was the seam of la Bescaina, near Pachuca, which laid the foundation of the fortune of the house of Aregla. The family of Fagoaga, well known for its beneficence, intelligence, and zeal for the public good, exhibits the ex- | 21 | ample ample of the greatest wealth which was ever derived from a mine. A single seam, which the family of the Marquis of Fagoaga possesses in the district of Sombrerete, left, in five or six months, all charges deducted, a net profit of 20 millions of francs.

From these data, one would suppose capitals in the Mexican families infinitely greater than what are really observed. The deceased Count de la Valenciana, the first of the title, sometimes drew from his mine alone, in one year, a net revenue of no less than six millions of livres. This annual revenue, during the last twenty-five years of his life, was never below from two to three millions of livres; and yet this extraordinary man, who came without any fortune to America, and who continued to live with great simplicity, left only behind him at his death, besides his mine, which is the richest in the world, ten millions in property and capital. This fact, which may be relied on, will not surprise those who are acquainted with the interior management of the great Mexican houses. Money rapidly gained is as rapidly spent. The working of mines becomes a game in which they embark with unbounded passion. The rich proprietors of mines lavish immense sums on quacks, who engage them in new undertakings in the most remote provinces. In a country where the works are conducted on such an extravagant scale, that the pit of a mine frequently requires two millions of francs to pierce,—the bad success of a rash project may absorb, in a few years, all that was gained in working the richest seams. We must add, that from the internal disorder which prevails in the greatest part of the great houses of both Old and New Spain, the head of a family is not unfrequently straitened with a revenue of half a million, though he display no other luxury than that of numerous yokes of mules.

In a country governed by whites, the families reputed to have the least mixture of negro or mulatto blood, are also naturally the most honoured. In Spain, it is almost a title of nobility to descend neither from Jews nor Moors. In America, the greater or less degree of whiteness of skin decides the rank which man occupies in society. A white, who rides barefooted on horseback, thinks he belongs to the nobility of the country. Colour establishes even a certain equality among men, who, as is universally the case where civilization ist either little advanced, or in a retrograde state, take a particular pleasure dwelling on the prerogatives of race and origin. When a common man disputes with one of the titled lords of the country, he is frequently heard to say, Do you think me not so white as yourself? This may serve to characterize the state and source of the actual aristocracy. It becomes consequently a very interesting business for the public vanity to estimate accurately the fractions of European blood which belong to the different casts. According to the principle sanctioned by usage, we have adopted the following proportions:—

Casts.	**Mixture of blood.**
Quarterons.........	1/4 negro 3/4 white.
Quinterons.........	1/8 negro 7/8 white.
Zambo..............	3/8 negro 1/4 white.
Zambo prieto.....	7/8 negro 1/8 white.

It often happens, that families suspected of being of mixed blood, demand from the high court of justice (*audiencia*), to have it declared that they belong to the whites. These declarations are not always corroborated by the judgment of the senses. We see very swarthy | 22 | mulattoes who have had the address to get themselves whitened, (this is the vulgar expression.) When the colour of the skin is too repugnant to the judgment demanded, the petitioner is contented with an expression somewhat problematical. The sentence then simply bears, that such or such individuals may consider themselves as whites; *(que a tengan por blancos.)*

Researches Concerning the Institutions and Monuments of the Ancient Inhabitants of America; with descriptions and views of some of the most striking scenes in the Cordilleras

Written in French by ALEXANDER DE HUMBOLDT.
And translated into English by
HELEN MARIA WILLIAMS.
In 2 vols. 8vo. Price 2l. 12s. 6d.

These volumes finish the extensive series of M. de Humboldt, which together form fourteen volumes in the French original. In these dissertations he exhibits the varied researches of the learned antiquary, physiologist, naturalist, astronomer, and geologist. The entire circle of literature exhibits nothing more profound or interesting than several of these essays. We have endeavored to introduce all the passages which bear separation from the context, to the notice and admiration of our readers; and we have no doubt but they will partake of our pleasure in the perusal.

OLD AND NEW WORLDS

Neither an attentive examination of the geological constitution of America, nor reflections on the equilibrium of the fluids, that are diffused over the surface of the globe, lead us to admit, that the new continent emerged from the waters at a later period than the old: we discern in the former the same succession of stony strata, that we find in our own hemisphere; and it is probable, that, in the mountains of Peru, the granites, the micaceous schists, or the different formations of gypsum, and gritstone, existed originally at the same periods as the rocks of the same denominations in the Alps of Switzerland. The whole globe appears to have undergone the same catastrophes. At a height superior to that of Mount Blanc, on the summit of the Andes, we find pe- | 609 | trified sea shells; fossil bones of elephants are spread over the equinoctial regions; and, what is very remarkable, they are not discovered at the feet of the palm trees in the burning plains of the Orinoco, but on the coldest and most elevated regions of the Cordilleras. In the new world, as well as in the old, generations of species long extinct have preceded those, which now people the earth, the waters, and the air.

MAN

There is no proof, that the existence of man is much more recent in America than in the other continent. Within the tropics, the strength of vegetation, the breadth of rivers, and partial inundations have presented powerful obstacles to the migration of nations. The extensive countries of the north of Asia are as thinly peopled as the savannahs of New Mexico and Paraguay; nor is it necessary to suppose, that the countries first peopled are those, which offer the greatest mass of inhabitants. The problem of the first population of America, is no more the province of history, than the questions on the origin of plants and animals, and on the distribution of organic germs, are that of natural science. History, in carrying us back to the earliest epochas, instructs us that almost every part of the globe is occupied by men who think themselves aborigines, because they are ignorant of their origin. Among a multitude of nations, who have succeeded, or have been incorporated with each other, it is impossible to discover with precision, the first basis of population, that primitive stratum beyond which the region of cosmogonical tradition begins.

The nations of America, except those which border on the polar circle, form a single race, characterized by the formation of the scull, the colour of the skin, the extreme thinness of the beard, and straight and glossy hair. The American race bears a very striking resemblance to that of the Mongul nations, which include the descendants of the Hiong-Nu, known heretofore by the name of Huns, the Kalkas, the Kalmucks, and the Burats. It has been ascertained, by late observations, that not only the inhabitants of Unalashka, but several tribes of South America, indicate, by the osteological characters of the head, a passage from the American to the Mongul race. When we shall have more completely studied the brown men of Africa; and that swarm of nations, who inhabit the interior and north-east of Asia, and who are vaguely described by systematic travellers under the name of Tartars and Tschoudes, the Caucasian, Mongul, American, Malay, and Negro races, will appear less insulated, and we shall acknowledge, in this great family of the human race, one single organic type, modified by circumstances which perhaps will ever remain unknown.

Though the nations of the new continent are connected by intimate ties, they exhibit, in the mobility of their features, in their complexions, tanned in a greater or less degree, and in their stature, a difference as remarkable as the Arabians, the Persians, and Sclavonians, who are all of the Caucasian race. The hordes who wander along the burning plains of the equinoctial regions have, however, no darker skins than the mountaineers of the temperate zone; whether it be that in the human race, and in the greater part of animals, there is a certain period of organic life, beyond which the influence of climate and food have no effect, or that the deviation from the primitive type becomes apparent only after a long series of ages. Besides, every thing concurs to prove, that the Americans, as well as the people of the Mongul race, have less flexibility of organization than the other nations of Asia and Europe.

LANGUAGES

The number of languages, which distinguish the different native tribes, appears still more considerable in the new continent than in Africa, where, according to the late researches of Messrs. Seetzen and Vater, there are above one hundred and forty. In this respect, the whole of America resembles Caucasus, Italy before the conquest of the Romans, Asia Minor when that country contained, on a small extent of territory, the Cilicians of Semitic race, the Phrygians of Thracian origin, the Lydians, and the Celts. The configuration of the soil, the strength of vegetation, the apprehensions of the mountaineers under the tropics of exposing themselves to the burning heat of the plains, are obstacles to communication, and contribute to the amazing variety of American dialects. This | 610 | variety, it is observed, is more restrained in the savannahs and forests of the north, which are easily traversed by the hunter, on the banks of great rivers along the coast of the ocean, and in every country where the Incas had established their theocracy by the force of arms.

When it is asserted, that several hundred languages are found in a continent, the whole population of which is not equal to that of France, we regard as different those languages, which bear the same affinity to each other, I will not say as the German and the Dutch, or the Italian and the Spanish, but as the Danish and the German, the Chaldean and the Arabic, the Greek and the Latin. In proportion as we penetrate into the labyrinth of American idioms, we discover, that several are susceptible of being classed by families, while a still greater number remain insulated, like the Biscayan among European, and the Japanese among Asiatic languages. This separation may, however, be only apparent; for we may presume that the languages, which seem to admit of no ethnographical classification, have some affinity, either with other languages which have been for a long time extinct, or with the idioms of nations which have never yet been visited by travellers.

The greater part of the American languages, even such as have the same difference with each other as the languages of Germanic origin, the Celtic and the Sclavonian, bear a certain analogy in the whole of their organization: for instance, in the complication of grammatical forms, in the modification of the verb according to the nature of its syntax, and in the number of additive particles (affixa et suffixa). This uniform tendency of the idioms betrays, if not a community of origin, at least a great analogy in the intellectual dispositions of the American tribes, from Greenland to the Magellanic regions.

Investigations made with the most scrupulous exactness, in following a method which had not hitherto been used in the study of etymologies, have proved, that there are a few words that are common in the vocabularies of the two continents. In eighty-three American languages, examined by Messrs. Barton and Vater, one hundred and seventy words have been found, the roots of which appear to be the same; and it is easy to perceive, that this analogy is not accidental, since it does not rest merely on imitative harmony, or on that conformity in the organs, which produces almost a perfect identity in the first sounds articulated by children. Of these one hundred and seventy words, which have this connexion with each other, three fifths resemble the Mantchou, the Tongouse, the Mongul, and the Samoyede; and two fifths the Celtic and Tschoud, the Biscayan, the Coptic, and the Congo languages. These words have been found by comparing the whole of the American languages with the whole of those of the old world; for hitherto we are acquainted with no American idiom, which seems to have an exclusive correspondence with any of the Asiatic, African, or European tongues. What some learned writers have asserted from abstract theories, respecting the pretended poverty of all the American languages, and the extreme imperfection of their numerical system, is as doubtful as the assertions which have been made respecting the weakness and stupidity of the human race throughout the new continent, the stunted growth of animated nature, and the degeneration of those animals, which have been transported from one hemisphere to the other.

ORIGIN

Though no traditions point out any direct connexion between the nations of North and South America, their history is not less fraught with analogies in the political and religious revolutions, from which dates the civilization of the Aztecks, the Muyscas, and the Peruvians. Men with beards, and with clearer complexions than the natives of Anahuac, Cundinamarca, and the elevated plain of Couzco, make their appearance without any indication of the place of their birth; and, bearing the title of high priests, of legislators, of the friends of peace and the arts, which flourish under its auspices, operate a sudden change in the policy of the nations, who hail their arrival with veneration. Quetzalcoatl, Bochica, and Manco Capac, are the sacred names of these mysterious beings. Quetzalcoatl, clothed in a black sacerdotal robe, comes from Panuco, from the shores of the Gulf of Mexico; Bochica, the Boudha of the Muyscas, presents himself on the | 611 | high plains of Bogota, where he arrives from the savannahs, which stretch along the east of the Cordilleras. The history of these legislators, which I have endeavoured to unfold in this work, is intermixed with miracles, religious fictions, and with those characters which imply an allegorical meaning. Some learned men have pretended to discover, that these strangers were shipwrecked Europeans, or the descendants of those Scandinavians, who, in the eleventh century, visited Greenland, Newfoundland, and perhaps Nova Scotia; but a slight reflection on the period of the Tolteck migrations, on the monastic institutions, the symbols of worship, the calendar, and the form of the monuments of Cholula, of Sogamozo, and of Couzco, leads us to conclude, that it was not in the north of Europe that Quetzalcoatl, Bochica, and Manco Capac framed their code of laws. Every consideration leads us rather towards Eastern Asia, to those nations who have been in contact with the inhabitants of Thibet, to the Shamanist Tartars, and the bearded Ainos of the isles of Jesso and Sachalin.

PYRAMID OF CHOLULA

Among those swarms of nations, which, from the seventh to the twelfth century of the christian era, successively inhabited the country of Mexico, five are enumerated; the Toltecks, the Cicimecks, the Acolhuans, the Tlascaltecks, and the Aztecks, who, notwithstanding their political divisions, spoke the same language, followed the same worship, and built pyramidal edifices, which they regarded as *teocallis*, that is to say, the houses of their gods. These edifices were all of the same form, though of very different dimensions; they were pyramids, with several terraces, and the sides of which stood exactly in the direction of the meridian, and the parallel of the place. The *teocalli* was raised in the midst of a square and walled enclosure, which, somewhat like the περιβολος of the Greeks, contained gardens, fountains, the dwellings of the priests, and sometimes arsenals; since each house of a Mexican divinity, like the ancient temple or Baal Berith, burnt by Abimelech, was a strong place. A great staircase led to the top of the truncated pyramid, and on the summit of the platform were one or two chapels, built like towers, which contained the colossal idols of the divinity, to whom the teocalli was dedicated. This part of the edifice must be considered as the most consecrated place; like the ναος, or rather the ςεκος, of the Grecian temples. It was there, also, that the priests kept up the sacred fire. From the peculiar construction of the edifice we have just described, the priest who offered the sacrifice was seen by a great mass of the people at the same time: the procession of the *teopixqui*, ascending or descending the staircase of the pyramid, was beheld at a considerable distance. The inside of the edifice was the burial place of the kings and principal personages of Mexico. It is impossible to read the descriptions, which Herodotus and Diodorus Siculus have left us, of the temple of Jupiter Belus, without being struck with the resemblance of that Babylonian monument to the teocallis of Anahuac.

The *teocalli* of Mexico was dedicated to Tezcatlipoca, the first of the Azteck divinities after Teotl, who is the supreme and invisible being; and to Huitzilopochtli, the god of war. It was built by the Aztecks, on the model of the pyramids of Teotihuacan, six years only before the discovery of America by Christopher Columbus. This truncated pyramid, called by Cortez the principal temple, was ninety-seven metres in breadth at its basis, and nearly fifty-four metres in height. It is not astonishing, that a building of these dimensions should have been destroyed a few years after the siege of Mexico. In Egypt there scarcely remains any vestiges of the enormous pyramids, which towered amidst the waters of the lake Mœris, and which Herodotus says were ornamented with colossal statues. The pyramids of Porsenna, of which the description seems somewhat fabulous, and four of which, according to Varro, were more than eighty metres in height, have equally disappeared in Etruria.

The greatest, most ancient, and most celebrated of the whole of the pyramidal monuments of Anahuac is the teocalli of Cholula. It is called in the present day the Mountain made by the hand of Man (*monte hecho a manos*). At a distance it has the aspect of a natural hill covered with vegetation.

The great teocalli of Cholula, called also the *Mountain of unbaked bricks* (tlalchihualtepec), had an altar on its top, dedicated to Quetzalcoatl, the god of | 612 | the air. This Quetzalcoatl, whose name signifies serpent clothed with green feathers, from *coatl*, serpent, and *quetzalli*, green feathers, is the most mysterious being of the whole Mexican mythology. He was a white and bearded man, like the Bochica of the Muyscus. He was high priest of Tula (Tollan,) legislator, chief of a religious sect, which, like the Sonyasis and the Bouddhists of Indostan, inflicted on themselves the most cruel penances. He introduced the custom of piercing the lips and the ears, and lacerating the rest of the body with the prickles of the agave leaves, or the thorns of the cactus; and of putting reeds into the wounds, in order that the blood might be seen to trickle more copiously.

Another very remarkable tradition still exists among the Indians of Cholula, according to which the great pyramid was not originally destined to serve for the worship of Quetzalcoatl. After my return to Europe, on examining at Rome the Mexican manuscript in the Vatican library, I found, that this same tradition was already recorded in a manuscript of Pedro de Los Rios, a Dominican monk, who, in 1566, copied on the very spot all the hieroglyphical paintings he could procure. "Before the great inundation which took place four thousand eight hundred years after the creation of the World, the country of Anahuac was inhabited by giants (tzocuillixeque). All those who did not perish were transformed into fishes, save seven, who fled into caverns. When the waters subsided, one of these giants, Xelhua, surnamed the architect, went to Cholollan; where, as a memorial of the mountain Tlaloc, which had served for an asylum to himself and his six brethren, he built an artificial hill in form of a pyramid. He ordered bricks to be made in the province of Tlamanalco, at the foot of the Sierra of Cocotl, and to convey them to Cholula he placed a file of men, who passed them from hand to hand. The gods beheld with wrath this edifice, the top of which was to reach the clouds. Irritated at the daring attempt of Xelhua, they hurled fire on the pyramid. Numbers of the workmen perished; the work was discontinued, and the monument was afterwards dedicated to Quetzalcoatl, the god of the air."

The size of the platform of the pyramid of Cholula, on which I made a great number of astronomical observations, is four thousand two hundred square metres. From it the eye ranges over a magnificant prospect; Popocatepetl, Iztaccihuatl, the peak of Orizaba, and the Sierra de Tlascalla, famous for the tempests which gather around its summit. We view at the same time three mountains higher than Mount Blanc, two of which are still burning volcanoes. A small chapel, surrounded with cypress, and dedicated to the Virgin de los Remedios, has succeeded to the temple of the god of the air, or the Mexican Indra. An ecclesiastic of the Indian race celebrates mass every day on the top of this antique monument.

No one of the ancient writers, neither Herodotus nor Strabo, Diodorus nor Pausanias, Arrian nor Quintus Curtius, asserts, that the temple of Belus was erected according to the four cardinal points, like the Egyptian and Mexican pyramids. Pliny observes only, that Belus was considered as the inventor of astronomy: *Inventor hic fuit sideralis scientiæ.* Diodorus relates, that the Babylonian temple served as an observatory to the Chaldeans. "It must be admitted," says he, "that this building was of an extraordinary height, and that here the Chaldeans made their observations on the stars, the rising and setting of which might be exactly perceived, on account of the elevation of the edifice." The Mexican priests, (*teopixqui*) made observations also on the stars from the summit of the teocallis; and announced to the people, by the sound of the horn, the hours of the night. These teocallis were built in the interval between the epocha of Mahomet and the reign of Ferdinand and Isabella; and we cannot observe without astonishment, that American edifices, the form of which is almost the same as that of one of the most ancient monuments on the banks of the Euphrates, belong to times so near our own.

VOLCANO OF COTOPAXI

Cotopaxi is the loftiest of those volcanoes of the Andes, which at recent epochs have undergone eruptions. Its absolute height is five thousand seven hundred and fifty-four metres (two thousand nine hundred and fifty-two toises); it is double that of Canigou; and consequently eight hundred metres higher than Vesuvius would be, | 613 | were it placed on the top of the Peak of Teneriffe. Cotopaxi is also the most dreadful volcano of the kingdom of Quito, and its explosions the most frequent and diastrous. The mass of scoriæ, and the huge pieces of rock, thrown out of this volcano, which are spread over the neighbouring valleys, covering a surface of several square leagues, would form, were they heaped together, a colossal mountain. In 1738 the flames of Cotopaxi rose nine hundred metres above the brink of the crater. In 1744 the roarings of the volcano were heard as far as Honda, a town on the borders of the Magdalena, and at the distance of two hundred common leagues. On the 4th of April, 1768, the quantity of ashes ejected by the mouth of Cotopaxi was so great, that in the towns of Hambato and Tacunga day broke only at three in the afternoon, and the inhabitants were obliged to use lanterns in walking the streets. The explosion which took place in the month of January, 1803, was preceded by a dreadful phenomenon, the sudden melting of the snows that covered the mountain. For twenty years before no smoke or vapour, that could be perceived, had issued from the crater; and in a single night the subterraneous fire became so active, that at sun-rise the external walls of the cone, heated, no doubt, to a very considerable temperature, appeared naked, and of the dark colour, which is peculiar to vitrified scoriæ. At the port of Guayaquil, fifty-two leagues distant in a straight line from the crater, we heard, day and night, the noises of the volcano, like continued discharges of a battery; we distinguished these tremendous sounds even on the Pacific Ocean, to the southwest of the island of Puna.

The form of Cotopaxi is the most beautiful and regular of the colossal summits of the high Andes. It is a perfect cone, which, covered with an enormous layer of snow, shines with dazzling splendor at the setting of the sun, and detaches itself in the most picturesque manner from the azure vault of heaven. This covering of snow conceals from the eye of the observer even the smallest inequalities of the soil; no point of rock, no stony mass, penetrates this coating of ice, or breaks the regularity of the figure of the cone. The summit of Cotopaxi resembles the Sugar-loaf (Pan de azucar) which terminates the Peak of Teyde; but the height of its con is six times the height of that of the great volcano of the Island of Teneriffe.

The greater the regularity in the form of the cone of this volcano the more we are struck in finding, on the side to the south-east, a small mass of rock, half concealed under the snow, studded with points, and which the natives call the head of the Inca. The origin of this singular denomination is very uncertain. A popular tradition prevails in the country that this isolated rock was heretofore a part of the top of Cotopaxi. The Indians relate that the volcano, at its first eruption, ejected far off a stony mass; which, like the cap of a dome, covered the enormous cavity that contains the subterraneous fire. Some pretend that this extraordinary catastrophe took place a short time after the invasion of the kingdom of Quito by the Inca Tupac Yupanqui; and that the rock to the left of the volcano, is called the head of the Inca, because its fall was the ominous presage of the death of the conqueror. Others, still more credulous, affirm that this mass of porphyry, with basis of pitchstone, was displaced in an explosion that happened at the very moment when the Inca Atahualpa was strangled by the Spaniards at Caxamarca.

LETTERS

The want of letters observed in the new continent, at the time of its second discovery by Christopher Columbus, leads to the idea that the tribes of the Tartar or Mongul race, which we may suppose to have passed from the east of Asia to America, were not in possession of alphabetical writing; or what is less probable, that, having relapsed into barbarism under the influence of a climate less favourable to the display of the understanding, they had lost this wonderful art, known only to a very small number of individuals. We shall not here examine the question, whether the Devanagari alphabet is of remote antiquity on the banks of the Indus and the Ganges; or whether, as Strabo asserts from Megasthenes, the Hindoos were ignorant of writing before the conquests of Alexander. Farther to the east and the north in the region of monosyllabic languages, as in that of the Tartarian, Samoiede, Ostiack, and Kamtschadale tongues, the use of letters, wherever it is at present found, | 614 | was introduced very late. It seems indeed probable that it was the Christian sect of Nestorians, who communicated the Stranghelo alphabet to the Oighours and the Manchou Tartars; an alphabet, which in the northern regions of Asia is still more recent than the Runic characters in the north of Europe. We need not therefore suppose the communications between eastern Asia and America to have been of very remote antiquity, in order to comprehend why this latter part of the world had not been instructed in an art, which for a long series of ages was unknown, except in Egypt, in the Phœnician and Grecian colonies, and in the small space lying between the Mediterranean, the Oxus, and the Persian gulf.

The first missionaries who visited America, Valades and Acosta, have already called the Azteck paintings a writing similar to that of the Egyptians. If Kircher, Warburton, and other learned men, have since contested the propriety of this expression, it is because they have not distinguished the paintings of a mixed kind, in which real hieroglyphics, sometimes curiological, sometimes tropical, are added to the natural representation of an action, from simple hieroglyphical writing, such as is found, not on the pyramidion, but on the great faces of the obelisks. The famous inscription of Thebes, cited by Plutarch, and by Clement of Alexandria, the only one, the explanation of which has reached us, expressed by the hieroglyphics of a child, an old man, a vulture, a fish, and a hippopotamus, the following sentence, "You who are born, and who are to die, know, that the Eternal hates impudence." A Mexican, to express the same idea, would have represented the great spirit, Teotl, chastising a criminal; certain characters placed above two heads would have been sufficient to indicate the age of the child, and that of the old man; he would have individualized the action, but the style of his hieroglyphical paintings would not have furnished him with the means of giving a general expression to the sentiment of hatred and vengeance.

When we compare the Mexican paintings with the hieroglyphics that decorated the temples, the obelisks, and perhaps even the pyramids of Egypt; and reflect on the progressive steps which the human mind appears to have followed in the invention of graphic means fitted to express ideas; we see that the nations of America were very distant from that perfection which the Egyptians had obtained. The Aztecks were indeed but little acquainted with simple hieroglyphics; they could represent the elements, and the relations of time and of place; but it is only by a great number of these characters, susceptible of being employed separately, that the painting of ideas becomes easy, and approximates to writing. We find among the Aztecks the germes of phonetic characters: they knew how to write names, by uniting certain signs which are associated with sounds; this contrivance might have led them to the beautiful discovery of giving an alphabetic form to their simple hieroglyphics; but ages would have elapsed before these nations of mountaineers, who adhered to their manners and customs with the same invincible obstinacy as the Chinese, the Japanese, and the Hindoos, could have raised themselves to the decomposition of words, the analysis of sounds, the invention of an alphabet.

Among the Mexican people the figures and symbolic characters were not traced on separate leaves. Whatever was the substance employed for manuscripts, they were seldom destined to form rolls, but were almost always folded in zigzag, in a particular manner, like the mounts of our fans. Two tablets of light wood were pasted at the ends, one at top, the other at bottom, so that before the painting was unfolded the manuscript had the most perfect resemblance with our bound books. By this arrangement, on opening a Mexican manuscript as we open our books, we can see only half of the characters at one time, those which are painted on the same side of the skin, or paper of maguey: to examine the whole of the pages, if the different folds of a band, which is often twelve or fifteen metres in length, can be called pages, we must extend the whole manuscript first from the left to the right, and then from the right to the left. In this respect the Mexican paintings are perfectly similar to the Siamese manuscripts, preserved in the public library at Paris, which are also folded in zigzag.

At Mexico the use of painting and of paper of maguey was extended far beyond the limits of the empire of | 615 | Montezuma, to the borders of the lake of Nicaragua, whither the Toltecks in their migrations had carried their language and their arts. In the kingdom of Guatimala, the inhabitants of Teochiapan had preserved traditions, that went back to the epocha of a great deluge; after which their ancestors, led by a chief called Votan, had come from a country lying toward the north. In the village of Teopixca there still existed in the sixteenth century descendants of the family of Votan, or Vodan, for these two names are the same, the Toltecks and the Aztecks not having the four consonants d, b, r, s, in their language. They who have studied the history of the Scandinavian nations in the heroic times must be struck at finding in Mexico a name which recalls that of Wodan or Odin, who reigned among the Scythians, and whose race, according to the very remarkable assertion of Bede, "gave kings to a great number of nations."

MEXICAN PAINTINGS

From the researches I have made, it appears that there exist at present in Europe only six collections of Mexican paintings, those of the Escurial, Bologna, Veletri, Rome, Vienna, and Berlin. The learned jesuit Fabrega, who is often cited in the works of Mr. Zoega, and whose manuscripts relating to the Azteck antiquities were communicated to me by the Chevalier Borgia, nephew to the Cardinal of that name, supposes that the archives of Simancas, in Spain, contain also some of these hieroglyphical paintings, which Robertson has so aptly denominated picture writings.

The collection preserved at the Escurial has been examined by Mr. Waddilove, chaplain to the English embassy at Madrid when Lord Grantham was ambassador. It has the form of a book in folio; which may lead us to suspect that it is only a copy of a Mexican manuscript, for the originals I have examined are all of the size of volumes in quarto. The objects represented seem to prove that the collection of the Escurial, like those of Italy and Vienna, are either astrological books or real rituals, which point out the religious ceremonies prescribed for particular days of the month. At the bottom of each page is an explanation in Spanish, which has been added since the conquest.

The collection of Bologna is deposited in the library of the Institute of Sciences of that city. We are unacquainted with its origin; but we read on the first page that this painting, which is 326 centimetres (eleven Roman palms) in length, was ceded, the 26th of December, 1665, by Count Valerio Zani to the Marquis of Caspi. The characters, which are traced on a thick and ill prepared skin, seem in a great measure to allude to the form of the constellations, and to astrological notions. There exists an engraved copy of this Codex Mexicanus of Bologna, in the Museum of Cardinal Borgia, at Veletri.

The collection of Vienna, which is sixty-five pages, is become celebrated since it fixed the attention of Dr. Robertson, who, in his classic work on the history of the New Continent, has published a few pages in outlines only, and without colouring. We read on the first page of this Mexican manuscript that it was sent by King Emanuel of Portugal to Pope Clement the Seventh, and that it has since been in the hands of the Cardinals Hippolito de Medicis and Capuanus.

The Codex Mexicanus of the Borgian museum at Veletri is the finest of the Azteck manuscripts that I have examined.

The collection preserved in the royal library at Berlin contains different Azteck paintings, which I purchased during my abode in New Spain. The twelfth plate gives two fragments of this collection; it contains the lists of tributes, genealogies, the history of the migrations of the Mexicans, and a calendar made at the beginning of the conquest, in which the simple hieroglyphics of the days are joined to figures of saints painted in the Azteck style.

The library of the Vatican at Rome possesses in the valuable collection of its manuscripts two Codices Mexicani, numbered 3738 and 3776, in the catalogue. These collections, as well as the manuscript of Veletri, were unknown to Dr. Robertson, when he enumerated the Mexican paintings preserved in the different libraries of Europe. Mercatus, in his description of the obelisks of Rome, relates that toward the end of the sixteenth century two collections of original paintings existed in the Vatican. It would seem that one of these collections is | 616 | entirely lost, unless it is that which is seen in the library of the Institute of Bologna; the other was found in 1785 by the jesuit Fabrega, after fifteen years search.

MEXICAN THEOLOGY

The ninety-sixth page of the Codex Vaticanus, represents on the left an adoration: the deity has on a helmet, the ornaments of which are very remarkable: he is seated on a small bench, called *icpalli*, before a temple, of which only the top, or small chapel placed on the upper part of the pyramid, is represented. The adoration consisted, at Mexico, as well as in the East, in the ceremony of touching the ground with the right hand, and carrying the left to the mouth. In the drawing, No. 1, the homage is rendered by a genuflexion; the attitude of the figure, which prostrates itself before the temple, is found in several paintings of the Hindoos.

The group, No. 2, represents the celebrated *serpentwoman,* Cihuacohuatl, called also Quilaztli, or Tonacacihua, *woman of our flesh*: she is the companion of Tonacateuctli. The Mexicans considered her as the mother of the human race; and, after the god of the *celestial Paradise,* Ometeuetli, she held the first rank among the divinities of Anahuac; we see her always represented with a great serpent. Other paintings exhibit to us a featherheaded snake, cut in pieces by the great spirit, Tezcatlipoca, or by the Sun personified, the god Tonatiuh. These allegories remind us of the ancient traditions of Asia. In the *woman and serpent* of the Aztecks we think we perceive the Eve of the Semetic nations: in the snake cut in pieces, the famous serpent *Kaliya*, or *Kalinaga*, conquered by Vishnu, when he took the form of Krishna. The Tonatiuh of the Mexicans appears also to be identical with the Krishua of the Hindoos, recorded in the Bhagavata Purana, and with the Mithras of the Persians. The most ancient traditions of nations go back to a state of things, when the earth, covered with bogs, was inhabited by snakes and other animals of gigantic bulk: the beneficent luminary, by drying up the soil, delivered the earth from these aquatic monsters.

Behind the serpent, who appears to be speaking to the goddess Cihuacohuatl, are two naked figures; they are of a different colour, and seem to be in the attitude of contending with each other. We might be led to suppose, that the two vases, which we see at the bottom of the picture, one of which is overturned, is the cause of this contention. The serpent woman was considered at Mexico as the mother of two twin children; these naked figures are perhaps the children of Cihuacohuatl; they remind us of the Cain and Abel of Hebrew tradition.

It is no way doubtful, that Nestorianism, mingled with the dogmata of the Bouddhistes and the Shamans, spread through Mantchou Tartary into the north-east of Asia: we may therefore suppose, with some appearance of reason, that Christian ideas have been communicated by the same means to the Mexican nations, especially to the inhabitants of that northern region, from which the Toltecks emigrated, and which we must consider as the *officina virorum* of the New World.

When we examine this question by the rules of the most rigid analysis, we find nothing among the Americans, which leads to the supposition, that the Asiatic nations migrated to the New Continent after the establishment of Christianity. I am very far from denying the possibility of these posterior communications; I am not ignorant that the Tchoutskis annually crossed Behring's Straits to make war on the inhabitants of the north-west coast of America; but I think I may affirm, from the knowledge we have acquired since the end of the last century of the sacred books of the Hindoos, that, in order to explain these resemblances of traditions, of which all the first missionaries speak, we have no need to recur to Western Asia, peopled by nations of the Semetic race; these same traditions, of high and venerable antiquity, are found both among the followers of Brahma, and among the Shamans of the eastern steppes of Tartary.

Notwithstanding these striking analogies existing between the nations of the New Continent, and the Tartar tribes who have adopted the religion of Bouddah, I think I discover in the mythology of the Americans, in the style of their paintings, in their languages, and especially in their external conformation, the descendants of a race of men, which, early separated | 617 | from the rest of mankind, has followed for a lengthened series of ages, a peculiar road in the unfolding of its intellectual faculties, and in its tendency towards civilization.

CHIMBORAZO AND CARGUAIRAZO

The group of Chimborazo and Carguairazo, has an absolute elevation of 2891 metres (1493 toises); it is only a sixth less elevated than the top of Etna. The summit of Chimborazo does not therefore surpass the height of this plain more than 3640 metres, which is 84 metres less than the height of the top of Mount Blanc above the priory of Chamonix; for the difference between Chimborazo and Mount Blanc is nearly equal to that which is observed between the elevation of the plain of Tapia, and the bottom of the valley of Chamonix. The top of the Peak of Teneriffe, compared with the level of the town of Oratava, is still more elevated than Chimborazo and Mount Blanc above Riobamba and Chamonix.

We distinguish three kinds of principal forms belonging to the high tops of the Andes. The volcanoes which are yet burning, those which have but a single crater of extraordinary size, are conic mountains, with summits truncated in a greater or less degree: such is the figure of Cotopaxi, of Popocatepec, and the Peak of Orizaba. Volcanoes, the summits of which have sunk after a long series of eruptions, exhibit ridges bristled with points, needles leaning in different directions, and broken rocks falling into ruins. Such is the form of the Altar, or Capac-Urcu, a mountain once more lofty than Chimborazo, and the destruction of which is considered as a memorable period in the natural history of the New Continent; such is the form also of Carguairazo, a great part of which fell in on the night of the 19th of July, 1698. Torrents of water and mud then issued from the opened sides of the mountain, and laid waste the neighbouring country. This dreadful catastrophe was accompanied by an earthquake, which, in the adjacent towns of Hambato, and Llactacunga, swallowed up thousands of inhabitants.

A third form of the high tops of the Andes, and the most majestic of the whole, is that of Chimborazo, the summit of which is circular; it reminds us of those paps without craters, which the elastic force of the vapours swells up in regions where the hollow crust of the globe is mined by subterraneous fires. The aspect of mountains of granite has little analogy with that of Chimborazo. The granitic summits are flattened hemispheres; the trappean porphyry forms slender cupolas. Thus on the shore of the South Sea, after the long rains of winter, when the transparency of the air has suddenly increased, we see Chimborazo appear like a cloud at the horizon; it detaches itself from the neighbouring summits, and towers over the whole chain of the Andes, like that majestic dome, produced by the genius of Michael Angelo, over the antique monuments, which surround the Capitol.

Travellers who have approached the summits of Mont Blanc and Mont Rose are alone capable of feeling the character of this calm, majestic, and solemn scenery. The bulk of Chimborazo is so enormous, that the part which the eye embraces at once near the limit of the eternal snows is seven thousand metres in breadth. The extreme rarity of the strata of air, across which we see the tops of the Andes, contributes greatly to the splendour of the snow, and the magical effect of its reflection. Under the tropics, at a height of five thousand metres,[1] the azure vault of the sky appears of an indigo tint. The outlines of the mountain detach themselves from the sky in this pure and transparent atmosphere, while the inferior strata of the air, reposing on a plain destitute of vegetation, which reflects the radiant heat, are vaporous, and appear to veil the middle ground of the landscape.

The elevated plain of Tapia, which extends to the east as far as the foot of the altar, and of Condorasto, is three thousand metres in height, nearly equal to that of Canigou, one of the highest summits of the Pyrennees. A few plants of schinus, molle, cactus, agave, and molina, are scattered over the barren plain: and we see in the foreground lamas (camelus lacma) sketched from nature, and groups of Indians going to the market of Lican. The flank of the mountain presents that gradation of vegetable life, which I have endeavoured to trace in my chart of the Geography of plants, and which may be followed on the western top of the Andes from the impenetrable | 618 | groves of palm trees to the perpetual snows, bordered by thin layers of lichens.

[1] 39,371 inches.

At three thousand five hundred metres absolute height, the ligneous plants with coriaceous and shining leaves nearly disappear. The region of shrubs is separated from that of the grasses by alpine plants, by tufts of nerteria, valerian, saxifrage, and lobelia, and by small cruciferous plants. The grasses form a very broad belt, covered at intervals with snow, which remains but a few days. This belt, called in the country the *pajonal*, appears at a distance, like a gilded yellow carpet. Its colour forms an agreeable contrast with that of the scattered masses of snow; and is owing to the stalks and leaves of the grasses burnt by the rays of the sun in the seasons of great draught. Above the *pajonal* lies the region of cryptogamus plants, which here and there cover the porphyritic rocks destitute of vegetable earth. Farther on, at the limit of the perpetual ice, is the termination of organic life.

MEXICAN CALENDAR

The civil year of the Aztecks was a solar year of three hundred and sixty-five days, and was divided into eighteen months, each of twenty days. After these eighteen months, or three hundred and sixty days, five complementary days were added, and the year began anew. The names of *Tonalpohualli* or *Cempohualilhuitl*, which distinguished this civil calendar from the ritual calendar, sufficiently indicated its principal characters. The first of these names signifies *reckoning of the Sun*, in opposition to the ritual calendar, called *reckoning of the Moon*, or *Metzlapohualli*: the second denomination is derived from *cempohualli*, twenty, and *ilhuitl*, festival; and it alludes, either to the twenty days contained in each month, or the twenty solemn festivals celebrated during the course of a civil year, in the teocallis, or *houses of the gods*.

The beginning of the civil day among the Aztecks was reckoned like that of the Persians, the Egyptians, the Babylonians, and the greater part of the nations of Asia, except the Chinese, from sunrising. It was divided into eight intervals, a division found among the Hindoos and the Romans; four of which were determined by the rising of the Sun, its setting, and its two passages across the meridian. The rising was called *yquiza tonatiuh*; noon, *nepantla tonatiuh*; the setting, *onaqui tonatiuh*; and midnight, *yohual-nepantla*. The hieroglyphic of the day was a circle divided into four parts. Although, under the parallel of the city of Mexico, the length of the day does not vary more than two hours twenty-one minutes, it is very certain, that the Mexican hours were originally unequal, like the *planetary hours* of the Jews, and all those which the Greek astronomers noted under the name of καιρικαι, in opposition to the ισημεριναι, *equinoxial hours.*

To correct the lunar year, and make it coincide with the solar year, eleven days, according to ancient custom, were added; which, by the edict of the Inca, were divided among the twelve moons. According to this arrangement, it was scarcely possible, that four equal periods, into which the lunar months should be divided, could be seven days each, and correspond to the phases of the Moon. The same historian, whose testimony is cited by M. Bailly in favor of the opinion, that the week of the Hindoos was known to the Americans, affirms, that, according to an ancient law of the Inca Pachacutec, there ought to be in each lunar month three days for festivals and for markets (*catu*); and that the people were to work, not seven, but eight consecutive days, and rest the ninth. This is undoubtedly a division of a lunar month, or a sideral revolution of the Moon, into three small periods of nine days.

We shall observe, on this occasion, that the Japanese, a nation of the Tartar race, are equally unacquainted with the small period of seven days; while it is in use among the Chinese, who seem also aborigines of the elevated plain of Tartary, but who have long had intimate communications with Hindostan and Thibet.

THEIR CHRONOLOGY

The Mexicans were in possession of annals, that went back to eight centuries and a half beyond the epocha of the arrival of Cortez in the country of Anahuac. We have already explained how these annals presented, in their subdivisions, sometimes a cycle of fifty-two years, at others a thalpilli of thirteen years, and at others a single year of two hundred and sixty days, contained in twenty small pe- | 619 | riods of thirteen days, according as the history was more or less minute. The Toltecks had disappeared four hundred and sixty-five years before the arrival of Cortez; the nation which the Spaniards found settled in the valley of Mexico was of the Azteck race: what he knew of the Toltecks he could have learnt only from paintings, which they had left in the country of Anahuac; or from some dispersed familes, who, restrained by the love of their native soil, had not thought proper to share the chances of the emigration.

Let us now examine the ingenious but very complicated methods, of which these people made use to denote the year and the day of a cycle of fifty-two years. This method is identic with that made use of by the

Hindoos, the Thibetans, the Chinese, the Japanese, and the Asiatic people of the Tartar race; who also distinguished the months and the years by the correspondence of several periodical series, the number of the terms of which is not the same. The Mexicans employ, for the cycle of years, the four following signs, which have the names of

Tochtli – a rabbit or hare,
Actal – a cane,
Tecpatl – a flint, or silex,
Calli – a house.

The same contrivance of the concordance of two periodical series was employed to distinguish the days of the same year. It appears, that originally among the Mexican nations, as well as among the Persians, each day of the month had a name, and a particular sign; these twenty signs recall to mind the *yogas*, which, in the astrological almanack of the Hindoos, we find added to the twenty-eight days of the lunar month. In the *Metztlapohualli,* or *reckoning of the Moon*, of the Aztecks, they were distributed among the small cycles of the half-lunations; so that a periodical series of thirteen terms, which were all ciphers, corresponded to a periodical series of twenty terms, which contained only hieroglyphical signs. It is in this series of days, that we find the four grand signs—rabbit, cane, flint, and house, by which, as we have just seen, the years of a cycle were denoted; sixteen other signs of an inferior order were so distributed, that in an equal number of four they separated the grand signs one from the other.

WODAN

Votan, or *Wodan*, an American, seems to be a member of the same family with the *Wods,* or *Odins*, of the Goths, and nations of Celtic origin. As Odin and Boudha, according to the learned researches of Sir William Jones, are probably one and the same person, it is curious to see the names of *Boudvar*, *Wodans dag*, (Wednesday), and *Votan*, denote in India, in Scandinavia, and Mexico, a day of a small period. According to the ancient traditions, collected by the Bishop Francis Nunnez de la Vega; "the Wodan of the Chiapanese was grandson of that illustrious old man, who, at the time of the great deluge, in which the greater part of the human race perished, was saved on a raft together with his family." Wodan co-operated in the construction of the great edifice, which had been undertaken by men to reach the skies; the execution of this rash project was interrupted; each family received from that time a different language, and the great spirit, *Teotl*, ordered Wodan, to go and people the country of Anahuac. This American tradition reminds us of the Menou of the Hindoos, the Noah of the Hebrews, and the dispersion of the Couschites of Singar [the Cushites of Shinar]. Comparing this tradition either with those of the Hebrews and Indians preserved in Genesis and the two sacred Pouranas, or with the fable of Xelhua the Cholulain, and other facts cited in the course of this work, it is impossible to avoid being struck with the analogy, which exists between the old memorials of the people of Asia, and those of the New Continent.

TRADITION OF THE FLOOD

A great inundation, which began the year *ce calli*, the day 4 water (*nahui atl*), destroyed mankind. This is the last of the great revolutions, which the world has undergone. Men were transformed into fish, except one man and one woman, who saved themselves in the trunk of an *ahahuète*, or cupressus disticha. The drawing represents the goddess of water, called *Matlalcueje*, or *Chalchiukcueje*, and considered as the companion of Tlaloc, descending towards the earth. Coxcox, the Noah of the Mexicans, and his wife *Xochiquetzal*, are seated in a trunk of a tree, | 620 | covered covered with leaves, and floating amidst the waters.

These four ages which are also designated under the name of *suns*, contain together eighteen thousand and twenty-eight years; that is to say, six thousand years more than the four Persian ages described in the ZendAvesta. I no where find how many years had elapsed from the deluge of Coxcox to the sacrifice of Tlalixco, or till the reform of the Azteck calendar; but, however near we may suppose these two periods, we still find that the Mexicans attributed to the world a duration of more than twenty thousand years. This duration certainly forms a contrast with the great period of the Hindoos, which consists of four millions three hundred and twenty thousand years; and still more with the cosmogonical fiction of the Thibetans, according to which mankind already compute eighteen revolutions, each of which has several *padu*, expressed by numbers of sixty-two ciphers. It is nevertheless remarkable that we find an American people, who, according to the same system of the calendar in use among them on the arrival of Cortez, indicate the days and the years in which the world underwent great catastrophes farther back than twenty ages.

The history begins by the Deluge of Coxcox, or the fourth destruction of the world, which, according to the Azteck cosmogony, terminates the fourth of the great cycles, *atonatiuh*, *the age of water*. This cataclysm took place, according to the two received chronological systems, one thousand four hundred and seventeen, or eighteen thousand and twenty-eight years after the beginning of the age of earth, *tlaltonatiuh*. The enormous difference of these numbers ought less to astonish us, when we recollect the hypotheses, which in our days have been advanced by Bailly, Sir William Jones, and Bentley, on the duration of the five Yougas of the Hindoos. Of the different nations that inhabit Mexico, paintings representing the deluge of Coxcox are found among the Aztecks, the Miztecks, the Zapotecks, the Tlascaltecks, and the Mechoacanese. The Noah Xisuthrus, or Menou of these nations, is called Coxcox, TeoCipactli, or Tezpi. He saved himself conjointly with his wife, Xochiquetzal, in a bark, or, according to other traditions, on a raft of *ahuahuete* (cupressus disticha). The painting represents Coxcox in the midst of the water, lying in a bark. The mountain, the summit of which, crowned by a tree, rises above the waters, is the Peak of Colhuacan, the Ararat of the Mexicans. The horn, which is represented on the left, is the phonetic hieroglyphic of Colhuacan. At the foot of the mountain appear the heads of Coxcox and his wife. The latter of these is known by the two tresses in the form of horns, which, as we have often observed, denote the female sex. The men born after the deluge were dumb: a dove, from the top of a tree, distributes among them tougues, represented under the form of small commas. We must not confound this dove with the bird which brings Coxcox tidings, that the waters were dried up. The people of Mechoacan preserved a tradition, according to which Coxcox, whom they called Tezpi, embarked in a spacious *acalli* with his wife, his children, several animals and grain, the preservation of which was of importance to mankind. When the great spirit, Tezcatlipoca, ordered the waters to withdraw, Tezpi sent out from his bark a vulture, the *zopilote* (vultur aura). This bird, which feeds on dead flesh, did not return on account of the great number of carcasses, with which the earth, recently dried up, was strewed. Tezpi sent out other birds, one of which, the humming bird, alone returned, holding in its beak a branch covered with leaves; Tezpi seeing that fresh verdure began to clothe the soil, quitted his bark near the mountain of Colhuacan.

These traditions, we here repeat, remind us of others of high and venerable antiquity. The sight of marine substances, found even on the loftiest summits, might give men, who have had no communication, the idea of great inundations, which for a certain time extinguished organic life on the earth: but ought we not to acknowledge the traces of a common origin, wherever cosmogonical ideas, and the first traditions of nations, offer striking analogies even in the minutest circumstances? does not the humming-bird of Tezpi remind us of Noah's dove, that of Deucalion, and the birds, which, according to Berosus, Xisuthrus sent out from his ark, to see whether the waters had run off, and whether he might erect altars to the protecting divinities of Chaldea. | 621 |

BOCHICA

In the description of the cataract of Tequendama, we have spoken of that marvellous personage, known in the American mythology under the name of Bochica, or Idacanzas, who opened a passage for the waters of the lake of Funzha, assembled the wandering tribes into a social state, introduced the worship of the Sun, and, like the Peruvian Manco-Capac, and the Mexican Quetzalcoatl, became the legislator of the Muyscas. These same traditions relate, that Bochica, son and emblem of the Sun, high priest of Sogamozo, or Iraca, seeing the chiefs of the different Indian tribes disputing for the supreme authority, advised them to choose for zaque, or sovereign, one among them called Huncahua, revered on account of his wisdom and justice. The advice of the high priest was universally adopted; and Huncahua, who reigned two hundred and fifty years, subdued the whole of the country that extends from the savannahs of San Juan de los Llanos to the mountains of Opon. Bochica, devoting himself to a life of severe penance, lived a hundred Muysca cycles, or two thousand years. He disappeared mysteriously at Iraca, to the east of Tunja. This town, which was then the most populous in the country, was founded by Huncahua, the first of the dynasty of the zaques of Cundinamurca; and took the name of Hunca, from its founder, which the Spaniards afterward changed into that of Tunca, or Tunja.

The form of government given by Bochica to the inhabitants of Bogota is very remarkable from its analogy with those of Japan and Thibet. The Incas of Peru united in their person the temporal and spiritual powers. The children of the Sun were both priests and kings. At Cundinamurca, at a period probably anterior to Manco-Capac, Bochica had constituted the four chiefs of tribes, Gameza, Busbanca, Pesca, and Toca, electors; and ordered, that, after his death, these electors, and their descendants, should have the right of choosing the high priest of Iraca. The pontiffs, or lamas, the successors of Bochica, were considered as heirs of his virtue and sanctity; and such as Cholula, in the time of Montezuma, was to the Aztecks, Iraca had been to the Muyscas. The people thronged in crowds to offer presents to the high priest, visiting those places which were consecrated by the miracles of Bochica; and amidst the horrors of the most sanguinary warfare, the pilgrims enjoyed the protection of those princes, through whose territories they passed to visit the sanctuary (*chunsua*), and prostrate themselves at the feet of the lama, who presided there. The temporal chief, called zaque of Tunja, to whom the *zippa*, or princes, of Bogota, paid an annual tribute, and the pontiff of Iraea, were consequently two distinct potentates, as the emperor and dairi are in Japan.

Bochica was not only considered as the founder of a new worship and lawgiver of the Muyscas; as emblem of the sun he regulated the seasons, and to him was attributed the invention of the calendar. He had prescribed also the order of the sacrifices, which were to be celebrated at the end of the smail cycles, on account of the fifth lunar intercalation.

The least division of time among the Muyscas was a period of three days. The week of seven days was unknown in America, as well as in part of eastern Asia. On the first day of this small period a great market was held at Turmequé.

Yellow Fever

We published, last week, some extracts from the personal narrative of Mesns Humboldt and Bompland, respecting the contagion of this disease M. Humboldt proceeds in his pertinent observations on this subject, by citing a series of pathological facts, for the judgment of his readers The well-known and acknowledged accuracy of this traveller, induces us to consider all his facts as authentic.

After mentioning the mystery attending the doctrine of contagion, he enquires how we can explore why during the eighteen years which preceeded the year 1794, there was not a single instance of Yellow-Fever at *Vera-Cruz*, though the concourse of unseasoned Europeans and of Mexicans from the interior was very considerable; though the sailors indulged in the same excesses with which they are still reproached, and though the town was not so clean as it has been since the year 1800?

The following is the series of pathological facts, considered in their greatest simplicity. When a great number of persons, born in a cold climate, arrive at the same time in a port of the torrid zone, not particularly dreaded by navigators, the typhus of America begins to appear. Those persons have not had the typhus during their passage; it manifests itself among them only on the very spot. Is the atmospheric constitution changed? or does a new form of disease display itself among individuals, whose irritability is highly increased?

The typhus soon begins to exert its ravages among the Europeans, born in more southern countries. If it propagate itself by contagion, it seems surprising, that in the towns of the equinoctial continent it does not attach itself to certain streets; and that immediate contact does not augment the danger, any more than seclusion diminishes it. The sick, when removed to the inland country, and especially to cooler and more elevated spots, to Xalappa, for instance, do not communicate the typhus to the inhabitants of those places, either because it is not contagious in it's nature, or that the predisposing causes are not the same as in the regions of the shore. When there is a considerable diminution of the temperature, the epidemy usually ceases, even on the spot where it first appeared. It again begins at the approach of the hot season, and sometimes long before; though during several months there has been no sick person in the harbour, and no ship has entered it.

The typhus of America appears to be confined to the shore, either because those persons who bring the disease disembark there, and goods supposed to be impregnated with deleterious miasmata are there accumulated; or because on the seaside gaseous emanations of a particular nature are formed. The aspect of the places where typhus exerts it's ravages seems often to exclude all idea of a local or endemical origin. It has been seen to prevail in the Canaries, the Bermudas, and among the smaller West India Islands, in dry places formerly distinguished for the great salubrity of their climate.— Examples of the propagation of the yellow fever in the inland parts of the torrid zone appear very doubtful; this malady may have been confounded with remitting bilious fevers. With respect to the temperate zone, in which the contagious character of the typhus of America is more decided, the disease has indubitably spread far from the shore, even into very elevated places, exposed to cool and dry winds, as in Spain, at Medina-Sidonia, at Carlotta, and the city of Murcia. That variety of phenomena, which the same epidemic exhibits, according to the difference of climates, the union of predisposing causes, it's shorter or longer duration, and the degree of it's *exacerbation*, should render us extremely circumspect in tracing the secret causes of the American typhus. An enlightened observer, who, at the time of the violent epidemics in 1802 and 1803, was chief physician to the colony of St. Domingo, and who has studied that disease in the Island of Cuba, the U. States and Spain, Mr. Bailly, thinks like me, 'that the typhus is very often, but not always, contagious.'

Since the yellow fever has made such cruel ravages in La Guayra, the want of cleanliness in that little town has been exaggerated, like that of Vera Cruz, and of the quays or wharfs of Philadelphia. In a place where the soil is extremely dry, destitute of vegetation, and where a few drops of water scarcely fall in seven or eight months, the causes, that produce what are called miasmata, cannot be very frequent. The streets of La Guayra appeared to me in general to be tolerably clean. with the exception of the quarter of the slaughter houses. The sea side has no beach, on which the remains of fuct and of Molluscæ are heaped up; but the neighboring coast, which stretches to the east towards Cape Codera, and consequently to the windward of La Guayra, is extremely unhealthy.

The yellow fever and the black vomit cease periodically at the Havana and Vera Cruz, when the north winds bring the cold air of Canada toward the Gulf of Mexico. But from the extreme equality of temperature, which characterizes the climates of Porto-Cabello, La Guayra, New Barcelona, and Comana, it may be feared, that the typhus will there become permanent, when ever, from a great concourse of strangers, it has acquired a high degree of exacerbation. Happily the mortality has diminished since the treatment of this epidemic has been varied, according to the character it presents in different years; and since the different stages of the disease have been better studied, which are recognized by symptoms of inflammation, and of *ataxy* or debility. It would, I think, be unjust to deny the success, which the new system of medicine has obtained over this terrible scourge; yet the persuasion of this success has not made much progress in the colonies. It is there said pretty generally, 'that the physicians now explain the course of the disease in a more satisfactory manner than they did formerly, but that they do not cure it better: that heretofore the patient was left to die slowly, taking no other remedy than an infusion of tamarinds; and that in our days a more active practice carries him to the grave in a more direct and expeditious manner.'

This opinion is not founded upon an accurate knowledge of what was done formerly in the West India Islands.— The voyage of Father Labat sufficiently demonstrates, that in the beginning of the 18th century the physicians of the West Indies did not suffer the sick to die so tranquilly, as seems to be supposed. They did not then kill by emetics, bark, and opium, employed in two large doses, and unseasonably;—but by frequent bleedings, and the abuse of purgatives. The physicians indeed seemed so well aware of the effects of their treatment that they had the candour 'to present themselves at the bed-side of the sick, accompanied at their first visit by a confessor and a notary.' At present, in neat and well conducted hospitals, they often succeed in reducing the number of deaths to eighteen or fifteen in a hundred, and even a little less. But whenever the sick are crowded together, the mortality increases to one half, or even to three quarters, of which the French army in St. Domingo afforded an example in 1802.

[Letter to the Editor of the *Moniteur*]

The *Moniteur* of the 11th contains a long letter to the Editor from M. De Humboldt, in which he brings a charge of plagiarism against Mr. Arrowsmith, a well-known English Geographer.

After some prefatory observations, M. De H. proceeds as follows:—

"I was engaged during four years in the construction of a map of Mexico, which was published at Paris, in September, 1809, in two large sheets, in my 'Geographical and Physical Atlas of the Kingdom of New Spain.' I had first drawn it up in Mexico, in 1803, and it was engraved at Paris by M. M. Aubert and Barriere: it was entitled 'General Map of the Kingdom of New Spain, drawn up from astronomical observations, and the whole of the materials which existed in Mexico, at the commencement of the year 1804, by Alexander De Humboldt.' The astronomical observations, the geodesic surveys, and the barometrical admeasurements of heights which I made in the course of my travels in the equinoctial regions of the new Continent, from 1799 to 1803, are to be found in the second volume of my "Collection of Astronomical Observations," published conjointly with M. Oltmanns. The numerous and unpublished materials which assisted in the construction of the 20 maps contained in the Mexican atlas, are indicated and discussed in an analysis prefixed to my Political Essay on the kingdom of New Spain.

"All these long and painful labours, however imperfect they appear to me, Mr. Arrowsmith has entirely appropriated to himself; he faithfully copied my general map of Mexico, and published it in London, before the appearance of the English translation of my Political Essay; and he substituted his own name instead of mine, under the title of "New map of Mexico, *compiled from original documents*, by Arrowsmith." The positions of the towns, of the villages and mines, the boundaries of the intendances, the ranges of mountains, the indications of the heights in toises, the notes on the migrations of the Azteques, and on the history of navigation, the small arrows annexed to a number of rivers, every thing, in short, is to be found in the copy of Mr. Arrowsmith. I was obliged to employ several new signs, for instance, two hammers placed crosswise to denote the chief station of a provincial council of the mines; and Mr. Arrowsmith, in adopting my signs has also adopted their explanation; he has copied with out translating, and without changing a single word on his English map he has engraved my explanations in French, always leaving out the sign pointing to the places where I made astronomical observations. If my name is looked for in the copy of the great map it is only to be found in one of the three squares which contain the same number of sketches borrowed from my Mexican Atlas. These small squares present hydrographic plans of the ports of Vera Cruz and Acapulco, and the chart of the valley of Mexico. To the words "Valléy of Mexico," Mr. Arrowsmith has thought fit to annex the following words, "from M. Humboldt's Map." But the only thing which Mr. Arrowsmith chooses to ascribe to me is not my own: it was a plan drawn up by D. Louis Martin, in 1804."

After a variety of other observations, M. de Humbolt adds: "This silence with regard to the sources of their information is become too common with geographers, particularly with those who do not accompany their maps with analytical memoirs on the subject of their construction, though a very different example has been given by the most distinguished of that class of literary men, such as d'Anville, Dalrymple, Fleuriu, and Rennell."

M. de H. concludes with expressing his hope, that the reclamations of a traveller will meet with some attention, when he proves that more copies of his labours are disseminated under the name of another person. The letter itself is written with every appearance of candeur, and deserves the attention of those who it concerns.

Extract from a report made to the class of phisico-mathematical sciences, on the 9th of Sept. 1811, by a committee, composed of Messrs. Hallé, Humboldt and Percy, on a memorial of Mr. Legallois, M. D. concerning the principle of action of the heart, and the seat of that principle.

M. Legallois has undertaken to prove, in the memoir which he presented to the class, that the principle of the action of the heart resides in the spinal marrow.

The committee, in the first place, give a succinct account of the divers systems suggested since the discovery of the circulation to explain the action of the heart. The several systems of Descartes, Sylvius, Delebre and Borelli, and that of Willis, as revised and corrected by Boerrhaave, the systems of Stahl, Ens and Stachelin, are noticed by them in their turn. But the one which Haller founded on irritability has particularly attracted their attention. As this last has consigned all the others to oblivion, it has given rise to lively discussions which have been continued to the present day. These disputes proceed from this fact; that Haller's system, as well as those that preceded it, are not sufficiently satisfactory to account for all the phenomena.

In taking a general view (say the committee) of the different systems invented before Haller, we may remark, that the nervous power, in one way or other, has always been considered as one of the essential conditions to produce the motions of the heart, and the seat of this power has constantly and uniformly been placed in the brain.

The cardiac nerves had of course a determinate use in all these systems, and it was easy to conceive how the heart was subjected to the influence of the passions; but they could not explain why the circulation continues in acephalous animals, nor why, in experimenting on animals, the interruption of all communication between the brain and the heart, does not stop the action of the latter organ. Since the days of | 343 | Haller, irritability has been the basis of every system: in considering this property as essential to the muscular fibre, and independent of the nervous power, the circulation in acephalous animals and the different phenomena observed in the experiments above alluded to, present no difficulty; but the use of the nerves of the heart, and the influence of the passions on that organ became inexplicable. The necessity of obviating these difficulties has split the advocates of irritability into two sects. On one side the zealous defenders of pure irritability have called to their assistance the most inconsistent hypotheses, and all their exertions tend only to prove how very difficult it is to defend the cause they have embraced. Their opponents have introduced the nervous power with irritability, and consider the latter as one of the functions of the former; but then they were necessitated to admit, either with regard to the seat, or the mode of existence of this nervous power, certain conditions, which as they themselves confess, are far from being demonstrated, on which they do not agree among themselves, and which, when applied to the action of the heart, do not entirely remove the old difficulties, or else give rise to new ones.

It is easy to perceive whence it is, that so little progress has been made in this long and important controversy. If we examine all that has been said on the subject since the time of Haller, we shall find nearly the same facts, the same experiments, and the same reasonings advanced by either party. The only new experiments are the application of Galvanism to stimulate the cardiac nerves, and even these are new only in appearance, as in the days of Haller electricity was employed for the same purpose and with the same results. It is evident that we have nothing more to hope for the advancement of this interesting part of physiology by following the track already beaten by so many celebrated men, for nearly sixty years. New paths must then be opened, and new methods discovered for the investigation of nature: above all, that degree of precision and strict logical reasoning must be introduced | 344 | into physiological researches, which has been already employed with such success in the other departments of natural philosophy. This is what M. Legallois has endeavoured to effect.

It was not, at first, the object of M. Legallois to search into the causes of the movements of the heart. He was attached to the theory of Haller, until experiments, undertaken with quite a different view, led him to conclude that it was still a mystery to him how the nervous power, if it interfered at all, was interested in aiding the functions of the heart.

That we may give a clear view of his investigations, we will relate on what occasion, and by what chain of facts and reasoning he was induced to engage in this research.

A particular case of midwifery gave him, some years ago, a wish to know how long a full grown fœtus may live without breathing, counting from the moment when, from any cause whatever, it has ceased to have any communication with the mother. This question, so very curious in itself, and so interesting to the practice of midwifery and medicine, had scarcely been glanced at by authors. M. Legallois undertook to solve it by direct experiments. And in order that the solution might extend to as great a number of cases as possible, he placed the fœtus of different animals in such different situations as might resemble the principle accidents that may happen to the human fœtus whilst its communication with the mother is intercepted. Among these accidents there is one which but too often takes place; this is separating the head from the body in the delivery of a *footling* case. The author wished to know what became of the fœtus in this case, whether it died at the moment of decapitation, and of what death it died? He discovered that the trunk remains living, and that, by preventing hemorrhage by passing ligatures around the vessels of the neck, it does not die except in the same space of time, and with the same phenomena, as if, without being separated from the head, respiration had been completely intercepted. And what evidently | 345 | proved to him that the decollated animal is only in a state of asphyxy, is, that he could at will prolong its existence by supplying respiration by means of the inflation of the lungs.

M. Legallois concludes from these facts, that decapitation does no more than arrest the inspiratory motions, and that consequently the principle of all these motions is in the brain, but that the principle of the life of the trunk resides in the trunk itself. Searching afterwards for the immediate seat of each of these principles, he found that the principle of the inspiratory motions resides in that part of the *medulla ablongata*, that gives origin to the nerves of the eighth pain, and that that of the life of the trunk has its source in the *medula spinalis*. It is not by the whole of the spinal marrow that each part of the body is animated, but only by that portion from which it receives its nerves: so that by destroying a portion of the spinal marrow, we deprive of life those parts only of the body that correspond with this portion. Besides, if we intercept the circulation of the blood in a portion of the spinal marrow, life languishes and soon becomes entirely extinct in all the parts that receive their nerves from that portion of marrow. There are then two means of destroying life in this or that part of an animal body; one by destroying the marrow from which the part receives its nerves, the other, by intercepting the circulation of the blood therein.

The result of the above was, that the maintenance of life in any part of the body essentially depended on two conditions; namely, the integrity of the corresponding portion of the spinal marrow, and the circulation of the blood: and consequently, that it was possible to continue life in any part of an animal as long as these two conditions could be made to subsist; that one might, for example, continue life to the anterior parts alone, after having deprived the hinder parts of life, by the destruction of the spinal marrow that corresponds with them, or to the hinder parts, after having, in like manner, killed the anterior parts.

M. Legallois wished to know if his experiments would | 346 | confirm these consequences, and if it were in reality possible to enable this or that portion of an animal to live, independently of the other parts.

He first submitted to experiment a rabbit twenty days old; in this animal he destroyed all the lumbar portion of the spinal marrow. This operation not immediately interfering with the rest of the marrow, and the circulation, according to Haller, not being necessarily affected by it, there was sufficient room to expect, in reasoning from preceding experiments, that the animal would survive a considerable length of time; but in less than two minutes respiration was suspended, and in less than four minutes it no longer gave signs of life. The same experiment frequently repeated was always attended with the same results, without the possibility of preventing them by any means whatsoever, and it was invariably true that a rabbit twenty days old could not survive the destruction of its lumbar marrow. This was the more surprizing as rabbits of that age can continue to live after decapitation, that is, after losing the whole of the brain. This is the fact that M. Legallois could not reconcile with his preceding experiments, and which has led him to discover that the principle of the powers of the heart resides in the spinal marrow.

He first of all assured himself that the destruction of either the dorsal or cervical portion of the marrow was mortal to rabbits of twenty days old, as also of the lumbar portion, and even in less than two minutes. He next discovered that the same experiments repeated on rabbits of different ages did not produce the same effects. In general, the destruction of the lumbar marrow is not suddenly mortal to these animals before the age of ten days; many even of fifteen days survived it. With rabbits above twenty days the effect is the same as with those of twenty days. Very young rabbits may continue to live after the destruction of either the dorsal or the cervical marrow, but a much shorter time, and in fewer cases when the latter is destroyed, than the former. Those that | 347 | have passed the age of fifteen days can survive the destruction of either the one or the other.

In all these partial injuries, even when death speedily ensues, it is never instantaneous, except in such parts as receive their nerves from the portion of marrow destroyed. It does not take place in the rest of the body until after a certain time; but this time is determinate, and no means can prolong it. This time, which is the same for animals of the same age and species, is longer in proportion as the animals are younger. For example when the cervical marrow is destroyed in rabbits, life is extinguished on the instant through all the neck, but it continues in the head, which we may ascertain from the spasmodic motions of the mouth. Life remains also in the posterior parts, from the shoulders downwards, as may be perceived by the sensations and voluntary motions they exhibit. On the day of birth, the motions of the mouth continue about twenty minutes, those of the rest of the body fifteen minutes; at the age of fifteen days, the spasmodic gaspings do not last longer than three minutes, nor are the motions of the body perceptible longer than two minutes and an half. In fine, at the age of thirty days all motions cease within a minute or a minute and an half. After the destruction of the dorsal marrow, it is the breast and not the neck that is struck with death; in the rest of the body, the same phenomena and duration may be observed. If the three portions of the marrow be simultaneously destroyed, the gapings, (the only signs of life that then are manifest,) have, in different ages, the duration above mentioned.

The author, who had so often practised decapitation on rabbits of different ages, had constantly remarked, that the head, when separated from the body, continued to gape, and that during a time proportioned to the age of each animal. Now this space of time was the same as after the destruction of the spinal marrow. But it is evident that after decapitation there is no longer any circulation in the head, and that the spasmodic gapings which take place in this case, do not | 348 | continue longer than life subsists in the brain after the circulation has totally ceased.

This was what first indicated to M. Legallois, that when the partial destruction of the spinal marrow extinguished life in the rest of the body, it was by suddenly arresting the circulation. To make himself certain of this, he cut out, at the base of the large vessels, the heart of rabbits from the age of five or six days to a month; and having attentively remarked the duration of the different signs of life, counting from the moment when the circulation was arrested by the above method, he found that the spaces of time were exactly the same with those he had observed after the destruction of the spinal marrow. He might have considered this correspondence of time as sufficient to decide the question. But he wished to determine to a certainty, whether the circulation is really intercepted at the very moment when the spinal marrow is destroyed. The absence of hemorrhage, and the vacuity of the arteries were the most evident signs he could have of the fact, and he found that in effect, the carotids are empty immediately after this operation, and no blood was yielded on the amputation of the members, even when it was performed close to the body, and before life is extinct in parts the marrow of which has not been destroyed. In a word, all the signs that serve to give a knowledge of the state of the circulation proved to him that whenever the destruction of any portion of the spinal marrow suddenly causes death in the rest of the body, it is by arresting that function. This last effect takes place, not because the action of the heart ceases on a sudden, but because it loses all its force, so as not to be able to impel the blood into the carotids.

The result of all this is, that it is from the spinal marrow that the heart derives the principle of its powers, and from the whole of this marrow, since the destruction of any one of the three portions can, according to the age arrest the circulation, and the destroying of the three portions arrests it in all ages. Another result is, that each portion of this marrow | 349 | has on life two modes of action very distinct from each other. One, by which it essentially constitutes life in all the parts that receive their nerves from it; the other, by which it serves to maintain life in all the parts of the body, by contributing to furnish all the organs that receive twigs from the great sympathetic, and especially the heart, with the principle of life and power, necessary for their functions.

It is plain then, that in order to enable the fore or hinder parts of an animal to live singly, (after having caused death in the other parts of the body by destroying the corresponding marrow,) it is necessary to be able to prevent this destruction from arresting the circulation. To effect this it is sufficient to diminish by ligatures passed round the arteries, the extent of the parts to which the heart has to distribute blood. We have seen, for instance, that the destruction of the lumbar marrrow is immediately mortal for rabbits that have attained or passed the age of twenty days; but they would not die of this operation, if, before we destroy the marrow, we were to commence by tying the abdominal aorta between the celiac and anterior mesenteric arteries.

The application of this principle to other parts of the body leads to a circumstance very singular in appearance; it is, that to be able to support life in rabbits of a certain age, after having destroyed the cervical marrow, we must begin by cutting off the head. They die inevitably, if without beheading them, we destroy this marrow. This fact ceases to surprise, when we remark that by decapitation we remove the entire head from the dominion of the circulation, and that by this means the heart having need of less force to continue its functions, we may weaken it by the destruction of the cervical marrow, without its ceasing to perform them. It is also easy to conceive that every other operation capable of suspending or considerably lessening the circulation in a certain extent of the body of an animal, must produce a similar effect, and afford an equal power of attacking without any | 350 | immediately fatal effects, a portion of the spinal marrow, the destruction of which would have been mortal without that preliminary.

This is what is obtained by the effect itself of destruction of the marrow. This destruction has two effects on the circulation: By the one it weakens the general circulation, by depriving the heart of the contingent powers it received from the marrow destroyed; by the other, without entirely arresting the circulation in the dead parts, it there diminishes it to a very great degree, so as to be to a certain point, equivalent to ligatures on the arteries of the parts.

But this effect does not plainly appear for many minutes after the destruction of the marrow. Thence it happens that the destruction of a first portion of the spinal marrow affords the means of destroying a second, this of a third, and so on. For instance, when, in beheading a rabbit of any age whatever, we are enabled to destroy the cervical marrow, the destruction of this marrow will, in a certain number of minutes, enable us to destroy the quarter of the dorsal marrow, and by continuing to operate thus at intervals on equal lengths of this same marrow, we at last succeed in destroying it entirely without arresting the circulation, which is then kept up by the lumbar marrow only.

We may gather from what we have said, that in rabbits any portion of the spinal marrow furnishes the heart with force sufficient to keep up the circulation in all the parts that correspond with this portion, and that consequently, if we were to cut a rabbit transversely, into several pieces, each piece might be made to live although insolated, if the lungs and heart, necessary to the formation and circulation of the arterial blood, could be included in the piece. But these can only form a part of the thorax; we might therefore very easily support life in the thorax alone, after having separated it from the rest of the body, and prevented hemorrhage by proper ligatures, and this would be the case in rabbits of thirty days and upwards. | 351 |

Such are the principle results of M. Legallois's investigations. These results, which have a mutual dependance on each other, are founded on direct experiments, performed with a degree of accuracy hitherto unknown in physiology. The author, say the committee, has repeated them in our presence; we have spent three sittings, each of several hours, in examining them, and to avoid all precipitation, as well as to give ourselves time to weigh all the facts at leisure, we have suffered the interval of a week to pass between each session.

All the experiments which were repeated before the committee are ranked, in the report, under two heads. One of which contains those that go to prove that the *primum mobile* of all the inspiratory motions resides in that part of the medulla ablongata which gives origin to the nerves of the eighth pair. The other contains those whose object is to shew that the principle of the powers of the heart is in the spinal marrow. As these details, however interesting, are too long to be inserted here, we shall only say that the author has made similar experiments on warm blooded and cold blooded animals, and it is astonishing to observe the uniformity which experiments made on animals so different, offer in the results, even with regard to the principle of the inspiratory motions. It is well known, that it was particularly after experiments made on frogs that the followers of Haller, and among others Fontana had maintained that the motions of the heart were independant of the nervous power. But experiments repeated with the greatest care on the same animals before the committee, demonstrated at once the inacuracy of this opinion, and of the experiments on which it was founded. The gentlemen of the committee conclude the recital of the experiments repeated before them, by declaring that they completely confirmed all the results above mentioned, and by adopting all the conclusions which the author has deduced from them.

They then continue in these words:

That we may here confine ourselves to the principle points we will say what we consider as demonstrated. | 352 |

1st. That the principle of all inspiratory motions has its seat in that part of the *medulla ablongata* which gives origin to the nerves of the eighth pair.

2d. That the principle which animates each part of the body resides in that part of the spinal marrow whence arise the nerves of that part.

3d. That it is, in like manner the spinal marrow which furnishes the principle of life and action; but in this the spinal marrow is throughout concerned, and not a particular portion of it.

4th. That the great sympathetic has its origin in the spinal marrow and that the particular function of this nerve is to bring all the parts to which it is distributed under the immediate influence of all the nervous power.

These results easily solve all the difficulties that have sprung up since the days of Haller, on the causes of the motions of the heart. The principle of these were, to explain for what purpose the heart receives nerves, why it is submitted to the empire of the passions; and why the circulation continues in acephalous animals and such as have been beheaded. No explication hitherto given was sufficient to reconcile all these points. But at present we can easily conceive why the heart receives nerves; and why it is so evidently subject to the empire of the passions, since it is animated by the whole of the spinal marrow. We may also see that when the circulation continues in acephalous animals, &c. it is because the motions of the heart do not depend on the brain, or at most depend on it but secondarily.

We ought to remark that this last point, on which M. Legallois has thrown so much light, presents nothing but confusion and error in the authors of the Hallerian, as well as of the new school. None of them have distinguished the motions of the heart that take place after decapitation, from those that are observed after the excision of this organ, or after the destruction of the spinal marrow.

The committee next proceed to testify that M. Legallois | 353 | is indebted to his own industry and zeal for his success in these researches. They then conclude in these terms:

In a word, it appears to us that we may say of the different authors who have considered the matters that M. Legallois has treated, what M. La Place so justly said on a like occasion. "One may find among them many facts, but these are almost always blended with numberless errors, and the discovery of the truth belongs only to him who separates it from the mass, and establishes it on a solid basis, by calculation or by observations."

The opinion of your committee is, that M. Legallois by the service he has done to physiology, merits all the encouragement the class can afford him; and that his researches are of more importance to this branch of knowledge than any that have been made since the days of Haller.

(Signed,)
DE HUMBOLT,
HALLE,
PERCY.

The class approves of the report, and admits the conclusions drawn in it.

The class moreover decrees that the report shall be printed in the records of the class, and that a committee shall consult with M. Legallois with regard to the expense incurred by him for the experiments already made, and the means of continuing them.

G. CUVIER,

Perpetual Secretary.

Description of the Ascent of the Peak of Teneriffe

From Humboldt's Travels in South America.

Towards three in the morning, by the sombrous light of a few fir torches, we began our expedition for the summit of the Piton. We scaled the volcano on the north-east, where the declivities are extremely steep; and we came, after two hours toil, to a small plain, which, on account of its isolated situation, bears the name of Alta Vista. It is the station also of the Neveros, those natives, whose occupation it is to collect ice and snow, which they sell in the neighbouring towns. Their mules, better practised in climbing mountains than those hired by travellers, reach Alta Vista, and the Neveros are obliged to transport the snow to this place on their backs. Above this point the Malpays begin, a term by which is designated here, as well as in Mexico, Peru, and every other country subject to volcanoes, a ground destitute of vegetable mould, and covered with fragments of lavas.

We observed during the twilight a phenomenon which is not unusual on high mountains, but which the position of the volcano that we were scaling rendered very striking. A layer of white and fleecy clouds concealed from us the sight of the ocean, and the lower region of the island. This layer did not appear above 1600 yards high; the clouds were so uniformly spread, and kept so perfect a level, that they wore the appearance of a vast plain covered with snow. The colossal pyramid of the peak, the volcanic summits of Lanzerota, of Fortaventura, and the isle of Palma, were like rocks amidst this vast sea of vapours, and their black tints were in fine contrast with the whiteness of the clouds.

I was anxious to make an exact observation of the instant of sunrising at an elevation so considerable as that we had reached on the Peak of Teneriffe. No traveller furnished with instruments had as yet taken such an observation. I had a telescope and a chronometer, of which I knew the great exactness. In the part where the sun was to appear the horizon was free from vapours. We perceived the upper limb at 4h. 48′ 55″ apparent time, and, what is very remarkable, the first luminous point of the disc was found immediately in contact with the limit of the horizon, consequently we saw the true horizon, that is to say, a part of the sea farther than 43 leagues. It is proved by calculation, that, under the same parallel in the plain, the rising would have begun at 5h. 1′ 50.4″, or 11′ 51.3″ later than at the height of the Peak. The difference observed was 12′ 55″, which arose no doubt from the uncertainty of the refraction for a zenith distance, of which observations are wanting.

We had yet to scale the steepest part of the mountain, the Piton, which forms the summit. The slope of this small cone, covered with volcanic ashes, and fragments of pumice stone, is so steep, that it would have been almost impossible to reach the top, had we not ascended by an old current of lava, the wrecks of which have resisted the ravages of time. These wrecks form a wall of scorious rocks, which stretches itself into the midst of the loose ashes. We ascended the Piton by grasping these half decomposed scoriæ, the sharp edges of which remained often in our hands. | 269 | We employed nearly half an hour to scale a hill, the perpendicular height of which is scarcely 180 yards.

When we gained the summit of the Piton, we were surprised to find scarcely room enough to seat ourselves conveniently. We were stopped by a small circular wall of porphyritid lava, with base of pitchstone, which concealed from us the view of the crater. The west wind blew with such violence that we could scarcely stand. It was eight in the morning, and we were frozen with the cold, though the thermometer kept a little above the freezing point.

The brink of the crater of the Peak bears no resemblance to those of the greater part of the other volcanoes which I have visited: for instance, the craters of Vesuvius, Jorullo, and Pichincha. In these the Piton preserves its conic figure to the very summit: the whole of their declivity is inclined the same number of degrees, and uniformly covered with a layer of pumice stone, very minutely divided; when we reach the top of these volcanoes, nothing obstructs the view of the bottom of the crater. The Peak of Teneriffe, and Cotopaxi, on the contrary, are of very different construction. At their summit a circular wall surrounds the crater; which wall, at a distance, has the appearance of a small cylinder placed on a truncated cone. On Cotopaxi this peculiar construction is visible to the naked eye at more than 2000 toises distance; and no person has ever reached the crater of this volcano. On the Peak of Teneriffe, the wall which surrounds the crater like a parapet, is so high, that it would be impossible to reach the caldera, if on the eastern side there was not a breach, which seems to have been the effect of a flowing of very old lava. We descended through this breach toward the bottom of the funnel, the figure of which is elliptic. Its greater axis has a direction from north-west to south-east, nearly N. 35° W. The greatest breadth of the mouth appeared to us to be 300 feet, the smallest 200 feet.

We descended to the bottom of the crater on a train of broken lava, from the eastern breach of the inclosure. The heat was perceptible only in a few crevices, which gave vent to aqueous vapours with a peculiar buzzing noise. Some of these funnels or crevices are on the outside of the inclosure, on the external brink of the parapet that surrounds the crater. We plunged the thermometer into them, and saw it rise rapidly to 68 and 75 degrees. It no doubt indicated a higher temperature, but we could not observe the instrument till we had drawn it up, lest we should burn our hands. M. Cordier found several crevices, the heat of which was that of boiling water. It might be thought that these vapours, which are emitted in gusts, contain muriatic or sulphurous acid; but, when condensed, they have no particular taste; and experiments, which several naturalists have made with reagents, prove, that the chimneys of the Peak exhale only pure water. This phænomenon, analogous to what I observed in the crater of Jorullo, deserves the more attention, as muriatic acid abounds in the greater part of volcanoes, and as M. Vauquelin has discovered it even in the porphyritic lavas of Sarcouy in Auvergne.

We prolonged in vain our stay on the summit of the Peak, to wait the moment when we might enjoy the view of the whole of the Archipelago of the Fortunate Islands. We discovered Palma, Gomera, and the Great Canary at our feet. The mountains of Lanzerota, free from vapours at sun-rise, were soon enveloped in thick clouds. On a supposition only of an ordinary refraction, the eye takes in, in calm weather, from the summit of the volcano, a surface of the globe of 5700 square | 270 | leagues, leagues, equal to a fourth of the surface of Spain. The question has often been agitated, if it were possible to perceive the coast of Africa from the top of this colossal pyramid; but the nearest parts of this coast are still farther from Teneriffe than 2° 49′, or 56 leagues. The visual ray of the horizon from the peak being 1° 57′, Cape Bojador can be seen only on the supposition of its height being 400 yards above the level of the ocean.

We could not withdraw our eyes, on the summit of the Peak, from beholding the colour of the azure vault of the sky. Its intensity at the zenith appeared to correspond to 41° of the cyanometer. We know by Saussure's experiment, that this intensity increases with the rarity of the air, and that the same instrument indicated at the same period 20° at the priory of Camouni, and 40° at the top of Mont-Blanc. This last mountain is 540 toises higher than the volcano of Teneriffe; and if, notwithstanding this difference, the sky is seen there of a less deep blue, we must attribute this phænomenon to the dryness of the African air, and the proximity of the torrid zone.

We saw on the summit of the Peak no trace of psora, lecidea, or other cryptogamous plants; no insect fluttering in the air. We found, however, a few hymenopteras adhering to masses of sulphur, moistened with sulphurous acid, and lining the mouths of the funnels. These are bees, which appear to have been attracted by the flowers of the spartium nubigenum, and which oblique currents of air had carried up to these high regions, like the butterflies found by M. Ramond at the top of Mont Perdu. The butterflies perished from cold, while the bees on the Peak were scorched, on imprudently approaching the crevices where they came in search of warmth.

Notwithstanding the heat we felt in our feet on the edge of the crater, the cone of ashes remains covered with snow during several months in the winter. It is probable, that under the cap of snow considerable hollows are found, like those we find under the glaciers of Switzerland, the temperature of which is constantly less elevated than that of the soil on which they repose. The cold and violent wind which blew from the time of sun-rise, engaged us to seek shelter at the foot of the Piton. Our hands and faces were frozen, while our boots were burnt by the soil on which we walked. We descended in the space of a few minutes the Sugar Loaf which we had scaled with so much toil; and this rapidity was in part involuntary, for we often rolled down on the ashes. It was with regret that we quitted this solitary place, this domain, where Nature towers in all her majesty; we soothed ourselves with the hope of once again visiting the Canary islands; but this, like many other plans which we then formed, has never been executed.

Not to interrupt the narrative of the excursion to the top of the Peak, I have said nothing of the geological observations I made on the structure of this colossal mountain, and on the nature of the volcanic rocks of which it is composed. Before we quit the Archipelago of the Canaries, I shall delay a moment, and bring into one point of view what relates to the physical picture of these countries.

Is the conical mountain of a volcano entirely formed of a liquified matter, heaped together by successive eruptions; or, does it contain in its centre a nucleus of primitive rocks covered with lavas, which are these same rocks altered by fire? What are the affinities which unite the productions of modern volcanoes with the basaltes, the phonolites, and those porphyries with basis of feldspar, which are without quartz, and which cover | 271 | the Cordilleras of Peru and Mexico, as well as the small groupes of the Monts d'Or, of Cantal, and of Mêzen in France? Has the central nucleus of volcanoes been heated in its primitive position, and raised up in a softened state by the force of the elastic vapours, before these fluids communicated, by means of a crater, with the external air? What is the substance which, for thousands of years, keeps up this combustion, which is sometimes so slow, and at other times so active? Does this unknown cause act at an immense depth; or does this chemical action take place in secondary rocks lying on granite?

The farther we are from finding a solution of these problems in the numerous works hitherto published on Etna and Vesuvius, the greater is the desire of the traveller to see with his own eyes. He hopes to be more fortunate than those who have preceded him; he wishes to form a precise idea of the geological relations the volcano and the neighbouring mountains bear to each other: but how often is he disappointed, when, on the limits of the primitive soil, enormous banks of tufa and puzzolana render every observation on the position and stratification impossible! We reach the inside of the crater with less difficulty than we at first expected, we examine the cone from its summit to its basis; we are struck with the difference in the produce of each eruption, and with the analogy which still exists between the lavas of the same volcano: but, notwithstanding the care with which we interrogate nature, and the number of partial observations which are presented at every step, we return from the summit of a burning volcano less satisfied than when we were preparing to go thither. It is after we have studied them on the spot, that the volcanic phænomena appear still more isolated, more variable, more obscure, than we figure them when consulting the narratives of travellers.

These reflections occurred to me on returning from the summit of the Peak of Teneriffe, the first unextinct volcano I had yet visited. They returned anew, whenever in South America, or in Mexico, I had occasion to examine volcanic mountains. If we reflect on the little progress, which the labours of mineralogists, and the discoveries in chemistry, have made toward the knowledge of the physical geology of mountains, we cannot help being affected with a painful sentiment; and this is felt still more strongly by those, who, questioning nature under different climates, are more occupied by the problems they have not been able to solve, than with the small number of results they have obtained.

The Peak of Teyde forms a pyramidical mass, like Etna, Tungurahua, and Popocatepetl. This physiognomic character is very far from being common to all volcanoes. We have seen some in the southern hemisphere, which, instead of having the form of a cone or bell, are lengthened in one direction, having the ridge sometimes smooth, and at others rough with small pointed rocks. This structure is peculiar to Antisana and

Pichincha, two burning mountains of the province of Quito; and the absence of the conic form ought never to be considered as a reason excluding a volcanic origin. It is here sufficient to observe in general, that the summits, which are still subject to eruptions of the greatest violence, and at the nearest periods to each other, are slender peaks of a conic form; that the mountains with lengthened summits, and rugged with small stony masses, are very old volcanoes, and near being extinguished; and that rounded tops, in the form of domes, or bells, indicate those problematic porphyries, which are supposed to have been heated in their primitive place, penetrated by vapours, and forced up in a softened state, without having ever flowed as real lithoidal lavas. To the first of these distinctions belong | 272 | Cotopaxi, the Peak of Teneriffe, and that of Orizava in Mexico. The second is common to Cargueirazo, and Pichincha, in the province of Quito; to the volcano of Puracey, near Popayan; and perhaps also to Hecla, in Iceland. The third and last is found in the majestic figure of Chimborazo, and, if it be permitted to place by the side of this colossus, a hill of Europe in the Great Sarcouy in Auvergne.

Personal Narrative of the Travels to the Equinoctial Regions of the New Continent, During the Years 1799–1804

[The jealous policy of the Spanish government having limited the travellers in its American colonies to privileged natives of Spain, no parts of the world are less known than the vast regions composing those colonies; at the same time that no countries are more interesting to naturalists, geologists, and philosophers. It would be superfluous to describe the superior pretensions of M. de Humboldt to gratify in the most ample manner the strongly-excited curiosity of the public on the various topics which these colonies present, his ardour in the pursuit of science having commanded our repeated eulogies, and excited the admiration of all his contemporaries. The extracts themselves will justify our taste in submitting so many of them to our readers.]

THE AUTHOR'S VIEWS

I had in view a double purpose in the travels, of which I now publish the historical narrative. I wished to make known the countries I had visited; and to collect such facts as are fitted to elucidate a science, of which we have possessed scarcely the outline, and which has been vaguely denominated *natural history of the world, theory of the Earth*, or physical *geography*. The last of these two objects seemed to me the most important. I was passionately devoted to botany, and certain parts of zoology, and I flattered myself that our investigations might add some new species to those which have been already described; but, preferring the connection of facts, which have been long observed, to the knowledge of insulated facts, although they were new, the discovery of an unknown genus seemed to me far less interesting than an observation on the geographical relations of the vegetable world, on the migration of the social plants, and the limit of the height which their different tribes attain on the flanks of the Cordilleras.

The natural sciences are connected by the same ties that link all the phenomena of nature. The classification of the species which we ought to consider as the fundamental part of botany, and the study of which is become more attractive and more easy by the introduction of natural methods, is to the geography of plants, what descriptive mineralogy is to the indication of the rocks which constitute the exterior crust of the globe. To comprehend the laws which are observed in the position of these rocks, and determine the age of their successive formations, and their identity in the most distant regions, the geologist ought to be previously acquainted with the simple fossils, which compose the mass of mountains, and of which the names and character are the object of oryctognostical knowledge. It is the same with that part of the natural history of the globe, that treats of the relations the plants have to each other, with the soil whence they spring, or the air which they inhale and modify. The progress of the geography of plants depends in a great measure on that of descriptive botany; and it would be injurious to the advancement of the sciences to attempt rising to general ideas, in neglecting the knowledge of particular facts.

Journeys by land are attended with considerable difficulty in the carriage of instruments and collections, but these difficulties are compensated by real advantages, which it would be useless to enumerate. It is not by | 588 | sailing sailing along the coast, that we can discover the direction of the chains of mountains, and their geological constitution, the climate of each zone, and its influence on the forms and the habits of organized beings. In proportion to the breadth of the continents, the greater is the display on the surface of the soil, of the richness of the animal and vegetable productions; the more distant the central chain of mountains from the shores of the ocean, the greater variety we find, in the bosom of the earth, of those stony strata, the regular succession of which unfolds to us the history of our planet. In the same manner, as every being considered apart is impressed with a particular type, we find the same impression in the arrangement of brute matter organized in rocks, in the distribution and mutual relations of plants and animals. The great problem of the physical description of the globe, is the determination of the form of these types, the laws of their relations with each other, and the eternal ties which link the phenomena of life, and those of inanimate nature.

HIS COLLECTIONS

The maritime war, during our abode in America, having rendered the communications with Europe very uncertain, we found ourselves compelled, in order to diminish the chance of losses, to form three different collections. Of these, the first was embarked for Spain and France, the second for the United States and England, and the third, which was the most considerable, remained almost constantly under our eyes. Towards the close of our expedition, this last collection formed forty-two boxes, containing an herbal of six thousand equinoctial plants, seeds, shells, insects, and, what had hitherto never been brought to Europe, geological specimens from the Chimborazo, New Grenada, and the banks of the river of the Amazons.

After the journey to the Orinoco we left a part of these objects at the island of Cuba, in order to take them on our return from Peru to Mexico. The rest followed us during the space of five years, on the chain of the Andes, across New Spain, from the shores of the Pacific Ocean to the West Indian seas. The conveyance of these objects, and the minute care they required, occasioned us such embarrassments as would scarcely be conceived, by those even who have traversed the most uncultivated parts of Europe. Our progress was often retarded by the threefold necessity of dragging after us, during expeditions of five or six months, twelve, fifteen, and sometimes more than twenty loaded mules, exchanging these animals every eight or ten days, and superintending the Indians who were employed in leading so numerous a caravan. Often, in order to add to our collections of new mineral substances, we found ourselves obliged to throw away others, which we had collected a considerable time before. These sacrifices were not less painful than the losses which we accidentally made. Sad experience taught us, but too late, that, from the sultry humidity of the climate, and the frequent falls of the beasts of burden, we could preserve neither the skins of animals too hastily prepared, nor the fishes and reptiles placed in phials filled with alcohol. I have thought proper to enter into these details, which, although little interesting in themselves, prove that we had no means of bringing back, in their natural state, many objects of zoology and comparative anatomy, of which we have published descriptions and drawings. Notwithstanding some obstacles and the expense occasioned by the carriage of these articles, I had reason to applaud the resolution I had taken before my departure, of sending to Europe the duplicates only of the productions we had collected. I cannot too often repeat, that, when the seas are infested with privateers, a traveller can be sure only of the objects in his own possession. A very small number of the duplicates, which we shipped for the ancient continent during our abode in America, were saved; the greater part fell into the hands of persons unknown to science. When a ship is condemned in a foreign part, boxes containing only dried plants or stones, far from being sent to the scientific men to whom they are addressed, remain consigned to oblivion. Some of our geological collections taken in the Southern Ocean had, however, a happier fate. We were indebted for their preservation to the generous activity of Sir Joseph Banks, President of the Royal Society of London, who, amidst the political agitations of Europe, has unceasingly laboured to strengthen the ties by which are united the scientific of all nations. | 589 |

CURRENTS OF THE ATLANTIC

The 9th of June, latitude 39° 50′, and longitude 16° 10′ west of the meridian of the observatory of Paris, we began to feel the effects of the great current, which from the Azores directs itself towards the straits of Gibraltar, and the Canary Islands. Comparing the place of our ship deduced from Berthoud's time-keeper with the pilot's reckoning, I was able to discover the smallest variations in the direction and velocity of the currents. From 37° to 30° of latitude, the vessel was sometimes carried, in twenty-four hours, from eighteen to twenty-six miles to the east. The direction of the current was at first E. by S. but nearer the straits it became due east. Capt. Mackintosh, and one of the most distinguished navigators of our time, Sir Erasmus Gower, have noticed the modifications of this movement of the waters at different seasons of the year. Several pilots who frequent the Canary Islands have found themselves on the coasts of Lancerotte, when they expected to make good their landing on the Isle of Teneriffe. M. de Bougainville, in his passage from Cape Finisterre to the Canary Islands, found himself in sight of the Isle of Ferro, 4° more to the east than his reckoning indicated.

The current which is felt between the Azores, the southern coasts of Portugal, and the Canary Islands, is commonly attributed to that tendency towards the east, which the straits of Gibraltar impress on the waters of the Atlantic Ocean. M. de Fleurieu, in notes added to the voyage of Capt. Marchant, observes even that the Mediterranean losing, by evaporation, more water than the rivers can supply, causes a movement in the neighbouring ocean, and that the influence of the Straits is felt at the distance of six hundred leagues. Without derogating from the sentiments of esteem which I owe to this celebrated navigator, from whose works I have derived much instruction, I may be permitted to consider this important object in a far more general point of view.

When we cast our eyes over the Atlantic, or that deep valley which divides the western coasts of Europe and Africa from the eastern coasts of the new continent, we distinguish a contrary direction in the motion of the waters. *Between the tropics, especially from the coasts of Senegal to the Caribbean sea, the general current, that which was earliest known to mariners, flows constantly from east to west.* This is called the equinoctial current. Its mean rapidity, corresponding to different latitudes, *is nearly the same in the Atlantic and in the Southern Ocean,* and may be estimated at nine or ten miles in twenty four hours, consequently from 0.59 to 0.65 of a foot every second! In those latitudes the waters run towards the west, with a velocity equal to a fourth of the rapidity of the greater part of the large rivers of Europe. *The movement of the ocean, in a direction contrary to that of the rotation of the globe, is probably connected with* this last phenomenon, only as far as the rotation changes the polar winds, which, in the low regions of the atmosphere, bring back the cold air of the high latitudes towards the equator, into trade winds. To this general impulsion, which these trade winds give the surface of the seas, we must attribute the equinoctial current, the force and rapidity of which are not sensibly modified by the local variations of the atmosphere.

THE GULF STREAM

The equinoctial current drives the waters of the Atlantic towards the coasts inhabited by the Mosquito Indians, and towards those of Honduras. The New Continent, stretching from south to north, forms an obstruction to this current. The waters are carried at first to the north-west, and, passing into the gulf of Mexico through the strait which is formed by False Cape and Cape St. Antonio, follow the bendings of the Mexican coast, from Vera Cruz to the mouth of the Rio del Norte, and thence to the mouths of the Mississippi, and the shoals to the west of the southern extremity of Florida. Having made this vast circuit to the west, the north, the east, and the south, the current takes a new direction towards the north, and throws itself with impetuosity into the Gulf of Florida. I there observed, in the month of May, 1804, in the 26th and 27th degrees of latitude, a celerity of eighty miles in twenty-four hours, or five feet every second, though at this period the north wind blew with great violence. At the end of the Gulf of Florida, in the parallel of Cape Cannaveral, the Gulf-stream, or current of Florida, runs to the north-east. Its rapidity resembles that of a torrent, and is sometimes five miles an hour. The pilot may judge, with | 590 | some some certainty, of the error of his reckoning, and of the proximity of his approach toward New York, Philadelphia, or Charles-town, when he reaches the edge of the stream; for the elevated temperature of their waters, their strong saltness, indigo-blue colour, and the shoals of sea weed which cover the surface, as well as the heat of the surrounding atmosphere, sensible even in winter, all indicate the Gulf-stream. Its rapidity diminishes towards the north, at the same time that its breadth increases, and the waters cool. Between Cayo Biscaino and the bank of Bahama, the breadth is only 15 leagues, whilst in the latitude of 28 degrees and a half it is 17, and in the parallel of Charles-town, opposite Cape Henlopen, from 40 to 50 leagues. The rapidity of the current is from three to five miles an hour, where the stream is the narrowest, and is only one mile as it advances towards the north. The waters of the Mexican Gulf, forcibly drawn to the north-east, preserve their warm temperature to such a point, that at 40 and 41 degrees of latitude I found them at 22.5° (18° R.), when, out of the current, the heat of the ocean at its surface was scarcely 17.5° (14° R.). In the parallel of New York and Oporto, the temperature of the Gulf-stream is consequently equal to that of the seas of the tropics in the 18th degree of latitude; as, for instance, in the parallel of Porto Rico, and the islands of Cape Verd.

To the east of the port of Boston, and on the meridian of Halifax, under 41° 25′ of latitude, and 67° of longitude, the current is near 80 leagues broad. From this point it turns suddenly to the east, so that its western edge, as it bends, becomes the western limit of the running waters, skirting the extremity of the great bank of Newfoundland, which Mr. Volney ingeniously calls the bar of the mouth of this enormous sea river. The cold waters of this bank, which according to my experiments are at the temperature of 8.7° or 10° (7° or 8° R.) present a striking contrast with the waters of the torrid zone, driven to the north by the Gulf-stream, the temperature of which is from 21° to 22.5° (7° to 8° R.).

From the bank of Newfoundland, or from the 52d degree of longitude to the Azores, the Gulf-stream *continues its course towards the east, and the east-south-east.* The waters still preserve a part of the impulsion they have received near a thousand leagues distance, in the straits of Florida, between the Isle of Cuba, and the shoals of Tortoise Island. This distance is double the length of the course of the river of the Amazons, from Jaen, or the Straits of Manseriche, to Grand-Para. On the meridian of the isles of Corvo and Flores, the most western of the group of the Azores, the breadth of the current is 160 leagues. When vessels, on their return from South America to Europe, endeavour to make these two islands to rectify their longitudes, they constantly perceive the motion of the waters to the south-east. At the 33d degree of latitude the equinoctial current of the tropics is in the near vicinity of the Gulf-stream. In this part of the ocean we may in a single day pass from waters that flow towards the west, into those which run to the south-east or east-south-east.

From the Azores the current of Florida turns towards the Straits of Gibraltar, the Isle of Madeira, and the group of the Canary Islands. The opening of the pillars of Hercules has no doubt accelerated the motion of the waters towards the east.

We may in this point of view assert, that the strait, by which the Mediterranean communicates with the Atlantic, produces its effects at a great distance; but it is probable also that without the assistance of this strait the vessels which sail to Teneriff would be driven to the south-east by a cause, which we must seek on the coasts of the New World. Every motion is the cause of another motion in the vast basin of the seas, as well as in the aerial ocean. Pursuing the currents to their most distant sources, and reflecting on their variable celerity, sometimes decreasing, as between the Gulf of Florida and the bank of Newfoundland; at other times augmenting, as in the neighbourhood of the Straits of Gibraltar, and near the Canary Islands, we cannot doubt but the same cause which drives the waters to make the circuitous sweep of the Gulf of Mexico, agitates them also near the Isle of Madeira.

VELOCITY OF THE CURRENTS

Between the parallels of 11 and 43 degrees, the waters of the Atlantic are drawn by the currents into a continual whirlpool. Supposing that a particle of water returns to the same place from which it departed, we can esti- | 591 | mate, mate, from our present knowledge of the swiftness of the currents, that its circuit of 3800 leagues is not terminated in less than two years and ten months. A boat, which may be supposed to receive no impulsion from the winds, would require thirteen months from the Canary islands to reach the coasts of Carracas, ten months to make the tour of the gulf of Mexico and reach Tortoise Shoals opposite the port of the Havannah, while forty or fifty days might be sufficient to carry it from the straits of Florida to the bank of Newfoundland. It would be difficult to fix the rapidity of the retrograde current from this bank to the coasts of Africa; estimating the mean velocity of the waters at seven or eight miles in twenty-four hours, we find ten or eleven months for this last distance. Such are the effects of this slow but regular motion, which agitates the waters of the ocean. Those of the river of the Amazons take nearly forty-five days to flow from Tomependa to Grand Para.

RECORDED PHENOMENA

In 1770 a small vessel laden with corn, and bound from the Island of Lancerotte to Santa Cruz, in Teneriff, was driven to sea, while none of the crew were on board. The motion of the waters from east to west carried it to America, where it went on shore at La Guayra, near Carracas.

Whilst the art of navigation was yet in its infancy, the gulf-stream furnished the genius of Christopher Columbus with certain indications of the existence of western regions. Two corpses, the features of which indicated a race of unknown men, were thrown on the coasts of the Azores, towards the end of the 15th century. Nearly at the same period, the brother-in-law of Columbus, Peter Correa, governor of Porto Santo, found on the strand of this island pieces of bamboo of an extraordinary size, brought thither by the western currents. These corpses and the bamboos attracted the attention of the Genoese navigator, who conjectured that both came from a continent situated towards the west; we know at present that in the torrid zone the trade-winds and the current of the tropics are in opposition to every motion of the waves in the direction of the earth's rotation. The productions of the new world cannot reach the old but by the very high latitudes, and in following the direction of the current of Florida. The fruit of several trees of the Antilles are often thrown on the coasts of the Isle of Ferro and Gomera. Before the discovery of America the Canarians considered these fruits as coming from the enchanted isle of St. Borondon, which, according to the reveries of the pilots, and certain legends, was placed towards the west in an unknown part of the ocean, buried, as was supposed, in eternal fogs.

The arm of the Gulf-stream, which in the 45th and 50th degrees of latitude, runs from the south-west to the north-east towards the coasts of Europe, deposits every year on the western coasts of Ireland and Norway the fruit of trees, which belong to the torrid zone of America. On the shores of the Hebrides we collect seeds of mimosa scandens, of dolichos urens, of guilandina bonduc, and several other plants of Jamaica, the isle of Cuba, and of the neighbouring continent. The current carries thither also barrels of French wine, well preserved, the remains of the cargoes of vessels wrecked in the West Indian seas. To these examples of the distant migration of the vegetable world, others no less striking may be added. The wreck of an English vessel, the Tilbury, burnt near Jamaica, was found on the coasts of Scotland. On these same coasts various kinds of tortoises are sometimes found, that inhabit the waters of the Antilles. When the western winds are of long duration, a current is formed in the high latitudes, which runs directly towards the east-south-east, from the coasts of Greenland and Labrador, as far as the north of Scotland. Wallace relates, that twice, in 1682 and 1684, American savages of the race of the Esquimaux, driven out to sea in their leathern canoes, during a storm, and left to the guidance of the currents, reached the Orcades. This last example is so much the more worthy of attention, as it proves at the same time how, at a period when the art of navigation was yet in its infancy, the motion of the waters of the ocean would contribute to disseminate the different races of men over the face of the globe.

THE CAUSES OF CURRENTS

In reflecting on the causes of the currents, we find that they are much more numerous than is generally believed; for the waters of the sea may
| 592 | be put in motion by an external impulse, by a difference in heat and saltness, by the periodical melting of the polar ice, or by the inequality of the evaporation, which takes place in different latitudes. Sometimes several of these causes concur to the same effect, and sometimes they produce effects that are contrary. Winds that are light, but which, like the trade winds, are continually acting on the whole of a zone, cause a real movement of transition, which we do not observe in the heaviest tempests, because these last are circumscribed within a small space. When, in a great mass of water, the particles placed at the surface acquire a different specific gravity, a superficial current is formed, which takes its direction towards the point where the water is coldest, or that which is most saturated with muriat of soda, sulphat of lime, and with muriate or sulphate of magnesia. In the seas of the tropics we find that at great depths the thermometer marks 7 or 8 centesimal degrees. Such is the result of the numerous experiments of Commodore Ellis and of M. Peron. The temperature of the air in those latitudes being never below 19 or 20 degrees, it is not at the surface that the waters can have acquired a degree of cold so near the point of congelation, and of the maximum of the density of water. The existence of this cold strata in the low latitudes is an evident proof of the existence of an inferior current, which runs from the poles towards the equator; it also proves that the saline substances, which alter the specific gravity of the water, are distributed in the ocean, so as not to annihilate the effect produced by the differences of temperature.

In fact, if the mean saltness of the sea was 0.005 greater under the equator than in the temperate zone, as several naturalists pretend, a current at the bottom, from the equator towards the pole, would be the result: for 0.005 produce a difference of density of 0.0017, while, according to the tables of Hallstrom, a refrigeration of 16 centesimal degrees, between the 20th and 4th of temperature, causes only a change of 0.00015 in the specific gravity. After attentive examination of the results of the experiments of Black, reduced by Mr. Kirwan to the temperature of 16°, I find on the average the density of the water of the sea,

from 0° to 14° latitude	at 1.0272
from 15° to 25°	1.0282
from 30° to 44°	0.0278
from 54° to 60°	1.0271

The proportion of salt corresponding to these four zones are, according to Bishop Watson, 0.0374, 0.0394, 0.0386, and 0.0372. Those numbers sufficiently prove that the experiments hitherto published do not in any way justify the renewed opinion, that the sea is saltier under the equator than under the 30th and 44th degrees of latitude. It is not therefore a greater quantity of saline substance held in solution, which opposes itself to this inferior current, by which the equinoctial ocean receives particles of water, which during the winter of the temperate zones have sunk towards the bottom of the sea, from the 30th to the 44th degree of southern and northern latitude. Baume has analysed the seawater collected by M. Pagès in different latitudes, and found in this water 0.005 less salt at 1° 16′ of latitude than between the 25th and 40th degrees.

FALLING METEORS

We were never wearied of admiring the beauty of the nights; nothing can be compared to the transparency and serenity of an African sky. We were struck with the innumerable quantity of falling stars, which appeared at every instant. The farther progress we made towards the south the more frequent was this phenomenon, especially near the Canaries. I have observed during my excursions that these igneous meteors are in general more common and luminous in some regions of the globe than in others; *I have never beheld them so multiplied as in the vicinity of the volcanoes of the province of Quito, and in the part of the Pacific Ocean which washes the volcanic coasts of Guatimala.* The influence which place, climate, and seasons, appear to have on the falling stars, distinguishes this class of meteors from those which give birth to stones that fall from the sky (aerolites), and which probably exist beyond the boundaries of our atmosphere. According to the corresponding observations of Messrs. Benzenberg and Brandes, *many of the falling stars seen in Europe were only sixty thousand yards high.* One was even measured, which did not exceed twenty-eight thousand yards, or fifteen miles. | 593 | These measures, which can give no result but by approximation, deserve well to be repeated. In warm climates, especially under the tropics, the falling stars leave a tail behind them, which remains luminous 12 or 16 seconds; at other times they seem to burst into sparks, and they are generally lower than those in the north of Europe. We perceive them only in a serene and azure sky; they have perhaps never been seen below a cloud. Falling stars often follow the same direction for several hours, which direction is then that of the wind. In the bay of Naples, M. Gay-Lussac and myself observed luminous phenomena, very nearly analogous to those which fixed my attention during a long abode at Mexico and Quito. These meteors are perhaps modified by the nature of the soil and the air, like certain effects of the looming, and of the terrestrial refraction peculiar to the coasts of Calabria and Sicily.

FLIGHT OF A SWALLOW

When we were forty leagues east of the island of Madeira, a common swallow came and perched on the topsail-yard. It was so fatigued that it suffered itself to be easily taken. What could engage a bird, in that season, and in calm weather, to fly so far? In the expedition of d'Entrecasteaux, a common swallow was seen at 60 leagues distance from Cape Blanc; but this was towards the end of October, and M. Labillardiere thought it had newly arrived from Europe. We crossed these latitudes in June, at a period when the seas had not for a long time been agitated by tempests. I dwell on this last circumstance, because small birds, and even butterflies, are sometimes forced out to sea by the impetuosity of the winds, as we observed in the southern ocean, when we were on the western coasts of Mexico.

APPARENT ATTRACTION

As we were prevented by the fall of the wind, and by the currents, from repassing the channel of Alegranza, we resolved on tacking during the night between the Isle Clara and the West Rock. This resolution had nearly proved fatal. A calm is very dangerous near this last rock, towards which the current drives with considerable force. We began to feel the effects of this current at midnight. The proximity of the stony masses, which rise perpendicularly above the water, deprived us of the little wind which blew: the sloop no longer obeyed the helm, and we dreaded striking every instant. It is difficult to conceive how a mass of basalt, insulated in the vast expanse of the ocean, can cause so considerable a motion in the waters. These phenomena, well worthy the attention of naturalists, are nevertheless well known to Mariners; they are extremely to be dreaded in the Pacific Ocean, particularly in the small Archipelago of the islands of Gallipagos. The difference of temperature which exists between the fluid and the mass of rocks cannot explain the direction which these currents take; and how can we admit that the water is engulfed at the base of these rocks, which often are not of volcanic origin; and that this continual engulphing determines the particles of water to fill up the vacuum that takes place?

TOP OF TENERIFFE

Towards three in the morning, by the sombrous light of a few fir torches, we began our expedition for the summit of the Piton. We scaled the volcano on the north-east, where the declivities are extremely steep; and we came, after two hours toil, to a small plain, which, on account of its isolated situation, bears the name of Alta Vista. It is the station also of the Neveros, those natives, whose occupation it is to collect ice and snow, which they sell in the neighbouring towns. Their mules, better practised in climbing mountains than those hired by travellers, reach Alta Vista, and the Neveros are obliged to transport the snow to this place on their backs. Above this point the Malpays begins, a term by which is designated here, as well as in Mexico, Peru, and every other country subject to volcanoes, a ground destitute of vegetable mould, and covered with fragments of lavas.

We observed during the twilight a phenomenon which is not unusual on high mountains, but which the position of the volcano that we were scaling, rendered very striking. A layer of white and fleecy clouds concealed from us the sight of the ocean, and the lower region of the island. This layer did not appear above 1600 yards high; the clouds were so uniformly spread, and kept so perfect a level, that they wore the appearance of a vast plain covered with snow. The colossal pyramid of the peak, the | 594 | volcanic summits of Lanzerota, of Fortaventura, and the isle of Palma, were like rocks amidst this vast sea of vapours, and their black tints were in fine contrast with the whiteness of the clouds.

I was anxious to make an exact observation of the instant of sunrising at an elevation so considerable as that we had reached on the Peak of Teneriffe. No traveller furnished with instruments had as yet taken such an observation. I had a telescope and a chronometer, of which I knew the great exactness. In the part where the sun was to appear the horizon was free from vapours. We perceived the upper limb at 4h 48′ 55″ apparent time, and, what is very remarkable, the first luminous point of the disc was found immediately in contact with the limit of the horizon, consequently we saw the true horizon, that is to say, a part of the sea farther than 43 leagues. It is proved by calculation that, under the same parallel in the plain, the rising would have begun at 5h. 1′ 50.4″ or 11′ 51.3″ later than at the height of the Peak. The difference observed was 12′ 55″, which arose no doubt from the uncertainty of the refraction for a zenith distance, of which observations are wanting.

We had yet to scale the steepest part of the mountain, the Piton, which forms the summit. The slope of this small cone, covered with volcanic ashes, and fragments of pumice stone, is so steep, that it would have been almost impossible to reach the top, had we not ascended by an old current of lava, the wrecks of which have resisted the ravages of time. These wrecks form a wall of scorious rocks, which stretches itself into the midst of the loose ashes. We ascended the Piton by grasping these half decomposed scoriæ, the sharp edges of which remained often in our hands. We employed nearly half an hour to scale a hill, the perpendicular height of which is scarcely 180 yards.

When we gained the summit of the Piton, we were surprised to find scarcely room enough to seat ourselves conveniently. We were stopped by a small circular wall of porphyritic lava, with base of pitchstone, which concealed from us the view of the crater. The west wind blew with such violence that we could scarcely stand. It was eight in the morning, and we were frozen with the cold, though the thermometer kept a little above the freezing point.

The brink of the crater of the Peak bears no resemblance to those of the greater part of the other volcanoes which I have visited: for instance, the craters of Vesuvius, Jorullo, and Pichincha. In these the Piton preserves its conic figure to the very summit: the whole of their declivity is inclined the same number of degrees, and uniformly covered with a layer of pumice stone, very minutely divided; when we reach the top of these volcanoes, nothing obstructs the view of the bottom of the crater. The Peak of Teneriffe, and Cotopaxi, on the contrary, are of very different construction. At their summit a circular wall surrounds the crater; which wall at a distance, has the appearance of a small cylinder placed on a truncated cone. On Cotopaxi this peculiar construction is visible to the naked eye at more than 2000 toises distance; and no person has ever reached the crater of this volcano. On the Peak of Teneriffe, the wall which surrounds the crater like a parapet, is so high, that it would be impossible to reach the Caldera, if on the eastern side there was not a breach, which seems to have been the effect of a flowing of very old lava. We descended through this breach toward the bottom of the funnel, the figure of which is elliptic. Its greater axis has a direction from north-west to south-east, nearly N. 35° W. The greatest breadth of the mouth appeared to us to be 300 feet, the smallest 200 feet.

We descended to the bottom of the crater on a train of broken lava, from the eastern breach of the enclosure. The heat was perceptible only in a few crevices, which gave vent to aqueous vapours with a peculiar buzzing noise. Some of these funnels or crevices are on the outside of the enclosure, on the external brink of the parapet that surrounds the crater. We plunged the thermometer into them, and saw it rise rapidly to 68 and 75 degrees. It no doubt indicated a higher temperature, but we could not observe the instrument till we had drawn it up, lest we should burn our hands. M. Cordier found several crevices, the heat of which was that of boiling water. It might be thought that these vapours which are emitted in gusts, | 595 | contain muriatic or sulphurous acid; but, when condensed, they have no particular taste; and experiments, which several naturalists have made with reagents, prove, that the chimneys of the Peak exhale only pure water. This phænomenon, analogous to what I observed in the crater of Jorullo, deserves the more attention, as muriatic acid abounds in the greater part of volcanoes, and as M. Vauquelin has discovered it even in the porphyritic lavas of Sarcouy in Auvergne.

We prolonged in vain our stay on the summit of the Peak, to wait the moment when we might enjoy the view of the whole of the Archipelago of the Fortunate Islands. We discovered Palma, Gomera, and the Great Canary at our feet. The mountains of Lanzerota, free from vapours at sunrise, were soon enveloped in thick clouds. On a supposition only of an ordinary refraction, the eye takes in, in calm weather, from the summit of the volcano, a surface of the globe of 5700 square leagues, equal to a fourth of the surface of Spain. The question has often been agitated, if it were possible to perceive the coast of Africa from the top of this colossal pyramid; but the nearest parts of this coast are still farther from Teneriffe than 2° 49′, or 56 leagues. The visual ray of the horizon from the peak being 1° 57′, Cape Bojador can be seen only on the supposition of its height being 400 yards above the level of the ocean.

We could not withdraw our eyes, on the summit of the Peak, from beholding the color of the azure vault of the sky. Its intensity at the zenith appeared to correspond to 41° of the cyanometer. We know by Saussure's experiment, that this intensity increases with the rarity of the air, and that the same instrument indicated at the same period 20° at the priory of Camouni, and 40° at the top of Mont-Blanc. This last mountain is 540 toises higher than the volcano of Teneriffe; and if, notwithstanding this difference, the sky is seen there of a less deep blue, we must attribute this phænomenon to the dryness of the African air, and the proximity of the torrid zone.

We saw on the summit of the Peak no trace of psora, lecidea, or other cryptogamous plants; no insect fluttered in the air. We found however a few hymenopteras adhering to masses of sulphur, moistened with sulphurous acid, and lining the mouths of the funnels. These are bees, which appear to have been attracted by the flowers of the spartium nubigenum, and which oblique currents of air had carried up to these high regions, like the butterflies found by M. Ramond, at the top of Mont Perdu. The butterflies perished from cold, while the bees on the Peak were scorched, on imprudently approaching the crevices where they came in search of warmth.

Notwithstanding the heat we felt in our feet on the edge of the crater, the cone of ashes remains covered with snow during several months in the winter. It is probable, that under the cap of snow considerable hollows are found, like those we find under the glaciers of Switzerland, the temperature of which is constantly less elevated than that of the soil on which they repose. The cold and violent wind which blew from the time of sun-rise, engaged us to seek shelter at the foot of the Piton. Our hands and faces were frozen, while our boots were burnt by the soil on which we walked. We descended in the space of a few minutes the Sugar Loaf which we had scaled with so much toil; and this rapidity was in part involuntary, for we often rolled down on the ashes. It was with regret that we quitted this solitary place, this domain where Nature towers in all her majesty; we soothed ourselves with the hope of once again visiting the Canary islands, but this, like many other plans which we then formed, has never been executed.

CANARY BIRDS

As we approached the town of Orotava, we met great flocks of canaries. These birds, well known in Europe, were in general uniformly green; some had a yellow tint on their backs; their note was the same as that of the tame canary. It is nevertheless remarked, that those which have been taken in the isle of the Great Canary, and in the islet of Monte Clara, near Lanzerota, have a stronger, and at the same time the most harmonious song. Under every zone, among birds of the same species, each flock has its peculiar note. The yellow canaries are a variety which has taken birth in Europe; and those we saw in cages at Orotava and Santa Cruz had been bought at Cadiz and in other ports of | 596 | Spain. But of all the birds of the Canary islands, that which has the most heart-soothing song is unknown in Europe; this is the capirote, which no effort has been able to tame, so sacred to his soul is liberty. I have stood in admiration at his soft and melodious warbling, in a garden at Orotava; but I have never seen him sufficiently near, to know to what family he belongs.

VOLCANIC PHILOSOPHY

Not to interrupt the narrative of the excursion to the top of the Peak, I have said nothing of the geological observations I made on the structure of this colossal mountain, and on the nature of the volcanic rocks of which it is composed. Before we quit the Archipelago of the Canaries, I shall delay a moment, and bring into one point of view what relates to the physical picture of these countries.

Is the conical mountain of a volcano entirely formed of a liquified matter, heaped together by successive eruptions; or does it contain in its centre a nucleus of primitive rocks covered with lavas which are these same rocks altered by fire? What are the affinities which unite the productions of modern volcanoes with the basaltes, the phonolites, and those porphyries with basis of feldspar, which are without quartz, and which cover the Cordilleras of Peru and Mexico, as well as the small groups of the Monts d'Or, of Cantal, and of Mézen in France? Has the central nucleus of volcanoes been heated in its primitive position, and raised up in a softened state by the force of the elastic vapours, before these fluids communicated, by means of a crater, with the external air? What is the substance which, for thousands of years, keeps up this combustion, which is sometimes so slow, and at other times so active? Does this unknown cause act at an immense depth; or does this chemical action take place in secondary rocks lying on granite?

The farther we are from finding a solution of these problems in the numerous works hitherto published on Etna and Vesuvius, the greater is the desire of the traveller to see with his own eyes. He hopes to be more fortunate than those who have preceded him; he wishes to form a precise idea of the geological relations the volcano and the neighbouring mountains bear to each other: but, how often is he disappointed, when, on the limits of the primitive soil, enormous banks of tufa and puzzolana render every observation on the position and stratification impossible! We reach the inside of the crater with less difficulty than we at first expected, we examine the cone from its summit to its basis; we are struck with the difference in the produce of each eruption, and with the analogy which still exists between the lavas of the same volcano: but, notwithstanding the care with which we interrogate nature, and the number of partial observations which are presented at every step, we return from the summit of a burning volcano less satisfied, than when we were preparing to go thither. It is after we have studied them on the spot, that the volcanic phænomena appear still more isolated, more variable, more obscure, than we figure them when consulting the narratives of travellers.

These reflections occurred to me on returning from the summit of the Peak of Teneriffe, the first unextinct volcano I had yet visited. They returned anew, whenever in South America, or in Mexico, I had occasion to examine volcanic mountains. If we reflect on the little progress, which the labours of mineralogists, and the discoveries in chemistry, have made toward the knowledge of the physical geology of mountains, we cannot help being affected with a painful sentiment; and this is felt still more strongly by those, who, questioning nature under different climates, are more occupied by the problems they have not been able to solve, than with the small number of results they have obtained.

The Peak of Teyde forms a pyramidal mass like Etna, Tungurahua, and Popocatepetl. This physiognomic character is very far from being common to all volcanoes. We have seen some in the southern hemisphere, which, instead of having the form of a cone or a bell, are lengthened in one direction, having the ridge sometimes smooth, and at others rough with small pointed rocks. This structure is peculiar to Antisana and Pichincha, two burning mountains of the province of Quito; and the absence of the conic form ought never to be considered as a reason excluding a volcanic origin. It is here sufficient to observe in general that the summits, which are still subject to eruptions of the greatest violence, and at the nearest periods to | 597 | each other, are slender peaks of a conic form; that the mountains with lengthened summits, and rugged with small stony masses, are very old volcanoes, and near being extinguished; and that rounded tops in the form of domes, or bells, indicate those problematic porphyries, which are supposed to have been heated in their primitive place, penetrated by vapours, and forced up in a softened state, without having ever flowed as real lithoidal lavas. To the first of these distinctions belong Cotopaxi, the Peak of Teneriffe, and that of

Orizava in Mexico. The second is common to Cargueirazo and Pichincha, in the province of Quito; to the volcano of Puracey, near Popayan; and perhaps also to Hecla, in Iceland. The third and last is found in the majestic figure of Chimborazo, and, if it be permitted to place by the side of this colossus a hill of Europe, in the Great Sarcouy in Auvergne.

SLOPES OF HILLS

If the slope of these three volcanoes were uniform from the summit to its basis, the Peak of Teyde would have an inclination of 12° 29′, Vesuvius 12° 41′, and Etna 10° 13′; a result which must astonish those who do not reflect on what constitutes an average slope. In a very long ascent, slopes of three or four degrees alternate with others which are inclined from twenty-five to thirty degrees; and the latter only strike our imagination, because we think all the slopes of mountains more steep than they really are.

The reader versed in the practice of levelling will not be astonished at the very gentle slope which these profiles seem to indicate. In nature an inclined plane of an angle of 35° appears to be 50°; we scarcely dare go down a hill of 22° slope in a carriage; and the parts of the volcanic cones that are inclined 40° or 42°, are almost inaccessible, though the foot may form steps by plunging it in the ashes.

In places where there were at the same time slopes covered with tufted grass and loose sands, I took the following measures:

> 5°, slope already of a very marked inclination. In France the high roads must not exceed 4° 46′ by law;
> 15°, slope extremely steep, and which we cannot descend in a carriage;
> 37°, slope almost inaccessible on foot, if the bottom be a naked rock, or a turf too thick to form steps. The body falls backwards, when the tibia makes a smaller angle than 53°, with the sole of the foot;
> 42°, the steepest slope that can be climbed on foot in a ground that is sandy, or covered with volcanic ashes.

When the slope is 44° it is almost impossible to scale it, though the ground permits the forming of steps by thrusting in the foot. The cones of volcanoes have a medium slope from 33° to 40°. The steepest parts of these cones, either of Vesuvius, the Peak of Teneriffe, the volcano of

Pichincha, or Jorullo, are from 40° to 42°. A slope of 55° is quite inaccessible. If seen from above it would be estimated at 75°.

GEOLOGICAL PHENOMENA

No enlightened observer has hitherto found at Teneriffe primitive strata, or even those trappean and ambiguous porphyries, which constitute the basis of Etna, and of several volcanoes of the Andes; but we must not conclude from this isolated fact, that the whole of the Archipelago of the Canaries is the production of submarine fires. The island of Gomera contains mountains of granite and mica-slate, and it is undoubtedly in these very ancient rocks, that we must here seek, as well as on all other parts of the globe, the centre of the volcanic action. Hornblende, sometimes pure and forming intermediate strata, at other times mixed with granite, as in the basanites or basalt of the ancients, may by itself furnish all the iron contained in the black and stony lavas. This quantity amounts in the basalt of the modern mineralogists only to 0.20, while in hornblende it exceeds 0.30.

Were these granites and these micaslates of Gomera anciently united to the chain of Atlas, as the primitive mountains of Corsica appear to be the central nucleus of Bochetta and the Apennines? This question can never be solved, till mineralogists shall have visited the islands that surround the Peak, and the mountains of Morocco covered with eternal snows.

Though Teneriffe belongs to a group of islands of considerable extent, the Peak exhibits nevertheless all the characters of a mountain placed on a solitary islet. As at St. Helena, the lead finds no bottom at a little distance from the ports of Santa Cruz, Orotava, | 598 | and and Garachico. The ocean, as well as the continents, has its mountains and its plains; and, if we except the Andes, the volcanic cones are formed every where in the regions of the globe.

As the Peak rises amid a system of basalts and old lava, and as the whole part which is visible above the surface of the waters exhibits burnt substances, it has been supposed, that this immense pyramid, is the effect of a progressive accumulation of lavas; or that it contains in its centre a nucleus of primitive rocks. Both of these suppositions appear to me improbable. I think that there as little existed mountains of granite gneiss, or primitive calcareous stone, where we at present see the tops of the Peak, of Vesuvius, and of Etna, as in the plains where almost in our own time has been formed the volcano of Jorullo, which is more than a third of the height of Vesuvius. On examining the circumstances, which accompanied the formation of the new island in the Archipelago of the Azores; on carefully reading the minute and ingenious narrative, which the Jesuit Bourguignon gave of the slow appearance of the islet of the little Kameni, near Santorino; we find, that these extraordinary eruptions are generally preceded by a swelling of the softened crust of the globe. Rocks appear above the waters before the flames force their way, and lava can issue from the crater: we must distinguish between the nucleus raised up, and the mass of lavas and scoriæ, which successively increase its dimensions.

It is true, in all the revolutions of this kind, which have taken place since the time that their history has been written, the perpendicular height of the stony nucleus appears never to have exceeded one hundred and fifty or two hundred toises; even taking into the account the depth of the sea, the bottom of which had been lifted up: but when we are considering the great effects of nature, and the intensity of its forces, it is not the bulk of the masses, that ought to stop the geologist in his speculations. Every thing indicates, that the physical changes of which tradition has preserved the remembrance, exhibit but a feeble image of those gigantic catastrophes, which have given mountains their present form, changed the positions of the rocky strata, and buried sea shells on the summit of the higher Alps. It was undoubtedly in those remote times, which preceded the existence of the human race, that the raised crust of the globe produced those domes of trappean porphyry, those hills of isolated basalt on vast elevated plains, those solid nuclei which are clothed in the modern lavas of the Peak, of Etna, and of Cotopaxi. The volcanic revolutions have succeeded each other after long intervals, and at very different periods: of this we see the vestiges in the transition mountains, in the secondary strata, and in those of alluvion. Volcanoes of earlier date than the sandstone and calcareous rocks have been for ages extinguished; those which are yet in activity are in general surrounded only with breccias and modern tufas; but nothing hinders us from admitting that the Archipelago of the Canaries may exhibit some real rocks of secondary formation, if we recollect that subterraneous fires have been there rekindled in the midst of a system of basalts and very ancient lavas.

I have observed in another place that the whole of the mountainous part of the kingdom of Quito may be considered as an immense volcano, occupying more than seven hundred square leagues of surface, and throwing out flames by different cones, known under the particular denominations of Cotopaxi, Tungurahua, and Pichincha. In like manner the whole group of the Canary Islands is placed, as it were, on the same submarine volcano. The fire makes its way sometimes by one and sometimes by another of these islands. Teneriffe alone contains in its centre an immense pyramid terminated by a crater, and throwing out, from one century to another, lava by its flanks. In the other islands the different eruptions have taken place in various parts; and we no where find those isolated mountains, to which the volcanic effects are restrained. The basaltic crust, formed by ancient volcanoes, seems every where undermined; and the currents of lava seen at Lanzerota and Palma, remind us by every geological affinity of the eruption, which took place in 1301 at the isle of Ischia, amid the tufas of Epomeo.

THEORY OF VOLCANIC FIRES

I might terminate these geological sketches by discussing the nature of the combustible, which feeds, for so many thousands of years, the fire of | 599 | the Peak of Teneriffe; I might examine whether it be sodium or potassium, the metallic basis of some earth, carburet of hydrogen, or pure sulphur combined with iron, that burns in the volcano; but, wishing to limit myself to what may be the object of direct observation, I will not take upon me to solve a problem, for which we have not yet sufficient data. We are ignorant whether we should conclude from the enormous quantity of sulphur contained in the crater of the Peak, that it is this substance which keeps up the heat of the volcano; or whether the fire, fed by a combustible of an unknown nature, effects merely the sublimation of the sulphur. What we learn from observation is, that in craters which are still burning sulphur is very rare; while all the ancient volcanoes finish by remaining true sulphur pits. We might presume that in the former the sulphur is combined with oxygen, while in the latter it is merely sublimed; for nothing hitherto authorises us to admit, that it is formed in the interior of volcanoes, like ammonia and the neutral salts. When we were yet unacquainted with sulphur, but as disseminated in the muriatiferous gypsum, and in the Alpine limestone, we were almost obliged to suppose that in every part of the globe the volcanic fire acted on rocks of floetz or secondary formation; but recent observations have proved that sulphur exists in great abundance in those primitive rocks, which so many phenomena indicate as the centre of the volcanic action. Near Alausi, on the summit of the Andes of Quito, I found an immense quantity in a bed of quartz, which formed a layer of micaslate; and this fact is so much the more important, as it is in strict conformity with the observation of those fragments of ancient rocks, which are thrown out untouched by the volcanoes.

PROGRESS OF VEGETATION

In the northern part of the temperate zone the cryptogamous plants are the first that cover the stony crust of the globe. The lichens and mosses, that display their foliage beneath the snows, are succeeded by gramina, and other phanerogamous plants. This order of vegetation is different on the borders of the torrid zone, and in the countries between the tropics. We there find, it is true, whatever some travellers may have asserted, not only on the mountains, but also in humid and shady places, almost on a level with the ocean, funaria, dicranum, and bryum; and these genera, among their numerous species, exhibit several which are common to Lapland, the Peak of Teneriffe, and the Blue Mountains of Jamaica. Nevertheless, in general, it is not by mosses and lichens that vegetation in the countries near the tropics begins. In the Canary islands, as well as in the Guinea, and on the rocky coasts of Peru, the first vegetables, that prepare the mould for others, are the succulent plants; the leaves of which, provided with an infinite number of orifices and cutaneous vessels, deprive the ambient air of the water it holds in solution. Fixed in the crevices of volcanic rocks, they form, as it were, that first layer of vegetable earth, with which the currents of lithoid lava are clothed. Wherever these lavas are scorified, and where they have a shining surface, as in the basaltic mounds to the north of Lanzerota, the unfolding of vegetation is extremely slow, and many ages may roll away before shrubs can take root. It is only when lavas are covered with tufa and ashes the volcanic islands lose that appearance of nudity which marks their origin, and deck themselves with a rich and brilliant vegetation.

In its present state the island of Teneriffe, the Chinerfe[1] of the Guanches, exhibits five zones of plants, which we may distinguish by the names of region of vines, region of laurels, region of pines, region of the retama, and region of grasses. These zones are arranged in stages, one above the other, and occupy, on the steep declivity of the Peak, a perpendicular height of 1750 toises; while fifteen degrees farther north, on the Pyrenees, the snows already descend to thirteen or fourteen hundred toises of absolute elevation. If the plants of Teneriffe do not reach the summit of the volcano, it is not because the perpetual snows, and the

[1] Of Chinerfe the Europeans have formed, by corruption, Tchineriffe and Teneriffe.

cold of the surrounding atmosphere, lay down limits which they cannot pass; it is the scorified lava of the Malpays, the powdered and barren pumice stone of the Piton, which impede the migration of the plants toward the brink of the crater.

The first zone, that of the vines, | 600 | extends from the sea-shore to four or six hundred yards of height; it is that which is most inhabited, and the only part carefully cultivated. The region of the vines exhibits, among its vegetable productions, eight kinds of arborescent euphorbia; mesembrianthema, which are multiplied from the Cape of Good Hope to the Peloponnesus; the cacalia kleinia, the dracæna, and other plants, which in their naked and tortuous trunks, in their succulent leaves, and their tint of blueish green, exhibit features distinguishing the vegetation of Africa. It is in this zone that the date tree, the plantain, the sugar cane, the India fig, the arum colocasia, the root of which furnishes the lower class with a nutritive fecula, the olive tree, the fruit trees of Europe, the vine, and corn, are cultivated. The wheat is reaped from the end of March to the beginning of May: and the culture of the bread-fruit tree of Otaheite, that of the cinnamon tree of the Moluccas, the coffee tree of Arabia, and the cocoa tree of America, have been tried with success.

The second zone, that of the laurels, contains the woody part of Teneriffe; this is the region of the springs, that rise up amidst a turf always verdant, and never parched with drought. Lofty forests crown the hills that lead to the volcano, and in them find four species of laurel, an oak nearly resembling the quercus Turneri of the mountains of Thibet, the visnea mocanera, the myrica faya of the Azores, a native olive (olea excelsa), which is the largest tree of this zone, two species of sideroxylon, the leaves of which are extremely beautiful, the arbutus callicarpa, and other evergreen trees of the family of myrtles.

The third zone begins at a mile of absolute height, where the last group of arbutus, of myrica faya, and that beautiful heath known to the natives under the name of texo, appears. This zone, half a mile in breadth, is entirely filled by a vast forest of pines.

The fourth and fifth zones, the regions of the retama and the gramina, occupy heights equal to the most inaccessible summits of the Pyrenees. It is the sterile part of the island, where heaps of pumice stone, obsidian, and broken lava, form impediments to vegetation. We have already spoken of those flowery tufts of alpine broom (spartium nub genum), that form oases amidst a vast sea of ashes. Two herbaceous plants, the scrofularia glabrata, and the viola cheiranthifolia, advance even to the Malpays. Just above a turf scorched by the heat of an African sun, an arid soil is overspread by the cladonia paschalis, to which the herdsmen often set fire, that rolls to considerable distances. Toward the summit of the Peak, the urceolarca, and other plants of the family of the lichens, labour at the decomposition of the schorified matter. By this unceasing action of organic forces the empire of Flora extends itself over islands ravaged by volcanoes.

TRADE WINDS

Our course was such as is taken by all vessels destined for the Antilles since the first voyage of Columbus. The latitude diminished rapidly, almost without gaining in longitude, from the parallel of Madeira to the tropic. When we reach the zone, where the trade winds are constant, we cross the ocean from east to west, on a calm and pacific sea, which Spanish sailors call the Ladies' Gulf, *el Golfo de las Damas*. We found, as all do who frequent those latitudes, that in proportion as we advance towards the west, the trade winds, which were at first east-north-east, fix to the east.

In the passage from Santa Cruz to Cumana, as in that from Acapulco to the Philippine islands, the sailors are scarcely ever under the necessity of touching the sails. We pass those latitudes as if we were descending a river, and we might deem it no hazardous undertaking, if we made the voyage in an open boat. Farther west, on the coast of St. Martha, and in the Gulf of Mexico, the trade wind blows impetuously, and renders the sea very stormy.

The wind fell gradually the farther we removed from the African coasts: it was sometimes smooth water for several hours, and these short calms were regularly interrupted by electrical phænomena. Black thick clouds with strong outlines rose on the east, and it seemed as if a squall would have forced us to hand our topsails; but the breeze freshened anew, there fell a few large drops of rain, and the storm was dispersed without our hearing any thunder. It was curious to observe, during this time, the effect of several black, isolated, and very low clouds, which passed the zenith. We felt the force of the wind aug- | 601 | ment or diminish progressively, according as small bodies of vesicular vapour approached or receded, while the electrometers, furnished with a long metallic rod and lighted match, showed no change of electric tension in the lower strata of the air.

Nothing equals the beauty and mildness of the climate of the equinoctial region on the ocean. While the trade wind blew strongly, the thermometer kept at 23 or 24 degrees in the day, and at 22 or 22·5 degrees during the night. To feel the full charm of these happy climates bordering on the equator, the passage from Acapulco or the coasts of Chili to Europe should be undertaken in a very rough season. What a contrast between the tempestuous seas of the northern latitudes, and the regions where the calm of nature is never disturbed! If the return from Mexico or South America to the coasts of Spain were as expeditious and as agreeable as the passage from the old to the new continent, the number of Europeans settled in the colonies would be much less considerable than it is at present. The sea which surrounds the Azores and the Bermuda islands, and which is traversed in returning to Europe by the high latitudes, is called by the Spaniards by the singular name of *Golfo de las Yeguas* (the Mares' Gulf). Colonists who are not accustomed to the sea, and who have led solitary lives in the forests of Guiana, the Savannahs of the Caraccas, or the Cordilleras of Peru, dread the neighbourhood of the Bermudas more than the inhabitants of Lima fear at present the passage round Cape Horn. They exaggerate the danger of a navigation which is perilous only in the winter. They defer from one year to another the execution of a project which appears hazardous, and death very often surprises them in the midst of the preparations which they make for their return.

SOUTHERN CONSTELLATION

From the time we entered the torrid zone, we were never wearied with admiring, every night, the beauty of the southern sky, which, as we advanced towards the south, opened new constellations to our view. We feel an indescribable sensation, when, on approaching the equator, and particularly on passing from one hemisphere to the other, we see those stars which we have contemplated from our infancy, progressively sink, and finally disappear. Nothing awakens in the traveller a livelier remembrance of the immense distance by which he is separated from his country, than the aspect of an unknown firmament. The grouping of the stars of the first magnitude, some scattered nebulæ, rivalling in splendor the milky way, and tracks of space remarkable for their extreme blackness, give a particular physiognomy to the southern sky. This sight fills with admiration even those, who, uninstructed in the branches of accurate science, feel the same emotion of delight in the contemplation of the heavenly vault, as in the view of a beautiful landscape, or a majestic site. A traveller has no need of being a botanist, to recognize the torrid zone on the mere aspect of its vegetation; and without having acquired any notions of astronomy, without any acquaintance with the celestial charts of

Flamstead and de la Caille, he feels he is not in Europe, when he sees the immense constellation of the Ship, or the phosphorescent clouds of Magellan, arise on the horizon. The heaven, and the earth, every thing in the equinoctial regions, assumes an exotic character.

We saw distinctly for the first time the Cross of the south only in the night of the 4th and 5th of July, in the sixteenth degree of latitude; it was strongly inclined, and appeared from time to time between the clouds, the centre of which, furrowed by uncondensed lightnings, reflected a silver light.

The pleasure we felt on discovering the southern Cross was warmly shared by such of the crew as had lived in the colonies. In the solitude of the seas, we hail a star as a friend, from whom we have long been separated. Among the Portuguese and the Spaniards peculiar motives seem to increase this feeling; a religious sentiment attaches them to a constellation, the form of which recalls the sign of the faith planted by their ancestors in the deserts of the new world.

The two great stars which mark the summit and the foot of the Cross having nearly the same right ascension, it follows hence, that the constellation is almost perpendicular at the moment when it passes the meridian. This circumstance is known to every nation that lives beyond the tropics, or in the southern hemisphere. It has been observed at what hour of the night, in different seasons, the Cross of the south is erect, or inclined. | 602 | It is a time-piece that advances very regularly near four minutes a day, and no other group of stars exhibits to the naked eye, an observation of time so easily made. How often have we heard our guides exclaim in the savannas of Venezuela, or in the desert extending from Lima to Truxillo, "Midnight is past, the Cross begins to bend!" How often those words reminded us of that affecting scene, where Paul and Virginia, seated near the source of the river of Lataniers, conversed together for the last time, and where the old man, at the sight of the southern Cross, warns them that it is time to separate.

TROPICAL HEATS

The extreme slowness with which the temperature increases during the passage from Spain to the New Continent, is highly advantageous to the health of Europeans, who go to settle in the colonies. At Vera Cruz and at Carthagena, the creoles who descend from the high savannas of Bogota, and the central elevated plain of New Spain, are more exposed on the coasts to the attack of the yellow fever, or *vomito*, than the inhabitants of the north, who arrive by sea. In travelling from Perote to Vera Cruz, the Mexicans descend in sixteen hours from the region of pines and oaks, from a mountainous country where the thermometer very often sinks at noon to four or five degrees, to a burning plain covered with cocoa trees, with mimosa cornigera, and other plants that vegetate only under the influence of a strong heat. These mountaineers feel a difference of temperature of eighteen degrees; and this difference produces the most fatal effects on the organs, by exciting their irritability. The European, on the contrary, crosses the Atlantic Ocean in thirty-five or forty days; he prepares himself gradually for the sweltering heat of Vera Cruz, which, without being the direct cause of the yellow fever, do not the less contribute to the rapidity of its progress.

A very sensible decrement of heat is observed on the Globe, whether we go from the equator to the poles, ascend from the surface of the earth into the highest regions of the air, or dive into the depth of the ocean. It is so much the more interesting to compare the rapidity of this threefold decrement, as this phænomenon has a great influence on the climatic distributions of vegetable and animal productions. The mean temperature of the lower strata of the air, which corresponds to the sixty-fifth, forty-eighth, and twentieth, degrees of north latitude, are, according to the most recent observations, 0·5°, 10·7°, and 25°; whence it results, that a centigrade degree corresponds nearly to a change of latitude of 1°, 45′. Now the decrement of caloric is one degree every ninety toises, when we raise ourselves perpendicularly into the atmosphere. It therefore follows, that under the tropics, where the lowering of the temperature is very regular on mountains of considerable height, 1000 yards of vertical elevation correspond to a change of latitude of 9° 45′. This result, conformable enough to those which other naturalists have adopted before me, is very important to the geography of plants; for, though in the northern countries the distribution of vegetables on the mountains and in the plains depends, like the height of the perpetual snows, more on the mean temperature of the months of summer than on that of the whole year, the latter does not less determine in southern countries the limits, which the species have not been able to pass in their distant migrations. The observation made by Tournefort on the summit of Ararat has been repeated by a great number of travellers. When we descend from a high chain of mountains and advance toward the poles, we find at first in plains of little height, and finally in the regions near the coast, the same arborescent plants which in the low latitudes cover only the heights near the perennial snows.

Table of the temperature of the Atlantic Ocean.

Latitude.	Temperature of the ocean.	Period.
0° 58′ S.	27·2°	Nov.
0 57 S.	27·7	April
0 33 S.	27·7	March
0 11 N.	28·0	Febr.
0 13 N.	27·1	May
25° 15′ N.	20·0°	June
25 29 N.	21·6	April
25 49 N.	20·7	March
27 40 N.	21·6	Jan.
28 47 N.	23·5	Octob.
42° 34′ N.	11·1°	Febr.
43 17 N.	15·5	May
43 58 N.	15 9	June
44 58 N.	12 7	Dec.
45 13 N.	15·5	Nov.
48 11 N.	14·3	June

| 603 |

Temperature of the air in both hemispheres.

		Mean temp of the Months.	
Latitude.	Correspondent months.	Southern hemisphere.	Northern hemisphere.
6°	December....	28·0°	....
15	June........	....	28·5
18	October	....	26·5
	April	27·5	....
22	January	....	19·3
26	July	22·5	....
	September ...	....	20·5
	March	20·8	....
34	December....	....	15·4
	June........	13·8	....

(*continued*)

(*continued*)

	February	....	17·0
	August	16·8	....
43	July	....	18·2
	January	15·2	....
48	June........	....	17·7
	December....	7	....
58	July	....	13·5
	January	6 2	....

HEIGHT OF TENERIFFE.

1. Geometrical measurements made on land.	
	Toises.
P. Feuillêe, in 1724 – – –	2213
the same result modified by Bouguer – – – – –	2062
Heberden and Cross, five operations, in 1752 – – – – –	2408
Hernandez, in 1742 – – – –	2658
Borda and Pingrê, in 1771 – –	1742
Borda, in 1776 – – – – –	1905
Made under sail.	
Mannevilette, in 1749 – – –	2000
Borda and Pingré, in 1771 – –	1701
Churruca, in 1788 – – – –	2193
Johnstone, – – – – – – –	2023
2. Barometric measurements calculated after the formula of La Place.	
	toises.
Feuillée and Verguin, in 1724	2025
Borda, in 1776 – – – – –	1976
Lamanon, in 1785 – – – – –	1902
Cordier, in 1803 – – – – –	1920

A toise is about two yards English.

The real height of the Peak of Teneriffe differs little probably from the mean between the three geometrical and barometrical measurements of Borda, Lamanon, and Cordier.

1905	toises
1902	
1920	
1909	

WEST INDIA LANDSCAPE

We anchored opposite the mouth of the river Manzanares on the 16th of July, at break of day: but we could not land till very late in the morning, because we were obliged to wait the visit of the officers of the port. Our eyes were fixed on the groups of cocoa-trees that border the river, and the trunks of which, more than sixty feet high, towered over the landscape. The plain was covered with tufts of cassias, capers, and those arborescent mimosas, which, like the pine of Italy, extend their branches in the form of an umbrella. The pinnated leaves of the palms were conspicuous on the azure of a sky, the clearness of which was unsullied by any trace of vapors. The sun was ascending rapidly toward the zenith. A dazzling light was spread through the air, along the whitish hills strewed with cylindric cactuses, and over a sea ever calm, the shores of which were peopled with alcatras, egrets, and flamingoes. The splendor of the day, the vivid colouring of the vegetable world, the forms of the plants, the varied plumage of the birds, every thing announced the grand aspect of nature in the equinoctial regions.

HORRIBLE RENCOUNTRE

We were conducted by the captain of the Pizarro to the governor of the province, Don Vincente Emparan, to present to him the passports which had been given us by the first secretary of state. He received us with that frankness, and that noble simplicity, which has at all times characterized the Biscayan nation. Before he was named governor of Portobello and Cumana, he had distinguished himself as a captain of a vessel in the royal navy. His name recalls to mind one of the most extraordinary and distressing events recorded in the history of maritime wars. At the time of the last rupture between Spain and England, two brothers of M. d'Emparan fought during a whole night before the port of Cadiz, taking each other's ship for an enemy's. The battle was so terrible that both vessels were sunk nearly at the same time. A very small part of the crew was saved, and the two brothers had the misfortune to recognize each other a little before they expired.

CITY OF CUMANA

The city, placed at the foot of a hill destitute of verdure, is command-
| 604 | ed by a castle. No steeple or dome attracts from afar the eye of the traveller, but only a few trunks of tamarind, cocoa, and date trees, which rise above the houses, the roofs of which are flat. The surrounding plains, especially those on the coasts, wear a melancholy, dusty, and arid appearance, while a fresh and luxuriant vegetation points out from afar

the windings of the river, which separates the city from the suburbs, the population of European and mixed race from the natives with a coppery tint. The hill of fort St. Antonio, solitary, white, and bare, reflects a great mass of light, and of radiant heat; it is composed of breccia, the strata of which contain pelagian petrifactions. In the distance, toward the south, a vast and gloomy curtain of mountains stretches along. These are the high calcareous Alps of New Andalusia, surmounted by sandstone, and other more recent formations. Majestic forests cover this Cordillera of the interior, and are joined by a woody vale to the open clayey lands and salt marshes of the environs of Cumana. A few birds of considerable size contribute to give a particular physiognomy to these countries. On the sea-shore, and in the gulf, we find flocks of fishing herons, and alcatras of a very unwieldy form, which swim, like the swan, raising their wings. Nearer the habitation of men, thousands of galinazo vultures, the true jackals of the winged tribe, are ever busy in uncovering the carcases of animals. A gulf, which contains hot and submarine springs, divides the secondary from the primary and schistose rocks of the peninsula of Araya. Each of these coasts is bathed by a tranquil sea, of an azure tint, and always gently agitated by the same wind. A bright and clear sky, with a few light clouds at sun-set, reposes on the ocean, on the peninsula destitute of trees, and on the plains of Cumana, while we see the storms accumulate and descend in fertile showers among the inland mountains. Thus on these coasts, as well as at the foot of the Andes, the earth and the skies offer the extremes of clear weather and fogs, of drought and torrents of rain, of absolute nudity and never ceasing verdure. In the New Continent, the low regions on the sea-coasts differ as widely from the inland mountainous districts, as the plains of Lower Egypt from the high lands of Abyssinia.

EARTHQUAKES

It is a very generally received opinion on the coasts of Cumana, and in the island of Margaretta, that the gulf of Cariaco owes its existence to a rent of the continent, attended by an irruption of the ocean. The remembrance of this great revolution was preserved among the Indians to the end of the fifteenth century; and it is related that, at the time of the third voyage of Christopher Columbus, the natives mentioned it as a very recent event. In 1530 the inhabitants were alarmed by new shocks on the coasts of Paria and Cumana. The lands were inundated by the sea, and the small fort, built by James Castellon at New Toledo, was entirely destroyed. At the same time an enormous opening was formed in the mountains of Cariaco, on the shores of the gulf that bears this name, when a great body of salt water, mixed with asphaltum, issued from the micaceous schist. Earthquakes were very frequent toward the end of the sixteenth century; and, according to the traditions preserved at Cumana, the sea often inundated the shores, rising from fifteen to twenty fathoms. The inhabitants fled to the Cerro of San Antonio, and to the hill where now stands the small convent of St. Francis. It is even thought that these frequent inundations induced the inhabitants to build that quarter of the town, which is backed by the mountain; and stands on a part of its declivity.

As no record exists at Cumana, and its archives, on account of the continual devastations of the termites, or white ants, contain no document that goes back farther than a hundred and fifty years, we are unacquainted with the precise dates of the ancient earthquakes. We only know that, in times nearer our own, the year 1776 was at the same time the most fatal to the colonists, and the most remarkable for the natural history of the country. A drought, like those which are felt at times in the islands of Cape Verd, had reigned during fifteen months, when, on the 21st of October, 1766, the city of Cumana was entirely destroyed. The remembrance of this day is every year renewed by a religious festival, attended with a solemn procession. The whole of the houses were overturned in the space of a few minutes, and the shocks were hourly repeated during fourteen months. In | 605 | several parts of the province the earth opened, and threw out sulphureous waters. These irruptions were very frequent in a plain extending toward Casanay, two leagues to the east of the town of Cariaco, and known by the name of the hollow ground, *tierra hueca*, because it appears entirely undermined by thermal springs. During the years 1766 and 1767, the inhabitants of Cumana encamped in the streets; and they began to rebuild their houses, when the earthquakes took place only once a month. What was felt at Quito, immediately after the great catastrophe of the 4th of February, 1797, took place on these coasts. While the ground was in a state of continual oscillation, the atmosphere seemed to dissolve itself into water. The rivers were swollen by these sudden torrents of rain, the year was extremely fertile, and the Indians, whose frail huts easily resist the strongest shocks, celebrated, from ideas of an old superstition, with feasting and dances, the destruction of the world, and the approaching epochs of its regeneration.

Tradition states, that in the earthquake of 1766, as well as in another very remarkable one in 1794, the shocks were mere horizontal oscillations;

it was only on the disastrous day of the 14th of December, 1797, that for the first time at Cumana the motion was felt by the raising up of the ground. More than four-fifths of the city were then entirely destroyed; and the shock, attended by a very loud subterraneous noise, resembled, as at Riobamba, the explosion of a mine at a great depth. Happily the most violent shock was preceded by a slight undulating motion, so that the greater part of the inhabitants could escape into the streets, and a small number only perished of those who had assembled in the churches. It is a generally received opinion at Cumana, that the most destructive earthquakes are announced by very feeble oscillations, and by a hollow sound, which does not escape the observation of persons habituated to this kind of phenomenon. In this fatal moment the cries of *misericordia, tembla, tembla*,[2] are every whcre heard; and it is very rarely that a false alarm is given by a native. Those who are most fearful attentively observe the motions of dogs, goats, and swine. The last of these animals, endowed with delicate olfactory nerves, and accustomed to turn up the earth, give warning of approaching danger, by their restlessness and their cries. We shall not decide, whether, placed nearer the surface of the ground, they are the first that hear the subterraneous noise; *or whether their organs receive the impression of some gaseous emanation, which issues from the earth*. We cannot deny the possibility of this latter cause. During my abode at Peru, a fact was observed in the inland country, which has an analogy with this kind of phænomenon, and which is not unfrequent. At the end of violent earthquakes, the herbs that cover the savannahs of Tucuman acquired noxious properties; an epidemic disorder took place among the cattle, and *a great number among them appeared stupified or suffocated by the deleterious vapours exhaled from the ground.*

At Cumana, half an hour before the catastrophe of the 14th of December, 1797, a strong smell of sulphur was perceived near the hill of the convent of St. Francis; and on the same spot a subterraneous noise, which seemed to proceed from the south east to the north-west, was heard the loudest. At the same time flames appeared on the banks of the Manzanares, near the hospice of the Capuchins, and in the gulf of Cariaco, near Mariguitar. This last phænomenon, so extraordinary in a country not volcanic, is pretty frequent in the Alpine calcareous mountains near Cumanacoa, in the valley of Bordones, in the island of Margaretta, and amidst the *Llanos*, or savannahs of New Andalusia. In these savannahs flakes of fire rise to a considerable height; they are seen for hours together

[2] Mercy! the earth trembles.

in the dryest places; and it is asserted, that, on examining the ground which furnishes the inflammable matter, no crevice is to be found. This fire, which resembles the sources of hydrogen, or *Salse*, of Modena, or what is called the Will-o'-the-wisp of our marshes, does not burn the grass; because, no doubt the column of gas, which developes itself, is mixed with azot, and carbonic acid, and does not burn at its basis. The people, although less superstitious here than in Spain, call these reddish flames by the singular name of the *soul of the tyrant Aguirre*; imagining, that the spectre of Lopez Aguirre, harassed by remorse, wan- | 606 | ders over these countries sullied by his crimes.

CAUSES

We can scarcely doubt, that the earth, when opened and agitated by shocks, spreads occasionally gaseous emanations through the atmosphere, in places remote from the mouths of volcanoes not extinct. At Cumana, as we have already observed, flames and vapors mixed with sulphurous acid spring up from the most arid soil. In other parts of the same province, the earth ejects water and petroleum. At Riobamba, a muddy and inflammable mass, which is called *moya*, issues from crevices that close again, and accumulates into elevated hills. At seven leagues from Lisbon, near Colares, during the terrible earthquake of the first of November, 1755, flames and a column of thick smoke were seen to issue from the flanks of the rocks of Alvidras, and, according to some witnesses, from the bosom of the sea. This smoke lasted several days, and it was the more abundant in proportion as the subterraneous noise, which accompanied the shocks, was louder.

I am inclined to think, that in the greater part of earthquakes nothing escapes from the agitated earth; and that, where gaseous emanations and vapours take place, they oftener accompany, or follow, than precede the shocks. This last circumstance explains a fact, which seems indubitable; I mean that mysterious influence, in equinoctial America, of earthquakes on the climate, and on the order of the dry and rainy seasons. If the earth generally acts on the air only at the moment of the shocks, we can conceive why it is so rare, that a sensible meteorological change becomes the presage of these great revolutions of nature.

The hypothesis according to which, in the earthquakes of Cumana, elastic fluids tend to escape from the surface of the soil, seems confirmed by the observation of the dreadful noise, which is heard during the shocks at the borders of the wells in the *plain of Charas*. Water and sand are sometimes thrown out twenty feet high. Similar phænomena have not escaped the observation of the ancients, who inhabited parts of Greece and Asia Minor abounding with caverns, crevices, and subterraneous rivers. Nature, in its uniform progress, every where suggests the same ideas of the causes of earthquakes, and the means by which man, forgetting the measure of his strength, pretends to diminish the effect of the subterraneous explosions. What a great Roman naturalist has said of the utility of wells and caverns, is repeated in the New World by the most ignorant Indians of Quito, when they show travellers the *guaicos*, or crevices of Pichincha.

The subterraneous noise, so frequent during earthquakes, is generally not in the ratio of the strength of the shocks. At Cumana it constantly precedes them, while at Quito, and for a short time past at Caracas, and in the West India Islands, a noise like the discharge of a battery was heard, a long time after the shocks had ceased. A third kind of phænomenon, the most remarkable of the whole, is the rolling of those subterraneous thunders, which last several months, without being accompanied by the least oscillating motion of the ground.

In every country subject to earthquakes, the point where, probably by a particular disposition of the stony strata, the effects are the most sensible is considered as the cause and the focus of the shocks. Thus at Cumana the hill of the castle of St. Antonio, and particularly the eminence on which the convent of St. Francis is placed, are believed to contain an enormous quantity of sulphur, and other inflammable matter. We forget, that the rapidity with which the undulations are propagated to great distances, even across the basin of the ocean, proves, that the centre of action is very remote from the surface of the globe. From this same cause, no doubt, earthquakes are not restrained to certain species of rocks, as some naturalists pretend, but all are fitted to propagate the movement. In order to keep within the limits of my own experience, I shall here cite the granites of Lima and Acapulco; the gneiss of Caracas; the mica-slate of the peninsula of Araya; the primitive thonschiefer of Tepecuacuilco, in Mexico; the secondary lime-stones of the Apennines, Spain, and new Andalusia; and finally the trappean porphyries of the provinces of Quito and Popayan. In these different places the ground is frequently agitated by the most violent shocks; but sometimes, in the same rock, the superior strata form invincible obstacles to the | 607 | propagation of the motion. Thus, in the mines of Saxony, we have seen workmen hasten up affrighted by oscillations, which were not felt at the surface of the ground.

If, in regions the most remote from each other, primitive, secondary, and volcanic rocks, share equally in the convulsive movements of the globe, we cannot but admire also, that, in ground of little extent, certain classes of rocks oppose themselves to the propagation of the shocks. At Cumana, for instance, before the great catastrophe of 1797, the earthquakes were felt only along the southern and calcareous coast of the gulf of Cariaco, as far as the town of this name; while in the peninsula of Araya, and at the village of Maniquarez, the ground did not partake of the same agitation. The inhabitants of this northern coast, which is composed of mica-slate, built their huts on a motionless earth; a gulf three or four thousand toises in breadth separated them from a plain covered with ruins, and overturned by earthquakes. This security, founded on the experience of several ages, has vanished; and, since the 14th of December, 1797, new communications appear to have been opened in the interior of the globe. At present the peninsula of Araya is not merely subject to the agitations of the soil of Cumana, the promontory of mica-slate is become in its turn a particular centre of the movements. The earth is sometimes strongly shaken at the village of Maniquarez, when on the coast of Cumana the inhabitants enjoy the most perfect tranquillity. The gulf of Cariaco nevertheless is only sixty or eighty fathoms deep.

The earthquakes of Cumana are connected with those of the West India islands; and it has even been suspected, that they have some connection with the volcanic phænomena of the Cordilleras of the Andes. On the 4th of November, 1797, the soil of the province of Quito underwent such a destructive commotion, that, notwithstanding the extreme feebleness of the population of that country, near 40,000 natives perished, buried under the ruins of their houses, swallowed up in the crevices, or drowned in lakes that were suddenly formed. At the same period, the inhabitants of the eastern Antilles were alarmed by shocks, which continued during eight months, when the volcano of Guadaloupe threw out pumice stones, ashes, and gusts of sulphureous vapors. This eruption of the 27th of September, during which very long-continued subterraneous noises were heard, was followed on the 14th of December by the great earthquake of Cumana. Another volcano of the West India islands, that of St. Vincents, has lately given a fresh instance of these extraordinary connections. This volcano had not emitted flames since 1718, when they burst forth anew, in 1812. The total ruin of the city of Caracas preceded this explosion thirty-five days, and violent oscillations of the ground were felt, both in the islands, and on the coasts of Terra Firma.

Every thing in earthquakes seems to indicate the action of elastic fluids seeking an outlet to spread themselves in the atmosphere. Often, on the coasts of the South Sea, the action is almost instantaneously communicated from Chili to the gulf of Guayaquil, a distance of six hundred leagues; and, what is very remarkable, the shocks appear to be so much the stronger, as the country is more distant from burning volcanoes. The granitic mountains of Calabria, covered with very recent breccia, the calcareous chain of the Apennines, the country of Pignerol, the coasts of Portugal and Greece, those of Peru and Terra Firma, afford striking proofs of this assertion. The globe, it may be said, is agitated with greater force, in proportion as the surface has a smaller number of funnels communicating with the caverns of the interior. At Naples and at Messina, at the foot of Cotopaxi and of Tunguragua, earthquakes are dreaded only when vapours and flames do not issue from the crater. In the kingdom of Quito, the great catastrophe of Riobamba, which we have before mentioned, has led several well-informed persons to think, that this unfortunate country would be less often desolate, if the subterraneous fire should break the porphyritic dome of Chimborazo; and if this colossal mountain should become a burning volcano. At all times analogous facts have led to the same hypothesis. The Greeks, who, like ourselves, attributed the oscillations of the ground to the tension of elastic fluids, cited in favour of their opinion the total cessation of the shocks at | 608 | the island of Eubœa, by the opening of a crevice in the Lelantine plain.

We have endeavoured to collect at the end of this chapter, the general phænomena of earthquakes under different climates. We have shown, that subterraneous vapours are subjected to laws as uniform as the mixture of gaseous fluids, which constitutes our atmosphere. We have abstained from all discussion of the nature of the chemical agents, which are the causes of the great derangements that the surface of the earth undergoes from time to time. It is sufficient here to observe, that these causes are concealed at immense depths; and that we must seek them in the rocks which we call primitive, perhaps even below the earthy and oxided crust of the globe, in the abysses that contain the metalloidal bases of silex, lime, soda, and potash.

SLAVE MARKET

The slaves exposed to sale were young men from fifteen to twenty years of age. Every morning cocoanut oil was distributed among them, with which they rubbed their bodies, to give their skin a black polish. The persons who came to purchase examined the teeth of these slaves, to judge of their age and health; forcing open their mouths as we do those of horses in a market. This degrading custom dates from Africa, as is proved by the faithful picture, which, in one of his dramatic pieces, Cervantes, released from his long captivity among the Moors, has drawn of the sale of the Christian slaves at Algiers. It is distressing to think, that even at this day there exist European colonists in the West Indies, who mark their slaves with a hot iron, to know them again if they escape. This is the treatment bestowed on those, "who save other men the trouble of sowing, tilling, and reaping, in order to live."

PHOSPHORESCENT INSECTS

The night was delightfully cool; swarms of phosphorescent insects glittered in the air, and over a soil covered with sesuvium, and groves of mimosa, that bordered the river. We know how common the glow-worm is in Italy, and in all the south of Europe; but the picturesque effect it produces cannot be compared to those innumerable, scattered, and moving lights, that embellish the nights of the torrid zone, and seem to repeat on the earth, along the vast extent of the savannahs, the spectacle of the starry vault of the sky.

Population of the Canaries

		Absolute Population.				
Islands.	**Surface in nautical square leagues.**	**1678.**	**1745.**	**1768.**	**1790.**	**Number of inhabitants to each square league in 1790.**
Teneriffe	73	49,112	60,218	66,354	70,000	958
Fortaventura	63	—	7,382	8,863	9,000	142
Grand Canary	60	20,458	33,864	41,082	50,000	838
Palma	27	13,892	17,580	19,195	22,600	837
Lanzerota	26	—	7,210	9,705	10,000	384
Gomera	14	4,373	6,251	6,645	7,400	528
Ferro	7	3,297	3,687	4,022	5,000	714
Total	270		136,192	155,866	174,000	644

Humboldt's Personal Narrative, i. 284.

Temperature of the Atlantic

The following table is given by Humboldt from observations made by him of the temperature of the Atlantic Ocean during his voyage to South America:—

North Latitude.	Longitude.	Temperature of the Surface of the Sea.
39° 10′	18° 38′	59°
34 30	19 15	61·34
32 16	19 24	63·86
30 36	19 14	65·48
29 18	19 00	66·74
26 51	21 33	68·00
20 8	31 11	70·16
17 57	35 34	72·32
14 57	47 00	74·66
13 51	52 3	76·46
10 46	63 14	78·44

Humboldt's Personal narrative, ii. 59.

Account of the Spermaceti Fishery on the Western Shore of America

The western coast of Mexico, especially that part of the great ocean, situated between the gulph of Bayonna, the three Mary Islands, and Cape St. Lucas, abounds in spermaceti whales, or cachalots, of which the fishery is one of the most important objects of mercantile speculation, on account of the extremely high prices given for spermaceti (adipocire) by the English and the inhabitants of the United States. Were it not for the cachalot fishery, and the trade in furs of sea otters, at Noutka, the great ocean would scarcely ever be frequented by the Anglo-Americans and Europeans. Of all the oily liquids which enter into trade, there are few so dear as spermaceti, or the particular substance contained in the enormous caverns of the snout of the cachalot. A single individual of these cetaceous giants yields as much as 125 English barrels (of 32 1/2 gallons each) of spermaceti. A tun, containing eight of these barrels, or 1024 pints of Paris, used to sell in London before the peace of Amiens at 70l. or 80l. and during the war at 95l. and 100l. sterling. 'Till 1788 the whale fishers only frequented the coasts of Chili and Peru. Only twelve or fifteen vessels then doubled Cape Horn annually for the cachalot fishery, while at the period when I was in the South Sea, there were more than sixty under the English flag alone.

The great ocean unites all the circumstances that render the cachalot fishery both easy and lucrative. It is richer in mol- | 402 | luscus fish, porpoises, tortoises, and sea calves of every species, and offers more nourishment to cetaceous animals than the Atlantic Ocean. Hence these last are there in greater numbers, as well fatter as larger. The calm, which prevails during so great a part of the year in the equinoctial region of the south sea, facilitates very much the pursuit of cachalots and whales. It is a general rule that the cachalot avoids shallows, whereas they are sought after by the whale.

The greatest number of English and Anglo-American vessels, which enter the great ocean, have the double object in view of carrying on the cachalot fishery and an illicit commerce with the Spanish colonies. The whale fishers can easily distinguish at a distance the cachalots from the whales, by the manner in which the former spout up the brine through their spiracles. The cachalots can remain longer under water, than the true whale. When they come to the surface, their respiration is more frequently interrupted. The female of the cachalot is four or five times smaller than the male; and its head does not yield more than 25 English barrels of adipocire. A great number of females go generally together, led by two or three males, which are perpetually describing circles round their flock. This fishery, to be profitable, must be conducted on the most economical principles. Vessels from 180 to 300 tons are employed in it, and the crew consists only of 16 or 24 individuals. The expences of equipment of a vessel of 180 tons, lined with copper, and provisioned for a voyage of two years, is estimated in London at 7000l. sterling. The fitter-out gives the sailors only their food, and a very small sum of money under the name of advance. Their pay depends on the produce of the fishery; for as the whole crew contribute to it, every individual has a right to the profit. It is not the want of hands which prevents the inhabitants of Mexico from applying to the cachalot fishery. In the present state of the Spanish colonies, the sloth of the inhabitants is inimical to the execution of similar projects; and it would be impossible to procure sailors willing to embrace so rude a business, and so miserable a life, as that of cachalot fishers. How could they be found in a country, where, according to the ideas of the common people, all that is necessary to happiness is, bananas, salted flesh, a hammock, and a guitar? The hope of gain is too weak a stimulus under a zone where beneficent nature provides man a thousand means of procuring an easy and peaceful existence without quitting his country, and without struggling with the monsters of the ocean.

Eye-Stones

—There are few prejudices of the vulgar more prevalent in the United States, than the belief of the efficacy of small calcareous substances, known by the name of eye-stones, to ex-tract foreign particles from the eyes. The origin of these substances was not unknown to naturalists, and the cause of their motions in vinegar was obvious to every attentive observer. The following extract from the personal narrative of Baron Humboldt will confirm the explanation which scientific men must doubtless have given to the movements and efficacy of these mysterious agents.

"Of all the productions on the coasts of Araya, that which the people consider as the most extraordinary, we may even say the most marvellous, is the stone of the eyes, *piedra de los ojos*. This calcareous substance is the subject of every conversation; according to the natural philosophy of the natives, it is both a stone and an animal. It is found in the sand, where it is motionless; but placed singly on a polished surface, for instance on a pewter or earthen plate, it moves when excited by lemon-juice. Placed in the eye, the pretended animal turns on itself, and expels every other foreign substance, that has been accidentally introduced. At the new salt works, and at the village of Maniquarez, the stones of the eyes were offered us by hundreds, and the natives were earnest to show us the experiment of the lemon-juice. They wished to put sand into our eyes, in order that we might ourselves try the efficacy of the remedy. It was easy to see, that these stones are thin and porous opercula, which have formed part of small univalve shells. Their diameter varies from one to four lines. One of their two surfaces is plain, and the other convex. These calcareous opercula effervesce with lemon-juice, and put themselves in motion in proportion as the carbonic acid is disengaged. By the effect of a similar re-action, loaves placed in an oven move sometimes in a horizontal plane; a phenomenon that has given | 83 | occasion in Europe to the popular prejudice of enchanted ovens. The *piedras de los ojos*, introduced into the eye, act like small pearls, and different round grains employed by the American savages to increase the flowing of tears. These explanations were little to the taste of the inhabitants of Araya. Nature has the appearance of greatness to man in proportion as she is veiled in mystery, and the philosophy of the people rejects every thing that bears a character of simplicity." — *New-Eng. Journ.*

Scenes at Cumana

Humboldt, in the "Personal Narrative of Travels to the equinoctial region of the New Continent," is particular in describing every thing around the city of Cumana, (now in possession of general Bolivar and the Patriots;) the geology of its vicinity; its climate, trees, fortresses, &c. We extract the following description of practices and customs which will be thought remarkable enough.

"The castle of St. Antonio is built at the eastern extremity of the hill, but not on the most elevated point, being commanded on the east by an unfortified summit. The tunal[1] is considered both here and every where in the Spanish colonies as a very important means of military defence, and when earthen works are raised, the engineers are eager to propagate the thorny opuntia, and promote its growth, as *they are careful to keep crocodiles in the ditches of fortified places*. Under a climate where organised nature is so powerful and active, man summons as auxiliaries in his defence the carnivorous reptile and the plant with its armour and formidable thorns."

"The banks of the Manzanares are very pleasant, and shadowed by mimosa, erythrinas, ceibas and other trees of gigantic growth. A river, the temperature of which, in the season of the floods, descends as low as 22 degrees, when the air is at 30 and 33 degrees is an inestimable benefit in a country where the heats are excessive during the whole year and where it is so agreeable to bathe several times in the day. The children pass as it were a part of their lives in the water; the whole of the inhabitants, even the women of the most opulent families, know how to swim; and in a country where man is so near the state of nature, one of the first questions asked at first meeting in the morning is whether the water is cooler than the preceding evening. The mode of bathing is various enough. We every evening visited a very respectable society in the suburb of the Guayquexias, (Indian inhabitants.) In a fine moon-light night, *chairs were placed in the water*; the men and women are lightly clothed, as in some baths of the north of Europe, and the family and strangers, assembled in the river, passed some hours in smoking segars, and in talking, according to the custom of the country, of the extreme dryness of the season, of the abundant rains in the neighboring districts, and

---

[1] A thicket of thorny trees, almost inpenetrable on account of the spikes or prickles of the cactus, &c.

particularly of the luxuries, of which the ladies of Cumana accuse those of the Caraccas and of the Havanna. The company were under no apprehensions from the bavas or small crocodiles, which are now extremely-scarce, and which approach men without attacking them. These animals are three or four feet long. We never meet them in Mazannares, but with a great number of dolphins, which sometimes ascend the river in the night, and frighten the bathers by spouting water."

Port of Cumana.—"The port of Cumana is a road capable of receiving all the navies of Europe. The whole of the Gulph of Carinco, which is thirty five miles long, und sixty eight miles broad, affords excellent anchorage. The great ocean is not more calm and pacific on the coasts of Peru than the seas of the Antilles from Porto Cabello, and especially from Cape Codera to the point of Paria. The hurricanes of the West Indies are never felt in these regions, the vessels of which are without decks."

The city, (N. lat. 10. 10.)—"The city of Cumana, properly speaking, occupies the ground that lies between the castle of St. Antonio & the small rivers of Manzanares and Santa Catalina. The Delta, formed by the bifurcation of the first of these rivers, is a plain covered with mammees, sopotas, plantains, and other plants, cultivated in the gardens or *chares* of the Indians. The town has no remarkable edifice, and the frequency of earthquakes forbids such embellishments. The heats are somewhat less oppressive on the side towards the sea shore, than in the old town when the reverberation of the calcarious soil, and the proximity of the mountain of St. Antonio raises the temperature to an extraordinary degree.

"The shore near the mouth of the small river Santa Catilina is bordered with mangrove trees, but these mangroves are not sufficiently spread to diminish the salubricy of the air of Cumana.

"On leaving the Indian suburbs, and asscending the river towards the south, we found a grove of cactus, a delightful spot, shaded by tamarinds, brazillettoes, bemcax, and other plants, remarkable for their leaves and flowers."

[N. Y. *Columbian*.]

On the Laws observed in the Distribution of vegetable Forms[1]

Botany, long confined to the simple description of the external forms of plants and their artificial classification, now presents several branches of study, which place it more on a footing with the other sciences. Such are the distribution of vegetables according to a natural method founded upon the whole part of their structure; physiology, which displays their internal organization; botanical geography, which assigns to each tribe of plants their height, limits, and climate. The terms *alpine plants, plants of hot countries, plants of the sea-shore*, are to be found in all languages, even in those of the most savage nations on the banks of the Oronoko. They prove that the attention of men has been constantly fixed on the distribution of vegetables, and on their connexion with the temperature of the air, the elevation of the soil, and the nature of the ground which they inhabit. It does not require much sagacity to observe, that on the slope of the high mountains of Armenia, vegetables of a different latitude follow each in succession, like the climates, superposed as it were upon each other. This idea of Tournefort, developed by Linnæus in two interesting dissertations, (*Stationes et Coloniæ Plantarum*,) nevertheless contains the first seeds of botanical geography. Menzel, the author of an unpublished Flora of Japan, strongly recommends to travellers researches as to the distribution of species in the different regions of the globe. He had even pointed out the result before by the name of the Geography of Plants. This appellation was again employed, and almost at the same time, about the year 1783, by the Abbé Giraud Soulavie, and by the celebrated author of the Studies of Nature; a work which, amid a great variety of very inaccurate ideas as to the *physique* of the globe, contains some profound and ingenious views as to the forms, relations, and habitudes of vegetables. Abbé Giraud Soulavie was occupied in preference with the plants already cultivated: he has distinguished the climates of the olive trees, the vines, and the chesnuts. He gives a vertical section of Mount Mezin, to which he has added the | 447 | barometrical heights, "because," as he says, "he has a great contempt for every result taken from barometrical measurement." His Geography of the Plants in the South of France was followed by the *Tentamen Historiæ geographicæ*

[1] Extracted from a paper read to the French Institute, Feb. 5, 1816.

Vegetabilium of the learned Professor Strohmayer, published in 1800 at Gottingen in the form of a dissertation; but this Tentamen exhibits rather the plan of a future work, and the catalogue of authors to be consulted, than information respecting the altitudes which spontaneous plants reach in different climates. The case is the same with the very philosophical views announced by M. Treviranus in his *Essai de Biologie*; we therein find general considerations, but no measurements of heights, and no thermometrical indications, which are the solid bases of the geography of plants. This study has not risen to the rank of a science, until men of science have perfected both the measures of heights by barometrical observations, and the determination of mean temperatures; or, what is more important for the development of vegetation, the determination of the differences between the temperature of summer and winter and between that of day and night. Few branches of study have in our day made more rapid progress; and a long time has not intervened between the first efforts and the present period, when by the united observations of a great number of travellers, we have succeeded in fixing the limits of vegetables in Lapland, the Pyrenees, the Alps, Caucasus, and the Cordilleras of America.

The vegetables which cover the vast surface of the globe present, when we study by natural classes or families, striking differences in the distribution of their forms: it is to the laws of this distribution that I have recently turned my attention. On limiting them to the countries in which the number of the species is exactly known[2], and by dividing this number by that of the *Glumaceæ*[3], the leguminous plants, the labiated, and the compound, we find numerical relations which form very regular series. We see certain forms become more common from the equator towards the pole, like the ferns, the glumaceæ, the ericineæ, and the rhododendrons. Other forms on the contrary increase from the poles towards the equator, and may be considered in our hemisphere as southern forms: such are the rubiaceæ, the malvaceæ, the euphorbia, the leguminous and the composite plants. Finally, others attain their maximum even in the temperate zone, and diminish also towards the equator | 448 | and the poles. Such are the labiated plants, the amentaceæ, the cruciferæ, and the umbelliferæ. Part of these data long since struck botanical travellers, and all those who have looked into herbals. It was known that the cruciferæ and umbelliferæ disappeared almost entirely in the plains of the torrid

[2] Lapland, France, England, &c. according to Messrs. Wahlenberg, Buch, Ramond, Decandolle, and Smith.

[3] The Glumaceæ contain the three families of *Grumineæ, Cyperaceæ*, and *Juncaceæ*.

zone, and that none of the malvaceæ were found beyond the polar circle. It is the same with the geography of the plants as with meteorology. The results of those sciences are so simple that in all ages general ideas have been formed of them: but it is only after laborious researches, and after having collected a great number of accurate observations, that numerical results were attained, and an acquaintance with the partial modifications undergone by *the law of the distribution of forms.* A table which we have drawn up exhibits this law with respect to sixteen families of plants distributed over the equatorial, temperate, and glacial zones. We there see with satisfaction mixed with surprise, how in organic nature, the forms present constant relations under the same *isothermal parallels, i.e.* on curves traced by points of the globe which receive an equal quantity of heat. The grasses form in England 1-12th, in France 1-13th, in North America 1-10th, of all the phanerogamous plants. The glumaceæ form in Germany 1-7th; in France 1-8th; in North America 1-8th; in New Holland, according to the researches of Mr. Brown, 1-8th; of the known phanerogamous plants. The composite plants increase a little in the northern part of the new continent; for, according to the new Flora of Pursch, there is between the parallels of Georgia and Boston 1-6th; whereas in Germany we find 1-8th; and in France 1-7th, of the total number of the species with visible fructification. In the whole temperate zone, the glumaceæ and the composite plants form together; nearly one-fourth of the phanerogamous plants; the glumaceæ, the compositæ, the cruciferæ and the leguminosæ, together nearly one-third. It results from these researches that the forms of organized beings are in a mutual dependence, and that the unity of nature is such that the forms are limited, the one after the other, according to constant laws easy of determination. When we know upon any point of the globe the number of species presented by one of the great families of the glumaceæ, the composite, the cruciferous, or the leguminous plants, we may estimate with considerable probability both the total number of the phanerogamous plants, and the number of species which compose the other vegetable families. It is thus that, by knowing under the temperate zone the number of the cyperaceæ or composite plants, we may guess at that of the gramineous or leguminous plants. | 449 |

The number of vegetable species described by botanists, or existing in European herbals, extends to 44000, of which 6000 are agamous. In this number we had already included 3000 new phanerogamous species enumerated by M. Bonpland and myself. France, according to M. Decandolle, possesses 3645 phanerogamous plants, of which 460 are glumaceæ,

490 composite, and 230 leguminous, &c. In Lapland there are only 497 phanerogamous plants; among which are 124 glumaceæ, 58 composite, 14 leguminous, 23 amentaceous, &c. See my Essay on the Geography of Plants published in 1806, and of which I am preparing a new edition.

In order to account for the differences which exist sometimes between the relations exhibited by Germany, North America, and France, we must take into consideration the more or less temperate climates of those regions. France extends from 42 1/2 ° to 51° of latitude, On this extent the mean annual heat is 16° 7′ to 11°: the mean heats of the summer months are 24° to 19°. Germany, comprised between 46° and 54° of latitude, presents at its extremities mean annual temperatures of 12° 5′ and 8° 5′. The mean heats of the summer months there are 21° and 18°. North America, in its immense extent, presents the most varied climates. Mr. Pursch has made us acquainted with 2000 phanerogamous plants which grow between the parallels of 35° and 44°; consequently under mean annual temperatures of 16° and 7°. The Flora of North America is a mixture of several Floras. The southern regions give it an abundance of malvaceæ and composite plants; the northern regions, colder than Europe under the same parallel, furnish to this Flora abundance of rhododendrons, amentaceæ, and coniferæ. The caryophylleæ, the umbelliferæ, and the cruciferæ are in general more rare in North America than in the temperate zone of the old continent.[4]

These constant relations observed on the surface of the globe, in the plains from the equator to the pole, are again traced in the midst of perpetual snows on the summits of mountains. We may admit, in general, that on the cordilleras of the torrid zone the boreal forms become more frequent. It is thus that we see prevail at Quito on the summit of the Andes, the ericineæ, the | 450 | rhododendrons, and the gramineous plants. On the contrary, the labiatæ, the rubiaceæ, the malvaceæ, and the euphorbiaceæ then become as rare as they are in Lapland. But this analogy is not supported in the ferns and the composite plants. The latter abound on the Andes, whereas the former gradually disappear when they rise above 1800 fathoms in height. Thus the climate of the Andes resembles that of northern Europe only with respect to the mean temperature of the

[4] For the sake of those who are not much conversant in descriptive botany, we shall here enumerate the plants which serve as a type to the forms or principal families: *Glumaceæ*, rushes, tares; *orchideæ*, orchis, satyrion, vanilla; *labiatæ*, sage; *ericineæ*, broom; *compositæ*, coltsfoot, tussilago; *rubiaceæ*, madder, quinquina; *umbelliferæ*, fennel; *cruciferæ*, radish, cabbage; *malvaceæ*, cotton; *leguminosæ*, furze, truffles, sensitive plant; *euphorbiaceæ*, milky thistle; *amentaceæ*, willow, oak; *coniferæ*, pine, yew, juniper.

year. The repartition of heat into the different seasons is entirely different, and powerfully influences the phænomena of vegetation. In general, the forms which prevail among the Alpine plants are, according to my researches, under the torrid zone, the *gramineæ* (ægopogon, podosæmum, deyeuxia, avena); the *compositæ* (culcitium, espeletia, aster, baccharis); and the *caryophylleæ* (arenaria, stellaria). Under the temperate zone, the *compositæ* (senecio, leontodon, aster); the *caryophylleæ* (cerastium, cherleria, silene); and the *cruciferæ* (draba, lepidium). Under the frozen zone, the *caryophylleæ* (stellaria, alsine); the *ericineæ* (andromeda) and the *ranunculaceæ*.

These researches into the law of the distribution of forms, naturally lead to the question whether there exist plants common to both continents? a question which inspires the more interest, as it belongs to one of the most important problems in Zoonomia. It has been long known, and it is one of the most interesting results from the *geography of animals*, that no quadruped, no terrestrial bird, and, as appears from the researches of M. Latreille, almost no insect, is common to the equatorial regions of the two worlds. M. Cuvier is convinced by precise inquiries that this rule applies even to reptiles. He has ascertained that the true boa constrictor is peculiar to America, and that the boas of the old continent were pytons. As to the regions beyond the tropics, Buffon has multiplied beyond measure the number of the animals common to America, to Europe, and the north of Asia. We are assured that the bison, the stag, and the goat of America, the rabbit and the musk rat, the bear, &c. &c. are species entirely different from those of Europe, although Buffon had affirmed the contrary. There remain only the glutton, the wolf, the white bear, the red fox, perhaps also the elau, which have not characters sufficient to entitle them to be specific. Among the plants, we must distinguish between the agamæ and the cotyledoneæ; and by considering the latter between the monocotylodens and the dicotyledons. There remains no doubt that many of the mosses and lichens are to be found at once in equinoctial America and in Europe: our herbals show this. But the case is not the same with the vascular agamæ as with the agamæ of a cellular texture. The ferus and the lycopodiaceæ | 451 | do not follow the same laws with the mosses and the lichens. The former, in particular, exhibit very few species universally to be found; and the examples cited are frequently doubtful. As to the phanerogamous plants (with the exception of the rhizophora, the avicennia, and some other littoral plants), the law of Buffon seems to be exact with respect to the species furnished with two cotyledons. It is absolutely false, although it has been often affirmed, that the ridges of the cordilleras of Peru, the climate of which has some analogy with the climate of France or Sweden, produce similar plants. The oaks, the pines, the yews, the ranunculi, the rose-trees, the alchemilla, the valerians, the stellaria, the draba of the Peruvian and Mexican Andes, have nearly the same physiognomy with the species of the same genera of North America, Siberia, or Europe. But all these alpine plants of the Cordilleras, without excepting one among three or four thousand which we have examined, differ specifically from the analogous species of the temperate zone of the old continent. In general, in that part of America situated between the tropies, the monocotyledontal plants alone, and among the latter almost solely the cyperaceæ and the gramineæ, are common to the two worlds. These two families form an exception to the general law which wc are here examining,—a law which is so important for the history of the catastrophes of our planet, and according to which the organized beings of the equinoctial regions differ essentially in the two continents. I have given in my *Prolegomena* a precise catalogue of those monocotyledontal plants common to the shores of the Oronoko, Germany, and the East Indies. Their number does not exceed 20 or 24 species, among which it is sufficient to cite the cyperus mucronatus, c. hydra, hypælyptum argenteum, poa eragrostis, andropogon, allioni, &c.

In North America placed beyond the tropics, we find nearly one-seventh of monocotyledontal and dicotyledontal plants common to the two continents. Of 2900 phanerogamous species collected in the New Flora of Pursch, 390 are European. It is true that we may hazard some doubts, as well with respect to the number of the plants which have accompanied Europeans from one hemisphere to the other, as upon those which, when better examined, will be recognised subsequently as new species: but it is impossible that this state of uncertainty should extend to all; and it is to be presumed that, even after a careful examination, the number of the species common to the temperate zones of the two worlds will still remain very considerably analogous. Mr. Brown recently undertook some researches on the plants of New Holland. A twenty-eighth part of all the monocotyledons hitherto found in the austral continent are common | 452 | to it with England, France, and Germany. Among the dicotyledons the ratio is only 1 in 200; which proves once more how, in the two hemispheres, the grasses and the cyperaceæ are the most diffused, on account of the extreme flexibility of their organization. It would be desirous that learned zoologists should endeavour to examine the analogous numerical ratios presented by the distribution of the different families of animals on the globe.

In the austral hemisphere the vegetable forms of the torrid zone advance more towards the pole than in the boreal hemisphere. The fern-trees in Asia and America are rarely to be found beyond the tropic of Cancer; whereas in the austral part of our globe the *Dicksonia antarctica*, the trunk of which is six metres in height, pushes its migrations as far as Van Diemen's Land under the latitude of 42°: it has even been found in New Zealand, in Dusky Gulph, under the parallel of Lyons.

Other forms not less majestic, and which were thought to belong exclusively to the equinoctial Flora, the parasite orchideæ (epidendra, dendrobia) are found mixed with the arborescent ferus far beyond the tropic of Capricorn, in the centre of the austral temperate zone. These phænomena of the geography of plants prove how vague is what has been generally said of the great diminution of temperature in the southern hemisphere, without distinguishing between the parallels more or less near the pole, and without any regard to the division of heat among the different seasons of the year. Those regions towards which the equinoctial forms extend, possess, on account of the immensity of the seas which surround them, a true *island climate*. From the tropic of Capricorn to the parallel of 34°, and perhaps still further, the mean heats of the year (*i e.* the quantity of heat received by any given point of the globe) do not differ considerably in the two hemispheres. On casting our eyes over the three continents, New Holland, Africa, and America, we find that the mean annual temperature of Port Jackson (lat. 33° 51′) is 19° 3′ of the centigrade thermometer: that of the Cape of Good Hope (lat. 33° 55′) 19° 4′; that of the town of Buenos Ayres (lat. 34° 36′) 19° 7′. We may be surprised at this great equality in the distribution of heat by the 34° of austral latitude. Meteorological observations, still more precise, prove that in the boreal hemisphere, under this very parallel of 34°, we find a mean temperature of 19° 8′. On advancing towards the antarctic pole, perhaps even to the parallel of 57°, the temperatures of the two hemispheres differ less in winter than summer. The Malouine Islands, situated in 51° and a half of south latitude, have less intense cold in winter than is experienced at London. The mean temperature of Van Diemen's Land seems to be 10°; it freezes | 453 | during winter, but not so much as to destroy the fern-trees and the parasite orchideæ. In the adjoining seas Capt. Cook, in 42° of austral latitude, did not see the thermometer fall below + 6°, 6 in the midst of winter (July). To these very mild winters, summers succeed remarkable for an extraordinary coolness. At the southern extremity of New Holland (lat. 42° 41′) the temperature of the air rarely rises in the midst of summer at noon-day higher than 12° or 14°;

and in Patagonia, as in the adjoining ocean (lat. 48° — 58°), the mean heat of the warmest month is only 7° — 8°; whereas in the boreal hemisphere at Petersburg and Umeo (lat. 59° 56′ and 63° 50′) this heat exceeds 17 — 19. It is this mild temperature of the islands, which the southern countries enjoy between 30° and 40° of latitude, which permits the vegetable forms to pass beyond the tropic of Capricorn. They embellish a great part of the temperate zone; and the genera which the inhabitant of the northern hemisphere regards as exclusively belonging to the tropical climates, present numerous species between the parallels of 35° and 38° of south latitude.

The Southern Constellations

From the time we entered the torrid zone, we were never wearied with admiring, every night, the beauty of the southern sky, which, as we advanced towards the south, opened new constellations to our view. We feel an indescribable sensation, when, on approaching the equator, and particularly on passing from one hemisphere to the other, we see those stars, which we have contemplated from our infancy, progressively sink, and finally disappear. Nothing awakens in the traveller a livelier remembrance, of the immense distance by which he is separated from his country, than the aspect of an unknown firmament. The grouping of the stars of the first magnitude, some scattered nebulæ, rivalling in splendour the milky way, and tracts of space remarkable for their extreme blackness, give a particular physiognomy to the southern sky. This sight fills with admiration even those, who, uninstructed in the branches of accurate science, feel the same emotion of delight in the contemplation of the heavenly vault, as in the view of a beautiful landscape, or a majestick site. A traveller has no need of being a botanist, to recognize the torrid zone on the mere aspect of its vegetation: and without having acquired any notions of astronomy, without any acquaintance with the celestial charts of Flamstead and de la Caille, he feels he is not in Europe, when he sees the immense constellation of the Ship, or the phosporescent clouds of Magellan, arise on the horizon. The heaven, and the earth, every thing in the equinoctial regions, assumes an exotic character.

The lower regions of the air were loaded with vapours for some days. We saw distinctly for the first time the cross of the south only in the night of the 4th and 5th of July, in the sixteenth degree of latitude; it was strongly inclined to, & appeared from time to time between the clouds, the centre of which furrowed by uncondensed lightnings, reflected a silver light. If a traveller may be permitted to speak of his personal emotions, I shall add that in this night I saw one of the reveries of my earliest youth accomplished.

When we begin to fix our eyes on geographical maps, & read the narratives of navigators, we feel for certain countries and climates a sort of predilection, for which we know not how to account at a more advanced period of life. These impressions, however, exercise a considerable influence over our determinations; and from a sort of instinct we endeavour to connect ourselves with objects, on which the mind has long been fixed as by a secret charm. At a period when I studied the heavens, not with the intention of devoting myself to astronomy, but only to acquire a knowledge of the stars, I was agitated by a fear unknown to those who love a sedentary life. It seemed painful to me to renounce the hope of beholding those beautiful constellations, which border the southern pole. Impatient to rove in the equinoctial regions, I could not raise my eyes towards the starry vault without thinking of the Cross of the South, and without recalling the sublime passage of Dante, which the most celebrated commentators have applied to this constellation:

> Io mi volsi a man destra e puosi mente
> All'altro polo e vidi quattro stelle
> Non viste mai fuor ch'alla prima gente.
> Goder pareva il ciel di lor fiammelle:
> O settentrional vedovo sito
> Poi che privato sei di mirar quelle!

The pleasure we felt on discovering the southern Cross was warmly shared by such of the crew as had lived in the colonies. In the solitude of the seas we hail a star as a friend, from whom we have long been separated. Among the Portuguese and the Spaniards peculiar motives seem to increase this feeling; a religious sentiment attaches them to a constellation, the form of which recalls the sign of the faith planted by their ancestors in the deserts of the new world.

The two great stars which mark the summit and the foot of the Cross having nearly the same right ascension, it follows hence, that the constellation is almost perpendicular at the moment when it passes the meridian. This circumstance is known to every nation, that lives beyond the tropics, or in the southern hemisphere. It has been observed at what hour of the night, in different seasons, the Cross of the south is erect or inclined. It is a time-piece that advances very regularly near four minutes a day, and no other group of stars exhibits to the naked eye, an observation of time so easily made. How often have we heard our guides exclaim in the savannas of Venezuela, or in the desert extending from Lima to Truxillo, "Midnight is past, the Cross begins to bend!" How often those words reminded us of that affecting scene, where Paul and Virginia seated near the source of the river of Lataniers, conversed together for the last time, and where the old man, at the sign of the southern Cross, warns them that it is time to separate.

To the Editor of the *Morning Chronicle*

Sir,

To prove how circumspect travellers ought to be in the explanation of hieroglyphic characters found among half civilised people, I have referred, in my *Researches on the Monuments of the New Continent, page* 62, to a fact observed among the Betjuans of Southern Africa. A traveller had seen children etching on a rock certain characters accidentally resembling the letters P and M of the Roman alphabet. This observation I have attributed to Mr. Trüter, and I have quoted the work from which it was taken—*The Geographical Ephemerides of Mr. Bertuch.* This fact, however, has exposed me to a severe criticism in the twenty-sixth number of the *Quarterly Review.* In the analysis of *Campbell's Missionary Travels* it is affirmed, that no such circumstance is related in Mr. Trüter's work, nor even any allusion of the kind; and after commencing with these words. "*If any credit were due to the authority of Mr. Humboldt,*" the article terminates with the following, "*yet this is a fact on which Mr. Humboldt hangs one of his numerous theories.*"

I do not pretend that my name is of great authority concerning facts not examined with my own eye. Few authors are more sensible of the imperfections of their own works than I am; but I believe that in what relates to the Betjuanas, the fault is not so great as this unfriendly critic would imagine. The fact in question, contained in the *Ephemerides of Mr. Bertuch, of May* 1807, *p. 32. l. 25.* is thus related:—"Among the Betjuanas' children of ten or twelve years of age, were occupied in engraving figures upon a rock with a pointed instrument, some of which so much resembled the M, H, and P of the Roman alphabet, that travellers might mistake them for true letters." This circumstance was noticed by me not for the purpose of establishing, but of condemning, theories, and particularly to prove, that a resemblance of Phœnician characters, with some signs traced on the rocks of the Orinoco, is merely accidental. I was only deceived in referring, as authority in this case, to Mr. Trüter, a name perfectly well known to me by the excellent *Travels of Mr. Barrow*. The observation, however, is that of Dr. Lichtenstein, sent by General Jansen, in 1803, to the country of the Betjuanas, and whose report is inserted in the *Ephemerides of Mr. Bertuch*. As Dr. Lichtenstein often compares his observations with those of Mr. Trüter, it so happened that I confounded the two travellers who had visited the same nation and examined the same objects.

ALEXANDER HUMBOLDT, F.R.S.
Paris, Jan. 24, 1816.

On Isothermal Lines, and the Distribution of Heat over the Globe[1]

The distribution of heat over the Globe belongs to that kind of phenomena, of which the general circumstances have been long known, but which were incapable of being rigorously determined or submitted to exact calculation, till experience and observation furnished data from which the theory might obtain the corrections and the different elements which it requires. The object of this memoir is to facilitate the collection of these data, to present results drawn from a great number of unpublished observations, and to group them according to a method which has not yet been tried, though its utility has been recognised for more than a century in the exposition of the phenomena of the variation and dip of the magnetic needle. As the discussion of individual observations will be published in a separate work, I shall at present limit myself to a simple sketch of the distribution of heat over the globe, according to the most recent and accurate data. Although we may not be able to refer the complex phenomena to a general theory, it | 2 | will be of considerable importance to fix the numerical relation by which a great number of scattered observations are connected, and to reduce to empirical laws the effects of local and disturbing causes. The study of these laws will point out to travellers the problems to which they should direct their principal attention, and we may entertain the hope, that the theory of the distribution of heat will gain in extent and precision, in proportion as observations shall be more multiplied, and directed to those points which it is of most importance to illustrate.

As the phenomena of geography and of vegetables, and in general the distribution of organised beings, depend on the knowledge of the three co-ordinates of Latitude, Longitude, and Altitude, I have been occupied for many years in the exact valuation of atmospherical temperatures; but I could not reduce my own observations without a constant

[1] As this interesting and valuable Memoir, the original of which was published in the *Memoires D'Arcueil*, tom. iii. p. 462, has never appeared in our language, and as it must be constantly referred to in all subsequent speculations on Meteorology, and should be familiar to every person who pursues this important study, we have resolved to present a translation of it to our readers. A small part of the memoir appeared in an English journal; but almost all the reductions from the Centigrade to Fahrenheit's scale were so erroneous, that the numbers cannot be trusted. We have added various notes, which will be distinguished from those of the Author by affixing Ed. to the former, and H to the latter.—Ed.

reference to the works of Cotte and Kirwan, the only ones which contain a great mass of meteorological observations obtained by instruments and methods of very unequal precision. Having inhabited for a long time the most elevated plains of the New Continent, I availed myself of the advantages which they present for examining the temperature of the superincumbent strata of air, not from insulated data, the results of a few excursions to the crater of a volcano, but from the collection of a great number of observations made day after day and month after month in inhabited districts. In Europe, and in all the Old World, the highest points of which the mean temperatures have been determined, are the Convent of Peissenberg in Bavaria, and the Hospice of St Gothard.[2] The first of these is placed at 3264, and the second at 6808 feet above the level of the sea. In America a great number of good observations have been made at Santa Fe de Bogota and at Quito, at altitudes of 8,727 and 9,544 feet. The town of Huancavelica, containing 10,000 inhabitants, and possessing all the resources of modern civilisation, is situated in the Cordilleras of the southern hemi- | 3 | sphere at 12,310 feet of absolute elevation; and the mine of Santa Barbara, encircled with fine edifices, and placed a league to the south of Huancavelica, is a place fit for making regular observations, at the height of 14,509 feet, which is double that of the Hospice of St Gothard.

These examples are sufficient to prove how much our knowledge of the higher regions of the atmosphere, and of the physical condition of the world in general, will increase, when the cultivation of the sciences, so long confined to the temperate zone, shall extend beyond the tropics into those vast regions, where the Spanish Americans have already devoted themselves with such zeal to the study of physics and astronomy. In order to compare with the mean heat of temperate climates, the results which M. Bonpland and I obtained in the equinoctial regions from the plains to the height of 19,292 feet, it was necessary to collect a great number of good observations made beyond the parallels of 30° and 35°. I soon perceived how vague such a comparison was, if I selected places under the meridian of the Cordilleras, or with a more eastern longitude, and I therefore undertook to examine the results contained in the most recent works. I endeavoured to find, at every 10° of latitude, but under different meridians, a small number of places whose mean temperature had been precisely ascertained, and through these, as so many fixed

[2] The mean temperature of the air at the Convent of the Great St Bernard, the height of which is 7,960 feet, is not determined. There are several villages in Europe placed at more than 5000 feet of altitude; for example, St Jacques de Ayas at 5,479, and Trinita Nuova, near Grasfoncy, at 5,315 feet.—H.

points, passed my *isothermal lines* or *lines of equal heat.* I had recourse, in so far as the materials have been made public, to those observations the results of which have been published, and I found, in the course of this easy, but long and monotonous labour, that there are many mean temperatures pointed out in meteorological tables, which, like astronomical positions, have been adopted without examination. Sometimes the results were in direct contradiction to the most recent observations, and sometimes it was impossible to discover from whence they were taken.

Many good observations were rejected, solely because the absolute height of the place where they were made was unknown. This is the case with Asia Minor, Armenia and Persia, and of almost all Asia; and while the equinoctial part alone of the New World presents already more than 500 points, the greater number of which are simple villages and hamlets, de- | 4 | termined by barometrical levelling, we are still ignorant of the height of Erzeroum, Bagdad, Aleppo, Teheran, Ispahan, Delhi and Lassa, above the level of the neighbouring seas. Notwithstanding the intimate relation in which we have lately stood with Persia and Candahar, this branch of knowledge has not made any progress in the last fifty years.

We are not authorised, however, on account of the decrease of temperature in the upper regions of the atmosphere, to confound the mean temperatures of places which are not placed on the same level. In the Old World, good observations, which can alone be used for establishing empirical laws, are confined to an extent between the parallels of 30 and 70 degrees of latitude, and the meridians of 30° east longitude, and 20° of west longitude. The extreme points of this region are the island of Madeira, Cairo, and the North Cape. It is a zone which is only a thousand nautical leagues, (1/7th of the circumference of the globe,) from east to west, and which, containing the Basin of the Mediterranean, is the centre of the primitive civilisation of Europe. The extraordinary shape of this part of the world, the interior seas and other circumstances, so necessary for developing the germ of cultivation among nations, have given to Europe a particular climate, very different from that of other regions placed under the same latitude. But as the physical sciences almost always bear the impress of the places where they began to be cultivated, we are accustomed to consider the distribution of heat observed in such a region, as the type of the laws which govern the whole globe. It is thus that, in geology, we have for a long time attempted to refer all volcanic phenomena to those of the volcanoes in Italy. In place of estimating methodically the distribution of heat, such as it exists on the surface of continents and seas, it has been usual to consider as real exceptions every thing which differs from the adopted type, or, by pursuing a method still more dangerous in investigating the laws of nature, to take the mean temperatures for every five degrees of latitude, confounding together places under different meridians. As this last method appears to exclude the influence of extraneous causes, I shall first discuss it briefly before I proceed to point out the method, essentially different, which I have followed in my researches. | 5 |

The temperature of the atmosphere, and the magnetism of the globe, cannot, like those phenomena which depend on one cause, or on a single centre of action, be disengaged from the influence of disturbing circumstances, by taking the averages of many observations in which these extraneous effects are mutually destroyed. The distribution of heat, as well as the dip and variation of thc nccdle, and the intensity of the terrestrial magnetism, depend, by their nature, on local causes, on the constitution of the soil, and on the particular disposition of the radiating surface of the globe. We must, however, guard against confounding under the name of extraneous and disturbing causes, those on which the most important phenomena, such as the distribution and the more or less rapid developement of organic life, essentially depend. Of what use would it be to have a table of magnetic dips, which, in place of being measured in parallels to the magnetic equator, should be the mean of observations made on the same degrees of terrestrial latitude, but under different meridians? Our object is to ascertain the quantity of heat which every point of the globe annually receives, and, what is of most importance to agriculture, and the good of its inhabitants, the distribution of this quantity of heat over the different parts of the year, and not that which is due to the solar action alone, to its altitude above the horizon, or to the duration of its influence, as measured by the semidiurnal arcs.

Moreover, we shall prove, that the method of means is unfit for ascertaining what belongs exclusively to the sun, (inasmuch as its rays illuminate only one point of the globe,) and what is due both to the sun and to the influence of foreign causes. Among these causes may be enumerated the mixture of the temperatures of different latitudes produced by winds;—the vicinity of seas, which are immense reservoirs of an almost invariable temperature;—the shape, the chemical nature, the colour, the radiating power and evaporation of the soil;—the direction of the chains of mountains, which act either in favouring the play of descending currents, or in affording shelter against particular winds;—the form of lands, their mass and their prolongation towards the poles;—the quantity of snow which covers them in winter, their temperature and their reflection in summer;— and, finally, the fields of ice, which form, as it were, circumpolar | 6 | continents, variable in their extent, and whose detached parts dragged away by currents modify in a sensible manner the climate of the temperate zone.

In distinguishing, as has long been done, between the *solar* and the *real climate*, we must not forget, that the local and multiplied causes which modify the action of the sun upon a single point of the globe, are themselves but secondary causes, the effects of the motion which the sun produces in the atmosphere, and which are propagated to great distances. If we consider separately (and it will be useful to do this in a discussion purely theoretical) the heat produced by the sun, the earth being supposed at rest and without an atmosphere, and the heat due to other causes regarded as disturbing ones, we shall find that this latter part of the total effect is not entirely foreign to the sun. The influence of small causes will scarcely disappear by taking the mean result of a great number of observations; for this influence is not limited to a single region. By the mobility of the aerial ocean, it is propagated from one continent to another. Every where in the regions near the polar circles, the rigours of the winters are diminished by the admixture of the columns of warm air, which, rising above the torrid zone, are carried towards the poles: Every where in the temperate zone, the frequent west winds modify the climate, by transporting the temperature of one latitude to another.[3] When we reflect, besides, on the extent of seas, on the form and prolongation of continents, either in the two hemispheres, or to the east and west of the meridians of Canton and of California, we shall perceive, that even if the number of observations on the mean temperature were infinite, the compensation would not take place.

It is, then, from the theory alone that we must expect to determine the distribution of heat over the globe, in so far as it depends on the immediate and instantaneous action of the sun. It does not indicate the degrees of temperature expressed by the dilatation of the mercury in a thermometer, but the ratios between the mean annual heat at the equator, at the parallel of 45°, and under the polar circle; and it determines the ratios between the solstitial and equinoctial heats in different zones. By comparing | 7 | the results of calculation, not with the mean temperature drawn from observations made under different longitudes, but with that of a single point of the earth's surface, we shall set out with that which is due to the immediate action of the sun, and to the whole of the

[3] Raymond, *Memoire sur la Formule Baromet.* p. 108 and 113.

other influences, whether they are solar or local, or propagated to great distances. This comparison of theory with experience will present a great number of interesting relations.

In the year 1693, previous to the use of comparable thermometers, and to precise ideas of the mean temperature of a place, Halley laid the first foundations of a theory of the heating action of the sun under different latitudes.[4] He proved that these actions might compensate for the effect of the obliquity of the rays. The ratios which he points out, do not express the mean heat of the seasons, but the heat of a summer day at the equator and under the polar circle, which he finds to be as 1.834 to 2.310.[5] According to Geminus,[6] Polybius among the Greeks had perceived the cause why there should be less heat at the equator than under the tropic. The idea also of a temperate zone, habitable and highly elevated in the midst of the torrid zone, was admitted by Eratosthenes, Polybius, and Strabo.

In two memoirs,[7] published at long intervals in 1719 and 1765, Mairan attempted to solve the problems of the solar action, by treating them in a much more extended and general manner. He compared, for the first time, the results of theory with those of observation; and as he found the difference between the heat of summer and winter much less than it ought to be by calculation, he recognised the permanent heat of the globe and the effects of radiation.

Without mistrusting the observations he employed, he conceived the strange theory of central emanations which increase the heat of the atmosphere from the equator to the pole. He supposes that these emanations decrease to the parallel of 74°, where the solar summers attain their maximum, and that they then increase from 74° to the pole. Lambert,[8] with that |8| sagacity which distinguishes all his mathematical researches, has pointed out in his *Pyrometrie* the error of Mairan's theory. He might have added, that this geometer confounds a quantity of heat which a point of the globe receives under the latitude of 60° during the three months of summer, with the maximum to which the inhabitants of these northern regions see their thermometers rising in a clear day. The mean temperatures of the summers, far from decreasing from the pole to the tropics, are under the equator, under the parallel of 45°, and under that

[4] *Phil. Trans.* 1693, p. 878.

[5] This should be 2.339.—Ed.

[6] *Isag. in Aratum*, cap. 13.; Strabo, *Geogr.* lib. ii. p. 97.

[7] *Mem. de l'Acad.* 1719, p. 133; and 1765, p. 145. and 210.

[8] *Pyrometrie oder Vom Maase des Feuers*, 1779, p. 342.

of Stockholm, Upsal, or St Petersburg, in the ratio of 81°.86; 69°.8; 61°.16 of Fahrenheit's scale. Reaumur had sent his new thermometers to the torrid zone, to Syria, and to the north. As it was then reckoned sufficient to mark the warmest days, an idea was formed of an *universal summer*, which is the same in all parts of the globe. It had been remarked, and with reason, that the extreme heats are more frequent, and even more powerful, in the temperate zone in high latitudes, than under the torrid zone. Without attending to the mean temperature of months, it was vaguely supposed, that in these northern regions the summers followed the ratio of the thermometrical extremes. This prejudice is still propagated in our own day, though it is well established, that in spite of the length of the days in the north, the mean temperatures of the warmest months at Petersburg, Paris, and the Equator, are 65°.66; 69°.44, and 82°.4. At Cairo, according to the observations of Nouet, the three months of summer are 84°.74, and consequently 19° warmer than at St Petersburg, and 15° warmer than at Paris. The summer heats of Cairo, are almost equal to those I have experienced at Cumana and La Guayra between the tropics.

With regard to the *central emanation* of the system of Mairan, or to the quantity of heat which the earth gives to the ambient air, it is easy to conceive that it cannot act in all seasons. The temperature of the globe at the depths to which we can reach, in general differs little from the mean annual temperature of the atmosphere. Its action is of great importance for the preservation of vegetables; but it does not become sensible in the air, unless where the surface of the globe is not entirely covered with snow, and during those months, whose mean temperature | 9 | is below that of the whole year. In the south of France, for example, the radiation of the earth may act upon the atmosphere in the five months which precede the month of April. We speak here of the proper heat of the globe, of that which is invariable at great depths, and not of the radiation of the surface of the globe, which takes place even at the summer solstice, and the nocturnal effects of which have furnished M. Prevost with an approximate measure of the direct action of the sun.[9]

Mairan had found, that in the temperate zone the heat of the solar summer is to that of the solar winter as 16 to 1. M. Prevost admits for Geneva 7 to 1. Good observations have given me for the mean temperature of the summers and the winters at Geneva 34°.7; 64°.94; and at St Petersburg 46°.94 and 62°.06. These numbers neither express ratios nor absolute quantities, but thermometrical differences considered as the total

[9] *Du Calorique rayonnant*, p. 271. 277. 292.

effect of the calorific influences; the ratios furnished by theory separate the solar heat from every other indirect effect. Euler was not more successful than Mairan in his theoretical essays on the solar heat. He supposes that the negative sines of the sun's altitude during the night give the measure of the nocturnal cooling, and he obtains the extraordinary result,[10] that under the equator the cold at midnight ought to be more rigorous than during winter, under the poles. Fortunately, this great geometer attached but little importance to this result, and to the theory from which it is deduced. The second memoir of Mairan, without adding to the problems which had been attempted since the time of Halley, has at least the advantage of containing some general views on the real distribution of heat in different continents. It is true, that the extreme temperatures are there constantly confounded with the mean temperatures; but previous to the works of Cotte and Kirwan, it was the first attempt to group the facts, and to compare the most distant climates.

Dissatisfied with the route followed by his predecessors, Lambert, in his Treatise on Pyrometry, directed his attention to two very different objects. He investigated analytical expressions for the curves, which express the variation of temperature in a | 10 | place where it had been observed, and he resumed in its greatest generality the theorem of solar action. He gives formulæ, from which we may find the heat of any day at all latitudes; but being perplexed with the determination of the nocturnal dispersion of the acquired heat, or the subtangents of the nocturnal cooling,[11] he gives tables of the distribution of heat under different parallels, and in different seasons,[12] which deviate so much from observation, that it would be very difficult to ascribe these deviations to the heat radiating from the globe, and to disturbing causes. We are struck with the slight difference which the theory indicates between the mean annual temperatures of places situated under the equator and the polar circle, and between the summers of the torrid zone and those of the temperate zone. It cannot be expected, indeed, that analysis is capable of determining the distribution of heat such as it exists on the surface of the globe. Without employing empirical laws, and deducing the data from actual observation, the theory can subject to calculation only a part of the total effect, or that which belongs to the immediate action of the solar rays; but after the recent successful applications of analysis to the phenomena of the radiation of surfaces, the transmission of heat through solid bodies, and

[10] *Comment. Petrop.* tom. ii. p. 98.
[11] *Pyrometrie*, p. 141, 179.
[12] *Id.* 318, 339.

the cooling of these bodies in media of variable density, we may still expect to be able to perfect the theory of solar action, and to compute the distribution of the heat received into the exterior crust of our planet.

In discussing what may be expected from the purely theoretical labours of Geometers, I have not spoken of a celebrated, but very concise Memoir of Mayer, the reformer of the Lunar Tables. This work, written in 1755, was published twenty years afterwards, in his *Opera Inedita.*[13] It is a method, and |11| not a theory: It is an essay essentially different from those we have quoted, and, as its learned author calls it, a determination of the mean heat found empirically by the application of coefficients furnished by observation. The method of Mayer is analogous to that which Astronomers pursue with so much success, when they correct by small steps the mean place of a planet, by means of the inequalities of its motion: It does not present the result of the solar action disengaged from the influence of foreign circumstances; but, on the contrary, it estimates the temperatures such as they are distributed over the globe, whatever be the cause of that distribution. The mean heat of two places situated under different latitudes being given, we find by a simple equation the temperature of every other parallel.[14] The calculations of Mayer, according to which the temperatures decrease from the equator to the poles, as the squares of the sines of the latitudes, give results sufficiently precise, when the place does not differ much in longitude from that of the regions where the empirical co-efficients have been obtained. But, even in the northern hemisphere, when we apply the formula to places situated 70° or 80° to the east or west of the meridian of Paris, the calculated results no longer agree with observation. The curve which

---

[13] *De Variationibus Thermometri accuratius definiendis*, (*Opera inedita*, vol. i. p. 3—10.) M. Daubuisson, in a note inserted in the Journal de Physique, tom. lxii. p. 449. has given a formula which accords better with observation than that of Mayer. He admits that the temperature increases from the pole to the equator, as the cosine of the latitude raised to the power of 2 (1/2) °; but he judiciously adds, that this formula is applicable only to a zone of the Old World, near the Northern Atlantic Ocean—H.

[14] The formula given by Mayer was $T = 24 \cos^2$ Lat.; or $T = 12 + 12 \cos 2$ Lat. for Reaumur's scale; and $T = 84—52 \sin^2$ Lat., or $T = 58 + 26 \cos 2$ Lat. for Fahrenheit's scale. Since the publication of Humboldt's memoir, M. Daubuisson has resumed the subject of the earth's temperature in his *Traité de Geognosie*, tom. i. p. 424. Paris, 1819. He gives the following formula, which is almost the same as that of Mayer, for finding the mean temperature, according to the Centigrade scale, viz. $T = 27° \cos^2$ Latitude. This formula, which is superior in accuracy to Mayer's, gives all the temperatures in *defect* for latitudes below 42°, and in *excess* for all the higher latitudes, as appears from Daubuisson's table. It is therefore obviously defective. M. Daubuisson, however, considers it as applicable principally between the parallels of 30° and 60° of N. lat. It ought to be remarked, that in the above formula, 27° has been *assumed as the mean temperature* of the equator, in order to make the results agree with observations made in the temperate regions, whereas the mean temperature of the equator, as ascertained by Humboldt, is 27°.5; and if this were used in Daubuisson's formula, it would make the differences still more in excess.— Ed.

passes through those points whose temperature is 32°, does not coincide with any terrestrial parallel. If, in the Scandinavian Peninsula, we meet with this curve under the 65th or 68th de- | 12 | gree of latitude, it descends, on the contrary, in North America, and Eastern Asia, to the parallel of from 53° to 58°. But the direction and the inflexions of this curve of 32° of temperature influences the neighbouring isothermal lines in the same manner as the inflexions of the magnetic equator modify the lines of inclination. To demand what is the mean temperature, or what is the magnetic inclination under a particular degree of latitude, is to propose problems equally indeterminate. Though, even in high latitudes the magnetic and the isothermal lines are not rigorously parallel to the magnetic equator, and to the curve of 32° of temperature; yet it is the distance of any place from this curve which determines the mean temperature, as the inclination of the needle depends on the magnetic latitude.

These considerations are sufficient to prove, that the empirical formulæ of Mayer require the introduction of a co-efficient, which depends upon the longitude, and consequently on the direction of the isothermal lines and their nodes with the terrestrial parallels. Mayer had no intention of disengaging the results which he obtained from the influence of all disturbing causes: He limited himself to the determination of the effects of altitude above the level of the sea, and those of the seasons, and the length of the day. He wished to point out the way which philosophers ought to pursue in imitating the method of astronomers. His Memoir was written at a time when we did not know the mean temperature of three points on the globe; and the corrections which I propose after tracing the isothermal lines, so far from being incompatible with the method of Mayer, are, on the contrary, among the number of those which this geometer seems to have indistinctly foreseen.

Kirwan, in his work on Climates, and in a learned Meteorological Memoir, inserted in the eighth volume of the Memoirs of the Irish Academy, attempted at first to pursue the method proposed by Mayer, but, richer in observations than his predecessors, he soon perceived, that after long calculations, the results agreed ill with observation.[15] In order to try a new method, he selected, in the vast extent of sea, those places | 13 | whose temperature suffered no change but from permanent causes. These were in the part of the great ocean commonly called the Pacific Ocean, from 40° of South to 45° of North latitude, and in the part of the Atlantic Ocean, between the parallels of 45° and 80°, from the coasts of

[15] Kirwan's *Estimate of the Temperature of the Globe*, chap. iii.

England to the Gulf Stream, the high temperature of which was first determined by Sir Charles Blagden. Kirwan tried to determine for every month the mean temperature of these seas at different degrees of latitude; and these results afforded him terms of comparison with the mean temperatures observed on the solid part of the terrestrial globe. It is easy to conceive, that this method has no other object, but to distinguish in climates that is in the total effect of *calorific influences*, that which is due to the immediate action of the sun on a single point of the globe. Kirwan first considers the earth as uniformly covered with a thick stratum of water, and he then compares the temperatures of this water at different latitudes, with observations, at the surface of continents indented with mountains, and unequally prolonged towards the poles.

This interesting investigation may enable us to appreciate the influence of local causes, and the effect which arises from the position of seas, on account of the unequal capacity of water and earth for absorbing heat. It is even better fitted for this object than the *Method of Means* deduced from a great number of observations made under different meridians; but in the actual state of our physical knowledge, the method proposed by Kirwan cannot be followed. A small number of observations made far from the coasts, in the course of a month, fixes, without doubt, the mean annual temperature of the sea at its surface, and, on account of the slowness with which a great mass of water follows the changes of the temperature of the surrounding air, the extent of variations in the course of a month is smaller in the ocean than in the atmosphere: But it is still greatly to be desired,[16] that we should be able to indicate by direct experience, for every parallel, and for every month, the mean temperature of the ocean under the temperate zone. The scheme which Kirwan has formed for the extent of the seas, that ought to | 14 | form the term of comparison, is founded only in a small degree upon the observations of navigators, and to a great degree on the theory of Mayer. He has also confounded experiments made on the superficial temperature of the ocean with the results of meteorological journals, or the indications of the temperature of the air which rests upon the sea: He has obviously reasoned in a circle, when he modified, either by theoretical suppositions or by observations made on the air upon the coasts of continents, the table of the temperature of the ocean, in order to compare afterwards with these same results, partly hypothetical, those which observation alone furnished in the interior of the earth. After the works of Kirwan, we must notice

[16] See my *Relation Historique*, tom. i.

those of Cotte, which are merely laborious, though useful, compilations, which, however, ought not to be used without much circumspection. A critical spirit has rarely presided over the reduction of the observations, and they are not arranged so as to lead to general results.

In detailing the actual state of our knowledge on the distribution of heat, I have shewn how dangerous it is to confound the results of observation with theoretical deductions. The heat of any point of the globe depends on the obliquity of the sun's rays, and the continuance of their action, on the height of the place, on the internal heat and radiation of the earth in the middle of a medium of variable temperature; and, in short, upon all those causes which are themselves the effects of the rotation of the earth, and the inequal arrangement of continents and seas. Before laying the foundation of a system, we must group the facts, fix the numerical ratios, and, as I have already pointed out, submit the phenomena of heat, as Halley did those of terrestrial magnetism, to empirical laws. In following this method, I have first considered whether the method employed by meteorologists for deducing the mean temperature of the year, the month and the day, is subject to sensible errors. Assured of the accuracy of the numerical averages, I have traced upon a map the isothermal lines, analogous to the magnetic lines of dip and variation. I have considered them at the surface of the earth in a horizontal plane, and on the declivity of mountains in a vertical plane. I have examined | 15 | the increase of temperature from the pole to the equator, which is inequal under different meridians; the distribution of the same quantity of heat over the different seasons, in the same isothermal parallel, and under different latitudes; the curve of perpetual snows, which is not a line of equal heat; the temperature of the interior of the earth, which is a little greater towards the north, and in high mountains, than the mean temperature of the atmosphere under the same parallel; and, lastly, the distribution of heat in the ocean, and the position of those bands, which may be called Bands of the warmest waters. As the limits of this extract will not permit me to enter, in a detailed manner, upon these different discussions, I shall confine myself solely to the principal results.

It was formerly the custom to take the *maximum* and *minimum* of temperature observed in the course of a year, and to consider half the sum as the mean temperature of the whole year. This was done by Maraldi, De la Hire, Muschenbröek, Celsius, and even Mairan, when they wished to compare the very warm year of 1718 with the excessively cold years of 1709 and 1740. De la Hire was struck with the identity between the uniform temperature of the caves of the Observatory of Paris, and the mean of the observed annual extremes. He appears to have been the first who had an idea of the mean quantity of heat which a point of the globe receives; and he adds, "We may regard the air of the caves as the mean state of the climate.[17]" Reaumur followed also the method of a *maximum*, though he confessed that it was incorrect.[18] He noticed the hours at which it was necessary to make observations; and after 1735, he published in the Memoirs of the Academy the extremes of daily temperature: he even compared the produce of two harvests with the sum of the degrees of heat to which during two consecutive years the crops had been exposed. When he treated, however, of the mean temperature of the month, he contented himself, as Duhamel did thirty years afterwards, with recording three or four thermometrical extremes. In order to have some idea of the errors of this imperfect method, I may | 16 | state, that even in 1777, the mean temperature of Toulon was estimated by Cotte[19] at 78°.08, though he afterwards found, by employing the whole mass of observations, that it was not more than 60°.26.

In order to diminish the errors of the method of annual extremes, it was perceived, though very late, that it was necessary to subdivide the curve which expresses the variation of temperature. Twenty-four extremes divided among twelve months of the year, give an annual mean more exact than the two extremes of all the observations. The ordinates do not increase uniformly and uninterruptedly up to the maximum of the year, and there are partial inflexions sufficiently regular. The more we subdivide, and the more we know the terms in the series, the more will these terms approximate, and the less error will there be in the supposition of an arithmetical progression, and in that of the equidistance of the different *maxima* and *minima* of temperature. These considerations enable us to appreciate the three methods according to which observations are at present made. 1. Observations are made *three times* a-day, at sunrise and sunset, and at two o'clock in the afternoon. This was done at Geneva during the three years 1796, 1797 and 1798. In the observations, the hour of mid-day was preferred to that of sunset. 2. Observations are made *twice* every day, at the two periods which are supposed to give the maximum and the minimum, namely, at sunrise, and at two o'clock in the afternoon. 3. Observations are made *once* a-day, at an hour which, in different seasons, has been found to represent the mean temperature of the day. It is thus that M. Raymond, by a judicious induction, has

[17] *Mem. de l'Acad.* 1719, p. 4.

[18] *Id.* 1735, p. 559.

[19] *Mem. de la Soc. Royale de Med.* 1777, p. 104.

proved, that the height of the barometer, at mid-day, gives, in our climates, the mean atmospherical pressure, corrected for the diurnal variation.

In calculating[20] a great number of observations made between the parallels of 46° and 48°, I have found, that a single observation at sunset, gives a mean temperature which differs only some tenths of a degree from that which is deduced from observations made at sunrise and at two o'clock. The devia- | 17 | tions of different months, do not exceed 1°.8, and they are very regularly positive or negative, according to the order of the seasons. M. Arago[21] has examined for seven years the observations of noon. They give for Paris 5°.4 more than the mean temperature of the whole year. Upon high mountains in the temperate zone, the difference is scarcely 1°.8[22]. By the application of coefficients, variable according to the latitude and the elevation, we may deduce the true mean temperatures from observations made at any particular period of the day, nearly in the same manner as we can ascertain the latitude of a place from altitudes of the sun, taken out of the meridian.

If we do not stop at two observations of the maximum and minimum, but add a third observation, we commit an error more or less serious, if we divide simply by three the sum of the observations, without attending to the duration of the partial temperatures, and to the place which the third observation occupies between the last terms of the series.[23] Experience proves, that the mean temperatures of the year, obtained by two or three observations, do not differ sensibly, if the intermediate observation is sufficiently distant (*four* or *five* hours) from the observation of the *maximum* and *minimum*. Whenever, therefore, we do not take into account the duration of the intermediate temperatures, we should prefer the two observations of the extreme temperature, which is the method most generally adopted. We shall content ourselves with pointing out the errors

---

[20] *De la Formule Barom.* p. 213.

[21] The mean of the observations at noon at Paris was 56°.84; at Clermont in Auvergne (elevation 1348 feet), 56°,30; at Strasburg (elevation 453 feet), 55°.22. *Bulletin de la Soc. Philom.* 1814, Oct. p. 95.—H.

[22] At the Hospice of St Gothard. *Ephem. Soc. Pal.* 1785, p. 47.

[23] *Example.*—On the 13th June, at 4^h in the morning, 46°.4; at 2^h in the afternoon, 55°.4; and at 11^h in the evening, 50°, (erroneously 46°.4, or 8° Centig. in the original). In calculating by the duration, we have

50°.9	the mean for	10^h	of interval,	= 509°.0
52.7		9		= 474.3
48.2		5		= 241.0

The true mean of which is 51°.0. The mean of the three observations, as commonly taken, is 50°.6. If we stop at the two extreme temperatures, we shall have for their half sum 50°.45.—H.

to which | 18 | it is liable. In our climates, the two extreme terms do not divide the series of twenty-fours into two equal parts. The maximum is an epoch nearly fixed: the rising of the sun retards or hastens it three hours. As we ought to take into account the duration of the partial temperature, in order to find the quantity of heat divided between the night and the day, we must couple the *maximum* of one day with the *minimum* of the day following, and not be content with taking half the sum of all the *maxima* and *minima* of a month. In the ordinary method, we determine only the mean temperature of the part of the day comprehended between the rising of the sun and two o'clock in the afternoon; and we take it for granted that the mean temperature is the same[24] from two o'clock to sunrise next day. This double error, of want of equi-distance and of the coupling of observations, does not in general produce errors of more than some tenths of degrees, sometimes in excess, and sometimes in defect, since the warm and cold days are mixed.[25]

All the calculated results will err in defect, if the 365 ordinates, through which the curve of the year passes, do not express an arithmetical progression, and if the partial irregularities do not sensibly compensate one another. It is only on this supposition that we can judge by the extreme terms of the series, of the sum of the terms, that is, of the partial temperatures. It is very obvious, that near the maximum, the increase ought to be more slow than in other points of the curve, and that this increase in the temperature of the air ought to depend on the sine of the sun's altitude, and on the emission of the radiant heat of the globe. | 19 |

It appeared to me very important to establish, by observations made at every hour, at different periods of the year, and under different latitudes, the degree of confidence that can be placed in those results which are called *Mean Temperatures*. I have selected from the registers of the Royal Observatory at Paris clear and calm days, which offered at least ten or twelve observations. Under the equator, I have spent whole days in determining the horary increments and decrements of temperature, in marking the thermometer both in the shade and in the sun, and also the

---

[24] *Example.*—At sunrise at 6^h, 50°; at 2 o'clock in the afternoon, 62°.6. At sunset, 51°.8; at 2^h, 66°.2; at sunrise, 50°. The true means will be for the first 24 hours 56°.9, and for the second 59°.0, for we shall have

For 8^h, 1/2 (50°.0 + 66°.2) × 8 = 450°.4 for 8^h = 472°.0

16 1/2 (51°.8 + 62°.6) × 16 = 915°.2 for 16^h = 929°.6

The method commonly employed gives 1/2 (50° + 62°.6) = 56°.3, and 1/2 (66°.2 + 51°8) = 58°.1. The errors being — 0°.6 and + 0,9, sometimes positive and sometimes negative.—H.

[25] The error disappears when days of equal temperature succeed one another. It amounts to 1°.8, if the mean temperatures of two successive days differ from 7° to 9°, which however very rarely happens.—H.

progress of evaporation and humidity; and in order to avoid calculation, I measured with a quadrant the altitude of the sun at each observation. I chose days and nights completely calm, and when the heavens were entirely free from clouds, because the mass of vesicular vapours interrupts the radiation from the earth. The result of these experiments has been very satisfactory, and proves, what had already been deduced from the coincidence between the temperature of the earth and the mean of daily observations, and from the regular progress of the mean temperatures of months in different years, that the effects of small disturbing causes may be compensated by a great number of observations.[26] I have obtained analogous results by taking, for several months, the mean of 9 o'clock in the morning, of sunrise and midnight. I have computed the temperatures by the distance of the *maximum* expressed in time, and on the supposition of an arithmetical progression. I have found that, under the Torrid Zone, the morning curve from sunrise to the *maximum*, differs very regularly from the even- | 20 | ing curve. In the morning the true mean temperature, such as we find by taking the duration into account, is a little greater

[26] On the 3d and 4th September 1811, lat. 48°.50′.

Sum of the temperatures during 24 hours.	True mean temperature, taking into account the duration.	Half sum of the two extreme temperatures.
625°.71 Fahr.	57°.92 Fahr.	58°.28 Fahr.
672.49	59.90	61.88
834.67	66.74	65.12
834.67	66.74	68.00
835.37	66.74	63.50
	—	—
	63.61 Mean.	63.35 Mean.

The three last days shew an equality of temperature, which is very surprising, and which does not appear but in the true means.—H.

than half the sum of the extremes.[27] In the evening the error is in a contrary direction, and the series of temperatures approaches more to a progression by quotients. The differences do not in general exceed half a degree, and calculation proves that the compensation is regular. It would be curious to examine the effect which the radiation of the earth has on these horary effects, as the changes of temperature at the surface do not follow the geometrical progression, in so far as they take place in a medium of uniform temperature.

In order to avoid the use of an arbitrary measure, astronomers express the diameters of the planets by taking that of the earth for unity. In like manner, I express the mean temperatures not in parts of the equatorial heat, but by the arithmetical ratios which subsist between this heat and that of the other parallels. This method frees us from the want of uniformity, which arises from the use of different thermometers. Instead of saying that in Europe, under the parallel of 45°, the mean temperature is 13°.4 Centigrade, or 56°.12 of Fahrenheit, we say that it is = 1.0°,487, and in lat. 55° = 1.0°,29. These arithmetical ratios inform us of what is most interesting in the theory of the distribution of heat, that in thermometers whose *zero* is the point of melting ice, the mean temperatures under the latitude of 45° and 55° are, in our regions, the *half* and the *third* nearly of the equatorial temperature, which I suppose to be 81°.5.

(To be Continued.)

[27] *Example.*—Latitude 10°.25′.

		Calculation of a true mean by the duration.	**Supposition of an arithmetical progression.**
Before the maximum,	11th September 1799,	70°.52 Fahr.	69°.44 Fahr.
	14th	69.26	68.00
	18th	71.24	70.34
After the maximum,	18th August,	68.72	69.80
	20th	70.16	71.24
	27th	68.72	69.26
Before the maximum,	17th August,	69.26	68.00
After the maximum,	17th August	65.48	66.02
		—	—
Total effect,	17th August,	67.37	67.01

H.

On Isothermal Lines, and the Distribution of Heat over the Globe. By Baron Alexander de Humboldt.

(Continued from Vol. III. p. 20.)

Having discussed the method of taking averages, and of reducing temperatures to general expressions, we shall now proceed to trace the course of the *Isothermal Lines* on the surface of the Globe, and at the level of the sea. From a slight attention to the difference of climates, it has been remarked, more than a century ago, that the temperatures are not the same under the same parallels; and that in advancing 70° to the east or the west, the heat of the atmosphere suffers a sensible diminution. In pursuance of our method, we shall reduce these phenomena to numerical results, and shew that places situated under the same latitudes do not differ, in America and Europe, by the same number of degrees of temperature, as has been vaguely stated. This assertion would make us suppose that the isothermal lines are parallel in the temperate zone.

		Lat.	Mean Temp.
I. Parallels of Georgia, of the State of Mississippi, of Lower Egypt, and Madeira.	*Natches*, –	31°28′	64° 8′
	Funchal, –	32 37	68 7
	Orotava, –	28 25	69 8
	Rome, –	41 53	60 4
	Algiers, –	36 48	70 0
		—	—
	Difference,	7 0	4 1
| 257 |			
II. Parallels of Virginia, Kentucky, Spain, and the South of Greece.	Williamsburg,	38° 8′	58° 0′
	Bourdeaux,	44 50	56 5
	Montpellier,	43 36	59 4
	Rome, –	41 53	60 4
	Algiers, –	36 48	70 0
		—	—
	Difference,	7 0	7 7
III. Parallels of Pennsylvania, Jersey, Connecticut, Latium, and Romelia.	*Philadelphia,*	39 56	54 9
	New-York,	40 40	53 8
	St Malo, –	48 39	54 5
	Nantes, –	47 13	54 7
	Naples, –	40 50	63 3
		—	—
	Difference,	7 0	9 5

(continued)

(continued)

		Lat.	Mean Temp.
	Ipswich, –	42 38	50 0
	Cambridge (Amer.)	42 25	50 4
	Vienna, –	48 13	50 5
	Manheim, –	49 29	51 3
	Toulon, –	43 7	62 1
	Rome, –	41 53	60 4
		—	—
	Difference,	6 30	11 0
IV. Parallels of Canada, Nova Scotia, France, and the South of Germany.	*Quebec,* –	46 47	41 9
	Upsal, –	59 51	41 9
	Padua, –	45 24	57 7
	Paris, –	48 50	51 4
		—	—
	Difference,	13 0	12 6
V. Parallels of Labrador, the South of Sweden, and Courland.	*Nains,* –	57 0	26 4
	Okak, –	57 20	29 8
	Umea, –	63 50	33 3
	Enontekies,	68 30	27 0
	Edinburgh,	55 58	47 8
	Stockholm,	59 20	42 3
		—	—
	Difference,[28]	11 0	17 1

This table[29] indicates the difference of climates, expressed by that of the mean temperature, and by the number of degrees in latitude which it is necessary to go northward in Europe, in | 258 | order to find the same quantity of annual heat as in America. As a place could not be found in the Old World, whose mean temperature was 48° the same as that of Williamsburg, I have supplied it with an interpolation between the latitudes of two points whose mean temperatures are 56°.5 and 59°.4. By an analogous method, and by employing only good observations, I have found that

1. The *isothermal line* of 32° (0° centig.) passes between Uleo and Enontekies in Lapland (lat. 66° to 68°; East long. from London 19° to 22°), and Table Bay in Labrador (lat. 54° 0′, west long. 58°.)

[28] The differences under the column of latitudes, is the difference of the latitude of a place in Europe and a place in America, which have the same mean temperature; and the differences under the column of mean temperatures, is the difference of the mean temperatures of a place in Europe and a place in America, which have the same latitude.—Ed.

[29] See *my Prolegomena de distributione geographica plantarum, secundum cœli temperiem et altitudinem montium.* p. 68.—H.

2. The *isothermal line* of 41° (5° centig.) passes by near Stockholm (lat. 60° east long. 18°) and the Bay of St George in Newfoundland (lat. 48°, and long. 59°.)

3. The *isothermal line* of 50° (10° centig.) passes by Belgium (lat. 51°, east long. 2°) and near Boston (lat. 42° 30′, west long. 70° 59.)

4. The *isothermal line* of 59° (15° centig.) passes between Rome and Florence (lat. 43° 0′, east long. 11° 40′) and near Raleigh in North Carolina (lat. 36° 0′, and west long. 76° 30′.)

The direction of these lines of equal heat, gives for the two systems of temperature, which we know by precise observations, viz. part of the middle and west of Europe, and that of the coast of America, the following differences:

Latitude.	**Mean Temp. of the west of the Old World.**	**Mean Temp. of the east of the New World.**	**Difference.**
30	70°.52	66°.92	3°.60
40	63.14	54.50	8.64
50	50.90	37.94	12.96
60	40.64	23.72	16.92

If we call the mean equatorial temperature 1, we shall have the half of this temperature in the Old World at 45°, and in the east of the New World, at 39° of lat.[30]

The mean temperatures decrease

Latitude.			**Temp.**		**Temp.**
From	0°—20°		3°.6		3°.6
	20 — 30		7.2		10.8
	30 — 40	In the	7.2	In the	12.6
	40 — 50	Old World,	12.6	New World,	16.2
	50 — 60		9.9		13.3
	0 — 60		40.5		56.5

In both continents, the zone in which the mean temperature decreases most rapidly is comprehended between the parallels of | 259 | 40° and 45°. Observation here presents a result entirely conformable to theory, for the variation of the square of the cosine, which expresses the law of the temperature, is a maximum towards 45° of latitude. This circumstance

[30] This observation relates to the Centigrade scale. If we count the temperatures from 32°, it applies also to Fahrenheit's scale.—Ed.

ought to have a favourable influence on the civilization and industry of the people who inhabit the regions under this mean parallel. It is the point where the regions of the vines touch those of the olives and the citrons. On no other part of the globe, in advancing from north to south, do we observe the temperatures increase more sensibly, and no where else do vegetable productions, and the various objects of agriculture, succeed one another with more rapidity. But a great difference in the productions of contiguous countries, gives activity to commerce, and augments the industry of the cultivators of the soil.

We have traced the direction of the isothermal lines from Europe to the Atlantic Provinces of the New World. We have seen them approach one another from parallelism towards the south, and converge towards the north, particularly between the thermometric curves of 41° and 50°: We shall now endeavour to pursue them to the west. North America presents two chains of mountains, extending from N. E. to S. W., and from N. W. to S. E. forming almost equal angles with the meridian, and nearly parallel to the coasts which are opposite to Europe and Asia, viz. the chain of the *Alleghanys* and the *Rocky Mountains*, which divide the waters of the Missouri and the Columbia. Between these chains stretch the vast basin of the Mississippi, the plains of Lousiana, and of the Tenessee, and the states of Ohio, the centre of a new civilization. It is generally believed in America that the climate is more mild to the west of the Alleghany Mountains, than under the same parallels in the Atlantic States.[31] Mr Jefferson, has estimated the difference at 3° of latitude; and the *Gleditsia monosperma*, the *Catalpa*, and the *Aristolochia Sypho*, and other vegetable productions, are found so many degrees farther to the north, in the basin of the Ohio, than on the coast of the Atlantic.[32] | 260 | M. Volney has endevoured to explain these phenomena by the frequency of the south-west winds, which drive back the warm air of the Gulf of Mexico towards these regions. A series of good observations, made for *seven* years by Colonel Mansfield at Cincinnati, on the banks of the Ohio, and recently published by Mr Drake, in an excellent treatise on American meteorology,[33] has removed the doubts which obscured this point. The thermometrical *means* prove that the isothermal lines do not rise again in the regions of the west. The quantity of heat which each point of the

[31] This is true also of the Columbian Valley. See Warden's *Account of the United States*, vol. iii. p. 169.—Ed.

[32] See my *Essai sur la Geographie des Plantes*, p. 154.

[33] *Natural and Statistical View or Picture of Cincinnati and the Miami Country*, 1 vol. 8vo. Cincinnati.—H. See Warden's *Account of the United States*, vol. ii. 236 for an abstract of Mr Drake's results.—Ed.

globe receives under the same parallels, is nearly equal on the east and the west of the Alleghany range, the winters being only a little milder to the west, and the summers a little warmer.[34] The migrations of vegetables towards the north are favoured in the basin of the Mississippi, by the form and the direction of the valley which opens from the north to the south. In the Atlantic Provinces, on the contrary, the valleys are transverse, and oppose great obstacles to the passage of plants from one valley to another.

If the isothermal lines remain parallel, or nearly so, to the equator, from the Atlantic shores of the New World to the east of the Mississippi and the Missouri, it cannot be doubted that they rise again beyond the *Rocky Mountains*, on the opposite coast of Asia, between 35° and 55° of latitude. To the con- | 261 | siderations which I pointed out in my work on Mexico,[35] are to be added the observations of Captain Lewis, and some other Anglo-American travellers, who have passed the winter on the banks of the Columbia. In New California, they cultivate with success the olive, along the canal of Santa Barbara, and the vine from Monterey to the north of the parallel of 37°, which is that of Chesapeake Bay. At Nootka, in the Island of Quadra and Vancouver, and almost in the latitude

[34] The following comparison of the mean temperatures has been deduced with great care.

Cincinnati. Lat. 36° 6′, west long. 84° 24′.	
Winter,	32°.9 Fahr.
Spring,	54.1
Summer,	72.9
Autumn,	54.9
	—
Mean,	53.7

Philadelphia. Lat. 39° 56′, west long. 75° 16′.	
Winter,	32°.2 Fahr.
Spring,	51.4
Summer,	73.9
Autumn,	56.5
	—
Mean,	53.5

I have taken for Philadelphia the means between the observations of Coxe and Rush. I have also referred for correction to the observations made by M. Legaux at Spring-Mill upon the Schuylkill, to the north of Philadelphia. As Cincinnati is 512 feet above the level of the sea, its mean temperature is 1°.4 too low.– H.

[35] *Essai Politique sur la Nouvelle Espagne*, tom. ii. p. 440, 478, 509.

of Labrador, the smallest rivers do not freeze before the month of January. Captain Lewis saw the first frosts near the embouchure of the Colombia, only on the 7th of January, and the rest of the winter was rainy. Through 122° 40′ of west long. the isothermal line of 50° Fahr. appears to pass almost as in the Atlantic part of the Old World, at 50° of lat. The western coasts of the two worlds resemble one another to a certain point.[36] But these returns of the isothermal lines do not extend beyond 60°. The curve of 32° Fahr. is already found to the south of the Slave Lake, and it comes still farther south in approaching Lakes Superior and Ontario.

In advancing from Europe towards the east, the isothermal lines again descend,[37] the number of fixed points being few. We can only employ those which are made in places whose known elevation allows us to reduce the mean temperatures to the level of the sea. The few good materials which we possess, have enabled us to trace the curves of 32° and 55°.4. We know even the nodes of the latter curve round the whole globe. It | 262 | passes to the N. of Bourdeaux, (lat. 45° 46′, W. long. 0° 37′,) near Pekin, (lat. 39° 54′, E. long. 116° 27′.,) and Cape Foulweather to the S. of the embouchure of the Colombia, (lat. 44° 40′, W. long. 104°.) Its nodes are distant at least 162° of longitude. We have here pointed out only the empirical laws, under which are ranged the general phenomena, and the variations of the temperature which embrace at once a vast extent of the globe. There are *partial inflexions* of the isothermal lines, which form, so to speak, particular systems modified by small local

[36] On account of the influence of west and south-west winds. See Dalton's *Meteor. Observ.* p. 125.

[37] In comparing places from the west to the east, and nearly under the same parallel, we find,

West.	**Lat.**	**Mean Temp.**
St Malo,	48° 39′	54°.5
Amsterdam,	52 21	53.4
Naples,	40 50	63.3
Copenhagen,	55 41	45.7
Upsal,	59 52	41.9
East.	**Lat.**	**Mean Temp.**
Vienna,	48° 13′	50°.5
Warsaw,	52 14	48.6
Pekin,	39 54	54.9
Moscow,	55 46	40.1
Petersburgh,	59 56	38.8

The elevation of Pekin is inconsiderable. That of Moscow is 984 feet. The absolute temperature of Madrid, to the west of Naples, is 59°; but the city is elevated 1978 feet above the level of the sea.—H.

causes; such as the strange inflexion of the thermometric curves on the shores of the Mediterranean, between Marseilles, Genoa, Lucca and Rome,[38] and those which determine the difference between the climate of the western coast and the interior of France. These last depend much less on the quantity of heat received by a part of the globe during the whole year, than upon the unequal distribution of heat between winter and summer. It will one day be useful to have upon particular charts the partial inflections of the isothermal lines, which are analogous to the lines of soundings or of equal declivity. The employment of graphical representations will throw much light upon phenomena, which are deeply interesting to agriculturists. If, in place of geographical charts, we possessed only tables containing the co-ordinates of latitude, longitude, and altitude, a great number of curious facts relative to the configuration and the superficial inequalities of continents would have remained forever unknown.

We have already found, that towards the north, the isothermal lines are neither parallel to the equator nor to one another; and it is on account of the want of parallelism, that we have, in order to simplify such complicated phenomena, traced round the whole globe the curves of equal heat. The position of the line of 32° acts like the magnetic equator, whose inflexions in the South Sea modify the inclinations at great distances. We may even believe that, in the distribution of climates, the line of 32° determines the position of the *curve of greatest heat*, | 263 | which is as it were the isothermal equator, and that in America and Asia through 78° of west, and 102° of east longitude, the torrid zone commences more to the south of the tropic of Cancer, or that it there presents temperatures of less intensity. An attentive examination of the phenomena proves that this is not the case. Whenever we approach the torrid zone below the parallel of 30°, the isothermal lines become more and more parallel to one another, and to the earth's equator. The great colds of Canada and Siberia do not extend their action to the equatorial plains. If we have long regarded the

38

	Lat.	Mean Temp.
Bologna,	44° 29′	56°.3
Genoa,	44 25	60.6
Marseilles,	43° 17′	58°.8
Rome,	41 53	60.4

H.

Old World as warmer between the tropics than the new world, it is, *first*, Because till 1760, travellers used thermometers of spirit of wine, coloured, and affected by light; *2d*, Because they observed it either under the reflection of a wall, or too near the ground, and when the atmosphere was filled with sand; and, *3d*, Because in place of calculating the true mean, they used only the thermometric *maximum* and *minimum*. Good observations give,

Old World.	**Lat.**	**Mean Temp.**
Senegambia,	15° 0	79°.07
Madras,	13 5	80 .42
Batavia,	6 12	80 .42
Manilla,	14 36	78 .08

New World.	**Lat.**	**Mean Temp.**
Cumana,	10° 27′	81°.86
Antilles,	17 0	81.05
Vera Cruz,	19 11	78.08
Havannah,	23 10	78.08

The mean temperature of the equator cannot be fixed beyond 81 1/2°. Kirwan values it at 84°, but only two places of the earth were known, viz. Chandernagor and Pondicherry, to which old travellers attributed annual temperatures above 81° 1/2. At Chandernagor, in latitude 21°.6, the mean temperature, according to Cotte, is 91°.9, but the Jesuite Boudier marked only the days when the thermometer was above 98°.6, and below 57°.2. And at Pondicherry, in latitude 11° 55′, the mean temperature, according to Cotte, is 85°.3, and according to Kirwan, 88°; but M. de Cossigny observed with a spirit-of-wine thermometer.

The distribution of heat over different parts of the year differs, not only according to the decrease of the mean annual temperatures, but also in the same isothermal line. It is this unequal division of the heat which characterises the two systems of climate of Europe and Atlantic America. Under the torrid zone, a small number of months are warmer in the Old World than in the New. At Madras, for example, according to Dr | 264 | Roxburgh, the mean temperature of June is 89°.4; at Abusheer 93°.2, but at Cumana I have found it only 84°.6.

With respect to the temperate zone, it has been long known, that from the parallel of the Canary Isles to the Polar Circle, the severity of the winter augments in a progression much more rapid than the summers diminish in heat. It is also known, that the climate of the islands and the coasts differs from that of the interior of continents, the former being characterised by mild winters and less temperate summers. But it is the heat of summer particularly which affects the formation of the amylaceous and saccharine matter in fruits, and the choice of the plants that ought to be cultivated. As the principal object of this memoir is to fix, after good observations, the numerical relations between the unequal quantities of heat distributed over the globe, we shall now compare the mean temperatures of three months of winter and summer under different latitudes, and shew how the inflections of the isothermal lines modify these relations. In following the curves of equal heat from west to cast, from the Basin of the Mississippi to the eastern coasts of Asia, through an extent of 4000 leagues, we are struck with the great regularity which appears in the variations of the winter temperature.

I. DIFFERENCES OF THE SEASONS FROM THE EQUATOR TO THE POLAR CIRCLE.

	Cisatlantic Region. Long. 1° W. and 17° E.			Transatlantic Region. Long. 58°—72° W.		
	Mean Temperature.			Mean Temperature.		
Isothermal Lines of	Winter.	Summer.	Diff.	Winter.	Summer.	Diff.
68°	59°.0	80°.6	21°.6	53°.6	80°.6	27°.0
59	44.6	73.4	28.8	39.2	78.8	39.6
50	35.6	68.0	32.4	30.2	71.6	41.4
41	24.8	60.8	36.0	14.0	66.2	52.2
32	14.0	53.6	39.6	1.4	55.4	54.0

This table shews the increase of the difference between the winters and summers from 28° and 30° to the parallels of 55° and 65°. The increase is more rapid in the Transatlantic Zone, where the isothermal lines of 32° and 50° approach one another very much; but it is remarkable, that in the two zones which form the two systems of different climates, the division of | 265 | the annual temperature between winter and summer is made in such a manner, that, upon the isothermal line of 32°, the difference of the two seasons is almost double of that which is observed on the isothermal line of 68°.

Cisatlantic Region.
Long. 31° E. and 22° W.

Places.	Latitude.	Mean Temperature.		
		Whole Year.	Winter.	Summer.
(Pondicherry)	11°.55	85°.3	77°.0	90°.5
Cairo,	30 02	72.7	57.7	84.7
Funchal,	32 37	68.7	63.9	72.5
Rome.	41 53	60.4	45.9	55.2
Bourdeaux,	44 50	56.5	42.1	70.7
Paris,	48 50	51.4	38.3	66.2
Copenhagen,	55 41	45.7	30.7	64.6
Stockholm,	59 20	42.3	25.5	61.9
Drontheim,	63 24	39.9	24.7	61.3
Umeo,	63 50	33.3	12.9	54.9

Transatlantic Region.
Long. 69° E. and 99° W

Places.	Latitude.	Mean Temperature.		
		Whole Year.	Winter.	Summer.
Cumana,	10°.27	81°.9	81°.7	83°.7
Havannah,	23.10	78.1	71.2	83.3
Natchez,	31.28	64.8	48.6	79.2
Cincinnati,	39.06	53.6	32.9	72.9
Philadelphia,	39 56	54.9	32.2	73.9
New York,	40.40	53.8	29.8	79.2
Cambridge,	42.25	50.4	34.0	70.5
Quebec,	46.47	41.9	14.2	68.0
Nain,	57.10	26.4	0.6	48.4
Fort Churchhill,	59.02	25.3	6.8	52.2

If, instead of the mean temperatures of the seasons, we consider, I do not say the days of the *maxima* and *minima* of the year, which are the ordinates of the concave and convex summits of the entire curve, but the mean temperatures of the warmest and the coldest month, the increase of the differences becomes still more perceptible. We request the reader to compare in the following Table only the places which belong to regions bounded by the same meridians, and consequently to the same system of climate; as for example, to the region of Eastern America to that of

Western Europe, and that of Eastern Asia. We must also attend to the changes of temperature produced by the monsoons in a part of the equinoctial regions, and | 266 | distinguish under the temperate zone between the climate of the interior, or the continental climate, and that of islands and coasts.

		Mean Temperature.			
Places.	**Lat.**	**Coldest Month.**	**Warmest Month.**	**Difference.**	**Observations.**
Cumana,	10° 27	80°.1	84°.4	4°.3	Uninterrupted trade winds.
Pondichcrry,	11 55	76.1	91.4	15.3	Monsoons. Radiation of the sands.
Manilla,	14 36	68.0	86.9	18.9	Monsoons.
Vera Cruz,	19 11	70.0	81.7	11.7	North winds in winter.
Cape Français,	19 46	77.0	86.0	9.0	Uninterrupted trade winds.
Havannah,	23 10	70.0	83.8	13.8	North winds in winter.
Funchal,	32 37	64.0	75.6	11.6	Insular climate.
Natchez,	31 28	46.9	78.8	31.9	Transatlantic region. Interior.
Cincinnati,	39 6	29.6	74.4	44.8	Same system of climate.
Pekin,	39 54	24.8	84.2	59.4	Region of eastern Asia.
Philadelphia,	39 56	29.8	77.0	47.2	Transatlant. region. Eastern coasts.
New York,	40 40	25.3	80.8	55.5	*Idem.*
Rome,	41 53	42.1	77.0	34.9	Cisatlantic region.
Milan,	45 28	33.8	55.2	21.4	Interior land.
Buda,	47 29	27.7	71.6	43.9	*Idem.*
Paris,	48 50	35.1	69.8	34.7	Nearer the western coast.
Quebec,	46 47	14.0	73.4	59.4	Transatlant. region. Eastern coasts.
Dublin,	53 21	37.6	60.3	22.7	Region of the west of Europe. Insular climate.
Edinburgh,	55 58	38.3	59.4	21.1	*Idem.*
Warsaw,	52 14	27.1	70.3	43.2	Interior land.
Petersburg,	59 56	8.6	65.7	57.1	East of Europe.
North Cape,	71 0	22.1	46.6	24.5	Climate of coasts and islands.

We may conclude in general, that for any given place in the curves which express the annual temperatures, the ordinates of the concave and convex summits differ the more from one another, as the temperatures diminish. In the New World, under 40° of latitude, we find a greater difference between the warmest and coldest months of the year than in the Old World, at Copenhagen and Stockholm under 56°—59° of latitude. At Philadelphia the thermometer descends to 50° or 59° below the freezing point, while under the same parallel in Europe it descends scarcely 30°.6 below it.

I have endeavoured to shew, in another work, how this circumstance which characterises the regions which Buffon indicates by the name of Excessive Climates, influences the physical constitution of the inhabitants. In the United States, the Europeans, and indeed all the natives, are, with great difficulty, inured to the climate. After winters that have been very rigorous, not from the general temperature, but from an extreme depression of the thermometer, the irritability of the nervous sys- | 267 | tem is prodigiously increased by the excessive heat of summer; and it is undoubtedly to this cause that we must, in a great measure, ascribe the difference in the propagation of the yellow fever, and the different forms of the marsh fever, under the equator, and in the temperate zone of the New World.[39]

On high mountains in islands of little extent, and along the shores, the lines of annual temperature take nearly the same form as in warm climates, having only a less degree of curvature. The difference between the seasons, too, becomes smaller. At the North Cape, in 71° of latitude, and in the isothermal line of 32°, it is almost 11° greater than at Paris, in 49° of latitude, and in the isothermal line of 50°. The sea-breezes and the fogs which render the winters so temperate, diminish at the same time the heats of summer.[40] The characteristic of any climate is not the difference between the winters, expressed in degrees of the thermometer;—it is this difference, compared with the absolute quantities indicated by the mean temperature of the seasons.

II. DIFFERENCE BETWEEN THE WINTERS AND SUMMER, IN FOLLOWING THE SAME ISOTHERMAL LINE FROM WEST TO EAST.

The differences between the seasons of the year are less great near the convex summits of the isothermal curves, where these curves rise again towards the North Pole, than near the concave summits. The same causes, which affect the inflexion or the greatest curvature of the isothermal lines, tend also to equalise the temperatures of the seasons.

[39] *Political Essay on the Kingdom of New Spain*, tom. iv. p. 528.

[40] Leopold von Buch's *Travels in Lapland*, tom. ii.

The whole of Europe, compared with the eastern parts of America and Asia, has an insular climate, and, upon the same isothermal line, the summers become warmer, and the winters colder, in proportion as we advance from the meridian of Mont Blanc towards the east or the west. Europe may be considered as the western prolongation of the old continent; and the western parts of all continents are not only warmer at equal latitudes than the eastern parts, but even in the zones of equal annual temperature, the winters are more rigorous, and the summers hotter on the eastern coasts than upon the western coasts of the two continents. The northern part of China, like | 268 | the Atlantic region of the United States, exhibits excessive climates, and seasons strongly contrasted, while the coasts of New California, and the embouchure of the Colombia, have winters and summers almost equally temperate. The meteorological constitution of these countries in the N. W. resembles that of Europe as far as 50° or 52° of latitude; and without wishing to ascribe the great revolutions of our species solely to the influence of climate, we may affirm that the difference between the eastern and western shores of continents, has favoured the ancient civilisation of the Americans of the west,—facilitated their migrations towards the south, and multiplied those relations with eastern Asia, which appear in their monuments, their religion, traditions, and the division of the year. In comparing the two systems of climates, the concave and convex summits of the same isothermal lines, we find at New York the summer of Rome and the winter of Copenhagen;—at Quebec the summer of Paris and the winter of Petersburg. In China, at Pekin for example, where the mean temperature of the year is that of the coasts of Britanny, the scorching heats of summer are greater than at Cairo, and the winters as rigorous as at Upsal.

The mean temperature of the year being equal to the fourth part of the winter, spring, summer and autumnal temperatures, we shall have upon the same isothermal line of 53°6′ (12° cent.)

At the *concave* summit in *America*, 74·40 West long.	$53{\cdot}6 = \frac{32° + 52{\cdot}3 + 75{\cdot}6 + 54{\cdot}5}{4}$
At the *convex* summit in *Europe*, 2·20 West long.	$53{\cdot}6 = \frac{40{\cdot}1 + 51{\cdot}8 + 68{\cdot}4 + 54{\cdot}1}{4}$

This analogy between the eastern coasts of Asia and America, sufficiently proves that the inequalities of the seasons, of which we have endeavoured to fix the numerical relations, depend on the prolongation and enlargement of continents towards the pole; of the size of seas in relation to their coasts, and on the frequency of the N. W. winds, which are the *Vents de Remous* of the temperate zone, and not on the proximity of some plateau or elevation of the adjacent lands. The great plateaus of Asia do not stretch beyond 52° of latitude; and in the interior of the New Continent, all the immense basin bounded by the Alleghany Range, and the rocky mountains, and covered with secondary formations, is not more than from 656 to 920 feet above | 269 | the level of the ocean, according to the levels taken in Kentucky, on the banks of the Monongahela, at Lake Erie.[41]

The following table indicates for all the habitable temperate zone the division of the same quantity of annual heat between the two seasons of winter and summer. The numbers which it contains, are either the result of direct observations, or of interpolations between a great number of observations made in neighbouring places, and situated under the same meridian. We have followed each isothermal curve from west to east, giving the preference to places situated near the summits of the curve, as presenting at the same time the greatest differences in the distribution of the annual heat. The longitudes are reckoned from the Observatory of Greenwich.

Isothermal Lines from 32°to 68°.

				Mean Temperature.	
	Long.	**Lat.**		**Winter.**	**Summer.**
Isoth. Line of 68°.	82·10 W.	29·30	*Florida,*	53·6	80·6
	16·56 W.	32·37	*Madeira,*	63·5	72·0
	3· 0 E.	36·48	*North Africa,*	59·0	80·6
Isoth. Line of 63·5.	89·40 W.	32·30	*Mississippi,*	46·4	77·0
	14·11 E.	40·50	*Italy,*	50·0	77·0
Isoth. Line of 59°.	84·10 W.	35·30	*Basin of the Ohio,*	39·2	77·9
	3°—4°E.	43·30	*Middle of France,*	44·6	75·2
Isoth. Line of 54·5.	84·40 W.	38·30	*America,* West of Alleghany,	34·7	75·2
	74·10 W.	40· 0	*America,* East of ditto,	32·5	77·0
	1·32 W.	47·10	*West of France,*	39·0	68·0
	9·20 E.	45·30	*Lombardy,*	34·7	73·4
	116·20 E.	40· 0	*East of Asia,*	26·6	82·4

(*continued*)

[41] Drake's *Nat. and Statist. View of Cincinnati*, p. 63.

(continued)

Isothermal Lines from 32°to 68°.

				Mean Temperature.	
	Long.	**Lat.**		**Winter.**	**Summer.**
Isoth. Line of 50°.	84·20 W.	41·20	*America, West of Al-leghany,*	31·1	71·6
	71·10 W.	42·30	*America, East of ditto,*	30·2	73·4
	6·40 W.	52·30	*Ireland,*	39·2	59·5
	0·40 W.	53·30	*England,*	37·4	62·6
	2·20 E.	51· 0	*Belgium,*	36·5	63·5
	19· 0 E.	47·30	*Hungary,*	31·1	69·8
	116·20 E.	40· 0	*Eastern Asia,*	23·0	78·8
Isoth. Line of 45 5.	71· 0 W.	44·42	*America,* East of Allegh-any,	23·9	71·6
	2·10 W.	57· 0	*Scotland,*	36·1	56·5
	12·35 E.	55·40	*Denmark,*	30·3	62·6
	21·20 E.	53· 5	*Poland,*	28·0	66·2
Isoth. Line of 41°.	71·10 W.	47· 0	*Canada,*	14·0	68·0
	9·20 E.	62·45	*West of Norway,*	24·8	62·6
	17·20 E.	60·30	*Sweden,*	24·8	60·8
	24·20 E.	60· 0	*Finland,*	23·0	63·5
	36·20 E.	58·30	*Central Russia,*	22·1	68·0
Isoth. Line of 36·5.	71·40 W.	50· 0	*Canada,*	6·8	60·8
	18· 5 E.	62·30	*West coast of Gulf of Bothnia,*	17·6	57·2
	22·20 E.	62·50	*East coast of ditto,*	16·5	59·0
Isoth. Line of 32°.	57·40 E.	53· 0	*Labrador,*	3·2	51·8
	19·50 E.	65· 0	*Sweden,*	11·3	53·6
	25·20 E.	71· 0	*North Extremity of Nor-way,*	23·9	43·7

| 270 |
When we consider that the annual temperature of a place is nothing more than the numerical expression of the mean of the ordinates, we may imagine an infinity of entirely dissimilar curves, in which the twelve ordinates of the months have exactly the same mean. This consideration should not lead us to believe, that a place which has the winter of the south of France, that is, where the mean temperature of winter is 44°.6, may, by the compensation of a summer and an autumn, much less warm, have the mean temperature of Paris. It is true, that the constant ratio which is observed in the same parallel, between the solstitial heights of

the sun and the semidiurnal arcs, is differently modified by the position of a place in the centre of a continent or upon the coast, by the frequency of certain winds, and by the constitution of an atmosphere more or less favourable to the transmission of light, and of the radiating caloric of the earth. But these variations, which travellers have often exaggerated, have a maximum which nature never oversteps. It is impossible to examine the preceding table without observing, that the division of the annual heat between summer and winter, follows on each isothermal line a determinate type; that the deviations of that type are contained between certain limits, and that they obey the same law in the zones which pass by the concave or convex summits of the isothermal lines, for example, by 58°—68° of West Long., and by 5°—7°, and 116° of East Longitude.

The following table shews the oscillations, or the *maxima* and *minima*, observed in the division of the heat between the seasons. I have added the means of the winters and summers found at different degrees of longitude, and under the same isothermal line.

		Oscillations observed in the Means.		Means calculated.	
Isoth. Lines.	**Degrees of Long. examined.**	**Winters.**	**Summers.**	**Winters.**	**Summers.**
32°	83	3·2 to 24·8	51·8 to 53·6	14·0	52·7
41	107	14·0 24·8	62·6 68·0	19·4	65·3
50	200	23·0 37·4	62·6 78·8	30·2	70·7
59	87	39·0 44·6	75·2 77·0	41·9	75·2
68	84	53·6 59·0	71·6 80·6	56·3	77·9

| 271 |

The deviations round the mean, that is, the inequality of the winters on the same isothermal line, increase in proportion as the annual heat diminishes, from Algiers to Holland, and from Florida to Pennsylvania. The winters of the curve of 68° are not found upon that of 51°, and the winters of 51° are not met with on the curve of 42°. In considering separately what may be called the same *system of climate*, for example, the European Region, the Transatlantic Region, or that of Eastern Asia, the limits of the variations become still more narrow. Wherever in Europe, in 40° of longitude the mean temperature rises

To 59·0	The winters are from	44·6 to 46·4	and the summers from	73·0 to 75·2
54·5	36·5 41·0	68·0 73·0		
50·0	31·1 37·4	62·6 69·8		
45·5	28·4 36·1	57·2 68·0		
41·0	20·3 26·8	55·4 66·2		

In tracing *five* isothermal lines between the parallels of Rome and Petersburg, the coldest winter presented by one of these lines is not found again on the preceding line. In this part of the globe, those places whose annual temperature is 54° 5, have not a winter below 32°, which is already felt upon the isothermal line of 50°. If, in place of stopping at the most rigorous winter which each curve presents, we trace the *lines of equal winter temperature*, (or the *Isocheimal lines*,) these lines, instead of coinciding with the lines of equal annual heat, oscillate round them. As the *Isocheimal lines* unite points placed on different isothermal lines, we may examine to what distance their summits extend. In considering always the same system of climates, for example, the European region, we shall find that the lines of equal winter cut isothermal lines, which are 9° distant. In Belgium[42] (in latitude 52°, and in isothermal latitude 51°8,)

[42] Throughout all Holland, 90 days of winter have a mean temperature of from 36°.7 to 38°.7. At Milan, at Padua, and at Verona, the same season is only from 34°.7 to 36°.7. The observations made in Belgium and Holland, offer also a very remarkable example of an equal quantity of heat distributed in the space of a year over a vast extent of territory. The mean temperatures scarcely vary from Paris to Franecker, over 3 1/2 degrees of latitude, which, in the interior of a continent, should produce a difference of 3 1/4 degrees of annual temperature. The canal of the Channel opens towards the north. The west winds blow, therefore, over a great part of the ocean, and during a long rainy winter, with the sky almost always clouded, the surface of the earth is less cooled by radiation than farther to the east, in the interior of the country, where the atmosphere is pure and dry.

Mean Temperature.

Lat.	**Year.**	**Winter.**	**Summer**	
Franecker,	52° 36′	51°.8	36°.7	67°.3
Amsterdam,	52 22	53.4	36.9	65.8
Hague,	52 3	51.8	38.3	65.5
Rotterdam,	51 54	51.1	36.9	64.9
Middelburg,	51 30	49.6	36.1	64.0
Dunkirk,	51 2	50.5	38.5	64.0
Brussels,	50 50	52.0	36.7	66.2
Arras,	50 17	49.4	35.8	63.3
Cambray,	50 10	52.0	39.0	66.6

and even in Scotland, (in latitude 57°, and isothermal latitude 45° 5,) the winters are more mild than at Milan, (in latitude | 272 | 45° 28′, and isothermal latitude 55° 8′,) and in a great part of Lombardy. Farther to the north, in the Scandinavian Peninsula, we meet with three very different systems of climate, viz. 1. The region of the west coasts of Norway to the west of the mountains. 2. The region of the eastern coasts of Sweden, to the east of the mountains. And, 3. The region of the west coasts of Finland, along the Gulf of Bothnia. Baron Von Buch has made us acquainted with the atmospherical constitution of these three different regions, in which the slowest increase of the winter cold is felt from Drontheim to the North Cape, on the west and north-west coasts. At the Isle of Mageroe, (in north lat. 82°,) at the northern extremity of Europe, under the parallel of 71°, the winters are still 7°.2 milder than at St Petersburg, (in north lat. 38°.8,) but the mean heat of the summers never reaches that of the winters of Montpellier, (in north lat. 59°.4). At the Faroe Isles, under 62° of north lat. the lakes are very seldom covered with ice, and to so temperate a winter succeeds a summer, during which snow often falls upon the plains. Nowhere without the tropics is the division of the annual heat among the seasons more equal. In the temperate zone, under parallels nearer to our own, Ireland presents an ex- | 273 | ample still more striking of the union of very mild winters, with cold and moist summers. Notwithstanding a difference of 4° of latitude, the winters there are as mild as in Britain, while the mean temperature of the summers is three degrees less. This is the true maritime climate. The month of August, which on the same isothermal line, in the east of Europe, (in Hungary)[43] has the temperature of 71°.6, reaches only 60°.8 at Dublin. The month of January, whose mean temperature at Milan, and in a great part of Lombardy, is only 35°.6, rises in Ireland to 5°.4, and

The mean duration of the observations at each place is from eight to nine years, and 52,000 partial observations have been employed to obtain nine mean temperatures. A similar harmony in the results is also found in Lombardy.

	Mean Temp.
Milan,	55°.8
Padua,	56.3
Verona,	55.8
Bologna,	56.3
Venice,	56.5

H.

[43] Wahlenberg, *Flora Carpath.* p. 90.

7°.2. On the coasts of Glenarm, also (in north lat. 54° 56′,) under the parallel of Konigsberg, the myrtle vegetates with the same strength as in Portugal.[44] It scarcely freezes there in winter, but the heat of summer is not capable of ripening the vine.

These examples are sufficient to prove, that the isocheimal lines deviate much more than the isothermal lines from the terrestrial parallels. In the system of European climates, the latitudes of two places that have the same annual temperature cannot differ more than from 4° to 5°, while two places whose mean winter temperature is the same, may differ more than 9° or 10 in latitude. The farther we advance to the east, the more rapidly do these differences increase.

The lines of equal summer, or *isotheral curves*, follow a direction exactly contrary to thc *isocheimal lines*. We find the same summer temperature at Moscow, in the centre of Russia, and towards the mouth of the Loire, notwithstanding a difference of 11° of latitude. Such is the effect of the radiation of the earth on a vast continent deprived of mountains. It is sufficiently remarkable, that the inflexions of the isothermal lines, and the division of lands and seas are such upon the globe, that every where in North America, in Europe, and in Eastern Asia, the mean temperature of the summers does not denote more than 36° in the parallels of from 45° to 47°. The same causes which in Canada, and in the north of China, sink the curves of equal annual heat, where the isothermal lines (those of | 274 | 51°.8, and 53°.6,) corresponding to the parallels of 45° and 47°, tend to raise the lines of equal summer or the *isotheral curves*.

However great be the influence of the unequal division of the heat between the seasons, on the physical condition of nations, on the developement of agricultural industry, and the selection of plants for culture, I would not recommend the tracing upon the same chart the isothermal lines, and the winter and summer curves. This combination would not be more fortunate than the lines of declination, inclination, and equal intensity of the magnetic forces, which, however, all depend upon one another. Instead of multiplying the intersection of the curves, it will be sufficient to add to the isothermal lines, near their summits, the indication of the mean temperatures of summer and winter. In this way, by following the line of 50°, we shall find marked in America, to the west of Boston, (30°.2/73°.4), in England (37°.4/62°.6), in Hungary), (31°.1/69°.8) and in China (23°.0/78°.8).

[44] *Irish Transactions*, tom. viii. p. 116, 203, 269.

(To be continued.)

On Isothermal Lines, and the Distribution of Heat Over the Globe. By Baron Alexander de Humboldt.

(Continued from Vol. III. p. 274.)

After what has already been stated respecting the limits between which the annual heat divides itself on the same isothermal curve, it will be seen how far we are authorised to say, that the *Coffee-tree*, the *Olive*, and the *Vine*, in order to be productive, require mean temperatures of 64°.4; 60°.8, and 53°.6 Fahr. These expressions are true only of the same system of climate, for example, of the part of the Old World which stretches to the west of the meridian of Mont Blanc; because in a zone of small extent in longitude, while we fix the annual temperatures, we determine also the nature of the summers and the winters. It is known likewise, that the olive, the vine, the varieties of grain, and the fruittrees, require entirely different constitutions of the atmosphere. Among our cultivated plants, some, slightly sensible of the rigours of winter, require very warm but not long summers; others require summers rather long than warm; while others, again, indifferent to the temperature of summer, cannot resist the great colds of winter. Hence, it follows, that, in reference to the culture of useful vegetables, we must discuss three things for each climate,—the mean temperature of the entire summer,—that of the warmest month,—and that of the coldest month. I have published the numerical results of this discussion in my *Prolegomena de Distributione Geographica Plantarum, secundum Cœli Temperiem*; and I shall confine myself at present to the limits of culture of the olive and the vine. The olive is cultivated in our continent between the parallels of 36° and 44°, wherever the annual temperature is from 62°.6 to 58°.1, where the mean temperature of the coldest month is not below from 41°.0 to 42°.8, and that of the whole summer from 71°.6 to 73°.4.[45] In the New World, the division of heat between the seasons is such, that on the isothermal line of 58°.1, the coldest month is | 24 | 35°.6, and that the thermometer

[45] In cases like the present, we have not used the round numbers of Fahrenheit, as is done in the original with the Centigrade scale, but have given the real value of the degrees used by the author, that his exact numbers may always be ascertained.—Ed.

sometimes sinks there even during several days from 14° to 10°.4. The region of potable wines extends in Europe between the isothermal lines of 62°.6 and 50°, which correspond to the latitudes of 36° and 48°. The cultivation of the vine extends, though with less advantage, even to countries whose annual temperature descends to 48°.2 and to 47°.48; that of winter to 33°.8, and that of summer to 66°.2 and 68°. These meteorological conditions are fulfilled in Europe as far as the parallel of 50°, and a little beyond it. In America, they do not exist farther north than 40°. They have begun, indeed, some years ago to make a very good red wine to the west of Washington, beyond the first chain of mountains, in the valleys which do not extend beyond 38° 54′ of Lat. On the Continent of Western Europe, the winters, whose mean temperature is 32°, do not commence till on the isothermal lines of 48°.2 and 50°, in from 51° to 52° of latitude; while in America, we find them already on the isothermal lines of from 51°.8 to 53°.6, under from 40° to 41° of latitude.

If, instead of considering the natural inflexions of the isothermal lines, that is to say, those that propagate themselves progressively at great intervals of longitude, we direct our attention to their partial inflexions, or to *particular systems of climates* occupying a small extent of country, we shall still find the same variations in the division of the annual heat between the different seasons. These partial inflexions are most remarkable,

1st, In the Crimea, where the climate of Odessa is contrasted with that of the S.W. shores of the Chersonesus, sheltered by mountains, and fit for the cultivation of the olive and the orange tree.

2dly, Along the Gulf of Genoa, from Toulon and the Hieres Isles to Nice and Bordighera, (*Annales du Museum*, tom. xi. p. 219.), where the small maritime palm-tree, *Chamærops*, grows wild, and where the date-tree is cultivated on a large scale, not to obtain its fruit, but the palms or etiolated leaves.

3dly, In England, on the coast of Devonshire, where the port of Salcombe has, on account of its temperate climate, been called the Montpellier of the North, and where (in South Hams) the Myrtle, the *Camellia Japonica*, the *Fuchsia coccinea*, and the | 25 | *Buddleia globose*,[46] pass the winter in the open ground, and without shelter.

4thly, In France, on the western coasts of Normandy and Brittany. In the Department of Finisterre, the arbutus, the pomegranate-tree, the

[46] Knight, *Trans. Hort. Soc.* vol. i. p. 32. In 1774, an Agave flowered at Salcombe, after having lived twenty-eight years without being covered in winter. On the coast of England, the winters are so mild, that orange trees are seen on espaliers, which are sheltered, as at Rome, only by means of a matting.—H.

Yucca gloriosa and *aloifolia*, the *Erica Mediterranea*, the *Hortensia*, the *Fuchsia*, the *Dahlia*, resist in open ground the inclemency of a winter which lasts scarcely fifteen or twenty days, and which succeeds to a summer by no means warm. During this short winter, the thermometer sometimes falls to 17°.6. The sap ascends in the trees from the month of February; but it often freezes even in the middle of May. The *Lavatera arborea* is found wild in the isle of Glenans, and opposite to this island, on the continent, the *Astragalus Bajonensis*, and the *Laurus nobilis*.[47]

From observations made in Brittany for twelve years, at St Malo, at Nantes, and at Brest, the mean temperature of the peninsula appears to be above 56°.3. In the interior of France, where the land is not much elevated above the sea, we must descend 3° of latitude in order to find an annual temperature like this.

It is known from the researches of Arthur Young,[48] that in spite of the great rise of the two isothermal lines of 53°.6 and 55°.4 on the western coast of France, the lines of culture (those of the olive, and of the maize and vine,) have a direction[49] quite opposite, from S.W. to N.E. This phenomenon has been ascribed,[50] with reason, to the low temperature of the summers | 26 | along the coast; but no attempt has been made to reduce to numerical expressions the ratios between the seasons in the interior and on the coast. In order to do this, I have chosen eight places, some of which lie under the same geographic parallels, and others in the prolongation of the same isothermal line. I have compared the temperatures of winter, of summer, and of the warmest months; for a summer of uniform heat excites less the force of vegetation, than a great heat, preceded by a cold season. The terms of comparison have been along the Atlantic; the coasts of Brittany, (from St Malo and St Brieux to Vannes and Nantes); the sands of Olonne; the Isle of Oleron; the embouchure of the Garonne and Dax, in the department of the Landes; and in the interior, corresponding to the same parallel, Chalons sur Marne, Paris, Chartres, Troyes, Poitiers, and Montauban. Farther south, from 44 1/2° of Lat. the comparisons become incorrect, because France, locked between the Ocean and the Mediterranean, presents, along this last basin, in the

[47] Bonnemaison, *Geogr. Botan. du Depart. du Finisterre*, (*Journal de Botan*. tom. iii. p. 118.).

[48] *Travels in France*, vol. ii. p. 91.

[49] The line which limits the cultivation of the vine, extends from the embouchure of the Loire and of the Vilaine, by Pontoise, to the confluence of the Rhine and the Moselle. The line of the olive trees commences to the west of Narbonne, passes between Orange and Montelimart, and carries itself to the N.E. in the direction of the Great St Bernard.—H.

[50] Decandolle, *Flor. Franç*. 3d edit. tom. ii. pl. viii. xi. Lequinio, *Voy. dans. le Jura*, tom. ii. p. 84.—91.

fine region of the olives, a system of climate of a particular kind, and very different from that of the western coast.

		Mean Temperature			
Places in the Interior.	**Latitude.**	**Of the Year.**	**Of Winter.**	**Of Summer.**	**Of the Warmest Month.**
		Fahr.	Fahr.	Fahr.	Fahr.
Chalons sur Marne, –	48°.57	50°.5	36°.1	66°.6	67°.5
Paris, – – –	48.50	51.1	38.7	65.3	67.5
Chartres, – –	48.26	50.7	37.0	64.6	65.7
Troyes, – –	48.18	52.2	38.3	67.3	68.4
Chinon, – –	47.26	53.4	38.7	69.1	70.2
Poitiers, – –	46.39	54.3	39.7	67.1	69.3
Vienne, – –	45.31	55.0	38.7	71.6	73.4
Montauban, –	44.01	55.6	42.6	69.3	71.4
Places on the Coast.					
St Malo, – –	48°.39′	55°.5	42°.4	66°.9	67°.5
St Brieux, – –	48.31	52.3	41.7	64.4	67.1
Vannes, – –	47.39	51.8	39.7	64.4	65.8
Nantes, – –	47.13	54.7	40.5	68.5	70.5
La Rochelle, – –	46.14	53.1	40.3	66.6	67.1
Oleron, – –	45.55	58.1	44.6	68.5	72.1
Bourdeaux, – –	44.50	56.5	42.1	70.9	71.4
Dax, – – –	43.52	54.1	44.4	67.3	68.9

| 27 |

These results are deduced from 127,000 observations, made with sixteen thermometers, of, no doubt, unequal accuracy. In supposing, on the theory of probabilities, that in such a number of observations, the errors, in the construction and exposure of the instruments, and in the hours of observation, will, in a great measure, destroy one another, we may determine, by interpolation, either under the same parallel, or upon the same isothermal line, the *mean winters* and *summers* of the interior and of the coast of France. This comparison gives,—

			Mean Winter.	**Mean Summer.**
I. Isothermal Lines of	52°.7	Coast, –	40°.6	65°.1
		Interior, –	38.5	68.0
	54°.7	Coast, –	41.4	67.3
		Interior, –	39.2	68.4

(continued)

(continued)

			Mean Winter.	Mean Summer.	Annual Temp.
I. Parallels of	47° to 49°	Coast,	41°.0	66°.7	53°.0
		Interior,	37.8	66.6	51.6
	45° to 46°	Coast,	42.3	67.8	55.8
		Interior,	39.2	69.3	54.7

As the isothermal lines rise again towards the western coasts of France; that is to say, as the mean temperature of the year becomes there greater than under the same latitude in the interior of the country, we ought to expect, that in advancing from east to west under the same parallel, the heat of the summers would not diminish. But the rising, again, of the isothermal lines, and the proximity of the sea, tend equally to increase the mildness of the winters; and each of these two causes acts in an opposite manner upon the summers. If the division of the heat between these seasons was equal in Brittany and in Orleannois, in the *climate of the coast*, and the *continental climates*, we ought to find the winters and summers warmer in the same latitude along the coast. In following the same isothermal lines, we readily observe, in the preceding table, that the winters are colder in the interior of the country, and the summers more temperate upon the coasts. These observations confirm in general the popular opinion respecting the climate of coasts; but in recollecting the cultivation and the developement of vegetation on the coasts and in the interior of France, we should expect differences of temperature still more considerable. It is surprising that these differences between the | 28 | winter and the summer should not exceed 1°.8, or nearly a quarter of the difference between the mean temperature of the winters or the summers of Montpellier and Paris. In speaking of the limits of the cultivation of plants upon mountains, I shall explain the true cause of this apparent contradiction. In the mean time, it may be sufficient to remark, that our meteorological instruments do not indicate the quantity of heat, which, in a clear and dry state of the air, the direct light produces in the more or less coloured parenchyma of the leaves and fruits. In the same mean temperature of the atmosphere, the developement of vegetation is retarded or accelerated, according as the sky is foggy or serene, and according as the surface of the earth receives only a diffuse light, during entire weeks, or is struck by the direct rays of the sun. On the state of the atmosphere, and the degree of the extinction of light, depend, in a great measure, those phenomena of vegetable life, the contrasts of which surprise us in islands, in the interior of continents, in plains, and on the summit of mountains. If we neglect these photometrical considerations, and do not appreciate the production of heat in the interior of bodies, and the effect of nocturnal radiation in a clear or a cloudy sky, we shall have some difficulty in discovering, from the numerical ratios of the observed summer and winter temperatures of Paris and London, the causes of the striking difference which appears in France and England in the culture of the vine, the peach, and other fruit-trees.[51]

When we study the organic life of plants and animals, we must examine all the stimuli or external agents which modify their vital actions. The ratios of the mean temperatures of the months are not sufficient to characterise the climate. Its influence combines the simultaneous action of all physical causes; and it depends on heat, humidity, light, the electrical tension of vapours, and the variable pressure of the atmosphere. It is the last cause which, on the tops of mountains, modifies the perspiration of plants, and even increases the exhaling organs. In making known the empirical laws of the distribution of heat | 29 | over the globe, as deducible from the thermometrical variations of the air, we are far from considering these laws as the only ones necessary to resolve all the problems of climate. Most of the phenomena of nature present two distinct parts, one which may be subjected to exact calculation, and another which cannot be reached but through the medium of induction and analogy.

Having considered the division of heat between winter and summer on the same isothermal line, we shall now point out the numerical ratios between the mean temperature of spring and winter, and between that of the whole year and the warmest month. From the parallel of Rome to that of Stockholm, and consequently between the isothermal lines of 60°.8 and 41°, the difference of the months of April and May is everywhere 10°.8 or 12°.6, and all the successive months are those which present the most rapid increase of temperature. But, as in northern countries, in Sweden, for example, the month of April is only 37°.4, the 10°.8 or 12°.6 which the month of May adds,[52] necessarily produces there a much

[51] Young's *Travels in France*, vol. ii. p. 195.

[52] In calculating for Europe, from 46° to 48° of Lat. for ten years the mean temperatures of every ten days, we find, that the decades which succeed one another, differ near the summits of the annual curve only 1°.44, while the differences rise in autumn from 3°.6 to 5°.4, and in spring from 5°.4 to 7°.2. —H.

greater effect on the developement of vegetation than in the south of Europe, where the mean temperature of April is from 53°.6 to 55°.4. It is from an analogous cause, that in passing from the shade to the sun, either in our climates in winter, or between the tropics on the back of the Cordilleras, we are more affected by the difference of temperature than in summer and in the plains, though in both cases the thermometrical difference is the same, for example from 5°.4 to 7°.2. Near the polar circle, the increase of the vernal heat is not only more sensible, but it extends equally to the month of June. At Drontheim, the temperatures of April and May, like those of May and June, differ not 10°.8 or 12°.6, but 14°.4 or 16°.2.

In distinguishing upon the same isothermal line the places which approach its concave or convex summits, in the same sys- | 30 | tem of climates in the northern and southern regions, we shall find,

1st, That the increase of the vernal temperature is great, (from 14°.4 or 16°.2, in the space of a month), and equally prolonged, wherever the division of the annual heat between the seasons is very unequal, as in the north of Europe, and in the temperate part of the United States.

2dly, That the vernal increase is great, (at least above 9° or 10°.8), but little prolonged, in the temperate part of Europe.

3dly, That the increase of the vernal temperature is small, (scarcely 7°.2), and equally prolonged, wherever there is an insular climate.

4thly, That in every system of climates, in the zones contained between the same meridians, the vernal increase is smaller, and less equally prolonged, in low than in high latitudes.

The isothermal zone from 53°.6 to 55°.4, may serve as an example for confirming these different modifications of spring. In Eastern Asia, near the concave summit, the differences of temperature between the four months of March, April, May and June, are very great, and very equal, (15°.7, 13°.3, and 13°.9). In advancing westward towards Europe, the isothermal line rises again, and in the interior of the country, near the convex summit, the increase is still greater, but little prolonged; that is to say, that of the four months which succeed one another, there are only two whose difference rises to 13°: they are 9°.4; 13°.3; 4°.1. Farther west, on the coasts, the differences become small and equal, viz. 3°.6; 6°.5; 5°.6. In crossing the Atlantic, we approach the western concave summit of the isothermal line of 53°.6. The increase of vernal temperature shews itself anew, and almost as great, and as much prolonged, as near the Arctic concave summit. The differences of the four months are 10°.4; 13°.9; and 10°.8. In the curve of annual temperature, the spring and autumn mark the transitions from the *minimum* and the *maximum*. The increments are naturally slower near the summits than in the intermediate part of the curve. Here they are greater, and of longer continuance, in proportion to the difference of the extreme ordinates. The autumnal decrease of temperature is less rapid than the vernal increase, because the sur- | 31 | face of the earth acquires the maximum of heat slower than the atmosphere, and because, in spite of the serenity of the air which prevails in autumn, the earth loses slowly, by radiation, the heat which it has acquired. The following Table will shew how uniform the laws are which I have just established.

Names of Places.	Latitude.	March.	April.	May.	June.	Differences of Temperature of the Four Months.			Mean Temp. Of the Year.
I. Group,—*Concave Summits in America.*									
Natchez, – – –	31° 28′	57°.9	66°.2	72°.7	79°.5	8°.3	6°.1	7°.2	64°.8
Williamsburg, – –	37 18	46.4	61.2	66.6	77.7	14.8	5.4	11.2	58.1
Cincinnati, – –	39 0	43.7	57.4	61.2	70.9	13.7	3.6	9.7	53.8
Philadelphia, – –	39 56	44.1	53.6	62.1	72.3	9.5	8.5	10.3	53.6
New York, – –	40 40	38.7	49.1	65.8	80.2	10.4	16.7	14.4	53.8
Cambridge, – –	42 25	34.5	45.5	56.8	70.2	11.0	11.3	13.3	50.4
Quebec, – – –	46 47	23.0	39.6	54.7	63.9	16.6	15.1	41.2	41.7
Nain, – – –	57 0	6.8	27.5	37.0	43.3	20.7	9.5	8.1	26.4
II. Group,—*Convex Summits in Europe.*									
1. *Continental Climate:*									
Rome, – – –	41 53	50.4	55.4	66.9	72.3	5.0	11.5	5.4	60.4
Milan, – – –	45 28	47.8	51.1	65.1	70.5	7.7	9.5	5.4	55.8
Geneva, – – –	46 12	39.6	45.5	58.1	62.2	6.1	12.4	4.1	49.3
Buda, – – –	47 29	38.3	49.1	64.8	68.4	10.8	15.7	3.6	51.1
Paris, – – –	48 50	42.3	48.2	60.1	64.4	8.5	11.9	4.3	51.1
Gottingen, – –	51 32	34.2	44.2	57.7	62.2	10.1	13.5	4.5	46.9
Upsal, – – –	59 51	29.5	39.7	48.7	57.9	10.3	9.0	9.2	41.9
Petersburg, – –	59 56	27.5	37.0	50.2	59.4	9.5	13.1	9.2	38.8
Umeo, – – –	63 50	23.0	34.2	43.7	55.0	11.2	9.5	11.3	33.3
Uleo, – – –	65 0	14.0	26.2	41.0	55.0	12.2	14.8	14.0	33.1
Enontekies, – –	68 30	11.5	26.6	36.5	49.5	15.1	9.9	13.0	27.0
2. *Climate of the Coast:*									
Nantes, – – –	47 13	50.0	53.6	60.1	65.7	3.6	6.5	5.6	54.7
London, – – –	51 30	44.2	49.8	56.5	63.1	5.6	6.7	6.7	51.6
Dublin, – – –	53 21	41.9	45.3	51.8	55.6	3.4	6.5	4.0	48.4
Edinburgh, – –	55 57	41.4	47.3	50.5	57.2	5.8	3.2	6.7	47.8
North Cape, – –	71 0	25.0	30.0	34.0	40.1	5.2	4.0	6.1	32.0
III. Group,—*Concave Summit of Asia.*									
Pekin, – – –	39 54	41.4	57.0	70.3	84.2	15.7	13.3	13.9	54.9 [32]

In all places whose mean temperature is below 62°.6, the revival of nature takes place in spring, in that month whose mean temperature reaches 42°.8 or 46°.4. When a month rises to,

41°.9, the Peach-tree (*Amygdalus Persica*) flowers.

46°.8, the Plum-tree (Prunus domestica) flowers.

51°.8, the Birch-tree[53] (*Betula alba*) pushes out its leaves.

At Rome, it is the month of March, at Paris the beginning of May, and at Upsal the beginning of June, that reaches the mean temperature of 51°.8. Near the Hospice of St Gothard, the birch cannot vegetate, as the warmest month of the year there scarcely reaches 46°.5. Barley, in order to be cultivated advantageously, requires, [54] during ninety days, a mean temperature of from 47°.3 to 48°.2. By adding the mean temperatures of the months above 51°.8, that is, the temperatures of those in which trees vegetate that lose their foliage, we shall have a sufficiently exact mean of the strength and continuance of vegetation. As we advance towards the north, vegetable life is confined to a shorter interval. In the south of France, there are 270 days of the year in which the mean temperature exceeds 51°.8; that is to say, the temperature which the birch requires to put forth its first leaves. At St Petersburgh, the number of these days is only 120. These two *cycles of vegetation*, so unequal, have a mean temperature which does not differ more than 5°.4; and even this want of heat is compensated by the effects of the direct light, which acts on the parenchyma of plants in proportion to the length of the days. If we compare, in the following *Table*, Eastern Asia, Europe, and America, we shall discover, by the increase of heat during the cycle of vegetation, the points where the isothermal lines have their concave summits. The exact knowledge of these cycles, will throw more light on the problem of Agricultural Geography, than the examination of the single temperatures of summer. | 33 |

[53] Cotte, *Meteorologie*, p. 448.—Wahlenberg, *Flor. Lap*. Pl. 51.

[54] Playfair, *Edin. Trans*, vol. v. p. 202.— Wahlenberg in Gilbert's *Annalen*, tom. xli. p. 282.

Equal Heat.	Names of Places.	Latitude.	Mean Temp. of the Year.	Sum of the Mean Temp. of the Months that reach 51°.8.	Number of those Months.	Mean Temperat. of the days which reach 51°.8.	Mean Temperature of the warmest Months.	Observations.
Isothermal Line of 59°.0,	Rome, –	41° 53′	60°.4	585°	9	64°.8	77°.0	Basin of the Mediterranean.
	Nismes,	43 50	60.3	593	9	65.8	78.3	*Idem.*
Isothermal Line of 53°.6,	Pekin, –	39 54	54.9	499	7	71.2	84.2	Eastern concave summit.
	Poitiers,	46 34	54.3	426	7	60.8	69.3	Convex summit.
	Nantes,	47 13	54.7	438	7	62.6	69.8	*Idem,* coasts.
	St Malo,	48 39	53.8	431	7	61.5	68.4	*Idem.*
	Philadelphia,	39 56	53.4	463	7	66.2	77.0	Western concave summit.
	Cincinnati,	39 6	53.8	458	7	65.5	74.3	*Idem.*
Isothermal Line of 50°.0,	London,	51 30	51.8	364	6	60.6	66.6	Insular climate.
	Paris, –	48 50	51.1	381	6	63.5	69.8	Near the coasts.
	Buda, –	47 29	51.1	323	5	64.6	72.0	Interior.
Isothermal Line of 48°.2,	Geneva,	46 12	49.3	311	5	62.2	66.6	Interior.
	Dublin,	53 21	48.7	282	5	56.5	60.8	Climate of the coasts.
	Edinburgh,	55 57	47.8	279	5	55.8	59.4	*Idem.*
Isothermal Line of 41°.0,	Upsal, –	59 51	41.9	229	4	57.2	61.9	Convex summit.
	Quebec,	46 47	41.7	318	5	63.7	73.4	Western concave summit.
Isothermal Line of 32°.0,	Petersburgh,	59 56	38.8	236	4	59.0	65.7	East of Europe.
	Umeo, –	53 50	33.3	118	2	59.0	62.6	E. Coast of Gulf of Bothnia.
	North Cape,	71 0	32.0	0	0	0	46.6	Interior climate.
	Enontekies,	68 30	27.0	116	2	58.1	59.5	Continental climate.

| 34 |

In the system of European climates, from Rome to Upsal, between the isothermal lines of 59° and 41°, the warmest month adds from 16°.2 to 18° to the mean temperature of the year. Farther north, and also in eastern Asia, and in America, where the isothermal lines bend towards the equator, the increments are still more considerable.

As two hours of the day indicate the temperature of the whole day, there must also be two days of the year, or two decades, whose mean temperature is equal to that of the whole year. From the mean of ten observations, this temperature of the year is found at Buda in Hungary from the 15th to the 20th of April, and from the 18th to the 23d of October. The ordinates of the other decades may be regarded as functions of the mean ordinates. In considering the temperatures of entire months, we find, that to the isothermal line of 35°.6, the temperature of the month of October coincides (generally within a degree) with that of the year. The following Table proves that it is not the month of April, as Kirwan affirms, (*Estimate,* &c. p. 166.), that approaches nearest to the annual temperature.

	Mean Temperature		
Names of Places.	**Of the Year.**	**Of October.**	**Of April.**
Cairo, –	72°.3	72°.3	77°.9
Algiers, –	69.8	72.1	62.6
Natchez,	65.0	68.4	66.4
Rome, –	60.4	62.1	55.4
Milan,	55.8	58.1	55.6
Cincinnati,	53.6	54.9	56.8
Philadelphia,	53.4	54.0	53.6
New York,	53.8	54.5	49.1
Pekin, –	54.7	55.4	57.0
Buda, –	51.1	52.3	49.1
London, –	51.8	52.3	49.8
Paris, –	51.1	51.3	48.2
Geneva, –	49.3	49.3	45.7
Dublin, –	48.6	48.7	45.3
Edinburgh,	47.8	48.2	46.9
Gottingen,	46°.9	47°.1	44°.4
Franeker,	52.3	54.9	50.0
Copenhagen,	45.7	48.7	41.0
Stockholm,	42.3	42.4	38.5
Christiania,	42.6	39.2	42.6
Upsal, –	41.7	43.3	39.7

(continued)

(continued)

	Mean Temperature		
Names of Places.	**Of the Year.**	**Of October.**	**Of April.**
Quebec,	41.9	42.8	39.6
Petersburgh,	38.8	39.0	37.0
Abo, –	41.4	4.0	40.8
Drontheim,	39.9	39.2	34.3
Uleo, –	33.1	37.9	34.2
Umeo, –	33.3	37.8	34.0
North Cape,	32.0	32.0	30.2
Enontekies,	27.0	27.5	26.6
Nain, –	26.4	33.1	27.5

As travellers are seldom able to make observations for giving immediately the temperature of the whole year, it is useful to know the constant ratios which exist in each system of climates, between the vernal and autumnal temperatures, and the annual temperature. | 35 |

The quantity of heat which any point of the globe receives, is much more equal during a long series of years than we would be led to believe from the testimony of our sensations, and the variable product of our harvests. In a given place, the number of days during which the N.E. or S.W. winds blow, preserve a very constant ratio, because the direction and the force of these winds, which bring warmer or colder air, depend upon general causes,—on the declination of the sun,—on the configuration of the coast,—and on the lie of the neighbouring continent. It is less frequently a diminution in the mean temperature, than an extraordinary change in the division of the heat between the different months, which occasions bad harvests. By examining between the parallels of 47° and 49° a series of good meteorological observations, made during ten or twelve years, it appears, that the annual temperatures vary only from 1°.8 to 2°.7; those of winter from 3°.6 to 5°.4; those of the months of winter from 9° to 10°.8. At Geneva, the mean temperatures of twenty years were as follows:

Years.	**Mean Temp.**	**Years.**	**Mean Temp.**
1796,	49°.3	1806,	51°.4
1797,	50.5	1807,	49.3
1798,	50.0	1808,	46.9
1799,	48.7	1809,	48.9

(continued)

(continued)

Years.	Mean Temp.	Years.	Mean Temp.
1800,	50.5	1810,	51.1
1801,	51.1	1811,	51.6
1802,	50.9	1812,	47.8
1803,	50.4	1813,	48.6
1804,	51.1	1814,	48.2
1805,	47.8	1815,	50.0
Mean of 20 Years, 49°.67			

If, in our climates, the thermometrical oscillations are a sixth part of the annual temperature, they do not amount to one twenty-fifth part under the tropics. I have computed the thermometrical variations, during eleven years, at Paris, for the whole year, the winter, the summer, the coldest month, the warmest month, and the month which represents most accurately the annual mean temperature; and the following are the results which I obtained: | 36 |

	Mean Temperature					
Observations of M. Bouvard.	**Of the Year.**	**Of Winter.**	**Of Summer.**	**Of January.**	**Of August.**	**Of October.**
Paris, 1803, –	51°.1	36°.7	67°.6	34°.3	67°.6	50°.5
1804, –	52.0	41.0	65.5	43.9	64.6	52.7
1805, –	49.5	36.0	63.1	34.9	64.8	49.3
1806, –	53.4	40.6	65.3	43.0	64.6	51.8
1807, –	51.4	42.3	67.8	36.1	70.5	54.3
1808, –	50.5	36.7	66.2	36.3	66.6	48.2
1809, –	50.9	40.5	62.4	40.8	64.2	49.6
1810, –	50.9	36.5	63.3	30.6	63.7	52.9
1811, –	52.7	39.2	65.1	26.6	63.7	57.6
1812, –	49.8	39.6	63.1	34.7	64.2	51.1
1813, –	49.8	36.1	61.7	32.5	62.6	53.1
Mean of these 11 years,	51°.1	38°.7	64°.0	36°.6	65°.1	51°.9

At Geneva, the mean temperatures of the summers were, from 1803 to 1809,—

Years.		**Mean Temp of Summers.**
1803,	–	67°.3
1804,	–	65.0
1805,	–	62.2
1806,		65.7
1807,	–	68.2
1808,	–	62.9
1809,	–	63.0
Mean of seven years,		64°.9

M. Arago has found, that in the two years 1815 and 1816, the last of which was so destructive to the crops in a great part of France, the difference of the mean annual temperature was only 2°, and that of the summer 3°.2. The summer of 1816 at Paris was 59°.9, 4°.7 below the mean of the former. From 1803 to 1813, the oscillations round the mean did not go beyond — 2°.9, and +3°4.

In comparing places which belong to the same system of climates, though more than eighty leagues distant, the variations seem to be very uniform, both in the annual temperature and that of the seasons, although the thermometrical quantities are not the same. | 37 |

Years.	Paris.		Geneva.		Paris.		Geneva.		Paris.		Geneva.	
	Mean Annual Temperature.	Difference between Mean Ann. Temp. and that for 12 years, 51°.1	Mean Annual Temperature.	Difference between Mean An. Temp. and that of 12 years, 49°.6.	Mean Temperature of Winter.	Difference with the mean Winter Temperature of 12 years, 38°.7.	Mean Temperature of Winter.	Difference with the Mean Winter Temperature of12 years, 34°.9.	Mean Temperature of Summer.	Difference with the Mean Temperature of Summer for 12 years, 64°.6.	Mean Temperature of Summer.	Difference with the Mean Temperature of Summer for 12 years, 64°.9.
1803,	51°.1	0	50°.4	+ 0.8	36°.7	— 2.0	32°.2	— 2°.7	67°.6	+ 3°.0	67°.6	+ 2.7
1804,	52.0	+ 0.9	51.1	+ 1.5	41.0	+ 2.3	38.3	+ 3.4	65.5	+ 0.9	66.2	+ 1.3
1805,	49.5	— 1.6	47.8	— 1.8	36.0	— 2.7	33.8	— 1.1	63.1	— 1.5	63.0	— 1.9
1806,	53.4	+ 2.3	51.4	+ 1.8	40.6	+ 1.9	38.5	+ 3.6	65.3	+ 0.7	64.6	— 0.3
1807,	51.4	+ 0.3	49.3	— 0.3	42.3	+ 3.6	35.8	+ 0.9	67.8	+ 3.2	68.2	+ 3.3
1808,	50.5	— 0.6	46.8	— 2.8	36.7	— 2.0	33.8	— 1.1	66.2	+ 1.6	63.5	— 1.4
1809,	50.9	— 0.2	48.7	— 0.9	40.5	+ 1.8	35.1	+ 0.2	62.4	— 2.2	63.1	— 1.8
1810,	50.9	— 0.2	51.1	+ 1.5	36.5	— 2.2			63.3	— 1.3		
1811,	52.7	+ 1.6	51.8	+ 2.2	39.2	+ 0.5			65.1	+ 0.5		
1812,	49.8	— 1.3	47.8	— 1.8	39.6	+ 0.9			63.1	— 1.5		
1813,	49.8	— 1.3	48.6	+ 1.0	36.1	— 2.6			61.7	— 2.9		

(To be continued in next Number.)

On Isothermal Lines, and the Distribution of Heat over the Globe. By Baron Alexander de Humboldt.

(Continued from Vol. IV. p. 37.)

All the ratios of temperature which we have hitherto fixed, belong to that part of the lower strata of the atmosphere which rests on the solid surface of the globe in the northern hemisphere. It now remains for us to discuss the *temperature of the southern hemisphere*. In few parts of natural philosophy, have naturalists differed so widely in opinion. From the beginning of the 16th century, and the first navigations round Cape Horn, an idea prevailed in Europe, that the southern was considerably colder than the northern hemisphere. Mairan and Buffon[55] combated this opinion by inaccurate reasonings of a theoretical nature. Æpinus[56] established it anew. The discoveries of Cook made known the vast extent of ice round the South Pole; but the inequality in the temperature of the two hemispheres was then exaggerated. Le Gentil, and particularly Kirwan,[57] had the merit of having first demonstrated, that the influence of the circumpolar ice extended much less into the temperate zone than was generally admitted. The less distance of the sun from the winter solstice, and his long continuance in the northern signs, act in an opposite manner[58] on the heat in the two hemispheres; and as (after the theorem of Lambert) the quantity of light which a planet receives from the sun, increases in proportion to the true anomaly, the inequality in the temperature of the two hemispheres is not the effect of unequal radiation. The southern hemisphere receives the same quantity of light; but the accumulation of heat in it is less,[59] on account of the emission of the radiant heat which takes place during a long winter. This hemisphere being also in a great measure covered with | 263 | water, the pyramidal extremities of the continents have there an irregular climate. Summers of a very low temperature are succeeded, as far as 50° of south latitude,

[55] *Theorie de la Terre*, tom. i. p. 312.—*Mémoires de l'Acad.* 1765, p. 174.
[56] *De Distributione Caloris*, 1761.
[57] *Estimate*, &c. p. 60.—Irish Trans. vol. viii. p. 423.—Le Gentil, *Voyage dans l'Inde*, vol. i. p. 73.
[58] Mairan, *Mem. Acad.* 1765, p. 166.—Lambert, *Pyrometrie*, p. 310.
[59] Prevost, *De la Chalcur Rayonnante*, 1809, p. 329. & 367. § 280,—306.

by winters far from rigorous. The vegetable forms also of the torrid zone, the arborescent ferns, and the orchideous parasites, advance towards 38° and 42° of S. Latitude. The small quantity of land[60] in the southern hemispheres, contributes not only to equalise the seasons, but also to diminish absolutely the annual temperature of that part of the globe. This cause is, I think, much more active than that of the small eccentricity of the earth's orbit. The continents during summer radiate more heat than the seas, and the ascending current which carries the air of the equinoctial and temperate zones towards the circumpolar regions, acts less in the southern than in the northern hemisphere. That cap of ice which surrounds the pole to the 71st and 68th degree of south latitude, advances more towards the equator, whenever it meets a free sea; that is, wherever the pyramidal extremities of the great continents are not opposite to it. There is reason to believe, that this want of dry land would produce an effect still more sensible, if the division of the continents was as unequal in the equinoctial as in the temperate zones.[61]

Theory and experience prove, that the difference of temperature between the two hemispheres, cannot be great near the limit which separates them.[62] Le Gentil had already observed, that the climate of Pondicherry is not warmer than that of Madagascar at the Bay of Antongel in 12° of S. Lat. Under the parallels of 20 the Isle of France has the same annual temperature, viz. 80°.1, as Jamaica and St Domingo. The Indian Sea between the east coasts of Africa, the Isles of Sonde and New Holland, form a kind of gulf which is shut up to the north by Arabia and Hindostan. The isothermal lines there appear to go back to the South Pole; for farther to the west in the open sea between Africa and the New World, the cold of the southern hemisphere already causes itself to be felt from the 22d degree, on account of its insulated mountains and particular loca- | 264 |
lities. I shall not mention the island of St Helena, Lat. 15° 55′ whose mean temperature, according to the observations of M. Beatson, at the sea side, does not exceed 71°6 or 73°4. It is the eastern coast of America, which, in the observations of a Portuguese astronomer, M. Bento Sanchez d'Orta,[63] present us with the S. Lat. of 22° 54′, almost at the limit of the equinoctial region with a plan, of which we know the climate by more than 3500 thermometrical and barometrical observations made every year,

[60] The dry lands in the two hemispheres are in the ratio of 3 to 1.

[61] The dry lands between the tropics, are in the two hemispheres as 5 to 4, and without the tropies as 13 to 1.

[62] Prevost, p. 343.

[63] *Mem. de l'Acad. de Lisbonne*, tome ii. p. 348. 369.

to ascertain the horary variations in the heat and pressure of the air. The mean temperature of Rio Janeiro is only 74°.3, whilst, notwithstanding the north winds which bring the cold air of Canada during winter into the Gulf of Mexico, the mean temperatures of Vera Cruz, (Lat. 19° 11′,) and of the Havannah, (Lat. 23° 10′,) are 77°.9. The differences of the two hemispheres become more sensible in the warmest months.

Rio Janeiro. Havannah.	Mean Temp.		Mean Temp.
June,	68°.0	December,	71°.8
July,	70 2	January,	70 2
January,	79 2	July,	83 3
February,	80 6	August,	83 8

The great equality in the division of the annual heat in 34° of N. and S. Lat. is very surprising. If we attend to the three continents of New Holland, Africa and America, we shall find, that the mean temperature of Port Jackson, (Lat. 33° 51′,) is, after the observations of MM. Hunter, Peron, and Freycinet, – – – 66°.7

That of the Cape of Good Hope, (Lat. 33° 53′,) 66 9

That of the city of Buenos Ayres, (Lat. 34° 36′,) 67 5

In the northern hemisphere 60°.8 or 69°.8 of annual temperature corresponds to the same latitude in the northern hemisphere, according as we compare the American system of climates[64] or the Mediterranean one;—the concave or the convex | 265 | parts of the isothermal lines. At Port Jackson, where the thermometer descends sometimes below the freezing point, the warmest month is 77°.4, and the coldest 56°.8. We find here the summer of Marseilles and the winter of Cairo. [65] In Louisiana

64

	Latitude.	Mean Temp.
Natchez,	31° 28′,	64°.8
Cincinnati,	39 06,	53.8.

65

	Latitude.	Mean Temp.
Cairo,	30° 2′	72°.3
Funchal,	—— 32 37	68 5
Algiers,	—— 36 48	70 0

2 1/2 of Lat. nearer the Equator, the warmest month is 79°.7, and the coldest 46°.9. In Van Diemen's Land, corresponding nearly in latitude to Rome, the winters are more mild than at Naples; but the coldness of the summers[66] is such, that the mean temperature of the month of February appears to be scarcely 64°.4, or 66°.2, whilst at Paris, under a latitude more distant from the Equator by 7°, the mean temperature of the month of August is also from 64°.4 to 66°.2′, and at Rome above 77°. Under the parallel of 51° 25′ south, the mean temperature of the Malouine Isles is well ascertained to be 47°.3. At the same Lat. N. we find the mean temperature in Europe from 50° to 51°.8, and in America scarcely from 35°.6 to 37°.4. The warmest and the coldest months are at London 66°.2 and 35°.6; at the Malouine Isles 55°.8 and 37°.4. At Quebec, the mean temperature of the water is 14°; at the Malouine Isles 39°.6, though those isles are 4° of Lat. farther from the Equator than Quebec. These numerical ratios prove, that, to the parallels of 40° and 50°, the corresponding isothermal lines are almost equally distant from the Pole in the two hemispheres; and that, in considering only the system of transatlantic climates between 70° and 80° of W. Long., the mean temperatures of the year, under the corresponding geographical parallels, are even greater in the southern than in the northern hemisphere.

The division of the heat between the different parts of the year, gives a particular character to southern climates. In the | 266 | southern hemisphere on the isothermal lines of 46°.4 and 50°.0, we find summers which in our hemisphere belong only to the isothermal lines of 35°.6 and 40°. The mean temperature is not precisely known beyond 51° of S. Lat. Navigators do not frequent those regions when the sun is in the northern signs, and it would be wrong to judge of the rigour of winter, from the low temperature of the summer. The eternal snows which in 71° of N. Lat. support themselves at the height of 2296 feet above the sea, descend even into the plains, both in South Georgia[67] and in Sandwich Land in 54° and 58° of S. Lat. But these phenomena, however striking

[66] In Van Diemen's Land the thermometer descends in February, in the morning, to 45°.5. The mean of mid-day is 60°.8. At Paris it is in August 73°.4. In Van Diemen's Land, in February the mean of the maxima is 78°8.; of the minima 54°.5. At Rome these means are 86° and 64°.5.—D'Entrecasteaux, *Voyage*, tom. i. p. 205. and 542.

[67] It is the more surprising to find in the Island of Georgia snow on the banks of the ocean, because 2° 39′ nearer the Equator at the Malouine Isles, the mean temperature of the summers is 53°.1, or 9° greater than at the point in our hemisphere in 71° of Lat. where the limit of perpetual snow exists at 2296 feet of absolute elevation. But we must recollect, 1st, That the Malouine Isles are near a continent which is heated in summer; 2d, That Georgia is covered with mountains, and is placed not only in a sea open to the north, but under the influence of the perennial ices of Sandwich Land; and, 3dly, That in Lapland, 20° of Lat. produce in certain local circumstances 10°.8 of difference in the temperatures of the summers.

they may appear, do not by any means prove that the isothermal line of 32° is 5° nearer the South Pole than the North Pole. In the system of transatlantic climates, the limit of eternal snow is not at the same altitude as in Europe; and in order to compare the two hemispheres, we must take into account the difference of longitude. Besides, an equal altitude of the snows, does not by any means indicate an equal mean temperature of the year. This limit depends parcularly[68] on the coldness of summer, and this again on the quick condensations of the vapour caused by the passage of the floating ice. Near the poles the foggy state of the air diminishes in summer the effect of the solar irradiation, and in winter that of the radiation of the globe. At the Straits of Magellan, MM. Churruca and Galeano have seen snow fall in 53° and 54° of S. Lat. in the middle of summer; and though the day was 18 hours long, the thermometer seldom rose above 42°.8 or 44°.6 and never above 51°.8.

The inequal temperature of the two hemispheres, which, as we have now proved, is less the effect of the eccentricity of the | 267 | earth's orbit, than of the unequal division of the continents, determines[69] the limit between the N. E. and S. E. Trade Winds. But as this limit is much more to the north of the Equator in the Atlantic Ocean, than in the South Sea, we may conclude that, in a region between 130° and 150° of W. Long. the difference of temperature between the two hemispheres, is less great than farther to the east in 20° or 50° of longitude. It is indeed under this region in the Great Ocean, that, as far as the parallel of 60°, the two hemispheres are equally covered with water, and equally destitute of dry land, which, radiating the heat during summer, sends the warm air towards the poles. The line which limits the N. E. and S. E. Trade Winds, approaches the Equator, whereas the temperature of the hemispheres is different; and if, without diminishing the cold of the southern atmosphere, we could increase the inflexion of the isothermal lines in the system of transatlantic climates, we should meet the S. E. winds in 20° and 50° of W. Long. to the north, and in 130° and 150° of W. Long. to the south of the Equator.[70]

The low strata of the atmosphere which rest upon the aqueous surface of the globe, receive the influence of the temperature of the waters. The sea radiates less absolute heat than continents; it cools the air upon the sea, by the effect of evaporation; it sends the particles of water cooled and heavier towards the bottom; and it is heated again, or cooled, by the

[68] Baron Von Buch's *Travels in Lapland*, vol. ii. p. 393,—420.

[69] Prevost, *Journ. de Phys*. tom. xxxviii. p. 369.—Irish *Trans*. vol. viii, p. 374.

[70] Humboldt's *Relat. Histor*. tom. i. p. 225, 237.

currents directed from the Equator to the Poles, or by the mixture of the superior and inferior strata on the sides of banks.

It is from these causes combined, that, between the tropics, and perhaps as far as 30° of Lat., the mean temperatures of the air next the sea, are 3°.6 or 5°.4 lower than that of the continental air. Under high latitudes, and in climates where the atmosphere is coolest in winter, much below the freezing point, the isothermal lines rise again towards the Poles, or become convex when the continents pass below the seas.[71]

With respect to the temperature of the ocean, we must distinguish between four very different phenomena. 1st, The tem- | 268 | perature of the water at the surface corresponding to different latitudes, the ocean being considered at rest, and destitute of shallows and currents. 2d, The decrease of heat in the superimposed strata of water. 3d, The effect of billows on the temperature of the surface water. 4th, The temperature of currents, which impell with an acquired velocity, the waters of our zone across the immoveable waters of another zone. The region of warmest waters no more coincides with the Equator, than the region in which the waters reach their maximum of saltness. In passing from one hemisphere to another, we find the warmest waters between 5° 45′ of N. Lat., and 6° 15′ of S. Lat. Perrins found their temperature to be 82°.3; Quevedo 83°.5; Churruca 83°.7, and Rodman 83°.8. I have found them in the South Sea to the east of the Galapagos Isles 84°.7. The variations and the mean result do not extend beyond 1°.3. It is very remarkable that in the parallel of warmest waters, the temperature of the surface of the sea is from 3°.6 to 5°.4 higher than that of the superincumbent air. Does this difference arise from the motion of the cooled particles towards the bottom, or the absorption of light, which is not sufficiently compensated by the free emission of the radiant coloric. As we advance from the Equator to the Torrid Zone, the influence of the seasons on the temperature of the surface of the sea becomes very sensible; but as a great mass of water follows very slowly the changes in the temperature of the air, the means of the months do not correspond at the same epochs in the ocean and in the air. Besides, the extent of the variations is less in the water than in the atmosphere, because the increase or decrease in the heat of the sea takes place in a medium of variable temperature, so that the *minimum* and the *maximum* of the heat which the water reaches, are modified by the atmospherical temperature of the months which follow the coldest of the warmest months of the year. It is from an analogous

[71] *Id.* p. 67, 230. 242.

cause, that in springs which have a variable temperature, for example, near Upsal,[72] the extent of the variations of temperature is only 19°.8, while the same extent in the air from the month of January to August, is 39°.6. In the parallel of the Canary Islands, | 269 | Baron Von Buch found the minimum of the temperature of the water to be 68°, and the maximum 74°.8. The temperature of the air in the warmest of the coldest months, is, in that quarter, from 64°.4 to 75°.2. In advancing towards the north, we find still greater differences of winter temperature between the surface of the sea and the superincumbent air. The cooled particles of water descend till their temperature reaches 39°.2; and hence in 46° and 50° of Lat. in the part of the Atlantic which is near Europe, the *maximum* and *minimum of heat* are

In the water at its surface,
68°.0 and 41°.9

In the air from the mean of warmest and coldest months,
66.2 and 35.6

The excess in the mean temperature of the water over that of the air, attains its maximum beyond the polar circle, where the sea does not wholly freeze. The atmosphere is cooled to such a degree in these seas, (from 63° to 70° of Lat., and 0° of Long.) that the mean temperature of several months of winter descend on the continents to 14° and 10°.4, and on the coasts to 23° and 21°.2, while the temperature of the surface of the sea is not below 32° or 30°.2. If it is true, that even in those high latitudes the bottom of the sea contains strata of water which, at the maximum of their specific gravity, have 39°.2 or 41° of heat, we may suppose that the water at the bottom contributes to diminish the cooling at the surface. These circumstances have a great influence on the mildness of countries in continents separated from the Pole by an extensive sea.

Hitherto we have attended to the distribution of heat on the surface of the globe at the level of the sea. It only remains for us to consider the variations of temperature in the higher regions of the atmosphere, and in the interior of the earth.

The decrease of heat in the atmosphere, depends on several causes, the principle of which, according, to Laplace and Leslie,[73] is the property

[72] Gilbert's *Annalen,* 1812, p. 129.

[73] *Essay on Heat and Moisture*, p. 11; and *Geometry*, p. 495.

of the air to increase its capacity for heat by its rarefaction. If the globe was not surrounded by a mixture of elastic and aëriform fluids, it would not be sensibly colder at the height of 8747 yards than at the level of the sea. As each | 270 | part of the globe radiates in every direction, the interior of a spherical envelope which would rest on the top of the highest mountains, would receive the same quantity of radiant heat as the lower strata of the atmosphere. The heat, it is true, will be spread over a surface a little greater; but the difference of temperature will be insensible, since the radius of the spherical envelope will be to that of the earth as 1.001 to 1.

Considering the earth as surrounded with an atmospherical fluid, it is obvious, that the air heated at its surface will ascend, dilate itself, and be cooled, either by dilatation, or, by a more free radiation across the other strata that are equally rarified. These are the ascending and descending currents, which keep up the decreasing temperature of the atmosphere.[74]

The cold of mountains is the simultaneous effect, 1st, Of the greater or less vertical distance of the strata of air at the surface of the plains and of the ocean. 2d, Of the extinction of light, which diminishes with the density of the superincumbent strata of air;[75] and, 3d, Of the emission of radiant heat, which is favoured by air very dry,[76] very cold, and very clear. The mean temperature of our present plains would be lowered, if the seas should experience a considerable diminution. The plains of continents would then become *plateaux*, and the air which rested on them would be cooled by the circumjacent strata of air, which, at the same level, would receive but a small portion of the heat emitted from the dry bottom of the seas.

The following Table contains the results of observations which I have made near the Equator, on the Andes of Quito, and towards the northern extremity of the torrid zone, in the Cordilleras of Mexico. These results are true means, given either by stationary observations made during several years, or by insulated observations. In these last, we have taken into account the hour of the day,—the distance of the solstices,—the direction of the wind,—and the reflection from the plains. | 271 |

[74] *Essay on Heat and Moisture*, p. 11.; and *Geometry*, p. 495.
[75] Humboldt on Refraction below 10°, *Observ. Astron.* tom. i. p. 126.
[76] Wells on *Dew*, p. 50.

	Cordilleras of the Andes. From 10° of North Lat. to 10° of South Lat.		Mountains of Mexico. From 17° of North Lat. to 21° of North Lat.	
Height above the Level of the Sea.	**Mean Temp. of the Year.**	**Examples, which may serve as a Type.**	**Mean Temp. of the Year.**	**Examples, which may serve as a Type.**
0 Toises. 0 Feet. (Comparative heights in Europe have been added for every 1000 metres.)	Fahr. 81°.5	*Cumana,* 33 *feet.* Temp. of day, 78°.8—86° —— night, 71.6—74.3 *Maximum,* 90.9 *Minimum,* 70.2 *Mean,* – 81.9	78°.80	*Vera Cruz,* 0 *feet.* Temp. of day, 80°.6—86° —— night, 78.26—82.4 in summer, —— night, 66.2—75.264.4—71.6 in winter, Mean Temp. 77.72
500 Toises. 3197 Feet. *Vesuvius* 3870 feet.	71°.24	*Caraccas,* 2906 *feet.* Temp. of day, 64°.4—73°4 ——night, *Maximum,* 78.3 *Minimum,* 54.5 Mean, – 69.4 *Guaduas,* 3769 feet. Temp. Mean, 67°.5	67°.64	*Xalapa,* 4330 *feet.* Temp. Mean in winter, 64°.76 —— of day, 57°.2—59 *Chilpantzingo,* 4523 *feet,* on a plateau which radiates. Temp. 69°.08 Mean –

(*continued*)

(continued)

	Cordilleras of the Andes. From 10° of North Lat. to 10° of South Lat.			Mountains of Mexico. From 17° of North Lat. to 21° of North Lat.		
		Popayan, 5815 *feet.*			*Valladolid de Mechoachan,* 6396 *feet.*	
		Temp. of day,	66°.2—75°.2		Mean Temp.	66°.2—68°.
		—— — night,	62.6—64.4			
		Mean, –	65.66		*Mexico,* 7468 *feet.*	
1000 Toises. 6394 Feet. *Hospice of St Gothard,* 6806 feet.	64°.4	*Santa Fé de Bogota, feet.*	8721	64°.4	Temp. of day,	60°.8—69°.8
		Temp. of day,	59°—64°.4			
		—— — night,	50—53.6		—— — night,	55.4—59
		Minimum,	36.5		Warmest months,	52.7—59
		Mean, –	57.74		Coldest months,	32—44.6
					Mean,	62°.6
		Quito, 9538 *feet.*			*Toluca,* 8823 *feet.*	
		Temp. of day,	60°.08-66°.74		Temp. Mean, –	59°
1500 Toises. 9591 Feet. >*Canigou,* 9118 feet.	57°.74	—— — night,	48.2--51.8	57°.2	*At the Nevado de Toluca,*	11,178 *feet.*
		Maximum,	71.6		Temp. of spring.	48°.2
		Minimum,	42.8			
		Mean, –	57.92			

	Cordilleras of the Andes. From 10° of North Lat. to 10° of South Lat.		Mountains of Mexico. From 17° of North Lat. to 21° of North Lat.	
2000 Toises. 12,789 Feet. *Peak of Teneriffe,* 12,169 feet.	44°.6	*Micuipampa,* 11,867 *feet.* Temp. of day, 41°—48°.2 —— — night, 35.6—31.28 *Les Paramos,* 11,480 feet. In February, at 9^h, Mean Temp. in gen. 47°.12	45°.5 50°.36	*At the Nevado* de Toluca, 12,178 *feet.* Temp. in Sept. at noon, 52°.7 *At Coffre de Perote,* 12,136 *f.*
2500 Toises. 15,985 Feet. *Mont Blanc,* 15,662 feet.	37°.7	*At the Inferior Limit of Perpetual Snows,* 15,774 *feet.* Temp. of day, 39°.2—46°.4 —— —night, 28.4—21.2 *Chimborazo,* 19,286 feet. In June, at 1 o'clock, I have seen the therm. at 29°.12.	33°.8	*At the Pic del Fraille,* 15,157 *feet.* I have seen the thermometer in September at 39°.74.

| 272 |

The means given by the Mexican observations are a little different from those given by the observations on the Cordilleras. When the differences and the coincidences amount to about a degree of Fahrenheit, they may be regarded as purely accidental. The length of the day is more unequal in the 20th degree of latitude, but the perpetual snows do not descend 656 feet lower than under the Equator. As the Cordilleras of New Granada, Quito, and Peru, present a great number of points where stationary observations have been made, I shall collect here the mean temperatures which M. Caldas[77] and I have determined with some certainty, and which all belong to a zone bounded by the parallels of 10° N. and 10° S. Lat.

	Alt. in Feet.	**Mean Temp.**
Coasts of Cumana,	0	80°.6
		82.4
Tomependa, Amazons R.	1279	78.44
Tocayma, – –	1581	81.5
Antioquia, – –	1666	77.00
Neiva, – –	1702	77.00
Caraccas, – –	2906	69.44
Caripe, – –	2959	65.3
Carthago, – –	3149	74.84
La Plata, – –	3437	74.66
Guaduas, – –	3772	67.46
La Meya, – –	4225	72.50
Medellin, – –	4858	68.9
Estrella, – –	5645	65.84
Popayan, – –	5815	65.66
Loxa, – –	6855	64.4
Almaguer, – –	7413	62.6
Pamplona, –	8016	61.16
Alausi, – –	7970	59°.00
Pasto, – –	8308	58.28
Santa Rosa, –	8459	57.74
Cuenca, – –	8633	60.08
Santa Fé de Bogota,	8721	57.74
Hambato, – –	8849	60.44
Caxamarca, –	9381	60.80
Llactacunga, –	9473	59.00
Riobamba Nuevo,	9482	61.16
Tunja, – –	9522	56.66

(continued)

[77] I have used the mean temperature and barometrical measurements published at Santa Fé de Bogota by MM. Caldas and Restrepo in the *Semanario del N. R. de Granada*, tom. i. p. 273.; tom. i. p. 93.—341.

(continued)

	Alt. in Feet.	Mean Temp.
Quito, – –	9538	57.92
Malbasa, – –	9971	54.50
Plateau de los Pastos,	10099	54.50
Les Paramos, –	11480	47.30
At the Inferior Limit of Perpetual tual Snow,	15744	34.88

These thirty-two points are not insulated points, as balloons would be if they were fixed in the atmosphere at a perpendicular height of 16,400 feet. They are stations taken on the declivity of mountains, upon that part of the solid mass of the globe which, in the form of a wall, rises into the higher regions of the atmosphere. These mountains, too, have at each height parti- | 273 | cular climates, modified by the radiation of the plateaus on which they stand,—upon the slope of the ground,—the nakedness of the soil,—the humidity of the forests,—and the currents which descend from the neighbouring summits.

Without knowing the localities themselves, the effect of disturbing causes will be readily seen, by comparing in the preceding Table the mean temperatures which correspond to the same elevations; and the discussion of these observations would prove, also, that the extent of the variations is much less than is generally believed. If we examine thirty-two temperatures, upon the hypothesis that a degree of cooling corresponds to an altitude of 200 metres (656 feet), we shall deduce the temperature of the plains (from 30°.6 to 82°.4) twenty-six times from that of elevated places. For the other six deductions, the temperatures differ only about 3°.6; and the errors of observation are here combined with the effects of localities. The air which rests on the plains of the Andes mixes itself with the great mass of the free atmosphere, in which there prevails under the torrid zone a surprising stability of temperature. However enormous be the mass of the Cordilleras, it acts but feebly on the strata of air which are unceasingly renewed. On the other hand, if the plains are heated during the day, they radiate as much during the night; for it is principally on the plains elevated 8856 feet above the sea, that the sky is most clear and uniformly serene. At Peru, for example, the magnificent plateau of Caxamarca, in which the wheat yields the eighteenth, and barley the sixtieth grain, has an extent of more than twelve square leagues: it is smooth like the bottom of a lake, and sheltered by a circular wall of mountains free from snow. Its mean temperature is 60°.8, yet the wheat is often frozen during the night; and in a season where the thermometer fell before sunrise to 46°.4, I have seen it rise in the day to 77° in the shade. In the vast plains of Bogota, which are 656 feet less elevated than that of Caxamarca, the mean temperature, as established by the fine observations of Mutis, is scarcely 57°.74.

In comparing towns situated on elevated plains with those which are placed on the declivity of mountains, I have found for the first an augmentation of temperature, which, on account of the | 274 | nocturnal radiation, does not exceed from 2°.7 to 4°.14. This augmentation is a little greater in the lower regions of the Andes, in those large valleys whose smooth bottoms reach the height of from 1312 to 1640 feet, principally in the valley of La Madaleine, between Neiva and Honda. It is singular to find in the middle of mountains heats which equal those of the plains, and which are more insupportable, as the air of the valleys is almost never agitated by the winds. If we compare, however, the mean temperatures of these same places with those of the strata of the true atmosphere, or on the declivity of mountains, we shall find them only from 3°.6 to 5°.4.

On these grounds, we may place some confidence on the four results which we have deduced from such a great number of observations, for the perpendicular heights of 1000, 2000, 3000, and 4000 metres. I have confined myself to a simple arithmetical mean, and to the fortuitous compensation of irregularities; for I could not have avoided employing an hypothesis on the decrease of heat, if I had wished to reduce to a standard height those heights which approach it the most. I have added the observations with which an intimate knowledge of localities has furnished me.

1. For 1000 Metres (3280 feet) of Elevation.	**Alt. in Feet.**	**Temp. Fahr.**
Convent of Caripe, (thick and damp forests), –	2959	65°.3
Caraccas, (a foggy sky, valley of small extent,) –	2906	69.44
La Plata, (very warm, valley communicating with that of L'Alto Magdalena,) – – – –	3437	74.66
Carthago, (very warm valley of Cauca), – –	3149	74.84
2. For 2000 Metres (6560 feet) of Elevation.		
Loxa, (a plateau of small extent), – –	6855	64.4
Almaguer, declivity covered with very thick vegetation), –	7413	62.6
Popayan, (small plateau, a little elevated above the valley of Cauca,) – – – – –	5815	65.66
3. For 3000 Metres (9840 feet) of Elevation.		
Caxamarca, (very extensive plateau, sky serene,) –	9381	60.80

(continued)

(continued)

1. For 1000 Metres (3280 feet) of Elevation.	Alt. in Feet.	Temp. Fahr.
Quito, (at the foot of Pinchincha, a narrow valley,) –	9538	57.92
Tunja, (mountains of New Grenada), – –	9522	56.66
Malvasa, (elevated plains, cooled by the snows of the volcano of Puracé,) – – – –	9971	54.50
Los Pastos, (very cold plateau, from which rise snow covered summits,) – – – – –	10099	54.50
Llactacunga, (temperate valley), – –	9473	59.0
Riobamba Nuevo, (arid plains of Tupia, covered with pumice-stone,)	9482	61.16

| 275 | Between the tropics, the Cordilleras form the centre of the civilization and industry of Spanish America. They are inhabited to the height of 4000 metres, (13,120 feet); and a small number of observations made on the back of the Andes, gives a sufficiently accurate idea of the mean temperature of the year. In Europe, on the contrary, in the temperate zone, the high mountains are in general little inhabited. The descent of the isothermal line of 32°, causes to cease the cultivation of crops of grain, at the point where they begin in the Cordilleras. Stationary habitations are rare above 2000 metres (6560 feet) of elevation; and in order to judge with any precision of the mean temperature of the superincumbent beds of air, we must unite at least 730 thermometrical observations made in the course of a year.[78] | 276 |

[78] Elevations of 400 metres (1312 feet) appear to have a very sensible influence on the mean temperature, even when great portions of countries rise progressively. In order to establish this point, I have examined the temperatures of places situated almost on the level of the sea, and under the same parallels.

	Lat.	Elevation in Feet.	Mean Temp.
Buda, – – –	47°.29	512	51°.08
Paris, – – –	48.50	116	51.08
Vienna, – – –	48.12	551	50.54
Manheim, – –	49.20	384	50.18

Whence, in the longitudes of Paris and Buda, and between the latitudes of 47° and 48°, and almost at the level of the sea, the mean temperature is from 50°.9 to 51°.44.
Under the same longitudes, we have,—

	Elevation in Feet.	Mean Temp.
Geneva, – – –	1177	49°.28
Zurich, – – –	1437	47.84
Munich, – – –	1711	50.74

	Elevations.		Mean Temperatures.		
Places situated between 46°—47° of North Lat.	**Metres.**	**Feet.**	**Of the Year.**	**Of the Coldest Months.**	**Of the Warmest Months.**
Level of the sea, –	0		53°.60	36°.32	69°.80
Geneva, – –	359	1177	49.64	34.16	66.56
Tegernsee, – –	744	2440	42.44	22.10	59.36
Peissenberg, –	995	3264	41.00	20.84	57.02
Chamouni, – –	1028	3372	39.20		55.40
Hospice de St Gothard,	2076	6809	30.38	15.08	46.22
Col de Géant, –	3436	11270	21.20		36.50

In comparing the mean temperature of superincumbent beds of air, I find that the isothermal line of 41°, which, in the parallel of 45°, is found at the height of 1000 metres, (3280 feet), makes the equatorial mountains of an absolute elevation of 4250 metres, (13,940 feet). It had, however, been long believed, after Bouguer, that the inferior limits of perpetual snows characterised every where a bed of air, whose mean temperature was 32°; but I have shewn in a Memoir read to the Institute in 1808,[79] that this supposition is contrary to experience. By uniting good observations, I have found, that at the limit of perpetual snows, the mean temperature of the air is,—

	Metres.	Feet.	Mean Temp. of Limit of Perpetual Snows.
At the Equator,	4800	15,744	34°.70
In Temperate Zone,	2700	8,856	25.34
In Frigid Zone, in Lat. 68°—69°,	1050	3,444	21.20

As the heat of the higher regions of the atmosphere depends on the radiation of the plains, we may conceive, that, under the same geographical

---

Berne, – – –	1755	49.28
Marschling, – –	1834	51.98*
Coire, – – –	1991	48.92**

* Heated by the winds of Italy.
** In spite of the winds of Italy.

By taking the means of these results, we cannot mistake the influence of *small elevations*, or of very *extensive plateaus*, on the decrease of the mean temperature.

[79] *Observations Astronomiques*, tom. i. p. 136.

parallels, we cannot find, in the transatlantic climates, (on the declivities of rocky mountains), the isothermal lines at the same height above the level of the sea as in European climates. The inflexions which these lines experience, when traced on the surface of the globe, necessarily influence their position in a vertical plane, whether we unite in the atmosphere points placed under the same meridians, or consider only those that have the same latitude.

Hitherto we have attempted to determine the mean temperatures which correspond under the Equator and in Lat. 45 and | 277 | 47° to beds of the atmosphere equally elevated. This determination is founded on stationary observations, and indicates the mean state of the atmosphere. General physics has its numerical elements, as well as the system of the world; and these elements, so important in the theory of barometrical measurements and in that of refractions, will be perfected in proportion as natural philosophers shall direct their attention to the study of general laws.

Height, in		**Equatorial Zone, from 0°——10°.**		**Temperate Zone, from 45°——47°.**	
Metres.	**Feet.**	**Mean Temp.**	**Diff.**	**Mean Temp.**	**Diff.**
0	0	81°.50	10°.26	53°.60	12°.60
974	3195	71.24	6.12	41.00	9.36
1949	6393	65.12	7.38	31.64	8.28
2923	9587	57.74	13.14	23.36	
3900	12792	44.60	9.90		
4872	15965	34.70			

This Table proves, in conformity with the deductions of theory, that in the mean state of the atmosphere, the heat does not decrease uniformly in an arithmetical progression. In the Cordilleras, (and the fact is extremely curious), we observe the decrease getting less and less between 1000 and 3000 metres, particularly between 1000 and 2500 metres of elevation, and then increasing anew from 3000 to 4000 metres. The strata, where the decrease attains its *maximum* and its *minimum*, are in the ratio of 1 to 2. From the height of the Caraccas to that of Popayan and Loxa, 1000 metres produce a difference of 6°.3. From Quito to the height of Paramos, the same 1000 metres change the mean temperature more than 12°.6. Do these phenomena depend only on the configuration of the Andes, or are they the effect of the accumulation of clouds in the aërial ocean? In considering that the Andes form an enormous mass, 3600 metres (11,808 feet) high, from which rise peaks or domes insulated and covered with snow, we may conceive how, from the point where the mass of the chain diminishes so rapidly, the heat decreases also with rapidity. It is not easy, however, to ex- | 278 | plain, by an analogous cause, why the progressive cooling diminishes between 1000 and 2000 metres. The great plateaus of the Cordilleras commence only at the height of 2600 or 2900 metres, (8528 or 9512 feet); and I am of opinion, that the slowness with which the heat decreases in the stratum of air between 1000 and 2000, is the triple effect of the extinction of light, or the absorption of the rays in the clouds,—of the formation of rain,—and the obstacle which the clouds oppose to the free passage of radiant heat. The bed of air of which we speak, is the region in which are suspended the large clouds which the inhabitants of the plains see above their heads. The decrease of temperature, which is very rapid from the plains to the region of clouds, becomes less rapid in that region; and if this change is less sensible in the temperate zone, it is no doubt because at the same height, the effect of radiation there is less sensible than above the burning plains of the equinoctial zone. In these zones, too, the cooling appears to follow the same law in the beds of air of equal temperature; but the force of radiation varies with the temperature of the radiating beds.

The results which we have now discussed, deserve the preference over those which are deduced from observations made during excursions to the tops of some lofty mountains. The first give for the

	Metres.	**Cent.**		**Fahr.**		**Metres.**
Equinoctial Zone,	0—4900	1°	or	1°.8	for	187[80]

[80] This is the mean result or the measure of the distribution of heat in the whole column of air. The partial results are from the back of the Andes.

Heights in Metres.	**Cent.**		**Fahr.**		**Metres.**
0—1000	1°	or	1°.8	for	170
1000—2000	1	or	1.8	for	294
2000—3000	1	or	1.8	for	232
3000—4000	1	or	1.8	for	131
4000—5000	1	or	1.8	for	180

In these numbers, we recognise, as in the above Table, the influence of the region of clouds upon the decrease of heat. In order to shew the utility of these numerical ratios, I shall give here the approximate calculation of the height of the plain of Thibet, deduced from the mean temperature of the month of October, which, according to the former, is 42°.26. As the latitude of Tissoolumbo 29°, gives 69°.8 for

Temperate Zone,	0—2900	1°	or	1°.8	for	174

| 279 |
The last give for the

	Cent.		Fahr.	Metres.
Equinoctial zone, –	1°	or	1°.8	190
Parallels of 45°—47°, –	1°	or	1°.8	160—172[81]

This agreement is no doubt very remarkable, and the more so, as, in comparing stationary with insulated observations, we confound the mean state of the atmosphere in the course of a whole year with the decrease which corresponds to a particular season, or a particular hour of the day. M. Gay-Lussac found, in his celebrated aëronautical voyage from 0 to 7000 metres, (0 to 22,960 feet), a centigrade degree for 187 metres, near Paris, at a period when the heat of the plains was nearly equal to that of the equinoctial region. It is on account of this observed equality in the decrease of heat, in reckoning from the standard temperature of the plains, that the astronomical refractions corresponding to angles below 10°, have been found the same under the equator and in temperate climates. This result, contrary to the theory of Bouguer, is confirmed by observations which I have made in South America, and by those of Maskelyne at Barbadoes, calculated by M. Oltmanns.

We have seen, that between the tropics, on the back of the Cordilleras, we find, at 2000 metres of elevation, I will not say the climate, but the mean temperature of Calabria and of Sicily. In our temperate zone, in 46° of Lat. we meet at the same elevation with the mean temperature of Lapland.[82] This comparison | 280 | leads us to an exact knowledge of

the mean temperature of the plain; and as at Mount St Gothard, the mean temperature of October is even a little below that of the whole year, it is probable that the height of the plain of Great Thibet exceeds from 2900 to 3000 metres.—See my Memoir on the Mountains of India in the *Ann. de Chim. et de Phys.* 1817.

Note by the Editor.

As the cold meridian of the globe passes through the plains of Great Thibet, we conceive that the mean temperature of Lat. 29° in that plain, when reduced to the level of the sea, will be about 65°, and therefore that the height of the plain of Great Thibet will not exceed 2800 metres or 9184 feet.—D. B.

[81] Saussure gives for the summer 160 metres, (525 feet); for winter 230, (754 feet); and for the whole year 195, (640 feet). M. Ramond gives 165, (538 feet).M. D'Aubuisson 173 metres, (567 feet).—See *Journ. de Phys.* tom. lxxi. p. 37.; *De la Formul. Barometr.* p. 189.; and my *Recueil d'Obs. Astron.* tom. i. p. 129.

[82] As the temperature varies very little in the course of a whole year in the equinoctial zone, we may form a pretty correct idea of the climate of the Cordilleras, by comparing them to the temperature of certain months in France or in Italy. We find in the plains of Orinoco the month of August of Rome;

the numerical ratios between the elevations and the latitudes, ratios which we find indicated with little precision in works on physical geography.

The following are the results which I have obtained from exact data in the temperate zone, from the plains to 1000 metres of elevation. Every hundred metres of perpendicular height, diminishes the mean temperature of the year, by the same quantity that a change of 1° of latitude does in advancing towards the Pole. If we compare only the mean temperature of summer, the first 1000 metres are equivalent to 0°.81 Fahr. From 40° to 50° of latitude, the mean heat of the plains in Europe decreases in Europe 12°.6 of Fahr.; and this same decrease of temperature takes place on the declivity of the Swiss Alps from 0 to 1000 metres of elevation.

Differences of Latitude, Compared with Differences of Elevation.	**Mean Heat of the Year.**	**Mean Heat of Summer.**	**Mean Heat of Autumn.**
I. At the Level of the Sea.			
a. Latitude, 40°, –	63°.14	77°.00	62°.60
b. Latitude, 50°, –	50.54	64.40	
II. On the Declivity of Mountains.			
a. At the foot in 46° of Lat.	53.60	68.00	51.80
b. At an elevation of 1000 metres,	41.00	58.46	42.80

These numerical ratios are deduced from observations made on the temperature of the air. We cannot measure the quantity of heat produced by the solar rays on the parenchyma of plants, or in the interior of fruits which receive their colour in ripening. The fine experiment of MM. Gay-Lussac and Thenard, the combustion of chlorine and hydrogen, proves what a powerful action direct light exercises on the molecules of bodies. But as the extinction of light is less upon the mountains in dry and rarified air, maize, fruit-trees, and the vine, still flourish at heights which, according to our thermometrical observations made in the air, and far from the ground, we ought to suppose | 281 | too cold for the cultivation of plants useful to man. M. De Candolle, indeed, to whom the geography of vegetables owes so many valuable observations, has seen the vine cultivated in the south of France at 800 metres (2624 feet) of absolute height, when, under the same meridian, this same cultivation went on

at Popayan, (2988 feet), the month of August of Paris; at Quito (4894 feet), the month of May; in the Paramos, (5904 feet), the month of March at Paris.

with difficulty at 4 of latitude farther north; so that if we consider only the ratios in France, an elevation of 100 metres, (328 feet), appears to correspond, not to 1°, but to half a degree of latitude.[83]

(To be concluded in next Number.)

On Isothermal Lines, and the Distribution of Heat over the Globe. By Baron Alexander de Humboldt.

(Concluded from Vol. IV. p. 281.)

I shall now conclude this Memoir by the enumeration of the most important results which have been obtained by Baron Von Buch, M. Wahlenberg, and myself, on the distribution of heat in the interior of the earth, from the Equator to 70° of N. Lat., and from the plains to 3600 metres (11,808 feet) of elevation. I shall limit myself to an enumeration of the facts. The theory by which these facts are connected, will be found in the fine analytical work with which M. Fourier will soon enrich natural philosophy.

The interior temperature of the earth is measured either by the temperature of subterraneous excavations, or by that of springs. This kind of observation is very liable to error, if the traveller does not pay the most minute attention to local circumstances, which are capable of altering the results.[84] The air, when cooled, accumulates in caverns, which communicate with the atmosphere by perpendicular openings. The humidity of rocks depresses the temperature by the effect of evaporation. Caverns that have little depth are more or less warmed, according to the colour, the density, and the moisture, of the strata of stone in which nature has hollowed them. Springs indicate too low a temperature, if they descend rapidly from a considerable height upon inclined strata. There are some under the torrid zone and in our climate, which do not vary in their

[83] See my *Prolegomena de Distributione Plantarum*, p. 151.-163. The small differences between the numbers given in the Prolegomena and in this Memoir, written subsequently, should be ascribed to the constant desire which I have had to perfect the mean results.

[84] Baron von Buch, in the *Bibl. Brit.* tom. xix. p. 263.; Saussure, *Voyages*, sect. 1418.; Wahlenberg, *De Veget. Helvet.* Pl. 77.—84.; Gilbert, *Annalen*, 1812, p. 150. 160. 277.; Lambert, *Pyrometrie*, p. 296. Dr Roebuck appears to have been the first who entertained exact notions on the temperature of springs, and upon their relation to the mean temperature of the air; Phil. Trans. 1775, vol. lxv. p. 461.

temperature throughout the whole year more than half a degree; and there are others which shew the | 29 | mean temperature of the earth only by observing them every month, and taking the mean of all the observations. From the Polar circle to the Equator, and from the tops of mountains towards the plains, the progressive increase of the temperature of springs diminishes with the mean temperature of the ambient air. The temperature of the interior of the earth is, at

	Lat.	**Temp. Fahr.**		**Lat.**	**Temp. Fahr.**
Vadso,	70°.0	35°.96	Paris,	48° 50′	53.°6
Berlin,	52.31	49.28	Cairo,	30 2	72.5

In equinoctial America, I have found it in the plains from 77° to 78°.8.

The following are examples of the decrease of temperature from the plains to the tops of mountains.

	Alt. in Feet.	**Temp. Fahr.**
Spring of Utliberg, near Zurich, –	1532	48°.92
Ditto of Rossbaden at St Gothard, –	7016	38.30

Between the Tropics I have found,

	Alt. in Feet.	**Temp. Fahr.**
Springs of Cumanacoa, – –	1,148	72°.5
Ditto Montserrate, above Santa Fé de Bogota,	10,680	59.9
Ditto in the Mine of Hualgayoc in Peru,	11,759	53.24

In the plains, and to the height of 3280 feet, between the parallels of from 40° to 45°, the mean temperature of the earth is nearly equal to that of the ambient air; but very accurate observations by Baron Von Buch and Wahlenberg tend to prove, that in high latitudes, towards the top of the Swiss Alps, for example, beyond the height of 1400 or 1500 metres (4592 or 4920 feet), the springs and the earth are 5°.4 warmer than the air.

Zone of 30°—55°.	Lat.	Mean Temp. of Air, Fahr.	Temp. of the Interior of the Earth.
Cairo, – –	30° 2′	72°.68	72°.50
Natchez, – –	31 28	64.76	64.94
Charlestown, – –	33 0	63.14	63.50
Philadelphia, – –	39 56	53.42	52.16
Geneva, – – –	46 12	49.28	50.74
Dublin, – – –	53 21	49.10	49.28
Berlin, – – –	52 31	47.30	49.28
Kendal, – –	54 17	46.22	47.84
Keswick, – –	54 33	48.02	48.56
Zone of 55°—70°.			
Carlscrona, – –	56 6	46.04	47.30
Upsal, – –	59 51	41.90	43.70
Umeo, – –	63 50	33.26	37.22
Vadso, – –	70 0	29.66	35.96

| 30 |

At Enontekies, in 68 1/2° of Lat. the difference between the mean temperatures of the earth and the air, is so great as 7°.74. Analogous differences are observed on the back of the Alps, at the altitude of 1400 metres (4592 feet).

In the following small table, I have added the mean temperature of the atmosphere, by supposing, with M. Ramond, that there is a decrease of 1° centigrade for 164 metres (1° Fahr. for 300 feet nearly), and by placing the temperature of 32° (according to observations made at the Hospice of St Gothard), at 1950 metres (6396 feet) of elevation.

		Temperature.	
	Alt. in Feet.	Springs.	Air.
Rigi Kaltebad, –	4717	43°.7	38°.12
Pilate, – –	4858	41.0	37.40
Blancke Alp, –	5786	37.4	35.78
Rossbaden, – –	7016	38.3	31.38

It may be objected, that in the Alps of Switzerland, the temperature of springs has only been observed from the beginning of June to the end of September, and that the differences between the air and the interior of the earth would almost entirely disappear, if we knew the temperature of the springs during the whole year. It must not be forgotten, however, that the springs of the Alps did not vary in the space of *four* months at the time of the observations of M. Wahlenberg;—that among the small number of scanty springs which indicate changes of temperature in different seasons, these variations amount from June to September to 11° or 15°;—and that several springs, particularly those which are very copious, do not vary during a whole year more than half a degree of Fahrenheit.

It appears to me, therefore, sufficiently certain, that where the earth is covered with a thick stratum of snow, while the temperature of the air descends to 15° or — 4° of Fahrenheit, the temperature of the earth is above the mean temperature of the air.

When we consider what a large portion of the globe is covered with the sea, and examine the temperature of the deepest waters,[85] we are constrained to admit, that in islands, along | 31 | coasts, and perhaps even in continents of small extent, thc interior heat of the earth is modified by the proximity of the strata of rocks on which the waters of the ocean rest.

I have considered successively in this memoir, the distribution of heat,

1. At the surface of the globe.
2. On the declivity of mountains.
3. In the ocean.
4. In the interior of the earth.

In explaining the theory of isothermal lines and their inflexions, which determine the different systems of climates, I have endeavoured to reduce the phenomena of temperature to empirical laws. These laws will appear much more simple, when we shall have multiplied and rectified by degrees the numerical elements which are the results of observation.

In the following general Table of the distribution of heat, the temperatures are expressed in degrees of Fahrenheit; the longitudes are reckoned from east to west of the meridian of the observatory of Greenwich. The mean temperatures of the seasons have been calculated, so that those of the months of *December, January*, and *February*, form the mean temperature of *Winter*. An asterisk (*) is prefixed to those places whose mean temperatures have been most accurately determined, and in general by means of 8000 observations. The isothermal lines have a convex summit in Europe, and two concave summits in Asia and Eastern America. | 32 | | 33 |

[85] At Funchal in Madeira, the temperature of caverns appears to be 61°.16, and consequently 7°.2 *below* that of the air.—*Phil. Trans*. 1778, p. 372.

Position.			Distribution of Heat in the different Seasons.				*Maximum* and *Minimum*.				
Isothermal Bands.	**Names of Places.**	**Lat.**	**Long.**	**Height in Feet.**	**Mean Temp. of the Year.**	**Mean Temp. of Winter.**	**Mean Temp. of Spring.**	**Mean Temp. of Summer.**	**Mean Temp. of Autumn.**	**Mean Temp. of Warmest Month.**	**Mean Temp. of Coldest Month.**
Isothermal	Nain, –	57° 8′	61°20′ W	0	26°.42	— 0°.60	23°.90	48°.38	33°.44	51°.80	—11°.20[1]
Bands from 32° to 41°.	* Enontekies,	68 30	20 47 E	1356	26.96	+ 0.68	24.98	54.86	27.32	59.54	— 0.58[2]
	Hospice de St Gothard,	46 30	8 23 E	6390	30.38	18.32	26.42	44.96	31.82	46.22	15.08 [3]
	North Cape,	71 0	25 50 E	0	32.0	23.72	29.66	43.34	32.08	46.58	22.10 [4]
	* Uleo, –	65 3	25 26 E	0	35.08	11.84	27.14	57.74	35.96	61.52	7.70 [5]
	* Umeo, –	63 50	20 16 E	0	33.26	12.92	33.80	54.86	33.44	62.60	11.48 [6]
	* St Petersburg,	59 56	30 19 E	0	38.84	17.06	38.12	62.06	38.66	65.66	8.60 [7]
	Drontheim,	63 24	10 22 E	0	39.92	23.72	35.24	61.24	40.10	64.94	19.58 [8]
	Moscow, –	55 45	37 32 E	970	40.10	10.78	44.06	67.10	38.30	70.52	6.08 [9]
	Abo, –	60 27	22 18 E	0	40.28	20.84	38.30	61.88	40.64	—	— [10]

(*continued*)

(continued)

		Position.			Distribution of Heat in the different Seasons.				*Maximum* and *Minimum.*		
Isothermal Bands from 41° to 50°.	* Upsal, –	59 51	17 38 E	0	42.08	24.98	39.38	60.26	42.80	62.42	22.46 [11]
	* Stockholm, –	59 20	18 3 E	0	42.26	25.52	38.30	61.88	43.16	64.04	22.82 [12]
	Quebec, –	46 47	71 10 W	0	41.74	14.18	38.84	68.00	46.04	73.40	13.81 [13]
	Christiania, –	59 55	10 48 E	0	42.8	28.78	39.02	62.60	41.18	66.74	28.41 [14]
	* Convent of Peyssenburg,	47 47	10 34 E	3066	42.98	28.58	42.08	58.46	42.98	59.36	30.20 [15]
	* Copenhagen,	55 41	12 35 E	0	45.68	30.74	41.18	62.60	48.38	65.66	27.14 [16]
	* Kendal, –	54 17	2 46 W	0	46.22	30.86	45.14	56.84	46.22	58.10	34.88 [17]
	Malouin Islands,	51 25	59 59 W	0	46.94	39.56	46.58	53.06	48.46	55.76	37.40 [18]
	* Prague, –	50 5	14 24 E	0	49.46	31.46	47.66	68.90	50.18	—	— [19]
	Gottingen, –	51 32	9 53 E	456	46.94	30.38	44.24	64.76	48.74	66.38	29.66 [20]
	* Zurich, –	47 22	8 32 E	1350	47.84	29.66	48.20	64.04	48.92	65.66	26.78 [21]
	* Edinburgh, –	55 57	3 10 W	0	47.84	38.66	46.40	58.28	48.56	59.36	38.30 [22]
	Warsaw, –	52 14	21 2 E	0	48.56	28.76	47.48	69.08	49.46	70.34	27.14 [23]
	* Coire, –	46 50	9 30 E	1876	48.92	32.36	50.00	63.32	50.36	64.58	29.48 [24]
	Dublin, –	53 21	6 19 W	0 *	49.10	39.20	47.30	59.54	50.00	61.16	35.42 [25]
	Berne, –	46 5	7 26 E	1650	49.28	32.00	48.92	66.56	49.82	67.28	30.56 [26]
	* Geneva, –	46 12	6 8 E	1080	49.28	34.70	47.66	64.94	50.00	66.56	34.16 [27]
	* Manheim, –	49.29	8 28 E	432	50.18	38.80	49.64	67.10	49.82	68.72	33.44 [28]
	Vienna, –	48 12	16 22 E	420	50.54	32.72	51.26	69.26	50.54	70.52	26.60 [29]

		Position.			Distribution of Heat in the different Seasons.				*Maximum* and *Minimum.*		
Isothermal Bands from 50° to 59°.	* Clermont, –	45 46	3 5 E	1260	50.00	34.52	50.54	64.40	51.26	66.20	28.04 [30]
	* Buda, –	47 29	19 1 E	494	51.08	33.98	51.08	70.52	52.34	71.60	27.78 [31]
	Cambridge, (U.S.)	42 25	71 3 W	0	50.36	33.98	47.66	70.70	49.82	72.86	29.84 [32]
	* Paris, –	48 50	2 20 E	222	51.08	38.66	49.28	64.58	51.44	65.30	36.14 [33]
	* London, –	51 30	0 5 W	0	50.36	39.56	48.56	63.14	50.18	64.40	37.76 [34]
	Dunkirk, –	51 2	2 22 E	0	50.54	38.48	48.56	64.04	50.90	64.76	37.75 [35]
	Amsterdam,	52 22	4 50 E	0	51.62	36.86	51.62	65.84	51.62	66.92	35.42 [36]
	Brussels, –	50 50	4 22 E	0	51.80	36.68	53.24	66.20	51.08	67.28	35.60 [37]
	* Franeker, –	52 36	6 22 E	0	51.80	36.68	51.08	67.28	54.32	69.08	32.90 [38]
	Philadelphia,	39 56	75 16 W	0	53.42	32.18	51.44	73.94	56.48	77.00	32.72 [39]
	New York,	40 40	73 58 W	0	53.78	29.84	51.26	79.16	54.50	80.78	25.34 [40]
	* Cincinnati, –	39 6	82 40 W	510	53.78	32.90	54.14	72.86	54.86	74.30	30.20 [41]
	St Malo, –	48 39	2 1 W	0	54.14	42.26	52.16	66.02	55.76	66.92	41.74 [42]
	Nantes, –	47 13	1 32 W	0	54.68	40.46	54.50	68.54	55.58	70.52	39.02 [43]
	Pekin, –	39 54	116 27 E	0	54.86	26.42	56.30	82.58	54.32	84.38	24.62 [44]
	* Milan, –	45 28	9 11 E	390	55.76	36.32	56.12	73.04	56.84	74.66	36.14 [45]
	Bourdeaux, –	44 50	0 34 W	0	56.48	42.08	56.48	70.88	56.30	73.04	41.00 [46]
Isothermal Band from 59° to 63°.	Marseilles, –	43 17	5 22 E	0	59.00	45.50	57.56	72.50	60.08	74.66	44.42 [47]
	Montpellier,	43 36	3 52 E	0	59.36	44.06	56.66	75.74	60.98	78.08	42.08 [48]
	* Rome, –	41 53	12 27 E	0	60.44	45.86	57.74	75.20	62.78	77.00	42.26 [49]
	Toulon, –	43 7	5 50 E	0	62.06	48.38	60.80	75.02	64.40	77.00	46.40 [50]
	Nangasacki,	32 45	129 55 E	0	60.80	39.38	57.56	82.94	64.22	86.90	37.40 [51]
	* Natchez, –	31 28	90 30 W	180	64.76	48.56	65.48	79.16	66.02	79.70	46.94 [52]

(*continued*)

(continued)

		Position.			Distribution of Heat in the different Seasons.				*Maximum* and *Minimum.*		
Isothermal Band from 68° to 77°.	* Funchal, –	32 37	16 56 W	0	68.54	64.40	65.84	72.50	72.32	75.56	64.04 [53]
	Algiers, –	36 48	3 1 E	0	69.98	61.52	65.66	80.24	72.50	82.76	60.08 [54]
Isothermal Bands above 77°.	* Cairo, –	30 2	31 18 E	0	72.32	58.46	73.58	85.10	71.42	85.82	56.12 [55]
	* Veracruz, –	19 11	96 1 W	0	77.72	71.96	77.90	81.50	78.62	81.86	71.06 [56]
	* Havannah, –	23 10	82 13 W	0	78.08	71.24	78.98	83.30	78.98	83.84	69.98 [57]
	* Cumana, –	10 27	65 15 W	0	81.86	80.24	83.66	82.04	80.24	84.38	79.16 [58]

| 34 |

[1] Coast of Labrador. Two years of observations. Floating ice towards the east. A transatlantic climate. Mean temp. of Oct. about 32°.72; Nov. 26°.68.
[2] Centre of Lapland. A European climate. Fine vegetation. June, 49°.46; July, 59°.54; Aug. 55°.94; Sept. 41°.74; Oct. 27°5; Nov. 12°.38. Inland situation. Specimen of a continental climate.
[3] Eleven years of observations, calculated afresh in decads by Wahlenberg. Thermometer verified by Saussure. Mean temp. of seven months of the year below 32°. Winds blow from Italy in the winter. *Minimum* observed in the winter—0°.4; in Aug. at noon, in the shade, *maximum* 54°.5; the nights in Aug. frequently from 33°.8 to 29°.3; the mean temp. of Oct. 31°.1 represents that of the whole year; at the Col de Géant, 10,598 feet high, the mean temp. of July is 36°.5. We find 32° to be the mean temp. in Europe in 45° of latitude, at 5,400 feet high; at the parallel of the Canaries, at 12,300 feet; in the Andes, under the Equator, at 16,500 feet.
[4] Buch, *Voy. en Norw.* ii. 416. Specimen of the climate of the islands and coasts in the north of Europe. April, 30°.02; May, 33°.98; Oct. 32°; Nov. 25°.88. At Alten, Lat. 70°, mean temp. of July, 63°.5; a continental climate.
[5] Finland, eastern coast. May, 40°.82; June, 55°.04; July, 61°.52; Aug. 56°.66; Sept. 46°.58; Oct. 38°.66; Nov. 24°.62. Julin and Buch.
[6] Eastern coast of Western Bothnia. Dr Nœzen. March, 23°.18; April, 33°.98; Oct. 38°.12; Nov. 24°.62.
[7] Euler. Mean temp. of the year, 37°.94. Inochodzow. *Acta. Petr.* xii. 519,—533.
[8] Two years. Berlin, in the *Mem. de l'Acad. de Drontheim*, iv. 216. April, 34°.34; May, 50°.74; Oct. 39°.2; Nov. 27°.68. Climate of the west coast of Europe.
[9] Four years. *Journal de Phys.* xxxix. 40. A continental climate. Winter colder, and summer warmer than at Petersburg. Eastern part of Europe; height as taken from Stritter. (Chamouni, Lat. 46° 1′; Long. 6° 18′ E.; height, 3,168 feet; mean temp. 39°.2.)
[10] Twelve years. Kirwan. Cotte, mean of the year, 41°.18; of the summer, 67°.46; too high. West coast of Finland. | 35 |
[11] Observations from 1774 to 1804, made by Mallet, Prosperin, Holmquist, and Schleling, calculated by M. De Buch, *Voy. de Norw.* ii. 309. It is, perhaps, the place the mean temp of which is the best determined. Winters more serene than at Stockholm; colder on account of the radiation of the ground and the air.
[12] Thirty-nine years of observations, 15 of which are very good. Wargentin. Cotte, mean temp. of the year, 44°.24. Five months below 32° as at Petersburg.
[13] Four years. A transatlantic climate.
[14] Buch, two years. Mean temp. of the winter often scarcely 31°.1. West coast.
[15] Alps of Bavaria. Six years' observations, calculated by M. Wahlenberg. Many fruit trees. Convent of Tegernsée, in Bavaria, height of, 2,292 feet; mean temp. of 1785, 42°.44; Peyssenberg, 41°.
[16] Bugge. Three months below 32°. Under the Equator, mean temp. of 44°.6, at an elevation of 18,000 feet.
[17] Dalton. West of England. Climate of islands; springs 47°.84. Keswick, Lat. 54° 33′, Long. 3° 3′ W.; mean temp. 48°.02; springs, 48°.56.
[18] Kirwan. Scarcely two years' observations. Southern latitude.
[19] Strnadt. Fifteen years. Climate of the continent of Europe.
[20] Maier.
[21] Six years' observations of M. Escher, calculated by Wahlenberg. The town is situated in a hollow, to which those warm winds cannot penetrate, that render the winters more temperate in the other parts of Switzerland.
[22] The calculation has been made from six years of excellent observations, by Professor Playfair; during this time the thermometer was never seen above 75°.74.* Vegetation continues from March 20. to Oct. 20.; mean temp. of these seven months is from 55°.76 to 50°.90, according as the years are more or less fruitful; wheat does not ripen if the mean temp. descends to 47°.66.
*See *Edinburgh Transactions*, vol. ix. p. 209.—Ed. | 36 |
[23] Guittard. Only three years. Mean temp. a little too high. Eastern part of Europe. A continental climate.
[24] Four years of observations, by M. de Salis Sewis, calculated by M. Wahlenberg. Mountains of the Grisons.
[25] Kirwan. *Irish Trans.* viii. 203. and 269. Specimen of the climate of the islands. Coldest days, 23°; interior of the ground, 49°.28. Hamilton.
[26] The climate of Berne is a continental climate, in comparison with that of Geneva; there is no lake near it.
[27] Seven years of observations. Saussure. Mean temp. 50°74. *Voy.* § 1418. I find the mean temp. from 1796—1815, 49°.7. Interior of the earth, 51°.98. Pictet, *Bibliotheque Brit.* 1817, iv. 109.
[28] Six years.
[29] Austria. Berlin, Lat. 52° 31′; mean temp. probably 46°.4 to 47°.3; according to Beguelin, 48°.74; springs, 49°.28. Ratisbon, Lat. 49°; height, 1,104 feet; mean temp. 47°.66. Munich, Lat. 48° 8′; height, 1,608 feet; mean temp. 50°.74.
[30] Ramond. Seven years of excellent observations. The mean of the months, at noon, well ascertained; winter, 39°.92; spring, 57°.02; summer, 70°.88; autumn, 57°.92. *Mem. Inst.* 1812, p. 49. Cotte, mean temp. 51°.26.
[31] Wahlenberg, *Flor. Carp.* p. 90. Continental climate. Height of the observatory, 474 feet.
[32] Two years, near Boston, in New England. Transatlantic climate. The thermometer sometimes descends to 0°.
[33] Eleven years (1803—1813) of observations made at the observatory. A greater number of years will, perhaps, give the mean temp. a little higher. Vaults, 53°.06. Kirwan finds for Paris, from seven years of observations of unequal value, 51°.62; he fixes upon 52°.7. Cotte, from 29 years of observations, (*Journ. de Phys.* 1782, July), 53°.24. Cotte, for 33 years, (1763—1781, *Mem. Instit.* iv. 266.), 52°.34. The extraordinary year of 1816 offers the mean temp. of 48°.74; winter, 37°.04; spring, 48°.92; summer, 59°.54; autumn, 50°: the preceding year, 1815, offers a mean temp. of 50°.74; winter, 37°.04; spring, 52°.7; summer, 62°.78; autumn, 50°.74. Arago. Mean temp. of Montmorency, for 33 years, 50°.74; height, 498 feet. Cotte, | 37 | Strasburg, Lat. 48° 34′; height, 480 feet; mean temp. 49°.28. Herrenschneider.
[34] Dr Young. Mean temp. varies from 47°.84 to 51°.62, (*Lectures*, ii. 453.) Cavendish, (*Trans.* 1788, p. 61.), 48°.74, Roebuck, Hunter, and Kirwan, 51°.62. Horsley, 51°.26. According to Kirwan, the four seasons in London are, 39°.56, 50°.9, 64°.76, 51°.98; at Paris, 36°.68, 51°.08, 65°.84, 52°.52; from which results, London, 51°.62; Paris, 51°.44. Cotte (*Journ. de Phys.* xxxix. 36.) thinks London is 51°.26, and Paris, 52°.34. The difference which we observe in cultivated plants depends less upon mean temp. than upon direct light, and the serenity of the atmosphere.

[35] Seven years. Cotte. Lisle, 48°.38; Rouen, 51°.44; Cambray, 51°.98; Soissons, 53°.42; Rethel, 53°.24; Metz, 52°.88; Nancy, 51°.98; Etampes, 51°.08; L'Aigle, 50°.9; Brest, 54°.14; Mayenne, 51°.98.

[36] Mohr, and Van Swinden. Five years.

[37] Thirteen years. Temperature rather too high?

[38] Eleven years. Van Swinden. From 1771—1783. Mean temp. 51°.26.

[39] Concave transatlantic summit. Seven years of observations give 54°.86; for the four seasons, 33°.98, 53°.06, 75°.2, 56°.12. Rush, 52°.52, (Drake's *View of Cincin.* p. 116.) Coxe, 54°.14. M. Legaux finds for 17 years, for Springmill on the Schuylkill, Lat. 40° 50′; mean temp. 53°.42. Springs, near Philadelphia, 54°.86, Warden.

[40] Two years only. Retif de la Serve. The thermometer sometimes descends to — 4° in the parallel of Naples! Springs, 54°.86. Ipswich, Lat. 42° 38′; mean temp. 50°. Williamsburg, in Virginia, 58°.1. Cotte and Kirwan. Transatlantic climates.

[41] Transatlantic climates west of the Alleghanys. Good observations, from 1806—1813. Col. Mansfield, (Drake, p. 93.) *Minimum* of the winter, from 5° to — 9°.4; Jan. 1797, as low as — 16°.6, for 39° latitude. *Maximum* 89°.6 to 107°.6 in the shade, without reflection; 1/3 of all the winds SW.; springs near Cincinnati, 54°.32. Little snow falls; but it is abundant between Lat. 40° and 42°.

[42] Three years only. Bougourd. Dijon, height, 810 feet; | 38 | Lat. 47° 19′; mean temp. 50°.9. Besançon, height, 804 feet; Lat. 47° 14′; mean temp. 51°.26.

[43] Six years. Duplessis, and Boudan. Temperature of the summer too high? Rochelle, 53°.06; Poitiers, 52°.7.

[44] Amyot. Six years. Concave. Asiatic summit. Three months below 32°, as at Copenhagen; the summer like that at Naples.

[45] One of the best determined points. The years 1789—1812 are calculated in decads of days. Observations of the Astronomer Reggio, April, 55°.76; Oct. 58°.1. The two decads which approach the nearest to the mean temp. of the year, are, the first of April, 53°.24; and the last of Oct. 54°.68. The mean temps. for January have varied in 10 years from 24°.98 to 38°.48; those of July, from 71°.42 to 78°.44; the mean of the years, from 54°.5 to 57°.2. (Reggio, taking only 24 *maxima* and *minima* in a year for 1763—1798; mean temp. 55.4, *Ephem. Mil.* 1779, p. 82.)

[46] Ten years. Guyot. Lyons, 528 feet, 55°.76. Mafra, near Lisbon, Lat. 38° 52′; height, 600 feet; mean temp. 56°.3, too small. *Mem. de Lisbonne*, ii. 105—158.

[47] Seven years, (1777—1782). St Jacques de Sylvabelle. The thermometer sometimes descends to 23°. Cotte, *Traité de Met.* ii. 420.) 34 years (Raymond, in *Mem. de la Soc. de Med.* 1777, p. 86.) give 62°.06. Cotte (*Journ. de Phys.* xxxix. 21.) fixes it at 58°.64. Kirwan, at 61°.24. The observations made at the Royal Observatory of Marseilles can alone decide.

[48] Ten years. Nismes, 60°.26; Perpignan, 59°.54; Tarascon, 59°.9; Arles, 59°; Rieux, 57°.2: Montauban, 55°.58; Tonains, 54°.86; Dax, 54°.14; Rodez, 57°.02; Aix, 56°.66. Under the equator, 57°.74, at 9,000 feet of elevation.

[49] William Humboldt. Calandrelli, 60°.08. The thermometer sometimes descends to 24°.5, and rises to 99°.5. Naples, 67°.1; Toaldo, probably 63°.5; Florence, 61°.52; Tartini, too high; Lucca, 60°.44; Genoa, 60°.26; Bologna, 56°.3; Verona, 55°.76; Venice, 56°.48; Padua, 56°.3. Kirwan regards it as an established fact, that in Europe the mean temp. in Lat. 40°, is 61°.88; in Lat. 50°, 52°.52.

[50] Only two years. Barberet, and d'Angos. Sheltered by mountains. Estimate a little too high. | 39 |

[51] Japan. A single year. *Voy. de Thunberg*, p. 121. Climate of islands. Under the Equator, 64°.4, at a height of 6000 feet.

[52] West of the Alleghanys, in Louisiana. Four years. Dunbar. Transatlantic climate.

[53] Madeira. Heberden. Climate of islands. St Croix, of Teneriffe, 71°.42. The remainder of the island of Teneriffe, in the plains, 69°.26. Buch.

[54] Old observations of Tartebout. They appear good. Bagdad, Lat. 33° 19′; according to Beauchamps, 73°.76. The four seasons, 50°.74; 74°.64; 92°.66; 77°; but there was reflection from a house. The thermometer falls to 29°.44. Under the Equator, at 3,000 feet high; mean temp. 71°.24.

[55] The calculations are made from the observations of Nouet, (*Decade*, ii. 213.) The following are the mean temps. of the 12 months: 58°.1; 56°.12; 64°.58; 77°.9; 78°.26; 83°.66; 85°.02; 85°.82; 79°.16; 72°.32; 62°.96; 61°.24. (Niebuhr, 72°.2.) Temp. of Joseph's Well, 72°.5. Catacombs of Thebes, 81°.5. Well of the great pyramid, surrounded by sand, 88°.16. Jomard. Bassora, on the Persian Gulf; mean temp. 77°.9; winter, 64°.04; summer, 90°.86; July, 93°.2.

[56] Orta. Humboldt, *Nouv. Esp.* iv. 516. Jamaica, coast, 80°.6. Blagden.

[57] Ferrer, 1810—1812. *Con. des Tems*, 1817, p. 338. Wells of 10 feet deep; air, 75°.92; water, 74°.48; in 1812, *maximum*, Aug. 14. 86°; *minimum*, Feb. 20. 61°.52. Grottos, 81°.5. Humboldt, *Observ. Astron.* i. 134.

[58] Humboldt. Pondicherry, 85°.1; Madras, 80°.42; Manilla, 78°.08; Isle de France, coast, 80°.42.

Characteristic sketches of american vegetables

Mr. Editor,

The following characteristic sketches of remarkable families of the vegetable kingdom, extracted from the magnificent botanical work of Messrs. Humboldt and Bonpland,[1] completed last year at Paris, will I trust be acceptable to many more of your readers than those who have made that department of nature the object of their particular study.

I. THE PALMS.

The family of the *Palms* thrives best between the tropics, in the plains, and to the height of 500 fathoms, in places where the mean temperature in summer is from 19° to 28° of the centigrade thermometer, and—excepting the island of Cuba, to which the north wind from Canada brings a colder air—where the cold in winter is not below 15°. Only very few of their mountain varieties grow in the equinoctial region on the acclivities of the Andes up to the height of 1300 fathoms and more, as the *Oreodoxa frigida* and the *Ceroxylon andicola*. Beyond the tropics the *Phœnix dactylifera*, the *Chamœrops humilis* and *Palmetta,* together with *Areca Novæ Zeelandiæ*, are found in tracts where the average warmth does not exceed 15° or 16°, and where the ground is sometimes covered for several days with snow.

In the time of Linnæus no more than fifteen varieties of the palm were known: to these Ruiz and Pavon have added eight; Willdenow and Bredemeyer, six; and Humboldt and Bonpland twenty more. From a list collected by Kunth, from the later botanists it appears that at least 87 kinds of palms from America alone were known at that time; and these added to the 50 species of the old continent discovered up to April, 1816, make the total family amount to 137 varieties. In Willdenow's *System of Botany*, 70 palms are enumerated; and Humboldt and Bonpland saw, during their travels, more than 45. We may, however, venture to assume, that in the whole torrid zone the number of the varieties of the palm must be much more extensive, if we consider that, on the one hand, the greatest

---

[1] *Nova Genera et Species Plantarum quas in Peregrinatione ad Plagam æquinoctialem Orbis novi colligerunt.* fol.

part of Africa, Asia, and New Holland, is yet unexplored; and that, on the other, according to authentic observations which Humboldt and Bonpland themselves had occasion to make during their five years' peregrinations, the plants of this family are confined within such narrow limits that almost every fifty leagues new species of them are met with.

In regard to port and form the palms present the greatest diversity. Some (*Kunthea montana, Aiphanes praga, Oreodoxa frigida*) have only a weak, reed-like stem; while that of others (*Jubæa spectabilis, Cocos butyracea*) measures from three to five feet in diameter. Some (*Mauritia flexuosa, Chamœrops humilis*) grow sociably together; while others (*Oreodoxa regia, Martinezia caryotæfolia*) occur but singly. This (*Actalea amygdalina*) has a very low trunk; that (*Ceroxylon andicola*) attains the height of 160 or 180 feet. These (*Corypha tectorum, Alfonsia oleifera*) inhabit the plains only of the tropical regions, or the mountain-slopes to the elevation of 300 fathoms, while others— which the travellers believe to have been wholly unknown previously to their visit to the Andes— approach the limits of everlasting snow. To the alpine and sub-alpine vegetables belong the following species:

Kunthia montana grows at the elevation of from 250 to 800 and even 1000 fathoms, in the temperate region of the heart-leafed cinchona, on the declivity of the Bater, in the neighbourhood of the valley of Hato Viejo and the village of St. Paul, whence the plant bears likewise the name of *Canna de San Pablo*, or, *de la Vibora*; in the woods of the Alps of Pastoens between Chilanquer and Barbacoas, as also on the mountains of the province of Mocoa, which have a moderate temperature.

Oreodoxa frigida, at the elevation of from 1000 to 1400 fathoms, on the Alps of Quito, in company with the walnuttree and *Podokarpos*.

Ceroxylon andicola, at 920 to 1500 fathoms, on the Alps of Tolimen and Mount St. John, in company with the | 398 | *Oreodoxa frigida* and the *Quercus granatensis*: it can endure a nocturnal cold of at farthest 6° to 8°. This beautiful palm, remarkable for having its whole stem overgrown with foliage, is most frequently met with on the eastern declivity of the Andes of Quito, between Nevados de San Juan and Tolima, in the vicinity of Passo del Machin, Alto de las Sepulturas and Los Galiegos. It is found also of very large size near El Inciencial, at an elevation of 1240 fathoms. In the environs of Los Volcanitos and La Garita del Paramo, the *Ceroxylon anaicola* was not found higher than 1500 fathoms; it grows in company with the *Escalloniæ* and *Thibaudiæ*.

Joseph de Caldas and George Thaddæus Lozano, two celebrated botanists of New Grenada, have found three other alpine palms on the mountains of Guana (*Paramo de Guanacas*) not far from the region of everlasting snow.

For a long time botanists knew of no other palms than those with six pistils. More recently we have been made acquainted with six varieties having many pistils; three of which (*Caryota, Manicuria*, and *Latania*,) belong to the old world; and the other three, described by Ruiz, Pavon, and Humboldt (*Ceroxylon, Jubœa*, and *Attalœa*) to the Andes. Those species the pith of which furnishes a kind of sago-flour that forius an excellent article of food for the inhabitants of the new world, deserve to be particularly recommended to the notice of future travellers in America.

It would lead me too far were I to attempt to enumerate the various uses to which the palm-trees are applied. All your readers know that they furnish articles of the most dissimilar nature—wine, oil, wax, flour, sugar, and salt.

In the tracts to the north of the line, between the mountains of Caraccas and the River of Amazons, traversed by Humboldt and Bonpland, they found the palms in general in blossom in January and February, but some of them flower at other seasons of the year.

If the family of the palms is distinguished by beauty of form and its large size from almost all the other families of vegetables, so likewise does it surpass them in fecundity and abundance of fruit; and the powerful action of the sap is developed not only in thick umbrageous foliage, but also in an innumerable multitude of flowers and fruit. Not only the fields planted with palm-trees, but even the soil that has never been cultivated by man is frequently covered to the depth of three inches with the fruit of the *Alfonsia, Cocos butyracea, Seje, Pihiguao*, and *Mauritia*.

Every flower capsule of the date-tree (*Phœnix dactylifera*) contains, according to Kämpfer (*Amœn. exot*. p. 699), above 12,000 male flowers. Humboldt and Bonpland have found a much greater number upon the *Alfonsia amygdalina*. In one single catkin-shaped bunch of flowers, of which each capsule contains from 110 to 120, they counted 1800 male flowers; so that the total number in each capsule may be computed at 207,000 flowers. According to a similar calculation every branch of the *Seje*, which furnishes an excellent aliment on the banks of the Oroonoko, produces 8000 fruits, many of which indeed drop off before maturity. In the palm-grounds planted with the *l'ihiguao* each tree yields annually upon an average 400 fruits of the shape of apples; and the Franciscans on the Oroonoko and Guainia know from experience that whenever there is an uncommonly abundant crop of palm fruit the Indians grow remarkably fat.

Selector.

(To be continued.)

Characteristic sketches of american vegetables. By Alexander Von Humboldt.

(Continued from Vol. VII. P. 398.)

II.—THE ORCHIDES.

The *monocotyledons*, or plants that grow with one seed-lobe only, present the greatest diversity as well in regard to the beauty of their forms as to the variety of their colours. In the family of the Orchides, the flower-sheath, which, in the grasses, the cyperus-grasses, the families of the junceæ and the palms, is pale, small, and nowise remarkable, is adorned with the most diversified colours, so as to surpass even the superb flowers of the amaryllis, the iris, and the banana genus. The orchides, in fact, are the chief ornament of the vegetable kingdom in the equinoctial regions; and if, according to the testimony of Brown, fewer plants of that family occur between the tropics in New Holland (where so many unusual and extraordinary things are met with), than between the 33d and 35th degree of latitude, this circumstance must be considered as an exception: but their preference of a damp and mild atmosphere explains the reason why the epidendronorchides extend in the southern hemisphere so far towards the south pole, since they there find in winter a moderate temperature similar to that of the vallies among the mountains of the tropical regions.

It is impossible to calculate—as the celebrated authors of the Flora of Peru also attest—how abundant a harvest of orchides the mild climate of the deep and shady vallies in the range of the Andes still reserve for future travellers; for it may be confidently asserted, that we are not acquainted with more than the twentieth part of them. In all Europe we find but from 70 to 80 species of orchis; whereas in the equinoctial regions of America, whose mountainous parts are yet so little known, 244 species have already been observed by botanists, and 61 of these are new ones discovered by Humboldt and Bonpland. The known orchides of the Old and New World do not amount to 700: Willdenow has enumerated only 394, of which 152 are American.

Though the plants of the orchis family are found scattered in both worlds from the sea-coast to the height of 1800 or 1900 fathoms; still we are justified in asserting, that, from the multitude of their varieties, from the profusion of their foliage and the brilliancy and fragrance of their flowers, the vallies of the Andes of Mexico, New Grenada, Quito and Peru may be properly called their native country, where they thrive best in the moist shade, in a mild air, at an elevation of between 800 and 1100 fathoms, and in a mean temperature of 17° to 19°.

The species of orchis with spur-shaped flowers are very rare in the torrid zone, and with few exceptions their forms in the temperate and frigid zones are essentially different, so that in the northern hemisphere we find the varieties of the *Orchis Habenaria, Cypripedium, Ophrys, Serapius, Epipactes, &c.*; but in the southern those of the *Satyrium, Pteripodium, Disperis, Corycium, Stilidium, Disa, Pterostylis, Acianthus, &c.* The orchides of the equinoctial regions belong chiefly to the epidendron genus. They are distinguished from the orchides of the temperate and frigid zones by a different appearance, and more especially by this circumstance, that the former almost always grow in society and upon trees, while the latter spring up singly without support. Very few species, which, if I may so express myself, display the physiognomy of the north, as the *Ophrys, Habenaria, Altensteinia*, occur between the tropics, and not merely on the ridges of the loftiest mountains, but sometimes also in the plains. The temperate zone of the two great continents possesses no more than four species in common: these are: *Satyrium viride, Orchis hyperborea,* | 21 | *Neottia repens, Neottia tortilis*. It is remarkable that the *Dendrobium polystachion*, which grows on the mountains of Jamaica and in the woods of Guiana, has been found also in the island of St. Maurice. (Schwartz, *Flor. Ind. occ.* T. 3. p. 1433). Whether it be actually the same species seems however to be doubtful. Most of the orchides of the torrid zone contain in their roots a white floury substance, which not unfrequently possesses alimentary properties; some, as the *Pleurothallis sagittifera* contain also in their bulbs a viscid juice which may be employed as glue.

The monocotyledons in general contain a large proportion of starch (*amylum*) partly in the fruit (the grasses, the unripe fruit of the *Musa*, the *Palma Pihiguao*) partly in the stem (*Sagus Mauritius*) partly in the roots (the arums, the orchides, the families of the Lilies and *Dioscoreœ, Maranta Indica*). The juice of the Grasses and of the Agaves, the Areng palm, and the ripe fruit of the Pisang on the other hand contain sugar. That starch and sugar should occur at once in the same plant, and often in the same parts of it, will appear the less extraordinary, when we consider, that modern chemists have found in both nearly the same proportions of oxygen, hydrogen and carbon, and that sweet beer is produced by sprouted barley. The ancients themselves seem to have had some notion of the conversion of the starch into sugar. Among others, Prosper Alpinus relates, that the Pisang (*Musa paradisiaca*) sprung from the engrafting of the sugar-cane upon the Colocasia: and in like manner Abd Allatif assures us that when a date-stone germinates in a Colocasia root the produce is a pisangplant. The glutinous juice secreted by nature in the bulbous roots of the orchides differs from the genuine viscid matter of corn, as does the latter from that made of bird-lime berries, and the bark of the holly. An aroma which excites the nerves is found in the flower of the lily, asphodel and narcissus, in the bulb of the crocus, in the fruit of the vanilla and cardamom, in the roots of the varieties of *Canna*, and in the whole family of the *Peperomiæ*. Acids, astringent matter, resins, camphor, poisons, tannin, and vegetable milk occur but very rarely among the monocotyledons. The *Colchica*, and some species of *amaryllis* alone furnish a poison, the only antidote to which is the juice of the palm that Humboldt calls *Kunthia*. The *Smilax*, and varieties of *Scilla* contain astringent matter; and resin is found in the *Aloe*. An astringent principle, which has not yet been duly examined, is contained in the *Dracæna Draco* and in the *Agave*, the juice of which is used as a caustic in wounds. It is inconceivable how the celebrated Fourcroy could assert that the monocotyledons, without exception, are destitute of oil, since the coco is planted in the tropical regions for the same purpose as our olive. Decandolle (*Essai sur les Proprietés medicales des Plantes*) has already made the ingenious remark, that the monocotyledons, because their returning sap vessels are distributed through the whole stem and not collected together into a mass of bark, want nearly all that which, as the proper cortical substance, is found in the plants that germinate with two-seed lobes (dicotyledons), as Knight's experiments (*Phil. Trans.* 1801, p. 337) unequivocally demonstrate.

Characteristic sketches of american vegetables. By Alexander Von Humboldt.

(Concluded from p. 21.)

III.—THE GRASSES.

Messrs. Humboldt and Bonpland were so fortunate as to meet with two specimens of the *Bamboo tree* in flower; the one on the bank of the river

Cassaquiar, which is a branch of the Oronoko, and the other at the village of El Muerto, in the province of Popayan. These reed-like trees, though they cover damp and swampy tracts of great extent, and attain the height of from 50 to 80 feet, are nevertheless very rarely seen in blossom in America. Various observant travellers who have visited the *guadales*—as the swamps occupied by bamboos are termed by the inhabitants—never had the good fortune to obtain a sight of the flowers or fruit either in Peru or New Granada. In the East Indies, on the other hand, the gigantic grasses blossom so abundantly, that the seeds of the bamboo mixed with honey are a common dish with the people of Mysore. According to a notion prevalent there, the plant does not bear fruit till it is fifteen years old, and soon afterwards dies. There too a distinction is made between the bamboo with a solid stem, which the natives call *chitter*, and which grows in a dry soil, and the bamboo with a hollow stem, termed *doda*, which grows more rapidly, and in moist situations.

The *guadua* bamboo was found by Messrs. Humboldt and Bonpland on mountains to the height of 860 fathoms; and what is still more remarkable, these mountain trees, though growing upon a soil equally damp, contained more water than those of the level country. In more elevated regions the tree grows separately in the thickets; but in the plains, and to the height of 400 fathoms, it forms extensive woods. The bamboo belongs to the plants that grow in society.

The American bamboos render the same services in the New World as those of the East Indies in the old. Whole houses are built of the *Bambus guadia*. The oldest and thickest trunks are employed for the walls; the roof is formed with the smaller, and these are covered with the young branches that have leaves. The doors and household furniture also are made of bamboo. The advantages which cause the Americans to prefer the use of bamboo to that of the hard wood of the lofty trees which every where grow in the vicinity of their habitations, consist in the facility with which it may be felled and wrought, in its durability and the coolness enjoyed in houses constructed with it, owing to the free current of air which they afford.

The water contained in the American bamboo has a somewhat saline but not disagreeable taste. The inhabitants assert that it acts powerfully on the urinary passages. Of the sweet bamboo-honey M. von Humboldt could discover no traces in the New World; but in the kingdom of Quito he met with the *tabaschir*, differing but little from that of the East Indies. A piece of it, given by him to M. Vauquelin, was found by that chemist to be composed of 70/100 of siliceous earth, and 30/100 of potash, lime and water. The American *tabaschir* is called by the Spaniards bamboo-fat. "I cannot conceive," says the traveller, "how writers who have made the sugar of the ancients a subject of their enquiry, can compare the *tabaschir*, which is white and brittle like starch, with honey. For my own part, I could not perceive any sweet taste in the *tabaschir* of Quito, though it was | 110 | yet moist and viscid: and I have some doubts whether the tree-like reeds of America contain a sweet juice at all. The *tabaschir*, indeed, before it becomes by desiccation as hard as stone, is clammy, white, and milky: when it has been kept five months it emits an extremely fetid animal odour. The same was observed by Russel respecting the salt of the Asiatic bamboo, and Garcias ab Orto, who long resided at Goa as physician to the viceroy, has alone ascribed a sweet taste to the juice of the bamboo. It appears, however, that the ancients mistook the *tabaschir* for real sugar; partly because both are the produce of reed-like plants, and partly because the Sanscrit word *scharkara*, which at the present day is used like the Persian *schakar*, and the Hindustanee *schukar* to denote our sugar, does not properly imply something sweet, but something stony or sandy, and even the urinary calculus. It is therefore probable that the word *scharkara* was at first used for the *tabaschir* alone, and in the sequel transferred to our sugar from the smaller sugar-cane on account of its similar figure. The word *bamboo* is derived from *mambu*, and from *canda* comes our term *sugar-candy*, as does *tabaschir* from the Persian word *schir*, which signifies milk.

"Under the name of *saccharum* Pliny certainly described the tabaschir of the bamboo as "a honey collected from the stems of reeds, which is white like gum, breaks between the teeth, is of the size of a nut, and is applicable to medicinal purposes alone." The ancients had, nevertheless, some knowledge of our sugar, which was said to be produced in India without bees. Several writers of antiquity believed that a sweet honeylike juice was expressed from the roots of lofty reeds, thus confounding the root with the stem, and the comparatively low sugar-cane with the tall bamboo-tree. Some of them even had a notion that the genuine cane-sugar was a dew which settled upon the leaves of those plants. The sugar-cane indeed grows wild and in a natural state near Almansura in the East Indies, on the banks of the Euphrates, and at Siraf; yet I presume that in the regions of Asia visited by the Greeks the juice of the cane was expressed only for the purpose of an immediate beverage, and that the ancients had not consequently any knowledge of solid sugar, so that wherever the term is employed by them it signifies the *tabaschir* of the bainboo.

"It is almost superfluous to observe that before the Spaniards opened the way to America, the inhabitants of that continent and the adjacent islands were strangers to the sugar cane as well as to our different kinds of corn and rice. The Spanish writers on America, indeed, give the name of *little rice* to the *chenopodium quino*, which is common in Bogota and Quito, as the Anglo-Americans term a species of *zizania*, wild rice of Canada. Maize or Turkey corn, like other plants of ancient cultivation, does not grow wild in any part of the new continent. It were to be wished that some future traveller would furnish more particular information respecting the *magu*-corn and *tuca*-barley, of which Molini makes very brief mention in his History of Chili, and of which the Araucanians formerly made their Covque bread. From maize and the agave the Americans prepared the honey not produced by bees, which, as Hernando Cortez informs us, was carried for sale to their markets."

Cavern of Guacharo

From the London Quarterly Review.

An object of great curiosity was pointed out to our travellers (Humboldt and Bonpland) at the head of the valley of Caripe; this was the grand cueva, or Cavern of Guacharro. Mr Humboldt observed, that in a country where they love the marvellous, a cavern which gives birth to a river and is inhabited by many thousands of nocturnal birds, the fat of which is employed in the Missions for dressing food, is an inexhaustable subject of conversation and discussion. There is nothing however very remarkable in this cavern, excepting its great length. The entrance is about eighty feet wide, by seventy two high, and it preserves the same direction, the same width, and nearly the same height for one thousand four hundred and fifty eight feet, which is said to be not one half of its whole length. The luxuriance of the vegetation near the mouth gave to it a character which, in a less favored climate, it would not have possessed; for, as our author very justly observes, it is with the openings of caverns as with the view of cascades, the character of the local scenery, and of the surrounding country, constitutes the principal charm. The bird of night which inhabits the Cueva de Guacharo is more curious than the cavern. It is a new genus, nearly allied to that of Caprimulgus, to which M. de Humboldt has given the significant name of Steatornis.

"It is difficult to form an idea of the frightful noise made by thousands of these birds in the dark part of the cavern. It can be compared only to that of our crows, which, in the fir forests of the north, live in society, and build their nests in trees which meet at the top. The shrill and piercing tones of the Guacharo reverberate from the arched roof, and echo repeats them in the depths of the cavern. The Indians by fixing torches to the end of a long pole, pointed out to us the nests of these birds; they were fifty or sixty feet above our heads, in funnel shaped holes, with which the whole roof of the grotto is riddled. The noise increased with our advance, and with the alarm of the birds at the flare of our copal torches. When it ceased for a few minutes, we heard distant moans from other branches of the cavern. The different flocks might be said to give alternate responses.

"The Indians go once a year into the Cueva del Guacharo, about mid summer, furnished with poles, with which they destroy the greater part of nests. At this time many thousand birds are killed, and the old ones, as if to protect their broods, hover over the heads of the Indians, uttering the most dreadful shrieks. The young that fall to the ground are ripped open immediately. The peritoneum is thickly loaded with an unctuous substance, and a layer of fat runs from the abdomen to the anus, forming a kind of cushion between the bird's thighs. This abundance of fat in frugivorous animals, not exposed to the light, and having few muscular motions, reminds us of the inclination to obesity long observed in geese and oxen. At the period commonly termed the *oil harvest*, the Indians construct little habitations of palm leaves close to the opening, and even in the mouth of the cavern. We saw some remains of such still standing. Here, over a fire of dry sticks, the grease of the young birds just killed, is melted and run into pots of white clay. This grease, known by the name of Guacharo butter, or oil, (*manteca* or *aceite*) is semi-liquid transparent and inodorous; and so pure, that it may be kept more than a twelve month without becoming rancid. At the convent of Caripe, no oil but that of the cavern was used in the monk's kitchen, and we never found it give to the dish either a disagreeable taste or smell."

Curious Circumstance

— [*Extracted from the Personal Narrative of Travels to the Equinoctial Regions of the New Continent, during the years 1799— 1804. By Alexander de Humboldt and Aimé Bonpland,* &c.]—"Near San Fernando is the village of Arenas, where the following singular story is related, which we choose to give in the author's own words rather than risk our credit by an abridgement. It relates to one Francisco Lozano, a white man, descended from Europeans.

"This man has suckled a child with his own milk. The mother having fallen sick, the father, to quiet the infant, took it into his bed, and pressed it to his bosom. Lozano, then 32 years of age, had never remarked, till that day, that he had milk; but the irritation of the nipple, sucked by the child, caused the accumulation of that liquid. The milk was thick and very sweet. The father, astonished at the increased size of his breast, suckled his child two or three times a-day during five months. The child had no other nourishment than the milk of his father. Lozano, who was not at Arenas during our journey in the *missions,* came to us at Cumana. He was accompanied by his son, who was then about 13 or 14 years of age. Mr. Bonpland examined with attention the father's breast, and found it wrinkled like those of women who have given suck. He observed that the left breast in particular was much enlarged; which Lozano explained to us from the circumstance that the two breasts did not furnish milk in the same abundance."

[From De Humboldt's Travels, part 2.]

The Mission of San Fernando, by the regularity of the town, the uniformity of the buildings, the sober and silent air of the inhabitants, and the extreme neatness of the houses, recalled to the recollection of our travellers the establishments of the Moravian brethren; and this is saying not a little in their favour. Every Indian family, besides its proper garden, assists in the cultivation of the common garden; *Conuco de la communidad* ; every adult working there one hour in the morning and one in the evening. The great square in the centre of the village contains the church, the dwelling of the missionary, and the humble edifice on which is bestowed the pompous apellation of *Casa del Rey*, the house of the king: a sort of caravansera, intended to give shelter to travellers, and of infinite service in a country where the hotel or inn is utterly unknown. The following is the portrait of the good father of San Fernando:

"The missionary of San Fernando was a Capuchin, a native of Aragon, very far advanced in years, but still vigorous and cheerful. His great corpulency, his sprightly disposition, and the interest which he took in battles and sieges, but ill accorded with the notion formed, in the northern countries, of the melancholic abstraction and contemplative life of a missionary. Though closely busied with a cow which was to be slaughtered the next morning, the old monk yet received us with good humour, and gave us leave to sling our hammocks in a gallery of his house. Seated, without employment, during the chief part of the day, in a great arm chair of reed wood, he complained bitterly of what he termed the idleness and ignorance of his countrymen. He asked us a thousand questions concerning the real motive of our travels which to him seemed hazardous and, at best, useless. Here, as on the Orinoco, we were harassed by that eager curiosity which, in the midst of the forests of America, Europeans retain respecting the wars and political storms of the Old World.

"In other respects, our missionary appeared to be satisfied with his situation. He treated the Indians mildly; he saw his mission proper; and he extolled with enthusiasm the water, the bananas, and the milk diet of the district. He smiled contemptuously at the sight of our instruments, books, and dried plants; and acknowledged, with a frankness peculiar to these climates, that, of all the enjoyments of life, not excepting sleep, none was to be compared with the pleasure of eating good beef, so true is it, that sensuality springs from the absence of mental occupation. Our host persuaded us repeatedly to visit a cow he had just purchased; and the next day, at sunrise, he insisted on our going to see the animal killed according to the custom of the country, namely, by cutting the hamstring, and plunging a large knife between the vertebrae of the neck. Disgusting as the operation was, we learnt from it the expertness of the Chaymas Indians, who, eight in number, cut the beast into small joints in less than twenty minutes. The cow had cost but seven piastres, yet this seemed to be considered a very high price. The same day the missionary paid eighteen piastres to a soldier of Cumana, for bleeding him in the foot. The fact, of little apparent importance, strikingly proves how much, in wild, uncultivated countries, the value of commodities differs from that of thc labour."

On the Geographical Distribution of Ferns

It will not escape the studious observer of the natural stations of plants, that in the cryptogamous[1] class of vegeta- | 58 | bles, the Order of Ferns distinguishes itself from those of Mosses, Liverworts, and Mushrooms, in this particular, viz. that the latter have many species which are owned in common as well by the north of Europe as by the tropical regions of both the Old and New Continents, and not merely as peculiar to the mountains, but likewise as inhabitants of their plains; such are FUNARIA *hygrometrica* (twisting cord moss,) DICRANIUM *glaucum* (white fork moss,) POLYTRICHUM *juniperinum* (juniper-leaved hair moss,) and the Lichens VERRUCARIA *Perella*, STICTA *crocata*, PARMELIA *perlata*, &c. all which are found as well on the trees and rocks of the West Indian Islands, the chain of the Andés, the East Indies, and that part of New Holland which stretches towards the North, as in the Forests of Sweden and Great Britain. While among ferns a far narrower range has been allotted to the species, for with very few exceptions those of the new and old continents, not only differ in the one from those in the other, but in the tropical regions are manifestly distinct from their co-ordinates of the temperate and frigid zones.

In tracing the native stations of the ferns, and the plan of their distribution over the face of the earth, I have ascertained that of the 1000 species that have been as yet observed, 760 belong to the torrid zone, and 240 to the temperate and frigid zones. But when we speak generally of the number of these plants, the subject presents itself three ways; one, regarding the relative numbers of the species among themselves, another regarding their numbers in relation to the phænogamous portion of vegetation, and a third in regard to the quantity each species may of itself exist in. Thus Wahlenberg reckons 19 species of fern in Lapland, Hoffmann, 40 in Germany, Swartz, 103 in Jamaica, making the proportions of the frigid, temperate, and torrid zones, as they relate to each other in this point, as 1, 2, and 5. But the amount of phænogamous plants | 59 | in these regions is greatly disproportionate; for in Lapland there are 514 known species, and in Germany 1884, making the proportions 1:25 and 1:48. Whether this relative disproportion increases towards the equator

[1] *Phanogamous* plants, are those in which the organs of fructification are apparent and determinate; *cryptogamous* plants, those where these are more or less clandestine and indeterminate; *agamous* plants, those in which those organs either are not present, or have not been detected. *K*

or not, is unknown to me; but if it were 1:50, and if the ferns have come incompletely within the view and observation of botanists as the phænogamous portion of the vegetable creation, (which has hardly been the case,[2]) the result would be, that 5000 phænogamous species have been already discovered in Jamaica, and 23,000 in the whole of the tropical extent of America. The degrees of abundance with which nature has sent forth particular species are manifestly various: at the extremity of Norway, on the shores of the frozen sea, ferns, although but few in species, cover nearly the whole face of the land.

Of the 1000 species of ferns recorded by Willdenow, 470 grow in the old world; viz. 170 in the temperate and frigid zones, 300 within the tropics; and 530 grow in the New World, viz. 70 in the temperate and frigid zones, and 460 within thc tropics.

I have elsewhere attempted to show, that if we represent the land within the tropics by = 1000, we are to allot it in the following proportions; to

Africa, ... 461
America 301
New Holland and the Islands in the Indian Sea, 124
Asia 114

leaving the proportion which belongs to the New World in relation to that of the Old as 5 is to 7. So that if we did not know the New World to be of a more humid constitution with a more mountainous surface than the Old, and that the interior of Africa and New Holland have been far less explored than the regions that lie between the Orinoco and the river of the | 60 | Amazons, we might be led to wonder why the portion of America, being considerably the least, should have by 1/3 the largest proportion of ferns.

I am fully aware that too much reliance must not be placed on the exactness of the amounts, since it has not been ascertained, what proportion of each of the various regions of the earth has been accurately explored; nor what proportion of the total of vegetation has escaped the notice of the botanist; nor whether all the different tribes of plants have been investigated with equal care. But independent of these doubts, it is clear that the number of species cannot at all events be less than that which I computed.

[2] Swartz, who attended diligently to ferns, has only recorded 764 phænogamous plants in Jamaica. In Michaux's Flora of North America 45 ferns and 1575 phænogamous plants have been described. The Isles of Bourbon and France comprehend 137 species of plants.

Although the phænogamous plants of tropical America are entirely different from the phænogamous ones of the Old World, yet the northern portion of America has many in common with Europe and the North of Asia. Whence it may be inferred, that at some period those continents have joined towards the North Pole; and it will be a matter of wonder, that so few European ferns are found in Canada, Pennsylvania, and California; for of such we do not hear of more than from 6 to 10 species, as for instance, OPHIOGLOSSUM *vulgatum*, (common adder's-tongue,) POLYPODIUM *calcareum*, (rigid three-branched polypody.) ASPIDIUM *thelypteris*, (marsh shield-fern, or Lady-fern,) *Acristatum*, (lesser crested shieldfern) PTERIS *aquilina,* (common brake,) &c.; but then Europe altogether does not contain more than 70 indigenous species of fern.

Towards the South Pole, the ferns, which grow at the extremities of the two continents, differ more than those of the northern temperate zone do from each other. The only instances of any common to all the continent in those regions, that I am aware of, are DAVALLIA *pinnata*, which grows in Chili and in the Philippine Islands; and OSMUNDA *barbara*, which grows in New Holland and at the Cape of Good Hope. Hence it is the more strange that ASPIDIUM *aculeatum*, (or common prickly shieldfern,) which is the only one among the ferns of the Old World that ranges through several of the zones, extending itself from England across Mount Atlas to the southernmost verge | 61 | of Africa, should not have yet been met with in America. According to Messrs. Forster and Brown, BOTRYCHIUM *Lunaria*, (common moonwort) familiar in every part of Germany, covers, in company with the PHLEUM *alpinum*, (alpine cat's-tail-grass) of our country, every rock in Tierra del Fuego. HYMENOPHYLLUM *tunbridgense*, (Tunbridge filmy-grass) grows in New Holland as well as in Ireland, Norway, and Italy.

The only instance of a species of the fern-tribe growing both in the Old and in the New continents, in the torrid as well as in the frigid zone, in a northern and a southern hemisphere, viz. in England, Jamaica, and the Isle of Bourbon, is the ADIANTUM *Capillus Veneris*, (true maiden-hair,) a plant which has been commemorated by Hippocrates, Theophrastus, and Dioscorides. It has however been surmised by some that the germs of this fern have adhered to the filtering stones used aboard ships in long voyages, and been thus carried along with them and disseminated over the world.

It is not quite clear, that any fern has been found that is common to the tropical regions of both the New and Old Continents. Authors have as yet only suggested three such instances, viz. ASPIDIUM *punctulatum*,

A. *coriaceum*, and ASPLENIUM *monanthemum*, which are said to be spontaneous in the West Indies, the Andés of Peru, Guinea, New Holland, and the Cape of Good Hope. To these may be added Asplenium falcatum, and BLECHNUM *caudatum*, supposed to belong to South America and the Islands of Magindanao as well as Ceylon. But I doubt whether the statements concerning these instances are so well authenticated, as those mentioned above of HYMENOPHYLLUM *tunbridgense*, and BOTRYCHIUM *Lunaria*.

More than half of all the ferns yet observed belong to four genera only, viz. to Polypodium, (Polypody,) Aspidium, (Shield-fern,) Pteris, (Brake,) and Asplenium, (Spleenwort.) Some types among them seem confined almost entirely to tropical lands, as Meniscium, Anemia, Hydroglossum, Mertensia, and Schizea: but all the generic types of the northern temperate zone are found likewise between the tropics. The New World alone has no genus of the Fern-tribe peculiar to itself, although many of its phænogamous genera are so; for example, Cactus, Calceolaria, Alstrœmeria, Bromelia, | 62 | and others. As for the fern-types Polybotrya, Pleopeltis, and Marattia, so few species of them have been found, that it is highly probable other congeners will be detected in our own continent, so that they can hardly be relied on as exceptions.

Ferns, which in the northern frigid zone grow along the ground in the shade, in the tropical regions, shoot up to the dimensions of trees, vying with the palms themselves in stature and comeliness. These tree-ferns constitute a principal ornament of the torrid zone, giving that peculiar character to the spots where they grow, which strikes so forcibly the European stranger by its novelty. The Greek and Roman writers who have treated of plants, mention in several places, that many which in Europe creep along the ground, in hotter climates become trees, making it the more remarkable, that none should speak of a tree-fern; particularly as Megasthenes, Aristobulus, and Nearchus, recount among the wonders of Indian and Ethiopian lands, that there are trees having leaves as large as a shield, a fig-tree that takes root at the ends of its branches, and palms too high for the flight of an arrow to pass over. Theophrastus, Dioscorides, and Pliny, who has trodden over again the ground of the two first, mention only eight or ten species of ferns, none above a yard high. And when a certain sort of fern brought from India is noticed, as being known to the philosopher Eresius, no observation is offered concerning the stature of the stem, attention being only bestowed on certain medicinal qualities imputed to the plant. So that tree-ferns must have been left unnoticed by the Romans and Greeks, either because in that portion of India, which the arms of Alexander had opened to them, or in Ethiopia and Lybia, which they visited in the view of commerce, none are to be found; or because their writers, as was usual in former times, left unnoticed all plants which did not recommend themselves to their attention by the fruit they bore, or the fragrance of their wood, or their medicinal virtues. Oviedo, the Spaniard, in his history of the West Indies, is the first who mentions a tree-fern; for the *Filix arborea* of Tragus is nothing more than a variety of our common forked spleenwort, transplanted into a more fertile land.

So late as the time of Linnæus scarcely 4 sorts of tree-fern | 63 | were known, and about 15 sorts of palms; at present we are acquainted with 25 of the former, and 100 of the latter. The tree-ferns of America are,

> CYATHEA *speciosa, C. arborea, C. Serra, C. muricata, C. multiflora, C. villosa, C. aspera*, PTERIS *aculeata, P. villosa*, MENISCIUM *arboreum*, ASPIDIUM *caducum, A. procerum, A. rostratum*, ASPLENIUM *arboreum*, cæt.

Of New Holland and the Islands of the Indian Sea;

> CYATHEA *affinis, C. medullaris, C. dealbata, C. extensa*, DICKSONIA *squarrosa, D. antarctica*, cæt.

Of Southern Africa, and the Islands of France and Bourbon.

> CYATHEA *excelsa, C. glauca, C. riparia*, cæt.

The tree-ferns of the East Indies, Cochinchina, the Island of Madagascar, and the Cape of Good Hope have not yet been accurately described. I have not included amongst the arborescent species, ASPIDIUM *Arbuscula*, LOMARIA *Boryana*, POLYPODIUM *rhizocaule, P. pruinatum*, PTERIS *marginata*, &c. because they are either climbers, or else shrubby or caulescent plants, with stems not more than three or four feet high. Of the five new species of tree-ferns observed by M. Bonpland and myself, CYATHEA *speciosa* is the one which makes the finest appearance, and has a stem 25 feet high. The *C. excelsa* of the Isle of Bourbon is said by Messrs. Du Petit Thouars and Bory St. Vincent to grow to the same height.

In general, the tropical plants are further advanced towards the South Pole than towards the North Pole, a fact which seems inconsistent with the received opinion of the cold being greater in the Southern Hemisphere. In North America and New Spain, the tree-ferns hardly ever grow beyond the limits of the tropic of Cancer, while only one species of palm (CHAMÆROPS *Palmetto*) advances to Carolina, or the latitude of 37°.

In the Southern Hemisphere, on the other hand, DICKSONIA *antarctica*, described by Labillardiére as having a stem near 20 feet high, grows in Van Diemen's Island; nay, another species of *Dicksonia* has been found at Dusky Bay in New Zealand, in the 46th degree of southern latitude, where in a parallel which corresponds with that of Lyons, the trees swarm with *Epidendra* and *Dendrobia*, vegetable parasites, which constitute the | 64 | most graceful ornaments of the tropical Flora. Phænomena, which excite the more wonder, when we find from the observations of the celebrated navigators Cook, Entrecasteaux, and Flinders, that the mean annual temperature of this zone is scarcely equal to 12,5 of the centigrade thermometer. But the great expanse of water in the Southern Hemisphere cools the heat in summer and assuages the cold in winter; so that in the 52d and 53d degrees of southern latitude, corresponding to that of Berlin in our hemisphere, the snow melts nearly every day of the winter, and the centigrade thermometer very seldom rises to 11° in the months of January and December, the hot summer months of these regions. Even in the 42d and 43d degrees of Southern latitude, owing to the currents of wind that blow from the South Pole, the summers are hardly hotter than at the sides of the mountains of Switzerland: these summers, on the other hand, are succeeded by winters even milder than those of Rome. Labillardiére never observed the centigrade thermometer in Van Diemen's Island rise higher at mid-day than 15-17°,5 in the months of January and February; and Cook in the same parallel, during July, a winter month of those regions; never found the degree of cold below 8° of the centigrade thermometer.

In tropical America, as I have shown in another work, the herbaceous ferns grow in all parts, from the border of the sea, and from the plains, up to the heights of the Andés, although but little below the limits of eternal snow. There are certain species peculiar to certain elevations, and each in its zone is unable to transgress that boundary which has been allotted to it. Thus in the mountains of Peru and New Spain, CHEILANTHES *marginatus*, ACROSTICHUM *muscosum*, and HEMIONITIS *rufa*, grow between the elevations of 1200 and 1600 fathoms: in the same way the PTERIS *crispa*, (rock-brake or curled stonefern) grows on Mount St. Gothard from above the region of Pines up to the elevation of 1,100 fathoms; and in Lapland, near Enontekies, to that of 300 fathoms; so that in Switzerland, its highest station is scarcely 280 fathoms removed from the limits of eternal snow, and in the North of Norway scarcely 100 fathoms. POLYPODIUM *hyperboreum*, (hairy alpine | 65 | polypody) in latitude 68° advances still higher, to beyond BETULA *nana*, (dwarf birch) DRABA *alpina* (mountain whitlow grass) and CAMPANULA *uniflora*, (one-flowered bell-flower.) I have elsewhere explained myself on the subject of the reason why plants approach nearer to the boundary of eternal snow towards the North Pole, than they do in the torrid zone.

In the kingdom of Quito, M. Bonpland and myself saw mountain-ferns reaching into the elevated plain which engirds Mount Antisana and along the sides of the Rucupichincha in the valley of Verdecuchu, both which spots are 2100 fathoms above the level of the sea; and also other herbaceous ferns on Mount Chimborazo, growing on porphyritic rock, as high up as 2300 fathoms. But at heights so great, where the cold is so severe, as well as in parched unshaded plains and spots, the quantity of ferns is plainly perceived to diminish from the want of moisture. Their main body is stationed in the temperate and sub-frigid zones, between the elevations of 300 and 1200 fathoms.

Tree-ferns in some regions occasionally grow down to the edge of the sea, and on lowlands in shady spots; but in those parts of tropical America where M. Bonpland and myself passed five years, they occupy a peculiar zone, in which the temperature is between 18 and 22 degrees of the centigrade thermometer. A delightful region, where vernal breezes prevail almost the year through, and which lies between the elevations of 400 and 800 fathoms from the level of the sea: though they sometimes descend as low as to the height of 200 fathoms. This zone is called by the natives in the Spanish dialect, *tierra templada de los helechos*, the temperate land of the ferns. The chief part of the tree-ferns we met with, was in New Andalusia, near the Convent of Caripa, and in New Granada near Ibague, Guaduas and Iconónzo, and in the valleys of Peru between Loxa and the river of the Amazons also in New Spain near Xalapa. The region of fern-trees is not far from that of the Cinchonas, or bark-trees; for we find CINCHONA *oblongifolia* and C. *multiflora*, however fond they are of heat, mingling themselves with the tree-ferns in the Andès of Peru, and of the mountains of Quito and New | 66 | Granada. In New Spain they grow in company with the oaks of that country; a strange association in the eye of the European.

Personal Narrative of Travels to the Equinoctial Regions of the New Continent, During the Years 1799–1804

EARTHQUAKE AT CUMANA

From the 28th of October to the 3d of November, the reddish fog was thicker than it had yet been. The heat of the night seemed stifling, though the thermometer rose only to 26°. The breeze, which generally cooled the air from eight or nine o'clock in the evening, was no longer felt. The atmosphere appeared as if it were on fire. The ground, parched and dusty, was cracked on every side. On the fourth of November, about two in the afternoon, large clouds, of an extraordinary blackness, enveloped the high mountains of the Brigantine and Tataraqual. They extended, by degrees, as far as the zenith. About four in the afternoon, thunder was heard over our heads, but at an im- | 591 | mense height, without rolling, and with a hoarse and often interrupted sound. At the moment of the strongest electric explosion, at 4h. 12m. there were two shocks of an earthquake, which followed at fifteen seconds distance from each other. The people in the streets filled the air with their cries. Mr. Bonpland, who was leaning over a table examining plants, was almost thrown on the floor. I felt the shock very strongly, though I was lying in a hammock. Its direction was from north to south, which is rare at Cumana. Slaves, who were drawing water from a well, more than eighteen or twenty feet deep, near the river Manzanares, heard a noise like the explosion of a strong charge of gunpowder. The noise seemed to come from the bottom of the well, a very singular phenomenon, though very common in the greater part of the countries of America exposed to earthquakes.

A few minutes before the first shock, there was a very violent blast of wind, followed by an electrical rain, in great drops. I immediately tried the atmospherical electricity by the electrometer of Volta. The small balls separated four lines; the electricity often changed from positive to negative, as is the case during storms; and, in the north of Europe, even sometimes in a fall of snow. The sky remained cloudy, and the blast of wind was followed by a dead calm, which lasted all night. The setting of the sun presented a scene of extraordinary magnificence. The thick veil of the clouds was rent asunder as in shreds, quite near the horizon:

the sun appeared at 12 degrees of altitude, on a firmament of indigo-blue. Its disk was enormously enlarged, distorted, and undulated toward the edges. The clouds were gilded; and fasciculi of divergent rays, which reflected the most brilliant colours of the rainbow, extended even to the midst of the heavens. There was a great crowd in the public square. This phenomenon, the earthquake, the clap of thunder which had accompanied it, the red vapour seen during so many days, all was regarded as the effect of the eclipse.

About nine in the evening there was a third shock, much slighter than the former two, but attended evidently with a subterraneous noise. The barometer was a little lower than usual; but the progress of the horary variations or small atmospheric tides, was no way interrupted. The mercury was precisely at the minimum of height at the moment of the earthquake; it continued rising till eleven in the evening, and sunk again till half after four in the morning, conformably to the law which regulates the barometrical variations In the night between the 3d and 4th of November, the reddish vapour was so thick, that I could not distinguish the place of the moon, except by a beautiful halo of 20° diameter.

It was scarcely twenty-two months since the town of Cumana had been almost totally destroyed by an earthquake. The people look on the vapours which darken the horizon, and the failure of the breeze during the night, as prognostics infallibly disastrous. We had frequent visits from persons, who wished to know if our instruments indicated new shocks for the next day. The inquietude was particularly great and general, when on the 5th of November, exactly at the same hour as the preceding day, there was a violent gust of wind, attended by thunder, and a few drops of rain. No shock was felt. The wind and storm returned for five or six days at the same hour, almost at the same minute. The inhabitants of Cumana, and of many other places between the tropics, have long ago, made the observation, that those atmospherical changes, which appear the most accidental, follow for whole weeks a certain type with astonishing regularity. The same phenomenon exists in summer, under the temperate zone; nor has it escaped the sagacity of astronomers, who often see clouds form in a serene sky, during three or four days together, at the same part of the firmament, take the same direction, and dissolve at the same height; sometimes before, sometimes after, the passage of a star over the meridian, consequently within a few minutes of the same point of apparent time.[1]

[1] Mr. Arago and I paid a great deal of attention to this phenomenon during a long series of observations made in the years 1809 and 1810, at the Observatory at Paris, to verify the declination of the stars.

The earthquake of the 4th of November, the first I had felt, made so much the more lively an impression on me, as it was accompanied with remarkable meteorological variations. It was, moreover, a real lifting-up, and not a shock by undulations. I did not then imagine, that, after a long abode on the table-lands of Quito, and the coasts of Peru, I should become almost as familiar with the abrupt movements of | 592 | the ground, as we are in Europe with the noise of thunder. We did not think of rising at night, in the city of Quito, when subterraneous rumblings (*bramidos*), which seem always to come from the volcano of Pichincha, announced, (two or three, and sometimes seven or eight minutes beforehand) a shock; the force of which is seldom in proportion to the intensity of the noise. The carelessness of the inhabitants, who recollect that, for three centuries past, their city has not been overwhelmed, communicates itself easily to the least intrepid traveller. In general, it is not so much the fear of the danger, as the novelty of the sensation, that strikes so forcibly, when the effect of the slightest earthquake is felt for the first time.

From our infancy, the idea of certain contrasts fixes itself in our minds: water appears to us an element that moves; earth, a motionless and inert mass. These ideas are the effect of daily experience; they are connected with every thing that is transmitted to us by the senses. When a shock is felt, when the earth is shaken on its old foundations, which we had deemed so stable, one instant is sufficient to destroy long illusions. It is like awakening from a dream; but a painful awakening. We feel, that we have been deceived by the apparent calm of nature; we become attentive to the least noise: we mistrust, for the first time, a soil on which we had so long placed our feet with confidence. If the shocks be repeated, if they become frequent during several successive days, the uncertainty quickly disappears. In 1784, the inhabitants of Mexico were accustomed to hear the thunder roll beneath their feet, as it is heard by us in the region of the clouds. Confidence easily springs up in the human breast, and we end by accustoming ourselves on the coast of Peru to the undulations of the ground, like the sailor to the tossings of the ship, caused by the motion of the waves.

METEORS

The night of the 11th of November, was cool and extremely beautiful. Toward the morning, from half after two, the most extraordinary luminous meteors were seen towards the east. Mr. Bonpland, who had risen to enjoy the freshness of the air in the gallery, perceived them first. Thousands of bolides and falling stars, succeeded each other during four hours. Their direction was very regularly from north to south. They filled a space in the sky extending from the true east 30° toward the north and south. In an amplitude of 60° the meteors were seen to rise above the horizon at E.N.E., and at E. describe arcs more or less extended, and fall toward the south, after having followed the direction of the meridian. Some of them attained a height of 40°, and all exceeded 25° or 30°. There was very little wind in the low regions of the atmosphere, and this blew from the east. No trace of clouds was to be seen. Mr. Bonpland relates, that from the beginning of the phenomenon, there was not a space in the firmament equal in extent to three diameters of the moon, that was not filled at every instant with bolides and falling stars. The first were fewer in number, but as they were seen of different sizes, it was impossible to fix the limit between these two classes of phenomena. All these meteors left luminous traces from five to ten degrees in length, as often happens in the equinoctial regions. The phosphorescence of these traces, or luminous bands, lasted seven or eight seconds. Many of the falling stars had a very distinct nucleus, as large as the disk of Jupiter, from which darted sparks of vivid light. The bolides seem to burst as by explosion; but the largest, those from 1° to 1° 15′ in diameter, disappeared without scintillation, leaving behind them phosphorescent bands (*trabes*) exceeding in breadth fifteen or twenty minutes. The light of these meteors was white, and not reddish, which must be attributed, no doubt, to the absence of vapours, and the extreme transparency of the air. For the same reason, under the tropics, the stars of the first magnitude have at their rising, a light evidently whiter than in Europe.

Almost all the inhabitants of Cumana were witnesses of this phenomenon, because they leave their houses before four o'clock, to attend the first morning mass. They did not behold these bolides with indifference; the oldest among them remembered, that the great earthquakes of 1766, were preceded by similar phenomena. The Guaiqueries in the Indian suburb, came out and asserted, "that the fire-work had begun at one o'clock; and that as they returned from fishing in the Gulf, they had already perceived very small falling stars toward the east." They affirmed at the same time, that igneous meteors were extremely rare on those coasts after two in the morning.

The phenomenon ceased by degrees after four o'clock, and the bolides and falling stars became less frequent; but | 593 | we still distinguished some toward the north-east by their whitish light, and the rapidity of their movement, a quarter of an hour after sun-rise. This circumstance will appear less extraordinary, when I bring to the reader's recollection, that in full day-light, in 1788, the interior of the houses in the town of Popayan was brightly illuminated by an aërolite of immense magnitude. It passed over the town when the sun was shining clearly, about one o'clock. Mr. Bonpland and myself, during our second residence at Cumana, after having observed on the 26th of September, 1800, the immersion of the first satellite of Jupiter, succeeded in seeing the planet distinctly with the naked eye, eighteen minutes after the disk of the sun had appeared in the horizon. There was a very slight vapour in the east, but Jupiter appeared on an azure sky. These facts prove the extreme purity and transparency of the atmosphere under the torrid zone. The mass of diffused light is so much less, as the vapours are more perfectly dissolved. The same cause, that weakens the diffusion of the solar light, diminishes the extinction of that which emanates either from a *bolis*, Jupiter, or the moon, seen on the second day after her conjunction.

The researches of Mr. Chladni having singularly fixed the attention of the scientific world upon the bolides and falling stars at my departure from Europe, we did not neglect, during the course of our journey from Caraccas to the Rio Negro, to enquire every where, whether the meteors of the 12th of November had been perceived. In a savage country, where the greater number of the inhabitants sleep out in the air, so extraordinary a phenomenon could not fail to be remarked, except when concealed by clouds from the eye of observation. The Capuchin missionary at San Fernando de Apura, a village situate amid the savannahs of the province of Varinas; the Franciscan monks stationed near the cataracts of the Oroonoko, and at Maroa, on the banks of the Rio Negro; had seen numberless falling stars and bolides illumine the vault of heaven. Maroa is south-west of Cumana, at one hundred and seventy-four leagues distance. All these observers compared the phenomenon to a beautiful firework, which had lasted from three till six in the morning. Some of the monks had marked the day upon their ritual; others had noted it by the nearest festivals of the church. Unfortunately, none of them could recollect the direction of the meteors, or their apparent height. From the position of the mountains and thick forests, which surround the Missions of the Cataracts and the little village of Maroa, I presume that the bolides were still visible at 20° above the horizon. On my arrival at the southern extremity of Spanish Guiana, at the little fort of San Carlos, I found some Portugueze, who had gone up the Rio Negro from the Mission of St. Joseph of the Marivitains; who assured me, that in that part of Brazil, the phenomenon had been perceived, at least as far as San Gabriel das Cachoeiras, consequently as far as the equator itself.

I was powerfully struck at the immense height, which these bolides muss have attained, to have been visible at the same time at Cumana, and on the frontiers of Brazil, in a line of two hundred and thirty leagues in length. But what was my astonishment, when, at my return to Europe, I learnt, that the same phenomenon had been perceived on an extent of the globe of 64° of latitude, and 91° of longitude; at the equator, in South America, at Labrador, and in Germany! I found accidentally during my passage from Philadelphia to Bordeaux (in the Memoirs of the Pennsylvanian Society) the corresponding observation of Mr. Ellicott (lat 30° 42′); and, upon my return from Naples to Berlin, I read the account of the Moravian Missionaries among the Eskimaux, in the Bibliothek of Gottingen. Several philosophers had already discussed at this period the coincidence of the observations in the north with those at Cumana, which Mr. Bonpland and I had published in 1800.

The following is a succinct enumeration of facts: 1st, the fiery meteors were seen in the east, and the east-north-east, to 40° of elevation, from 2 h. to 6 h. at Cumana (lat. 10° 27′ 52″, long. 66° 30′); at Porto Cabello (lat. 10° 6′ 52″, long. 67° 5′); and on the frontiers of Brazil, near the equator, in the longitude of 70° west of the meridian of Paris. 2d, In French Guiana (lat. 4° 56′, long. 54° 35′), "the northern part of the sky was seen all on fire. Innumerable falling stars traversed the heavens during an hour and a half, and diffused so vivid a light, that those meteors might be compared to the blazing sheaves shot out from a fire-work." The knowledge of this fact rests upon highly respectable testimony, that of the Count of Marbois, at that time transported to Cayenne, a victim to his love of justice and of rational constitutional liberty. 3d, Mr. Ellicott, | 594 | astronomer to the United States, having terminated his trigonometric operations for the rectification of the limits on the Ohio, being on the 12th of November in the Gulf of Florida, in the latitude of 25°, and longitude 81° 50′, saw, in all parts of the sky, "as many meteors as stars, moving in all directions: some appeared to fall perpendicularly; and it was expected every minute that they would drop into the vessel." The same phenomenon was perceived upon the American continent as far as the latitude of 30° 42′. 4th, In Labrador, at Nain (lat. 56° 55′), and Hoffenthal (lat. 58° 4′); in Greenland, at Lichtenau (lat. 61° 5′), and at New Herrenhut (lat. 64° 14′, long. 52° 20′); the Eskimaux were frightened at the enormous quantity of bolides, that fell during twilight toward all points of the firmament, and "some of which were a foot broad. 5th, In

Germany, Mr. Zeissing, vicar of Itterstadt near Weimar (lat. 50° 59′, long. 9° 1′ east), perceived, on the 12th of November, between the hours of six and seven in the morning, when it was half after two at Cumana, some falling stars, which shed a very white light. "Soon after, toward the south and south-west, luminous rays appeared from four to six foot long; they were reddish, and resembled the luminous track of a sky-rocket. During the morning twilight, between the hours of seven and eight, the southwest part of the sky was seen, from time to time, strongly illuminated by white lightning, which ran in serpentine lines along the horizon. At night the cold increased, and the barometer had risen." It is very probable, that the meteor might have been observed more to the east, in Poland and in Russia. If Mr. Ritter had not taken a particular account of them from the Vicar of Itterstadt's papers, we should have supposed, that the bodies had not been visible beyond the limits of the New Continent.

The distance from Weimar to the Rio Negro, is 1800 sea leagues; and from Rio Negro to Herrenhut in Greenland, 1300 leagues. Admitting that the same fiery meteors were seen at points so distant from each other, we must also admit, that their height was at least 411 leagues. Near Weimar, the appearance like sky-rockets was seen in the south, and south-east; at Cumana, in the east, and in the east-north-east. We may therefore conclude, that numberless aërolites must have fallen into the sea, between Africa and South-America, to the west of the Cape-Verde Islands. But, since the direction of the bolides was not the same at Labrador and at Cumana, why were they not perceived in the latter place towards the north, as at Cayenne? We can scarcely be too cautious on a subject, on which good observations made in very distant places are still wanting. I am rather inclined to think, that the Chayma Indians of Cumana did not see the same bolides as the Portugueze in Brazil, and the missionaries in Labrador; but, at the same time, it cannot be doubted, and this fact appears to me very remarkable, that in the New World, between the meridians of 46° and 82°, between the equator and 64° north, at the same hour, an immense number of bolides and falling stars were perceived; and that those meteors had every where the same brilliancy throughout a space of 921,000 square leagues.

The scientific men, who have lately made such laborious researches on falling stars and their parallaxes, considered them, as meteors belonging to the farthest limits of our atmosphere, between the region of the Aurora Borealis and that of the lightest clouds. Some have been seen, which had not more than 14,000 toises, or about five leagues, of elevation.

The highest do not appear to exceed thirty leagues. They have often more than a hundred feet diameter; and their swiftness is such, that they dart, in a few seconds, over a space of two leagues. Some of these have been measured, the direction of which was almost perpendicularly upward, or forming an angle of 50° with the vertical line. This extremely remarkable circumstance has led to the conclusion, that falling stars are not aërolites, which, after having hovered about a long time in space, take fire on entering accidentally into our atmosphere, and fall toward the earth.

Whatever may be the origin of these luminous meteors, it is difficult to conceive any instantaneous inflammation taking place in a region, where there is less air than in the vacuum of our air-pumps; and where (at 25,000 toises high) the mercury in the barometer would not rise to 0 012 of a line. We have ascertained the uniform mixture of atmospheric air to 0 003 nearly, only to an elevation of 3000 toises; consequently not beyond the last stratum of fleecy clouds. It might be admitted, that, in the first revolutions of the globe, gaseous substances, which yet remain unknown to us, may have risen toward that region, | 595 | through which the falling stars pass: but accurate experiments, made upon mixtures of gasses which have not the same specific gravity, prove, that we cannot admit a superior stratum of the atmosphere entirely different from the inferior strata. Gaseous substances mix and penetrate each other with the least motion, and a uniformity of their mixture would have taken place in the lapse of ages, unless we suppose in them the effects of a repulsive action unexampled in those substances which we can subject to our observations. Farther, if we admit the existence of particular aërial fluids in the inaccessible regions of luminous meteors, falling stars, bolides, and the Aurora Borealis; how can we conceive why the whole stratum of those fluids does not at once take fire, but that the gaseous emanations, like the clouds, occupy only limited spaces? How can we suppose an electrical explosion without some vapours collected together, capable of containing unequal charges of electricity, in air, the mean temperature of which is, perhaps, 25° below the freezing point of the centigrade thermometer, and the rarefaction of which is so considerable, that the compression of the electrical shock could scarcely disengage any heat? These difficulties would, in great part, be removed, if the direction of the motion of falling stars allowed us to consider them as bodies with a solid nucleus, as cosmic phenomena (belonging to space beyond the limits of our atmosphere), and not as telluric phenomena (belonging to our planet only).

Supposing that the meteors of Cumana were only at the usual height, at which falling stars in general move, the same meteors were seen above the horizon in places more than 310 leagues distant from each other. Now, what an extraordinary disposition to incandescence must have reigned on the 12th of November, in the higher regions of the atmosphere, to have furnished, during four hours, myriads of bolides and falling stars, visible at the equator, in Greenland, and in Germany.

Mr. Benzenberg judiciously observes, that the same cause, which renders the phenomenon more frequent, has also an influence on the largeness of the meteors, and the intensity of their light. In Europe, the nights when there are the greatest number of falling stars, are those in which very bright ones are mixed with very small ones. The periodicalness of the phenomenon augments the interest which it excites. There are months, in which Mr. Brandes has reckoned in our temperate zone, only sixty or eighty falling stars in one night; and in other months their number has risen to two thousand. Whenever one is observed, which has the diameter of Sirius or of Jupiter, we are sure of seeing so brilliant a meteor, succeeded by a great number of smaller meteors. If the falling stars be very frequent during one night, it is very probable, that this frequency will continue during several weeks. It would seem, that, in the higher regions of the atmosphere, near that extreme limit where the centrifugal force is balanced by gravity, there exists, at regular periods, a particular disposition for the production of bolides, falling stars, and the Aurora Borealis. Does the periodicalness of this great phenomenon depend upon the state of the atmosphere? or upon something which this atmosphere receives from without, while the earth advances in the ecliptic? Of all this we are still ignorant, as men were in the days of Anaxagoras.

With respect to the falling stars themselves, it appears to me, from my own experience, that they are more frequent in the equinoctial regions than in the temperate zone; more frequent over the continents, and near certain coasts, than in the middle of the ocean. Do the radiation of the surface of the globe, and the electric charge of the lower regions of the atmosphere, which varies according to the nature of the soil, and the positions of the continents and seas, exert their influence as far as those heights, where eternal winter reigns? The total absence even of the smallest clouds, at certain seasons, or above some barren plains destitute of vegetation, seem to prove, that this influence can be felt at least as far as five or six thousand toises high. A phenomenon analogous to that of the 12th of November, was observed thirty years before, on the table land of the Andes, in a country studded with volcanoes. At the city of Quito, there was seen, in one part of the sky, above the volcano of Cayambo, so great a number of falling stars, that the mountain was thought to be in flames. This singular sight lasted more than an hour. The people assembled in the plain of Exico, where a magnificent view presents itself of the highest summits of the Cordilleras. A procession was already on the point of setting out from the Convent of St. Francis, when it was perceived, that the blaze on the horizon was caused by fiery meteors, which ran along the skies in all | 596 | directions, at the altitude of twelve or thirteen degrees.

LA GUAYRA

I landed in the port of La Guayra, and the same evening made preparations for transporting my instruments to Caraccas. The persons for whom I had recommendation, advised me not to sleep in the town, where the yellow fever had only ceased a few weeks, but in a house on a little hill, above the village of Maiquetia, more exposed to fresh winds than La Guayra.

La Guayra is rather a roadstead than a port. The sea is constantly agitated, and the ships suffer at once by the action of the wind, the tideways, the bad anchorage, and the worms. The landing is taken in with difficulty, and the height of the swell prevents embarking mules here, as at New Barcelona and Porto Cabello. The free Mulattoes and Negroes, who carry the cacao on board the ships, are a class of men of very remarkable muscular strength. They go up to their middles through the water; and, what is well worthy of attention, they have nothing to fear from the sharks, which are so frequent in this harbour. The fact seems connected with what I have often observed between the tropics, relatively to other classes of animals, that live in society; for instance, monkeys and crocodiles. In the missions of the Oroonoko, and the river of Amazons, the Indians who catch monkeys to sell them, know very well, that they can easily succeed in taming those which inhabit certain islands; while monkeys, of the same species, caught on the neighbouring continent, die of terror or rage, when they find themselves in the power of man. The crocodiles of one pool in the Llanos are cowardly, and flee even in the water; while those of another attack with extreme intrepidity. It would be difficult to explain this difference of manners and habits, by the aspect of their respective localities. The sharks of the port of La Guayra seem to furnish an analogous example. They are dangerous and blood-thirsty, at the island opposite the coast of Caraccas, at the Roques, at Bonayre, and at Curassao: while they forbear to attack persons swimming in the ports of La Guayra and Santa Martha. The people, who, in order to simplify the explanation of natural phenomena, have always recourse to the marvellous, affirm, that, in both places, a bishop gave his benediction to the sharks.

The situation of La Guayra is very singular, and can only be compared to that of Santa-Cruz in Teneriffe. The chain of mountains, that separates the port from the high valley of Caraccas, descends almost directly into the sea; and the houses of the town are backed by a wall of steep rocks. There scarcely remains one hundred or one hundred and forty toises breadth of flat ground between the wall and the ocean. The town has six or eight thousand inhabitants, and contains only two streets, running parallel to each other east and west. It is commanded by a battery of *Cerro colorado*; and its fortifications along the seaside are well disposed, and kept in repair. The aspect of this place has something solitary and gloomy; we seemed not to be on a continent, covered with vast forests, but in a rocky island, destitute of mould and vegetation. With the exception of Cape Blanco, and the cocoa-trees of Maiquetia, no view meets the eye but that of the horizon, the sea, and the azure vault of heaven. The heat is stifling during the day, and most frequently during the night. The climate of La Guayra is justly considered as more ardent than that of Cumana, Porto Cabello, and Coro; because the sea-breeze is less felt, and the air is heated by the radiant caloric, which the perpendicular rocks emit from the time the sun sets.

The examination of the thermometric observations, made during nine months at La Guayra by a distinguished physician, enabled me to compare the climate of this port, and that of Cumana, the Havannah, and Vera Cruz.

These four places are considered as the hottest on the shores of the New World. The mean of the observations made at noon, from the 27th of June to the 16th of November, were, at La Guayra, 31 6° of the centigrade thermometer; at Cumana, 29 3°; at Vera Cruz, 28 7°; at the Havannah, 29 5°. The daily difference, at the same hour, scarcely exceeded 0 8° or 1 4°. During this period it rained but four times, and then only for seven or eight minutes. At this season, the yellow fever prevails; which usually disappears at La Guayra, as at Vera Cruz, and the island of St. Vincent, when the temperature of the day descends below twenty-three or twenty-four degrees. The mean temperature of the hottest month was, at La Guayra, nearly 29 3°; at Cumana, 29 1°; at Vera Cruz, 27 7°; at Cairo, according to Nouet, 29 9°; at Rome, 25°. From | 597 | the 16th of November to the 19th of December, the mean temperature of noon, at La Guayra, was only 24 3°; at night, 21 6°. This is the time at which the sufferings from heat are the least; and yet I do not believe, that the thermometer ever falls (and it is lowest a little before sunrise) below 21°. It sometimes descends at Cumana, to 21 2°; at Vera Cruz, to 16°; at the Havannah (always when the north wind blows), to 8°, and even lower. The mean temperature of the coldest month is in these four places, 23 2°, 26 8°, 21 1°, 21 0°. At Cairo, it is 13 4°. The mean of the whole year, according to good observations carefully calculated, is, at La Guayra nearly 28 1°; at Cumana, 27 7°; at Vera Cruz, 25 4°; at the Havannah, 25 6°; at Rio Janeiro, 23 5°; at Santa Cruz in Teneriffe, in the latitude of 28° 28′, but backed like La Guayra, by a wall of rocks, 21 9°: at Cairo, 22 4°; at Rome, 15 8°.

From the whole of these observations it follows, that La Guayra is one of the hottest places on the earth,[2] that the quantity of heat which it receives in the course of a year, is a little greater than that felt at Cumana; but that, in the months of November, December, and January, (at equal distance from the two passages of the sun through the zenith of the town), the atmosphere cools more at La Guayra.

At the time of my abode at La Guayra, the scourge of yellow fever, or *calentura amarilla*, had been known only two years; and the mortality had not been considerable, because the confluence of strangers on the coast of Caraccas was less than at the Havannah and Vera Cruz.

I found the latitude of La Guayra to be 10° 36′ 19″; and the longitude 69° 26′ 13″. The dip of the needle on the 24th of January, 1800, was 42 20°; and its variation 4° 20′ 35″ east. The intensity of the magnetic forces was found proportional to 237 oscillations.

CARACCAS

After having described the scenery, and the atmospheric constitution of La Guayra, we shall now leave the coasts of the Caribbean Sea. The road that leads from the port to Caraccas, the capital of a government of near 90,000 inhabitants, resembles, as I have already observed, the passages over the Alps, the road of St. Gothard, and of the Great St. Bernard. Taking the level of the road had never been attempted before my arrival in the province of Venezuela. No precise idea had even been formed of the elevation of the valley of Caraccas. It had indeed been long observed, that the descent was much less from La Cumbre and La Vueltas, which

---

[2] In Asia, the mean temperature of Abushar, of Madras, and of Batavia, are not above 25° and 27°; but the hottest month at Madras rises to 32°, according to Roxburgh; and at Abushar, on the Persian Gulf, according to Mr. Jukes, to 33 90°; which is from two to four degrees higher than at Cairo.

is the culminating point of the road toward the Pastora at the entrance of the valley of Caraccas, than toward the port of La Guayra: but the mountain of Avila, having a very considerable bulk, the eye cannot discover, at the same time, the points to be compared. It is even impossible to form a precise idea of the elevation of Caraccas from the climate of the valley. The air in it is cooled by the descending currents of air; and by the fogs, which envelop the lofty summit of the Silla during a great part of the year. I have often gone on foot from La Guayra to Caraccas; and I sketched a profile of the road, founded on twelve points, the heights of which were determined by barometric measurements.

When in the season of the great heats we breathe the burning atmosphere of La Guayra, and turn our eyes toward the mountains, we are strongly affected by the idea, that, at the direct distance of five or six thousand toises, a population of forty thousand souls, assembled in a narrow valley, enjoys all the coolness of spring, of a temperature which, at night, descends to 12° of the centesimal thermometer. This near approach of different climates is common in the Cordilleras of the Andes; but every where, at Mexico, at Quito, in Peru, and in New Grenada, a long journey must be made into the interior, either by the plains, or by proceeding up the rivers, in order to reach the great cities, which are the centres of civilization. The height of Caraccas is but a third of that of Mexico, Quito, and Santa Fe de Bogota; yet, among all the capitals of Spanish America, which enjoy a cool and delicious climate in the midst of the torrid zone, Caraccas stands nearest to the coast. What a privilege, to possess a sea-port at three leagues distance, and to be situate among mountains, on a table-land, which would produce wheat, if the cultivation of the coffee-tree were not preferred!

The road from La Guayra to the Val- | 598 | ley of Caraccas, is infinitely finer than that from Honda to Santa Fe, or that from Guayaquil to Quito. It is even kept in better order than the ancient road, which led from the port of Vera Cruz to Perote, on the eastern declivity of the mountains of New Spain. With good mules, it requires but three hours to go from the port of La Guayra to Caraccas; and only two hours to return. With loaded mules, or on foot, the journey is from four to five hours. The ascent begins with a ridge of rocks extremely steep, and stations that bear the name of Torrequemada, Curucuti, and Salto, to a large inn (La Venta) built at six hundred toises above the level of the sea. The denomination of the Burnt Tower indicates the sensation that is felt in descending toward La Guayra. A suffocating heat is reflected by the walls of rock, and still more by the barren plains, on which the traveller looks down.

When I passed for the first time that table-land, on my way to the capital of Venezuela, I found several travellers assembled round the little inn of Guayavo, to rest their mules. They were inhabitants of Caraccas, and were disputing on the efforts toward independence, which had been made a short time before. Joseph Espana had perished on the scaffold; and his wife groaned in a prison, because she had given an asylum to her husband when a fugitive, and had not denounced him to the government. I was struck with the agitation which prevailed in every mind, and the bitterness with which questions were debated, on which men of the same country ought not to have differed in opinion. While they descanted on the hatred of the Mulattoes against the free negroes and whites, on the wealth of the monks, and the difficulty of holding slaves in obedience, a cold wind, that seemed to descend from the lofty summit of the Silla of Caraccas, enveloped us in a thick fog, and put an end to this animated conversation. We sought for shelter, in the Venta del Guayavo. When we entered the inn, an old man, who had spoken with the most calmness, reminded the others how imprudent it was, in a time of denunciation, on the mountain as well as in the city, to engage in political discussions. These words, uttered in a spot of so wild an aspect, made a lively impression on my mind; which was often renewed during our journeys in the Andes of New Grenada and Peru. In Europe, where nations decide their quarrels in the plains, we climb the mountains in search of solitude and liberty. In the New World, the Cordilleras are inhabited to the height of twelve thousand feet; and thither men carry with them their political dissentions, and their little and hateful passions.

Caraccas is the capital of a country, which is nearly twice as large as Peru at present, and which yields little in extent to the kingdom of New Grenada. This country, which the Spanish government designates by the name of *Capitania General de Caraccas*, or of the (united) provinces of Venezuela, has nearly a million of inhabitants, among whom are sixty thousand slaves. It contains, along the coast, New Andalusia, or the province of Cumana (with the island of Margaretta), Barcelona, Venezuela or Caraccas, Coro, and Maracaybo; in the interior, the provinces of Varinas and Guiana, the first along the rivers of Santo Domingo and the Apure, the second along the Oroonoko, the Casiquiare, the Atabapo, and the Rio Negro. In a general view of the seven united provinces of Terra Firma, we perceive, that they form three distinct zones, extending from east to west.

Caraccas is the seat of an *audiencia* (high court of justice) and one of the eight archbishoprics, into which Spanish America is divided. Its population in 1800, according to the researches I made into the number of births, was nearly 40,000; the best informed inhabitants believed it even to be 45,000, of which 18,000 are whites, and 27,000 free men of colour. Computations made in 1778, had already given from 30,000 to 32,000. All the direct numberings have remained a quarter, and more, below the effective number. In 1766, the population of Caraccas, and the fine valley in which that city is placed, suffered immensely by a severe attack of the small-pox. The mortality rose in the town to six or eight thousand. Since that memorable period, inoculation is become general, and I have seen it practised without the aid of physicians. In the province of Cumana, where the communications with Europe are less frequent, there had not been, in my time, one instance of the small-pox during fifteen years; while, at Caraccas, that cruel malady was constantly dreaded, because it always showed itself sporadically on several points at a time. I say sporadically, for in Equinoctial America, where the changes of the atmosphere, and the phenomena of organic | 599 | life, seem subject to a remarkable periodicalness, the small-pox, before the benevolent introduction of the vaccine disease, exerted its ravages only, if we may place confidence in general belief, every fifteen or twenty years. Since my return to Europe, the population of Caraccas has continued to augment. It amounted to 50,000 souls; when, at the great earthquake of the 26th of March, 1812, twelve thousand inhabitants perished beneath the ruins of their houses. The political events, which have succeeded this catastrophe, have reduced the number of inhabitants to less than 20,000; but these losses will soon be repaired, if the fertile and commercial country, of which Caraccas is the centre, should have the happiness of enjoying repose, and a wise administration, for a few years.

The town is seated at the entrance of the plain of Chacao, which extend three leagues east towards Caurimare and the Cuesta de Auyamas, and which is two leagues and a half in breadth. This plain, through which runs the Rio Guayra, is four hundred and fourteen toises in height above the level of the sea. The ground, which the town of Caraccas occupies, is uneven, and has a steep slope from N.N.W. to S.S.E. In order to form an exact idea of the situation of Caraccas, we must recollect the general disposition of the mountains of the coast, and the great longitudinal valleys, by which they were traversed. The Guayra rises in the group of primitive mountains of Higuerota, which separates the valley of Caraccas from that of Aragua. It is formed near Las Ajuntas by the junction of the little rivers of San Pedro and Macarao, and runs first to the east as far as the Cuesta of Auyamas, and then to the south, to unite its waters with those of Rio Tuy, below Yare. The Rio Tuy is the only considerable river in the northern and mountainous part of the province.

The climate of Caraccas has often been called a perpetual spring. It is found every where, half way up the Cordilleras of Equinoctial America, between four hundred and nine hundred toises of elevation, unless the great breadth of the valley, joined to an arid soil, causes an extraordinary intensity of radiant caloric. What indeed can we imagine more delightful than a temperature, which in the day keeps between 20° and 26°; and at night, between 16° and 18°, which is equally favourable to the plantain (cambury), the orange-tree, the coffee-tree, the apple, the apricot, and corn? A national writer compares the situation of Caraccas to the terrestrial Paradise, and recognizes in the Anauco and the neighbouring torrents, the four rivers of the Garden of Eden.

It is to be regretted, that such a temperate climate is generally inconstant and variable. The inhabitants of Caraccas complain of having several seasons in the same day; and of the rapid change from one season to another. In the month of January for instance, a night, of which the mean temperature is 16°, is followed by a day, when the thermometer, during eight successive hours, keeps above 22° in the shade. In the same day, we find the temperature of 24° and 18°.

The cool and delightful climate we have been describing, agrees also with the culture of equinoctial productions. The sugar-cane is cultivated with success, even at heights exceeding that of Caraccas; but in the valley, on account of the dryness of the climate, and the stony soil, the cultivation of the coffee-tree is preferred; which there yields little fruit indeed, but of the finest quality. When the shrub is in blossom, the plain extending beyond Chacao presents a delightful aspect. The banana-tree, which is seen in the plantations near the town, is not the great *platano harton*; but the varieties *camburi* and *dominico*, which require less heat. The great plantations are brought to the market of Caraccas from the haciendos of Turiamo, situate on the coast between Burburata and Porto Cabello. The highest flavoured pine-apples are those of Baruto, of Empedrado, and of the heights of Buenavista, on the road to Victoria. When a traveller ascends, for the first time, to the valley of Caraccas, he is agreeably surprised to find the culinary plants of our climate, the strawberry, the vine, and almost all the fruit-trees of the temperate zone, growing by the side of the coffee and banana-tree. The apples and peaches esteemed the best come from Macaroa, or from the western extremity of the valley. There, the quince-tree, the trunk of which attains only four or five feet in height, is so common, that it has almost become wild.

Preserved apples and quinces, particularly the latter, are much used in a country, where it is thought, that, to drink water, thirst must previously be excited by sweetmeats. In proportion as the environs of the town have been cultivated with coffee; and the establishment of | 600 | plantations, which dates only from the year 1795, has increased the number of agricultural negroes; the apple and quince trees, scattered in the savannahs, have given place, in the valley of Caraccas, to maize and pulse.

I remained two months at Caraccas, where Mr. Bonpland and I lived in a large and nearly solitary house, in the highest part of the town. From a gallery, we could survey at once the summit of the Silla, the serrated ridge of the Gallipano, and the charming valley of the Guayra, the rich cultivation of which, formed a pleasing contrast with the gloomy curtain of the surrounding mountains. It was the season of drought, and, in order to improve the pasturage, the savannahs, and the turf that covers the steepest rocks, were set on fire. These vast conflagrations, viewed from a distance, produce the most singular effects of light. Wherever the savannahs, following the undulating slope of the rocks, have filled up the furrows hollowed out by the waters, the inflamed land appears, in a dark night, like currents of lava suspended over the valley. Their vivid but steady light assumes a reddish tint, when the wind, descending from the Silla, accumulates streams of vapour in the low regions. At other times, and this aspect is still more solemn, these luminous bands, enveloped in thick clouds, appear only at intervals, where it is clear; and as the clouds ascend, their edges reflect a splendid light. These various phenomena, so common under the tropics, become still more interesting from the form of the mountain, the disposition of the slopes, and the height of the savannahs covered with alpine grasses. During the day, the wind of Petare, blowing from the east, drives the smoke toward the town, and diminishes the transparency of the air.

Civilization has in no other part of Spanish America, assumed a more European physiognomy. The great number of Indian cultivators, who inhabit Mexico and the interior of New Grenada, have impressed a peculiar, I might also say an exotic character, on those vast countries. Notwithstanding the increase of the black population, we seem to be nearer Cadiz and the United States at Caraccas and the Havannah, than in any other part of the New World.

I found in several families at Caraccas a taste for instruction, a knowledge of the master-pieces of French and Italian literature, and a particular predilection for music, which is cultivated with success, and which, as it always happens in the pursuit of the fine arts, serves to bring the different classes of society nearer to each other. The mathematical sciences, drawing, painting, cannot here boast of any of those establishments, with which royal munificence, and the patriotic zeal of the inhabitants, have enriched Mexico. In the midst of the marvels of nature, so rich in productions, no person on this coast was devoted to the study of plants and minerals. In a convent of St. Francis alone I met with a respectable old gentleman, who calculated the almanac for all the provinces of Venezuela, and who possessed some precise ideas on the state of modern astronomy.

Remarks on the natural Family of the Grasses

I have, in a former part of this work, classed the Orders *Gramineæ* (Grasses,) *Cyperaceæ* (Club-rushes,) and *Junceæ* (Rushes,) in one natural family, under the denomination of *Glumaceæ* (the plants with a chaffy flower,) and shall now proceed to a summary notice of the species, in regard to number, configuration, and geographical distribution. We may form a conception of the richness of America in plants of this nature, and of the smallness of the proportion of those which had come to the knowledge of the botanists of Europe, when we find, that of 343 species, observed by M. Bonpland and myself in the course of our travels, scarcely a fifth or sixth part had been recorded. In casting up the *Glumaceæ*, enumerated in Persoon's Synopsis Plantarum, those found by Mr. Brown in New Holland and Van Diemen's Island, and the new ones published by myself and fellow-traveller, we find that we are now acquainted with about 1200 *Gramineæ*, 900 *Cyperaceæ*, and 100 *Junceæ*, forming a total of 2200 *Glumaceæ*. But, although this may appear a considerable number, and one that has, of late, received very extensive additions, it turns out, in fact, to be a proof of the negligence of botanists in respect to these three orders. It has been already shown, that these three orders form one-tenth part of the Phænogamous vegetation of the earth; so that, in the 30,000 Monocotyledonous and Dicotyledonous species, which have been recorded, we ought to have met with, at least, 3000 *Glumaceæ*, if they had received from travellers an equal share of attention, with that which has been bestowed upon *Compositeæ* and *Leguminosæ*.

All over the world the *Glumaceæ* are found to increase their number in a wonderful proportion, either as you recede from the line, towards the poles, or as you ascend the mountain from the level of the sea. But then this augmentation of number takes place in a far smaller ratio, from the line to the temperate zone, than that in which it is found to take place from the latitudes of France and Germany, towards the polar circle. In Lapland, for instance, there are three times more *Glumaceæ* than *Compositæ*; while, in the temperate parts of Europe, the families are nearly equal. On the other hand, in North America, from the 32d to the 45th degree of latitude, the *Compositæ* are already found to exceed the *Glumaceæ* by a fourth: a proportion which becomes still greater in the tropical regions of that continent. I have purposely taken the *Glumaceæ* and *Compositœ* for points of comparison, as being the two families which, in every part of the world, comprise the largest portion of vegetable species, and display the greatest variety of configuration. Next, in point of numbers, to the *Glumaceœ* and *Compositœ*, as far as I am able to judge, are the *Caryophylleœ, Amentaceœ*, and *Ericinœ*, in the frozen zone; the *Leguminosœ, Cruciferœ*, and *Labiatœ*, in the temperate zone; the *Leguminosœ, Rubiaceœ*, and *Malvaceœ*, in the torrid zone.

In considering, separately, the three natural orders which compose the family of the *Glumaceœ*, we shall find, that the respective relations of the *Gramineœ, Cyperaceœ*, and *Junceœ*, under the line, are, to each other, nearly as 25.7.1; in the temperate latitudes of the old world, as 7.5.1; under the polar circle, as 2 2/5. 2 3/5. 1. So that it is only in Lapland that the *Cyperaceœ* are equal, in number of species, to the *Gramineœ*; but through the temperate zone to the tropics, the quantity of *Cyperaceœ* and *Junceœ* diminishes, in a far greater proportion, in the northern hemisphere, than that of the *Gramineœ*: insomuch that the *Junceœ* disappear almost entirely in the torrid zone. The *Cyperaceœ*, on the other hand, seem better qualified to support every degree of climate; and it is specially among them that we find the plants which are common to both the new and the old continents, such as *Kyllingia monocephala*, Cyperus *monostachyus*, Chætospora *aurea,* and other species, which we have enumerated elsewhere. So New Holland and South America produce, in common, Scirpus *triqueter*, Scirpus *capitatus*, and Fuirena *umbellata*; Europe and Australasia, Scirpus *fluitans*, Scirpus *supinus*, Scirpus setaceus, Scirpus *lacustris*, Scirpus *triqueter*, Schœnus *mariscus*, Carex *cœpitosa*, Carex *PseudoCyperus*, Juncus *maritimus*, and Juncus *effusus*. In general the countries which lie within the tropic of Capricorn appear to abound in the Cyperaceæ; for, of the 456 Glumaceæ of New Holland, described by Mr. Brown, 214 are ranked in the *Gramineœ*, and 200 in the *Cyperaceœ*; which proportion, if it be admitted as the true one of the relative distribution of these plants, is widely different from that which is exemplified in the tropic of Cancer.

As to what I have to offer in regard to the secondary groups or tribes, into which the *Glumaceœ* have been divided according to natural affinity, I shall make use of an extract from the writings of Mr. Kunth: "Some of the tribes of the *Gramineœ* are represented by numerous species in the tropical regions, while in Europe they have not a single species, or at least, such only as are very rare: for instance, the *Paniceœ, Stipaceœ, Chlorideœ, Saccharinœ, Orizeœ, Olyreœ*, and *Bambusaceœ*. Europe does not produce a single species of *Paspalum*, only five species of the *Stipaceœ*, very few of the *Saccharinœ*, but one of the *Oryzeœ*, (Leersia *oryzoides*,) and not one of the *Chlorideæ, Olyreæ*, and *Bambusaceæ*. On the other hand, the *Agrostideæ, Avenaceæ, Arundinaceæ*, and *Bromeæ*, are peculiar to our temperate latitudes. In a like manner the *Hordeaceæ*, (an order which comprehends the principal part of our corn-plants,) seem specially adapted to the warmer regions of Europe and to Asia, while the alpine grasses of both the new and old continents belong principally to the *Agrostideæ, Avenaceæ*, and *Bromeæ*. The genus *Cyperus*, is almost entirely tropical; for of the 150 species, which we know at present, scarcely 20 belong to Europe and the northern part of America. There is not a Mariscus, nor a Kyllingia in all Europe; nor is there any species of the true *Cyperaceæ*, (those with two ranked imbricated glumes,) in the whole European continent. The *Scirpeæ* seemed to be dispersed indiscriminately over every part of the globe, and of all the monocotyledonous plants, are those among which we meet with the greatest number of instances of species, which are common both to the new and old continents."

In regard to Bamboos, M. Bonpland and myself had the good fortune to meet with them in bloom twice, once on the banks of the Cassiquiare, (a branch of the Oronoco,) and again near the plantation of El Muerto, in the province of Popayan, between Bugus and Quilichao. For though these tree-like canes cover the marshy lands of the new world to a great extent, and often attain the height of 50 or 60 feet, yet they very seldom flower there. Neither Mutis, who had explored so many *Guadales*, (as the marshes covered with Bamboos are called by the Creoles,) nor our friend Tafalla, who had accompanied Ruiz and Pavon in their well-known botanical expedition through Peru, had either of them ever been able to procure the fruit of the flower of a Bamboo. On the other hand, in the East-Indies these gigantic grasses are known to flower in such abundance, that according to Dr. Buchanan, the seed mixed with honey, is a common article of food in the Mysore country. The plant is there believed to bear fruit as soon as it has attained the age of 15 years, and to die immediately afterward. It is distinguished by the natives of those parts into two sorts, one with a solid cane, called *Chittu*, and which grows in dry places, and the other with a hollow cane, called *Doda*, which comes quicker to maturity, and grows in watery places.

Having procured some canes of a Bamboo (Bambusa *Guaduæ*) at Guaduas, in New Granada, we saw at once how deficiently and incorrectly the genus had been characterized in all our botanical works, and made it our first concern to take the description of the plant on the spot where it grew, and attended in particular to the deeply three-parted style, and the three scales which surround the parts of fructification, and were then denominated by us its triphyllous nectary. Loureiro is almost the only author who has described the style correctly in the Asiatic species, (for example in Bambusa *verticellata*.)

The Bamboo is not so general in the wet lands of South America as has been usually thought. They are rather scarce, both in the Caraccas and New Andalusia, (if we except the valleys that lie between Cumanacoa and the town of San Fernando,) and the marshy woods of Guayana, that line the banks of the Cassiaquiare and Atabasso. There are hardly any other on the shores of the Apure which runs through the province of Varinas, or on those of the Guainia, or Rio Negro. In all the parts of America which were visited by Bonpland and myself, we only found them common in those places which lay exposed to the setting sun. They abound principally in New Granada, where they constitute vast forests, and grow both in the sultry lowlands, between Turbaco, and Mahates, as well as in the highland valleys, where the climate is more temperate; for instance, between the towns of Guaduas and Villeta, in Santa Fè de Bogotà; on the western declivities of the Quindu Andès, near Buenavista and Carthagena; on the bank of the Cauca (between Bugas and Quilichao, in Popayan;) and, lastly, at the back of the volcanic mountain, Rucu-Pichincha, near the city of Quito; where a wide marshy level, covered with a close rank vegetation, extends through the province of Esmeralda, to the shores of the Pacific Ocean.

We found the Bambusa Guaduæ from on a level with the sea, up to the height of 860 fathoms; and what struck us was, that the highland plants of this species always contained more water than those on the plains, although both grew in soils of equal degrees of humidity. In the highest stations (as from 600 to 900 fathoms) we only found them dispersed about in small bushes; but, in the level country, even as high up as 400 fathoms, they formed extensive forests. In general the plants which belong to the Bamboo tribe, may be reckoned among the gregarious plants (*plantœ sociatœ*.) The Nastus[1] of the Isle of Bourbon, according

[1] The genera Bambusa and Nastus, which most botanists of the present day have blended together, are to be distinguished from each other by the following characteristics:—In Bambusa the long subcylindrical ears comprise a great number of bipaleaceous (double-chaffed) flowers, of which the lower only are male ones. Each of these bipaleaceous flowers is enclosed by a calyx of two glumes, (husks.) The manner of the inflorescence, and the form of the chaffs are nearly the same as in the Poæ, from which, however, the Bambusæ are sufficiently distinguishable by an arborescent haulm, by having six stamens, a deeply trifid style, and three scales that surround the parts of fructification. In the genus Nastus, on the other hand, the ear is oblong, compressed, and comprises a fixed number of chaffs, which overlay each other in two rows, nearly as in the *Cyperaceœ*. Of these glumes, only the two upper ones enclose a flower like that of Bambusa, viz. one with a trifid style, six stamens, and three nectaries. Judging from analogy, the two lower glumes may stand for the calyx in Bambusa, the others may be looked upon as neutral

to Bory St. Vincent, is a true subalpine grass, and never descends into the plains lower than to an elevation of 600 fathoms above the sea.

The water which is found secreted in the hollow of the American Bamboos has a somewhat brackish taste, but is not unpalatable. It is said by the natives to have an injurious effect upon the urinary passages. I could never detect any secretion from the American Bamboos, that gave me any idea of the sweetness of honey; but I met with the true *Tabasheer* or *Tabaxir* in the kingdom of Quito, and it differed but very slightly from that of the East Indies. I brought home a specimen of it, which was analyzed by the celebrated Vauquelin. It is called by the Creoles *manteca de Guaduas*, (Guaduas butter,) and contains 0,70 of siliceous earth, and 0,30 of potash, lime, and water.

I cannot account for the *Tabasheer*, which is a white substance, and friable like starch, having been compared to honey by those who have treated on the subject of the sugar of the ancients. I could perceive no sweet taste in the Quito Tabasheer, not even when it was in the state of a mucilage, and before it was hardened by drying: and strongly suspect that none of the arborescent canes in all America contain any sweet liquid whatever. As to the Tabasheer, before it coagulates into its wonted stony hardness, it is a viscid, white, and milky substance. Kept for five months, it exhales a strongly fetid animal smell. The same property was observed by Dr. Patrick Russell, the oriental traveller, in what he terms the *salt of the Asiatic Bamboo*, while Garcias del Huerto, who resided for a long time at Goa, in quality of physician to the viceroy, is the only author who speaks of a sweet juice from the Bamboo. The ancients seem to have been led to confound true sugar with Tabasheer in the first place, from both being the produce of a cane, and in the second from the Sanscrit word *sharkara*, which at this day (like the Persian *shaker*, and the Hindustanée *schukur*) is used for our sugar, not properly meaning something which is sweet, but something that is lapideous and granulated, as we learn from Boppius, on the authority of Amarasinha. It is probable that the word *scharkara* originally meant only Tabasheer, (*saccar mombu*,) but was subsequently transferred from similitude of appearance to our sugar from the smaller cane (*ikschut, kandekschu, kanda.*) The word Bamboo is derived from *mombu*; and from *kanda* we get candy,

flowers with only one valve. The species which belong to Bambusa are, *arundinacea et stricta* Roxb. *verticillata* Willd. *latifolia et Guadœ*, Bonpl. and an unpublished one of the Isle of Bourbon. To the genus Nastus belong *Calumet des hauts de Bourbon*, and a Madagascar, species of which the specimen is preserved in the Herbarium of M. Du Petit Thouars. Bory St. Vincent has described the style and nectaries of Nastus correctly, but has blended the genus with Bambusa.

(*sugar-candy.*) In tabasheer we trace the Persian word *scher*, which means milk, in Sanscrit *kschiram*.

Pliny, as has been ingeniously demonstrated by Salmasius, in opposition to Scaliger, clearly describes tabasheer of the Bamboo under the name of sugar, when he speaks of it as "a honey formed in canes, white like gum, crumbling between the teeth, found in lumps about the size of a filbert, and only used medicinally." Yet it is manifest from the verses of Varro, quoted by Isidorus, from passages of Theophrastus, Lucan, Solinus, and especially Strabo, on the authority of Eratosthenes, that the ancients had derived some notion of our sugar from the East Indies, and affirmed that it was obtained from canes without the assistance of the bee. But then they believed that a liquid of the sweetness of honey was expressed from the roots of the large kind of canes, confounding the root with the haulm and the humble sugar cane (Saccharum *officinale*) with the Bamboo, "a joint of which, when split asunder, they described as large enough to form a navigable bark." Others among them, as Seneca, Dioscorides, and Alexander Aphrodisæus, believed true sugar to be the morning dews of the canes, collected on the foliage of the plant. It is certain that the sugar-cane is indigenous in the neighbourhood of Almangar, in the East Indies, on the banks of the Euphrates, and at Siraf; but I suspect that in that part of Asia which was frequented by the Greeks, the plant was only expressed for the purpose of procuring a beverage for immediate use, and that the juice never could have been exported, owing to its tendency to ferment. So that I conclude, that hard sugar was unknown to the ancients,[2] and that when they speak of a solid sugar, they mean tabasheer or scharkara of Bamboo.

It may be hardly thought necessary that we should mention in this place, that before America was discovered by the Spaniards, the inhabitants of that continent and the adjacent islands, were entirely unacquainted either with the sugar canes, or with any of our corn plants, (which last are indigenous of the country between the Kur and the Terek, in Persia and Armenia,) or with rice. The Spanish writers on the subject of America, give the name of *small-rice* (*Oryza parva,*) to the Chenopodium *Quinoa*, which is common in Santa Fè de Bogotá, and in the kingdom of Quito, in the same way as the Anglo-Americans did that of *Canada rice* to a species of Zizania. Turkey corn, (Zea Mays,) like many of the plants

[2] The art of making sugar from the cane was not mentioned by any writers until the fifth century, and as the learned Sprangell has first shown, Moses Chorenensis, in describing the beauties of the province of Chorasan, mentions the valley of *Gundi-Saporem* "as place where the precious sugar was made." I have proved in another place, that the manufacture of sugar was of the highest antiquity in China.

which have been in general cultivation from remote periods, is not found growing wild in any part of America. We have to lament that travellers have not made us correctly acquainted with the characters of the plants mentioned cursorily by Molina in his history of Chili, under the names of *Secale Magu* and *Hordeum Tuca*, of which the Araucanos formerly made a bread called *coque*. We know from Cortes that the Agave and Turkey corn afforded the Americans a honey independently of the bee, and that he saw it in market at Tenochlitl.

On the Milk extracted from the Cow Tree (l'Arbre de la Vache), and on Vegetable Milk in general

We have given (say the editors of the Annales) the extract of a memoir read by M. Humboldt in one | 425 | of the last sittings of the Academy of Sciences. The readers who may desire a more extended detail on a subject so interesting in vegetable chemistry, will find it in the sixth volume of the *Relation Historique*, by the author, which will be very shortly published.

"We had for several weeks past heard mention made of a certain tree in the valleys of Aragua, the juice of which was said to be a nourishing milk; the tree, indeed, was called the cow tree, and we learnt that the negroes of the place, who drank it abundantly, looked upon it as a very wholesome species of aliment. This account excited surprise in our minds, since the general characters attached to the milky juices of plants, are those of being acrid, bitter, and more or less poisonous. But since our residence in Barbula (province of the Caraccas), we have found that in this account of the *Palo de vaca* there was no exaggeration. The tree thus named, is one of a very beautiful appearance. Its flowers we had not an opportunity of seeing; the fruit of it is rather pulpy, and encloses sometimes one and sometimes two kernels. When incisions are made into the trunk of this tree, which appears to belong to the sapota tribe of plants, it gives out an abundance of a glutinous thick kind of milk, void of all acrimony, and exhaling an odour by no means unpleasant. We drank considerable quantities of it, both in the evening before we retired to rest and early in the morning, without experiencing the slightest inconvenience: it is only the glutinous nature of the fluid in question that occasions its taste to be at all unpleasant. Both the slaves and others employed upon the plantations drink it freely, and mix it with the maize and capada plant. The master of the plantation assured us that the slaves always thrived and gained flesh during the season in which the *Palo de vaca* furnished them with milk. When this milk is exposed to the air, its surface becomes covered with a strongly animalized substance of a yellowish hue, and fibrous stringy appearance, resembling a cheesy matter. This change in the juice is probably produced by an absorption of oxygen from the air. When the membranous substance of which we speak is separated from the more fluid part of the milk, it proves to be nearly as elastic as caoutchouc; but it undergoes in the course of time a like putrefaction with gelatine. The people of the place call this matter cheese. It becomes sour in about five or six days, according to the observation which I made on some portions of it which I took with me to Orenoque. The milk, enclosed in a well stopped vial, had thrown down a small quantity of *coagulum*, which, far from being fetid, constantly exhaled a sort of balsamic odour. The fresh milk scarcely coagulated at all upon being mixed with cold water; | 426 | but a separation of this viscous matter took place when I heated it with nitric acid. We sent two bottles of this milk to M. Fourcroy, at Paris. The one bottle contained the juice pure, in the other it was mixed with a quantity of carbonate of soda.

This remarkable production, the Palo de vaca, appears to be confined to the Cordilleras (la Cordillière du Littoral), especially near the lake of Marakabo. It is found also near the town of San Mateo; and, according to M. Bredmeyer (whose voyages have so much contributed to enrich the beautiful greenhouses of Schônbrun and Vienna), it is to be seen in the vale of Caucagua, three days' journey east of the Caraccas. This naturalist, like ourselves, found the vegetable juice in question of an agreeable taste and aromatic odour. The natives of Caucagua call the tree which furnishes it, the milk tree *arbol de loche*.

The inhabitants of the Andes had been in the practice of fabricating wax lights from the wax which is found on the trunk of the palm tree long before the chemists of Europe had discovered quantities of wax in the pollen of flowers, varnish in leaves, and farina in fruits; in like manner the *caseum*, the basis of cheese, has but recently been detected in the emulsions of almonds; whereas we find that ages ago, in the mountains of Venezuela, the milk of a tree, and the cheese which separates itself from such milk, were used as aliment. How are we to account for these singularities in regard to the development of our knowledge of nature's productions? How can we explain the fact, that the people of another hemisphere have discovered and applied properties which had for so long a time escaped the penetration of men whose very occupation it is to search Nature's laws, and penetrate her mysterious operations?

It would appear that the fact is to be explained partly from the circumstance of the elements and principles of plants being distributed among so many orders and families of the vegetable creation, partly from that difference of quantity, with respect to their essential principles, which is observed in the vegetable world, according as the particular plant is a native of equatorial, or of cold and temperate latitudes, and partly by an ingenuity derived from necessity, which impels uncultivated man to seek for his sustenance in the natural productions by which he is surrounded. Thus the juices, the bark, the roots, and the fruits of trees, become the subjects naturally of instinctive investigation; and when poisonous productions are combined with those that are wholesome and nutritive, man is taught by the same necessity to separate the one principle from the other, as in the arum, tacca, &c. The American savage, as well as the inhabitants of the South Sea islands, have thus | 427 | learnt the art to prepare the fecula of plants, by compressing it and separating its juice. In the milk of plants, as also in the milky emulsions, materials considerably nutrient, such as albumen, the caseous and the saccharine principles are intermixed with caoutchouc, and deleterious ingredients, such as morphine and hydrocyanic acid. These combinations vary not only in the different tribes of plants, but also in the respective species of the same genera. Sometimes it is the morphine or narcotic principle, which predominates in vegetable milk; in other instances the caoutchouc is the characterizing ingredient; and lastly, as we have seen in the juice of the papaya, and cow tree, the caseous is the main principle.

The lactiferous plants belong principally to the three families of the Euphorbia, the Urticaria, and the Apocyna; and as upon investigating the different distribution of vegetable growth in the several parts of the world, we find that the species of these orders of plants are most numerous in the tropical regions, we infer that a very high temperature is necessary for the proper elaboration of the milky juices, as well as to the complete formation of the caoutchouc of albumen, and of the caseous principle. The juice of the *Palo de vaca* certainly presents one of the most striking examples of a vegetable milk in which the acrid and deleterious principles are not united to albumen, caseum, or caoutchouc; but the euphorbium and asclepias genera, so generally known by their caustic properties, had before furnished us with some species of which the juice is bland and innocent, as in the instance of the tabayba dulce, (euphorbia balsamifera) of the Canary Isles, and the asclepias lactifera of Ceylon. Bruman has told us, that the inhabitants of Ceylon make use of this last in lieu of milk, and that they mix its leaves in cookery with those articles of food that are generally prepared with milk. It is to be wished that Mr. John Davy may pay attention to this particular during his stay at Ceylon; for it appears probable, as has been suggested by M. Decandolle, that it is only the juice which exudes from the young plant, which is used for the purposes in question, viz. that which flows from the vegetable before the development of the acrid principle. Indeed, in some countries, the first shoots of even the apocyna are eaten.

I have thus endeavoured to institute a general resemblance between the juices which circulate in plants, and the lactiferous emulsions which are yielded by the almond and palm tribe; and I may here add a few remarks respecting some experiments upon the carica papaya, which I made during my stay in the valleys of Aragua; not, of course, of so satisfactory a nature as they might have been, had I been furnished with the several reactives necessary to conduct investigations of this kind. I ascertained | 428 | that the younger the fruit of the plant, the greater was the quantity of milk which it furnished: this juice, indeed, was found even in the germ of the fruit; and as it proceeded to maturity, its milk became not merely less abundant, but also more aqueous. In it was found less of the animal coagulable matter (matière animale coagulable), and it is conceivable that as the plant advances in growth, this matter, abstracted from the fruit, comes to constitute its pulpy or fleshy parts. The coagulum from the very young fruit, heated with nitric acid, becomes viscous, and exhales a waxy odour, similar to that which I have observed in muscular flesh and in small mushrooms (morilles), when the acid is added to them in the same manner. Following the hints of Fourcroy and Vauquelin, I mixed the milk of the papaya with a solution of carbonate of soda. The mixture did not run into coagulated lumps, even when water was added to it. The membranous substance did not show itself, until, by the addition of an acid to the solution, the soda became neutralized. The coagulum formed by the nitric acid, lemon juice, and hot water, I caused to disappear, by adding to the mixture the carbonate of soda. By this addition the juice resumed its milky and liquid appearance; but the experiments only answered when the addition was made to recently formed coagula.

In conclusion, it may be remarked, that a comparison of the milky juices of the papaya with those of the cow-tree and the hevea, offers a striking analogy between the juices which abound in the caseous, and those that contain the caoutchouc principle. All the productions in which the latter is predominant, as likewise the impenetrable garments (manteaux imperméables) that are fabricated in Spanish America, are made to exhale a like animal and disagreeable odour, by placing the milk of the hevea between two layers of them; a circumstance which would seem to argue that the caoutchouc in coagulating, attracts to itself the caseum, which is no other, perhaps, than albumen in a different form. The fruit of the bread-tree is not any more bread than are the bananas before their maturity, or the tuberous and amilaceous roots of the cassada, the convolvulus battatas, or the potatoe. On the contrary, the milk of the *Palo de vaca* contains the caseous principle in the same manner as does the milk from the mammalia. In pursuing these analogies between the animal and vegetable world, we may, with M. Gay-Lussac, consider the caoutchouc as the oily part, the butter as it were of the vegetable milk. In a word, we may find in vegetable milk, caseum and caoutchouc; in animal milk, caseum and butter. The two principles vary in their proportions in different animals, as they do in the various species of lactiferous plants. In these last, they are more generally mingled with | 429 | other materials which are not alimentary; but from such materials they may, perhaps, be always separated by chemical processes. Vegetable milk becomes wholesome and nourishing when it is deprived of its acrid and narcotic principle, and when it abounds more in caseous matter than in caoutchouc.

Baron Humboldt's Last Volume. Extract from Baron Humboldt's Personal Narrative of Travels to the Equinoctial Regions of the New Continent. Vol. 4

On leaving the Rio Apure, we found ourselves in a country of a totally different aspect. An immense plain of water stretched before us like a lake, as far as we could see. White-topped waves rose to the height of several feet, from the conflict of the breeze and the current. The air resounded no longer with the piercing cries of the herons, the flamingoes, and the spoonbills, crossing in long files from one shore to the other. Our eyes sought in vain those water fowls, the inventive snares of which vary in each tribe. All nature appears less animated.— Scarcely could we discover in the hollows of the waves a few large crocodiles, cutting obliquely, by the help of their long tails, the surface of the agitated waters. The horizon was bounded by a zone of forests, but these forests no where reached so far as the bed of the river. A vast beach, constantly parched by the heat of the sun, desert and bare as the shores of the sea, resembled, at a distance, from the effect of the *mirage*, pools of stagnant water. These sandy shores, far from fixing the limits of the river, rendered them uncertain, by approaching or withdrawing them alternately, according to the variable action of the inflected rays.

In these scattered features of the landscape, in this character of solitude and of greatness, we recognize the course of the Oroonoko, one of the most majestic rivers of the New World. The water, like the land, displays every where a characteristic and peculiar aspect.— The bed of the Oroonoko resembles not the bed of the Meta, the Guaviare, the Rio Negro, or the Amazon. These differences do not depend altogether on the breadth or the velocity of the current: they are connected with a multitude of impressions, which it is easier to perceive upon the spot, than to define with precision. Thus the mere form of the waves, the tint of the waters, the aspect of the sky and the clouds, would lead an experienced navigator to guess, whether he were in the Atlantic, in the Mediterranean, or in the equinoctial part of the Great Ocean.

The wind blew fresh from eastnortheast. Its direction was favourable for our sailing up the Oroonoko, toward the mission of Encaramada; but our canoes were so ill calculated to resist the shocks of the waves, that, from the violence of | 139 | the motion, those who suffered habitually at sea were incommoded on the river. The short broken waves are caused by the conflict of the waters at the junction of the two rivers. This conflict is very violent, but far from being so dangerous as Father Gumilla asserts.[1] We passed the Punta Curiquima, which is an isolated mass of quartzose granite, a small promontory composed of rounded blocks. There, on the right bank of the Oroonoko, Father Rotella founded, in the time of the Jesuits, a mission of Palenka and Viriviri or Guire Indians. At the period of inundations, the rock Curiquima and the village placed at its foot, were surrounded every where by water. This serious inconvenience, and the sufferings of the missionaries and Indians from the innumerable quantity of moschettoes and niguas,[2] led them to forsake this humid spot. It is now entirely desert, while opposite to it, on the right bank of the river, the little mountains of Coruato are the retreat of wandering Indians, expelled either from the missions, or from tribes that are not subjected to the government of the monks. Struck with the extreme breadth of the Oroonoko, between the mouth of the Apure and the rock Curiquima, I ascertained it by means of a base measured twice on the western beach. The bed of the Oroonoko, in its present state of low water, was 1906[3] toises broad; but this breadth attains 5517[4] toises, when, in the rainy season, the rock Curiquimi, and the farm of Capuchino near the hill of Pocopocori, become islands. The swelling of the Oroonoko is augmented by the impulse of the waters of the Apure, which far from forming, like other rivers, an acute angle with the upper part of that into which it flows, meets it at right angles. The temperature of the waters of the Oroonoko, measured in several parts of its bed, was in the middle of the *thalweg*, or deepest part of the channel, where the current has the most swiftness, 28.3, and towards the banks 29.2.

We went up first towards the southwest, as far as the shore of the Guaricoto Indians, on the left bank of the Oroonoko, and then towards the south. The river is so broad, that the mountains of Encaramada appear to rise from the water, as if they were seen above the horizon of the sea. They form a continued chain from east to west. As you approach them, the aspect of the country becomes more picturesque. These mountains are composed of enormous blocks of granite, cleft and piled one upon

---

[1] *Orinoco illustrado*, vol. i, p. 47.

[2] The *chego, pulex* penetrans, which penetrates under the nails of the toes in men and monkeys, and there deposits its eggs.

[3] 3714 metres, or 4441 varas, supposing 1 metre = 0.51307 of a toise = 1.19546 vara.

[4] 10753 metres, or 12855 varas.

another. Their division into blocks is the effect of decomposition. What contributes above all to embellish the scene at Encaramada is the force of vegetation, that covers the sides of the rocks, leaving bare only their rounded summits. They look like ancient ruins rising in the midst of a forest. The mountain immediately at the back of the mission, the *Tepu-pano*[5] of the Tamanack Indians, is covered by three enormous granitic cylinders; two of which are inclined, while the third, worn away at its basis, and more than eighty feet high, has preserved a vertical position. This rock, which calls to mind the form of the *Schnarcher* in the Hartz, or that of the *Organs of Aetopan* in Mexico,[6] composed for- | 140 | merly a part of the rounded summit of the mountain. In every zone it is the property of unstratified granite, to separate by decomposition into blocks of prismatic, cylindric, or columnar figures.

Opposite the shore of the Guaricotoes, we drew near another heap of rocks, which is very low, and three or four toises long. It rises in the midst of the plain, and has less resemblance to a *tumulus* than to those masses of granitic stones, which in the north of Holland and of Germany bear the name of *Huenenbette, beds* (or tombs) *of heroes*. The shore at this part of the Oroonoko, is no longer of pure and quartzose sand, but is composed of clay and spangles of mica, deposited in very thin strata, and most frequently with a dip of forty or fifty degrees. It looks like decomposed mica-slate. This change in the geological constitution of the shore extends far beyond the mouth of the Apure. We had begun to observe it in this latter river as far off as Algodonal and Canodel Manati. The spangles of mica come no doubt from the granite mountains of Curiquima and Encaramada; since farther to the north and to the east we find only quartzose sand, sandstone, compact limestone, and gypsum. Alluvial earth carried successively from the south to the north need not surprise us in the Oroonoko; but to what shall we attribute the same phenomenon in the bed of the Apure, seven leagues west of its mouth? In the present state of things, notwithstanding the swellings of the Oroonoko, the waters of the Apure never retrogade so far; and, to explain this phenomenon, we are forced to admit, that the micaceous strata were deposited at a time when the whole of the very low country that lies

[5] *Tepu-pano*, "place of stones," in which we recognize *tepu*, "stone rock," as in *tepuiri*, mountain. We here again perceive that Lesgian Oigour-Tatar root *tep*, (stone) found in America among the Mexicans, in *tepetl*; among the Caribbees, in *tebou*; among the Tamanacks, in *tepuiri*; a striking analogy between the languages of Caucasus and Upper-Asia and those of the banks of the Oroonoko.

[6] In Captain Tuckey's Voyage on the River Congo, we find represented a granitic rock, the *Taddi Enzazi*, which bears a striking resemblance to the mountain of Encaramada.

between Caycara, Algodonal, and the mountains of Encaramada, formed the basin of an inland lake.

We stopped some time at the port of Encaramada; it is a sort of *embarcadere*, a place where boats assemble. A rock of forty or fifty feet high forms the shore. It is composed of the same blocks of granite, heaped one upon another, as at the Schneeberg in Franconia, and in almost all the granitic mountains of Europe. Some of these detached masses have a spheroidal form; they are not balls, however, with concentric layers, as we have elsewhere described; but merely rounded blocks, nuclei separated from their envelopes by the effect of decomposition. This granite is of a grayish lead-colour, often black, as if covered with oxide of manganese; but this colour does not penetrate one fifth of a line into the rock, which is of a reddish white within, coarse-grained, and destitute of hornblende.

The Indian names of the mission of *San Luis del Encaramada*, are *Guaja* and *Caramana*.[7] This small village was founded in 1749, by Father Gili, the Jesuit, author of the *Storia dell Orinoco*, published at Rome. This missionary, learned in the Indian tongues, lived in this solitude during eighteen years, till the expulsion of the Jesuits. To form a precise idea of the savage state of these countries, we must recollect, that Father Gili speaks of Carichana, which is forty leagues | 141 | from Encaramada, as of a spot far distant; and that he never advanced so far as the first cataract of the river, of which he ventured to undertake the description.

In the port of Encaramada we met with some Caribbees of Panapana. A Cacique was going up the Oroonoko in his canoe, to join in the famous fishing of turtles' eggs. His canoe was rounded towards the bottom like a *bongo*, and followed by a smaller boat called *curiara*. He was seated beneath a sort of tent, *toldo*, constructed, as well as the sail, of palm leaves. His cold and silent gravity, the respect with which he was treated by his attendants, every thing denoted him to be a person of importance. He was equipped, however, in the same manner as his Indians. They

[7] All the missions of South America have names composed of two words, the first of which is necessarily the name of a saint, the patron of the church, and the second an Indian name, that of the nation, or the spot where the establishment is placed. Thus we say, San Jose de Maypures, Santa Cruz de Cachipo, San Juan Nepomuceno de los Atures, &c. These compound names appear only in the official documents; the inhabitants adopt but one of the two names, and generally, provided it be sonorous, the Indian. As the names of saints are several times repeated in neighbouring places, great confusion in geography arises from these repetitions. The names of San Juan, San Diego, and San Pedro, are scattered in our maps as if by chance. It is pretended, that the mission of *Guaji* affords a very rare example of the composition of two Spanish words. The word *Encaramada* means things raised one upon another, from *encaramar, attollere*. It is derived from the figure of Tepupano and the neighboring rocks perhaps it is only an Indian word, *caramana*, in which, as in *manati*, from a love for etymology, a Spanish signification was believed to be discovered.

were all equally naked, armed with bows and arrows, and covered with *onoto*, which is the colouring fecula of the bixa orellana. The chief, the domestics, the furniture, the boat, and the sail, were all painted red. These Caribbees are men of an almost athletic stature; they appeared to us much taller than the Indians we had hitherto seen. Their smooth and thick hair, cut upon their forehead like that of choristers, their eyebrows painted black, their look at once gloomy and animated, give their physiognomy a singular hardness of expression. Having till then seen only the skulls of some Caribbees of the West India Islands preserved in the collections of Europe, we were surprised to find that these Indians, who were of pure race, had the forehead much more rounded than it has been described. The women, very tall, but disgusting from their want of cleanliness, carried their infants on their backs, having their thighs and legs bound at certain distances by broad strips of cotton cloth. The flesh, strongly compressed beneath the ligatures, was swelled in the interstices. It is generally to be observed, that the Caribbees are as attentive to their exterior, and their ornaments, as it is possible for men to be, who are naked and painted red. They attach great importance to certain forms of the body; and a mother would be accused of culpable indifference towards her children, if she did not employ artificial means to shape the calf of the leg after the fashion of the country. As none of our Indians of Apure understood the Caribbee language, we could obtain no information from the Cacique of Panama respecting the encampments that are made at this season in several islands of the Oroonoko for collecting turtles' eggs.

Near Encaramada, a very long island divides the river into two branches. We passed the night in a rocky creek, opposite the mouth of the Rio Cabullare, which is formed by the Payara and the Atamacia, and is sometimes considered as one of the branches of the Apure, because it communicates with this river by the Rio Arichuna. The evening was beautiful. The moon illumined the tops of the granitic rocks. The heat was so uniformly distributed, that, notwithstanding the humidity of the air, no twinkling of the stars was remarked, even at four or five degrees above the horizon. The light of the planets was singularly dimmed; and if, on account of the smallness of the apparent diameter of Jupiter, I did not suspect some error in the observation, I should say, that here, for the first time, we thought we distinguished the disk of Jupiter with the naked eye. Toward midnight, the northeast wind became extremely violent. It brought no clouds, but the vault of the sky was covered more and more with vapours. Strong gusts were felt, and made us in fear for the safety of our canoe. During this whole day we had seen very few crocodiles, but all of an extraordinary size, from twenty to twenty-four feet. The Indians assured us, that the young crocodiles prefer the marshes, and the rivers | 142 | that are less broad, and less deep. They crowd together, particularly in the *Canos*, and we might be tempted to say of them, what Abd-Allatif says of the crocodiles of the Nile, that "they swarm like worms in the shallow waters of the river, and in the shelter of uninhabited Islands."

Baron Humboldt's Last Volume. Extract from Baron Humboldt's Personal Narrative of Travels to the Equinoctial Regions of the New Continent. Vol. 4

DEPARTURE FROM CARACCAS

To take the shortest road from Caraccas to the banks of the Oroonoko, we should have crossed the southern chain of mountains between Baruta, Salamanca, and the savannahs of Ocumare; traversed the steppes or *Llanos* of Orituca; and embarked at Cabruta, near the mouth of the Rio Guarico. But this direct road would have deprived us of the opportunity of surveying the finest and most cultivated part of the province, the valleys of Aragua; of taking the level of an important part of the chain of the coast by means of the barometer; and of descending the Rio Apure as far as its junction with the Oroonoko. A traveller, who has the intention of studying the configuration and natural riches of a country, is not guided by distances, but by the interest which the regions he may traverse excites in his mind. This powerful motive led us to the mountains of Los Teques, to the thermal springs of Mariara, to the fertile banks of the lake of Valencia, and through the immense savannahs of Calabozo to San Fernando de Apure, in the eastern part of the province of Varinas. Following this road, our first direction was to the west, then to the south, and finally to the east-southeast, to enter the Oroonoko by the Apure in the latitude of 7° 36′ 23′.

The day that we quitted the capital of Venezuela, subsequently swallowed up by terrible earthquakes, we reached the foot of the woody mountains, that close the valley toward the southwest, where we slept. We followed the right bank of the Rio Guayra as far as the village of Antimano, by a very fine road, partly scooped out of the rock: and passed by La Vega, and Carapa. The church of La Vega displays itself in a very picturesque manner on a range of hills covered with thick vegetation. Scattered houses are surrounded with date-trees, and seem to proclaim the easy circumstances of their inhabitants. A chain of low mountains separates the little river Guayra from the valley of La Pascua,[1] so much

[1] Valley of Cortes, or Easter Valley, so called because Diego de Losada, after having defeated the Teques Indians, and their cacique Guaycapuro, in the mountains of San Pedro, spent his Easter there in 1567, before entering the valley of San Francisco, where he founded the city of Caraccas (Oviedo, p. 252.)

celebrated in the history of the country, and from the ancient gold-mines of Baruta and Oripoto. Ascending toward Carapa, we enjoy once more the sight of the Silla, which appears like an immense dome with a cliff toward the sea. This rounded summit, and the ridge of Galipano crenated like a wall, are the only objects that, in this basin of gneiss and mica-slate, impress a character on the landscape. The other mountains are of a uniform and dull monotonous aspect.

Near Antimano all the orchards were full of peach-trees loaded with flowers. This village, the *Valle*, and the banks of the Macarao, furnish great abundance of peaches, quinces, and other European fruits, for the market of Caraccas. Between Antimano and Ajuntas we crossed the Rio Guayra seventeen times. The road is very fatiguing; yet, instead of constituting a new one, it would perhaps be better to change the bed of the river, which loses a great quantity of water by the combined effects of filtration and evaporation. Each sinuosity forms a marsh more or less extensive. This loss of water is to be regretted in a province, the whole cultivated region of which, with the exception of the land between the sea and the littoral chain of Mariara and Niguatar, is extremely dry. The rains are much less frequent and less violent in this place than in the inte- | 210 | rior of New Andalusia, at Cumanacoa, and on the banks of the Guarapiche. Many of the mountains of Caraccas enter the region of the clouds; but the strata of primitive rocks dip at an angle of 70° or 80°, and generally toward the northwest, so that the waters are either lost in the interior of the earth, or gush out in copious springs, not toward the south, but to the north of the mountains of the coast of Niguatar, Avila, and Mariara. The rising of the gneiss and mica-slate strata toward the south appears to me to explain, in great part, the extreme humidity of the coast. In the interior of the province we meet with spaces of land, two or three leagues square, quite destitute of springs. The sugar-cane, indigo, and coffee, can grow only in places where running waters can be made to supply the artificial irrigations necessary during very dry weather. The first colonists very imprudently destroyed the forests. Evaporation is enormous on a stony soil surrounded with rocks, that radiate heat on every side. The mountains of the coast appear like a wall, extending east and west from Cape Codera toward Point Tucacas. They prevent the humid air of the shore, those inferior strata of the atmosphere resting immediately on the sea, and dissolving the largest proportion of water, from penetrating to the inlands. There are few apertures, few ravines, which, like those of Catia, or of Tipe, lead from the coast to the high longitudinal valleys. No bed of a great river, no gulf allowing the sea to flow into the lands, spreads moisture by an abundant evaporation. In the eighth and tenth degrees of latitude, in regions where the clouds do not glide along the soil, many trees are stripped of their leaves in the months of January and February; not on account of the sinking of the temperature, as in Europe, but because the air, at this season, the farthest from that of rains, has nearly attained its maximum of dryness. The plants with very tough and glossy leaves alone resist this absence of humidity. Beneath the fine sky of the tropics, the traveller is struck with the aspect almost hibernal of the country; but the freshest verdure again appears, when he has reached the banks of the Oroonoko, where another climate prevails; and the great forests preserve by their shade a certain quantity of moisture in the soil, which they shelter from the devouring ardour of the sun.

Beyond the small village of Antimano the valley becomes much narrower. The river is bordered with *lata*, that fine gramineous plant with distich leaves, which sometimes reaches the height of thirty feet, and which we have described under the name of *gynerium*. Every hut is surrounded with enormous trees of persea, at the foot of which the aristolochiæ, paullinia, and other creepers, vegetate. The neighbouring mountains, covered with forests, seem to spread humidity over the western extremity of the valley of Caraccas. We passed the night in a plantation of sugar-canes, before our arrival at Las Ajuntas. A square house contained nearly eighty negroes; they were lying on skins of oxen spread upon the ground. In each apartment of the house were four slaves; it looked like a barrack. A dozen fires were burning in the farm-yard, where people were employed in dressing victuals. We were again struck with the noisy mirth of the blacks, which almost prevented us from sleeping. The clouds hindered me from observing the stars; the moon appeared only at intervals. The aspect of the landscape was dull and uniform, and all the surrounding hills were covered with aloes. Workmen were employed at a small canal, which was intended to convey the waters of the Rio San Pedro to the farm, at a height of more than seventy feet. According to a barometic calculation, the site of the *hacienda* is only fifty toises above the bed of the Rio Guayra at La Noria, near Caraccas. | 211 |

COFFEE-TREE

The soil of these countries has been found but little favorable to the cultivation of the coffee-tree, which in general is less productive in the valley of Caraccas, than was imagined when the first plantations were made near Chacao. In order to form a general idea of the importance of this branch of commerce, we must re- member, that the produce of the whole province of Caraccas in its greatest prosperity, before the revolutionary wars of 1812, amounted to fifty or sixty thousand quintals of coffee. This quantity, almost equal to the produce of Guadaloupe and Martinico both together, must appear so much the more considerable, as it is only since 1781 that a respectable citizen, Don Bartholomew Blandin, endeavoured to introduce this branch of culture on the coasts of Terra Firma. Mr. Depons, in his statistical account of the Capitania-general of Venezuela, having been able to give information respecting the state of commerce and agriculture only as far as the year 1804, it will not be uninteresting to add some of a more recent date, and not less exact. The finest coffee-plantations are now found in the savannah of Ocumare, near Salamanca, and at Rincon, in the mountainous countries of Los Mariches, San Antonio Hatillo, and Los Budares. The coffee of the last three places, to the east of Caraccas, is of a superior quality, but the trees bear a smaller quantity, which is attributed to the height of the spot, and the coolness of the climate. The greater plantations of the province of Venezuela, as Aguacates, near Valencia, and Rincon, yield in good years a produce of three thousand quintals. In 1796 the total exportation of the province was only four thousand eight hundred quintals, and in 1804, ten thousand; yet it began in 1789. The prices varied from six to eighteen piastres per quintal. At the Havana it has sunk as low as three piastres; but at this period, so disastrous for the planters, in 1810 and 1812, more than two millions of quintals of coffee, amounting in value to ten millions sterling, were accumulated in the warehouses of England.

The extreme predilection entertained in this province for the culture of the coffee-tree is partly founded on the circumstance, that the berry can be preserved during a great number of years; whereas, notwithstanding every possible care, cocoa spoils in the warehouses after ten or twelve months. During the long dissensions of the European powers, at a time when the metropolis was too weak to protect the commerce of the colonies, industry was directed in preference towards productions, of which the sale was less urgent, and could wait the chances of political and commercial events. I remarked, that, in the coffee-plantations, the nurseries are formed not so much by collecting together those young plants which accidentally rise under trees that have yielded a crop, as by exposing the seeds of coffee to germination during five days, in heaps between plantain leaves. These seeds are taken out of the pulp, but yet retaining a part of it adherent to them. When this seed has germinated, it is sown, and produces plants that can bear the ardour of the sun better than those that spring up in the shade in the coffee-plantations. In this country five thousand three hundred coffee-trees are generally planted in a *vanega* of ground, amounting to five thousand four hundred and seventy-six square toises. This land, if it be capable of artificial irrigation, costs five hundred piastres in the northern part of the province. The coffee-tree bears flowers only the second year, and its flowering lasts only twenty-four hours. At this time the shrub has a charming aspect; seen from afar, it appears covered with snow. The produce of the third year becomes very abundant. In plantations well weeded and watered, and recently cultivated, | 212 | we find trees bearing sixteen, eighteen, and even twenty pounds of coffee. In general, however, a produce of more than a pound and an half or two pounds cannot be expected from each plant; and even this is superior to the mean produce of the West-India Islands. Rains at the time of flowering, the want of water for artificial irrigations, and a parasitic plant, a new species of loranthus, which clings to the branches, are extremely injurious to the coffee-trees. When, in plantations of eighty or a hundred thousand shrubs, we consider the immense quantity of organic matter contained in the pulpy berry of the coffee-tree, we may be astonished that no attempts have been made to extract a spirituous liquor from them.

FIG-TREE

We were here struck with those ligneous excrescences, which in the form of ridges, or ribs, augment in so extraordinary a manner, and as far as twenty feet above the ground, the thickness of the trunk of the fig-trees of America. I found trees twenty-two feet and a half in diameter near the roots: These ligneous ridges sometimes separate from the trunk at a height of eight feet, and are transformed into cylindrical roots two feet thick. The tree looks as if it were supported by buttresses. This scaffolding however does not penetrate very deep into the earth. The lateral roots wind at the surface of the ground, and when at twenty feet distance from the trunk, they are cut with a hatchet: we see the milky juice of the fig-tree gush out, which, when deprived of the vital influence of the organs of the tree, is altered and coagulates. What a wonderful combination of cells and vessels exists in these vegetable masses, in these gigantic trees of the torrid zone, which without interruption, perhaps during a thousand years, prepare nutritious fluids, raise them to the height of one hundred and eighty feet, convey them down again to the ground, and conceal beneath a rough and hard bark, under the inaminate layers of ligneous matter, all the movements of organic life!

Baron Humboldt's Last Volume. Personal Narrative of Travels to the Equinoctial Regions of the New Continent. Vol. 4. London, 1819

For ten years past, Baron Humboldt has been engaged in publishing, in detached parts, his learned and interesting observations and discoveries in the interior regions of South-America, and New-Mexico.

In Europe his various works have been eagerly sought after, and the learned and curious inquirers of France, England, and Germany, have complimented this celebrated traveller with the highest eulogiums. The calculating, and to us, the tedious manner in which the Baron continues to yield to the eager curiosity of the world his scientific researches, and his adventures, may have the effect of continually refreshing our memory with the recollection of his former works, and keeping alive the interest derived from his new and successive publications; but, the lapse of twenty years that has occurred since his travels in those regions, and the dreadful revolutions that have since | 66 | convulsed, and are still continuing to ravage those devoted regions, are diverting our minds to the interest that would be excited by more recent description. Of the former volumes of this narrative, two only have been printed in this country, and the remainder, we presume, will not be republished until the termination of the series can be discovered, as it can hardly be expected that any bookseller will venture again to begin a work of such indefinite extent.

Taking his departure from Caraccas, he describes the mountainous regions of San Pedro, and of Los Teques, La Victoria, the Valleys of Aragua, the Lake of Tacarigua, the Hot Springs of Mariara; then descends toward the coasts of Porto Cabello; describes the mountains that separate the valleys of Aragua from the Llanos of Caraccas, Villa de Cura, Parapara, the great Llanos, or Plains, Calabozo, San Fernando De Apure; descends the river Apure to its confluence with the great Oroonoko; and ascends the Oroonoko beyond the mountains of Encuramada, Uruana, Baraguam, the mouth of the Meta, and above the Cataracts of the Oroonoko, where the volume abruptly terminates. The closing part of this interesting voyage will undoubtedly be in due time laid before the public, who can only await with exemplary patience the time that may suit the learned traveller to publish the remainder, or the conclusion of his observations. In the mean time, we propose to present our readers with such portions of his recently published and most interesting volume, as we may find suitable to our pages.

We left Caraccas on the 7th of February, in the cool of the evening, and began our journey to the Oroonoko. The remembrance of that departure is more painful to us now than it was some years ago. Our friends have perished in the sanguinary revolutions, which have successively given liberty to those distant regions, and deprived them of it. The house which we inhabited is now a heap of ruins. Tremendous earthquakes have changed the surface of the soil. The city, which I have described, has disappeared; and on the same spot, on the ground fissured in various directions, another city is slowly rising. Already those heaps of ruins, the grave of a numerous population, are become anew the habitation of men.

In retracing changes of so general an interest, I shall be led to notice events that took place long after my return to Europe. I shall pass over in silence the popular commotions, and the modifications which the state of society has undergone. Modern nations, careful of their own remembrance, snatch from oblivion the history of human revolutions, which is that of ardent passions, and inveterate hatred. It is not the same with respect to the revolutions of the physical world. These are described with the least accuracy when they happen to coincide with the period of civil dissentions. Earthquakes, and the eruptions of volcanoes, strike the imagination by the evils which are their necessary consequence. Tradition seizes in preference whatever is vague and marvellous; and amid great public calamities, as in private misfortunes, man seems to shun that light which leads us to discover the real causes of events, and recognise the circumstances by which they are attended. I have thought proper to record in this work all I have been able to collect, with certainty, respecting the earthquake of the 26th of March, 1812, which destroyed the town of Caraccas; and by which more than twenty thousand persons perished, almost at the same instant, in the province of Venezuela. The intercourse which I have kept up with | 67 | persons of all classes has enabled me to compare the description of many eye-witnesses, and to interrogate them on objects that may throw light on physical science in general. The traveller, being the historian of nature, should authenticate the dates of great catastrophes, examine their connection and mutual relations, and mark, in the rapid course of ages, in this continual progress of successive changes, those fixed points with which other catastrophes may one day be compared. All epochas approach each other in the immensity of time comprehended in the history of nature. Years passed away seem but a few instants; and if the physical descriptions of a country do not excite a very powerful and general interest, they have, at least, the advantage of never becoming old. Similar considerations, no doubt, led Mr. De la Condamine to describe, in his *Voyage à l'Equateur*, the memorable eruptions[1] of the volcano of Cotopaxi, which took place long after his departure from Quito. Following the example of this celebrated traveller, I think I shall the less deserve blame, as the events which I am going to relate will serve to elucidate the theory of *volcanic reactions*, or the influence of a *system of volcanoes* on a vast space of circumjacent country.

At the time that Mr. Bonpland and myself inhabited the provinces of New Andalusia, Nueva Barcelona, and Caraccas, a general opinion prevailed, that the easternmost parts of these coasts were the most exposed to the destructive effects of earthquakes. The inhabitants of Cumana dreaded the valley of Caraccas, on account of its damp and variable climate, and its gloomy and foggy sky; while the inhabitants of that temperate valley considered Cumana as a town where only a burning air was breathed, and where the soil is periodically agitated by violent commotions. Forgetful of the overthrow of Riobamba, and other very elevated towns; ignorant that the peninsula of Araya, composed of micaslate, has partaken of the commotions of the calcareous coast of Cumana; well-informed persons thought they perceived motives of security in the structure of the primitive rocks of Caraccas, as well as in the elevated situation of this valley. Religious ceremonies celebrated at La Guayra, and even in the capital, in the middle of the night[2] recalled, no doubt, to their memory, that the province of Venezuela had been subject, at intervals, to earthquakes; but dangers that seldom recur, are slightly feared. Cruel experience destroyed, in 1811, the charm of theory, and of popular opinions. Caraccas, situate in the mountains, three degrees west of Cumana, and five degrees west of the volcanoes of the Caribbee islands, has felt greater shocks than were ever experienced on the coast of Paria or New Andalusia.

At my arrival in Terra Firma, I was struck with the connection of two physical events, the destruction of Cumana on the 14th of December, 1797, and the eruption of the volcanoes in the smaller West India islands.[3]

---

[1] Those of the 30th of November, 1744, and of the 3d of September, 1750. (*Introd. Hist.* p. 156 and 160.)

[2] For instance, the nocturnal procession of the 21st of October, instituted in commemoration of the great earthquake which took place on that day of the month, at one in the morning, in 1778. Other very violent shocks were those of 1641, 1703, and 1802.

[3] See vol, ii. p. 231.

This connection has been again manifested in the destruction of Caraccas on the 26th of March, 1812. The volcano of Guadaloupe seemed to have reacted, in 1797, on the coasts of Cumana. Fifteen years after, it was a volcano situate nearer the continent, that of St. Vincent's, which appeared to have extended it's influence as far as Caraccas and the | 68 | banks of the Apure. It is possible that, at these two epochas, the centre of the explosion was at an immense depth, equally distant from the regions toward which the motion was propagated at the surface of the globe.

From the beginning of 1811 till 1813, a vast extent of the earth,[4] limited by the meridian of the Azores, the valley of the Ohio, the Cordilleras of New Grenada, the coasts of Venezuela, and the volcanoes of the smaller West India islands, has been shaken almost at the same time by commotions; which may be attributed to subterraneous fires. The following series of phenomena seems to indicate communications at enormous distances. On the 30th of January, 1811, a submarine volcano appeared near the island of St. Michael, one of the Azores. At a place where the sea was sixty fathoms deep, a rock appeared above the surface of the waters. The heaving up of the softened crust of the globe appears to have preceded the eruption of flames by the crater[5], as had already been observed at the volcanoes of Jorullo in Mexico, and at the apparition of the little island of Kameni near Santorino. The new islet of the Azores was at first nothing more than a shoal; but on the 15th of January an eruption, which lasted six days, enlarged its extent, and carried it progressively to the height of fifty toises above the surface of the sea. This new land, of which Captain Tillard hastened to take possession in the name of the British government, calling it *Sabrina island*, was nine hundred toises in diameter. It has again, it seems, been swallowed up by the ocean. This is the third time that submarine volcanoes have presented this extraordinary spectacle near the island of St. Michael; and, as if the eruptions of these volcanoes were subject to a regular period, owing to a certain accumulation of elastic fluids, the island raised up has appeared at intervals of ninety-one or ninety-two years.[6] It is to be regretted that, notwithstanding the proximity of the spot, no European government, or learned society, has sent natural philosophers and geologists to the Azores,

---

[4] Between the latitudes of 5° and 36° North, and the meridians of 31° and 91° West from Paris.

[5] See vol. i. p. 240.

[6] *Malte Brun, Geogra. Univ.* vol. iii. p. 177— 180 There remains, however, some doubt respecting the eruption of 1628, which some place in 1638. The rising always happened near the island of St. Michael, though not identically on the same spot. It is remarkable that the small island of 1720 reached the same elevation as the island of Sabrina in 1811. See above, vol. i chap. i. p. 95.

to investigate a phenomenon which would throw so much light on the history of volcanoes, and on that of the globe in general.

At the time of the appearance of the new island of Sabrina, the smaller West India islands, situate eight hundred leagues to the south-west of the Azores, experienced frequent earthquakes. More than two hundred shocks were felt from the month of May, 1811, to April, 1812, in the island of St. Vincent; one of the three where there are still active volcanoes. The commotion did not remain circumscribed to that insular portion of Eastern America. From the 16th of December, 1811, the earth was almost incessantly agitated in the valleys of the Mississippi, the Arkansas, and the Ohio. The oscillations were more feeble on the east of the Alleghanies than to the west of these mountains, in Tennessee and Kentucky. They were accompanied by a great subterraneous noise, coming from the south-west. At the spots between New Madrid and Little Prairie, as at the Saline, north of Cincinnati, in latitude 37° 45′, the shocks were felt every day, nay, almost every hour, during several months. The whole of these phenomena lasted from the 16th of | 69 | December, 1811, till the year 1813. The commotion, confined at first to the south, in the valley of the lower Mississippi, appeared to advance slowly toward the north.[7]

At the same period, when this long series of earthquakes began in the *Transalleghanian States*, in the month of December, 1811, the town of Caraccas felt the first shock in calm and serene weather. This coincidence of phenomena was probably not accidental; for we must not forget that, notwithstanding the distance which separates these countries, the low grounds of Louisiana, and the coasts of Venezuela and Cumana, belong to the same basin, that of the Gulf of Mexico. This *Mediterranean sea, with several outlets*, runs from the south-east to the northwest; and an ancient prolongation of it seems to be found in those vast plains, rising gradually thirty, fifty, and eighty toises[8] above the level of the ocean, covered with secondary formations, and watered by the Ohio, the Missouri, the Arkansas, and the Mississippi. When we consider geologically *the basin of the Caribbean sea, and of the Gulf of Mexico*, we find it bounded on the south by the chain of the coast of Venezuela and the Cordilleras of Merida and Pamplona; on the east by the mountains of the West India islands, and the Alleghanies; on the west by the Andes

[7] See the interesting description of these earthquakes, given by Mr. Mitchill, in the *Trans. of the Liter. and Phil. Soc. of New York*, vol. i. p. 281—308; and by Mr. Drake, in the *Nat. and Stat. View of Cincinnati*, p. 232—238.

[8] Cincinnati, situated on the Ohio, in latitude 39° 6 min has only eighty-five toises absolute elevation.

of Mexico, and the Stony Mountains;[9] and on the north by very inconsiderable elevations which separate the Canadian lakes from the rivers that flow into the Mississippi. More than two-thirds of this basin are covered with water. It is bordered by two ranges of active volcanoes; to the east, in the Caribbee islands, between the latitudes of 13° and 16°; and to the west in the Cordilleras of Nicaragua, Guatimala, and Mexico, between 11° and 20°. When we reflect that the great earthquake at Lisbon, of the 1st of November, 1755, was felt almost at the same moment on the coasts of Sweden, at lake Ontario, and at the island of Martinico, it will not appear too daring to suppose, that all this basin of the West Indies, from Cumana and Caraccas as far as the plains of Louisiana, may be simultaneously agitated by commotions proceeding from the same centre of action.

An opinion very generally prevails on the coasts of Terra Firma, that earthquakes become more frequent when electric explosions have been very rare during some years. It is thought to have been observed, at Cumana and Caraccas, that the rains were less frequently attended with thunder from the year 1792; and the total destruction of Cumana in 1797, and the commotions felt[10] in 1800, 1801, and 1802, at Maracaibo, Porto Cabello, and Caraccas, have not failed to be attributed to "an accumulation of electricity in the interior of the earth." It would be difficult for a person who has lived a long time in New Andalusia, or in the low regions of Peru, to deny that the season the most to be dreaded, from the frequency of earthquakes, is that of the beginning of | 70 | the rains, which is, however, the time of thunder storms. The atmosphere, and the slate of the surface of the globe, seem to have an influence unknown to us on the changes produced at great depths; and I believe, that the connection which some persons pretend to recognise between the absence of thunder storms and the frequency of earthquakes, is rather a physical hypothesis, framed by the half-learned of the country, than the result of long experience. The coincidence of certain phenomena may be favoured by chance. The extraordinary commotions felt almost

[9] It is with regret I use this vague and improper denomination, which is given to the northern prolongation of the mountains of New Mexico. I should prefer the name of Chippewan Range, which Mr. Drake (Stat. View of Cincinnati, p. 91.) and other geographers of the United States, begin to substitute for the received denomination of Stony Mountains; but nations almost of the same name, very distant from each other, and speaking different languages, the Chippeways of the sources of the Mississippi, and the Chepewyans of the Slave Lake, described by Pike and Mackenzie, may occasion those mountains to the south and south-west of the great Canadian lakes, which lie east and west, to be confounded with the Stony Mountains, which run north and south.

[10] Depons, vol. i. p. 125.

continually during two years on the borders of the Mississippi and the Ohio, and which coincided in 1812 with those of the valley of Caraccas, were preceded at Louisiana by a year almost exempt from thunder storms.[11] Every mind was again struck with this phenomenon. We cannot deem it strange that in the country of Franklin a great predilection is retained for explanations founded on the theory of electricity.

The shock felt at Caraccas in the month of December, 1811, was the only one that preceded the horrible catastrophe of the 26th of March, 1812. The inhabitants of Terra Firma were ignorant of the agitations of the volcano in the island of St. Vincent on one side, and on the other, of those that were felt in the basin of the Mississippi, where, on the 7th and 8th of February, 1812, the earth was day and night in perpetual oscillation. A great drought prevailed at this period in the province of Venezuela. Not a single drop of rain had fallen at Caraccas, or in the country ninety leagues round, during the five months which preceded the destruction of the capital. The 26th of March was a remarkably hot day. The air was calm, and the sky unclouded. It was Holy Thursday, and a great part of the population was assembled in the churches. Nothing seemed to presage the calamities of the day. At seven minutes after four in the afternoon the first shock was felt; it was sufficiently powerful to make the bells of the churches toll; it lasted five or six seconds, during which time the ground was in a continual undulating movement, and seemed to heave up like a boiling liquid. The danger was thought to be past, when a tremendous subterraneous noise was heard, resembling the rolling of thunder, but louder, and of longer continuance, than that heard within the tropics in time of storms. This noise preceded a perpendicular motion of three or four seconds, followed by an undulatory movement somewhat longer. The shocks were in opposite directions, from north to south, and from east to west. Nothing could resist the movement from beneath upward, and undulations crossing each other. The town of Caraccas was entirely overthrown. Thousands of the inhabitants (between nine and ten thousand) were buried under the ruins of the houses and churches. The procession had not yet set out; but the crowd was so great in the churches, that nearly three or four thousand persons were crushed by the fall of their vaulted roofs. The explosion was stronger toward the north, in that part of the town situate nearest the mountain of Avila, and the Silla. The churches of La Trinidad and Alta Gracia, which were more than one hundred and fifty feet high, and the naves of which were

---

[11] *Trans. of New York*, vol. i. p. 285; *Drake*, p. 210.

supported by pillars of twelve or fifteen feet diameter, left a mass of ruins scarcely exceeding five or six feet in elevation. The sinking of the ruins has been so considerable that there now scarcely remain any vestiges of pillars or columns. The barracks, called *El Quartel de* | 71 | *San Carlos*, situate farther north of the church of the Trinity, on the road from the custom-house de la Pastora, almost entirely disappeared. A regiment of troops of the line, that was assembled under arms, ready to join the procession, was, with the exception of a few men, buried under the ruins of this great edifice. Nine tenths of the fine town of Caraccas were entirely destroyed. The walls of the houses that were not thrown down, as those of the street San Juan, near the Capuchin Hospital, were cracked in such a manner that it was impossible to run the risk of inhabiting them. The effects of the earthquake were somewhat less violent in the western and southern parts of the city, between the principal square and the ravine of Caraquata. There, the cathedral, supported by enormous buttresses, remains standing.[12]

Estimating at nine or ten thousand the number of the dead in the city of Caraccas, we do not include those unhappy persons who, dangerously wounded, perished several months after, for want of food and proper care. The night of Holy Thursday presented the most distressing scene of desolation and sorrow. That thick cloud of dust which, rising above the ruins, darkened the sky like a fog, had settled on the ground. No shock was felt, and never was a night more calm or more serene. The Moon, nearly full, illumined the rounded domes of the Silla, and the aspect of the sky formed a perfect contrast to that of the earth, covered with the dead, and heaped with ruins. Mothers were seen bearing in their arms their children, whom they hoped to recall to life. Desolate families wandered through the city, seeking a brother, a husband, a friend, of whose fate they were ignorant, and whom they believed to be lost in the crowd. The people pressed along the streets, which could no more be recognized but by long lines of ruins.

All the calamities experienced in the great catastrophes of Lisbon, Messina, Lima, and Riobamba, were renewed on the fatal day of the 26th of March, 1812. "The wounded, buried under the ruins, implored, by their cries, the help of the passers by, and nearly two thousand were dug out. Never was pity displayed in a more affecting manner; never had it been seen more ingeniously active than in the efforts employed to save the miserable victims whose groans reached the ear. Implements

[12] *Sur le Tremblement de Terre de Venezuela, en* 1812, *par M. Delpeche.* MS.

for digging, and clearing away the ruins, were entirely wanting; and the people were obliged to use their bare hands to disinter the living. The wounded, as well as the sick who had escaped from the hospitals, were laid on the banks of the small river Guayra. They found no shelter but the foliage of trees. Beds, linen to dress the wounds, instruments of surgery, medicines, and objects of the most urgent necessity, were buried under the ruins. Every thing, even food, was wanting during the first days. Water became alike scarce in the interior of the city. The commotion had rent the pipes of the fountains; the falling in of the earth had choaked up the springs that supplied them; and it became necessary, in order to have water, to go down to the river Guayra, which was considerably swelled; and then vessels to convey the water were wanting.

There remained a duty to be fulfilled toward the dead, enjoined at once by piety, and the dread of infection. It being impossible to inter so many thousand corpses, half buried under the ruins, commissaries were appointed to burn the bodies: and for this purpose funeral piles were erected between the heaps of ruins. This ceremony | 72 | lasted several days. Amid so many public calamities, the people devoted themselves to those religious duties, which they thought were the most fitted to appease the wrath of Heaven. Some, assembling in processions, sung funeral hymns; others, in a state of distraction, confessed themselves aloud in the streets. In this town was now repeated what had been remarked in the province of Quito, after the tremendous earthquake of 1797: a number of marriages were contracted between persons, who had neglected for many years to sanction their union by the sacerdotal benediction. Children found parents, by whom they had never till then been acknowledged; restitutions were promised by persons, who had never been accused of fraud; and families, who had long been enemies, were drawn together by the tie of common calamity. If this feeling seemed to calm the passions of some, and open the heart to pity, it had a contrary effect on others, rendering them more rigid and inhuman. In great calamities vulgar minds preserve still less goodness than strength: misfortune acts in the same manner, as the pursuits of literature and the study of nature; their happy influence is felt only by a few, giving more ardour to sentiment, more elevation to the thoughts, and more benevolence to the disposition.

Shocks as violent as those, which in the space of one minute[13] overthrew the city of Caraccas, could not be confined to a small portion

---

[13] The duration of the earthquake, that is to say, the whole of the movements of undulation and rising (*undulacion y trepidacion*,) which occasioned the horrible catastrophe of the 26th of March, 1812, was estimated by some at 50″, by others at 1′ 12″.

of the continent. Their fatal effects extended as far as the provinces of Venezuela, Varinas, and Maracaybo, along the coast; and still more to the inland mountains. La Guayra, Mayquetia, Antimano, Baruta, La Vega, San Felipe, and Merida, were almost entirely destroyed. The number of the dead exceeded four or five thousand at La Guayra, and at the town of San Felipe, near the copper mines of Aroa. It appears, that it was on a line running East-North-East, and West-South-West, from La Guayra and Caraccas to the lofty mountains of Niquitao and Merida, that the violence of the earthquake was principally directed. It was felt in the kingdom of New Grenada from the branches of the high Sierra de Santa Martha[14] as far as Santa Fe de Bogota and Honda, on the banks of the Magdalena, one hundred and eighty leagues from Caraccas. It was every where more violent in the Cordilleras of gneiss and micaslate, or immediately at their foot, than in the plains, and this difference was particularly striking in the savannahs of Varinas and Cassanara. (This is easily explained according to the system of those geologists, who admit, that all the chains of mountains, volcanic and not volcanic, have been formed by being raised up, as if through crevices.) In the valleys of Aragua, situate between Caraccas and the town of San Felipe, the commotions were very weak: and La Victoria, Maracay, and Valencia, scarcely suffered at all, notwithstanding their proximity to the capital. At Valecillo, a few leagues from Valencia, the earth, opening, threw out such an immense quantity of water, that it formed a new torrent. The same phenomenon took place near Porto-Cabello.[15] On the other hand, the lake of Maracaybo diminished sensibly. At Coro no commotion was felt, though | 73 | the town is situate upon the coast, between other towns which suffered from the earthquake."[16] Fishermen, who had passed the day of the 26th of March in the island of Orchila, thirty leagues North-East of La Guayra, felt no shock. These differences, in the direction and propagation of the shock, are probably owing to the peculiar arrangement of the stony strata.

Having thus traced the effects of the earthquake to the West of Caraccas, as far as the snowy mountains of Santa Marta, and the table land of Santa Fe de Bogota, we will proceed to consider their action on the country East of the capital. The commotions were very violent beyond Caurimare, in the valley of Capaya, where they extended as far as the

[14] As far as Villa de Los Remedios, and even to Carthagena.

[15] It is asserted, that in the mountains of Aros, the ground, immediately after the great shocks, was found covered with a very fine and white earth, which appeared to have been projected through crevices.

[16] *Apuntamientos sobre las principales Circunstancias del Terremoto de Caracas, por Don Manuel Palacio Fajardo*. MS.

meridian of Cape Codera: but it is extremely remarkable, that they were very feeble on the coasts of Nueva-Barcelona, Cumana, and Paria; though these coasts are the continuation of the shore of La Guayra, and formerly known to have been often agitated by subterraneous commotions. Admitting that the destruction of the four towns of Caraccas, La Guayra, San Felipe, and Merida, may be attributed to a volcanic focus placed under or near the island of St. Vincent, it may be conceived, that the motion might have been propagated from North-East to North-West[17] in a line passing through the islands of Los Hermanos, near Blanquilla, without touching the coasts of Araya, Cumana, and Nueva-Barcelona. This propagation of the shock might even have taken place, without the intermediate points at the surface of the globe, the Hermanos Islands for instance, having felt any commotion. This phenomenon is frequently remarked at Peru and Mexico, in earthquakes which have followed during ages a determinate direction. The inhabitants of the Andes say with simplicity, speaking of an intermediary ground, which is not affected by the general motion, "that it forms a bridge" (*que hace puente*:) as if they meant to indicate by this expression, that the undulations are propagated at an immense depth under an inert rock.

Fifteen or eighteen hours after the great catastrophe, the ground remained tranquil. The night, as we have already observed, was fine and calm; and the commotions did not recommence till after the 27th. They were then attended with a very loud and long continued subterranean noise (*bramido.*) The inhabitants of Caraccas wandered into the country; but the villages and farms having suffered as much as the town, they could find no shelter till they were beyond the mountains of Los Teques, in the valleys of Aragua, and in the Llanos or Savannahs. No less than fifteen oscillations were often felt in one day. On the 5th of April there was almost as violent an earthquake as that which overthrew the capital. During several hours the ground was in a state of perpetual undulation. Large masses of earth fell in the mountains; and enormous rocks were detached from the Silla of Caraccas. It was even asserted, and this opinion prevails still in the country, that the two domes of the Silla sunk fifty or sixty toises; but this assertion is founded on no measurement whatever. I am informed, that in the province of Quito also, the people, at every period of great commotions, imagine that the volcano of Tunguragua is diminished in height.—It has been affirmed, in many descriptions published of the destruction of Caraccas, "that the mountain of the Silla is

[17] Nearly in a line directed South, 64° west.

an extinguished volcano; that a great quantity of volcanic substances are found on the road from La | 74 | Guayra to Caraccas;[18] that the rocks do not present any regular stratification; and that every thing bears the stamp of the action of fire." It is even added, "that, twelve years before the great catastrophe, Mr. Bonpland and myself, from our physical and mineralogical researches, had considered the Silla as a very dangerous neighbour to the city, because that mountain contained a great quantity of sulphur, and that the commotions must come from the North-East." It is seldom that natural philosophers have to justify themselves for an accomplished prediction; but I think it my duty to combat ideas that are too easily adopted on the *local causes* of earthquakes.

[18] See the account given by Mr. Drouet of Guadaloupe, translated in the *Trans. of New York*, vol. i. p. 303 The author, in giving to the Silla nine hundred toises of *absolute* height, has confounded the height of the mountain, in my measurement, above the level of the sea, with its height above the valley of Caracas, which makes a difference of four hundred and sixty toises.

Baron Humboldt's Last Volume. Personal Narrative of Travels to the Equinoctial Regions of the New Continent. Vol. 4. London, 1819

Consejo.—We left the plantation of Manterola on the 11th of February, at sunrise. The road follows the smiling banks of the Tuy; the morning was cool and humid, and the air seemed embalmed by the delicious odour of the Prancatium undulatum, and other large liliaceous plants. In our way to La Victoria, we passed the pretty village of Mamon, or of Consejo, celebrated in the country for a miraculous image of the Virgin. A little before we reached Mamon, we stopped at a farm belonging to the family of Monteras. A negress more than a hundred years old was seated before a small hut, constructed with earth and reeds. Her age was known, because she was a creole slave. She seemed still to enjoy very good health. "I hold her to the Sun" (*la tengo al sol*) said her grand-son; "the heat keeps her alive." The means appeared to us rather violent, for the Sun darted its rays almost perpendicularly. The nations with a brown skin, Blacks well seasoned, and Indians, attain a happy old age in the torrid zone. I have mentioned elsewhere the history of a native of Peru, who died at the age of one hundred and forty three years, after having been ninety years married.

Don Francisco Montera, and his brother, a young and enlightened ecclesiastic, accompanied us, in order to conduct us to their house at La Victoria. Almost all the families, with whom we had lived in friendship at Caraccas, the Ustarizes, the Tovars, and the Toroes, were assembled in the fine valleys of Aragua. Proprietors of the richest plantations, they contended with each other in whatever could render our stay agreeable. Before we plunged into the forests of the Oroonoko, we enjoyed once more all the advantages of an advanced civilization.

The road from Mamon to La Victoria runs toward the south and the south-west. We soon lost sight of the river Tuy, which, turning to the east, forms an elbow at the foot of the high mountains of Guaraima. As we drew nearer to Victoria, the ground became smoother; it looked like the bottom of a lake, the waters of which had been drained off. We might have fancied ourselves in the valley of Hasli, in the canton of Berne. The neighbouring hills, only one hundred and forty toises in height, are composed of calcareous tufa; but their abrupt declivities project like promontories on the plain. The form indicates the ancient shore of the lake. The eastern extremity of this valley is parched and uncultivated. No advantage has been derived from the ravines, that water the neighbouring mountains; but a fine cultivation is commencing in the proximity of the town. I say of the town, though in my time Victoria was considered only as a village (*pueblo*).

La Victoria.—The environs of La Victoria present a very remarkable aspect, with regard to agriculture. The height of the cultivated ground is from two hundred and seventy to three hundred toises above the level of the ocean, and yet we there find fields of corn mingled with plantations of sugar canes, coffee, and plantains. Excepting the interior of the island of Cuba, we scarcely find any where else in the equinoctial regions European corn cultivated in large quantities in so low a region. The fine fields of | 267 | wheat in Mexico are between six hundred and twelve hundred toises of absolute elevation; and it is rare to see them descend to four hundred toises. We shall soon perceive, that the produce of grain augments sensibly, from high latitudes toward the equator, with the mean temperature of the climate, in comparing spots of different elevations. The success of agriculture depends on the dryness of the air; on the rains distributed among different seasons, or accumulated in one rainy season; on winds blowing constantly from the east, or bringing the cold air of the north into very low latitudes, as in the gulf of Mexico; on mists, which for whole months diminish the intensity of the solar rays; in short, on a thousand local circumstances, which have less influence on the mean temperature of the whole year, than on the distribution of the same quantity of heat among the different parts of the year. It is a striking spectacle, to see the grain of Europe cultivated from the equator as far as Lapland, in the latitude of 69°, in regions where the mean heat is from 22° to —2°, in every place where the temperature of summer is above 9° or 10°. We know the minimum of heat requisite to ripen wheat, barley, and oats; we are less certain in respect to the maximum, which these species of grain, accommodating as they are, can support. We are even ignorant of all the circumstances, which favour the culture of corn between the tropics at very small heights. La Victoria and the neighbouring village of San Matheo yield an annual produce of four thousand quintals of wheat. It is sown in the month of December, and the harvest is reaped on the seventieth or seventy-fifth day. The grain is large, white, and abounding in gluten: its pellicle is thinner, and not so hard as that of the wheat of the very cold table-lands of Mexico. An acre near Victoria generally yields from three thousand to three thousand two hundred pounds weight of wheat. The average produce is consequently here, as at Buenos Ayres, three or four times as much as that of northern countries. Nearly sixteen times the quantity of the seed is reaped; while, according to Lavoisier, the surface of France yields on a mean only five or six for one; or from one thousand to twelve hundred pounds per acre. Notwithstanding this fecundity of the soil, and this happy influence of the climate, the culture of the sugar cane is more productive in the valleys of Aragua, than that of corn.

La Victoria is traversed by the little river Calanchas, running not into the Tuy, but into the Rio Aragua. It thence results, that this fine country, producing at once sugar and corn, belongs already to the basin of the lake of Valencia, to a system of interior rivers which do not communicate with the sea. The quarter of the town West of the Rio Calanchas is called *la otra banda*; it is the most commercial part; merchandize is every where exhibited, and ranges of shops form the streets. Two commercial roads pass through La Victoria; that of Valencia, or of Porto Cabello, and the road of Villa de Cura, or of the plains, called *camino de los Llanos*. We here find more white in proportion than at Caraccas. We visited at sunset the little hill of Calvary, where the view is extremely fine and extensive. We discover on the west the smiling valleys of Aragua, a vast space covered with gardens, cultivated fields, clumps of wild trees, farms, and hamlets. Turning toward the south and south-east, we see, extending as far as the eye can reach, the lofty mountains of La Palma, Guayraima, Tiara, and Guiripa, which conceal the immense plains or steppes of Calabozo. This interior chain stretches to the west | 268 | along the lake of Valencia, toward the Villa de Cura, the Cuesta de Yusma, and the denticulated mountains of Guigue. It is very steep, and constantly covered with that light vapour, which in hot climates gives a vivid blue tint to distant objects, and far from concealing their outlines, renders them more strongly marked. It is believed, that, among the mountains of the interior chain, that of Guayraima reaches an elevation of twelve hundred toises. I found, in the night of the eleventh of February, the latitude of La Victoria 10° 13′ 35″, the magnetic dip 40.8°, the intensity of the forces equal to 236 oscillations in 10′ of time, and the variations of the needle 4.4° to the north-east.

We proceeded slowly on our way by the villages of San Matheo, Turmero, and Maracay, to the Hacienda de Cura, a fine plantation belonging to Count Tovar, where we did not arrive till the fourteenth of February, in the evening. The valley which gradually widens, is bordered with hills of calcareous tufa, called here *tierra blanca*. The scientific men of the country have made several attempts to calcine this earth; mistaking it for the porcelain earth proceeding from the decomposition of strata of feldspar. We stayed some hours with a family equally enlightened and respectable, the Ustarizes, at Concession. The house, which contains a collection of choice books, is placed on an eminence, and surrounded with plantations of coffee and sugar canes. A grove of balsam trees (*balsamo*) gives coolness and shade to this spot. We observed with a lively interest the great number of scattered houses in the valley, inhabited by freedmen. In the Spanish colonies, the laws, the institutions, and the manners, are more favourable to the liberty of the blacks, than in other European settlements.

San Matheo, Turmero, and Maracay, are charming villages, where every thing announces persons in the most easy circumstances. We seemed to be transported to the most industrious parts of Catalonia. Near San Matheo we find the last fields of wheat, and the last mills with horizontal hydraulic wheels. A harvest of twenty for one was expected: and, as if the produce were but moderate, I was asked, whether corn produced more in Prussia and in Poland. It is an error that generally prevails under the tropics, to consider grain as plants which degenerate in advancing toward the equator; and to believe, that the harvests are more abundant in the northern climates. Since calculations have been made on the progress of agriculture in the different zones, and the temperatures under the influence of which corn will flourish, it has been found, that, beyond the latitude of 45°, the produce of wheat is no where so considerable, as on the northern coast of Africa, and on the table-lands of New Grenada, Peru, and Mexico. Without comparing the mean temperature of the whole year, but only the mean temperature of the season which embraces the *cycle of vegetation* of corn, we find for three months of summer, in the north of Europe, from 15° to 19°; in Barbary, and in Egypt, from 27° to 29°; within the tropics, between fourteen and six hundred toises of height, from 14° to 25.5° of the centigrade thermometer.

The fine harvests of Egypt and of the kingdom of Algiers, those of the valleys of Aragua and the interior of the island of Cuba, sufficiently prove, that the augmentation of heat is not prejudicial to the harvest of wheat and other alimentary grain, unless it is attended with an excess of drought or moisture. To this circumstance, no doubt, we must attribute the apparent anomalies, that are sometimes observed between | 269 | the tropics, in the *inferior limit of corn*. We are astonished to see the east of the Havannah, in the famous district of Quatro Villas, this limit descend almost to the level of the ocean; while to the west of the Havannah, on the slope of the mountains of Mexico and Xalappa, at six hundred and seventy-seven toises of height, the luxury of vegetation is such, that wheat does not form ears. At the beginning of the conquest, the corn of Europe was cultivated with success in several regions, which are now thought too hot, or too damp, for this branch of agriculture. The Spaniards recently removed to America were less accustomed to live on maize; they still adhered to their European customs; they did not calculate whether corn would be less profitable than coffee or cotton; they tried seeds of every kind; they interrogated nature with more boldness, because their reasonings were less founded on false theories. The province of Carthagena, crossed by the chain of the mountains Maria and Guamoco, produced wheat till the sixteenth century. In the province of Caraccas, this culture is very ancient among the mountainous lands of Tocuyo, Quibor, and Barquisimeto, which connect the littoral chain with the Sierra Nevada of Merida. It is still happily practised there, and the environs of the town of Tocuyo alone export annually more than eight thousand quintals of excellent flour. But, though the province of Caraccas, in its vast extent, presents several spots very favourable to the cultivation of European corn, I believe, that in general this branch of agriculture will never become of great importance there. The most temperate valleys are not sufficiently wide; they are not real table-lands; and their mean elevatation above the level of the sea is not considerable enough, for the inhabitants to avoid perceiving, that it is more their interest to establish plantations of coffee, than to cultivate corn. Flour now comes to Caraccas either from Spain, or from the United States. When, in circumstances more favourable to industry and public tranquillity, the road from Santa Fe de Bogota to the port of Pachaquiaro is laid open, the inhabitants of Venezuela will receive the flour of New Grenada by way of the Rio Meta and the Oroonoko.

Turmero.—The village of Turmero is four leagues distant from San Matheo. The road leads through plantations of sugar, indigo, cotton, and coffee. The regularity, which we observed in the construction of the villages, reminded us, that they all owe their origin to monks and missions. The streets are straight and parallel; they cross each other at right angles; and the church is erected in the great square, situate in the centre. The church of Turmero is a sumptuous edifice, but overloaded with architectural ornaments. Since the missionaries have been replaced by vicars, the whites have mixed their habitations with the Indians. The latter gradually disappear as a separate race; that is to say, they are represented in the general statement of the population by the Mestizoes and the Zamboes, whose numbers daily increase. I still found, however, four thousand tributary Indians in the valleys of Aragua. Those of Turmero and Guacara are the most numerous. They are little, but less squat than the Chaymas; their eyes announce more vivacity and intelligence, owing perhaps less to a diversity in the race, than to a superior state of civilization. They work like freemen by the day; they are active and laborious during the short time they allot to labour; but what they earn in two months is spent in one week, in buying strong liquors, at the small inns, of which, unhappily, the numbers daily increase. | 270 |

We saw at Turmero the remains of the assembled militia of the country, and their appearance alone sufficiently indicated, that these valleys had enjoyed for ages undisturbed peace. The captain-general, in order to give a new impulse to the military service, had ordered a grand review; and the battalion of Turmero, in a mock fight, had fired on that of La Victoria. Our host, a lieutenant of the militia, was never weary of describing to us the danger of these manœuvres. "He had been surrounded with muskets, which might have burst at every instant; he had been kept four hours in the Sun, and his slaves were not permitted to hold an umbrella over his head." With what rapidity do nations, who appear the most pacific, acquire military habits! I then smiled at a timidity avowed with such simple frankness; and twelve years afterward those valleys of Aragua, those peaceful plains of La Victoria and Turmero, the defile of Cabrera, and the fertile banks of the lake of Valencia, have become the theatre of the most obstinate and bloody conflicts between the natives and the soldiers of the mother country.

Upon quitting the village of Turmero, we discover, at a league distant, an object, which appear at the horizon like a round hillock, or *tumulus*, covered with vegetation. It is neither a hill, nor a group of trees close to each other, but one single tree the famous *zamang del Guayre*, known throughout the province for the enormous extent of its branches, which form a hemispheric head five hundred and seventy-six feet in circumference. The zamang is a fine species of mimosa, the tortuous branches of which are divided by bifurcation. Its delicate and tender foliage displayed itself agreeably on the azure of the sky. We stopped a long time under this vegetable roof. The trunk of the zamang del Guayre, which is found on the road from Turmero to Maracay, is only sixty feet high, and nine thick; but its real beauty consists in the form of its head. The branches extended like an immense umbrella, and bend toward the ground, from which they remain at a uniform distance of twelve or fifteen feet. The circumference of this head is so regular, that, having traced different diameters, I found them one hundred and ninety-two and one hundred and eighty-six feet. One side of the tree was entirely stripped of its foliage, owing to the drought; and on the other side there remained at once leaves and flowers. Tillandsias, lorantheæ, cactus pitahayas, and other parasite plants, cover its branches, and crack the bark. The inhabitants of these villages, but particularly the indians, hold in veneration the zamang del Guayre, which the first conquerors found almost in the same state in which it now remains. Since it has been observed with attention, no change has appeared in its thickness or height. This zamang must be at least as old as the Orotava dragon-tree. There is something solemn and majestic in the aspect of aged trees, and the violation of these monuments of nature is severely punished in countries, that are destitute of monuments of art. We heard with satisfaction, that the present proprietor of the zamang had brought an action against a farmer, who had had the temerity to cut off a branch. The cause was tried, and the tribunal condemned the farmer. We find near Turmero and the hacienda de Cura other zamangs, the trunks of which are larger than that of Guayre, but their hemispherical head is not of equal extent.

The culture and population of the plains augment as we advance toward Cura and Guacara, on the northern side of the lake. The valleys of Aragua contain more than | 271 | 52,000 inhabitants, on a space of ground thirteen leagues long, and two wide. This is a relative population of two thousand souls on a square league, which equals almost that of the most populous parts of France. The village, or rather the small town of Maracay, was heretofore the centre of the indigo plantations, when this branch of colonial industry was in its greatest prosperity. In 1795, seventy tradesmen had established shops in a population of six thousand inhabitants. The houses are all of masonry, and every court contains cocoa-trees, which rise above the habitations. The aspect of general wealth is still more striking at Maracay, than at Turmero. The anil, or indigo, of these provinces, has always been considered in commerce as equal, and sometimes superior to that of Guatimala. This branch of culture has, since 1772, followed that of cacao, and preceded the cultivation of cotton and coffee. The predilection of the colonists has been alternately fixed on each of these four productions; but the cacao and coffee are now the only important branches of commerce with Europe. In the most prosperous times the fabrication of indigo has almost equalled that of Mexico; it rose in the province of Venezuela to 40,000 arrobas, or a million of pounds, the value of which exceeded 1,250,000 piastres.

Famine

The effects of *famine* (says Humboldt) are common to almost all equinoctial countries. In South America, in the province of New Andalusia, I have seen the inhabitants of villages, compelled by famine, disperse themselves, from time to time, thro' uncultivated regions, to seek nourishment from wild plants. The missionaries employ in vain their authority to prevent this dispersion. In the province ot *Los Pastos* the Indians, when the potatoes (*pommes de terre*) fail, flee to the highest elevation of the Cordillere to sustain life by using the *achupiallas*, a plant related to the *genus pircarina*. The Otomaques at *Uruara*, on the borders of the Oronoque, eat for several months, a clayey earth, to absorb the gastric-juice and to diminish, in some measure, the torments of hunger. In some Islands of the South Sea, on a fertile soil, in the midst of the beauties of nature (*au sein d'une grande et belle nature*) famine drives the inhabitants to the dreadful necessity of being canibals. Under the *torrid zone*, man, carelessly, improvident and phlegmatic, experiences periodically that want of nourishment which the industry of civilized people banishes from the more sterile regions of the *north*.

Scenery on the Rio Apure

During the whole of my voyage from San Fernando to San Carlos del Rio Negro, and thence to the town of Angostura, I confined myself to writing day by day, either in the boat, or where we disembarked at night, what appeared to me worthy of observation. Violent rains, and the prodigious quantity of moschettoes with which the air is filled on the banks of the Oroonoko and the Cassiquiare, necessarily occasioned some breaks in this labour; which I supplied by notes taken a few days after. The following pages are extracts from | 554 | my journal. Whatever is written while the objects we describe are before our eyes bears a character of truth, I had almost said of individuality, which gives attraction to things the least important.

In order to avoid useless repetitions, I have sometimes added to this journal the notions I afterward acquired respecting the objects I had described. The more nature appears great and awful in forests traversed by immense rivers, the more we should preserve in our pictures of the scenery that character of simplicity, which constitutes the principal and often the sole merit of a first sketch.

March the 31st. A contrary wind obliged us to remain on shore till noon. We saw a part of some cane-fields laid waste by the effect of a conflagration, which had spread from a neighbouring forest. The wandering Indians every where set fire to the forest where they have encamped at night; and during the season of drought, vast provinces would be the prey of these conflagrations, if the extreme hardness of the wood did not prevent the trees from being entirely consumed. We found trunks of desmanthus, and mahogany (*cahoba*), that were scarcely charred two inches deep.

Having passed the Diamante, we entered a land inhabited only by tigers, crocodiles, and *chiguires*, a large species of the genus cavia of Linneus. We saw flocks of birds, crowded so close together, as to appear against the sky like a dark cloud, that every instant changed its form. The river widens by degrees. One of its banks is generally barren and sandy from the effect of inundations: the other is higher, and covered with lofty trees. Sometimes the river is bordered by forests on each side, and forms a straight canal a hundred and fifty toises broad. The manner in which the trees are disposed is very remarkable. We first find bushes of *sauso*, forming a kind of hedge four feet high; and appearing as if they had been clipped by the hand of man. A copse of cedars, brazillettoes, and lignum vitæ, rises behind this hedge. Palm-trees are rare; we saw only a few scattered trunks of the thorny piritu and corozo. The large quadrupeds of those regions, the tigers, tapirs, and pecaris, have made openings in the hedge of *sausos* which we have just described. Through these the wild animals pass, when they come to drink at the river. As they fear but little the approach of a boat, we had the pleasure of viewing them pace slowly along the shore, till they disappeared in the forest, which they entered by one of the narrow passes left here and there between the bushes. I confess that these scenes, which were often repeated, had ever for me a peculiar attraction. The pleasure they excite, is not owing solely to the interest, which the naturalist takes in the objects of his study; it is connected with a feeling common to all men, who have been brought up in the habits of civilization. You find yourself in a new world, in the midst of untamed and savage nature. Now it is the jaguar, | 555 | the beautiful panther of America, that appears upon the shore; and now the hocco with it's black plumage and its tufted head, that moves slowly along the sausoes. Animals of the most different classes succeed each other. "*Esse como en el Paradiso*," said our pilot, an old Indian of the missions. Every thing indeed here recalls to mind that state of the primitive world, the innocence and felicity of which ancient and venerable traditions have transmitted to all nations: but, in carefully observing the manners of animals between themselves, we see that they mutually avoid and fear each other. The golden age has ceased; and in this Paradise of the American forests, as well as every where else, sad and long experience has taught all beings, that benignity is seldom found in alliance with strength.

When the shore is of considerable breadth, the hedge of *sauso* remains at a distance from the river. In this intermediate ground we see crocodiles, sometimes to the number of eight or ten, stretched on the sand. Motionless, the jaws opened at right angles, they repose by each other without displaying any of those marks of affection, observed in other animals that live in society. The troop separates as soon as they quit the shore. It is, however, probably composed of one male only, and many females; for, as Mr. Descourtils, who has so much studied the crocodiles of Saint Domingo, observed before me, the males are rare, because they kill one another in fighting during the season of their loves. These monstrous reptiles are so numerous, that throughout the whole course of the river we had almost at every instant five or six in view.

The crocodiles of the Apure find abundant nourishment in the chiguires (the thick-nosed tapir of naturalists), which live fifty or sixty together in troops on the banks of the river. These unfortunate animals, as large as our pigs, have no weapons of defence; they swim somewhat better than they run: yet they become the prey of the crocodiles in the water, as of the tigers on land. It is difficult to conceive, how, persecuted by two powerful enemies, they can become so numerous; but they breed with the same rapidity as the cobayas, or little guinea-pigs, which come to us from Brazil.

Near the Joval nature assumes an awful and savage aspect. We there saw the largest tiger we had ever met with. The natives themselves were astonished at it's prodigious length, which surpassed that of all the tigers of India I had seen in the collections of Europe. The animal lay stretched beneath the shade of a large zamang. It had just killed a chiguire, but had not yet touched it's prey, on which it kept one of it's paws. The zamuroes, a species of vulture which we have compared above to the perenopterus of Lower-Egypt, were assembled in flocks to devour the remains of the jaguar's repast. They afforded the most curious spectacle, by a singular mixture of boldness and timidity. | 556 | They advanced within the distance of two feet from the jaguar, but at the least movement the beast made they drew back. In order to observe more nearly the manners of these animals, we went into the little boat, that accompanied our canoe. Tigers very rarely attack boats by swimming to them; and never but when their ferocity is heightened by a long privation of food. The noise of our oars led the animal to rise slowly, and hide itself behind the *sauso* bushes, that bordered the shore. The vultures tried to profit by this moment of absence to devour the chiguire: but the tiger, notwithstanding the proximity of our boat, leaped into the midst of them; and in a fit of rage, expressed by his gait and the movement of his tail, carried off his prey to the forest. The Indians regretted, that they were not provided with their lances, in order to go on shore, and attack the tiger. They are accustomed to this weapon, and were right in not trusting to our musquets, which, in an air so excessively humid, often miss fire.

Continuing to descend the river, we met with the great herd of chiguires, which the tiger had put to flight, and from which he had selected his prey. These animals saw us land with great tranquillity; some of them were seated, and gazed upon us, moving the upper lip like rabbits. They seem not to be afraid of men, but the sight of our great dog put them to flight. Their hind legs being longer than their fore legs, their pace is a slight gallop, but with so little swiftness, that we succeeded in catching two of them. The chiguire, which swims with the greatest agility, utters a short moan in running, as if it's respiration were impeded. It is the largest of the family of gnawing animals. It defends itself only at the last extremity, when it is surrounded and wounded. Having great strength in it's grinding teeth, particularly the hinder ones, which are pretty long, it can tear the paw of a tiger, or the leg of a horse, with it's bite. It's flesh has a smell of musk somewhat disagreeable; yet hams are made of it in this country, which almost justifies the name of *water hog*, given to the chiguire by some of the older naturalists.

[Crocodiles of South America]

It appears from Humboldt's Travels, that the crocodiles of South America are of the same species as those of the Nile, and that at San Fernando scarcely a year passes without several persons, particularly women, being drowned by them. The following anecdote is extracted from this interesting work:—

"They related to us the history of a young girl of Uritucu, who by singular intrepidity and presence of mind saved herself from the jaws of a crocodile. When she felt herself seized, she sought the eyes of the animal, and plunged her fingers into them with such violence, that the pain forced the crocodile to let her loose, after having bitten off the lower part of her left arm. The girl, notwithstanding the enormous quantity of blood she lost, happily reached the shore, swimming with the hand she had still left. In those desert countries, where man is ever wrestling with nature, discourse daily turns on the means, that may be employed to escape from a tiger, a boa or *traga venado*, or a crocodile; every one prepares himself in some sort for the dangers that await him. "I knew," said the young girl of Uritucu coolly, "that the *cayman* lets go his hold if you put your fingers into his eyes." Long after my return to Europe I learned, that in the interior of Africa the negroes know and practise the same means. Who does not recollect with a lively interest *Isaaco*, the guide of the unfortunate Mungo Park, seized twice, near Boulinkombuo, by a crocodile, and twice escaping from the jaws of the monster, having succeeded in placing his fingers under water in both his eyes? The African Isaaco and the young American, owed their safety to the same presence of mind, and the same combination of ideas."

On the Temperature of the different Mines in America[1]

MINES OF NEW SPAIN

Guanaxuato is situated in latitude 21° 0′ 15″, and the height of the plain is about 1100 toises above the level of the sea. The mean temperature of the air is probably 60°.8 of Fahrenheit. The mine of Valenciana is so warm, that in the deepest parts the miners are constantly exposed to a temperature of 91°.4.

	Temperature.	Depth in Feet.
In September the air without the mine was	66°.7	
Between *Despacho del tiro Nuevo* and *Boveda de San Pablo*,	74°.7 81°.7	273 546
In the *Levels (Plancs) of San Bernardo*,	92°.8	1638

The spring which comes from the same veins at the depth of 1638 feet, has a temperature of 98°.2, and is 5°.4 warmer than the air of the levels (*planes*) in which the miners work.[2]

The mine of Rayas, near that of Valenciana, is erroneously regarded by the miners as much warmer than the levels of San Bernardo. I have found the thermometer,

	Temperature.	Depth in Feet.
Near the *Boca de la Mina*, –	69 .4	0
In the levels, – – –	92°.7	630

In the mines of Villalpando, three leagues to the north of Guanaxuato, on a plateau of 1330 toises, I have found the thermometer,

[1] These observations have been just published in the *Annales de Chim.* tom. xiii. p. 207, from the unpublished journals of M. Humboldt.

[2] In the *Nivellement Barometrique des Andes*, No. 336, the temperature of the spring is said to be 29°.3 Cent., whereas it is 36°.8 Cent. or 98°.2 of Fahrenheit.

	Temperature.	Depth in Feet.
In the open air, – – –	72°.3	0
In the levels, – – –	84 .9	436

It was in the mines of Guanaxuato that there was heard, in 1784, a subterranean noise, unattended with any shock. The volcano of Jorullo, which broke out of the earth on the 14th | 287 | September 1759, is fifty leagues distant from the metalliferous transition rocks of Guanaxuato. There are warm springs around Guanaxuato which issue from a basaltic conglomerate. Those of Comangillas which I have examined, have 205°.1 of temperature.

The mine of Cabrera, near Moran, (latitude 20° 10′ 4″) has an altitude of 1331 toises, and a mean annual temperature probably of 60°.4. I found

	Temperature.	Depth in Feet.
The external air, – – – –	50° to 53°.2	0
In the Gallery *del Conde de Regla*, without any traces of metal in the transition porphyry,	70°.2	164
Subterranean water at this place, – –	62°.8	164

In the village of Tehuilotepec near Tasco, where the latitude is 18° 35′ 0″, its altitude 919 toises, and the mean annual temperature, probably 68°, I have found the thermometer,

	Temperature.	Depth in Feet.
Out of the mine in the day-time, – –	80°	0
Ditto, during the night, – – –	61°.7	0
In the Gallery of *San Ignacio*, where there was no miner, und no current of air,	75°.7	354
In the water of the mines at the same place, –	68°	354

At Moran, the waters of the mines was 7°.2 colder, and at 7°.7 colder than the air of the mines.[3]

[3] In the mines of St Ana, in the kingdom of New Granada, I have found the air of the interior every where 70°.5, when the air without was in the daytime 72°.5, and in the night 65°.7, but the depth of

MINES OF PERU

The only observations made at great depths, but in parts of the globe raised more than 1800 toises above the level of the sea, are those of Hualgayoc, near Micuipampa, on the back of the Andes of Chota, in south latitude 6° 43′ 38″. The height of the plateau is 1816 toises, and the mean annual temperature of the air at this height is probably 46°. The metalliferous mountain of Hualgayoc, which is insulated in the plateau, appears to be more than 2100 toises above the level of the ocean. I found the thermometer to stand thus: | 288 |

	Temperature.	Depth in Feet.
In the open air, – –	9° to 10°.8	0
In the *Mina de Guadalupe*, –	57 .7	unknown.
Water of the mine, – –	52 .2	

In the *Mina Purgatorio*, which is extremely dry, the air was 67°.3. Such a temperature in the interior of the earth, and almost at the height of the Peak of Teneriffe, is undoubtedly very remarkable. The thermometer in these countries stands at from 41° to 48° in the day-time, and from 32°.7 to 35°.6 in the night. The point where I measured the temperature at Purgatorio, is nearly thirty toises lower than that of Guadalupe. It is almost impossible to determine the depth relative to the surface of the earth, because the insulated mountain in which the mines are wrought, has a very irregular declivity. It is sufficient to recollect that the levels of the two mines are a little above the level of the Plateau of Micuipampa, and from 250 to 300 toises below the summit of the Mountain Hualgayoc.

Two points in the interior of this mountain of 1840 toises of absolute height, have therefore 57°.7 and 67°.3 of temperature. When the surrounding air has a mean temperature of 46°, the water of the mines is in the same places 52°.2.

---

the mine was scarcely thirty-five toises. It is situated in 5° 10′ of north latitude, has 500 toises of height, and is in a region whose mean annual temperature is probably from 69°.8 to 71°.6.

On the Nocturnal Increase in the Intensity of Sound

From the Annales de Chimie et de Physique.

The increase in the intensity of sound during the night, is a problem which is not resolved in any of the works on natural philosophy. I shall endeavour to give a solution derived from the latest researches on the theory of sonorous undulations; but before treating of the cause of this phenomenon, I shall mention the circumstances under which it takes place.

It has been remarked from the earliest antiquity that the intensity of sound increases during the night. Aristotle has spoken of it in his Problems, and Plutarch in his Dialogues. It is only the increase of sound when the air is calm that we are now treating of, not that which accompanies a change of wind in the night, and which is modified by the relation that exists between the direction of the wind, and that of the sonorous ray. Under the same latitude (between the tropics for example), the nightly increase of sound has appeared to me greater on the plains than on the heights of the Andes, 3000 metres above the level of the sea; and also, when on a low level, more considerable in the middle of continents than in the open sea. This estimation is founded on the comparative noise of two volcanos, that of Guacamayo, and of Cotopaxi, which I have frequently heard both by day and by night, one in a level plain between the city of Quito and the savannahs of Chillo, and the other on the Pacific Ocean, ten leagues west of the coast of Peru.

The roarings of the volcanos of the Cordilleras usually succeed each other with great uniformity every five minutes: they are not attended with any visible explosion | 376 | above the edges of the craters; and resemble, sometimes distant thunder, at other times, repeated discharges of heavy cannon. In places in which the ground is covered in winter with snow in the vicinity of cascades, it would be interesting to examine whether the nocturnal increase of sound is not less in winter than in summer, when the soil is strongly heated by the rays of the sun. It is to be observed, that in what I have mentioned of the difference between the daily and nightly sound from sonorous objects in the high and low levels of South America, I only consider the comparative intensities under the same barometric pressure. I do not compare the absolute intensity at different altitudes, but the difference between the diurnal and nocturnal intensity on high levels and on plains. The variation of absolute intensity observed at different heights of the atmosphere, is a problem which the mathematical theory of sound has long resolved. M. Poisson has even arrived at this remarkable result, that the intensity of sound from below upwards or from above downwards, whether the sonorous rays are propagated vertically or obliquely, depends only on the density of the stratum of air whence the sound has proceeded. One must not therefore confound totally distinct problems.

When the noise of the great cataracts of the Orinoco is heard, at more than a league distance, in the plain that surrounds the mission of Alures, the observer thinks he is close to a coast surrounded with reefs and breakers. The noise is three times as great by night as by day, and gives an inexpressible charm to those solitary regions. What can be the cause of this nocturnal increase of sound in a desert, in which nothing seems to break in upon the silence of nature? The diminution of temperature, so far from increasing, would retard the velocity of sound. The intensity of sound diminishes in a wind | 377 | blowing in an opposite direction: it also diminishes by the expansion of the air; it is weaker in the higher than in the lower regions of the atmosphere, where the particles of the agitated air have more density and more elasticity in the same ray. The intensity is the same in a dry air as in one full of vapour, but it is weaker in carbonic acid than in the mixtures of azote and oxygen. On considering these facts (the only ones that we know with certainty), it is difficult to explain a phenomenon which is observed near every cascade, and has been the subject of common remark both in Europe and in America. Near the Orinoco the temperature of the air is lower by about five degrees (Fahrenheit) by night than by day; whilst the visible moisture increases at night, and the haze round the cataracts becomes denser. We have just seen that the hygroscopic state of the air has no influence on the propagation of sound, whilst the velocity of the latter is diminished by the cooling of the atmosphere.

It might be supposed that, even in places remote from the habitation of man, the buzzing of insects, the singing of birds, and the rustling of leaves in the gentlest winds, produce, by day, a confused noise; which, being uniform, may fill the ears without materially distracting the attention, and thus imperceptibly lessen the effect of any louder sound. But this reasoning, whatever be its force, can hardly apply to the forests of the Orinoco, when the air is constantly filled with an innumerable multitude of moskitoes, so that the buzzing of insects is much stronger by night than by day; and where also the breeze, when it occurs at all, springs up after sunset.

I am inclined to think, that the presence of the sun affects the propagation and intensity of sound by the obstacles opposed to its transmission by currents of air of | 378 | different densities and partial undulations, the result of the unequal heating of various parts of the earth's surface. In air at rest, whether it be dry, or mixed with elastic vapours equally distributed through it, the sonorous undulation is propagated without difficulty. But when this air is crossed in every direction by small currents of a warmer temperature, the sonorous undulation divides into two waves, at the spot where there is the most sudden change in the density of the medium; thus producing partial echoes, which weaken the body of sound, because one of the sonorous waves is reflected back upon itself. The theory of these partitions of sonorous waves has been lately explained by M. Poisson, with the intelligence which distinguishes all his labours.[1] It is not, therefore, the motion of the passage of the particles of air from below upwards, nor the small oblique currents of this fluid that we consider as opposing, by impulse, the propagation of the sonorous waves. A *stroke* or impulse impressed on the surface of the liquid will form circles around the impinging centre, even when the liquid is in agitation. Several kinds of waves may cross in air, as well as in water, without interfering with each other; but the true cause of the less intensity of sound in the day time appears to be the want of homogeniety in the elastic medium. There is at this time a sudden change of density throughout, produced by small currents of air, of a high temperature, rising from portions of the earth's surface that are unequally heated. The sonorous waves are then divided in the same manner as luminous rays are refracted, and form a *mirage* of sound wherever strata of air of unequal density are contiguous. A distinction must be kept between the *intensity* of sound or of light, and the *direction* of the sonorous or luminous | 379 | wave. When these waves are propelled across strata of different densities two simultaneous effects will be produced, there will be a change in the direction of the wave, and extinction of light or sound. The reflexion that accompanies each refraction weakens the intensity of light; the separation of the sonorous wave causes partial echoes, and that portion which returns on itself becomes insensible to our ear, in weak noises, at the spot where the density of the medium suddenly changes.

---

[1] Annales de Chimie, T. 7.

In the *mirage* with double images, that which has undergone refraction contiguous to the earth is always weaker than the direct image. Strata of fluids, of very different density, may so alternate, that the primitive direction of the luminous or the sonorous ray will remain the same, but the intensity of the ray will be not the less weakened on that account. During the night the surface of the earth cools; the parts covered with grass, or with sand take the same temperature; the atmosphere is no longer crossed by currents of hot air, rising obliquely or vertically in every direction. The medium being now become more homogeneous, the sonorous wave passes with less difficulty, and the intensity of sound increases, as the separations of the sonorous waves and echoes become less frequent.

To give a precise idea of the cause of these currents of hot air, which rise in the day-time from a soil unequally heated; I will mention some observations which I have made under the Tropic. In the *Llanos*, or *steppes* of Venzuela, I have found the sand at two o'clock heated to 126° Fahrenheit, and sometimes as high as 140°. The air under the shade of a Bombax was 97°; in the sun, eighteen inches from the ground, 109°. At night the sand was only 82°, having lost at least 46°. Near the falls of the Orinoco, the ground being covered with | 380 | grasses, its temperature rose in the day only to 86°, that of the air being 79°, but at the same time the temperature of beds of granite lying on the surface was 118°. I have published a great number of similar observations in my account of experiments on the *mirage* at Cumana, at the same period that Dr. Wollaston was employed on the same subject in Europe.

If I have given the true cause of the nightly increase of sound, it will not be surprising that, under the torrid zone, this increase is greater within land than on the open sea; greater in the level plains than on the sides of the Cordilleras. The surface of the equatorial seas heats uniformly, and not beyond 85° Fahrenheit, whilst that of the continents, being variously coloured, and composed of materials of very different powers of radiation, acquires a heat of from 86 to 126°. Under the tropics the earth usually remains warmer during the night than the air: in the temperate zones, on the other hand, in calm nights, the surface of the ground cools down to seven or eight degrees below that of the air. In Europe, the temperature, instead of diminishing as the distance from the ground recedes, presents a regular progressive increase to the height of fifty or sixty feet. It is not surprising, therefore, that the terrestrial refractions are sometimes in the temperate zone almost as great by night as by day. There will always be strata of air, of different densities, resting horizontally one upon the other; but the streams of warm air that cross the atmosphere obliquely, will be more rare at night than in the day. At 3000 metres in height, in that part of the Andes which is situated beneath the equator, the mean temperature of the air is only 57°, and the force of radiation towards a cloudless sky, through a dry and very pure air, hinders the soil from heating considerably in the day time. | 381 |

I shall not dwell farther on these local circumstances, it is sufficient to have made the general deduction of the nocturnal increase of sound from the theory of sonorous undulations and their divisions. The whole depends on the want of homogeniety in the vertical columns of the atmosphere, which (according to the ingenious application that M. Arrago has made of the interference and neutralisation of the rays) is also the true cause of the scintillation of the stars. Besides, it is well known that the propagation of sound is sensibly altered when a stratum of hydrogen gas is made to rest above one of common air, in a tube closed at one extremity.

The mountaineers of the Alps, as well as those of the Andes, regard an unusual increase of sound in calm nights as a sure prognostic of a change in the weather. "It is about to rain," say they, "for the noise of the torrent comes nearer." M. Deluc has endeavoured to explain this phenomenon by a change in barometric pressure, and by an increased number of bubbles of air breaking on the surface of the water. This explanation is forced, and but little satisfactory. I will not attempt another hypothesis, but I will suggest the analogy that exists between the prognostic derived from an increase of sound, and from that of light. The inhabitants of the mountains also expect a change of weather, when, in calm weather, the mountains covered with snow appear suddenly to approach nearer to the observer, and when their outlines are unusually well defined upon the blue sky that surrounds them. Whatever be the constitution of the atmosphere that causes these appearances, it is not a little curious to observe in it a new analogy between the motions of the sonorous and the luminous undulations.

Fish Flour

"The Indians in all the Upper Oroonoko fry fish, dry them in the sun, and reduce them to powder without separating the bones. I have seen masses of fifty or sixty pounds of this flour, which resembles that of cassava. When it is wanted for eating, it is mixed with water and reduced to a paste. In every climate the abundance of fish has led to the invention of the same means of preserving them. Pliny and Diodorus Siculus have described the fish bread of the Ichthyophagous nations, that dwelt on the Persian Gulf, and the shores of the Red Sea."— Humboldt.

Humboldt's Personal Narrative

M. Humboldt relates some remarkable particulars concerning the serpents in South America, and the following will show to what perils, from them, the traveller is exposed. On taking up a night's lodging, we are told — "Before we took possession of the deserted hut, the Indians killed two great *mapanare* serpents. These grow to four or five feet long. They appeared to me to be the same species as those I described in the Rio Magdalena. It is a beautiful animal, but extremely venomous, white below the belly, and spotted with brown and red on the back. As the inside of the hut was filled with grass, and as we lay upon the ground, there being no means of suspending our hammocks, we were not without inquietude during the night. In the morning a large viper was found on lifting up from the ground the jaguar skin, upon which one of our domestics had slept. The Indians say, that these reptiles, slow in their movements when they are not pursued, creep near a man because they are fond of heat. In fact, on the banks of the Magdalena a serpent entered the bed of one of our fellow travellers, where he remained a part of the night, without doing him any harm. Without wishing here to take up the defence of vipers and rattlesnakes, I believe it may be affirmed, that, if these venomous animals had such a disposition for offence as is supposed, the human species would certainly not have resisted their numbers in some parts of America; for instance, on the banks of the Oroonoko, and the humid mountains of Choco."

After this the wanderers entered the Rio Negro, and the view here taken is both geographically good and morally affecting. —"After all we had endured (observes our interesting author), I may be permitted, perhaps, to speak of the satisfaction we felt in having reached the tributary streams of the Amazon, having passed the isthmus that separates two great systems of rivers, and in being sure of having fulfilled the most important object of our voyage, the determining astronomically the course of that arm of the Oroonoko, which falls into the Rio Negro, and of which the existence has been alternately proved and denied during half a century. In proportion as we draw near to an object we have long had in view, its interest seems to augment. The uninhabited banks of the Cassiquiare, covered with forests, without memorials of times past, then occupied my imagination, as do now the banks of the Euphrates, or the Oxus, celebrated in the annals of civilized nations. In that interior part of the New Continent we almost accustomed ourselves to regard men as not being essential to the order of nature. The earth is loaded with plants, and nothing impedes their free developement. An immense layer of mould manifests the uninterrupted action of organic powers. The crocodiles and the boas are masters of the river; the jaguar, the pecari, the dante, and the monkeys, traverse the forest without fear, and without danger; there they dwell as in an ancient inheritance. This aspect of animated nature, in which man is nothing, has something in it strange and sad. To this we reconcile ourselves with difficulty on the ocean, and amid the sands of Africa; though in these scenes, where nothing recalls to mind our fields, our woods, and our streams, we are less astonished at the vast solitude through which we pass. Here, in a fertile country adorned with eternal verdure, we seek in vain the traces of the power of man; we seem to be transported into a world different from that which gave us birth. These impressions are so much the more powerful, in proportion as they are of longer duration. A soldier, who had spent his whole life in the missions of the Upper Oroonoko, slept with us on the bank of the river. He was an intelligent man, who, during a calm and serene night, pressed me with questions on the magnitude of the stars, on the inhabitants of the Moon, on a thousand subjects of which I was as ignorant as himself. Being unable, by my answers, to satisfy his curiosity, he said to me in a firm tone: 'With respect to men, I believe there are no more above, than you | 442 | would have found, if you had gone by land from Javita to Cassiquiare. I think I see in the stars, as here, a plain covered with grass, and a forest (*mucho monte*) traversed by a river.' In citing these words, I paint the impression produced by the monotonous aspect of those solitary regions."

We have now conducted our readers to the 8th book, and, though this notice is very brief, find this a convenient place to rest for the present.

Humboldt's Personal Narrative. Dirt Eaters

"The inhabitants of Uruana belong to those *nations of the savannahs* (*Indios Andantes*), who, more difficult to civilize than the *nations of the forest*, (*Indios del monte*), have a decided aversion to cultivate the land, and live almost exclusively on hunting and fishing. They are men of a very robust constitution; but ugly, savage, vindictive, and passionately fond of fermented liquors. They are omnivorous *animals* in the highest degree; and therefore the other Indians, who consider them as barbarians, have a common saying, 'nothing is so disgusting that an Otomac will not eat it.' While the waters of the Oroonoko and its tributary streams are low, the Otomacs subsist on fish and turtles. The former they kill with surprising dexterity, by shooting them with an arrow, when they appear at the surface of the water. When the rivers swell, which in South America, as well as in Egypt and in Nubia, is erroneously attributed to the melting of the snows, and which occurs periodically in every part of the torrid zone, fishing almost entirely ceases. It is then as difficult to procure fish in the rivers which are become deeper, as when you are sailing on the open sea. It often fails the poor missionaries, on fast-days as well as flesh-days, though all the young Indians are under the obligation of 'fishing for the convent.' At the period of these inundations, which last two or three months, the Otomacs swallow a prodigious quantity of earth. We found heaps of balls in their huts, piled up in pyramids three or four feet high. These balls were five or six inches in diameter. The earth which the Otomacs eat is a very fine and unctuous clay, of a yellowish grey colour; and, being slightly baked in the fire, the hardened crust has a tint inclining to red, owing to the oxid of iron which is mingled with it. We brought away some of this earth, which we took from the winter provision of the Indians; and it is absolutely false, that it is steatitic, and contains magnesia. Mr. Vauquelin did not discover any traces of this earth in it; but he found that it contained more silex than alumin, and three or four per cent. of lime.

"The Otomacs do not eat every kind of clay indifferently; they choose the alluvial beds or strata that contain the most unctuous earth, and the smoothest to the feel. I inquired of the missionary, whether the moistened clay were made to undergo, as Father | 199 | Gumilla asserts, that peculiar decomposition, which is indicated by a disengagement of carbonic acid and sulphuretted hydrogen, and which is designated in every language by the term of *putrefaction*; but he assured us, that the natives neither cause the *clay* to *rot*, nor do they mingle it with flour of maize, oil of turtles' eggs, or fat of the crocodile. We ourselves examined, both at the Oroonoko and after our return to Paris, the balls of earth which we brought away with us, and found no trace of the mixture of an organic substance, whether oily or farinaceous. The savage regards every thing as nourishing that appeases hunger; when, therefore, you inquire of an Otomac on what he subsists during the two months when the river is the highest, he shows you his balls of clayey earth. This he calls his principal food; for at this period he can seldom procure a lizard, a root of fern, or a dead fish swimming on the surface of the water. If the Indian eat earth from want during the two months, (and from three quarters to five quarters of a pound in twenty-four hours), he does not the less regale himself with it during the rest of the year. Every day in the season of drought, when fishing is most abundant, he scrapes his balls of *poya*, and mingles a little clay with his other aliment. What is most surprising is, that the Otomacs do not become lean by swallowing such quantities of earth; they are, on the contrary, extremely robust, and far from having the belly tense and puffed up. The missionary Fray Ramon Bueno asserts, that he never remarked any alteration in the health of the natives at the period of the great risings of the Oroonoko.

"The following are the facts in all their simplicity which we were able to verify. The Otomacs, during some months, eat daily three quarters of clay slightly hardened by fire, without their health being sensibly affected by it. They moisten the earth afresh when they are going to swallow it. It has not been possible to verify hitherto with precision how much nutritious vegetable or animal matter the Indians take in a week at the same time; but it is certain that they attribute the sensation of satiety which they feel, to the clay, and not to the wretched aliments which they take with it occasionally. No physiological phenomenon being entirely insulated, it may be interesting to examine several analogous phenomena, which I have been able to collect.

"I observed every where within the torrid zone, in a great number of individuals, children, women, and sometimes even full-grown men, an inordinate and irresistible desire of swallowing earth; not an alkaline or calcareous earth, to neutralize (as it is vulgarly said) acid juices, but a fat clay, unctuous, and exhaling a strong smell. It is often found necessary to tie the childrens' hands, or to confine them, to prevent their eating the earth, when the rain ceases to fall. At the village of Banco, on the bank of the river Magdalena, I saw the Indian women who make pottery, continually swallowing great pieces of clay."

The author goes at some length into analogies and reasoning on them, but we confine our quotation principally to the facts.

"The negroes on the coast of Guinea delight in eating a yellowish earth, which they call *caouac.* The slaves who are taken to America try to procure for themselves the same enjoyment: but it is constantly detrimental to their health. They say, that the earth of the West Indies is not so easy of digestion as that of their country.'"

"In the Indian Archipelago, at the island of Java, Mr. Labillardière saw, between Surabaya and Samarang, little square and reddish cakes exposed to sale. These cakes, called *tanaampo*, were cakes of clay, slightly baked, which the natives eat with appetite. The attention of physiologists, having been powerfully fixed on these phenomena of *geophagy*, Mr. Leschenault (one of the naturalists of the expedition to the Southern Lands under the command of Captain Baudin) has published some curious details on the *tanaampo*, or *ampo*, of the Javanese. 'The reddish and somewhat ferruginous clay,' he says, 'which the inhabitants of Java are fond of eating occasionally, is spread on a plate of iron, and baked, after | 200 | having been rolled into little cylinders in the form of the bark of cinnamon. In this state it takes the name of *ampo*, and is sold in the public markets. This clay has a peculiar taste, which is owing to the torrefaction; it is very absorbent, and adheres to the tongue, which it dries. In general it is only the Javanese women who eat the *ampo*, either in the time of their pregnancy, or in order to grow thin; the want of plumpness being a kind of beauty in this country. The use of this earth is fatal to health; the women lose their appetite imperceptibly, and no longer take without disgust a very small quantity of food: but the desire of becoming lean, and of preserving a slender shape, can brave these dangers, and maintains the credit of the *ampo*.' The savage inhabitants of New Caledonia also, to appease their hunger in times of scarcity, eat great pieces of a friable *lapis ollaris*. Mr. Vauquelin analysed this stone, and found in it, besides magnesia and silex in equal portions, a small quantity of oxid of copper. Mr. Goldberry had seen the negroes in Africa, in the islands of Bunck and Los Idolos, eat an earth of which he had himself eaten, without being incommoded by it, and which also was a white and friable steatite."

"When we reflect on the whole of these facts, we perceive that this disorderly appetite for clayey magnesian, and calcareous earth, is most common among the people of the torrid zone; that it is not always a cause of disease; and that some tribes eat earth from choice, while others (the Otomacs in America, and the inhabitants of New Caledonia, in the Pacific Ocean), eat it from want, and to appease hunger."

"The observations, which I made on the banks of the Oroonoko, have been recently confirmed by the direct experiments of two distinguished young physiologists, Messrs. Hippolyte Cloquet and Breschet. After long fasting, they ate as much as five ounces of a silvery green and very flexible laminar talc. Their hunger was completely satisfied, and they felt no inconvenience from a kind of food, to which their organs were unaccustomed. It is known, that great use is still made in the East of the bolar and sigillated carths of Lemnos, which are clay mingled with oxid of iron. In Germany, the workmen employed in the quarries of sandstone worked at the mountain of Kiffhæuser spread a very fine clay upon their bread, instead of butter, which they call *steinbutter*,[1] stone butter; and they find it singularly filling, and easy of digestion."

[1] "This *steinbutter* must not be confounded with the *mountain butter, bergbutter,* which is a saline substance, owing to a decomposition of aluminous schists."

Humboldt's Travels[1]

There are some men whose names seem to irradiate the age in which they are born, whose every step in life forms an epoch in science, and who, as if Nature herself were sedulous to guard them as her historians, escape unhurt through perils that would alike appal the mind, and overwhelm the bodies, of less enthusiastic, less gifted individuals. Such is Humboldt, every addition to whose travels is an addition to our stock of knowledge. In him all the qualities that are requisite for a philosopher and a man of science are most happily combined, whilst the energies of his mind seem to transform themselves into physical powers of more than natural strength, to enable him to follow whither his ardour leads him. The termination of his "Personal Narrative" has made its appearance, and the last part is no way inferior to the first, in vigour of research, truth of inference, and beauty of moral reflection. It is not easy for persons who stay quietly at home, to imagine the exceeding energy of mind which must be called up, to bear some of the torments, the privations, the perplexities, of a man exposed to every variation of climate, and to peculiarities attendant on each, of which he may be totally ignorant. The moschettoes, for instance, just beyond the mouth of the Rio Aranca, assume the appearance of an evil so formidable as to forbid all the attempts of man towards civilization in the quarter which they infest, in a degree greater than is known in any other part of the habitable globe.

"The lower strata of air, from the surface of the ground to the height of fifteen or twenty feet, are filled with venomous insects like a condensed vapour. If in an obscure spot, for instance in the grottoes of the cataracts formed by superincumbent blocks of granite, you direct your eyes toward the opening enlightened by the sun, you see clouds of moschettoes more or less thick, according as these little animals, in their slow and regular movements to their own music [*mouvemens lents et cadencés*], form into groups, or spread themselves abroad. At the mission of San Borja, the suffering from moschettoes is greater than at Carichana; but in the *raudales*, at Atures, and above all at Maypures, this suffering may be said to attain its *maximum*. I doubt whether there be a country upon earth, where man is exposed to more cruel torments in the rainy season. Having

[1] "Personal Narrative of Travels to the Equinoctial Regions of the New Continent, during the years 1799—1804. By Alexander De Humboldt and Aimé Bonpland; with maps, plans, &c. Written in French by Alexander de Humboldt, and translated into English by Helen-Maria Williams." Vol. V.

passed the fifth degree of latitude, you are somewhat less stung; but on the Upper Oroonoko the stings are more painful, because the heat, and the absolute want of wind, render the air more burning, and more irritating in its contact with the skin."

No wonder that in the missions of the Oroonoko, in the villages placed on the banks of the river, surrounded by immense forests, stations to which the unfortunate priests of Spain are condemned for twenty years together, the *plaga de los moscos*, the plague of the flies, | 315 | affords an inexhaustible subject of conversation, and that when two persons meet in the morning the first questions they address to each other are "How did you find the zancudoes during the night? How are we to-day for the moschettoes?" Some of these poor missionaries shewed their legs to Mr. Humboldt, which were so discoloured by the repeated stinging of the different species of these tormentors (for their name is legion) that it was difficult to see a trace of the whiteness of the skin between the spots of coagulated blood. The different species do not congregate together, but occupy the air in succession, and sometimes there is a short interval of quiet between the changes, which is enjoyed with an ecstasy that must make the recommencement of the attack still more insupportable. It is a mistake to imagine that the Indians are less susceptible of the bites of these insects than the Europeans. "How comfortable must people be in the moon," said a Saliva Indian to father Gumilla; "she looks so beautiful and so clear that she must be free from moschettoes." Mr. Humboldt's account of this plague of the deserts is minute and highly interesting; as is also his description of the missions in general, and of those whom God has ordained, as the monks despairingly express themselves, to inhabit them. But our limits forbid us to enter into particulars: instead of which we will present our readers with the following characteristic sketch of a part of the travelling suite of Mr. Humboldt and his companion Bonpland, in their passage up the Rio Negro.

"In one of the huts of the Pacimonales we made the acquisition of two large fine birds, a toucan (*piapoco*),[2] approaching the ramphastos erythrorhynchos, and an *ana*, a species of macaw, seventeen inches long, having the whole body of a purple colour, like the p. macao. We had already in our canoe seven parrots, two manakins (pipra), a motmot, two *guans*, or *pavas de monte*, two manaviris (cercoleptes or viverra caudivolvula), and eight monkeys, namely, two ateles, [3] two titis,[4] one

[2] *Kiapoco*, or *aviapoco*.

[3] *Marimonda* of the Great Cataracts, simia belzebuth, Brisson.

[4] Simia sciurea, the saimiri of Buffon. (See my "Rec. d'Observ. de Zoologie," vol. i. p. 327, 334, 353, and 357.)

viudita,[5] two douroucoulis or nocturnal monkeys,[6] and the cacajao with a short tail.[7] Father Zea whispered some complaints at the daily augmentation of this ambulatory collection. The toucan resembles the raven in its manners and intelligence. It is a courageous animal, but easily tamed. Its long and stout beak serves to defend it at a distance. It makes itself master of the house, steals whatever it can come at, and loves to bathe often and fish on the banks of the river. The toucan we had bought was very young; yet it took delight, during the whole voyage, in teazing the cusicusis, or nocturnal monkeys, which are sad and passionate. I did not observe what has been related in some works of natural history, that the toucan is forced, from the structure of its beak, to swallow its food by throwing it up into the air. It raises it indeed with some difficulty from the ground, | 316 | but, having once seized it with the point of its enormous beak, it has only to lift it up by throwing back its head, and hold it perpendicularly as long as it is in the act of swallowing. This bird makes extraordinary gestures when preparing to drink. The monks say, that it makes the sign of the cross upon the water; and this popular belief has obtained for the toucan, from the creoles, the singular name of *diostede* (God grant it thee.)

"Most of our animals were confined in small willow cages; others ran at full liberty all over the boat. At the approach of rain, the macaws sent forth frightful cries, the toucan wanted to gain the shore to fish, and the little monkeys, the titis, went in search of father Zea, to take shelter in the large sleeves of his Franciscan habit. These scenes were often repeated, and made us forget the torment of the *moschettoes*."

Of many tribes of Indians being in the constant habit of eating human flesh, M. Humboldt does not appear to entertain the smallest doubt; and, indeed, it would be folly to disbelieve what the missionaries, though unwillingly, admit, and the Indians themselves boast of. Cocuy, a celebrated warrior, and chief of the Manitivitanoes, used to devour the finest and fattest of his wives, an instance of epicurism which scandalized no little the worthy missionaries, who converted him to christianity; though, without working that change in his appetite which they did in his belief. "It is a *bad habit*," says father Gili, of these people of Guyana, "in other respects *so good and so mild*."

"'You cannot imagine,' said the old missionary of Mandavaca, 'all the perversity of this *familia de Indios*. You receive men of a new tribe

[5] Simia lugens. (Ib., p. 319).
[6] Cusicusi or simia trivirgata. (Ib. p. 307 and 358.) This is the aotus of Illiger.
[7] Simia melanocephala, *mono feo*. (Ib. p. 317.) These last three species are new.

into the village; they appear to be mild, good, and laborious; but, suffer them to take part in an incursion (*entrada*) to bring in the natives, and you can scarcely prevent them from murdering all they meet, and hiding some portions of the dead bodies.' In reflecting on the manners of these Indians, we are almost terrified at that combination of sentiments, which seem mutually to exclude each other; that faculty of nations to become but partially humanized; that preponderance of customs, prejudices, and traditions, over the natural reflections of the heart. We had a fugitive Indian from the Guaisia in our canoe, who had become sufficiently civilized in a few weeks, to be useful to us in placing the instruments necessary for our observations at night. He was no less mild than intelligent, and we had some desire of taking him into our service. What was our regret, when, talking to him by means of an interpreter, we learned, 'that the flesh of the marimonde monkeys, though blacker, appeared to him to have the taste of human flesh.' He told us 'that his *relations* (that is, the people of his tribe) preferred the inside of the hands in man, as in bears.' This assertion was accompanied with gestures of savage joy. We inquired of this young man, so calm and so affectionate in the little services which he rendered us, whether he still felt sometimes a desire to eat of a Cheruvichahena. He answered without discomposure, that, living in the mission, he would only eat what he saw was eaten by *los Padres*. Reproaches addressed to the natives on the abominable practice, which we here discuss, produce no effect; it is as if a Bramin of the Ganges, travelling in Europe, reproached us with our habit of feeding on the flesh of animals. In the eyes of the Indian of the Guaisia, the Cheru-
| 317 | vichahena was a being entirely different from himself; and whom he thought it was no more unjust to kill, than the jaguars of the forest. It was merely from a sense of propriety, that, as long as he should remain in the mission, he would only eat the same food as *los Padres*. The natives, if they return to their tribe (*al monte*), or find themselves pressed by hunger, soon resume their ancient habits of anthropophagy. And why should we be so much astonished at this inconstancy in the tribes of the Oroonoko, when we are reminded, by terrible and well ascertained examples, of what has passed among civilized nations in times of great scarcity? In Egypt, in the thirteenth century, the habit of eating human flesh pervaded all classes of society; extraordinary snares were spread for physicians in particular. They were called to attend persons, who pretended to be sick, but who were only hungry; and it was not in order to be consulted, but devoured. An historian of great veracity, Abd-Allatif,

has related, how a practice, which at first inspired dread and horror, soon occasioned not the slightest surprise.[8]"

Next to the contemplation of this feast of the anthropophagi, we recommend the following for its singularity, to the imagination of our readers, to look upon with what appetite they may.

"The hut, where the natives were assembled, displayed during several days a very singular aspect. There was neither table nor bench, but large roasted monkeys, blackened by smoke, were ranged in order resting against the wall. These were the *marimondes* (ateles belzebuth), and those bearded monkeys called *capuchins*, which must not be confounded with the weeper, or sai (simia capucina of Buffon). The manner of roasting these anthropomorphous animals contributes singularly to render their appearance disagreeable in the eyes of civilized man. A little grating, or lattice, of very hard wood, is formed, and raised one foot from the ground. The monkey is skinned, and bent into a sitting posture; the head generally resting on the arms, which are meagre and long; but sometimes these are crossed behind the back. When it is tied on the grating, a very clear fire is kindled below. The monkey, enveloped in smoke and flame, is broiled and blackened | 318 | at the same time.[9] On seeing the natives devour the arm or leg of a roasted monkey, it is difficult not to believe, that this habit of eating animals, that so much resemble man in their physical organization, has, in a certain degree, contributed to diminish the horror of anthropophagy among savages. Roasted monkeys, particularly those that have a very round head, display a hideous resemblance to a child; the Europeans therefore, who are obliged to feed on quadrumanes, prefer separating the head and the hands, and serve up only the rest of

[8] *Account of Egypt by Abd-Allatif, Physician of Bagdad, translated into French by M. Silv. de Sacy*, p. 360—374. "When the poor began to eat human flesh, the horror and astonishment caused by repasts so dreadful were such, that these crimes furnished the never ceasing subject of every conversation. But at length the people became so accustomed to it, and conceived such a taste for this detestable food, that people of wealth and respectability were found to use it as their ordinary food, to eat it by way of regale, and even to lay in a stock of it. This flesh was prepared in different ways, and the practice being once introduced, spread into the provinces, so that examples of it were found in every part of Egypt. It then no longer caused any surprise; the horror, it had at first inspired, vanished; and it was mentioned as an indifferent and ordinary thing. This fury of devouring one another became so common among the poor, that the greater part perished in this manner. These wretches employed all sorts of artifices to seize men by surprise, or decoy them into their houses under false pretences. This happened to three physicians among those who visited me; and a bookseller, who sold me books, an old and very corpulent man, fell into their snares, and escaped with great difficulty. All the facts which we relate as ocular witnesses fell under our observation accidentally, for we generally avoided seeing spectacles, which inspired us with so much horror."

[9] "Soon after my return to Europe, an engraving was published at Weimar, from a drawing composed with great spirit by Mr. Schick at Rome, representing one of our resting-places on the banks of the Oroonoko. In the foreground some Indians are occupied in roasting a monkey."

the animal at their tables. The flesh of monkeys is so lean and dry, that Mr. Bonpland has preserved in his collections at Paris an arm and hand, which had been broiled over the fire at Esmeralda; and no smell arises from them after a great number of years."

Whilst we are on the subject of *taste*, which is allowed to be mere matter of opinion, not of dispute, we will make mention of another diet, not exactly either animal or vegetable, but which we suppose may combine the flavours of both to those who indulge in it. We speak of the Otomacs, a tribe in the rudest state, who inhabit the picturesque mission of Uruana, and who present one of the most extraordinary physiological phenomena.

"The Otomacs," says Mr. Humboldt, "eat earth; that is, they swallow every day, during several months, very considerable quantities, to appease hunger, without injuring their health." And his arguments on this singular custom are so curious, that though our limits forbid us from pursuing them further, we cannot but recommend them, with the whole of the work, to the attention of our readers, as very highly interesting.

Indian Rubber

—The following curious account of the mode in which Indian Rubber is procured, is extracted from *Humboldt's Travels*:— "Here (says Mr. H. at the mission of St. Balthasar on the Atabapo) we saw, for the first time, that white and fungous substance, which I have made known by the name of *dapicho* and *zapis*. We immediately perceived, that it was analogous to the *elastic resin*: but, as the Indians made us understand by signs, that it was found under ground, we were inclined to think, till we arrived at the mission of Javita, that the *dapicho* was a fossil caoutchouc, though different from the elastic bitumen of Derbyshire. A Poimisano Indian, seated by the fire, in the hut of the missionary, was employed in reducing the *dapicho* into black caoutchouc. He had spitted several bits on a slender stick, and was roasting them like meat. The *dapicho* blackens in proportion as it grows softer, and gains in elasticity. The resinous and aromatic smell, which filled the hut, seemed to indicate, that this coloration is the effect of the decomposition of a carburet of hydrogen, and that the carbon appears in proportion as the hydrogen burns at a low heat. The Indian beat the softened and blackened mass with a piece of brazil wood, ending in form of a club; he then kneaded the *dapicho* into balls of three or four inches in diameter, and let it cool. These balls exactly resemble the caoutchouc of the shops, but their surface remains in general slightly viscous. They are used at San Balthasar in the Indian game of tennis, which is so celebrated among the inhabitants of Uruana and Encara-
| 523 | mada; they are cut into cylinders, to be used as corks, and are far preferable to those made of the bark of the cork-tree."—Soon after, the travellers obtained precise information respecting this substance:—it was shown them at the depth of two or three feet, in a marshy soil, "between the roots of two trees known by the name of the *jacio* and the *curvana*. The first is the hevea of Aublet, or siphonia of the modern botanists, known to furnish the caoutchouc of commerce in Cayenne and the Grand Para; the second has pinnate leaves, and its juice is milky, but very thin, and almost destitute of viscosity. The *dapicho* appears to be the result of an extravasation of the sap from the roots. This extravasation takes place more especially when the trees have attained a great age, and the interior of the trunk begins to decay. The bark and alburnum crack; and thus is effected naturally, what the art of man performs to collect in abundance the milky juices of the hevea, the castilloa, and the caoutchouc fig tree."

Memoir of Don Jose Celestino Mutis

Translated from the French of the Baron de Humboldt, by Colonel R. Wright, late Governor of the Province of Loxa, Consul General for the Republic of the Equator.

Don Jose Celestino Mutis, director of the *Botanical Expedition* of the kingdom of New Granada,[1] and astronomer royal at Santa Fé de Bogota, was born in Cadiz in 1732, of a family in easy circumstances. He is best known in Europe through his vast knowledge and researches in Botany. Linnæus terms him, "*phytologorum Americanorum princeps*;" but the services he has rendered to all the branches of natural history,—the discovery of the *quinquinas* in regions where their existence was unknown, the beneficent influence which he has exercised upon the civilisation and enlightenment of the Spanish colonies, assign to him a distinguished station amongst the men who have illustrated the new world.

After having applied himself with uncommon ardour to the study of mathematics, Mutis was forced by his parents to direct his attention to practical medicine, and pursued his courses at the college of San Fernando de Cadiz; he took his degrees at Sevilla, and, in 1757, was named substitute of a chair of anatomy at Madrid. During a sojourn of three years in the Spanish capital, he took more delight in botanical excursions than in visiting the hospitals, and he had the rare good fortune of acquiring the acquaintance of the celebrated naturalist of Upsal, who was desirous of adding to his herbal collection the plants of the Peninsula. This correspondence of Mutis with Linnæus became of still greater importance to science, inasmuch as the viceroy, Don Pedro Mexia de la Cerda engaged the former to accompany him to America in the quality of physician, 1750. Our young botanist had been named by the ministry amongst the persons selected to terminate their studies at Paris, Leyden and Bologna, but he did not hesitate a moment in sacrificing the prospect of visiting the most renowned universities of Europe, to the advantages and attractions of a distant expedition.

On his arrival at New Granada, he was vividly struck with the natural richness of a country where different climates are encountered in

[1] Now the republic.

progressive succession, one above another, with the regularity of a flight of stairs. After remaining for a long period in Carthagena, Turbaco, and Honda, (chief port of the river Magdelena), Mutis accompanied the viceroy on his journey to Bogota, situate on the table land[2] 1365 toises above the level of the sea, and possessing a temperature similar to that of Bordeaux. Between Honda and Bogota he passed through forests containing precious species of the cinchona (*quinquina*), but until the year 1772, he had not given his attention to this useful production.

Named Professor of Mathematics in the "*Colegio Mayor de nuestra Senora del Rosario*," he circulated at Sante Fé de Bogota, the first notions of the true planetary system. The Dominican Friars did not behold without inquietude that "*the Heresies of Copernicus*," already professed by Bouguer, Godin, and La Condamine at Quito, should penetrate into New Granada; but the viceroy protected Mutis against the monks, who would have the earth remain immoveable; they, however, slowly accustomed themselves to what they yet call the "hypotheses of the new philosophy."

Mutis, anxious to examine the plants of the warm region, and to visit the argentiferous mines of New Granada, quitted the table land of Bogota. He made a long stay, first at La Montuosa, between Giron and Pamplona, and subsequently (from 1777 to 1782) at Real del Zapo, and Mariquita, situate at the foot of the Quindio Andes and Paramo of Herveo. It was at La Montuosa that he commenced the "*Grand Flore de la Nouvelle Grenade*," a botanical work at which he laboured without intermission for the space of forty years, and which we have reason to fear, never will be published in a complete form. Linnæus, in his supplement of *Species Plantarum*, and in his *Mantissa*, has named a great number of rare species which Mutis remitted him from La Montuosa, but by a strange, and for the geography of plants, most unfortunate error, he has noted them as coming from Mexico.

The trifling sums which our traveller earned by the practice of his art, sometimes at the mines, he expended in the formation of a botanic library, in the purchase of barometers, geodætical instruments, and glasses to observe the occultations of Jupiter's satellites. He engaged artists to draw the most curious plants, and who painted in oil the indigenous animals almost all as large as life. The author of this article has seen a part of this collection, formed before Mutis had become the object of his sovereign's munificence. It was also during his sojourn at Real del Zapo (1786), that he made the important discovery of a quicksilver mine near

---

[2] Of the Andes.

Ibague Viejo, between the Nevada de Tolima and the river Saldane. So much useful exertion met at length with honourable encouragement. The court of Madrid, at the instance of the viceroy, Archbishop Don Antonio Caballero y Gongora, resolved (1782) upon founding, first at Mariquita, and afterwards (1790) at Santa Fé de Bogota, a grand establishment of natural history, under the name of "Expedicion Real Botanica," at the head of which was placed Don Celestino Mutis. An immense edifice of the capital was destined for this establishment. He included the herbaria, school of design, and library, one of the richest and most beautiful ever consecrated to such a purpose even in Europe, in one sole branch of natural history. Mutis had embraced the ecclesiastical state since the year 1772, and was named canon of the metropolitan church of Santa Fé and Confessor to a convent of nuns. Zealous in the discharge of the duties he had imposed upon himself, he found no leisure to extend his excursions beyond the vicinity of the capital, but he despatched the artists attached to the "Expedicion," to the warm and temperate regions which surround the table land of Bogota. Some Spanish artists, whose talents had been perfected by the councils of Mutis, formed, in a very few years, a school of young painters. The Indians, coloured, and indigenous natives of mixed blood, shewed an extraordinary disposition to imitate the form and colour of the plants. The drawings of the *Flore de Bogota* were executed on the largest sized paper; they chose the branches which were most laden with flowers; the analysis or anatomy of the fructified parts was subjoined at the foot of the drawing—generally, each plant was represented on three or four large sheets in colours and in black at the same time. The colours were extracted partly from indigenous colouring matter unknown in Europe. Never was a collection of drawings executed with greater splendour, nor, it may be said, on a larger scale. Mutis had taken for his model the most admired botanical works of his time, those of Jacquin, of L'Heritier, and the Abbé Cavanilles. The aspect of the vegetation and physiognomy of the plants, were pourtrayed with astonishing fidelity. Modern botanists who study the affinity of vegetables according to the insertion and adherence of the organs | 220 | would have wished for a more detailed analysis of the fruit and seed. When MM. de Humboldt and Bonpland were at Bogota in 1801, and enjoyed the noble hospitality of Mutis, the latter estimated the number of drawings already finished at 2000, amongst which 43 species of the *passiflora* and 123 species of *orchideous* plants, deserved particular admiration. Those travellers were the more surprised at the richness of Mutis's botanic collections (formed by himself and his estimable pupils MM. Valenzuela, Zea, Caldas, and his chief painters MM. Rizo and Mathis), as the most fertile countries of New Granada, the plains of Tolu and San Benito Abad, the Andes of Quindio, the provinces of Santa Martha, Antioquia, and Choco, remained yet unexplored by any botanist. The larger the mass of materials gathered by this indefatigable savant grew, the greater the difficulties he encountered to publish the fruits of his labours. He had copies taken of the drawings in his *Flore de Bogota*, with the view of sending one to Spain, and of preserving the rest at Santa Fé. But how could it be hoped the scientific world should enjoy this immense work, when the *Flora Peruviana et Chilensis* of Ruiz and Pavon, notwithstanding the pecuniary succour of the Government and Colonies, hardly advanced at an extremely slow pace? Mutis was too much attached to the establishments he had founded—he was too fond of a country which had become his second home, to undertake, at the age of 76, a return to Europe.

He continued, till his death, to accumulate materials for his work without deciding on the mode of its publication. Accustomed to overcome apparently insurmountable obstacles, he dwelt with pleasure on the idea of establishing some day a press in his house, and of instructing his young indigenous pupils, who had learned to paint so successfully, in the art of engraving. Notwithstanding his advanced age, he undertook, in 1802, the construction of an observatory in the centre of his garden. It is an octagon tower, 72 feet in height, which contained, in 1808, a *gnomon* of 37 feet, a quarter circle by Sisons, Graham's pendulum which La Condamine had left at Quito, two chronometers by Emery, and a few of Dollond's telescopes.

Mutis had the good fortune not to witness the bloody revolutions which have devastated those lovely countries: he was overtaken by death 11th September, 1808, at a period when he was in the enjoyment of all the happiness that can be derived from the esteem of honest men, from literary glory, and from the certainty of having largely contributed to an improved social state in the New World, by his instruction, example, and the practice of every virtue.

We have given a brief sketch of the life of Mutis; we shall proceed to a summary mention of his labours, which embrace nearly all the branches of natural science.

Nothing remains of him but a few dissertations, printed in the Memoirs of the Royal Academy of Stockholm, for the year 1769, and in an excellent journal published at Santa Fé in 1794, under the title of *Papel Periodico*; but Linnæus's Supplement—the works of the Abbé Cavanilles, and of M. de Humboldt—the *Seminario del Nuevo Reyno de*

Granada, edited by M. Caldas, in 1808-9, have published a part of his observations. We are ignorant of the fate of the manuscripts which this celebrated man left to the care of his friends and nearest relatives. M. Caldas, the director of the observatory of Santa Fé,—the favourite pupil of Mutis, Don Salvader Rizo, chief painter to the *Botanic Expedition*, and the greater part of the inhabitants distinguished for their talents and acquirements, were all put to death[3] during the unhappy reaction of the metropolis. The precious collection of paintings was sent to Spain, where had been previously remitted the inedited matter of the *Flore du Perou et du Mexique*. We hope that when political agitation shall cease in the Peninsula, and in the colonies, the labours of Mutis may not remain consigned to oblivion, like those of Sesse and of Mocino.

It was the communications which Mutis had made to Linnæus that gave him celebrity in Europe, long before the works he was preparing were known to exist. Most of the genera, *alstonia, vallea, barnadesia, escallonia, manettia, acæna, brathys, myroxylon, befaria, telipogon, brabejum, gomozia*, and many others, published in Linnæus's supplement, are due to the sagacity of the botanist of Santa Fé.

Speaking of the genus *mutisia*, Linnæus adds, *nomen immortale quod nulla œtas unquam delebit.*

It was Mutis who first made known the true characters of the genus cinchona. As this last has become of great importance, we shall endeavour to call to mind the opinions heretofore entertained respecting the *quinquinas* of the New World. La Condamine and Joseph de Jussieu had examined, in 1738, the trees of the forests of Loxa, which yield the febrifuge *quina,* or bark. The first-named has published a description and drawing of the Peruvian quinquina in the *Memoires de l'Academie*, the same species made known by MM. Humboldt and Bonpland, under the name *cinckona condaminea*, and which botanists have long confounded with many other kinds, under the vague denomination of *cinchona officinalis*. This cinchona condaminea, also called *cascarilla fina de Loxa*,[4] *de Caxanuma*, and *de Urituzinga*, is the species most rare, most precious, and without doubt the kind originally employed. There are but 100 quintals exported annually from Guayaquil, a port of the South Sea.[5] The whole

[3] Note by Tr.—By the Spanish general, Morillo, Caldas requested a respite for a few days to regulate "papers of interest to the world,"—the stern tyrant refused.

[4] The cascarilla fina of Loxa (orange coloured and primitive), called also by the natives of Loxa, *cascarilla de Urituzinga*, that is, in the Quichua tongue, *bear's nose*, the highland forest in which it grows, having a profile when viewed from the vallies, somewhat resembling that of a bear.—Tr.

[5] A Port of the Republic of the Equator on the Pacific Ocean.

exportation of the different species of quina from America is annually 14,000 quintals. Linnæus had formed, since 1742, his genus *cinchona*, a name derived from a vice-queen of Peru. He could only have founded the *genus* on the imperfect description of La Condamine. In 1753, an intendant of the mint of Bogota (Don Miguel de Santestevan), visited the forests of Loxa, and discovered on his route, between Popayan and Quito, trees of quinquina in many quarters, especially near the village of Guanacas and the Sitio de los Corrales. He sent samples of cinchona to Mutis, and it was upon those samples that the latter made the first exact description of the genus. He lost no time in transmitting to Linnæus the flower and fruit of the yellow quinquina (cinchona cordifolia,) but the great naturalist of Upsal, on publishing the observations of Mutis, (Syst. Nat. ed. 12, page 164,) confounded the yellow quinquina with that which La Condamine had described. Up to that time, Europe only received the febrifuge bark by the ports of the South Sea. The tree which furnishes this precious production was not yet known north of the parallel of 2 1/2° lat. boreal. In 1772, Mutis discovered the quina within 6 leagues of Santa Fé de Bogota, in the woods of Tena. This important discovery was quickly followed, 1773, by another of the same plant on the route from Honda to Villeta, and at the Mesa de Chinga. We have entered into some details on this head, because the quinquina of New Granada, exported by Carthagena, and consequently by a port of the Caribbean Sea, nearer to Europe, has had the most beneficial influence on colonial industry, and on the diminution of the price of febrifuge bark in the markets of the old world. Mutis was right when he attached great importance to this discovery, for which he was never recompensed by his government. An inhabitant of Panama, Don Jose Lopez Ruiz, who himself acknowledges, in *Informes al Rey.*, never to have known the *quinquinas* of Honda until 1774, passed a long time for the original discoverer of the quinas of Santa Fé. He enjoyed as such a pension of 10,000 francs, until, in 1775, the viceroy of Gongona made known to the court the priority of Mutis's rights. About the same period, (1776) Don Francisco Renjifo found the quinquina in the southern hemisphere on the slopes of the Peruvian Andes of Guanuco. It is now to be found in the whole length of the Cordilleras, at an elevation of from 700 to 1500 toises, upon an extent of above 600 leagues, from La Paz and Chuquisaca, to the mountains of Santa Martha and Merida.

Mutis has the merit of having been the first to distinguish the different species of Cinchona, some of which, with downy petals, are much more active than others whose surface is smooth. He has proved, that the active species, *whose properties vary with the organic structure and form, should not be indiscriminately employed.* The *Quinologia* of Mutis, which is about to be published at Madrid by M. Lagasca, and of which a fragment only has been inserted in the *Papel Periodico de Santa Fé de Bogota*, Feb. 1794, comprises the whole of his medical and botanical researches. That work describes a preparation of fermented quinquina, in great esteem at Bogota, Quito, and Lima, under the name of cerveza de quina (beer of quina.)

Amongst the useful plants in medicine and commerce, originally described by Mutis, we must reckon the *psychotria emetica*, or ipecacuanha, (*raisilla*) of the river Magdalena, the *toluifera*, and the *myroxylon*, which yield the balm of Tolu and of Peru—the *winteria granadensis*, near neighbour of the *canella alba* of our pharmacopœias, and the *alstonia theæformis*, which furnishes the tea of Bogota, the infusion of which cannot be too strongly recommended to travellers who remain any length of time exposed to the rain of the tropics.

At Mariquita, in a temperate and delicious climate, Mutis formed a small plantation of quinquina, cinnamon, (*laurus cinnamomoides*,) which abounds in the missions of Andaquies, and of indigenous nutmeg (*myristica otova*.) The name of this celebrated botanist is also attached to a discovery which has greatly interested the public mind in America. It was known that the Indians and negroes who worked at the *lavaderos* (washings) of gold and platinum, in the province of Choco, were possessed of what they called the secret of a plant, the most powerful antidote against the bite of venemous serpents. Mutis was enabled to discover the mystery, and make the plant known: it is of the family of the compositæ, and is known in its native country by the name | 221 | of *vejuco del guaco*.[6] MM. Humboldt and Bonpland were the first to represent it, (*v. Mikania Guaco*, in the *Plantæ Æquinoctiales*, vol. 2, p. 85, pl. 105.) This plant has a nauseous odour, which appears to affect the olfactory organs of the vipers. The scent of the guaco mixes itself, no doubt, with the cutaneous transpiration of man: one is considered out of danger after a space of time, more or less, when the patient has been *curado*, that is, when inoculation of the juice of the guaco into the dermoid system has been effected. Bold experiments made in the house of Mutis by MM. Zea, Vargas, and Mathis, and during which they were seen to handle with impunity the most venemous vipers, have been described in the *Seminario de Agricultura* of Madrid, 1798, vol. 4, p. 397. As the guaco

[6] Found chiefly in Zaruma, a canton of the province of Loxa, abounding in venemous reptiles.—Tr.

has been found in several of the warm vallies of the Andes, from Peru down to Carthagena, and on the mountains of Varinas, a great number of persons owe their recovery to this valuable discovery of Mutis. It is to be regretted that this plant, which has been frequently confounded with the *Ayapana*, loses its virtue when the leaves and trunk are preserved in alcohol. The guaco is not found everywhere that serpents abound.

We know but little of the zoological and physical labours of Mutis; but we are aware he studied for a long time the habits of the ants and termites, which in America, as in Senegal, construct mounds from five to six feet in height. He had painted a number of the mammiferous species, of the birds and fishes of New Granada. He has described, according to the Linnæan method, in the *Memoires de l'Academie de Stockholm*, of which he was a member, a new species of marten, (*viverra mapurito*.)

The manuscripts of Mutis also contained a great number of valuable observations on the *atmospheric currents*, which appear much more distinctly under the tropics than in the temperate climates, by the horary variations of the barometer. This instrument rises and falls, at the level of the sea as on the highest table land, four times in 24 hours under the torrid zone, with such regularity, that one may tell, to within a quarter of an hour, the time of day, by the simple inspection of the column of mercury. It appears that this curious observation, which has so much excited the attention of scientific men, and the discovery of which La Condamine (*Voyage à l'Equatcur*, p. 50), has so falsely attributed to Godin, had been already made at Surinam in 1722, (*Journal Litteraire de la Haye* for the year 1722, p. 234.) Father Bondier (1742), employed it at Chandernagor; Godin (1737), at Quito; Thibault de Chanvalon (1751), at Martinique; Lamanon, in 1786, in the South Sea: Mutis avers having found that the moon exercises a sensible influence on the period and on the extent of the horary variations, (Caldas, *Seminario de la nuevo Reyno de Granada*, vol. 1, p. 55 and 361, No. 3.)

The man who has displayed such astonishing activity during forty-eight years of labour in the New World, was endowed by nature with a most happy physical constitution. He was tall in stature; his features bore the imprint of nobleness; his air was grave, and his manners easy and polite. His conversation was as varied as the objects of his studies. If he frequently spoke with warmth, he also loved to practice that art of listening to which Fontenelle attached so much value, and which he found so rare in his days. Although much occupied by a science which requires the most attentive study of organization, Mutis never lost sight of the great problems respecting the *physique* of the universe. He travelled over the Cordilleras barometer in hand. He determined the medium temperature of those table lands, which appear, like islets, in the middle of the aerial ocean. He had been struck with the aspect of the vegetation, which varies according as we descend towards the vallies, or climb towards the frozen summits of the Andes. Every question connected with the geography of plants was to him of the most lively interest, and he sought to delineate the limits more or less, betwixt which, on the mountain declivities, grow the various species of Cinchona.

This taste for the physical sciences—this active curiosity directed to the development of the phenomena of organisation and meteorology, maintained the empire of his mind to the last moment of his life. Nothing proves better the superiority of his talent, than the enthusiasm with which he received the intelligence of any important discovery. He had not seen a chemical laboratory since 1760; but the perusal of the works of Lavoisier, of Guyton de Morveau and Fourcroy, had offered him the most valuable knowledge of the state of modern chemistry. Mutis encouraged with great interest the young people who manifested a disposition for study; he furnished them with books and instruments, and sent many to travel at his proper cost

After having spoken of his liberality, and of the daily sacrifices he made to science, it is unnecessary to extol his disinterestedness. He enjoyed all his life the confidence of the viceroys, who exercised unlimited sway in those countries; but he only employed his credit for the benefit of science—to make known modest merit, or to plead with courage the cause of the unfortunate. His only ambition was to aid in the triumph of justice and of truth; he fulfilled with zeal, it may be said with austerity, the duties he had imposed on himself, in embracing the ecclesiastical profession, but his piety sought not the eclat of renown—it was mild and sincere, as it ever is, when found combined with feeling and elevation of character.

New Inquiries into the Laws which are observed in the Distribution of Vegetable Forms

By Baron Alexander Humboldt[1]

The numerical proportions of vegetable forms may be viewed in two perfectly distinct lights. If we consider plants, grouped together in natural families, without having regard to their geographical distribution, we enquire what are the types of organization, according to which the greatest number of species are formed. Are there more Glumaceæ than Compositæ on the globe? Do these two tribes of vegetables constitute together the fourth part of Phænogamous plants? What is the proportion of Monocotyledones to the Dicotyledones? These are questions of general Phytology—of the sciences which examine the organization and mutual connection of vegetables. If we view the species which we associate according to the analogy of their form, not in an abstract manner, but with regard to their climacteric relations, or their distribution over the surface of the globe, the questions which arise afford an interest highly varied. What are the families of plants which predominate over the other phænogamous vegetables more within the torrid zone than under the polar circle? Are the Compositæ more numerous, either in the same geographical latitude, or on the same isothermal band, in the new Continent than in the old? Do the types which predominate less in advancing from the | 274 | equator to the pole, follow the same law of decrease, in proportion as they rise toward the summit of the equatorial mountains? Do not the mutual proportions of families vary, on isothermal lines of the same denomination, in the temperate zones, to the north and to the south of the equator? These questions belong to the geography of plants, properly so called: they are connected with the most important problems belonging to meteorology and the natural history of the globe in general. Upon the preponderance of certain families of plants, depends also the character of the landscape, the aspect of a country, whether of a beautiful or majestic nature. The abundance of the Gramineæ, which form vast

[1] A separate copy of this memoir was sent to us by the author for insertion in our Journal, through Dr Marcet.

savannahs, and that of the Palms or the Coniferæ, are much influenced by the social state of the people, by their manners, and by the more or less perfect development of the economical arts.

In considering the geographical distribution of forms, we may attend to species, to genera, and to natural families, (Humboldt, *Proleg. in Nov. Gen.* vol. i. pp. xiii. li. & 33.). Often a single species of plants, especially among those which I have named social plants, covers a vast extent of country. Of this kind, in the north, are the heaths and forests of pines; in equinoctial America, the associations of Cactus, Croton, Bambusa, and Brathys of the same species. It is curious to examine the proportions of organic multiplication and development. We may demand what species in a given zone produces the greatest number of individuals;—we may point out the families to which in different climates belong the species which predominates over the others. Our imagination is peculiarly struck with the preponderance of certain plants, which we consider, on account of their easy reproduction, and the great number of individuals which present the same specific characters, as the more common plants of this or that zone. In a northern region, where the Compositæ and the Ferns are to the phænogamous plants in the proportion of 1:13, and of 1:25, (that is to say, where we find these proportions on dividing the total number of the phænogamous plants by the number of the species of the Compositæ and Ferns,) a single species of Fern may occupy ten times as much space as the whole species of Compositæ together. In this case, the Ferns predominate over the Compositæ by the | 275 | mass,—by the number of *individuals* belonging to the same species of Pteris or of Polypodium; but they do not predominate, if we compare with the total sum of the species of phænogamous plants the different forms which compose the two groups of Ferns and Compositæ. As the multiplication of all the species does not follow the same laws, since they do not all produce the same number of individuals, the quotient obtained in dividing the total number of phænogamous plants by the number of species of different families, does not of itself decide the aspect, I might almost say kind of monotony of nature in the different regions of the globe. If the traveller is struck with the frequent repetition of the same species,—with the sight of those which predominate by their mass,—he is not less so with the paucity of individuals of some other species useful to man. In the countries where the Rubiaceæ, the Leguminosæ, or the Terebinthaceæ, compose the forests, we are surprised to see how rare are the trees of certain species of Cinchona, Hæmatoxylon, and the Balsamiferæ.

In turning our attention to species, we may also, without having regard to their multiplication, and to the greater or smaller number of individuals, compare in each zone, in an absolute manner, the species which belong to different families. This interesting comparison has been made in the great work of M. Decandolle, (*Regni Vegetabilis Systema Naturæ*, vol. i. p. 128. 396. 439. 464. 510.). M. Kunth has attempted it with more than 3300 Compositæ already known up to the present day, (*Nova Genera*, vol. iv. p. 238). It does not point out what family predominates in the same degree above the other indigenous phænogamous plants, either by the mass of individuals, or by the number of species; but it presents the numerical proportions between the species of the same family belonging to a different country. The results of this method are generally more precise, because they are obtained without valuing the total mass of phænogamous plants, after being freed with care from the study of each isolated family. The forms which are the most varied, the Ferns, for example, are found under the tropics: it is in the mountainous, temperate, humid and shady parts of the equatorial regions, that the family of Ferns produces the greatest number of species. In the tem- | 276 | perate zone, there are not so many as under the tropics: their absolute number still diminishes as we advance toward the pole, but since the cold region, for example Lapland, produces species of Ferns which resist the cold better than the great mass of phænogamous plants, the Ferns, by the number of species, predominate more over the other plants in Lapland than in France or Germany. The numerical proportions presented in the table which I have published in my *Prolegomena de Distributione Geographica Plantarum*, and which appears again here perfected by the great labours of Mr Robert Brown, differ entirely from the proportions given by an absolute comparison of the species which grow in the different zones. The variation which we observe in proceeding from the equator to the poles is not consequently the same in the result of the two methods. In this, two of the fractions used by Mr Brown and myself are variable, since, in changing the latitude, or rather the isothermal zone, the total number of phænogamous plants is not seen to vary in the same proportion as the number of species which constitute the same family.

When from *species* or *individuals* of the same form which are reproduced according to constant laws, we pass to divisions of the *natural method*, we may turn our attention to genera, to families, or to sections still more general. There are some genera and some families which belong exclusively to certain zones, to a particular association of climacteric conditions; but there is a great number of genera and of families which have representatives in all zones, and at all heights. The first researches which have been made regarding the geographical distribution of forms, those of M. Treviranus, published in his ingenious work on Biology, (vol. ii. p. 47. 63. 83. 129.), have for their object the dispersion of genera over the globe. That method is less proper for presenting general results than this, which compares the number of species of each family, or the large groups of the same family, with the total mass of phænogamous plants. In the frigid zone, the variety of generie forms does not diminish in the same degree as the variety of species: we find more genera, with a smaller number of species, (Decandolle, *Théorie Elément.* p. 190.; Humboldt, *Nova Gen.* vol. i. p. xvii. & l.). It is nearly the same on the summit of the lofty mountains, | 277 | which receive colonists of a great number of genera, which we suppose to belong exclusively to the vegetation of the plains.

I have deemed it necessary to show the different points of view from which the laws of the distribution of vegetables may be seen. It is in confounding them that we think the contradictions are to be found, which are not otherwise than apparent, and which are erroneously attributed to the uncertainty of observations. (*Berliner Jahrbücher der Gewächskunde*, Bd. i. p. 18. 21. 30.) When the following expressions are used: "this form or this family loses itself toward the frigid zone;" "it has its true native country in such and such a parallel;" "it is a southern form;" "it abounds in the temperate zone;" we must expressly mention, if we consider the absolute number of species, the increase or decrease of their absolute frequency with the latitudes, or if we speak of families which predominate in the same degree over the rest of the phænogamous plants. These expressions are correct: they afford a precise signification, if we distinguish the different methods according to which we consider the variety of forms. The Island of Cuba (to use an analogous case taken from political economy) contains a much greater number of individuals of the African than of the Martinique race; and yet the mass of these individuals predominates much more over the number of whites in this latter island than in that of Cuba.

The rapid progress which the geography of plants has made within these twelve years, by the united labours of Messrs Brown, Wahlenberg, Decandolle, Leopold de Buch, Parrot, Ramond, Schouw and Hornemann, are owing in a great measure to the advantages of the natural method of M. de Jussieu. In following, I shall not say the artificial classifications of the sexual system, but the families founded upon vague and erroneous principles, (*Dumosœ*, *Corydales, Oleraceœ*,) we no longer perceive the great physical laws in the distribution of vegetables on the globe. It was Mr Robert Brown, who, in a celebrated memoir on the vegetation of

New Holland, first made known the true proportions between the great divisions of the vegetable kingdom, the Acotyledonous, Monocotyledonous, and Dicotyledonous plants. (Brown, in *Flinders' Voyage to Terra* | 278 | Australis, vol. ii. p. 538; *and Observations systematic and geographical on the Herbaries of the Congo* , p. 3.) I made an attempt, in 1815, to pursue this kind of research, in extending it to the different orders or natural families. The natural history of the globe is, in its numerical elements, like the system of the world, and can be brought to perfection only by the joint efforts of botanical travellers, to discover the true laws of the distribution of vegetables. The collection of facts is not of itself sufficient: in order to obtain the most accurate approximations, (and we do not pretend to give any thing but approximations,) the different circumstances under which the observations have been made must be discussed. I think with Mr Brown, that we ought to prefer in general to calculations made upon incomplete lists of all the plants published, the examples taken from countries of considerable extent, and whose Flora is well known, such as France, England, Germany, and Lapland. It would be desirable to have still a complete Flora of two countries of 20,000 square leagues, destitute of lofty mountains and of platforms, and situated between the tropics in the Old and in the New Worlds. Until this shall be accomplished, we must be contented with the great herbaries formed by travellers, who have resided for some time in the two hemispheres. The habitations of plants are so vaguely and incorrectly pointed out, in the vast compilations known under the names of *Systema Vegetabilium*, and *Species Plantarum*, that it would be very dangerous to use them in an absolute manner. I have not employed these lists otherwise than in a subsidiary manner, to control and modify a little the results obtained by the Floras and the partial herbaries. The number of equinoctial plants which M. Bonpland and I have brought to Europe, and of which our learned colleague M. Kunth will have soon finished the publication, is perhaps numerically greater than any of the herbaries formed between the tropics; but it is composed of the vegetables of the plains and elevated platforms of the Andes. The alpine plants are even much more considerable than in the Floras of France, of England, and of the Indies, which associate also the productions of different climates belonging to the same latitude. In France, the number of species which vegetate exclusively at above 500 toises of height, does not appear to be more | 279 | than 1/9th of the entire mass of phænogamous plants. (Decandolle, in the *Mem. de Arcueil*, vol. iii. p. 295.).

It will be useful to consider at a future period the vegetation of the tropics and that of the temperate region between the parallels of 40° and

50°, according to two different methods, either in searching the numerical proportions in the whole of the plains and the mountains, which nature presents over a great extent of country, or in determining these proportions in the plains alone of the temperate zone and of the torrid zone. As our herbaries are the only ones that point out, according to barometrical measurement, in more than 4000 plants of the equinoctial region, the height of each station above the level of the sea, the numerical proportions of the table which I have already published may be rectified, when our work, the *Nova Genera*, shall be finished, by taking away from the 4000 phænogamous plants which M. Kunth has employed in this work (*Prolegom*. p. 16.), the plants which grow at above 1000 toises, and by dividing the total number of plants which are not alpine, of each family, by that of plants which live in the cold and temperate regions of equinoctial America. This mode of proceeding should affect more, as we shall show by and bye, the families which abound in alpine species, for example, the Gramineæ and the Compositæ. At 1000 toises of elevation, the mean temperature of the air is still, on the back of the equatorial Andes, 17° centigr., which is equal to that of the month of July at Paris. Although on the platform of the Cordilleras, we find the same annual temperature as in the high latitudes, (because the *isothermal line* of 8°, for example, is the track marked in the plains by the intersection of the *isothermal surface* of 8°, with the surface of the earth's spheroid,) it is not too much to generalise these analogies of the temperate climates of the equatorial mountains, with the low regions of the circumpolar zone. These analogies are not so great as might be thought; they are modified by the influence of the partial distribution of the heat in the different parts of the year. (*Proleg*. p. 54., and my *Mémoire sur les Lignes Isothermes,*[2], p. 137.) The quotients do | 280 | not change, however, in ascending from the plains toward the mountains, in the same manner as they change in approaching the pole: this is the case with the Monocotyledones, considered in a general view, as well as with the Ferns and Compositæ. (*Proleg*. p. 51. and 52.; Brown on *Congo*, p. 5.)

It may further be remarked, that the development of vegetables of different families, and the distribution of forms, depend not on isothermal latitudes, nor on geographical latitudes alone; but that the quotients are not always similar on the same isothermal line of the temperate zone, in the plains of America and of the Old Continent. There exists, under the

[2] A translation of this valuable Memoir will be found in this *Journal*, vol. III. pp. 1, 256.; IV. pp. 23, 262.; V. p. 28.

tropies, a very remarkable difference between America, India, and the west coasts of Africa. The distribution of organic beings on the globe, depends not only on very complicated climatic circumstances, but also on geological causes, with which we are entirely unacquainted, because they are connected with the original state of our planet. The great Pachydermata are wanting at the present day in the New World, although we find them still in abundance in analogous climates, in Africa, and in Asia. In the equinoctial zone of Africa, the family of palms is far from numerous, compared with the great number of species of equinoctial America. These differences, far from deterring us from the scrutiny of the laws of nature, ought to excite us to study these laws in all their complications. The lines of equal heat are not parallel to the equator. They have, as I have tried to prove elsewhere, convex summits, and concave summits, which are distributed with great regularity over the globe, and form different systems along the eastern and western coasts of the two worlds, in the centre of continents, and in the neighbourhood of the ocean. It is probable, that when philosophical botanists have travelled over a larger extent of the globe, we shall find, that often the lines of the *maxima* of *agroupment* (the lines taken from the points where the fractions are reduced to the smallest denominator,) become isothermal lines. In dividing the globe by longitudinal bands comprehended between two meridians, and in comparing the numerical proportions under the same isothermal latitudes, we perceive the existence of different systems of *agroupment.* We can already, with the actual state of our knowledge, distinguish | 281 | four systems of vegetation, those of the New Continent, of Western Africa, of India, and of New Holland. Since, notwithstanding the regular increase of the mean heat from the pole to the equator, the *maximum* of heat is not identical in the different regions at different degrees of longitude, there exist also places where certain families attain a more perfect development than at any other: this is the case with the family of Compositæ in the temperate region of North America, and especially at the southern extremity of Africa. These partial accumulations determine the physiognomy of the vegetation, and are what we call vaguely the characteristic features of the landscape.

In the whole temperate zone, the Glumaceæ and Compositæ form together more than the fourth part of the phænogamous plants. We find, from the same inquiries, that the forms of organised beings have a mutual dependence. The unity of nature is such, that the forms are universally limited according to constant and immutable laws. When we know at any point of the globe the number of species which a great family presents, (for example, that of the Glumaceæ, the Compositæ, or the Leguminosæ,) we can estimate with much probability both the total number of phænogamous plants, and the number of species which compose the other vegetable families. It is thus, that, on knowing the number of Cyperaceæ or of Compositæ in the temperate zone, we can form an estimate of that of the Gramineæ or Leguminosæ. These estimates enable us to see in what tribes of vegetables the Floras of a country are still deficient: they are so far from being uncertain, as to enable us to avoid confounding the quotients which belong to the different systems of vegetables. The labour which I have bestowed upon plants, will no doubt one day be applied with success to the different classes of vertebral animals. In the temperate zones, there are nearly five times as many birds as mammalia, and the latter increase much less toward the equator than the birds and reptiles.

The geography of plants may be considered as a part of the natural history of the globe. If the laws which nature has followed in the distribution of vegetable forms, should prove to be more complicated than they appear at first sight, still, we ought not on this account to be deterred from submitting them to ac- | 282 | curate investigation. We do not relinquish the tracing of a map, when we perceive the sinuosities of rivers, and the irregular form of coasts. The laws of magnetism become intelligible to him who has commenced with tracing the lines of equal inclination and declination, and who has compared a great number of observations, which, at first sight, might seem contradictory. He who thinks that it is not yet time to search the *numerical elements* of the geography of plants, forgets the progressive march by which the physical sciences have elevated themselves to determinate results. In examining a complicated phenomenon, we commence with a general scrutiny of the circumstances by which it is determined or modified; but, before discovering certain proportions, we find, that the first results to which we attend, are not sufficiently free from local influence: it is then that we modify and correct the numerical elements, and discover the regularity in the very effects of partial disturbances. Criticism exercises itself on whatever has been prematurely announced as a general result; and the spirit of criticism once excited, becomes favourable to the investigation of truth, and accelerates the progress of human knowledge.

Acotyledones. Cryptogamous plants (fungi, lichens, mosses, and ferns); *Agames celluleuses et vasculaires* of M. Decandolle. On uniting the plants of the plains with those of the mountains, we have found them to be under the tropics 1/9; but their number ought to be much greater. Mr Brown has rendered it very probable, that, in the torrid zone, the proportion is in the plains 1/15, on the mountains 1/5[3]. (*Congo*, p. 5.) In the temperate zone, the agamous plants are generally to the phænogamous as 1:2; in the frigid zone they attain the same number, and often exceed it considerably.

On dividing the agamous plants into three groups, we observe that the ferns are more frequent (the denominator of the fraction being less) in the frigid zone than in the temperate zone, (*Berl. Jahrb.*, bd. i. p. 32.), and the lichens and mosses also in- | 283 | crease towards the frigid zone. The geographical distribution of ferns depends upon the union of the local circumstances of shade, humidity, and moderate heat. Their *maximum* (that is to say, the place where the denominator of the normal fraction of a group becomes the least possible,) is found in the mountainous parts of the tropics, particularly in islands of small extent, where the proportion rises to 1/3, and upwards. When the plains and mountains were not separated, Mr Brown found the ferns of the torrid zone to be 1/20. In Arabia, India, New Holland, and Western Africa, (between the tropics,) they are 1/26: our herbaries of America do not give more than 1/38; but the ferns are rare in the great valleys and on the dry platforms of the Andes, where we were forced to remain a long time. (*Congo*, p. 43., and *Nova gen.*, vol. i. p. 33.) In the temperate zone, the ferns are 1/70; in France, 1/75; in Germany, according to late inquiries, 1/71. (*Berl. Jahrb.*, b. i. p. 26.) The group of ferns is extremely rare on the Atlas mountains, and almost completely disappears in Egypt. In the frigid zone, the ferns appear to rise to 1/25.

Monocotyledones. The denominator becomes progressively smaller in proceeding from the equator toward the 62d degree of North Latitude; it increases again in the regions still farther north, on the coast of Greenland, where the Gramineæ are very scanty. (*Congo*, p. 10.) In the different parts of the tropics, the proportion varies from 1/5 to 1/6. In 3880 phænogamous plants of equinoctial America, found by M. Bonpland and me, in flower and in fruit, there are 654 monoctyledonous, and 3226 dicotyledonous; hence the great division of the Monocotyledones would be 1/6 of the phænogamous plants. According to Mr Brown, the proportion in the Old Continent (in India, equinoctial Africa, and New Holland,) is 1/5; in the temperate zone we find 1/4. (France, 1:4 5/5; Germany, 1:4 1/2; North America, according to Pursh, 1:4 1/2; Kingdom of Naples, 1:4 1/5; Switzerland, 1:4 1/4; British Isles, 1:3 5/4); in the frigid zone, 1/3.

[3] In this article, the fractions 1/9, 1/15, 1/5, indicate the proportions between the species of a family and the total number of Phænogamous plants which vegetate in the same country. The abbreviations *Trop. Temp. Frig.*, signify *Tropies* or *Torrid Zone, Temperate Zone, Frigid Zone.*

Glumaceæ. (The three families of Junceæ, Cyperaceæ, and Gramineæ, united). = *Trop., 1/11. Temp., 1/8. Frig., 1/4.* | 284 |

The increase toward the north is owing to the Junceæ and Cyperaceæ being very rare compared with the other Phænogamous plants, in the temperate and torrid zones. On comparing the species belonging to the three families, we find that the Gramineæ, the Cyperaceæ, and the Junceæ, are under the tropics, as 25, 7, 1; in the temperate region of the Old Continent, as 7, 5, 1; in the polar circle, as 2 2/5, 2 5/5, 1. In Lapland the Gramineæ and Cyperaceæ are equal: toward the equator the Cyperaceæ and Junceæ diminish much more than the Gramineæ; the junceal form disappears almost entirely under the tropics, (*Nova Gen.* vol. i. p. 240.).

Junceæ alone. = Trop., 1/400. Temp., 1/90. Frig., 1/25, (Germany, 1/94; France, 1/86.)

Cyperaceæ alone. = Trop. America, nearly 1/57; Western Africa, 1/18; India, 1/25; New Holland, 1/14. (*Congo*, p. 9). Temp., probably 1/20. (Germany, 1/16; France, according to the works of M. Decandolle, 1/27; Denmark, 1/16.) Frig. 1/9. This is the proportion found in Lapland, and as far as Kamtschatka.

Gramineæ alone. Trop. I have allowed as much as 1/15. Mr Brown found in Western Africa, 1/12; in India, 1/12. (*Congo* , p. 41.) M. Hornemann fixed this part of Africa at 1/10. (*De Indole Plant. Guineensium*, 1819, p. 10.) Temp. Germany, 1/15; France, 1/15. Frig. 1/10.

Compositæ. On blending the plants of the plains with those of the mountains, we have found in equinoctial America 1/6 and 1/7; but, of 534 Compositæ of our herbaries, there are only 94 which grow to 500 toises above the plains, (the height at which the mean temperature is still 21°.8; equal to that of Cairo, of Algiers, and of the Island of Madeira.) From the equatorial plains to 1000 toises of height (where we have still the mean temperature of Naples), we have collected 265 Compositæ. This last result gives the proportion of Compositæ, in the regions of equinoctial America, beneath 1000 toises, from 1/9 to 1/10. This result is very remarkable, because it proves, that between | 285 | the tropics in the lowest and warmest region of the New Continent, there are fewer Compositæ, in the subalpine and temperate regions more, than under the same circumstances in the Old World. Mr Brown found on the Rio-Congo, and in SierraLeone, 1/25; in India and New Holland 1/16. (*Congo*, p. 26; *Nova Gen.* vol. iv. p. 239.) In the temperate zone, the Compositæ are in America, 1/6, (this is probably also in equinoctial America the proportion of the Compositæ of the highest mountains to the whole mass of alpine phænogamous plants); at the Cape of Good Hope, 1/5; in France, 1/7 (correctly 2/15); in Germany 1/8. Under the frigid zone the Compositæ are, in Lapland, 1/13; in Kamtschatka, 1/13. (*Hornemann*, p. 18.; *Berl. Jahrb*. b. i. p. 29.)

Leguminosæ. = Trop. America, 1/12; India, 1/9; New Holland, 1/9; Western Africa, 1/8, (*Congo*, p. 10.) Temp. France, 1/16; Germany, 1/20; North America, 1/19; Siberia, 1/14, (*Berl. Jahrb*. b. i. p. 22.) Frig. 1/35.

Labiatæ. = Trop. 140. Temp. North America, 1/40; Germany, 1/26; France, 1/24. Frig. 1/70. The rarity of Labiatæ and Cruciferæ in the temperate zone of the New Continent is a very remarkable phenomenon.

Malvaceæ. = Trop. America, 1/47; India and Western Africa, 1/54, (*Congo*, p. 9.); on the coast of Guinea alone, 1/20, (*Hornemann*, p. 20.) Temp. 1/200. Frig. 0.

Cruciferæ. Almost wanting under the tropics, on taking away the mountains to within from 1200 to 1700 toises. (*Nova Gen*. p. 16.) France, 1/19; Germany, 1/18; North America, 1/62.

Rubiaceæ. Without dividing the family into sections, we find beneath the tropics, in America, 1/29; in Western Africa, 1/14: under the temperate zone in Germany, 1/70; in France, 1/73: under the frigid zone, in Lapland, 1/80. Mr Brown separates the great family of Rubiaceæ into two groups, which present very distinct climatic proportions. The group of Stellatæ without interposed stipules, belong chiefly to the temperate | 286 | zone: it almost disappears between the tropics, excepting at the summit of the mountains. The group of Rubiaceæ with opposite leaves and stipules, belong very peculiarly to the equinoctial region. M. Kunth has divided the great family of Rubiaceæ into eight groups, one of which, that of the Coffeaceæ, forms in our herbaries a third part of the whole Rubiaceæ of equinoctial America. (*Nov. Gen*. vol. iii. p. 341.)

Euphorbiaceæ. = Trop. America, 1/35; India and New Holland, 1/30; Western Africa, 1/28. (*Congo*, p. 25.) Temp. France, 1/70; Germany, 1/100. Frig. Lapland, 1/500.

Ericeæ and Rhododendra. = Trop. America, 1/130. Temp. France, 1/125; Germany, 1/90; North America, 1/36. Frig. Lapland, 1/25.

Amentaceæ. = Trop. America, 1/800. Temp. France, 1/50; Germany, 1/40; North America, 1/25. Frig. Lapland, 1/20.

Umbelliferæ. = almost none under the tropics to the height of 1200 toises; but, on taking both the plains and high mountains in equinoctial America, 1/100; under the temperate zone much more numerous in the Old than in the New Continent. France, 1/34; North America, 1/57; Lapland, 1/60.

On comparing the two worlds, we find in general in the New World, under the equatorial zone, fewer Cyperaceæ and Rubiaceæ, and more Compositæ; under the temperate zone, fewer Labiatæ and Cruciferæ, and more Compositæ, Ericæ, and Amentaceæ, than in the corresponding zones of the Old World. The families which increase from the equator toward the pole (according to the *fractional method*), are the Glumaceæ, the Ericeæ, and the Amentaceæ. The families which decrease from the pole toward the equator are the Leguminosæ, the Rubiaceæ, the Euphorbiaceæ, and the Malvaceæ. The families which appear to attain the *maximum* under the temperate zone, are the Compositæ, the Labiatæ, the Umbelliferæ, and the Cruciferæ.

I have thrown together the principal results of this work in one table; but I enjoin naturalists to have recourse to illustra- | 287 | tions of the several families, whenever any of the numbers seem doubtful. The quotients of the tropics are modified in such a manner that they are proportioned to regions whose mean temperature is from 28° to 20°, (from the level of the sea to 750 toises of height). The quotients of the temperate zone are adapted to the central part of that zone, between 13° and 10° of mean temperature. In the frigid zone, the mean temperature is from 0° to 1°. To this table of quotients or fractions, which indicate the proportions of each family to the total mass of phænogamous plants, might be added another, in which the absolute number of species might be compared. We here present a fragment which comprehends only the temperate and frigid zones.

	France.	**North America.**	**Lapland.**
Glumaceæ, – –	460	365	124
Compositæ, – –	490	454	38
Leguminosæ, – –	230	148	14
Cruciferæ, – –	190	46	22
Umbelliferæ, –	170	50	9
Caryophylleæ, –	165	40	29
Labiatæ, – –	149	78	7

(*continued*)

(*continued*)

Rhinantheæ, – –	147	79	17
Amentaceæ, – –	69	113	23

These absolute numbers are taken from the works of Messrs Decandolle, Pursh, and Wahlenberg. The mass of plants described in France is to that of North America in the proportion of 1 1/2:1; to that of Lapland in the proportion of 7:1.[4] | 288 |

Groups, founded on the Analogy of Forms.	Proportions to the whole mass of Phænogamous Plants. Equatorial Zone, Lat. 0°—10°.	Temperate Zone, Lat. 45°—52°.	Frigid Zone, Lat. 67°—70°.	Signs indicating the direction of Increase.
Agamous Plants, Ferns, Lichens, Mosses, Fungi,	Plains, – – – 1/15 Mountains, – – – 1/5	1/2	1/1	↗
Ferns alone.	Country slightly Mountainous, 1/20 Country very Mountainous, 1/3—1/8	1/70	1/25	← →
Monocotyledonous Plants.	Old Continent, – – 1/5 New Continent, – – 1/6	1/4	1/3	↗
Glumaceæ, Junceæ, Cyperaceæ, Gramineæ,	1/11	1/8	1/4	↗
Junceæ alone.	1/400	1/90	1/25	↗
Cyperaceæ alone.	Old Continent, – – 1/22 New Continent, – – 1/50	1/20	1/9	↗
Gramineæ alone.	1/14	1/12	1/10	↗
Compositæ.	Old Continent, – – 1/18 New Continent, – – 1/12	Old Continent, – 1/8 New Continent, – 1/5	1/13	→ ←
| 289 | Leguminosæ.	1/10	1/18	1/35	↙

(*continued*)

[4] A series of additional observations on this subject, so highly interesting to the philosophical botanist and the geologist, also by Baron Humboldt, will appear in our next Number.

(*continued*)

Rubiaceæ.	Old Continent, – – 1/14 New Continent, – – 1	1/60	1/80	↙
Euphorbiaceæ.	1/32	1/80	1/500	↙
Labiatæ.	1/40	America, – – 1/40 Europe, – – 1/25	1/70	→ ←
Malvaceæ.	1/35	1/200	0	↙
Ericeæ and Rhododendra.	1/130	Europe, – – 1/10 America, – – 1/36	1/25	↙
Amentaceæ.	1/800	Europe, – – 1/45 America, – – 1/25	1/20	↙
Umbelliferæ.	1/500	1/40	1/60	→ ←
Cruciferæ.	1/800	Europe, – – 1/18 America, – – 1/60	1/24	→ ←

Explanation of the Signs: ↗ The denominator of the fraction diminished from the Equator toward the North Pole: ↙ The denominator diminished toward the Equator: ← → The denominator diminished toward the Equator and toward the North Pole: → ← The denominator diminished from the North Pole, and from the Equator toward the Temperate Zone.

New Inquiries into the Laws which are observed in the Distribution of Vegetable Forms. By Baron Alexander Humboldt.[5] (Continued from Vol. VI. p. 289.)

It is with the distribution of organic beings as with all the other phenomena of the physical world. In the midst of apparent disorder, which seems to arise from the influence of a multitude of local causes, we discover the immutable laws of Nature, as soon as we cast our eyes over a great extent of country, or employ a mass of facts with which the partial disturbances mutually agree. I have had the satisfaction to see this work completed by a minute examination made in Germany, England, Italy, and of late in Denmark. One of the greatest botanists of our times, and indeed of

[5] This continuation is taken from a Memoir read before the Academy of Sciences on the 19th February 1821.

any age, Mr Robert Brown, has compared each of the numerical results with those given by the rich herbaries which he has had the opportunity of consulting. Many of the numbers have been rectified, while the others have been confirmed by an agreement some- | 48 | what unexpected. The mass of facts is increased by the very desire of invalidating or corroborating the results to which I have applied myself. It is thus that, in the progress of the physical sciences, general ideas, which at first have only been deduced from a small number of facts, compel observers to multiply the partial results. Enriched by these materials, profiting daily by whatever truth or utility is contained in the most severe criticism of my works, I have been enabled to give to the numerical results of which the table of vegetable forms consists, a degree of exactness which I have not till now been able to attain. It is in the nature of these researches, that we are not able to rectify the *co-efficients* otherwise than progressively, in proportion as the observations accumulate. I have attended here only to the general development of principles. As this kind of botanical arithmetic demands more minute discussions of the proportions of each family of vegetables to the whole mass of phænogamous plants, I have thrown together these discussions in the notes which I have published separately.[6]

We may foresee, that the labour which I have bestowed on the families of plants, will one day be applied with success to most of the classes of vertebral animals. The immense collections which are to be seen at Paris, in the Museum of Natural History, show that we are already acquainted with nearly 56,000 species of Plants, cryptogamous and phænogamous; 44,000 Insects , 2500 Fishes; 700 Reptiles; 4000 Birds; and 500 species of Mammalia. Agreeably to the inquiries which M. Valenciennes and I have made, there exist in Europe alone nearly 80 Mammalia, 400 Birds, and 30 Reptiles: it follows, that, in this northern temperate zone, there are five times as many species of Birds as of Mammalia, as there are (in Europe) five times as many Compositæ as Amentaceæ and Coniferæ; five times as many Leguminosæ as Orchideæ and Euphorbiaceæ. The beautiful collections lately brought from the Cape of Good Hope by M. De Lalande prove, (if we compare them with the works of MM. Temminik and Le Vaillant), that in that part of the southern temperate zone, the Mammalia are to the Birds in the | 49 | proportion of 1:4,3. Such an agreement between two opposite zones is abundantly striking. The Birds, and especially the Reptiles, increase towards the

[6] See *Dictionnaire des Sciences Naturelles*, arranged by the Professors of the Garden of Plants, vol. xviii. p. 423,—436.

equatorial zone in a much more considerable degree than the Mammalia. According to the discoveries of M. Cuvier regarding fossil bones, we may be induced to believe, that these proportions have not been the same in all ages, and that, amid the ancient catastrophes of our planet, a much greater number of Mammalia has been lost than of Birds. M. Latreille, in an excellent memoir on the Geographical Distribution of Insects, has not compared the number of articulated animals with the number of plants, and with that of the different classes of vertebral animals which inhabit the same climates; but he has exhibited, in an interesting manner, a view of the increase or diminution of the particular groups of insects, in advancing from the pole towards the equator. I pass over in silence the laborious researches of M. Illiger on the Geography of Birds.[7] The author has discussed the habitation of upwards of 3800 species; but he has contented himself with viewing them according to their distribution among the five divisions of the world,—a method by no means philosophical, and altogether improper for investigating the influence of climate over the development of organized beings. All the continents, with the exception of Europe, extend from the temperate to the equatorial zone: the laws of Nature cannot, therefore, manifest themselves, when we group the phenomena according to divisions which are arbitrary, and which depend, so to speak, simply upon the difference of meridians. It is not my intention to push to a greater extent these considerations regarding the numerical proportions of animals of different classes. I am satisfied with having called the attention of the learned to a branch of natural philosophy which has appeared to me very worthy of investigation. We conceive how, in a given space of country, the *individuals* belonging to different tribes of plants and of animals may be *numerically limited*,—how, after an obstinate struggle, and long wavering, a state of equilibrium is esta- | 50 | blished, which results from the necessity of food, and the habitudes of life; but the causes which have *limited the forms*, are hidden under the impenetrable veil which conceals from our sight all that extends from the origin of things to the first development of organic life.

On examining in detail all that we already know regarding the proportion of monocotyledonous to dicotyledonous plants, we find that the denominator becomes gradually smaller (and this with a very great degree of regularity), in proceeding from the equator toward the 62d degree of N. Lat. It increases probably anew in regions still more northern, on the coast of Greenland, where the Gramineæ have become very rare

[7] *Memoirs of the Academy of Berlin* for the years 1812 and 1813, p. 221,— 237.

(*Congo*, p. 4.). The proportion varies from 1/5 to 1/6 in the different parts of the tropics. In upwards of 3880 phænogamous plants of Equinoctial America, found by M. Bonpland and myself in flower and in fruit, there were 654 monocotyledones, and 3226 dicotyledones: hence the great division of monocotyledones would be 1/6 of the phænogamous plants. According to Mr Brown, the proportion is generally in the Old Continent (in India, Equinoctial Africa, and New Holland,) 1/5.

In the temperate zone, we find (according to my *Proleg.* p. xii. and the partial results published by M. Decandolle, *Dict. des Sciences Nat.* vol. xviii. p. 594,—597.) that the monocotyledones are to the dicotyledones,

In Barbary, – – = 1:4,8
Egypt. – – = 1:5,0
Caucasus and Crimea, = 1:6,0
Kingdom of Naples, = 1:4,7
The State of Venice, = 1:4,0
In France, – – – = 1,4,7
Germany, – – = 1:4,0
Switzerland, – – = 1:4,3
The British Isles, – = 1:3,6
North America, – = 1:4,6

In the frigid zone, the proportion is,

In Lapland, – – – = 1:2,8
Iceland, – – = 1:2,8

We see, that, from the tropics to the pole, the relative increase of the monocotyledones is very regular. As the monocotyledones affect moisture, they are more numerous in the British Isles, and more rare in Egypt and on the dry mountains of Caucasus. I have already observed, that in the Alps of Switzerland, up to the region of the Rhododendra, the monocotyledones are to the | 51 | phænogamous plants as 1:7, while in the plains, they are, at the foot of the Alps, as 1:4,3, (*Prolegomena*, p. lii.)

In the most fertile part of Europe, at the centre of the temperate zone, an extent of country of 30,000 square leagues produces about 6000 species of plants, of which 2200 are acotyledonous or cryptogamous, and 3800 phænogamous. Among the last there are nearly 500 Compositæ, 300 Gramineæ, (exclusive of the Cyperaceæ and Junceæ), 250 Leguminosæ, and 200 Cruciferæ; but only 70 Amentaceæ, 50 Euphorbiaceæ, and 25 Malvaceæ. The great families constitute from 1/7 to 1/20, the small ones about 1/50 of the total mass of phænogamous plants. This is, so to speak, the mean state of the vegetation in Europe, in the fertile countries, between 42° and 50° of N. Lat. To convince the most incredulous of the reality of *fixed proportions*, or of the regularity to be observed in Europe in the distribution of forms in a particular zone, I shall here present the proportions given by two neighbouring countries, France and Germany. The figures marked in the following Table may be regarded as the coefficients of each family; for, on multiplying the number of phænogamous plants of the temperate zone of Europe by 0.076 or 0.053, the number of species which compose the families of Gramineæ or Cruciferæ may be found.

Compositæ,
Fr. 1/7.4=*0.135*
Ger. 1/8=*0.125*
↙

Glumaceæ,
Fr. 1/7.9=*0.127*
Ger. 1/7.1=*0.141*
↗

Gramineæ alone,
Fr. 1/13=*0.077*
Ger. 1/13=*0.077*
↗

Leguminosæ,
Fr. 1/16=*0.063*
Ger. 1/18=*0.056*
↙

Cruciferæ,
Fr. 1/19=*0.052*
Ger. 1/18=*0.056*
↙

Umbelliferæ,
Fr. 1/21=*0.048*
Ger. 1/22=*0.046*
↙

Labiatæ,
Fr. 1/24=*0.042*

Ger. 1/26=*0.038*
↙

Cyperaceæ alone,
Fr. 1/27=*0.037*
Ger. 1/18=*0.056*
↗

Amentaceæ,
Fr. 1/50=*0.020*
Ger. 1/40=*0.025*
↗

Orchideæ,
Fr. 1/67=*0.015*
Ger. 1/43=*0.023*
↙

Boragines,
Fr. 1/74=*0.014*
Ger. 1/72=*0.014*
↗

Rubiaceæ.
Fr. 1/75=*0.014*
Ger. 1/70=*0.014*
↙

Euphorbiaceæ,
Fr. 1/70=*0.014*
Ger. 1/100=*0.010*
↙

Junceæ,
Fr. 1/85=*0.012*
Ger. 1/94=*0.011*
↗

Ericeæ,
Fr. 1/125=*0.008*
Ger. 1/90=*0.011*
↗

Malvaceæ,
Fr. 1/140=*0.007*

Ger. 1/250=*0.004*
↙

Coniferæ,
Fr. 1/192=*0.005*
Ger. 1/269=*0.004*
↗ | 52 |

This agreement in the greater part of the results is the more striking, that the coefficients have been obtained from very unequal masses of plants. In France 3645, in Germany only 1844 phænogamous plants have been employed, in order to determine the particular proportions of the families. Since the two countries border on each other, the species should in like manner be much the same. The agreement of the results within limits thus marked (the greater part commonly within 1/8 of difference), prove two facts equally remarkable: 1st, That the 1700 to 1800 species of phænogamous plants which the French catalogue has more than the excellent catalogue of M. Schrader employed for Germany, are distributed among the different families almost in the same proportions which we observe among the plants common to the two countries: 2d, That the species of Leguminosæ, of Cruciferæ, and of Umbelliferæ, which Germany would appear to possess exclusively, are substituted in France by a nearly equal number of species belonging to the same families. In general, where the greatest discrepancies are observed, they may be attributed to the circumstance, that Germany is situate more to the north than France is. We know that the Cyperaceæ and the Ericeæ increase so rapidly towards the north pole, that, while in the temperate zone the Cyperaceæ are 1/20, and the Ericeæ 1/100, in the frigid zone there is 1/9 of Cyperaceæ, and 1/25 of Ericeæ. On the other hand, the proportions of the Orchideæ, of the Malvaceæ, and of the Euphorbiaceæ, increase with equal rapidity towards the south. On comparing the preceding table with the table of the three zones, (the torrid, the temperate, and frigid), we observe the same laws. I have added to this comparative table of the vegetation of France and Germany the arrows which, in the general table, indicate the *directions of increase*, from the pole to the equator, and from the equator to the pole. It is a very remarkable circumstance, also, that the coefficients of the families do not change so much, if, instead of examining extensive countries which have from 2600 to 3800 species of phænogamous plants, we confine our researches to an extent of some square leagues; for example, to the Flora of Berlin, which, according to the work of M. Kunth, comprehends only 900 species. In that small extent of land, the Leguminosæ are 1/19, | 53 | (in the whole of France 1/16, in the whole of Germany 1/18); the Glumaceæ 1/6, (in France 1/7.9, in Germany 1/7,) of the whole mass of phænogamous plants.

In the same way that the *system of climates* of the New Continent differs essentially from that of the Old, on account of the unequal distribution of the heat among the different parts of the year, so also the *system of agroupment* of American plants has its peculiar features. It is to the new researches of *botanical arithmetic* that we owe the knowledge of these contrasts between the temperate zones of the two worlds. I have thrown together in the following Table the results of the American Flora of Pursh and of the French Flora of M. Decandolle. I have added certain coefficients of the European frigid region, in order to show how much of a *boreal character* the American temperate region presents in the five families of Ericeæ (and Rhododendra), Coniferæ, Amentaceæ, Umbelliferæ, and Labiatæ.

Temp. America.	**France.**	**Lapland.**	
Compositæ,	1/6	1/7	0
Glumaceæ,	1/8	1/7.9	0
Gramineæ alone,	1/10	1/15	0
Junceæ alone,	1/125	1/85	0
Cyperaceæ alone,	1/40	1/27	0
Cruciferæ,	1/62	1/19	0
Leguminosæ,	1/19	1/16	0
Malvaceæ,	1/125	1/140	0
Labiatæ,	1/40	1/24	1/70
Ericeæ and Rhododendra,	1/36	1/125	1/25
Umbelliferæ,	1/57	1/20	1/25
Amentaceæ,	1/25	1/50	1/51
Coniferæ,	1/103	1/200	1/160

The differences which are manifested in this table between the two continents, bear, not merely upon the five last families, which might be called *boreal forms*, but also upon the Cruciferæ, the Junceæ, and the Cyperaceæ, which are equally rare in the torrid zone and in the temperate zone of the New Continent.

It is conceived that the inquiries regarding the numerical proportions of vegetable families, will present results much more interesting when the Floras of different countries shall be circumscribed within more precise geographical limits, and when botanists shall attend more particularly to the principles according to which varieties and species ought to be distinguished. The catalogues which we see under the vague name of

Flora of the United States of America , comprehend countries placed in very different climates, from 18° to 9° of mean temperature. Here we have a difference of climate as great as in Europe be- | 54 | tween Calabria and Austria. When the vegetation of South Carolina, of Pennsylvania, and of New England, shall be described individually, and with the same degree of exactness, we shall notice a regular increase and decrease in the numerical proportions of families from south to north. At present, we only know the *general mean* of these partial proportions. Many countries seem richer in plants, because the botanists have inconsiderately elevated varieties to the rank of species. In another country, the travellers often neglect the plants which they believe to be the same as those of their own country. But when we attend to the great divisions, and when the number of species which we compare is very considerable, our researches are favoured with happy compensations. It is thus that the new Floras, much more complete of America and Lapland, published by Messrs Pursh and Wahlenberg, have not sensibly altered the numerical proportions which we find on confining ourselves to the old Floras of Michaux and Linnæus, (*Berl. Jahrb. der Gew*. b. i. s. 24.). Whatever may be the corrections which shall be made in my work, I am persuaded beforehand, that the more that exact observations are accumulated, and the more we look beyond the same hemisphere, the same *system of agroupment*, the partial variations of the coefficients will be found not to be made by sudden starts, but according to invariable laws. It may be that the *tropical proportion* of the Malvaceæ is 1/32 or 1/33, instead of 1/35, but it is not the less certain, that the Leguminosæ and the Malvaceæ increase toward the equator, while the Junceæ and the Ericeæ increase toward the pole. One may recall in doubt the *quantities* of the variations, the *rapidity* of the increase, but not its *direction*.

On comparing the coefficients which belong to the same families, in different zones, we find, in the rapidity of the increase, contrasts strongly marked. In the Old Continent, the proportions of the Gramineæ, the Leguminosæ, and the Euphorbiaceæ, change much less from the temperate zone to the equator, than from the temperate zone to the pole.

Those who are accustomed to consider each phenomenon in an absolute and irrelative point of view,—who regard the mean temperature of places, the laws which are observed in the variations of terrestrial magnetism, with the proportions between births | 55 | and deaths, as daring hypotheses, and as vague, theoretical speculations,—will probably look with contempt on the discussions which form the principal object of this memoir. Those, on the other hand, who delight to contemplate the mutual concatenation of organized beings,—who are aware that the numerical results correct themselves by the accumulation and the careful investigation of particular facts,—will cherish a kind of research, which throws light upon the economy of Nature,—on the connection which is observed between the climates and the form of beings,—on the distribution of plants and of animals in the different regions of our planet. It is only by the numerical examination and the comparison of species, that we can form a correct idea of the state of the vegetation of a given country;—of the general influence which the temperature exercises over the frequency of certain forms, near the equator, in the mean parallel, and towards the polar circle;—of the characteristic features which distinguish, in isothermal zones, the two systems of agroupment of the Old and the New World.[8]

---

[8] For the use of such as have not made descriptive botany their particular study, and who are yet desirous of acquiring a knowledge of the investigations which have been attempted in the different branches of the natural sciences, we shall add in this place the names of some of the most common plants which characterise, if we may so speak, the tribes or families which are frequently the subject of discussion in this memoir. (The translation of the following is modified so as to give an English reader an idea of the families). *Junceæ* (Rushes); *Cyperaceæ* (Hard or Moor Grasses, Cotton-Grass); *Gramineæ* (Corn, Grasses); *Compositæ* (Dandelion, Thistles, Sunflower); *Leguminosæ* or *Papilionaceæ* (Vetches, Pease, Clover); *Rubiaceæ* (Rennet, Madder); *Euphorbiaceæ* (Sun-spurge, Dogs' Mercury); *Labiatæ* (Mint, Thyme, Rosemary); *Malvaceæ* (Mallows, Hollyhock); *Umbelliferæ* (Carrot, Hemlock, Chervil, Carraway); *Cruciferæ* (Mustard, Cresses, Radish, Turnip). The great mass of plants which cover the globe is divided by botanists into *Phænogamous* (those having visible flowers), and *Cryptogamous* or *Agamous* (Ferns, Lichens, Mushrooms).

Personal Narrative of Travels to the Equinoctial Regions of the New Continent, during the years 1799–1804

By Alexander de Humboldt, and Aimé Bonpland, &c. &c. London, 1821, 8vo. 2 Vols. pp. 864.

These volumes, translated by H. Maria Williams, terminate the second volume (in quarto) of M. Humboldt's personal narrative; and belong to a work so universally celebrated, that we need only say, they are, if possible, more thickly studded with pieces of valuable information and curious matter, than the parts which have preceded them.

We never take up Humboldt but he reminds us of Othello, who
—Spake of most disastrous chances,
Of moving accidents, by flood, and field;
Of hair-breadth 'scapes—
And portance in his travel's history;
Wherein of antres vast, and deserts idle,
Rough quarries, rocks, and hills whose heads
touch heaven,—
And of the cannibals that each other eat;
The Anthropophagi, and men whose heads
Do grow beneath their shoulders,

he told the marvellous stories. Our author is hardly a trace behind him; and, like the fair Desdemona, we, with greedy ear, devour up his discourse; whence, without further preface, we shall now proceed to draw for the benefit of our readers.

The natives near the cataracts or raudales of the Oroonoko, up which river M. de Humboldt made his way to a height little known to Europeans, are distinguished by several remarkable prejudices, among which, none are more fatal than those narrated in the following:— "Among the causes of the depopulation of the Raudales, I have not reckoned the small pox; that malady which, in other parts of America, makes such cruel ravages, that the natives, seized with dismay, burn their huts, kill their children, and renounce every kind of society.[1] This scourge is almost unknown on the banks of the Oronoko. What depopulates the Christian settlements is, the repugnance of the Indians for the regulations of the missions, the insalubrity of a climate at once hot and damp, bad nourishment, want of care in the diseases of children, and the guilty practice of mothers of preventing pregnancy by the use of deleterious herbs. Among the barbarous people of Guyana, as well as those of the half-civilized islands of the South Sea, young wives will not become mothers. If they have children, their offspring are exposed, not only to the dangers of savage life, but also to the dangers arising from the strangest popular prejudices. When twins are born, false notions of propriety and family honour require, that one of them should be destroyed. 'To bring twins into the world, is to be exposed to public scorn; it is to resemble rats, opossums, and the vilest animals, which bring forth a great number of young at a time.' Nay more: 'two children born at the same time cannot belong to the same father.' This is an axiom of physiology of the Salivas; and in every zone, and in different states of society, when the vulgar seize upon an axiom, they adhere to it with more stedfastness than the better informed men, by whom it was first hazarded. To avoid a disturbance of conjugal tranquillity, the old female relations of the mother, or the *mure japoicnei* (midwives,) take care, that one of the twins shall disappear. If the new-born infant, though not a twin, have any physical deformity, the father instantly puts it to death. They will have only robust and well-made children, for deformities indicate some influence of the evil spirit Ioloquiamo, or the bird *Tikitiki*, the enemy of the human race. Sometimes children of a feeble constitution undergo the same fate. When the father is asked, what is become of one of his sons, he will pretend, that he has lost him by a natural death. He will disavow an action, that appears to him blameable, but not criminal. 'The poor *mure*,'[2] he will tell you, 'could not follow us; we must have waited for him every moment; he has not been seen | 476 | again, he did not come to sleep where we passed the night.' Such is the candour and simplicity of manners, such the boasted happiness of man in the *state of nature*! He kills his son, to escape the ridicule of having twins, or to avoid journeying more slowly; in fact, to avoid a little inconvenience."

Amid the prodigality and magnificence of nature, such are the moral evils which deform the scene; and we are often compelled to leave the

[1] As the Mahas in the plains of the Missoury, according to the accounts of the American travellers, Clark and Lewis.

[2] In Tamanack *mure* signifies a child; *emuru*, a son.

author's glowing descriptions of superb landscape in the torrid zone, to vex our spirits with similar details. But, the able manner in which distant objects and remote similitudes are brought to bear on almost every subject discussed, is the great charm of this work; and we have so vast a quantity of intelligence combined with so rich a fund of amusing anecdote, that the mind never tires. It has been alleged, that Mr. H. is too prone to this sort of classification, and to theories built upon it; but however that may be in a philosophical point of view, as a popular performance, it wonderfully enhances the attractions of his narrative. He is, in truth, the very Jacques of travellers; and his way is delectable, "compounded of many simples, extracted from many objects; and, indeed, the sundry contemplation of his travels, in which his often rumination wraps him in a most humorous sadness." He morals on every thing; for example:—

"The inhabitants of Atures and Maypures, whatever the missionaries may have asserted in their works, are not more struck with deafness by the noise of the great cataracts, than the catadupes of the Nile. When this noise is heard in the plain that surrounds the mission, at the distance of more than a league, you seem to be near a coast skirted by reefs and breakers. The noise is three times as loud by night as by day, and gives an inexpressible charm to these solitary scenes. What can be the cause of this increased intensity of sound in a desert, where nothing seems to interrupt the silence of nature? The velocity of the propagation of sound, far from augmenting, decreases with the lowering of the temperature. The intensity diminishes in air, agitated by a wind, which is contrary to the direction of the sound; it diminishes also by dilatation of the air, and is weaker in the higher than in the lower regions of the atmosphere, where the number of particles of air in motion is greater in the same radius. The intensity is the same in dry air, and in air mingled with vapours; but it is feebler in carbonic acid gas, than in mixtures of azot and oxygen. From these facts, which are all we know with any certainty, it is difficult to explain a phenomenon observed near every cascade in Europe, and which, long before our arrival in the village of Atures, had struck the missionary and the Indians. The nocturnal temperature of the atmosphere is 3° less than the temperature of the day; at the same time the apparent humidity augments at night, and the mist that covers the cataracts becomes thicker. We have just seen, that the hygroscopic state of the air has no influence on the propagation of the sound, and that the cooling of the air diminishes its swiftness.

"It may be thought, that, even in places not inhabited by man, the hum of insects, the song of birds, the rustling of leaves agitated by the feeblest winds, occasion, during the day, a confused noise, which we perceive the less because it is uniform, and constantly strikes the ear. Now this noise, however slightly perceptible it may be, may diminish the intensity of a louder, noise; and this diminution may cease, if during the calm of the night the song of birds, the hum of insects, and the action of the wind upon the leaves, be interrupted. But this reasoning, even admitting its justness, can scarcely be applied to the forests of the Oroonoko, where the air is constantly filled by an innumerable quantity of moschettoes, where the hum of insects is much louder by night than by day, and where the breeze, if ever it be felt, blows only after sunset.

"I rather think, that the presence of the sun acts upon the propagation and intensity of the sound by the obstacles which they find in the currents of air different density, and the partial undulations of the atmosphere caused by the unequal heating of different parts of the soil. In calm air, whether it be dry, or mingled with vesicular vapours equally distributed, the *sonorous undulation* is propagated without difficulty. But when the air is crossed in every direction by small currents of hotter air, the sonorous undulation is divided into two undulations; where the density of the medium changes abruptly, partial echoes are formed, that weaken the sound, because one of the streams comes back upon itself; and those divisions of undulations take place, of which Mr. Poisson has recently developed the theory with great sagacity. It is not therefore the movement of the particles of air from below to above in the ascending current, or the small oblique currents, that we consider as opposing by a shock the propagation of the sonorous undulations. A shock, given to the surface of a liquid, will form circles around the center of percussion, even then the liquid is agitated. Several kinds of undulations may cross each other in water, as in air, without being disturbed in their propagation; little movements may *ride over each other*, and the real cause of the less intensity of sound during the day appears to be the interruption of homogeneity in the elastic medium. During the day, there is a sudden interruption of density, where ever small streamlets of air of a high temperature rise over parts of the soil unequally heated. The sonorous undulations are divided, as the rays of light are refracted, and form the *mirage* (looming), wherever strata of air of unequal density are contiguous. The propagation of sound is altered, when a stratum hydrogen gas is made to rise in a tube closed at one end above a stratum of atmospheric air; and Mr. Biot has well explained, by the interposition of bubbles of carbonic acid gas, why a glass filled with Champagne wine is little sonorous so long as the gas is evolved, and continues to pass through the strata of the liquid."

This hypothesis is well worth further investigations; but we must surrender it to the scientific journals, and continue our more mixed career.

"The Indians of Atures," says Mr. H., "are mild, moderate, and accustomed, from the effects of their idleness, to the greatestprivations. Formerly, excited to labour by the Jesuits, they did not want for food. The fathers cultivated | 477 | maize, French beans, (*frisoles*), and other European vegetables; they even planted sweet oranges and tamarinds round the villages ; and they possessed twenty or thirty thousand head of cows and horses, in the savannahs of Atures and Carichana. They had at their service a great number of slaves and servants (*peones*), to take care of their herds. Nothing is now cultivated but a little cassava, and a few plantains. The fertility of the soil however is such, that at Atures I counted on a single branch of musa 108 fruits, 4 or 5 of which would almost suffice for the daily nourishment of a man. The culture of maize is entirely neglected, and the horses and cows have disappeared. Near the *raudal*, a part of the village still bears the name of *Passo del ganado* (ford of the cattle), while the descendants of those very Indians, whom the Jesuits had assembled in a mission, speak of horned cattle as of animals of a race that is lost. In going up the Oroonoko, toward San Carlos del Rio Negro, we saw the last cow at Carichana. The fathers of the Observance, who now govern these vast countries, did not immediately succeed the Jesuits. During an interregnum of eighteen years, the missions were visited only from time to time, and by Capuchin monks. The agents of the secular government, under the title of *Commissioners of the King*, managed the *hatos* or farms of the Jesuits with culpable negligence. They killed the cattle in order to sell the hides. Many heifers were devoured by tigers, and a greater number perished in consequence of wounds made by the bats of the *raudales*, which are much less, but far bolder than the bats of the Llanos. At the time of the expedition of the boundaries, the horses of Encaramada, Carichana, and Atures, were conveyed as far as San Jose of Maravitanos, where, on the banks of the Rio Negro, the Portugueze could only procure them after a long passage, and of a very inferior quality, by the river Amazon and Grand Para. Since the year 1795, the cattle of the Jesuits have entirely disappeared. There now remains in testimony of the ancient cultivation of these countries, and the industrious activity of the first missionaries, only a few trunks of the orange and tamarind in the savannahs, surrounded by wild trees.

"The tigers, or jaguars, which are less dangerous for the cattle than the bats, come into the village at Atures, and devour the pigs of the poor Indians. The missionary related to us a striking instance of the familiarity of these animals, upon the whole so ferocious. Some months before our arrival, a jaguar, which was thought to be young, though of a large size, had wounded a child in playing with him; I use confidently this expression, which may seem strange, having on the spot verified facts which are not without interest in the history of the manners of animals. Two Indian children, a boy and a girl, about eight and nine years of age, were seated on the grass near the village of Atures, in the middle of a savannah, which we have often traversed. At two o'clock in the afternoon, a jaguar issued from the forest, and approached the children, bounding around them; sometimes he hid himself in the high grass, sometimes he sprang forward, his back bent, his head hung down, in the manner of our cats. The little boy, ignorant of his danger, seemed to be sensible of it only when the jaguar with one of bis paws gave him some blows on the head. These blows, at first slight, became ruder and ruder; the claws of the jaguar wounded the child, and the blood flowed with violence. The little girl then took a branch of a tree, struck the animal, and it fled from her. The Indians ran up at the cries of the children, and saw the jaguar, which retired bounding, without the least show of resistance.

"The little boy was brought to us, who appeared lively and intelligent. The claw of the jaguar had taken away the skin from the lower part of the forehead, and there was a second scar at the top of the head."

[*To be continued.*] | 486 |

Humboldt's Narrative. Personal Narrative of Travels to the Equinoctial Regions of the New Continent, during the years 1799–1804. (Continued from page 477.)

"Among the monkeys," the author continues, "which we saw at the mission of the Atures, we found one new species of the tribe of *sais* and *sajous*, which the Creoles vulgarly call *machis*. It is the *ouavapavi* with grey hair and a bluish face. It has the orbits of the eyes and forehead as white as snow, which at first sight distinguish it from the *simia capucina*, the *simia apella*, the *simia trepida*, and the other weeping monkeys hitherto so confusedly described. This little animal is as gentle as it is ugly. Every day in the court-yard of the missionary it seized a pig, upon which it remained from morning till night, traversing the savannahs. We have also seen it upon the back of a large cat, which bad been brought up with it in father Zea's house.

"It was among the cataracts that we began to hear of the hairy man of the woods, called salvaje, that carries off women, constructs huts, and sometimes eats human flesh. The Tamanacks call it *achi*, and the Maypures *vasitri*, or *great devil*. The natives and the missionaries have no doubt of the existence of this anthropomorphous monkey, which they singularly dread. Father Gili gravely relates the history of a lady in the town of San Carlos, who much praised the gentle character and attentions of the man of the woods. She lived several years with one in great domestic harmony, and only requested some hunters to take her back, 'because she was tired, she and her children (a little hairy also), of living far from the church and the sacraments.' The same author, notwithstanding his credulity, confesses, that he had not been able to find an Indian, who asserted positively that he had seen the *salvaje* with his own eyes. This fable, which the missionaries, the European planters, and the negroes of Africa, have no doubt embellished with many features taken from the description of the | 487 | manners of the ourang-outang, the gibbon, the jocko or chimpanzee, and the pongo, pursued us during five years from the northern to the southern hemisphere; and we were every where blamed, in the most cultivated class of society, for being the only persons to doubt the existence of the great anthropomorphous monkey of America. We shall first observe, that there are certain regions, where this belief is particularly prevalent among the people; such are the banks of the Upper Oroonoko, the valley of Upar near the Lake of Maracaybo, the mountains of Santa Martha and of Merida, the provinces of Quixos, and the banks of the Amazon near Tomependa. In all these places, so distant one from the other, it is repeated, that the *salvaje* is easily recognized by the traces of its feet, the toes of which are turned backward. But if there exist a monkey of a large size in the New Continent, how has it happened, that during three centuries no man worthy of belief has been able to procure the skin of one? Several hypotheses present themselves to the mind, in order to explain the source of so ancient an error or belief. Has the famous *capuchin* monkey of Esmeralda, the canine teeth of which are more than six lines and a half long, the physiognomy much more like man's than that of the ourang outang, and which, when irritated, rubs its beard with its hand, give rise to the fable of the *salvaje*? It is not so large indeed as the coaita (*simia paniscus*); but when seen at the top of a tree, and the head only visible, it might easily be taken for a human being. It may be also (and this opinion appears to me the most probable), that the man of the woods was one of those large bears, the footsteps of which resemble those of a man, and which is believed in every country to attack women.

The animal killed in my time at the foot of the mountains of Merida, and sent, by the name of *salvaje*, to Colonel Ungaro, the governor of the province Varinas, was in fact a bear with black and smooth fur."

These extraordinary accounts are succeeded by a detailed history of the Moschettoes of this region; perhaps the most remarkable of all its animal phenomena.

"Persons who have not navigated the great rivers of equinoctial America, for instance, the Oroonoko and the Rio Magdalena, can scarcely conceive, how without interruption, at every instant of life, you may be tormented by insects flying in the air, and how the multitude of these little animals may render vast regions wholly uninhabitable. However accustomed you may be to endure pain without complaint, however lively an interest you may take in the objects of your researches, it is impossible not to be constantly disturbed by the moschettoes, *zancudoes, jejens* and *tempraneroes*, that cover the face and hands, pierce the clothes with their long sucker in the form of a needle, and, getting into the mouth and nostrils, set you coughing and sneezing whenever you attempt to speak in the open air. In the missions of the Oroonoko, in the villages placed on the banks of the river, surrounded by immense forests, the *plaga de las moscas*, the plague of the flies, affords an inexhaustible subject of conversation. When two persons meet in the morning, the first questions they address to each other are, 'How did you find the zancudoes during the night? How are we to day for the moschettoes.' These questions remind us of a Chinese form of politeness, which indicates the ancient state of the country where it took birth. Salutations were made heretofore in the *celestial empire*, in the following words, *vou-touhou*, 'Have you been incommoded in the night by the serpents?' We shall soon see, that on the banks of the Tuamini, in the river Magdalena, and still more at Choco, the country of gold and platina, the Chinese compliment on the serpents might be added to that of the moschettoes."

Other curious facts are recorded, and illustrate this subject. Mr. H. says—

"At Mandavaca we found an old missionary, who told us with an air of sadness, that he bad spent *his twenty years of moschettoes* in America. He desired us to look well at his legs, that we might be able to tell one day, 'poor alla (beyond sea), what the poor monks suffer in the forests of Cassiquiare.' Every sting leaving a small darkish brown point, his legs were so speckled, that it was difficult to recognize the whiteness of his skin through the spots of coagulated blood. If the insects of the simulium genus abound in the Cassiquiare, which has *white waters*, the culices, or *zancudoes*, are so much the more rare; you scarcely find any there, while on the rivers of *black waters*, in the Atabapo and the Rio Negro, there are generally some *zancudoes* and no *moschettoes*."

(*To be continued*).

HUMBOLDT'S TRAVELS, (Continued.)

Our last number left us in the midst of a swarm of moschettoes, and, however painful it has been to remain a week among these pestiferous insects, there we are still. Proceeding with his anecdotes to illustrate their natural history, our author says:

"I have just shown, from my own observations, how much the geographical distribution of venomous insects varies in this labyrinth of rivers with white and black waters. It were to be wished, that a learned entomologist could study on the spot the specific differences of these noxious insects, which in the torrid zone, in spite of their littleness, act an important part in the economy of nature. What appeared to us very remarkable, and is a fact, known to all the missionaries, is, that the different species do not associate together, and that at different hours of the day you are stung by a distinct species. Every time that the scene changes, and to use the simple expression of the missionaries, other insects 'mount guard,' you have a few minutes, often a quarter of an hour, of repose. The insects that disappear have not their places instantly supplied in equal numbers by their successors. From half after six in the morning till five in the afternoon, the air is filled with moschettoes; which have not, as we find related in some travels, the form of our gnats, but that of a small fly. They are simuliums of the family nemoceræ of the system of Latreille. Their sting is as painful as that of *stomoxes*. It leaves a little reddish-brown spot, which is extravasated and coagulated blood, where their proboscis has pierced the skin. An hour before sun-set a species of small gnats, called *tempraneros*, because they appear also at sun-rise, take the place of the moschettoes. Their presence scarcely lasts an hour and a half; they disappear between six and seven in the evening, or, as they say here, after the *Angelus* (*a la oracion*). After a few minutes repose, you feel yourself stund by *zancudoes*, another species of gnat (culex) with very long legs. The zancudo, the proboscis of which contains a sharp pointed sucker, causes the most acute pain, and a swelling that remains | 488 | several weeks. Its hum resembles that of our gnats in Europe, but is louder and more prolonged. The Indians pretend to distinguish 'by their song' the *zancudoes* and the *tempraneroes*; the latter of which are real *twilight insects*, while the *zancudoes* are most frequently *nocturnal insects*, and disappear towards sun-rise.

"The *culices* of South America, have generally the wings, corselet, and legs of an azure colour, annulated, and variable from a mixture of spots of a metallic lustre. Here, as in Europe, the males, which are distinguished by their feathered antennæ, are extremely rare; you are seldom stung except by females. The preponderance of this sex explains the immense increase of the species, each female laying several hundred eggs. In going up one of the great rivers of Ameriva, it is observed, that the appearance of a new species of *culex* denotes the approximity of a new stream flowing in.'

"The whites born in the torrid zone walk barefoot with impunity in the same apartment where a European recently landed is exposed to the attack of the *niguas* or *chegoes* (*pulex penetrans*). These animals, almost invisible to the eye, get under the nails of the feet, and there acquire the size of a small pea by the quick increase of its eggs, which are placed in a bag under the belly of the insect. The *nigua*, therefore, distinguishes, what the most delicate chemical analysis could not distinguish, the cellular membrane and blood of a European from those of a Creole white. It is not so with the moschettoes.

"In the day, even when labouring at the oar, the natives, in order to chase the insects, are continually giving one another smart slaps with the palm of the hand. Rude in all their movements, they strike themselves and their comrades mechanically during their sleep. The violence of their blows reminds us of the Persian tale of the bear, that tried to kill with his paw the insects on the forehead of his sleeping master. Near Maypures we saw some young Indians seated in a circle and rubbing cruelly each others backs with the bark of trees dried at the fire. Indian women were occupied with a degree of patience, of which the copper-coloured race alone are capable, in extirpating by means of a sharp bone the little mass of coagulated blood, that forms the centre of every sting, and gives the skin a speckled appearance. One of the most barbarous nations of the Oroonoko, that of the Otomacs, is acquainted with the use of moschetto curtains *(mosquiteros*) formed of a tissue of fibres of the palm tree, *murichi*. We had lately seen, that at Higuerote, on the coast of Caraccas, the people of a copper colour sleep buried in the sand. In the villages of the Rio Magdalena the Indians often invited us to stretch ourselves with them on ox-skins, near the church, in the middle of the *plaza grande*, where they had assembled all the cows in the neighbourhood. The proximity of cattle give some repose to man. The Indians of the Upper Oroonoko and the Cassiquiare, seeing that Mr. Bonpland could not prepare his herbal, on account of the continual torment of the moschettoes, invited him to enter their oveus, *(hornitos*). Thus they call little chambers, without doors or windows, into which they creep horizontally through a very small opening. When they have driven away the insects by means of a fire of wet brush-wood, which emits a great deal of smoke, they close the opening of the oven. The absence of moschettoes is purchased clearly enough by the excessive heat of stagnant air, and the smoke of a torch of *copal*, which lights the oven during your stay in it. Mr. Bonpland, with courage and patience well worthy of praise, dried hundreds of plants, shut up in these *hornitos* of the Indians.

"It is difficult not to smile at hearing the missionaries dispute on the size and voracity of the moschettoes at different parts of the same river. In the centre of a country ignorant of what is passing in the rest of the world, this is the favorite subject of conversation. 'How I pity your situation!' said the missionary of the raudales to the missionary of Cassiquiare, at our departure: 'you are alone, like me, in this country of tigers and monkeys; with you fish is still more rare, and the heat more violent; but as for my flies, (*mia moscas*) I can boast, that with one of mine I would beat three of yours.'

"This voracity of insects in certain spots, the rage with which they attack man, the activity of the venom varying in the same species, are very remarkable facts; which find their analogy, however, in the classes of large animals. The crocodile of Angostura pursues men, while at Nueva Barcelona, in the Rio Neveri, you may bathe tranquilly in the midst of the carnivorous reptiles. The jaguars of Maturin, Cumanacoa, and the isthmus of Panama, are cowardly in comparison to those of the Upper Oroonoko. The Indians well know, that the monkey of some valleys can easily be tamed, while others of the same species, caught elsewhere, will rather die of hunger, than submit to slavery."[3]

By this time we fancy our readers are as well acquainted with the habits of the moschettoes, as if they had been bitten by them all over; and further knowledge being unnecessary, we shall advance to other subjects.

---

[3] "I might have added the example of the scorpion of Cumana, which it is very difficult to distinguish from that of the island of Trinidad, Jamaica, Carthagena, and Guayaquil; yet the former is not more to be feared than the *scorpio europæus* (of the south of France), while the latter produces consequences far more alarming than the *scorpio occitanus* (of Spain and Barbary). At Carthagena and Guayaquil, the sting of the scorpion (*alacran*) instantly causes the loss of speech. Sometimes a singular torpor of the tongue is observed for fifteen or sixteen hours. The patient, when stung in the legs, stammers as if he had been struck with apoplexy."

Above the cataract of Atures, at the mouth of the Rio Calaniapo, Mr. Humboldt gives the following account of an extinct tribe:

"We were shown at a distance, on the right of the river, the rocks that surround the cavern of Ataruipe; but we had not time to visit the cemetry of the destroyed tribe of the Atures. We regretted this so much the more, as father Zea was never weary of talking to us of the skeletons painted with anotta, which this cavern contained; of the large vases of baked earth, in which the bones of separate families appeared to be collected; and of many other curious objects, which we proposed to examine at our return from the Rio Negro." | 489 |

At Maypure, higher up, we hear more of the pottery of the Indians:

"In every part of the forests, far from any human habitation, on digging the earth, fragments of pottery and delft are found. The taste for this kind of fabrication seems to have been common heretofore to the natives of both Americas. To the north of Mexico,—on the banks of the Rio Gala— among the ruins of an Azteck city—in the United States—near the *tumuli* of the Miamis; in Florida—and in every place where any trace of ancient civilization could be found, the soil covers fragments of painted pottery; and the extreme resemblances of the ornaments they display is striking. Savage nations, and those civilized people, who are condemned by their political and religious institutions always to imitate themselves, strive, as if by instinct, to perpetuate the same forms, to preserve a peculiar type or style, and to follow the methods and processes which were employed by their ancestors. In North America, fragments of delft have been discovered in places where lines of fortification are found, and the walls of towns constructed by an unknown nation, now entirely extinct. The paintings on these fragments have a great similitude to those which are executed in our days on earthenware by the natives of Louisiana and Florida. Thus too the Indians of Maypure often painted before our eyes the same ornaments as we had observed in the cavern of Ataruipe, on the vases containing human bones. They are real *grecques*, meandrites, and figures of crocodiles, of monkeys, and of a large quadruped, which I could not recognize, though it has always the same squat form."

Above Maypure this is indeed a "New World,": Mr. Humboldt says:—

"When the traveller has passed the great cataracts, he feels as if he were in a new world; and had overstepped the barriers which nature seems to have raised between the civilized countries of the coast and the savage and unknown interior. Toward the east, in the bluish distance, appeared for the last time, the high chain of the Cunavami mountains. Its long horizontal ridge reminded us of the Mesa of Bergantin, near Cumana; but it terminates by a truncated summit. The peak of Calitamini (the name given to this summit) glows at sun-set as with reddish fire. This appearance is every day the same. No one ever approached the summit of this mountain, the height of which does not exceed six hundred toises. I believe this splendor, commonly reddish and sometimes silvery, to be a reflexion produced by large plates of talc, or by gneiss passing into mica-slate. The whole of this country contains granitic rocks, on which here and there, in little plains, an argillaceous grit-stone immediately reposes, containing fragments of quartz, and of brown iron ore.

"In going to the *embarcadere*" he continues, "we caught on the trunk of a hevea, a new species of tree frog, remarkable for its beautiful colours; it had a yellow belly, the back and head of a fine velvetty purple, and a very narrow stripe of white from the point of the nose to the hinder extremities. This frog was two inches long, and allied to the *rana tinctoria*, the blood of which, it is asserted, introduced into the skin of a parrot, in places where the feathers have been plucked out, occasions the growth of frizzled feathers of a yellow or red colour.

But this is not only the region of real wonders; it has its fictions also.

"The forests of Sipapo are altogether unknown, and there the missionaries place the nation of *Rayas*, who have their mouth in the navel. An old Indian, whom we met at Carichano, and who boasted of having often eaten human flesh, had seen these *acephali* 'with his own eyes.' These absurd fables are spread as far as the Llanos, where you are not always permitted to doubt the existence of the *Raya* Indians. In every zone intolerance accompanies credulity; and it might be said, that the fictions of ancient geographers had passed from one hemisphere to the other, did we not know, that the most fantastic productions of the imagination, like the works of nature, furnish every where a certain analogy of aspect and form."

Religious Opinions of American Indians

"The nations of the Upper Oroonoko, the Atabapo, and the Inirida, like the ancient Germans and the Persians, have no other worship than that of the powers of nature. They call the good principle Cachimana; it is the Manitou, the Great Spirit, that regulates the seasons and favors the harvests. By the side of Cachimana there is an evil principle, Iolokiamo, less powerful, but more artful, and in particular more active. The Indians of the forest, when they visit occasionally the missions, conceive with difficulty the idea of a temple or an image. "These good people," said the missionary, "like only processions in the open air. When I last celebrated the patron festival of my village, that of San Antonio, the Indians of Inirida were present at mass. 'Your God,' said they to me, 'keeps himself shut up in a house, as if he were old and infirm; ours is in the forest, in the fields, and on the mountains of Sipapu, whence the rains come.'" Among the more numerous, and on this account less barbarous tribes, religious societies of a singular kind are formed. Some old Indians pretend to be better instructed than others in what regards the divinity; and to them is confined the famous *botuto*, of which I have spoken, and which is sounded under the palm-trees, that they may bear abundance of fruit. On the banks of the Oroonoko there exists no idol, as among all the nations who have remained faithful to the first worship of nature, but the *botuto*, the sacred trumpet, is become an object of veneration. To be initiated into the mysteries of the botuto, it is requisite to have pure manners, and to have lived single. The initiated are subjected to flagellations, fastings, and other painful exercises. There are but a small number of these sacred trumpets. The most anciently celebrated is that upon a hill near the confluence of the Tomo and the Guainia. It is pretended that it is heard at once on the banks of the Tuamini and at the mission of San Miguel de Davipe, a distance of ten leagues. Father Cereso assured us that the Indians speak of the *botuto* of Tomo as an object of worship common to many surrounding tribes. Fruit and intoxicating liquors are placed by the sacred trumpet. Sometimes the great spirit (*Cachimana*) himself make the *botuto* resound; sometimes he is content to manifest his will by him to whom the keeping of the instrument is entrusted. These juggleries being very ancient, (from the fathers of our fathers, say the Indians) we must not be surprised that some incredulous are already to be found; but these express their disbelief of the mysteries of the *botuto* only in whispers. Women are not permitted to see this marvellous instrument, and are excluded from all the ceremonies of this worship. If a woman have the misfortune to see the trumpet, she is put to death without mercy."

Shirt Trees

"We saw on the slope of the Cerra Duida," says M. Humboldt, "*shirt trees* fifty feet high. The Indians cut off cylindrical pieces two feet in diameter, from which they peel the red and fibrous bark, without making any longitudinal incision. This bark affords them a sort of garment, which resembles sacks of a very coarse texture, and without a seam. The upper opening serves for the head, and two lateral holes are cut to admit the arms. The natives wear these shirts of marima in the rainy season: they have the form of the *ponchos* and *ruanas* of cotton, which are so common in New Granada, at Quito, and in Peru. As in these climates the riches and beneficence of nature are regarded as the primary causes of the indolence of the inhabitants, the Missionaries do not fail to say in showing the shirts of marima, 'In the forests of the Oroonoko, garments are found ready made on the trees.' We may add to this tale of the shirts, the pointed caps, which the spathes of certain palm trees furnish, and which resemble coarse net work."

Stories of Crocodiles

Our latter extracts from this publication have been as desultory as the curious nature of the author's inquiries seemed to require, without servilely following him through all his topographical details, and philosophical *generalizations*. In the same spirit, we shall now conclude our notice of these volumes with a brief sequel relating to the crocodiles of the Oroonoko.

"When the waters (says Mr. H.) are high, the river inundates the keys; and it sometimes happens, that even in the town imprudent men become the prey of crocodiles. I shall transcribe from my journal a fact, that took place during Mr. Bonpland's illness. A Guaykeri Indian, from the island de la Margaretta, went to anchor bis canoe in a cove, where there were not three feet of water. A very fierce crocodile, that habitually haunted that spot, seized him by the leg, and withdrew from the shore, remaining on the surface of the water. The cries of the Indian drew together a crowd of spectators. This unfortunate man was first seen seeking with astonishing courage for a knife in the pocket of his pantaloons. Not being able to find it, he seized the head of the crocodile, and thrust his fingers into its eyes. No man in the hot regions of America is ignorant, that this carnivorous reptile, covered with a buckler of hard and dry scales, is extremely sensible in the only parts of his body which are soft and unprotected, such as the eyes, the hollow underneath the shoulders, the nostrils, and beneath the lower jaw, where there are two glands of musk. The Guaykeri Indian had recourse to the same means which saved the negro of Mungo Park, and the girl of Urituco, whom I have mentioned above; but he was less fortunate than they had been, for the crocodile did not open its jaws, and lose hold of its prey. The animal, yielding to the pain, plunged to the bottom of the river; and, after having drowned the Indian, came up to the surface of the water, dragging the dead body to an island opposite the port. I arrived at the moment when a great number of the inhabitants of Angostura had witnessed this melancholy spectacle.

"As the crocodile, on account of the structure of its larynx, of the hyoid bone, and of the folds of its tongue, can seize, though not swallow, its prey under water; a man seldom disappears without the animal being perceived some hours after near the spot where the misfortune has happened, devouring its prey on a neighbouring beach. The number of individuals who perish annually, the victims of their own imprudence and of the ferocity of these reptiles, is much greater than it is believed to be in Europe. It is particularly so in villages, where the neighbouring grounds are often inundated. The same crocodiles remain long in the same places. They become from year to year more daring, especially, as the Indians assert, if they have once tasted of human flesh. These animals are so wary, that they are killed with difficulty. A ball does not pierce their skin, and the shot is only mortal when directed at the throat, or beneath the shoulder. The Indians, who know little of the use of fire-arms, attack the crocodile with lances, after it is caught with large pointed iron hooks, baited with pieces of meat, and fastened by a chain to the trunk of a tree. They do not approach the animal till it has struggled a long time to disengage itself from the iron fixed in the upper jaw. There is little probability that a country in which a labyrinth of rivers without number brings every day new bands of crocodiles from the eastern back of the Andes, by the Meta and the Apure, towards the coast of Spanish Guyana, should ever be delivered from these reptiles. All that will be gained by civilization will be, to render them more timid, and more easily put to flight.

"Affecting instances are related of African slaves, who have exposed their lives to save those of their masters, who had fallen into the jaws of the crocodile. A few years ago, between Uritucu and the *Mission* | 663 | *de Abazo*, a negro, hearing the cries of his master, flew to the spot, armed with a long knife, (*machette*,) and plunged into the river. He forced the crocodile, by putting out his eyes, to let go his prey, and hide himself under the water. The slave bore his expiring master to the shore, but all succour was unavailing to restore him to life. He died of suffocation, for his wounds were not deep; the crocodile, like the dog, appears not to close its jaws firmly while swimming. It is almost superfluous to add, that the children of the deceased, though poor, gave the slave his freedom."

Upon the whole, this portion of Mr. Humboldt's work is equally entertaining with what has gone before, and throws much light on Physics and Geography.

War Poison of the Indians

Esmeralda is the most celebrated spot on the Oroonoko for the fabrication of that active poison which is employed in war, in the chase, and, what is singular enough, as a remedy for gastric obstructions. The poison of the ficunas of the Amazon, the upastieute of Java, and the *curare* of Guyana, are the most deleterious | 232 | substances that are known. Raleigh, toward the end of the sixteenth century, had heard the name of *curare* pronounced as being a vegetable substance, with which arrows were envenomed; yet no fixed notions of this poison had reached Europe. The missionaries Gumilla and Gili had not been able to penetrate into the country where the curare is manufactured. Gumilla asserts that this preparable was enveloped in great mystery; that its principal ingredient was furnished by a subterraneous plant, by a tuberose root, which never puts forth leaves, and which is called the root by way of eminence, *raiz de si misma*; that the venomous exhalations, which arise from the pots, cause the old women (the most *useless*) to perish who are chosen to watch over this operation; finally, that these vegetable juices never appear sufficiently concentrated, till a few drops produce *at a distance* a repulsive action on the blood. An Indian wounds himself slightly; and a dart dipped in the liquid *curare* is held near the wound. If it make the blood return to the vessels without having been brought into contact with them, the poison is judged to be sufficiently concentrated. I shall not stop to refute these popular tales collected by Father Gumilla.

When we arrived at Esmeralda, the greater part of the Indians were returning from an excursion which they had made to the east beyond the Rio Padamo, to gather *juvias*, or the fruit of the Bertholletia, and the liana which yields the curare. Their return was celebrated by a festival, which is called in the Mission *la fiesta de las juvias*, and which resembles our harvest homes and vintage feasts. The women had prepared a quantity of fermented liquor, and during two days the Indians were in a state of intoxication. Among nations that attach great importance to the fruits of the palm-trees and of some others useful for the nourishment of man, the period when these fruits are gathered is marked by public rejoicings, and time is divided according to these festivals, which succeed one another in a course invariably the same. We were fortunate enough to find an old Indian less drunk than the rest, who was employed in preparing the *curare* poison from freshly gathered plants. He was the chemist of the place. We found at his dwelling large earthen pots for boiling vegetable juice, shallower vessels to favour the evaporation by a larger surface, and leaves of the plane-tree rolled up in the shape of our filters, and used to filtrate the liquors more or less loaded with fibrous matter. The greatest order and neatness prevailed in this hut, which was transformed into a chemical laboratory. The Indian who was to instruct us, is known throughout the mission by the name of the *master of poison* (*amo del curare*): he had that self-sufficient air and tone of pedantry, of which the pharmacopolists of Europe were formerly accused; "I | 233 | know," said he, "that the whites have the secret of fabricating soap, and that black powder which has the effect of making a noise and killing animals, when they are wanted. The *curare*, which we prepare from father to son, is superior to any thing you can make *down yonder* (beyond sea). It is the juice of an herb which *kills silently* (without any one knowing whence the stroke comes)."

This chemical operation, to which the *master of the curare* attached so much importance, appears to us extremely simple. The liana (*bujuco*), which is used at Esmeralda for the preparation of the poison, bears the same name as in the forest of Javita. It is the *bejuco de mavacure*, which is gathered in abundance east of the Mission, on the left bank of the Oroonoko, beyond the Rio Amaguaca, in the mountains and granatic lands of Guanaya and Yumariquin.

The juice of the liana, when it has been recently gathered, is not regarded as poisonous; perhaps it acts in a sensible manner only when it is strongly concentrated. It is the bark, and a part of the alburnum, which contains this terrible poison. Branches of the *mavacure* four or five lines in diameter, are scraped with a knife; and the bark that comes off is bruised, and reduced into very thin filaments, on the stone employed for grinding cassava. The venomous juice being yellow, the whole fibrous mass takes this colour. It is thrown into a funnel nine inches high, with an opening four inches wide. This funnel was, of all the instruments of the Indian laboratory, that of which the *master of poison* seemed to be most proud. He asked us repeatedly, of *por alla* (*down yonder*, that is in Europe) we had ever seen any thing to be compared to his *empudo*. It was a leaf of a plantain tree rolled up in the form of a cone, and placed in another stronger cone made of the leaves of the palm-tree. The whole of this apparatus was supported by slight frame work made of the petioli and ribs of palm-leaves. A cold infusion is first prepared by pouring water on the fibrous matter, which is the ground bark of the *mavacure*. A yellowish water filters during several hours, drop by drop, through the leafy funnel. This filtered water is the venomous liquor, but it acquires strength only when it is concentrated by evaporation, like molasses in a large earthen pot. The Indian from time to time invited us to taste the liquid; its taste, more or less bitter, decides when the concentration by fire has been carried sufficiently far. There is no danger in this operation, the *curare* being deleterious only when it comes into immediate contact with the blood. The vapours, therefore, that are disengaged from the pans, are not hurtful, notwithstanding what has been asserted on this point by the Missionaries of the Oroonoko. Fontana, in his fine experiments on the poison of the | 234 | *ticunas* of the river of Amazons, long ago proved, that the vapours arising from this poison when thrown on burning charcoal, may be inhaled without apprehension; and that it is false, as M. de la Condamine has announced, that Indian women, when condemned to death, have been killed by the vapours of the poison of the *ticunas*.

The juice is thickened with a glutinous substance to cause it to stick to the darts, which it renders mortal; but taken internally, the Indians consider the *curare* to be an excellent stomachic. Scarcely a fowl is eaten (adds our author) on the banks of the Oroonoko, which has not been killed with a poisoned arrow. The Missionaries pretend, that the flesh of animals is never so good as when these means are employed. Father Zea, who accompanied us, though ill of a tertian fever, caused every morning the live fowl allotted for our repast to be brought to his hammock, together with an arrow. Notwithstanding his habitual state of weakness, he would not confide this operation, to which he attached great importance, to any other person. Large birds, a guan (*pava de monte*) for instance, or a curassoa (*alsetor*,) when wounded in the thigh, perish in two or three minutes; but it is often ten or twelve before a pig or a pecari expires.[1]

[1] M. Humboldt does not seem to be acquainted with any certain antidote, if such exists, to this fatal poison. Sugar, garlic, the muriate of soda, &c. are mentioned doubtingly.

[The famous Ahauete]

The Intendancy of Puebla (in New Spain) presents to the curious traveller one of the most ancient monuments of vegetation. The famous *Ahauete*, or the Cypress of the village of Atlixco, is 23 metres and 3 tenths, viz. 73 feet in circumference. The hollow of its trunk is 15 feet diameter. This Cypress of Atlixco is, within a few feet, as large as the Baobab (*Adansonia digitata*) of Senegal; — Humboldt's New Spain.

[The Guahibi mother]

M. Humboldt, in the account of his travels in South America, relates a most affecting story: where the Atabapo enters the Rio Temi, the narrative says, "Before we reached its confluence, a granitic hummock, that rises on the western bank, near the mouth of the Guasacavi, fixed our attention; it is called the *Rock of the Guahiba woman*, or the Rock of the Mother, *Piedra de la Madre*. We inquired the cause of so singular a denomination. Father Zea could not satisfy our curiosity; but some weeks after, another missionary, one of the predecessors of this ecclesisastic, whom we found settled at San Fernando as president of the missions, related to us an event, which I recorded in my journal, and which excited in our minds the most painful feelings. If, in these solitary scenes, man scarcely leaves behind him any trace of his existence, it is doubly humiliating for an European to see perpetuated by the name of a rock, by one of those imperishable monuments of nature, the remembrance of the moral degradation of our species, and the contrast between the virtue of a savage, and the barbarism of civilized man! In 1797, the missionary of San Fernando had led his Indians to the banks of the Rio Guaviare, on one of those hostile incursions, which are prohibited alike by religion and the Spanish laws. They found, in an Indian hut, a Guahiba mother with three children, two of whom were still infants: she was occupied in preparing the flour of Cassava. Resistance was impossible; the father was gone to fish, and the mother tried in vain to flee with her children. Scarcely had she reached the savannah, when she was seized by the Indians of the mission, who go to hunt men, like the whites and the negroes in Africa. The mother and her children were bound, and dragged to the bank of the river. The monk, seated in his boat, waited the issue of an expedition, of which he partook not the danger. Had the mother made too violent a resistance, the Indians would have killed her, for every thing is permitted when they go to the conquest of souls (*à la conquista espiritual*,) and it is children in particular they seek to capture, in order to treat them, in the mission, as *poitos* or slaves of the Christians. The prisoners were carried to San Fernando, in the hope that the mother would be unable to find her way back to her home by land. Far from those children who had accompanied their father on the day in which she had been carried off, this unhappy woman showed signs of the deepest despair. She attempted to take back to her family the children who had been snatched away by the missionary, and fled with them repeatedly from the village of San Fernando, but the Indians never failed to seize her anew; and the missionary, after having caused her to be mercilessly beaten, took the cruel resolution of separating the mother from the two children who had been carried off with her. She was conveyed alone toward the missions of the Rio Negro, going up the Atabapo. Slightly bound, she was seated at the bow of the boat, ignorant of the fate that awaited her; but she judged, by the direction of the sun, that she was removed farther and farther from her hut and her native country. She succeeded in breaking her bonds, threw herself into the water, and swam to the left bank of the Atabapo. The current carried her to a shelf of rock, which bears her name to this day. She landed, and took shelter in the woods, but the president of the missions ordered the Indians to row to the shore, and follow the traces of the Guahiba. In the evening she was brought back. Stretched upon the rock (*la Piedra de la Madre*) a cruel punishment was inflicted on her with those straps of manatee leather, which serve for whips in that country, and with which the alcades are always furnished. This unhappy woman, her hands tied behind her back with strong stalks of mavacure, was then dragged to the mission of Javita. She was there thrown into one of the caravanseras that are called *Casa del Rey*. It was the rainy season, and the night was profoundly dark. Forests, till then believed to be impenetrable, separated the mission of Javita from that of San Fernando, which was twenty-five leagues distant in a straight line. No other part is known than that of the rivers; no man ever attempted to go by land from one village to another, were they only a few leagues apart. But such difficulties do not stop a mother, who is separated from her children. Her children are at San Fernando de Atabapo; she must find them again, she must execute her project of delivering them from the hands of Christians, of bringing them back to their father on the banks of the Guaviare. The Guahiba was carelessly guarded in the caravansera. Her arms being wounded, the Indians of Javita had loosened her bonds, unknown to the missionary and the alcades. She succeeded by the help of her teeth in breaking them entirely; disappeared during the night; and at the fourth rising sun was seen at the mission of San Fernando, hovering around the hut where her children were confined. 'What that woman performed,' added the missionary who gave us this sad narrative, 'the most robust Indian would not have ventured to undertake. She traversed the woods at a season when the sky is constantly covered with clouds, and the sun during whole days appears but for a few minutes. Did the course of the waters direct her way? The inundations of the rivers forced her to go far from the banks of the main stream, through the midst of woods where the movement of the waters is almost imperceptible. How often must she have been stopped by the thorny lianas, that form a network around the trunks they entwine! How often must she have swum across the rivulets, that run into the Atabapo! This unfortunate woman was asked how she had sustained herself during the four days? She said, that, exhausted with fatigue, she could find no other nourishment than those great black ants, called rachacos, which climb the trees in long bands to suspend on them their resinous nests.' We pressed the missionary to tell us, whether the Guahiba had peacefully enjoyed the happiness of remaining with her children; and if any repentance had followed this excess of cruelty. He would not satisfy our curiosity; but at our return from the Rio Negro we learnt, that the Indian mother was not allowed time to cure her wounds, but was again separated from her children, and sent to one of the missions of the Upper Oroonoko. There she died, refusing all kind of nourishment, as the savages do in great calamities.

"Such is the remembrance annexed to this fatal rock, *Piedra de la Madre*."

View of the various Productions of the different Countries of the Earth

Every hemisphere produces plants of a different species; and it is not by the diversity of climates that we can attempt to explain why equinoctial Africa has no laurineæ, and the New World no heaths; why the calceolariæ are found only in the Southern hemisphere; why the birds of the continent of India glow with colours less splendid than the birds of the hot parts of America; finally, why the tiger is peculiar to Asia, and the ornithorhincus to New Holland. In the vegetable, as well as in the animal kingdom, the causes of the distribution of the species are among the number of mysteries which natural philosophy cannot reach. This science is not occupied in the investigation of the origin of beings, but of the laws according to which they are distributed on the globe. It examines the things that are, the coexistence of vegetable and animal forms, in each latitude, at different heights, and at different degrees of temperature; it studies the relations under which particular organizations are more vigorously developed, multiplied or modified; but it approaches not problems, the solution of which is impossible, since they touch the origin, the first existence of a germ of life. We may add, that the attempts that have been made to explain the distribution of various species on the globe, by the sole influence of climate, date at a period when physical geography was still in its infancy; when, recurring incessantly to pretended contrasts between the two worlds, it was imagined, that the whole of Africa and of America resembled the deserts of Egypt, and the marshes of Cayenne. At present, when men judge of the state of things not from one type arbitrarily chosen, but from positive knowledge, it | 82 | is ascertained that the two continents, in their immense extent, contain countries that are altogether analogous. There are regions of America, as barren and burning as the interior of Africa. The islands that produce the spices of India are scarcely remarkable for their dryness; and it is not on account of the humidity of the climate, as it has been affirmed in recent works, that the New Continent is deprived of those fine species of laurineæ and myristicæ, which are found united in one little corner of the earth, in the archipelago of India. For some years past, the real cinnamon has been cultivated with success in several parts of the New Continent; and a zone that produces the coumarouna, the vanilla, the pucheri, the pine-apple, the myrtus pimenta, the balsam of tolu, the myroxylon peruvianum, the crotons, the citrosmas, the pejoa, the *incienso* of the silla of Caraccas, the *quereme*, the pancratium, and so many majestic liliaceous plants, cannot be considered as destitute of aromatics. Besides, a dry air favours the developement of the aromatic, or exciting properties, only in certain species of plants. The most cruel poisons are produced in the most humid zone of America: and it is precisely under the influence of the long rains of the tropics that the American pimento, capsicum baccatum, the fruit of which is often as caustic and fiery as Indian pepper, vegetates best.

From the whole of these considerations, it follows, 1st, that the new continent possesses spices, aromatics, and very active vegetable poisons that are peculiar to itself, differing specifically from those of the ancient world; 2dly, that the primitive distribution of species in the torrid zone cannot be explained by the influence of climate solely, or by the distribution of temperature, which we observe in the present state of our planet; but that this difference of climate leads us to perceive, why a given type of organization developes itself more vigorously in such or such local circumstances. We can conceive, that a small number of the families of plants, for instance the musaceæ and the palms, cannot belong to very cold regions, on account of their internal structure and the importance of certain organs; but we cannot explain why no one of the family of melastomas vegetates north of the parallel of thirty degrees, or why no rose tree belongs to the southern hemisphere. Analogy of climates is often found in the two continents, without identity of productions.—*Personal Narrative*, vol. v. p. 180.

Description of the Muschelkalk and Quadersandstein

MUSCHELKALK

A formation which varies but slightly, and whose vague denomination of *shelly limestone* has caused it to be confounded out of Germany, with the lower or upper strata of the oolite formation (with the lias or forest marble). It is well characterized by its more simple structure, by the prodigious quantity of shells, that are partly broken, which it contains, and by its position above the Nebra sandstone (new red sandstone), and beneath the quadersandstein which separates it from the oolite formation. It covers a vast portion of northern Germany (Hanover, Heinberg near Gottingen, Eichfeld, Cobourg, Westphalia, Pyrmont, and Bielfeld), where it is much thicker than the zechstein or Alpine limestone (magnesian limestone).

It extends in southern Germany throughout the platform between Hanau and Stutgard. In France, where, notwithstanding the labours of Omalius d'Halloy, the secondary formations beneath the chalk have been so long neglected, | 314 | Messrs. de Beaumont and Boué have recognised it round the Vosges chain. The muschelkalk generally possesses pale, whitish, grey or yellow tints: its fracture is compact and dull, but the mixture of small laminæ of calcareous spar, arising perhaps from fossil remains, sometimes renders it granular and brilliant. Many beds are marly, arenaceous, or passing into the oolite structure. (Seeberg near Gotha; Weper near Gottingen; Preussisch-Minden; Hildesheim. Chert (hornstein) passing into flint and jasper (Dransfeld, Kandern, Saarbrück), are either disseminated in nodules in the muschelkalk, or form beds of small continuity. The inferior strata of this formation alternate with the new red sandstone (between Bennstedt and Kelme), or insensibly pass into the sandstone, by becoming charged with sand, clay, and even (to the E. of Cobourg) with magnesia (magnesian beds of the muschelkalk).

Subordinate beds. The Marls and clays so frequent in the oolite formation, the new red sandstone, and the zechstein (magnesian limestone) are rare in the muschelkalk. In Germany, this rock contains hydrate of iron, a little fibrous gypsum (Sulzbourg near Naumbourg), and coal (lettenkohle of Voigt; at Mattstedt and Eckardsberg near Weimar) mixed with aluminous schist and carbonized fruits (coniferæ?). The nearer coal is found to the tertiary formations, the more do at least some of its strata approach the state of lignite and aluminous earth.

Fossils. From the researches of M. von Schlotheim, and rejecting the beds which do not belong to the muschelkalk, the fossils are: Chamites striatus, Belemnites paxillosus, Ammonites amalteus, A. nodosus, A. angulatus, A. papyraceus, Nautilites binodatus, Buccinites gregarius, Trochilites lævis, Turbinites cerithius, Myacites ventricosus, Pectinites reticulatus, Ostracites spondyloides, Terebratulites fragilis, T. vulgaris, Gryphites cymbium, G. suillus, Mytulites socialis, Pentacrinites vulgaris, Encrinites liliiformis, &c. Some isolated beds of the oolite formation perhaps contain more fossils than the muschelkalk; but in no secondary formation do organic remains so uniformly abound as in that | 315 | which now occupies us. An immense quantity of shells, partly broken, and partly well preserved, but strongly adhering to the stone (entrochi, turbinites, strombites, mytulites) is accumulated in many strata from 20 to 25 millimetres (about 1 in.) thick, which occur in the muschelkalk. Many species occur united in families (belemnites, terebratulites, chamites). Between these very shelly strata are disseminated ammonites, turbinites, some terebratulites, with their nacreous shells, the Gryphæa cymbium, and superb pentacrinites. Corals, echinites, and pectinites are rare. From the abundance of entrochi in the muschelkalk, this formation has received the name of entrochite limestone (trochitenkalk) in some parts of Germany. As a bed of entrochi often also characterizes the zechstein, and separates it from the coal measures, this name may lead us to confound two very distinct formations. The denomination of gryphite limestone (calaire à gryphites of the zechstein and of the oolite formation) and all those which allude to fossils, without indicating the species, expose us to the same danger. It is stated that the muschelkalk contains the bones of large animals (oviparous? quadrupeds, Friesleben, T. 1, p. 74; T. iv, p. 24, 305) and birds (ornitholithes of the Heimberg; Blumenbach, *Naturgesch*; 3te. Aufl. p. 663); but these bones may belong, as also the teeth of fish, to the breccias and marls resting on the muschelkalk.

Messrs. Buckland and Conybeare, during their tour in Germany, considered the muschelkalk of Werner as identical with the lias. I am inclined to think that there is rather a parallelism than an identity of formation. The muschelkalk occupies the same place as the lias, it equally abounds in ammonites, terebratulæ, and encrinites; but the fossil species differ, and its structure is much more simple and uniform. The muschelkalk strata are not separated by the blue clays which abound in the lias.

The middle strata of the latter possess a dull compact and even fracture, much more resembling the lithographic varieties of the oolite formation than the muschelkalk of Gottingen, Jena, and Eichsfeld. M. Boué has recognised the muschelkalk in | 316 | France, in the platform of Burgundy, near Viteaux and Coussy-les-Forges, near Dax, in the commune of St. Pan de Lon, &c. I have not observed this formation in the equinoxial part of America.

QUADERSANDSTEIN (SANDSTONE OF KÖNIGSTEIN)

A very distinct formation (Banks of the Elbe, above Dresden, between Pirna, Schandau, and Königstein; between Nuremberg and Weissenburg; Staffelstein in Franconia; Heuscheune, Adersbach; Teufelsmauer at the foot of the Hartz; valley of the Moselle and near Luxembourg; Vic in Lorraine; Nalzen in the Pays de Foy, and Navarreins, at the foot of the Pyrenees) characterized by M. Hausmann, and for a long time confounded either with the quartzose varieties of the new red sandstone, the sandstone of the plastic clay, or with the sandstone of Fontainebleau, above the calcaire grossier of Paris: it is the white sandstone of M. de Bonnard, and the third formation sandstone of M. d'Aubuisson. Preferring geographical names, I often call this formation the Königstein sandstone, the new red sandstone the Nebra sandstone, and the muschelkalk the Gottingen limestone.

The quadersandstein has a white, yellowish, or grey colour, with very fine grains, which are agglutinated together by a nearly invisible argillaceous or quartzose cement. Mica is not abundant in it, and is always silvery and disseminated in isolated plates. It neither contains the included oolite beds, nor the flattened lenticular masses of clay (thongallen) which characterise the new red sandstone. It is never schistose; but is divided into very thick beds, which are cut at a right angle by fissures, and of which some easily decompose into a very fine sand. It contains hydrate of iron (Metz) disposed in nodules. The organic remains disseminated in this formation present, according to Messrs. von Schlotheim, Haussmann, and Raumer, an extraordinary mixture of sea shells (very analogous to those of the muschelkalk) and | 318 | dicotyledonous phytolithes. In it have been found mytulites, tellinites, pectinites, turritellæ, and ostreæ, (with cerithia, but no ammonites; Habelschwerd, Alt-Lomnitz in Silesia), and at the same time the wood of palms, the impressions of leaves belonging to the class of the dicotyledons, and small deposites of coal

(Deister, and Wefersleben near Quedlinbourg), very well described by Messrs. Rettberg and Schulze, and passing into lignite.

M. von Raumer had observed that the quadersandstein is separated from the new red sandstone by the muschelkalk; it is placed between this limestone and the Jura limestone, and consequently beneath the oolite formations of England and the continent. In this position we cannot consider it, with M. Keferstein (see his Essay on the mineral geography of Germany, T. 1. p. 12. and 48.), as parallel to the molasse of Argovy (mergelsandstein), which represents the plastic clay beneath the chalk. The nature of the vegetable remains contained in the quadersandstein, and its resemblance to the plänerkalk which belongs to the chloritous and sandy strata of the chalk, have caused it to be regarded by many celebrated geologists as a formation posterior to the oolite formation: thus Messrs. Buckland, Conybeare, and Phillips place it between the chalk and the upper beds of the oolites. But, according to the observations of M. Boué and many other celebrated German geologists, the quadersandstein, sometimes alternating with marly and conglomerate beds, rests immediately on gneiss near Freyberg, on the coal measures in Silesia, and in Bohemia; on the new red sandstone near Nuremberg in Franconia; on the muschelalk between Hildesheim and Dickholzen near Helmstädt, and near Schweinfurt on the Mein. It is covered by the oolite formation, and alternates with marly beds of this limestone in Westphalia, between Osnabrück, Bielfeld, and Bückebourg.

Visit of the ruins of Ataruipe

We climbed with difficulty, and not without some danger, a steep rock of granite, entirely bare. It would have been almost impossible to fix the foot on its smooth and sloping surface, if large crystals of feldspar, resisting decomposition, did not stand out from the rock, and furnish points of support. Scarcely had we attained the summit of the mountain, when we beheld with astonishment the singular aspect of the surrounding country. The foamy bed of the waters is filled with an archipelago of islands covered with palm-trees. Toward the west, on the left bank of the Oroonoko, stretch the savannahs of the Meta and the Casanare. They resembled a sea of verdure, the misty horizon of which was illumined by the rays of the setting sun. Its | 81 | orb, resembling a globe of fire, suspended over the plain; and the solitary Peak of Uniana, which appeared more lofty from being wrapped in vapours that softened its outline; all contributed to augment the majesty of the scene. Near us the eye looked down into a deep valley, enclosed on every side. Birds of prey and goatsuckers winged their lonely flight in this inaccessible circus. We found a pleasure in following with the eye their fleeting shadows as they glided slowly over the flanks of the rock.

A narrow ridge led us to a neighbouring mountain, the rounded summit of which supported immense blocks of granite. These masses are more than forty or fifty feet in diameter; and their form is so perfectly spherical, that, appearing to touch the soil only by a small number of points, it might be supposed, at the least shock of an earthquake, they would roll into the abyss. I do not remember to have seen any where else a similar phenomenon, amid the decompositions of granitic soils. If the balls rested on a rock of a different nature, as it happens in the blocks of Jura, we might suppose, that they had been rounded by the action of water, or thrown out by the force of an elastic fluid; but their position on the summit of a hill alike granitic makes it more probable, that they owe their origin to the progressive decomposition of the rock.

The most remote part of the valley is covered by a thick forest. In this shady and solitary spot, on the declivity of a steep mountain, the cavern of Ataruipe opens itself; it is less a cavern than a jutting rock, in which the waters have scooped a vast hollow, when, in the ancient revolutions of our planet, they attained that height. We soon reckoned in this tomb of a whole extinct tribe near six hundred skeletons well preserved, and so regularly placed, that it | 82 | would have been difficult to make an error in their number. Every skeleton reposes in a sort of basket, made of the petioles of the palm-tree. These baskets, which the natives call *mapires*, have the form of a square bag. Their size is proportioned to the age of the dead; there are some for infant's cut off at the moment of their birth. We saw them from ten inches to three feet four inches long, the skeletons in them being bent together. They are all ranged near each other, and are so entire, that not a rib, or a phalanx is wanting. The bones have been prepared in three different manners, either whitened in the air and the sun; dyed red with onoto, a colouring matter extracted from the bixa orellana; or, like real mummies, varnished with odoriferous resins, and enveloped in leaves of the heliconia or of the plantain tree. The Indians related to us, that the fresh corpse is placed in damp ground, in order that the flesh may be consumed by degrees; some months after, it is taken out, and the flesh remaining on the bones is scraped off with sharp stones. Several hordes in Guyana still observe this custom. Earthen vases half-baked are found near the *mapires*, or baskets. They appear to contain the bones of the same family. The largest of these vases, or funeral urns, are three feet high, and five feet and a half long. Their colour is greenish gray; and their oval form is sufficiently pleasing to the eye. The handles are made in the shape of crocodiles, or serpents; the edge is bordered with meanders, labyrinths, and real grecques, in straight lines variously combined. Such paintings are found in every zone, among nations the most remote from each other, either with respect to the spot which they occupy on the globe, or to the degree of civilization which they have attained. The inhabitants of the little mission | 83 | of Maypures still execute them on their commonest pottery; they decorate the bucklers of the Otaheiteans, the fishing implements of the Eskimoes, the walls of the Mexican palace of Mitla, and the vases of ancient Greece. Every where a rhythmic repetition of the same forms flatters the eye, as the cadenced repetition of sounds soothes the ear. Analogies founded on the internal nature of our feelings, on the natural dispositions of our intellect, are not calculated to throw light on the filiation and the ancient connexions of nations.

We withdrew in silence from the cavern of Ataruipe. It was one of those calm and serene nights, which are so common in the torrid zone. The stars shone with a mild and planetary light. Their scintillation was scarcely sensible at the horizon, which seemed illumined by the great nebulæ of the southern hemisphere. An innumerable multitude of insects spread a reddish light on the ground, loaded with plants, and resplendent with these living and moving fires, as if the stars of the firmament had sunk down on the savannah. On quitting the cavern, we stopped several times to admire the beauty of this singular scene. The odoriferous vanilla, and festoons of bignonia, decorated the entrance; and above, on the summit of the hill, the arrowy branches of the palm-trees waved murmuring in the air.

Extent and population of America

The following estimate of the extent, in square leagues, and population, of the Continent of America in 1822, was lately transmitted by Baron de Humboldt from Paris, to President Bolivar, with the accompanying letter: —

	Square Leagues.	**Population.**
Mexico, or New Spain	75,830	6,800,000
Guatemala	16,740	1,600,000
Cuba and Puerto Rico	4,430	800,000
Colombia	33,700	900,000
Venezuela	58,250	1,800,000
New Granada		
Peru	42,150	1,400,000
Chile	14,240	1,100,000
Buenos Ayres	126,770	2,000,000
Total formerly Spanish	372,110	16,400,000
United States	125,440	10,200,000
Brazil	256,990	4,000,000

"The above surfaces have been calculated with great care, on maps rectified by astronomical observations. The calculations have been several times repeated by M. Mathieu, Member of the Board of Longitude and of the Institute, as well as by myself. The results differ from those published in 1809, in the Political Essay on Mexico. Then the inhabited parts alone had been calculated, without including the desert regions, over which the independent tribes of the indigenous inhabitants wander. Now the whole extent of each country, to its furthest limits, has been measured.

"The population of the different parts of *ci-devant* Spanish America is very uncertain, nevertheless each portion has been calculated according to the latest data which have reached me. The present estimate ought to be considered in the same light as all my other works on America; they are no other than essays, in which every thing will have to be done over again. Statistical calculations can only be rendered' perfect by degrees, in like manner as the elements of meteorological and astronomical tables.

"Comparison.—Spain contains 16,094 square leagnes—Europe contains 304,710 square leagues—and the whole of Southern America 581,891 square leagues.

(Signed) "HUMBOLDT."

On Petrifactions, or Fossil Organic Remains

Of the various proofs of the identity of formations in the most distant regions of the globe, one of the most striking, and for which we are indebted to the assistance of zoology, is the identity of the organic bodies contained in strata having a similar geological position. The researches which lead to this kind of proofs, have exercised the skill of the learned, since Messrs De Lamarck and Defrance began to determine the fossil shells of the neighbourhood of Paris, and since Messrs Cuvier and Brongniart published their memorable works on the fossil bones and tertiary formations. As the greater portion of the formations which compose the crust of our planet contain no organic remains, and as these remains are of very rare occurrence in the transition formations, and are often broken and difficult to detach from the rock in the older secondary formations, the study of fossil organic bodies comprehends but a small part of geognosy,—a part, however, very worthy of the attention of the philosopher. The problems which present themselves in this department, are numerous: They bear relation to the geography of animals, whose races are extinct, and which, for this reason, already belong to the history of our planet: They render necessary the discussion of the zoological characters, by which we would distinguish the different formations. To remain faithful to the plan which I have proposed to myself, of only considering the various objects in the most general way, I shall now mention those questions of *Geognostical Zoology*, which appear the most important in the present state of science, and whose solution has been tried with more or less success. What are the Genera, and (if the state of preservation and degree of attachment to the rock permit a more complete determination) what are the Species, to which the fossil remains may be referred? Does an exact determination of the species discover any which are identical with the plants and animals of the present world? What are the classes, the orders, and families of organized beings, which present the greatest number of these analogies? In what proportion does the number of identical genera and species increase, with the newness of rocks or of earthy deposits? Is the | 21 | order observed in the superposition of transition, secondary, tertiary, and alluvial formations, every where in harmony with the increasing analogy which the types of organization present? Do these types succeed each other from below upwards, passing from the grey-wackes and black transition limestone, through the coal sandstone, Alpine limestone, Jura limestone, chalk, the tertiary gypsum, to the fresh water and modern alluvial formations, in the same order which we have adopted in our systems of natural history, by arranging animals according as their structure becomes more complicated, and as other systems of organs are added to those of nutrition? Does the distribution of fossil organic bodies indicate a progressive development of vegetable and animal life upon the globe,—a successive appearance of acotyledonous and monocotyledonous plants, of zoophytes, crustacea, mollusca; (cephalopoda, acephala, gasteropoda), of fishes; oviparous quadrupeds; of dicotyledonous plants; of marine mammifera and of terrestrial mammifera? Considering fossil organic bodies, not in their relation with such or such a rock in which they have been discovered, but simply with regard to their climacteric distribution, is an appreciable difference to be observed between the species which predominate in the old and in the new continent, in the temperate climates, and under the torrid zone, in the northern and in the southern hemisphere? Is there a certain number of tropical species which are found every where, and which seem to announce, that, independently of a distribution of climates similar to the present, they have experienced, in the first age of the world, the high temperature which the cracked crust of the globe, strongly heated in its interior, had given to the ambient atmosphere? Can the fresh-water shells be distinguished with certainty, by means of precise characters, from those of marine origin? Is the determination of the genus sufficient for this purpose, or are there not (as among fishes) certain genera, the species of which live both in the sea and in rivers? Although, in some of the tertiary rocks, the fluviatile shells occur mixed (for example, at the mouths of our rivers) with the pelagic shells, do we not observe, that the first form particular deposits, characterizing formations, the examination of which has been hitherto neglected, and which are of very recent origin? Have fresh-water shells ever been discovered under the Jura | 22 | limestone, among reputed fluviatile fishes, in the bituminous slate of the Alpine limestone? Do identical species of petrifactions or fossils occur in the same formations in different points of the globe? Are they capable of furnishing zoological characters for distinguishing the different formations? Or, should it not rather be admitted, that species which the zoologist is entitled to consider as identical, following the methods usually adopted, penetrate across several formations; that they are even seen in those which are not in immediate contact? Should not the zoological characters be taken both from the total absence of certain species, and from their relative frequency *or predominance*, as well as from their constant association with a certain number of other species? Is it right to divide a formation, whose unity has been determined from relations of position, and from the identity of the beds, which are equally interposed between upper and lower strata, merely because the first of these strata contain fresh-water shells, while the last contain marine ones? Is the total absence of organic bodies in certain masses of secondary and tertiary formations, a sufficient reason for considering these masses as particular formations, if other geognostical relations do not justify this separation?

Some of these problems engaged the attention of geologists at an early period. Lister had already maintained, upwards of 150 years ago, that each rock was characterized by different fossil shells,— (*Philosophical Trans*., No. 76, p. 2283.) With the view of proving that the shells of our seas and lakes are specifically different from fossil shells (*lapides sui generis*), he affirms, "that the latter, for example, those of the Northamptonshire quarries, have all the characters of our *murices, tellinæ*, and *trochi*; but that naturalists, who are not accustomed to content themselves with a vague and general perception of things, will find the fossil shells *specifically* different from those of the present world." Nearly at the same period, Nicolas Stenon, (*De solido intra solidum contento*, 1669, p. 2, 17, 28, 63, 69, fig. 20, 25,) first distinguished "the (primitive) rocks anterior to the existence of plants and animals upon the globe, and consequently never containing organic remains, and the (secondary) rocks superimposed upon the first, and abounding with these remains (*turbidi maris sedimenta sibi invicem imposita*)." | 23 | He considered each bed of secondary rock "as a sediment deposited by an aqueous fluid;" and, exposing a system perfectly similar to that of Deluc, "On the formation of valleys by longitudinal sinkings, and on the inclination of beds originally all horizontal," he admits, with regard to Tuscany, in the manner of our modern geologists, "six great epochs of nature (*sex distinctæ Etruriæ facies, ex præsente facie Etruriæ collectæ*), according as the sea periodically inundated the continent, or retired within its ancient limits." In those times, when the observation of nature gave rise in Italy to the first ideas regarding the relative age and succession of the primitive and secondary strata, zoology and geognosy could not yet be of mutual assistance, because the zoologists of those days were not acquainted with the rocks, and the geognosts were entirely strangers to the natural history of animals. Naturalists confined themselves to vague perceptions; they regarded as specifically identical all that presented any analogy of form; but, at the same time, and this was a step made in the true road, they were attentive to the fossils found in such or such rocks. Hence the denominations *limestone with gryphites, limestone with trochites, slates with ferns*, and *slates with trilobites*, (Gryphiten und TrochitenKalk, Kräuter und Trilobiten-Schiefer), were long ago employed by the mineralogists of Germany. The determination of genera whose characters are derived from the teeth, and other parts of the hinge, and from the opening of the shell, is much more difficult in the oldest secondary rocks than in the tertiary formations, the former being in general harder, and adhering more closely to the body of the shell. This difficulty increases, when it is attempted to distinguish the species; it becomes almost insurmountable in some limestone-rocks of the Transition class, and in the shell-limestone or muschelkalk, which contains broken shells. Could the zoological characters of a certain number of formations be taken from very distinct genera, and did the trilobites and orthoceratites belong exclusively to the transition formations, the gryphites to the alpine limestone (zechstein), the pectinites to the variegated sandstone or *bunter* sandstone (grès de Nebra), the trochites and mytilites to the shell-limestone (muschelkalk), the tellinæ to the quadersandstein, the ammonites and turritellæ to the Jura limestone | 24 | and its marls, the ananchytes and spatangi to the chalk, the cerithia to the coarse limestone; the knowledge of these genera would prove an easy and effectual aid in the determination of rocks: it would no longer be requisite to examine the superposition of formations on the spot, for they could be distinguished without stirring from one's cabinet, by the mere examination of specimens. But it is far from being the case that nature has rendered the study of the shelly masses which compose the crust of the earth so easy to man. The same types of organization are repeated at very different epochs. The same genera occur in different formations. There are orthoceratites in the transition-limestones, the Alpine limestones, and the variegated sandstone; terebratulites in the Jura limestone and shell-limestone (muschelkalk); trilobites in the transition clayslates (thonschiefer), in the bituminous marl-slate, and, according to an excellent geognost, M. de Schlottheim, even in the Jura limestone. There are pentacrinites in the transition clay-slate and the newest shell-limestone (muschelkalk). The ammonites penetrate across many of the limestone and marl formations, from the greywackes to the lower beds of the chalk. Trunks of monocotyledonous plants are found both in the red sandstone and in the marls of the fresh-water gypsum, formed at a period when the world was already filled with monocotyledonous vegetables.

But at a period when naturalists are no longer satisfied with vague and uncertain notions, it has been discovered that the greater number

of those petrifactions or fossils (gryphites, terebratulites, ammonites, trilobites, &c.) imbedded in different formations, are not specifically the same; and that a great number of species which have been determined with precision, vary with the superimposed rocks. The fishes which are observed in the transition-slates (Glaris), in the zechstein or bituminous marlslate, in the Jura limestone, in the tertiary limestone with cerithia of Paris and Monte Bolca, and in the gypsum of Montmartre, are distinct species, partly pelagic and partly fluviatile. Might it be concluded, from these facts, that all the formations are characterised by particular species; that the fossil-shells of the chalk, of the shell-limestone (muschelkalk), of the Jura limestone, and of the alpine limestone, are all different from
| 25 | each other? I am of the opinion that this would be pushing the induction too far; and even M. Brongniart, who was so well acquainted with the value of zoological characters, has limited their absolute application to cases where the superposition (the circumstances of relative situation) do not oppose it. I might mention the cerithia of the coarse limestone, which occur (near Caen) beneath the chalk, and which seem, like the repetition of the clays with lignites above and below the chalk, to shew a certain connection between formations which at first sight might be considered entirely distinct. I might adduce other species of shells, which belong at once to several tertiary formations, and observe, that should species, which at the present day are believed to be identical, be separated at some future time by means of characters not readily observable, and by slight shades of difference, even the delicacy of these distinctions would not prove too great an inducement to believe in the universality, otherwise so desirable, of zoological characters in geognosy. Another objection, taken from the influence exercised by climate even upon the pelagic animals, appears to me more important still. Although the seas, from well-known physical causes, present at great depths the same temperature under the equator and in the temperate zone, we find, however, that, in the present state of our planet, the shells of the Tropics (among which the univalves predominate, as among the fossil testacea) differ much from those of the temperate climates. The greater number of these animals live upon reefs and shoals; whence it follows that the specific differences are often very sensible, under the same parallel, on opposite coasts. Now, were the same formations repeated and extended, so to speak, to prodigious distances, from east to west, and from north to south, from one hemisphere to the other, is it not probable, whatever may have been the complicated causes of the ancient temperature of our globe, that variations of climate have modified, in ancient times as at the present,

the types of organisation, and that one and the same formation (that is to say, a rock of the same nature placed, in the two hemispheres, between homogeneous formations) may have enveloped distinct species? It no doubt often happens, that superimposed beds present a very striking contrast of fossil bodies; but should it be concluded from this, that, after a deposit was | 26 | formed, the beings which then inhabited the surface of the globe were all destroyed?

It is incontestible, that generations of different types have succeeded one another. The ammonites, which scarcely occur among the transition rocks, attain their *maximum* in the strata which on different points of the globe represent the shell-limestone (muschelkalk) and Jura limestone; they disappear in the upper beds of the chalk, and above this formation. The echinites, which are very rare in the alpine limestone, and even in the shell-limestone (muschelkalk), become, on the contrary, very common in the Jura limestone, in the chalk and the tertiary formations. But nothing proves to us that this succession of different organic types, this gradual destruction of genera and of species, necessarily coincides with the periods when each formation was deposited. "The consideration of similarity, or of discrepancy between organic remains, is not of great importance," says M. Beudant (*Voyages Min.* v. iii. p. 278.), "when we compare deposits formed in countries very remote from each other; but it is of much importance, when we compare those which are very near."

In combating the two absolute conclusions which might be attempted to be derived from the validity of zoological characters, I am far from denying the important services which the study of fossil bodies renders to geognosy, if we consider this science under a philosophical point of view. Geognosy does not limit itself to the searching for diagnostical characters; it embraces the aggregate of relations under which each formation may be considered: 1st, Its relative position; 2dly, Its oryctognostic constitution (that is to say, its chemical composition, and the particular mode of aggregation, more or less crystalline, of its molecules); 3dly, The association of different organic bodies which occur imbedded in it. If the superposition of heterogeneous rocky masses reveals to us the successive order of their formation, is it not also interesting to know the state of organic nature at the different periods when these deposits were formed? Can it be questioned, that, over a surface of many thousands of square leagues (in Thuringia, and in all the northern part of Germany), nine superimposed formations, those, namely, of the transition limestone, greywacke, red sandstone, | 27 | zechstein with bituminous marl-slate, (muriatiferous gypsum,) oolitic sandstone, (argillaceous gypsum,) shell limestone or muschelkalk, and white sandstone (quadersandstein), have been recognised as distinct, without having any recourse to geological characters; but it does not follow from thence, that the most minute investigation of these characters, or, to speak more correctly, the most intimate knowledge of the petrifactions or fossils contained in each of the formations, is not indispensable for presenting a complete and truly geognostical picture. It is with the study of formations as with that of organic beings. Botany and zoology, which are now considered under a higher point of view, are no longer confined to the investigation of some external characters distinctive of the species; these sciences investigate the whole of vegetable and animal organization. The characters taken from the forms of the shell, suffice to distinguish the different species of acephalous testacea; but, for all this, could the knowledge of the animals which inhabit these shells be regarded as superfluous? Such is the connexion of phenomena, and of their natural relations, that, if some be neglected, there is produced not only an incomplete image, but also a false one.

In the case of the conformity of position, there may be identity of mass (that is to say, of mineralogical composition) and diversity of petrifactions, or diversity of mass and identity of petrifactions. The rocks β and β′, placed at great horizontal distances, between two identical formations α and γ, either belong to the same formation, or are of parallel formations. In the former case, their mineral composition is similar; but, from the distance of the places, and the effects of climate, the organic remains which they contain may differ considerably. In the other case, the mineralogical composition is different, but the organic remains may be analogous. I am of opinion that the words *identical formations, parallel formations*, vindicate the conformity or non-conformity of mineralogical composition; but that they determine nothing with regard to the identity of petrifactions. If it is pretty probable that the deposits β and β′, placed at great horizontal distances between the same rocks α and γ, have been formed *at the same period*, because they contain the same fossils, and an analogous mass; it is not equally probable, on the other hand, that when the fossils are distinct, the periods | 28 | of formation have been very remote from each other. We may conceive that, under the same zone, in a country of but little extent, generations of animals have succeeded each other, and have characterised, as by particular types, the *periods* of formation; but, at great horizontal distances, beings of very different forms, may, in different climates, have simultaneously occupied the surface of the globe, or the basin of the ocean. Further still, the position of β, between α and γ, proves that the formation of β is anterior to that of γ, and posterior to that of α; but nothing gives us the precise measure of the interval between the epochal limits, and different isolated deposits of β cannot be simultaneous.

There seems to result from the facts which the zeal and sagacity of naturalists have united within these few years, that, if we should not always expect to find, as Lister pretended, in each different formation different organic remains, yet that most commonly formations which are shewn to be distinct by their relative situation and composition, contain in the most remote countries of the globe associations of entirely similar species. M. Brongniart, whose labours, together with those of other celebrated mineralogists, have so much advanced the study of *subterranean conchology*, has lately made known the striking analogies which certain formations of Europe and of North America present with regard to fossil organic bodies. He has endeavoured to prove that a formation is sometimes so disguised, that it is only by its zoological characters that it can be recognised, (Brongniart, *Hist. Nat. des Crust. Fossiles*, p. 57, 62). In the study of formations, as in all the descriptive physical sciences, it is the assemblage of many characters which should alone guide us in our search after truth. The specific description of the remains of plants and animals contained in the different formations, presents us, so to speak, their *Flora* or *Fauna*. Now, in the primordial, as in the present world, the vegetation and the animal productions of different portions of the globe, appear to have been less characterised by some isolated forms of an extraordinary aspect, than by the association of many forms specifically different, but analogous to each other, notwithstanding the distance of the places. On discovering a new land near Torres Straits, it would not be easy to determine, from a small | 29 | number of productions, whether this land were contiguous to New Holland, or to one of the Molucca Islands, or to New Guinea . To compare formations with relation to petrifactions, is to compare *Floras* and *Faunas* of different countries and of different epochs; it is to resolve a problem so much the more complicated, that it is modified at the same time by space and by time.

Among the zoological characters applied to geognosy, the absence of certain fossils often afford a better characteristic than their presence. This applies particularly to the transition rocks; in general there only occur in them, madrepores, encrinites, trilobites, orthoceratites, and shells of the family of terebratules, that is to say shells of which some species, not identical but analogous, are met with in very modern secondary beds; but these transition rocks are destitute of many other organic remains, which appear abundantly above the red sandstone. The judgment which we form regarding the absence of certain species, or the total absence of fossil organic bodies, may, however, be founded upon an error which it will be of importance to explain here. On examining the formations which contain imbedded shells in a general manner, we observe that the organic bodies are not always equally distributed in the mass; but, 1st, That strata entirely destitute of shells alternate with others which are provided with them; 2dly, That, in the same formation, particular associations of fossils characterize certain strata, which alternate with other strata having distinct fossils. This phenomenon, which has been long since observed, occurs in the shell limestone (muschelkalk,) and in the alpine limestone (zechstein), which are often separated by a bed of trochites from the coal sandstone, (Buch, Beob. v. i. p. 135, 146, 171); it is found also in the Jura limestone, and in several tertiary formations. On examining the chalk of the neighbourhood of Paris only, it might almost be thought that the univalve shells are entirely awanting in this formation. The polythalamous univalves, however, the ammonites, as we have already mentioned, are very common in England, in the oldest beds of the chalk. Even in France, (coast of St Catherine near Caen), the tufaceous and chloritic chalk, contains many fossils which are not met with in the white chalk. (Brongniart, Caractères Zool. p. 12.) | 30 | As in different countries the formations are not equally developed, and shreds of formations may be taken for complete and entire ones, those which are destitute of shells in one region may present them in another. This consideration is important in obviating the pretty general tendency of multiplying formations in too great a degree; for when, in a particular point of the globe, a formation (sandstone for example) abounds in its lower part with petrifactions, while its upper part is entirely destitute of them, this absence of petrifactions or fossils does not of itself justify the splitting of the deposit into two distinct formations. In the geological description of the neighbourhood of Paris, M. Brongniart has very properly united the millstones without shells, with those which are in a manner studded with freshwater shells.

We have seen that a formation may contain in different strata, petrifactions specifically different, but that most commonly some species of the lower stratum mingle with the great mass of heterogeneous species which occur united in the superimposed stratum. When this difference relates to genera, of which some are pelagic and others fresh-water shells, the problem of the unity or indivisibility of a formation becomes still more puzzling. Two cases must first be distinguished, that in which some fluviatile shells occur, mixed with a great mass of marine shells, and that where marine and fluviatile shells may alternate bed by bed. Messrs Gilet de Laumont and Beudant, have made interesting observations regarding the admixture of marine and freshwater productions, in one and the same bed. M. Beudant has proved, by ingenious experiments, how many of the fluviatile mollusca gradually habituate themselves to live in a fluid which has all the saltness of the ocean. The same geologist has, together with M. Marcel de Serres, examined certain species of paludinæ, which, preferring brackish waters, are found near our coasts, sometimes with pelagic shells, sometimes with fluviatile ones. (*Journ. de Phys*. t. lxxxiii. p. 137; t. lxxxviii. p. 211; Brongniart, *Geogr. Min*. p. 27, 54, 89.) To these curious facts are joined others, which I have published in the account of my *Journey to the Equatorial Regions*, (v. i. p. 535, and v. ii. p. 606), and which seem to explain what has formerly taken place upon the globe, according to what we still see taking place | 31 | at the present day. On the coast of Terra Firma, between Cumana and Nueva Barcelona, I have seen crocodiles advance far into the sea. Pigafetta has made the same observation regarding the crocodiles of Borneo. To the south of the Island of Cuba, in the Gulph of Xagua, there are lamantines in the sea, at a point where springs of fresh water issue in the midst of the salt water. When we reflect upon these facts, we become less surprised at the mixture of some terrestrial productions with many others indisputably marine. The second case which we have adverted to, that of alternation, is never present, I believe, in a manner so distinct as the alternation of clay-slate and of black limestone, in the same transition formation, or (to adduce a fact relative to the distribution of organic bodies) as the alternation of two great marine formations, (limestone with cerithia, and the sandstone of Romainville), with two great freshwater formations, (gypsum and millstone of the plain of Montmorency). The result of an attentive examination of the superpositions is merely this; that there are alternating beds of gypsum and marl placed between two marine formations, and containing at the centre (in their largest mass), terrestrial and fresh-water productions, and towards the upper and inferior limits, as well in the gypsum as in the marls, marine productions. Such is the geological constitution of the gypsum of Montmartre. The specific variation in the petrifactions, the mixture observed at Pierrelaie, and the phenomenon of alternation which Montmartre presents, are not sufficient to justify the division of the same formation into fragments. The marls and gypsum which contain marine shells, (No. 26. of the third mass), cannot be geognostically separated from the marls and gypsums which contain fresh-water productions. Nor have Messrs Cuvier and Brongniart hesitated to consider the general mass of these marine and fresh-water marls and gypsums as of one and the same formation. These geologists have even cited this union of alternating beds as one of the clearest examples of what the word formation should be intended to express. (*Geogr. Miner.*, p. 31, 39, 189). In fact, different systems of beds may be contained in the same formation: they are groups, subdivisions, or as the geognosts of the Freyberg School express it, | 32 | members more or less developed of the same formation, (Freiesleben, *Kupf.* v. i. p. 17. v. iii. p. 1).

Notwithstanding the mixture of pelagic and fluviatile shells which is sometimes observed at the point of contact of two formations of different origin, the name of *Marine Limestone*, or of *Marine Sandstone*, may be given to the one of these formations, when the denomination of the rocks could only be taken from the species which constitute the greatest mass and the centre of the beds. This terminology recalls a fact which bears relation, so to speak, to geogony, to the ancient history of our planet; it fixes (and perhaps a little too much) the alternation of the fresh and of the salt waters. I do not dispute the utility of the denominations *marine limestone*, or *marine sandstone*, with regard to local descriptions; but, according to the principles which I have proposed to follow, in the general classification of formations, characterized, according to the place which they occupy, as terms of a series, I have deemed it necessary to avoid them with care. Are all the formations beneath the chalk, or even beneath the limestone, with cerithia (coarse limestone of the basin of Paris), without exception, *marine limestones* and *sandstones*? or do the monitors and fishes of the bituminous copper-slate in the alpine limestone of Thuringia; the ichthyosauri of Mr Home, placed beneath the Oxford and Bath oolites, in the lias of England, (which, on the Continent, is represented by a part of the Jura limestone); the crocodiles of Honfleur, contained in clays, with limestone-beds above the oolites of Dive and Isigny limestone, and consequently superior to the Jura limestone,—do all these prove, that there are already beneath the chalk, between that formation and the red sandstone, small fresh-water formations superadded to the great *marine formations*? Do not the coals containing ferns, situated beneath the red-sandstone and beneath the secondary porphyry, present an evident example of a very ancient formation, not of marine origin? These circumstances show the necessity of much caution in the present state of science, when, from characters purely zoological, it is attempted to divide formations, whose unity has appeared confirmed by the alternation of the same beds, and by other phenomena of relative position. (Engelhard und Raumer, *Geogn. Vers.* p. 125.–133.) This caution is so much the more necessary, that, according to the testimony of a mine- | 33 | ralogist who has long attended to this matter, M. Brongniart, "there exists a kind of transition between the marine limestone, and the fresh water gypsum which follows this limestone; and that these two formations do not present, in an abrupt manner, the separation which is seen, in the same places, between the chalk and coarse limestone; that is to say, between two marine formations. It cannot be doubted," adds the same observer, "that the first beds of gypsum have been deposited in a liquid analogous to the sea, while the following ones have been deposited in a liquid analogous to fresh water." (*Geogr. Min.*, p. 168. and 193.)

In announcing the motives which prevent me from generalizing a terminology founded upon the contrast between freshwater productions and marine ones, I am far from disputing the existence of a fresh-water formation, superior to all the other tertiary formations, and which contains only bulimæ, limneæ, cyclostomata and potamides. Recent observations have demonstrated, that this formation is more general than was at first believed. It is a new and last term to add to the geognostical series. We are indebted for the knowledge of this fresh-water limestone, to the useful labours of M. Brongniart. The phenomena which the fresh-water formations present, whose existence was formerly known only by the tufas of Thuringia, and by the Travertine of the plains of Rome, (Reuss, *Geogn.* v. ii. p. 642.; Buch, *Geog. Beob.* v. ii. p. 21.–30.), are conformable, in the most satisfying manner, to the admirable laws discovered by M. Cuvier, in the position of the bones of viviparous quadrupeds. (Brongniart, *Annales du Museum*, vol. xv. p. 357.–581.; Cuvier, *Rech. sur les Ossemens fossiles*, vol. i. p. 54.)

The distinction between fluviatile and marine shells is the object of very delicate inquiries; for it may happen, when the remains of organic bodies are, with difficulty, detached from the siliceous limestone in which they are contained, that ampullariæ may be confounded with naticæ, potamides with cerithia, and so forth. In the family of Conchæ, the cyclades and cyrenæ, the venuses and lucinæ, cannot be separated with certainty, but by the examination of the teeth of the hinge. The work which M. de Ferussac has undertaken, on the terrestrial and fluviatile shells, | 34 | will throw much light on this important subject. Besides, when a genus of pelagic shells is supposed to be seen in the midst of a genus of fresh water ones, it may be asked, if, in fact, the same generic types might not occur in the lakes and in the seas. We are already acquainted with the example of a true fluviatile mytilus. Perhaps the ampullariæ and corbulæ, may be found to present analogous mixtures of marine forms and of fresh-water ones. (See a Memoir of *M. Valenciennes*, inserted in my *Recueil d'Obs. de Zoologie et d'Anatomie comparée*, v. ii. p. 218).

It results from these general considerations regarding the zoological characters and examination of fossil bodies, that, notwithstanding the beautiful and old works of Camper, of Blumenbach and Sömmering, the exact specific determination of species, and the examination of their relations with very modern beds, was only commenced about twenty years ago. I am of opinion, that this examination of fossil bodies applied to all the other secondary and transition strata by geognosts, who, at the same time, consult the relative situation and mineral composition of rocks, far from overturning the system of formations already established, would rather serve to confirm it, to perfect it, and to complete the vast picture which it presents. The geognostical knowledge of formations may, without doubt, be considered under very different points of views, according as the preference is given to the superposition of mineral masses, to their composition (that is to say, their chemical and mechanical analysis), or to the fossils which are found contained in many of those masses. The denominations, *geognosy of situation*, or *of superposition, oryctognostic geognosy* (analysing the texture of masses), *geognosy of fossils*, designate, I shall not say branches of one and the same science, but different classes of relations which are attempted to be isolated for their more particular examination. *This unity of the science, and the vast field which it comprehends, were well known to Werner, the creator of positive geognosy.* Although he did not possess the necessary means for giving himself up to a rigorous determination of fossil species, he did not cease, in his lectures, to fix the attention of his pupils upon the relations which exist between certain petrifactions and the formations of different ages. I have been witness of the | 35 | lively satisfaction which he experienced, when, in 1792, M. de Schlottheim, one of the most distinguished geognosts of the Freyberg School, began to make these relations the principal object of his investigation. Positive geognosy enriches itself with all the discoveries which have been made regarding the mineral constitution of the globe; it contributes the most precious materials to another science, improperly called *Theory of the Earth*; and which embraces the first history of the catastrophes of our planet. It reflects more light upon this science than it receives from it in return; and, without calling in question the ancient fluidity or softening of all the rocky beds (a phenomenon which is manifested by fossil organic remains, by the crystalline aspect of the masses, and by the rolled pebbles or fragments imbedded in the transition and secondary rocks), positive geognosy does not pronounce on the nature of those liquids in which the deposits are alleged to have been formed, on those *waters of granite, porphyry and gypsum*, which, according to hypothetical geology, have successively covered the same points of the globe.[1]

[1] Translated from Humboldt's "Essai Geognostique sur le Gisement des Roches dans les deux hemispheres." 8vo, Paris, 1823.

On the Constitution and Mode of Action of Volcanoes, in different Parts of the Earth[1]

When we consider the influence which scientific travels into distant regions, and a more extended geographical knowledge, have for some centuries past exerted upon the study of nature, we soon discover how this influence has varied according to the objects of inquiry, which have been, on the one hand, the forms of the organic world, and, on the other, the inanimate formation of the earth;—the knowledge of rocks, their relative ages, and origin. Different forms of plants and animals enliven the earth in every zone, as well in the plains, where the heat of the atmosphere is determined by the geographical latitude and the different inflexions of the isothermal lines, as where it changes suddenly on the steep declivities of the mountains. Organic nature gives a peculiar physiognomical character to every zone, which is not the case with the inorganic world where the solid crust of the earth is divested of its vegetable covering. The same rocks approaching | 122 | to and receding from each other in groups occur in both hemispheres, from the equator to the poles. On a distant island, surrounded by strange plants, under a sky where the well-known stars do not shine, the sailor recognises, often with glad surprise, the clayslate which is the common rock of his native country.

This independence of the geognostical relations of places on the present constitution of their climate, does not diminish, but only gives a particular direction to the favourable effect upon the progress of geology and physical geognosy, which is produced by numerous observations made in foreign countries. Every expedition enriches natural history with new plants, and new genera of animals; at one time they are organic forms ranging themselves with well-known types, and representing to us, in its original perfection, a regularly woven, though often apparently interrupted texture of animated creatures; at another, they are forms which appear to be isolated, as vestiges of genera which have been destroyed, or as surprising members of groups still to be discovered. Such a variety is not presented by the examination of the solid crust of the earth; it rather reveals to us an agreement, which excites the admiration of the

[1] Read before the Royal Academy of Sciences of Berlin, Jan. 24, 1823.

geognost, between the parts of which it is composed, in the superposition of masses of different natures, and in their periodical repetition. In the chain of the Andes, as well as in the central mountains of Europe, one formation seems, as it were, to occasion the existence of another; masses of the same character assume similar forms:[2] mountains are formed by basalt and dolerite; steep declivities by dolomite, porphyry, and quader-sandstein; bell-shaped eminences and high-vaulted domes by vitreous trachyte rich in felspar.

In the most distant zones, larger crystals, as it were by internal evolution out of the more compact texture of the greater mass, aggregate into subordinate beds, and thus frequently announce the vicinity of a new and independent formation. Thus is the whole inorganic world reflected, more or less clearly, in every mountain of considerable extent; but in order to ascertain completely the most important phænomena respecting the composition, the relative age, and the origin of the different species of rocks, observations from the most distant parts of the earth must be compared together. Problems which had appeared enigmatical to the geognost in his mother country are solved near the equator. If distant zones do not furnish new species of rocks, that is to say, unknown arrangements of simple substances, as has already been remarked, they yet teach us how to discover the great laws which are every where the same, and according to which, the different strata of the earth support each | 123 | other, appear in the form of veins, or are elevated by elastic powers.

We need not be surprised, that, notwithstanding the great assistance which our geological information derives from inquiries, having whole countries for their object, an extensive class of phænomena (with which I venture to entertain this assembly), has been treated, during so long a period, in a confined manner; the points of comparison being more difficult, and, I might say, more troublesome to find. Whatever we believed we knew, until the end of the last century, respecting the form of volcanoes, and the action of their subterraneous forces, had been derived from two mountains of the south of Italy,—from Ætna, and from Vesuvius. The first being more accessible, and having, like all low volcanoes, more frequent eruptions, has served for a type, according to which a whole distant world,—the powerful volcanoes of Mexico, South America, and

[2] In an imperfect translation of this paper, which has been forwarded to the Editor from the Continent, a word here occurs which cannot be decyphered; and on account of other inaccuracies which it has been necessary to correct, unaided by the original, the translation, as now given, is not to be regarded as exact in every particular.

the Asiatic Islands, has been considered. Such a method recalls to our remembrance the shepherd of Virgil, who expected his narrow cottage to contain the ideal of the eternal city, imperial Rome.

A careful examination of the whole Mediterranean, and principally of its easterly islands and shores, where mankind first awakened to mental culture, and to noble feelings, might certainly have dispelled such a narrow idea of nature. Out of the deep bed of the sea, among the Sporades, rocks of trachyte have arisen, like the Azoric island, which has thrice reappeared during three centuries, the intervening periods being almost equal. Between Epidaurus and Troezene, near Methone, the Peloponnesus has a Monte Nuovo which has been described by Strabo, and seen by Dodwell, higher than the Monte Nuovo of the Campi Phlegræi, near Baia; perhaps higher than the new volcano of Xorullo in the plains of Mexico, which I have found among a thousand basaltic cones, raised out of the earth, and still smoking. In the bason of the Mediterranean Sea also, the volcanic fire bursts forth, and not only from permanent craters, from isolated mountains which preserve a lasting communication with the interior of the earth, like Stromboli, Vesuvius, and Ætna; —on Ischia, near the Epomæus, and also, as it would appear from the reports of the ancients, near Chalcis in the Lelantic plains, has lava flowed out of fissures which have suddenly opened. Besides these phænomena, which have taken place in the period of history within the narrow limits of certain traditions, and which Ritter will collect and explain in his masterly Geography, the shores of the Mediterranean contain abundant remains of more ancient igneous effects. The south of France shows, in Auvergne, a range of hills, in which *bells* of trachyte occur alternately with cones of eruption, from which currents of lava have descended. The Lombardic plain, which forms the innermost bay of the Adriatic Sea, surrounds the trachyte of the Euganean Hills, where domes of granular trachyte, of obsidian, and of pearlstone, rise, which, passing into each other, break through the Jura | 124 | limestone, but never occur in narrow streams which have flowed. Similar evidences of former revolutions may be found in many parts of the Grecian continent, and in Asia Minor, countries which will afford the geognost copious subjects for examination, when the light once returns to the land whence it first beamed over the western world—when tormented mankind ceases to sink under the savage lethargy of the Ottoman.

I mention the geographical neighbourhood of so many phænomena, in order to prove, that the bed of the Mediterranean, with all its chains of islands, might have afforded to the attentive observer, every thing that has been discovered, in latter periods, under the most varied forms, in South America, on Teneriffe, or on the Aleutian islands, near the polar regions. There were accumulated objects for observation, but tours into distant regions, and the comparison of large tracts of country within and beyond Europe, were necessary, in order to discover what was common to all these phænomena, and to learn, clearly, their dependence on each other.

By the usage of language, which often gives stability and respect to the first erroneous views of things, but often, as it were, by instinct, distinguishes the truth, we apply the term volcanic to all eruptions of subterranean and melted matter; to columns of smoke and steam, which rise sporadically out of rocks, as at Colares after the great earthquake at Lisbon; to Salsæ, or conical hills of clay which emit mud, asphaltum, and hydrogen, as those near Girgenti, in Sicily, and near Turbaco, in South America; to hot Geyser springs which rise by the pressure of elastic vapours; and, in general, to all violent powers of nature which have their seat deep in the interior of our planet. In the Spanish main of America, and in the Philippine islands, the inhabitants make a distinction between igneous and aqueous volcanoes, *vulcanes de agua y de fuego*: they apply the first name to mountains, which, during violent earthquakes, from time to time, eject subterraneous water, and with a dull noise.

Without denying the connexion between the different phænomena just mentioned, it seems advisable to give a distinct language to the physical as well as to the oryctognostic branch of geognosy; and not to apply the term volcano in one instance to a mountain that terminates in a permanent crater; and in another, to every subterranean cause of volcanic phænomena.

In the present state of the earth, the most common from of volcanic eminences is that of isolated cones; such are Vesuvius, Ætna, the Peak of Teneriffe, Tunguragna, and Cotopaxi. I have seen them of every magnitude, from the lowest hills to mountains rising to the height of 17,700 feet above the level of the sea. Besides these conical mountains, there are other craters, permanently communicating with the interior of the earth, situated upon lengthened craggy ranges of mountains, not always in the middle of their wall-like summits, but towards the end, and near their declivities. Such is Pichincha which rises | 125 | between the Pacific Ocean and the town of Quito, and which has become celebrated by Bouguer's earliest formula for the barometer; such also are the volcanoes that rise in the plain de los Pastos, at the elevation of 10,000 feet.

All these differently formed summits consist of trachyte, or trap-porphyry, a granular rock, full of cracks and fissures, and composed of glassy felspar and hornblende, but often containing in addition, augite, mica, laminar felspar, and quartz.

Where the evidence of the first eruption, and where the first scaffolding, I might say, has been entirely preserved, the isolated conical hills are surrounded by a high wall of rocks forming a circus, consisting of superposed strata; such walls, or annular surrounding masses, are called *craters of elevation*; of these very important phænomena, Leopold von Buch, the first geognost of our times, from whose works I have taken several views contained in this paper, read a remarkable account, five years ago.

The volcanoes which communicate with the atmosphere by means of craters, and the conical hills of basalt and bell-shaped trachytic hills without craters, the latter either low like Sarcouy, or high like Chimborazo, form different groups. A geographical comparison shows, in one place, small Archipelagi, or, as it were, classed systems of mountains, either with craters and currents of lava, as in the Canaries and Azores, or devoid of craters and real currents of lava in the Euganeans, and the Siebengebirge near Bonn; or it shows, in other places, single and double chains of volcanoes, connected with each other, and forming tracts of many hundred miles in length, which are either parallel to the direction of the mountains, as in Guatimala, Peru, and Java, or in directions perpendicular to their axis, as in the land of the Aztekes, where none but volcanic trachytemountains attain the limits of eternal snow, and those, probably, have been thrust out of a fissure nearly 500 miles in length, which divides the whole continent, from the Pacific Ocean to the Atlantic.

This aggregation of volcanoes either in single round groups, or in double ranges, affords the most determinate proof that volcanic effects do not depend upon slight causes existing near the surface of the earth, but that they are great and deeply founded phænomena. The whole eastern part of the American continent, which is poor in metals, is at present without craters, without trachyte, probably even without basalt. All the volcanoes are situated in the part opposite to Asia, in the meridian line of the Andes chain, 1800 geographical miles long; the whole of the elevated district of Quito is nothing but a single volcanic hearth, the summits of which are Pichincha, Cotopaxi, and Tunguragua. The volcanic fire now bursts forth from one, and then from another of these apertures, which we are accustomed to consider as separate volcanoes.

The progressive motion of the fire here, in the space of three centuries, turned from north to south. The earthquakes with | 126 | which this part of the world is so terribly visited, furnish remarkable evidences of the existence of subterraneous communication, not only between countries without volcanoes, as was known long ago, but even between craters which are far distant from each other. Thus the volcano of Pasto, situated to the east of the river Guaytara, uninterruptedly vomited a high column of smoke, during three months of the year 1797; and this column disappeared at the very moment, when, at the distance of nearly 300 miles, the great earthquake of Riobamba and the mud eruption of the Moya, killed from 30,000 to 40,000 Indians. The sudden appearance of the Azoric island Sabrina, on the 30th of January, 1811, was the forerunner of those dreadful shocks, which, further to the west, shook, almost uninterruptedly, from the month of May, 1811, to that of June, 1813, first the Antilles, afterwards the plains of the Ohio and the Mississippi, and at last the opposite coast of Venezuela. Thirty days after the complete destruction of the town of Caraccas, the eruption of the volcano of St. Vincent in the neighbouring Antilles took place; at the same moment when this explosion happened, on the 30th of April, 1811, a subterranean noise was heard throughout a country of 2200 geographical square miles, or 47,900 English square miles, in extent.

The inhabitants near the Apure, where it is joined by the Rio Nula, as well as those of the most distant part of the coast, compared this noise to that of artillery. From where the Rio Nula falls into the Apure, through which river I came into the Orinoco, to the volcano of St. Vincent, the distance, in a direct line, is 731 English miles. The noise just alluded to, which certainly was not communicated through the air, must, therefore, have had a deep internal cause. Its intensity on the coast of the Antillic sea was scarcely greater than in the interior of the country.

It would be useless to augment the number of examples, but for the purpose of recalling to memory a phenomenon which has become historically interesting to Europe, I will mention the earthquake at Lisbon. At the same time with this, on the 1st of November, 1755, not only were the Swiss lakes, and the sea on the Swedish shores violently agitated, but even in the easterly Antilles, around Martinique, Antigua, and Barbadoes, where the tide never exceeds 28 inches, it suddenly rose to 20 feet. All these phænomena prove, that the subterranean powers act either dynamically, by producing tension and vibration, as in earthquakes; or chemically, by producing or altering substances, as in volcanoes. They prove, likewise, that these powers do not act from superficial causes, from the exterior crust of the earth; but from deeply-seated causes, from the interior of our planet; extending their simultaneous effects to the most distant parts of the earth, through fissures and empty veins.

The more different the structure of volcanoes; that is to say, of those raised masses which surround the canal through which | 127 | the melted substances proceed from the interior of the earth to its surface, the more important is it to become thoroughly acquainted with that structure, by exact measurement. The interest attached to this measurement, which has been a particular object of my examination in another part of the world, is heightened by the consideration, that that which is to be measured is a variable magnitude. The physiognomy of nature consists in the change of phænomena tending to connect the present with the past. In order to ascertain a periodical return, or the laws of progressive natural changes in general, certain fixed points are necessary; and observations carefully made at stated periods, may serve for numerical comparison. Had the mean temperature of the atmosphere in different latitudes been observed for a few thousand years, and the mean height of the barometer at the level of the sea, we might now know in what proportion the heat of different climates has increased, or diminished, and whether the height of the atmosphere has undergone any changes. Similar points for comparison are required, for the variation and the declination of the magnetic needle, and for the intensity of the electromagnetic power, upon which two excellent philosophers of this Academy have thrown so much light. If it be a praiseworthy undertaking of learned societies to inquire assiduously into the changes of temperature undergone by the globe, into those which take place in the pressure of the atmosphere, and in the magnetic variation,— it is the duty of a travelling geognost, in ascertaining the inequality of the earth's surface, to consider, principally, the variable height of the volcanoes. What I formerly attempted on the mountains of Mexico, on the Toluca Nauhiampatepetl and Xorullo, and in the Andes of Quito, on the Pichincha, I have found opportunity, since my return to Europe, to repeat at different periods on Vesuvius. Saussure measured this mountain in 1773, at the time when both sides of the crater, the south-eastern and north-western, appeared to be of equal altitude; he found their height to be 609 toises (3894 feet) above the level of the sea. The eruption of 1794 occasioned a fall on the south side, which even the unaccustomed eye discovers at a great distance. In 1805, I measured Vesuvius three times, in conjunction with M. von Buch, and M. Gay-Lussac; we found the elevation of the northern edge, opposite to Monte Somma, la Rocca del Palo, to be exactly the same as Saussure had before determined it; the southern edge we found 71 toises (454 feet) lower than it was in 1773; the total height of the volcano on the side opposite Torre del Greco (towards which side the fire seems to have acted the most powerfully, during the last 30 years), had diminished one-ninth part.

The cone of ashes on Vesuvius bears the proportion of one-third to the height of the whole mountain, that on Pichincha is as 1 to 10, and that on the Peak of Teneriffe as 1 to 22; Vesuvius has, therefore, the largest cone of ashes in proportion, | 128 | because, probably, as a low volcano, it has acted principally through its summit. A few months ago, I succeeded not only in repeating my former measurements on Vesuvius, but also in ascertaining the elevation of all the edges of the crater. This work, perhaps, deserves some consideration, for the periods at which it was executed include those of the great eruptions from 1805 to 1822, and it is, perhaps, the only admeasurement yet published of any volcano which may be compared in all its parts. It proves that the edges of the craters, not only where they evidently consist of trachyte, as in the volcanoes of the Andes, but likewise everywhere else, are much more constant phænomena than has hitherto been believed. Simple angles of elevation ascertained from the same points are more proper for these examinations than barometrical and trigonometrical measurements. According to my last determination, the north-western edge of the crater of Vesuvius has not changed its form in the least since Saussure's time, a period of 49 years. The south-eastern edge towards Bosche tre Case, which became about 450 feet lower in 1794, has sunk very little since that time.

If in the description of great eruptions, in the public papers, the completely changed form of Vesuvius has frequently been mentioned, if this opinion often seems to be corroborated by the picturesque views of the mountain made at Naples, the cause of this mistake may be found in the circumstance, that the outlines of the edges of the crater have been confounded with those of the cone of eruption which is accidentally formed in the middle of the crater, upon a bottom that has been raised by vapours. Such a cone of eruption, consisting of rapilli and slags loosely heaped together, has become visible over the south-eastern edge of the crater, since 1816 and 1818. The eruption of February, 1822, had so much increased it that it had become from 70 to 80 feet higher than the north-eastern edge of the crater, Rocca del Palo. This remarkable cone, which, at Naples, they were accustomed to consider as the true summit of Vesuvius, fell in with a tremendous noise, during the eruption of the 22d of October, so that the bottom of the crater, which had been uninterruptedly accessible from the year 1811, now lies 850 English feet beneath the northern edge, and about 213 feet deeper than the southern edge of the volcano. The variable form and relative situation of the crater of eruption, the opening of which must not be taken for the real crater of the volcano, as frequently has been done, gives, at different times, a peculiar physiognomy to Vesuvius; and the historiographer of that volcano, from the mere outline of the summit, and the relative height of the northern or southern side of the mountain, as it is drawn in Hackert's Views in the palace of Portici, would guess the year in which the artist made the sketch of his picture.

In the night between the 23d and 24th of October, one day after the fall of the cone of slags 400 feet in height, when small | 129 | but numerous currents of lava had already flowed, the fiery eruption of ashes and rapilli began. It continued uninterruptedly for twelve days, but was most violent during the first four. During this time the detonations in the interior of the volcano were so violent, that the mere concussion of the air (no earthquake had been observed) caused the roofs to burst in the palace of Portici. In the surrounding villages of Resina, Torre del Greco, Torre del Annonciata, and Bosche tre Case, an interesting phænomenon was observed; the atmosphere was so thickly filled with ashes, that the most intense darkness overspread the whole country for several hours in the middle of the day. The people walked in the streets with lanterns, as is often done at Quito when Pichincha is in eruption. The flight of the inhabitants was never more general; currents of lava were less feared than a fall of ashes, a phenomenon which was unknown there with such violence, and in consequence of the relations respecting the destruction of Herculaneum, Pompeii, and Stabiæ, filled the minds of the people with frightful images.

The hot steam which rose from the crater during the eruption and passed into the atmosphere, formed on cooling a thick mass of clouds, around the column of ashes and fire, 9000 feet in height. This sudden condensation of steam, and, as Gay-Lussac has shown, the very formation of the clouds, increases the electric tension. Lightnings burst forth in all directions from the column of ashes, and the rolling thunder might clearly be distinguished from the interior noise of the volcano. At no former eruption had the play of electric charges been so surprising.

On the morning of the 26th of October, a singular account was circulated, that a current of boiling water had issued from the crater, and rushed down from the cone of ashes. Monticelli, the zealous and learned observer of the volcano, soon discerned that the rumour had been occasioned by an optical deception. The supposed current of water was nothing but a dry mass of ashes, which flowed down, like quicksand, from a fissure in the superior edge of the crater. A drought, which had completely desolated the fields, preceded the eruption, but the volcanic thunderstorm occasioned, towards its termination, a very heavy and continued rain. Such a phenomenon characterizes the conclusion of an eruption in every zone. On account of the cone of ashes being generally covered with clouds during this time, and likewise because the torrents of rain are heaviest in its neighbourhood, currents of mud flow down on all sides. The affrighted peasant considers it to be water which has risen from the interior of the crater, and the deceived geognost conceives that he recognizes in it either sea-water, or mud-like volcanic productions, which are called *eruptions boueuses*, or, as the old French systematic writers termed them, products of a fiery-aqueous liquefaction. | 130 |

When the summits of volcanoes (as is generally the case in the chain of the Andes), extend into the region of eternal snow, or even to double the height of Ætna, the melted snow renders the inundations amazingly frequent and destructive. They are phænomena meteorologically connected with volcanic eruptions, and are multifariously modified by the altitude of the mountains, the extent of their summits covered with eternal snow, and the calefaction of the sides of the cone of ashes; but they should never be considered as real volcanic phænomena. Subterranean lakes, in connexion with alpine rivers, are formed both on the slopes and at the feet of the mountains. When the earthquakes which precede every eruption in the chain of the Andes, shake with mighty force the entire mass of the volcano, the subterranean vaults are opened, and emit, at the same time, water, fishes, and tufa-mud. This is the singular phenomenon that furnishes the fish *pimelodes cyclopum*, which the inhabitants of the high lands of Quito call *preñadilla*, and which was described by me soon after my return. When the summit of the mountain Carguairazo, to the north of Chimborazo, and 18,000 feet high, fell, in the night between the 19th and 20th of June, 1698, the surrounding fields, to the extent of about 43 English square miles, were covered with mud and fishes. The fever which raged in the town of Ibarra, seven years before, had been ascribed to a similar eruption of fishes from the volcano Imbaburu. I recur to these facts, because they throw some light on the difference between the eruption of ashes, and that of mud-like masses of tufa and trass, which contain wood, coal, and shells.

The quantity of ashes ejected by Vesuvius in the late eruptions, like all other things which are connected with great and appalling phænomena, has been enormously exaggerated in the public papers; and two Neapolitan chemists, Vincenzo Pepe, and Giuseppe di Nobili, have affirmed, that they contain gold and silver, notwithstanding the contradiction of Monticelli and Covelli.[3] According to my examination, the stratum of ashes which had fallen in twelve days, towards Bosche tre Case, on the slope of the cone, where rapilli were mixed with it, was only three feet in thickness, and in the plain, it did not exceed from 15 to 18 inches. Measurements of this kind must not be made in places where the ashes have been drifted by wind, like snow, or sand, nor in those where they have been accumulated by water. The times are past in which we sought only for the marvellous in volcanic phænomena, and, like Ctesias, made the ashes of Ætna fly to the Indian peninsula. Some of the Mexican gold and silver mines are certainly in trachytic porphyry, but in the ashes of Vesuvius which I collected, and which, at my desire, have been analyzed by Henry Rose, of | 131 | Berlin, an excellent chemist, no traces of either metal, could be discovered.

However great may be the discrepancy between the results that I have here given, but which agree with Monticelli's more exact observations, and those which have been circulated during several months past, yet the eruption of ashes from Vesuvius, from the 24th to the 28th of October, still remains the most remarkable of which we have any certain account since the death of the elder Pliny. Its quantity, perhaps, was three times as great as that of all the ashes, collectively, which have been observed to fall, during the time in which volcanic phænomena have been attentively considered. A stratum of from 15 to 18 inches in thickness, seems at first view unimportant, if compared to the mass with which we find Pompeii to be covered; but without speaking of the torrents and inundations which certainly may have increased this mass for centuries, without renewing the violent dispute concerning the cause of the destruction of the Campanian towns, which has been carried on with so much scepticism on the other side of the Alps, it may be affirmed that the eruptions of one and the same volcano at distant periods can by no means be compared with respect to their intensity. All conclusions founded on analogy are insufficient, when the question is about quantitative proportions,—the quantity of ashes and lava, the height of the column of smoke, or the violence of the detonation.

From the geographical description of Strabo, and from an opinion of Vitruvius concerning the volcanic origin of pumice, we see that until the year in which Vespasian died, that is to say, until the eruption which overwhelmed Pompeii, Vesuvius was more like an extinguished volcano than a solfatara.

---

[3] See Annals, v. 236.

When after long rest the subterranean powers suddenly open new passages, and again break through beds of primitive rocks and of trachyte, effects must necessarily take place, for which all the phænomena subsequently observed do not afford any standard of comparison. It may be clearly seen from the well-known letter in which the younger Pliny announces the death of his uncle to Tacitus, that the recommencement of the eruptions, I might say, the awakening of the dormant volcano, began with an eruption of ashes. The same circumstance was observed at Xorullo, in Sept. 1759, when the new volcano, breaking through beds of syenite and trachyte, suddenly arose in the plain. The peasants fled, because they found in their huts, ashes that had been ejected from the fissures of the earth, which was burst in every place. Every partial eruption, in the periodical general eruptions of volcanoes terminates with a shower of ashes.

There is a passage in Pliny's letter, which shows, that the dry ashes which had fallen from the air had already attained a height of from four to five feet, in the commencement of the eruption, | 132 | and without the effect of accumulation by water. "The court which led to his [uncle's] apartment," he says, "being now almost filled with ashes and pumice, it would have been impossible for him, if he had continued there any longer, to have made his way out." In the narrow space of a court, the wind could not have had any great effect in accumulating the ashes.

I have ventured to interrupt my comparative view of volcanoes by observations solely on Vesuvius, partly on account of the great interest which the last eruption has excited, and partly because every great fall of ashes almost involuntarily reminds us of the classic ground of Pompeii and Herculaneum.[4] We have hitherto considered the form and the effects of those volcanoes which are in permanent communication with the interior of the earth, by means of a crater. Their summits are raised masses of trachyte and lava, intersected by numerous veins; the duration of their effects causes us to believe that they have a very stable and undisturbed structure. They possess, I may say, a more individual character, which remains the same during long periods. Neighbouring mountains often furnish completely different products, leucite-lava, and felspar-lava; obsidian, with pumice, and basaltic masses containing olivine. They belong to the newer phænomena of the earth, pass generally through all the strata of secondary rocks, and their eruptions and currents of lava

[4] The author here mentions a paper on the data of his measurements at Vesuvius, which was unsuitable for reading; and then proceeds to notice a collection of minerals that he brought with him, and which will be added to the Royal Museum at Berlin.

are of later origin than our valleys. Their life, if I may use that expression, depends upon the manner and duration of their connexion with the interior of the earth. They often rest for centuries, suddenly take fire again, and terminate as solfataras, which emit steam, gases, and acids. Sometimes, as on the Peak of Teneriffe, their summit has already become such a depository of reproduced sulphur, while mighty currents of lava flow from the sides of the mountain, like basalt below, and above, where the pressure is less, like obsidian with pumice.

Independently of these with permanent craters, volcanic phænomena of another kind exist, which have been observed less frequently, but are principally interesting in geognosy, and remind us of the primitive world; that is to say, of the earliest revolutions of our earth. Mountains of trachyte suddenly open, eject lava and ashes, and close again, perhaps, for ever: thus was it with the mighty Antisana; and thus with the Epomæus, in Ischia, in 1302. Such an eruption sometimes takes place even in the plain, as in the high lands of Quito; in Iceland, far from Hecla; and in Eubœa, in the Lelantic fields. Many of the islands which have been raised up are owing to these temporary phænomena. In these cases the communication with the interior of the earth is not permanent, and the effect ceases as soon as the fissure, which is the communicating channel, is | 133 | closed again. The veins of basalt, dolerite, and porphyry, which, in different parts of the world, pass through every formation; and those of syenite, augite-porphyry, and amygdaloid, which are characteristic of the newest strata of the transition formation, and of the oldest rocks of the secondary strata, have probably been formed in a similar manner. In the first age of our planet, the yet liquid substances penetrated through the crust of the earth, which was every where intersected by fissures, and assumed the form of granular rocks, either in veins, or spreading over and expanding themselves in strata. The rocks strictly volcanic which the primitive ages have afforded us, have not flowed in currents like the lava of our insulated conical hills; the same mixture of augite, titaniferous iron, glassy felspar, and hornblende, may have existed at different periods, but at one time it may have approached nearer to basalt, and at others to trachyte; the chemical substances may have combined in a crystalline form, in distinct proportions, as we are taught by M. Mitscherlich's new and important labours, and by the analogy of artificial products of fire: we find that substances similarly formed have arrived at the surface of the earth in very different ways; they have either been merely raised, or protruded by temporary fissures through the older strata; that is to say, through the already oxidized surface of the earth; or they have flowed, as currents of lava, from conical hills with a permanent crater. By confounding such different phænomena together, the geognosy of volcanoes is carried back to that darkness from which a great number of comparative observations are beginning to extricate it.

The question has often been asked, What is it that burns in volcanoes? What was it that excited the heat by which earths and metals were melted? Modern chemistry answers, that the substances which melt are the metals of the earths and alkalies. The solid crust of the earth, already oxidized, separated the surrounding air with its oxygen, from the combustible unoxidized substances of the interior of our planet. The observations which have been made in mines and caves in every zone, and which, in conjunction with M. Arago, I have collected in a particular paper, demonstrate that the heat of the mass of the earth is yet much greater than the mean temperature of the atmosphere at the same place. Such a remarkable and almost generally proved fact, is closely connected with those which are proved by volcanic phænomena. Laplace has even gone so far as to endeavour to calculate the depth at which the body of the earth may be considered to be a melted mass. Whatever doubts may be entertained, notwithstanding the veneration due to so great a name, with respect to the numerical certainty of such a calculation, thus much remains probable; that all volcanic phænomena originate in a very simple cause, in | 134 | a permanent or in a variable communication between the interior and the exterior of our planet.

The pressure of elastic vapour forces the melted substances upwards through deep fissures while they are undergoing oxidation; volcanoes, if I may so speak, are intermitting springs of the earth; the liquid mixtures of metals, alkalies, and earths, which on cooling become currents of lava, flow quietly when they are raised, and find a vent. The ancients imagined, according to Plato's Phædon, that all volcanic currents of fire flowed, in a similar way, from the Periphlegeton.

It may be permitted me, perhaps, to add to these considerations one which is still more hazardous. In this interior heat of the earth, indicated by experiments with the thermometer, and by observations on volcanoes, the cause, perhaps, may be found, of one of the most wonderful phænomena which the examination of fossils presents to us. Tropical forms of animals, arboriform ferns, palms, and bamboo-like plants, lie interred in the cold north. The primitive world every where shows a distribution of organic forms at variance with the then existing nature of the climate. In order to solve this important problem, several hypotheses have been invented; as the neighbourhood of a comet, the altered inclination of the ecliptic, the increased intensity of the solar light. Neither of these has been sufficient to satisfy at once the astronomer, the natural philosopher, and the geognost. For my part, I leave the axis of the earth unaltered, as well as the light of the solar disc, by the spots on which, a celebrated astronomer has explained both the fertility and the unfruitfulness of the fields; but I believe, that in every planet, independently of its relation to a central body, and of its astronomical situation, various causes exist of the production of heat; oxidation, precipitation, and a change in the capacity of bodies; by increase of electromagnetic charge, by the opening of a communication between the interior and the exterior part of the earth.

Where the deeply cleft crust of the earth in the primitive world radiated heat from its fissures, whole countries, perhaps, could produce for centuries, palms and arborescent ferns, and sustain all the animals of the torrid zone. According to this view, to which I have already alluded in a work just published, "Essai Géognostique sur le Gissement des Roches dans les deux Hemispheres," the temperature of volcanoes would be that of the interior of the earth itself, and the same cause which now occasions such dreadful destruction, would once have occasioned, on the newly oxidated crust of the earth, upon the deeply cleft strata of rocks, the most luxuriant growth of plants in every zone.

Even if any one should be inclined to suppose, in order to explain the marvellous distribution of tropical forms in their | 135 | ancient graves, that shaggy animals of the elephant tribe now imbedded in icebergs, were once peculiar to a northern climate, and that similar forms belonging to the same primary types, like lions and lynxes, could live in very different climates, such an explanation could not, however, be extended to the products of vegetation. For reasons which the physiology of plants explains, palms, and arboriform monocoty ledones cannot sustain the northern cold, and in the geological problem we here speak of, it seems difficult to me to separate plants and animals. The same explanation must be applied to both.

Towards the end of this paper, I have combined uncertain hypothetical suppositions with facts collected from the most different parts of the world. The philosophical knowledge of nature rises above a mere description of nature. It does not consist in a sterile aggregation of isolated observations. It may sometimes be allowed, therefore, to the curious and ever-active mind of man, to look back upon the past, to imagine what cannot be clearly known, and to amuse himself with the ancient, and, under many forms, returning mysteries of geogony.

Analysis of the Water of the Rio Vinagre, in the Andes of Popayan, by M. Mariano de Rivero; with geognostic and Physical Illustrations of some Phenomena which are exhibited by Sulphur, Sulphuretted Hydrogen, and Water, in Volcanoes

"In compliance with the desire of M. de Humboldt, I procured some of the water of Rio Vinagre. It was sent to me by M. Torrés, who takes an interest in all that can contribute to scientific researches. This water has yielded per litre: sulphuric acid, 1.080; muriatic acid, 0.184; alumine, 0.240; lime, 0.160; and some indications of iron.[1] The presence of muriatic acid confirms the observations made on the vapours and the stony productions of Vesuvius and of several other volcanoes."

I had made known, at the time of my return from America, the presence of the sulphuric and muriatic acids in the water of the Rio Vinagre, which the aborigines call Pusambio. (See Views of the Cordilleras, and Monuments of the People of America, vol. ii. p. 166; Barometric Levelling of the Andes, No. 126; *Caldas, Samanario del Nuevo Reyno de Granada*, t. i. p. 265); but not being furnished with the salts of barytes. I had engaged MM. Rivero and Boussingault, when they departed from Bogota, to verify these facts. The analysis, which we owe to one of these expert chemists, is the first which has been attempted on the water of the Rio Vinagre. I shall give some extracts from the journal of my travels, in great part still unpublished, explanatory of the local circumstances.

The town of Popayan is situated in the beautiful valley of the Rio Cauca, on the Bogota road to Quito, at the foot of the two great volcanoes of Puracé and Sotarà. | 461 | These volcanoes, almost extinct, and exhibiting only the phenomena of solfataras, form part of the central chain of the Andes of New Granada. At 1° 55′ and 2° 20′ of north latitude, the group of mountains which incloses the sources of the Magdalena is

[1] It cannot be doubted that the indications are by grammes and fractions of grammes; a litre of the water of the Rio Vinagre includes 1.080 grammes of sulphuric acid, and 0.184 grammes of muriatic acid. This proportion of sulphuric acid is nevertheless very sensible to the taste, and is manifest by an abundant precipitate with the salts of barytes. [The litre being 2.113 pints, the contents of the water in English grains will be as follows: sulphuric acid 16.68; muriatic acid 2.84; alumine 3.7; lime 2.47. Edit.]

divided into three branches, of which the eastern is continued towards Timana and the Nevados of Chita and of Merida; the intermediate and central one towards the Paramos of Guanacas and of Quimdiú; the western towards the platiniferous district of the Choco and the isthmus of Panama. In ascending from the town of Popayan to the summit of the volcano of Puracé, M. Bonpland and I found at the height of 8672 feet a little plain (Llane del Corazon), inhabited by poor Indian husbandmen. This plain is separated from the rest of the acclivity, with which it would otherwise be continuous, by two ravines extremely deep: it is at the edge of these precipices that the village of Puracé is built. Springs rise every where from the trachytic rock; each garden is surrounded with a quickset hedge of narrow-leaved euphorbium (*lechero*) of the most delicate green. This beautiful verdure contrasts in a striking manner with the background of black and arid mountains, which surround the volcano, and which are rent by the effects of the earthquakes.

The site of the village is celebrated in the country on account of three beautiful cascades [*choreras*] of the river of Pusambio, whose water is acid, and which the people, who know no other acid than vinegar, call *Rio Vinagre*, sometimes *Gran Vinagre*. This river takes its rise at the height of nearly 10,871 feet, in a very inaccessible spot. Although the temperature of the water be little different in the lower cascades from that of the surrounding atmosphere, it is no less certain that the sources of the Pusambio or Venagre are very hot. This fact was attested to me by the natives and by the missionary of the village of Puracé. In going to the summit of the volcano I saw a column of smoke rise at the place where the acid waters make their appearance. I have drawn the second of the falls of the Vinagre (plate xxx of the Views of the Cordilleras): the water which opens itself a passage across a cavern, is precipitated more than 383 feet in depth. The fall has a very picturesque effect; but the inhabitants of Popayan would be better pleased if the river, instead of throwing itself into the Rio Cauca, became engulfed in some other crevice; for such is the delicacy of constitution of animals which breath by gills, and which absorb the oxygen dissolved | 462 | in the water, that the Cauca, during a course of four leagues, is destitute of fish, on account of the mixture of its waters with those of the Rio Vinagre,[2] which are charged both with oxide of iron and with sulphuric and muriatic acid. After staying a considerable time on the craggy wall of rock which

---

[2] M. Caldas has even attributed to this mixture, doubtless with little reason, the absence of goitres in the valley of Rio Cauca.—*Semanario*, t. l. p. 265. See my Memoir on the Goitres in the Cordilleras. (*Magendie, Jour. de Physiol.* t. iv, p. 109)

borders the cascade, a pricking sensation is felt in the eyes from the minute spray in the atmosphere. Fish re-appear in the Rio Cauca at the point where it becomes enlarged by the influx of the Pindamon and of the Placé.[3]

A little to the north of the sources of the Pusambio rise two other rivulets charged in like manner with free sulphuric acid, which the people call the Little Vinegars (*los dos Vinagres chicos*): they throw themselves into the Rio de San Francisco, which is itself but a tributary of the Gran Vinagre. During my stay at Popayan it was an opinion generally received that all these acid waters contained some iron dissolved by a great quantity of carbonic acid. When it was merely remembered that the sources of the Vinagre are very hot, this opinion ought to have been abandoned. I boiled some water taken from the cascade; and I found, after the ebullition, the same acid taste and the precipitates as in the unboiled water. At this period I had very few re-agents left.

The nitrate of silver[4] gave a white and milky precipitate, indicating the presence of muriates. The presence of iron was shown by the prussiate of lime, that of lime by the oxalate of potash. When the water was weighed with great care in the office of the mint of Popayan, the weight of an equal quantity of the water of the Vinagre was found to be to that of distilled water as 2735 1/2 gr. to 2731 gr.; that is to say that the specific gravity of the water of the cascade was 1.00 5.

The waters which I describe, and of which M. Rivero has given the first analysis, must not be confounded with those of the two *subterraneous lakes* which we have found near the summit of the volcano; one is 14,356 feet high, the other, above the snows, 15.475 feet. This volcano of Puracé is a dome | 463 | of semivitreous trachyte of a bluish grey, and having a conchoidal fracture. It does not present a great crater at its summit, but several little mouths. It differs very much from the neighbouring volcano, the Sotarà, which is of a conical form, and which has thrown out an immense quantity of obsidians. These masses, covering the plains of Julumito, are balls or tears of obsidian, the surface of which is often tubercular. They present, what I have seen no where else in the two hemispheres, all the shades of colour, from deep black to that of an artificial glass entirely colourless. It may appear surprising to see that this deprivation of colour has not been accompanied by any inflation or

[3] *Journal de Physique*, t. lxii, p. 61.

[4] The conjoint presence of the sulphuric and muriatic acids has also been observed by M. Vauquelin in the water which M. Leschenault had taken from the crater-lake of Mount Idienne in Java (*Journal de Physique*, t. lxv. p. 406.) See Phil. Mag. vol. xlii. pp. 126, 182.

porosity. The obsidians of Sotarà are mixed with fragments of enamel which resemble the porcelain of Réaumur, and adhering to which I have found masses of felspar which have resisted fusion.

Here, as in the Andes of Quito, as at Mexico, and at the Canary Islands, the system of basaltic rocks lies far from the trachytes which form the volcano of Puracé and of Sotarà. The basalts of the Tetilla of Julumito belong only to the left bank of the Cauca. They rise from transition porphyries free from augite, containing some hornblende, a very little quartz in small crystals embedded in the mass, and a felspar which passes from the common to the vitreous variety. This porphyry is covered, near to Los Serillos, with a blackish-grey limestone, traversed by veins of carbonate of lime, and so much overcharged with carbon that in some parts it stains the fingers like an aluminous schist, or like the Lydian[5] stone of Steeben in the Fichtelgebirge.

The trachytic dome of Puracé which gives birth to the little river of sulphuric acid, rises out of a porphyritic syenite (with common felspar), which in its turn is superimposed on transition granite abounding in mica. This observation,[6] very important for the position of volcanic rocks, may be made near to Santa Barbara in ascending from Popayan to the village of | 464 | Puracé. The volcano, like the most part of the great volcanoes of the Andes, presents layers or mantles of melted stony matter, not real currents of lava. Some fragments of granular limestone, probably magnesian, which I found at more than 12,790 feet high, seem to have been thrown up through crevices which have since become closed. They are like those of the *Fosso Grande* of Vesuvius, which owe their granular texture to volcanic fire. It is not possible to go on horse back further than the cascades of the Rio Vinagre. From thence we were eight hours in mounting on foot to the summit of the volcano and in descending from it. The weather was dreadful; snow and hail fell. I had a great deal of difficulty in lighting the tinder the point of the conductor of Volta's electrometer; the balls of elder pith separated from 5 to 6 lines, and the electricity passed often from positive to negative without there being any other symptom of storm: for thunder and lightning are (according to my

[5] M. Vauquelin has recently proved by a direct analysis the presence of carboa in the purest lydian stones. I had found, in a series of experiments made on the galvanic exciters in 1798, that the lydian stones of the transition schists of Steeben produced jointly with zinc the same effect as graphite or carburet of iron. I have since made some trials to prove chemically the presence of carbon in several varieties of lydian stone. See my Experiments on the Nervous and Muscular Fibre (in German), t. ii. p. 163.

[6] See an account of the whole of these phenomena of the volcanoes of Popayan in my *Essai sur le Gisement des Roches*, 1823, pp. 129, 139, 340.

experience) generally very rare when we are above 12,800 or 14,000 feet high. The hail was white;[7] the hailstones, from five to seven lines in diameter, composed of layers varying in translucency. They were not only much flattened towards the poles, but so much increased in their equatorial diameter, that rings of ice separated themselves on the least shock. I had already twice observed and described this phenomenon, in the mountains of Bareuth, and near Cracow, during a journey in Poland. Can it be admitted that the successive layers which are added to the central nucleus are in a state of fluidity so great that the rotary motion can cause the flattening of the spheroids?

When the barometer indicated that we were come very near the limit of perpetual snow, we found the masses of sulphur disseminated in imperfectly columnar trachytic rocks augmented. This phenomenon struck me the more, as I knew how rare sulphur is on the sides of inflamed volcanoes: a column of yellowish smoke and a frightful noise informed us of the neighbourhood of one of the mouths (*bocas*) of the volcano. We had some trouble to approach its edge; the declivity of the mountain being very steep, and the crevices only covered by a crust of sulphur, of whose thickness we | 465 | were ignorant. We believed we might rate the extent of this crust, which is often interrupted by rocks, at more than 12,000 square feet. These little ridges of trachytic rocks act strongly on the magnet. I tried to keep at as much distance from them as possible to determine the inclination of the needle. It was at the town of Popayan (height 5,825 English feet) 23°.05, centesimal division; at the village of Puracé (height 8671 feet) 21°.81; near the summit of the volcano of Puracé (height 14,542 feet) 20°.85. The intensity of the magnetic force varied very little at Popayan and at the village of Puracé; and the diminution of the inclination is certainly not the effect of the height, as is proved by so many other observations which I have made on the summit of the Andes, but the effect of local attractions depending on certain centres of action in the trachytes.

The mouth of the volcano of Puracé is a perpendicular cleft, the visible opening of which is only 6 feet long and 3 broad. It is covered in form of a vault by a layer of very pure sulphur, which is 18 inches thick, and which the force of the elastic vapours has split on the north side. At the distance of 12 feet from the mouth we felt an agreeable heat.

[7] I have already remarked elsewhere in the *Ann. de Chimie*, that at Paramo de Guanacas, where the road from Bogota to Popayan passes to the height of 14,700 feet, there has been seen fall, no snow, but red hail. Did it inclose those same germs of vegetable organization which have been discovered above the polar circle?

The centigrade thermometer, which had kept till then at 6°.2 (43° F.) (a cold not at all considerable in a time of hail, and at a height of 14,356 feet), rose to 15° (61° F). Placed in such a manner as not to be incommoded by the vapours, we had the pleasure of drying our clothes. The frightful noise which is heard near this opening has almost always the same intensity: it can only be compared to that which would be caused by several steam-engines together, were the dense steam suffered to escape from all at the same moment. We threw great stones into the crevice, and we discovered on this occasion that the opening communicated with a basin full of boiling water. The vapours which escape with so much violence are of the sulphurous acid, which is indicated by their suffocating smell. We shall soon see that the water of the subterraneous lake is charged with sulphuretted hydrogen; but the odour of this gas is not smelt at the summit of the volcano, because it is disguised by the much stronger smell of the sulphurous acid vapours. I had not any means of determining the temperature of these vapours, which seem to undergo a prodigiously strong pressure in the interior of the volcano. The Indians pretend that the opening has several compartments which are not all filled with water, and that the noise which is heard at times in the interior of the | 466 | crevice is the forerunner of flames. I introduced by means of a long pole, some papers coloured with the tincture of violets, under the vault, where I could be sure of not touching the surface of the water. Drawing back the pole I found the papers strongly reddened, but not at all inflamed, as was easy to be foreseen.

We succeeded after several vain attempts in obtaining some water from the crevice: this was by tying a *tutuma* (the fruit of the *Cresentia Cujete*) to a stick 8 feet long. The water was directly poured into a bottle and hermetically stopped. We examined it on our return to the village of Puracé; it exhaled a strong smell of sulphuretted hydrogen; it had no acid taste, but some weak precipitates caused by the nitrate of silver showed the presence of muriatic acid. The crust of sulphur which forms above the mouth arises without doubt from the contact of vapours of sulphurous acid with the sulphuretted hydrogen which the subterraneous lake disengages. Even the water of this lake is covered with a coat of sulphur which disappeared in the places where we threw the stones. It results from these observations, that only the presence of the muriatic acid, or of combinations of this acid with salifiable bases, indicates a feeble analogy between the waters of Rio Vinagre and those of the lakes. The first, which spring much lower, at the declivity of the volcano of Puracé, are charged with free sulphuric acid; the others, which are found at the summit of the volcano, contain sulphuretted hydrogen. As the upper mouths are found at very different heights above the level of the sea, it may be supposed that their subterraneous waters are owing to the melting of the snows, and that they do not communicate. The Rio Vinagre receives its acid in the interior of a volcano which abounds in sulphur, and the temperature of which appears extremely elevated, although for centuries no luminous phenomenon has been perceived at its summit.

[*To Be Continued.*]

Analysis of the Water of the Rio Vinagre, in the Andes of Popayan, by M. Mariano de Rivero; with geognostic and Physical Illustrations of some Phenomena which are exhibited by Sulphur, Sulphuretted Hydrogen, and Water, in Volcanoes

(Concluded from p. 466.)

The good curate of the village of Puracé thought to render a great service to his parishioners, as well as to the inhabitants of the town of Popayan, in causing, as he said, the *chimneys of the volcano* to be cleaned now and then. He ordered the Indians to take away the crust of sulphur which rises in form of a dome above the crevice. This crust has acquired sometimes, as they affirm, a thickness of as much as four feet in less than two years. It lessens without doubt the opening by which the vapours of sulphurous acid escape; but it may be conceived that the elastic force of these vapours is such, that, if the opening were entirely stopped up for some moments, it would sooner break the new arch than produce commotions by acting against the rocky sides of the volcano. For several years the lakes, which represent in miniature the *crater-lakes* of our extinguished volcanos, seem each to preserve the same level of their line of water; which proves that the evaporation is equal to the infiltration of the waters of snow and rain: but this equilibrium has not always been equally steady. About the year 1790 the *Boca grande* caused partial inundations. I dwell on this phenomenon, because | 514 | it seems to throw some light on a problem of the geology of volcanos, which has not been sufficiently examined: I mean the ejections of water and mud. At Vesuvius these ejections are only apparent, and come neither from the interior of the crater, nor from the lateral crevices. An immense electric tension manifests itself in the atmosphere which surrounds the summit of the volcano at the time of great eruptions.

Flashes of lightning cleave the air; the aqueous vapours thrown out by the crater are cooled; thick clouds envelop the summit during the continuance of this storm, confined to a little space; the water descends in torrents, and is mixed with the tufaceous substances which it drags with it.[8] These effects, purely meteorological, have given rise to the traditions about boiling waters that issued from the crater of Vesuvius in 1631; fabulous traditions, which are perpetuated by an inscription at Portici.

In the volcanos of the Andes, which exceed the limit of perpetual snow, the causes of inundations are very different from those which we have just indicated. As the eruptions of these colossal summits take place only after long intervals, (every thirty or forty years, or still more rarely) banks of snow of an enormous thickness accumulate on the sides of the mountains. The snows do not melt at the time of the explosion only, but sometimes several days before. Thus in February 1803, during my stay at Guayaquil, the inhabitants of the province of Quito were frightened at the appearance of the cone of Cotopaxi, which lost a great part of its snows in a single night, and showed plainly the black colour of its burnt rocks. Whatever idea may be formed of the power of the volcanic forces, and of the intensity of the subterraneous fires in the Andes, it cannot be admitted that the thick sides of a cone could be uniformly warmed, and transmit the heat with such rapidity (by the conductibility of their mass) to | 515 | the outside. The sudden melting of the snows, when, in the Cordilleras, it precedes the eruptions, is probably owing only to an infinity of little *fumaroles*, which disengage hot vapours through the fissured rock of the cone. These vapours, according to what I have had opportunity of observing in the craters of Vesuvius, the Peak of Teneriffe, and the volcano of Jorullo in Mexico, are more frequently pure water, which does not act at all on the most sensible re-agents; at other times they contain muriatic acid. It is remarkable that the same crevice gives at very

[8] M. de la Condamine (*Mémoires de l'Académie* 1754, p. 18) had already expressed very precise ideas on the cause of these phenomena, which are found equally well explained in the *Storia dell' incendio del* 1737, published by the Academy of Naples. I saw in my last journey to Naples, (December 1822) the ravages caused by the torrents of water from the side of Ottajano, at the foot of Vesuvius. They had transported into the plain, not only mud, but masses of lava 48 feet in circumference and 25 feet high. See the excellent description of these phenomena by M M. Monticelli and Covelli. (*Storia del Vesuvio degli anni* 1821-23, p. 91-98) Phil. Mag. vol. lxiii. p. 46. By the mixture of the rain and the volcanic cinders, there is formed in the air *(l. c.* p. 94) a kind of pisolites with concentric layers, which I also found on the plain of Hambato, among the ancient ejections of the Carguairazo. The inhabitants of the province of Quito call these pisolites *earth hailstones*.

near epochs, distilled (pure) water and very acid waters. The artificial spring which M. Gimbernat has had the ingenious idea of forming at the summit of Vesuvius, by the condensation of the vapours in a glass tube, has sometimes shown these variations; they prove either the change of chemical action in the interior of the volcano, or the accidental opening of some new communications. In the Andes of Quito, as in Iceland, and in the eruptions of Etna of March 23, 1536, and March 6, 1755, the sudden melting of the banks of snow produced great devastations.[9]

At other times, by slow infiltrations, the snow waters are accumulated in the lateral cavities of the volcano; shocks of violent earthquakes, which do not always coincide with the epoch of the fiery eruptions, open these cavities; and waters long kept in, which support little fish of the genus *Pimelodes*, carry with them pulverized trachytes, pumice-stones, tufas, and other incoherent matters. These liquid ejections spread sterility over the plains for centuries. Muddy clays (*lodazales*) covered a space of more than four square leagues, when, in the night of the 19th of June 1698, the Peak of Carguairazo, the actual height of which exceeds 15,700 feet, sunk down with a noise. The lakes of sulphureous water that we found at the summit of Puracé, explain what the inhabitants of Quito report of the fetid smell of the waters which descend sometimes from the sides of the volcanos during great eruptions. Struck with the novelty of these phœnomena, which we only mention here, the Spanish *Conquistadores* have, since the sixteenth century, distinguished two sorts of volcanos,— the *fire volcanos* and the *water volcanos* (*volcanes de fuego y de agua*). This last denomination, which one might say was invented to bring near to each other the *volcanists* and the *neptunists*, and to put an end to the famous schism of dog-matical | 516 | geology, has been applied especially to the mountains of Guatin ala and of the Archipelago of the Philippines. The *Volcan de agua*, placed between the volcano of Guatamala,[10] and that of Pocaya, ruined, by torrents of water, and stones which it sent forth the 11th of September 1541, the town of Almolonga, which is the ancient capital of the country. This mountain does not attain the limit of perpetual snow, but it remains covered with snow several months of the year. When we call to mind the confusion of the accounts that are found in our own days in the public papers of Europe, every

[9] Ferrara, *Campi Flegrei*, 1810, p. 165.—Idem, *Descriz. dell' Etna*, 1818, p. 89, 116—120.

[10] Juarros, *Compendio de la Historia de Gautemala*, 1809, t. i. p. 72; t. ii. p. 351. Remesal. *Hist. de la Provincia de San Vincente*, lib. iv. cap. 6.— Also in the great eruption of the volcano of the province of Sinano in Japan (July 27, 1783, boiling waters were mixed with the *rapilli. Mémoire sur la Dynastie regnante des Djogouns*, 1820, p 182.

time that Etna or Vesuvius are in action, we cannot complain of the uncertainty in which the chroniclers of Spanish America and the *Conquistadores* of the sixteenth century, leave us respecting the phœnomena of *volcanic inundations*, so worthy of engaging the attention of natural philosophers. During the eruption of Etna in 1792, there opened on the declivity of the volcano, three miles from the crater, a gulf[11] from which issued for several weeks water mixed with ashes, scoriæ, and clays. These liquid ejections, which must not be confounded with the phœnomenon of the Salses,[12] or *air volcanos*, were very thick. It is easily conceivable that in the equinoctial zone, even very low mountains may, by a particular disposition of their subterraneous cavities, and by the excessive abundance of the tropical rains, be subject to cause frightful inundations each time that they undergo shocks of earthquakes. Furthermore, the phœnomena which we have been describing are repeated from time to time far from the volcanos, in secondary mountains, in the centre of Europe. Sad examples have proved in our days that in the Alps of Switzerland, where no shocks of earthquakes are felt, a simple hydrostatic pressure lifts up and breaks with violence banks of rocks, throwing them to a great distance, as if they were projected by elastic forces.

The trachytes of Puracé contain sulphur like those of Mont- | 517 |
Dore in Auvergne, of Budoshegy in Transylvania, of the Isle of Montserrat in the Little Antilles, and of the Antisana in the Andes of Quito. It is still formed daily in the clefts around the gulfs of Puracé, either by a very slow sublimation, or by the contact of the sulphurous acid vapours with the sulphuretted hydrogen of the lake. The volcano labours in its interior like the *solfataras*; but it presents nothing in its form that resembles the places which are designated by that name, and which I have visited; for example, the *solfataras* of Puzzuoli, the Peak of Teneriffe, and the volcano of Jorullo in Mexico. These last three are craters which have vomited lava; they show that their first state was very different to that in which we see them at present. With very elevated temperatures, the chemical products of a volcano are not the same as with a very low temperature. If the appellation *solfatara* be given indefinitely to every place where sulphur is formed or deposited, this denomination may also

[11] *Ferrara, Descr. dell' Etna*, p. 132. As this phenomenon seems to have some relation to that of the *Maya de Pelileo*, which contains the carburets of hydrogen, and which I made known at my return from America, I obtained very lately an explanatory manuscript note from the learned Sicilian geologist, M. Ferrara, on the muddy eruption of Etna, observed March 25, 1792.

[12] There is only the muddy torrent (*fiume di fango*) of Santa-Maria-Nascemi (March 18, 1790) in the Val di Noto, which seems to me to belong to the actions of the Salses.

be applied to a district which I shall describe here, and which contrasts singularly with the trachytes of volcanos. In crossing the Cordilleras of the Andes of Quindiu, between the basins of the Cauca and of the Magdalena (lat. 4° 50′—4° 45′) I saw an immense formation of gneiss and of micaceous schist resting immediately on an ancient granite. The layers of micaceous schist which alternate with strata of gneiss are free from garnets, whilst the gneiss contains many. But in these same primitive micaceous schists, a little to the west of the station of the Moral, at the height of 6800 feet above the level of the sea, in the *Quebrada del Azufral*, some *decayed veins* extremely full of crevices abound in sulphur,[13] and exhale a sulphureous vapour, the temperature of which rose to 47° 8′ centessimal (118° F.), when the surrounding air was at 20° 2′ (63° F.) Here then is repeated on a small scale, in the clifts of a primitive rock, the phœnomena of the trachytic *solfatara* of Budoshegy in Transylvania, which has been recently examined by M. Bouè. The micaceous schist of Quindiu, which surrounds the *open veins*, is decomposed, and the sulphur is formed in masses considerable enough to become the object of a sulphur-work which supports a family settled in the ravine of the Azufral. The rock contains some decomposed pyrites; but I much doubt whether these pyrites perform the important part in nature which has been so long ascribed to them in geological treatises. In the midst of the granitic rocks of Quindiu rise the trachytes of the volcano of Tolima. | 518 | a truncated cone, which reminds us of the form of the Cotopaxi, and which, according to a geodesic measurement made by me at the west of Ibagué, is the highest summit of the Andes in the northern hemisphere.[14] A rivulet which emits considerably the smell of sulphuretted hydrogen descends from the Peak of Tolima, and proves that the trachytes which have penetrated the granitic rocks also contain sulphur. Two learned travellers, M M. Rivero and Boussingault, have recently visited this little *solfatara in the micaceous schist* of Quindiu: they have sent some specimens to the cabinet of the *Ecole des Mines* at Paris, which contains the most complete and instructive series of geognostic specimens. Following the Cordillera of the Andes southwards, these same alternations of primitive formations, and of porphyritic and trachytic, are found:—but what was my surprise, when beyond the equator I ascertained that the celebrated *mountain of sulphur of Ticsan* (S. lat. 2° 10′), between Quito and Cuenca, is neither composed of trachyte, nor of chalk or of gypsum, but of micaceous schist.

[13] See my Barometric and Geognstic Levelling of the Cordilleras, No. 102.
[14] Height 18,321 feet; N. lat. 14° 46′.

This mountain of sulphur, which the Indians call *Quello*, is situated, according to my barometric measurement, at the height of 8000 feet above the level of the ocean. It is entirely composed of primitive micaceous schist (*glimmerschiefer*), which is not even anthracitic, as are the varieties of this rock peculiar to transition countries. In some very deep ravines between Ticsan and Alausia, the micaceous schist is seen resting on gneiss. The sulphur is contained in a stratum of quartz which is more than 12,00 feet thick: it lies in a tolerably regular direction, N. 18° E., and inclined like the micaceous schist from 70° to 80° to the north-west. The bed of quartz, which passes sometimes into the hornstone, is wrought in an open working. The declivity of the *Cerro Quello*, on which the works were begun some centuries since, is opposite to the south-south-east; and the bed of quartz appears to be prolonged towards the north-north-west, that is to say, towards the coast of the Pacific Ocean. It is however asserted that the sulphur has not been found on the surface of the ground in this direction to the distance of 2000 toises from Ticsan. All is covered there with a thick vegetation. Towards the end of the eighteenth century, masses of sulphur were still worked, which were from 2 to 3 feet in diameter. At present they are working some quartzose strata much less | 519 | rich, in which the sulphur is only dispersed in nodules from 3 to 4 inches thick. It is observed that the quantity of sulphur increases with the depth; but the working has been so unskilfully directed that the lower strata are nearly inaccessible. The quartz in which the sulphur is dispersed presents neither great fissures nor cavities, or *druses*; nor have I been able to find any specimen of crystallized sulphur.

The mineral which is the object of the working of the *Cerro Quello* does not form a mass of complication of veins, as might be supposed: the sulphur is disseminated without any continuity by little masses in the quartz which traverses the micaceous schist in a direction parallel to its strata. The clefts that have perhaps formerly united these masses are no longer visible, but all the quartz seems to have undergone an extraordinary change. It is tarnished, often brittle, and breaks in some parts on the least shock; which indicates an imperceptible cleavage. The temperature of the rock did not differ from that of the exterior air. The inhabitants like to attribute the violent earthquakes to which their country has been sometimes exposed to concavities which they suppose to exist under the mountain of sulphur. If this hypothesis be well founded, it must be admitted that the cause which it indicates acts but locally. In the great catastrophe of the 4th of February 1797, which destroyed so many thousand Indians in the province of Quito—the three places where there is the most sulphur, the Cerro Quello, the Azufral of Cuesaca near to the villa of Ibarra, and the Machay of Saint Simon, near the volcano of Antisana, were but very feebly agitated; but at a much earlier period there has been experienced, even on the bed of quartz which includes the sulphur near Ticsan, an explosion similar to that of a mine.

The bed of quartz appears at the surface on the two sides of the little river of Alausi; and facing the Cerro Quello is found a little plain, where, in the seventeenth century, was situated the village of Ticsan. The ruins of the church of *Pueblo Viejo* are still seen. An earthquake wholly local (for its effects were confined to a very small space of country) made the surrounding hills sink down: a part of the village sunk: another part was thrown into the air, as happened at Riobamba, where I found the bones of the unfortunate inhabitants of the town thrown on the Cerro de la Culca, to a height of several hundred feet. The Indians of Ticsan who survived this catastrophe constructed their habitations more to the north, far from the mountain of sulphur whose neigh- | 520 | bourhood they dreaded. It may be that the coincidence of these phœnomena of explosion and of the position (*gisement*) of a substance easy to be converted into elastic vapours has only been accidental, but it may be also that ancient communications with the interior of the globe, those upon which is formed by sublimation the immense deposit of sulphur, became re-established from time to time, and allow the volcanic forces to shake the surface of the soil. Near the ruins of *Pueblo Viejo* of Ticsan I found a hill of gypsum lying above the micaceous schist: as this hill is not covered by other formations, it is difficult to decide whether the gypsum, partly fibrous and mixed with clay, is primitive, like that of Val Canaria, or transition, like the gypsum of the Tarentaise.

The abundance of sulphur in primitive countries is a very important geological fact, in relation to the study of volcanos and of rocks, through which the subterraneous fire has opened itself a passage. Before I had visited the Andes of Quito and the mountain of Ticsan, sulphur was known only in the transition limestone and gypsum; in the gypsums, marles and muriatiferous clays of secondary countries, and in the rocks exclusively called *volcanic*. These different geological situations, to which may be added the tertiary districts, very ill explained the frequency of the sulphureous vapours exhaled by the mouths of the volcanos, whose centre of action was placed (and doubtless with propriety) very much below the secondary and intermediate rocks. In proportion as we become acquainted with a greater part of the globe, we not only see positive geognosy, that is to say, the view of the formations and of the geological

positions extended; but even *geogony*, or systematic geognosy, the conjectural science which investigates the causes of phœnomena, begins to be founded on the analogy of more certain facts. We may have been struck for some time past with the little masses of native sulphur which are disseminated in some metalliferous veins, and which traverse granitic rocks; for example, in Schwarzwald, near Riepoldsau. The mountain of Ticsan which I have made known leaves no further doubt respecting the existence of sulphur in the primitive districts. It has also been lately found in Brazil, that the chloritic quartz formation which covers, in the Capitania de Minas Geraes, the primitive clayslate contains both gold and sulphur. Laminæ of this rock strongly heated burn with a blue flame. Near to Villarica, in the district called Antonio Pereira, a schist, of the same age as that on which is superposed the intacolumite or chlo- | 521 |
itic quartz, contains a calcareous bed traversed by veins of quartz, which the Baron d'Eschwege (director of the gold and diamond mines of these countries) has found filled with little nodules of pulverulent sulphur. All these phenomena increase in interest, when we reflect that this learned geologist, and also another German traveller (M. Pohl), incline to the opinion that gold, micaceous iron, diamonds, euclases, platina, and palladium, which are peculiar to the alluvial districts of Brazil, have been derived either from the destruction of the great formation of chloritic quartz, or from that of a ferruginous bed (*itabarite*), which is placed above this formation.

Scenery of the Upper Oroonoko

To take in at one view the grand character of these stupendous scenes, the spectator must be stationed on the little mountain of Manimi, a granitic ridge, that rises from the Savannah, north of the church of the mission, and is itself only a continuation of the steps of which the cataract of Manimi is composed. We often visited this mountain, for we were never weary of the view of this astonishing spectacle, concealed in one of the most remote corners of the earth. Arrived at the summit | 53 | of the rock, the eye suddenly takes in a sheet of foam, extending a whole mile. Enormous masses of stone, black as iron, issue from its bosom. Some are paps grouped in pairs, like basaltic hills; others resemble towers, strong castles, and ruined buildings. Their gloomy tint contrasts with the silvery splendour of the foam. Every rock, every islet, is covered with vigorous trees collected in clusters. At the foot of those paps, far as the eye can reach, a thick vapour is suspended over the river, and through this whitish fog the tops of the lofty palm-trees shoot up. This majestic plant, the trunk of which is more than eighty feet high, has a leafy plume of a brilliant lustre, which rises almost straight toward the sky. At every hour of the day the sheet of foam displays different aspects. Sometimes the hilly islands and the palm-trees project their broad shadows; sometimes the rays of the setting sun are refracted in the humid cloud that shrouds the cataract. Coloured arcs are formed, and vanish, and appear again alternately; light sport of the air, their images wave above the plain.

Such is the character of the landscape discovered from the top of the mountain of Manimi, which no traveller has yet described. I do not hesitate to repeat, that neither time, nor the view of the Cordilleras, nor any abode in the temperate vallies of Mexico, have effaced from my mind the powerful impression of the aspect of the cataracts. When I read a description of those places in India that are embellished by running waters and a vigorous vegetation, my imagination retraces a sea of foam and palm-trees, the tops of which rise above a stratum of vapour. The majestic scenes of nature, like the sublime works of poetry and the arts, leave remembrances that are incessantly awakening, and through the whole of life mingle with all our feelings of what is grand and beautiful. | 54 |

The calm of the atmosphere, and the tumultuous movement of the waters, produce a contrast peculiar to this zone. Here no breath of wind ever agitates the foliage, no cloud veils the splendour of the azure vault of heaven; a great mass of light is diffused in the air, on the earth strewn with plants with glossy leaves, and on the bed of the river, which extends far as the eye can reach. This appearance surprizes a traveller born in the north of Europe. The idea of wild scenery, of a torrent rushing from rock to rock, is linked in his imagination with that of a climate, where the noise of the tempest is mingled with the sound of the cataracts; and where, in a gloomy and misty day, sweeping clouds seem to descend into the valley, and rest upon the tops of the pines. The landscape of the tropics in the low regions of the continents has a peculiar physiognomy; something of greatness and repose, which it preserves even where one of the elements is struggling with invincible obstacles. Near the equator, hurricanes and tempests belong to islands only, to deserts destitute of plants, and to those spots where parts of the atmosphere repose upon surfaces, from which the radiation of heat is very different.

Humboldt.

South America—missions in the interior

It was at five o'clock in the morning on the 4th of September (says M. de Humboldt) that we commenced our journey to the Missions of the Chaymas Indians, and towards a group of high mountains which cross New Andalusia. We soon met with cabins inhabited by Mestizos, in the Ravine of Los Frailes. In Europe we judge of the number of inhabitants from the extent of cultivation: under the Tropics, on the contrary, in the hottest and most humid parts of South America, the most populous provinces appear to be deserted; because men, for the sake of nourishment alone, will only cultivate a small proportion of land. Without neighbours, almost without intercourse with men, each family of these Colonists forms an isolated people. It is said, that in the New World man appears not as an absolute master who changes the surface of the soil at his pleasure, but rather as a passing guest, who tranquilly enjoys the bounties of Nature. In fact, in the neighbourhood of the most populous cities, the land retains the inheritance of its forests, or is covered with a thick underwood, which has never been touched with the axe. The natural productions of the earth so exceed in magnitude and number those introduced by cultivation, as to characterise the aspect of the country. It is to be presumed that this state of things cannot be changed but in a long process of time.

The forests of the New World present a singular appearance. Every where the trunks of the trees are hidden under a species of verdant tapestry, and if due care were taken, in transplanting, to preserve the ligaments, a vast extent of ground might be covered. The same climbers which at present spread rampantly over the soil, might be made to ascend the trees, passing from one to another to the height of 100 feet.

We travelled for some hours under the shade of these arbours, which hardly left an opening through which we could view the azure sky. It appeared to me of an indigo colour, and remarkably dark, as the green of the equinoctial plants is generally deep, and verging towards a brown.

Near San Fernando, the evaporation caused by the action of the sun was so great, that being slightly clothed, we felt ourselves as damp as if coming from a shower bath.

A path bordered by bamboos brought us to the small village of San Fernando. This was the first Mission we had seen in America. The houses or rather huts of the Chaymas Indians, which are separated from all others, have no gardens round them. Every Indian family cultivates at some distance from the village, besides his private garden, the *conuco* of the commune. It is in the latter of these that the adults of both sexes are accustomed to labour one hour each morning and evening. In those Missions nearest the coast, the garden of the community is generally a sugar or indigo plantation under the superintendence of the Missionary, the produce of which can only be employed in the support of the Church and the purchase of sacerdotal ornaments. The great square of San Fernando is situated in the middle of the village, inclosing the Church, the Missionary dwelling, and that unaspiring edifice which is called the King's House. It is truly a Caravansary, appropriated to the shelter of travellers, and is infinitely precious in a country where even the word Inn is unknown.

These King's houses are to be met with in all the Spanish colonies, and one might believe that they are, in imitation of the Tambos of Peru, established by the laws of Manco-Capac.

The Mission of San Fernando was founded at the end of the 17th century, at the junction of the small rivers of Manzanares and Lucalparez. The resident families have increased to about a hundred, and the Missionary here remarked that the custom of the young folks marrying at the ages of thirteen or fourteen had greatly contributed to the rapid growth of the population. The government of these Indian Communes is, in other respects, most complicated; they have their governor, their alguazil major, their commandants of militia, all of whom carry their bows and arrows. The company of the archers have flags, and, habited in white, exercise the bow and arrow. These constitute the National Guards of the country.

I have already proved (says M. de Humboldt), in my work on Mexico, how very wrong it was to state, generally, the assumed fact of the destruction or diminution of the Indians in the Spanish Colonies. There still exists in the two Americas a colored population of above six million of souls; and although a great number of the tribes and languages has been extinguished or confounded, the number of Indians has not greatly diminished since the discovery by Columbus.

The religious orders have formed their establishments between the territory of the Colonists and the Three Indians. The Missions should be considered as a species of intermediate States; they have encroached, no doubt, upon the liberty of the natives, but almost everywhere they have been subservient to the increase of population, which is incompatible with the wandering life of the independent Indians. In proportion as the religious orders approach the forests, and gain on the natives, the white colonists in their turn endeavour to invade on the opposite side the settlements of the Missionaries; after an unequal contest the Missionaries are gradually supplanted by the regular clergy. The whites and the mulattoes, favored by the Corregidors, are establishing themselves in the midst of the Indians. The residences of the Missionaries become Spanish villages, and the natives lose even the recollection of their national idiom. Such has been the progress of civilisation from the coasts to the interior.

Emendations

The following errors, which distort the meaning of Humboldt's texts, have been emended.

Extract of a Letter from Mr. Humboldt to Mr. Blumenbach, containing new Experiments on the Irritation caused by the Metals with Respect to their different Impressions on the Organs of Animals

P. 9, L. 9: Schalleru → Schallern
P. 13, L. 14: feven → seven

A Letter from M. de Humboldt to M. Pictet, on the Magnetic Polarity of a Mountain of Serpentine

P. 17, L. 39: sussiciently → sufficiently
P. 18, L. 2: sussicient → sufficient
P. 18, L. 4: sorm → form

A Letter from Mr. Von Humboldt to M. H. Van Mons on the Chemical Process of Vitality; together with the Extract of a Letter from Citizen Fourcroy to Citizen Van Mons on the same Subject

P. 21, L. 33: aud → and
P. 22, L. 12: persorms → performs

On the Effects of Oxygen in accelerating Germination

P. 25, L. 20: vegetation → vegetation.
P. 25, Fn. 1: Physiology → Physiology,
P. 25, Fn. 2: Pfass → Pfaff
P. 26, L. 7: Vander → Van der

Extract of a letter from M. Von Humboldt to Lalande

P. 35, L. 25: depend that → depend on that

Sketch of a Geological Delineation of South America

P. 44, L. 17: desarts → deserts
P. 45, L. 26: Loango → Loango,
P. 47, L. 17: Tonto → Torito
P. 61, L. 16: Atarnipe → Ataruipe
P. 62, L. 32: Guigne → Guigue
P. 64, L. 11: Guigne → Guigue

Letter from M. Humboldt to C. Delambres, one of the perpetual Secretaries of the National Institute

P. 80, L. 2: Delambres → Delambre
P. 85, L. 4: Chumgat → Chuchungat

Letter from Mr. Humboldt to Prof. Cavanilles

P. 88, L. 19: I → "I
P. 89, L. 14: Popayau → Popayan

Account of the Travels of M. A. de Humboldt in South America, extracted from some of his Letters

P. 98, L. 7: re spiration → respiration
P. 99, L. 2: Oratuen → Uritucu

Baron Humboldt

P. 104, L. 19: Papayan → Popayan
P. 104, L. 31: in in → in
P. 105, L. 8: Montutar → Montufar
P. 105, L. 36: Saen → Jaen
P. 106, L. 20: asstonished → astonished
P. 107, L. 16: or or → or

Baron Humboldt's Travels in America
P. 109, L. 21: *a* → *à*

On the Variations of the Terrestrial Magnetism in different Latitudes
P. 110, L. 17: Berda → Borda

Experiments on the Eudiometrical Means, and on the Proportion of the Constituent Principles of the Atmosphere
P. 164, L. 22: combussion → combustion

Memoir on a new Species of Monkey found in the eastern Declivity of the Andes
P. 179, L. 5: albo-cœrulescens → albo-cærulescens

Memoir on a new Species of Pimelodus thrown out of the Volcanoes in the Kingdom of Quito; with some Particulars respecting the Volcanoes of the Andes
P. 184, L. 19: Posc → Posse

On the Cinchona Forests in South America
P. 216, L. 6: Selfkabet → Selskabet

Cow Pock in Mexico
P. 236, L. 13: *Valmis* → *Balmis*
P. 236, L. 20: *Valmis* → *Balmis*
P. 236, L. 26: Valmis → Balmis
P. 236, L. 31: Valmis → Balmis
P. 237, L. 20: viccination → vaccination
P. 238, L. 4: sever → fever
P. 238, L. 8: principle → principal

On the Volcanoes of Jorullo
P. 243, L. 1: Colluet → Collnet

Account of the Character and present Condition of the different Classes of Inhabitants in Mexico, or New Spain
P. 245, L. 2 (Table): Indiaus → Indians
P. 248, L. 12: Europe. → Europe,
P. 249, L. 8: paintiug → painting
P. 249, L. 29: ars → are
P. 251, L. 9: particalar → particular
P. 252, L. 21: delos → de los

Yellow Fever
P. 281, L. 7: tystem → system

Personal Narrative of the Travels to the Equinoctial Regions of the New Continent, During the Years 1799–1804
P. 300, L. 14: cotemporaries → contemporaries
P. 304, L. 10: wtth → with
P. 305, L. 28: R.) → R.).
P. 307, L. 14: situate → situated
P. 309, L. 10: salter → saltier
P. 312, L. 5: were → where
P. 324, L. 34: aspec → aspect
P. 326, L. 21: productions, → productions.
P. 331, L. 38: epocha → epochs
P. 334, L. 32: thouschiefer → thonschiefer
P. 335, L. 27: Gaudaloupe → Guadaloupe

Account of the Spermaceti Fishery on the Western Shore of America
P. 341, L. 3: adipocire → adipocire.

Scenes at Cumana
P. 343, L. 10: timal → tunal
P. 343, L. 18: crythrtias → erythrinas
P. 344, L. 7: water. → water."
P. 344, L. 19: bifurication → bifurcation

The Southern Constellations
P. 354, L. 6: couuld → could
P. 354, L. 10: posi → puosi
P. 354, L. 13: pare lo ceil → pareva il ciel
P. 354, L. 13: fammelle; → fiammelle:

P. 354, L. 14: settrentrional → settentrional
P. 354, L. 15: se' → sei

To the Editor of the Morning Chronicle
P. 355, L. 25: *Bostuch* → *Bertuch*
P. 355, L. 37: *Bostuch* → *Bertuch*

On Isothermal Lines, and the Distribution of Heat over the Globe
P. 358, L. 14: Begota → Bogota
P. 362, L. 14: Esatosthenes → Eratosthenes
P. 375, L. 4: America → America.
P. 384, L. 25 (Table): coasts → coasts.
P. 387, L. 31 (Table): *America* → *America,*
P. 387, L. 33 (Table): *America* → *America,*
P. 388, L. 14 (Table): *America* → *America,*
P. 391, L. 26: (in (Hungary) → (in Hungary)
P. 391, Fn. 43: Wahlenberg → Wahlenberg,
P. 410, L. 32: Benito → Bento
P. 410, L. 33: Dorta → d'Orta
P. 419, L. 3 (Table): *Nivado* → *Nevado*
P. 422, L. 11: Houda → Honda
P. 424, L. 7 (Table): Tegernsée → Tegernsee
P. 429, Fn. 83: p. 151. 163. → p. 151.–163.

Cavern of Guacharo
P. 448, L. 23: darks → dark
P. 449, L. 14: *aceiba* → *aceite*
P. 449, L. 18: tast → taste

Curious Circumstance
P. 450, L. 4: *Boupland* → *Bonpland*

[From De Humboldt's Travels, part 2.]
P. 451, L. 9: there- → there

Personal Narrative of Travels to the Equinoctial Regions of the New Continent, During the Years 1799–1804
P. 466, L. 36: reseaches → researches
P. 470, L. 5: SantaCruz → Santa-Cruz

Remarks on the natural Family of the Grasses
P. 483, L. 34: Bagotá → Bogotá

On the Milk extracted from the Cow Tree (l'Arbre de la Vache), and on Vegetable Milk in general
P. 487, L. 25: at → as

Baron Humboldt's Last Volume. Extract from Baron Humboldt's Personal Narrative of Travels to the Equinoctial Regions of the New Continent. Vol. 4
P. 493, L. 7: granatic → granitic

Baron Humboldt's Last Volume. Personal Narrative of Travels to the Equinoctial Regions of the New Continent. Vol. 4. London, 1819
P. 501, L. 24: extent → extent.
P. 508, L. 27: recal → recall
P. 508, L. 34: 1812 → 1812.
P. 509, L. 9: fouutains → fountains
P. 509, Fn. 13: *undulaceon* → *undulacion*

Baron Humboldt's Last Volume. Personal Narrative of Travels to the Equinoctial Regions of the New Continent. Vol. 4. London, 1819
P. 513, L. 10: Victora → Victoria
P. 513, L. 31: Victora → Victoria
P. 514, L. 7: (pueblo) → (*pueblo*)
P. 515, L. 5: mien → mean
P. 519, L. 13: monnuments → monuments
P. 519, L. 15: tried,, → tried,

Scenery on the Rio Apure
P. 521, L. 30: it's → its

P. 521, L. 31: it's → its
P. 522, L. 17: it's → its

Humboldt's Personal Narrative
P. 535, L. 32: Oroonnko → Oroonoko

Humboldt's Personal Narrative. Dirt Eaters
P. 537, L. 1: **Personaln Arrative** → **Personal Narrative**
P. 538, L. 5: mingl, → mingle

Memoir of Don Jose Celestino Mutis
P. 549, L. 26: Mantissa → *Mantissa*

New Inquiries into the Laws which are observed in the Dirstribution of Vegetable Forms
P. 558, L. 6: *Pen.* → *Gen.*
P. 559, L. 6: 238.) → 238).
P. 559, L. 30: latude → latitude
P. 560, L. 17: *Jährbücher* → *Jahrbücher*
P. 560, L. 38: *Dumosœ* → *Dumosæ*
P. 562, Fn. 2: thie → this
P. 563, L. 16: nd → and
P. 565, L. 20: *Novagen.* → *Nova gen*
P. 573, L. 16: 1.4,7 → 1,4,7
P. 578, L. 18: *Jarhb.* → *Jahrb.*
P. 578, L. 18: *Pew.* → *Gew.*

Personal Narrative of Travels to the Equinoctial Regions of the New Continent, during the years 1799–1804.
P. 580, L. 19: desarts → deserts
P. 581, Fn. 1: planes → plains
P. 582, L. 13: humourous → humorous
P. 587, L. 26: works → words
P. 587, L. 27: serpents? → serpents?'
P. 588, L. 23: guard. → guard,
P. 589, L. 11: in.j → in.
P. 591, L. 31: Ataroipe → Ataruipe
P. 591, L. 32: ineaudrites → meandrites
P. 592, L. 13: hevea → hevea,

Shirt Trees
P. 595, L. 3: Dnida → Duida
P. 595, L. 10: *ruanos* → *ruanas*

War Poison of the Indians
P. 599, L. 9: accused;" I → accused; "I

[The famous Ahauete]
P. 601, L. 4: *Akauete* → *Ahauete*

[The Guahibi mother]
P. 602, L. 18: sivilized → civilized
P. 603, L. 25: villiage → village

On Petrifactions, or Fossil Organic Remains
P. 620, L. 12: of → of the
P. 627, L. 25: xv → xv.
P. 627, L. 26: p. 54 → p. 54.)

On the Constitution and Mode of Action of Volcanoes, in different Parts of the Earth
P. 638, L. 17: phœnomenon → phænomenon

Analysis of the Water of the Rio Vinagre, in the Andes of Popayan, by M. Mariano de Rivero; with geognostic and Physical Illustrations of some Phenomena which are exhibited by Sulphur, Sulphuretted Hydrogen, and Water, in Volcanoes
P. 653, L. 25: *de agua*) → *de agua*).

South America—missions in the interior
P. 663, L. 5: Indians - → Indians.

Sources

Titles are listed according to the chronological order of the respective original publications.

All the texts of this edition are English-language translations of Alexander von Humboldt's original publications in German and French, published during his lifetime—with one exception: His first article ever (from 1789), has been translated by Rex Clark for this edition.

"Letter to the Author of this Journal; on the Bohon-Upas, by a Young Gentleman of This City," translated by Rex Clark.
Original publication: "Lettre à L'Auteur de cette Feuille; sur le Bohon-Upas, par un jeune Gentilhomme de cette ville." *Gazette littéraire de Berlin* 1270 (5 January 1789): 4–8; 1271 (12 January 1789): 11–13.

"Extract of a Letter from Mr. Humboldt to Mr. Blumenbach, containing new Experiments on the Irritation caused by the Metals with Respect to their different Impressions on the Organs of Animals." *A Journal of Natural Philosophy, Chemistry, and the Arts* 1 (September 1797): 256–60.
Original publication: "Neue Versuche über den Metallreitz, besonders in Hinsicht auf die verschiedenartige Empfänglichkeit der thierischen Organe." *Neues Journal der Physik* 3, no. 2 (1796): 165–84.

"A Letter from M. de Humboldt to M. Pictet, on the Magnetic Polarity of a Mountain of Serpentine." *A Journal of Natural Philosophy, Chemistry, and the Arts* 1 (June 1797): 97–100.

"A Letter from Mr. Von Humboldt to M. H. Van Mons on the Chemical Process of Vitality; together with the Extract of a Letter from Citizen Fourcroy to Citizen Van Mons on the same Subject." *A Journal of Natural Philosophy, Chemistry, and the Arts* 1 (November 1797): [359]–63.
Original publication: "Lettre de M. Von Humboldt à M. Van-Mons, sur le procédé chimique de la vitalité"/"Addition à la Lettre précédente." *Annales de chimie* 22 (11 Floréal year 5, 30 April 1797): 64–76.

"On the Effects of Oxygen in accelerating Germination." *The Philosophical Magazine* 1, no. 3 (1798): 309–11.
Original publication: "De la germination." *Journal de physique, de chimie, d'histoire naturelle et des arts* 4, no. 1 [= 47, no. 1] (Messidor year 6 [June/July 1798]): 63–65.

"Remarks on the above Observations." *The Philosophical Magazine* 1 (1798): 419–20.
Original publication: "Note de Frédéric von-Humboldt, sur les Observations précédentes." *Journal de physique, de chimie, d'histoire naturelle et des arts* 4, no. 3 [= 47, no. 3] (Fructidor year 6 [August/September 1798]): 205–6.

"Von Humboldt's Expedition to Spanish America." *The Philosophical Magazine* 6 (February 1800): 94–95.
Original publication: "Gelehrte Reisen." *Intelligenzblatt der Allgemeinen Literatur-Zeitung* 163 (21 December 1799): 1322–24.

"Extract of a letter from M. Von Humboldt to Lalande." *The Philosophical Magazine* 11, no. 44 (1802): 355–361.
Original publication: "Lettre du baron de Humboldt (de Berlin), à Jerôme Lalande." *Magasin encyclopédique, ou Journal des Sciences, des Lettres et des Arts* 5, no. 6 (Germinal year 8 [March/April 1800]): 376–91.

"Humboldt's travels through Spanish America." *The Philosophical Magazine* 7, no. 27 (August 1800): 275–79.
Original publication: "Sur plusieurs objets d'histoire naturelle et de chimie." *Annales de chimie* 35 (30 Messidor year 8 [19 July 1800]): 102–11.

"Sketch of a Geological Delineation of South America." *The Philosophical Magazine* 17, no. 68 (January 1804): 347–57; 18, no. 69 (February 1804): 26–36; 18, no. 70 (March 1804): 172–79.
Original publication: "Esquisse d'un tableau géologique de l'Amérique méridionale." *Journal de physique, de chimie, d'histoire naturelle et des arts* 53, no. 1 (Messidor year 9 [June/July 1801]): 30–60, plate.

"Letter from C. Humboldt to C. Fourcroy, Member of the French National Institute." *The Philosophical Magazine* 10, no. 37 (June 1801): [3]–8.
Original publication: "Extrait d'une lettre de M. Humboldt, au C. Fourcroy." *Bulletin des Sciences, par la Société Philomatique de Paris* 3, no. 50 (Floréal year 9 [April/May 1801]): 9–11.

"Letter from M. A. Humboldt to C. Delambre, Member of the French National Institute." *The Philosophical Magazine* 9, no. 6 (May 1801): 365–70.

Original publication: "Lettre de M. A. Humboldt, au cit. Delambre, membre de l'institut national." *Gazette Nationale ou le Moniteur Universel* 214 (4 Floréal year 9 [4 April 1801]: 898–99.

"Letter from M. Humboldt to C. Delambre, one of the perpetual Secretaries of the National Institute." *The Philosophical Magazine* 16, no. 62 (July 1803): 165–72.

Original publication: "Copie d'une lettre de M. Humboldt, adressée au C. Delambre, l'un des secrétaires perpétuels de l'Institut national (datée de Lima le 25 novembre 1802)." *Magasin encyclopédique, ou Journal des Sciences, des Lettres et des Arts* 8, no. 6 (Germinal year 11 [March/April 1803]): 537–49.

"Letter from Mr. Humboldt to Prof. Cavanilles." *Annals of Botany* 1 (1806): 573–76.

Original publication: "Extracto de la carta que el Baron de Humboldt escribió desde México en 22 de Abril de 1803 á D. Antonio Josef Cavanilles." *Anales de ciencias naturales* 6, no. 18 (October 1803): 281–87.

"Account of the Travels of M. A. de Humboldt in South America, extracted from some of his Letters." *The Philosophical Magazine* 16, no. 63 (August 1803): 237–47.

Original publication: "Extrait de plusieurs lettres de M. A. de Humbold." *Magasin encyclopédique, ou Journal des Sciences, des Lettres et des Arts* 9, no. 2 (Thermidor year 11, 1803 [July/August]): 241–58.

"Baron Humbold." *The Literary Magazine and American Register* 2, no. 10 (July 1804): 321–27.

"Baron Humboldt's Travels in America." *The Medical Repository* 2 (May–July 1804): 95–96.

"On the Variations of the Terrestrial Magnetism in different Latitudes." *The Philosophical Magazine* 22, no. 87 (August 1805): 248–57; 22, no. 88 (September 1805): 299–308.

Original publication: "Sur les variations du magnétisme terrestre à différentes latitudes," with Jean-Baptiste Biot. *Journal de physique, de chimie, d'histoire naturelle et des arts* 59, no. 6 (Frimaire year 13 [November/December 1804]): 429–50, 2 plates.

"Experiments on the Torpedo, by Messrs. Humboldt and Gay-Lussac. Extracted from a Letter from M. Humboldt to M. Berthollet, dated Rome, 15 Fructid. Year 13." *The Philosophical Magazine* 23, no. 92 (January 1806): 356–60.

Original publication: "Expériences sur la torpille," with Louis Joseph Gay-Lussac. *Annales de chimie* 56, no. 1 (30 Vendémiaire year 14 [22 October 1805]): 15–23.

"Experiments on the Eudiometrical Means, and on the Proportion of the Constituent Principles of the Atmosphere," with Gay-Lussac. *The Repertory of Arts, Manufactures, and Agriculture* 8, no. 45 (February 1806): 231–39; 8, no. 46 (March 1806): 303–18; 8, no. 47 (April 1806): 365–81; 8, no. 48 (May 1806): 449–61.

Original publication: "Expériences sur les moyens eudiométriques, et sur la proportion des principes constituans de l'atmosphère," with Louis Joseph Gay-Lussac. *Journal de physique, de chimie, d'histoire naturelle et des arts* 60, no. 2 (Pluviôse year 13 [January/February 1805]): 129–67, table.

Corresponding book publication: Alexander von Humboldt. "Untersuchungen über die eudiometrischen Mittel und über das Verhältniß der wesentlichen Bestandtheile der Atmosphäre, von Alexander v. Humboldt und J. L. Gay-Lussac. (Gelesen in der ersten Classe des Instituts am 21. Januar 1805, abgedruckt im Journal de Physique T. LX. An. XIII (1805) p. 129–167)." In *Kleinere Schriften. Geognostische und physikalische Erinnerungen,* vol. 1, 315–70. Stuttgart: Cotta, 1853.

"Universal Diffusion of Life." *Boys' and Girls' Magazine and Fireside Companion* (1 April 1850): 87.

Original publication: "Fragment aus der am 30sten Jan. 1806 in der öffentlichen Sitzung der Königl. Akademie gehaltenen Vorlesung: Ideen zu einer Physiognomik der Gewächse." *Der Freimüthige oder Ernst und Scherz* 4:1:31 (13 February 1806): 121–23.

Corresponding book publications: Alexander von Humboldt. *Ansichten der Natur,* 157–278. Tübingen: Cotta, 1808.

Alexander Humboldt. *Ansichten der Natur*, Zweite verbesserte und vermehrte Ausgabe, vol. 2, 1–47. Stuttgart and Tübingen: Cotta, 1826.

Alexander von Humboldt. *Ansichten der Natur*, Dritte verbesserte und vermehrte Ausgabe, vol. 2, 39. Stuttgart: Cotta, 1849.

"Memoir on a new Species of Monkey found in the eastern Declivity of the Andes." *The Philosophical Magazine* 24, no. 96 (February–May 1806): 339–41, plate 8.

Corresponding book publication: Alexander von Humboldt, "Mémoire sur une nouvelle espèce de singe, trouvée sur la pente orientale des Andes." In *Recueil d'observations de zoologie et d'anatomie comparée, faites dans l'Océan Atlantique, dans l'intérieur du Nouveau Continent et dans la Mer du Sud pendant les années 1799, 1800, 1801, 1802 et 1803,* vol. 1, 14–16. Paris: F. Schœll/G. Dufour, 1811 [1812]; J. Smith/Gide, [1813–]1833.

"Memoir on a new Species of Pimelodus thrown out of the Volcanoes in the Kingdom of Quito; with some Particulars respecting the Volcanoes of the Andes." *The Philosophical Magazine* 24, no. 96 (February–May 1806): 333–39, plate 8.
Corresponding book publication: Alexander von Humboldt. "Mémoire sur une nouvelle espèce de pimelode, jetée par les volcans du royaume de Quito." In *Recueil d'observations de zoologie et d'anatomie comparée, faites dans l'Océan Atlantique, dans l'intérieur du Nouveau Continent et dans la Mer du Sud pendant les années 1799, 1800, 1801, 1802 et 1803,* vol. 1, 21–25. Paris: F. Schœll/G.EL Dufour, 1811 [1812]; J. Smith/Gide, [1813–]1833.

"Memoir on the Eremophilus and Astroblepus, two new Genera of the Order of Apodes." *The Philosophical Magazine* 24, no. 96 (February–May 1806): 329–32, plate 8.
Corresponding book publication: Alexander von Humboldt. "Mémoire sur l'eremophilus et l'astroblepus, deux nouveaux genres de l'ordre des apodes." In *Recueil d'observations de zoologie et d'anatomie comparée, faites dans l'Océan Atlantique, dans l'intérieur du Nouveau Continent et dans la Mer du Sud pendant les années 1799, 1800, 1801, 1802 et 1803,* vol. 1, 17–20. Paris: F. Schœll/G.EL Dufour, 1811 [1812]; J. Smith/Gide, [1813–]1833.

"Baron Humboldt's Description of the Mode of Catching the Gymnotus, or Electrical Eel." *The Zoological Magazine, or Journal of Natural History* (1833): 76–80.
Original publication: "Jagd und Kampf der electrischen Aale mit Pferden." *Annalen der Physik* 25, no. 1 (1807): 34–43.

"On the Cinchona Forests of South America." In Aylmer Bourke Lambert, *An Illustration of the Genus Cinchona; Comprising Descriptions of All the Officinal Peruvian Barks, Including Several New Species*, [19]–59. London: Longman, Hurst, Rees, Orme, and Brown, 1821.
Original publication: "Über die Chinawälder in Südamerika." *Der Gesellschaft naturforschender Freunde zu Berlin Magazin für die neuesten Entdeckungen in der gesammten Naturkunde* 1 (1807): 57–68, 104–20.

"Account of the Ottomacs, a People who eat Clay." *The Edinburgh Medical and Surgical Journal* 6 (1 July 1810): 316–19.
Original publication: "Ueber die erdefressenden Otomaken." *Morgenblatt für gebildete Stände* 241 (8 October 1807): 961–62.
Corresponding book publication: Alexander von Humboldt. *Ansichten der Natur*, 142–53. Stuttgart: Cotta, 1808.

Alexander von Humboldt. *Ansichten der Natur*, Dritte verbesserte und vermehrte Ausgabe, vol. 1, 231–38. Stuttgart: Cotta, 1848.

"The Cataracts of Orinoco." *The Carlisle Patriot* 33, no. 1779 (10 November 1849): [4].

Original publication: "Ansichten der Natur mit wissenschaftlichen Erläuterungen, von Alex. v. Humboldt. Erster Band. 16. Tübingen, in der J. G. Cotta'schen Buchhandlung. Ueber die Wasserfälle des Orinoco bey Atures und Maypures." *Morgenblatt für gebildete Stände* 49 (26 February 1808): 193–95; 50 (27 February 1808): 197–99.

Corresponding book publication: Alexander von Humboldt. *Ansichten der Natur*, 279–334.Tübingen: Cotta, 1808.

"Scenes in the Pampas." *Reading Mercury Oxford Gazette, Newbury Herald, and Berks County Paper,* 127, October 13, 1849, p. [4].

Original publication: "Ansichten der Natur, mit wissenschaftlichen Erläuterungen von Alexander von Humboldt. Erster Band. Tübingen in der Cottaschen Buchhandlung." *Zeitung für die elegante Welt.* 89, June 3, 1808, col. 705–9; 90, June 6, 1808, col. 713–18.

Corresponding book publication: Alexander von Humboldt, *Ansichten der Natur, mit wissenschaftlichen Erläuterungen*, Zweite verbesserte und vermehrte Ausgabe, vol. 1, 29–40. Stuttgart: Cotta, 1826.

"Observations upon the Face of the Country of Spain and its Climate. In Alexandre de Laborde, *A View of Spain; Comprising a Descriptive Itinerary, of Each Province, and a General Statistical Account of the Country*, vol. 1, clxiii–clxxiv. London: Longman, Hurst, Rees, and Orme, 1809.

Original publication: "Notice sur la configuration du sol de l'Espagne et son climat." In Alexandre de Laborde, *Itinéraire descriptif de l'Espagne, et Tableau élémentaire des différentes branches de l'administration et de l'industrie de ce royaume*, vol. 1, cxlvii–clvi. Paris: H. Nicolle et Lenormant, 1808.

"Cow Pock in Mexico." *The Hull Advertiser, and Exchange Gazette*, 16, no. 865, December 15, 1810, p. [4].

Original publication: [Extrait de l'Essai politique sur la Nouvelle-Espagne.] *Journal de médecine, chirurgie, pharmacie, etc.* 16 (October 1808): 291.

Corresponding book publication: Alexander von Humboldt. *Essai politique sur le royaume de la Nouvelle-Espagne. Avec un atlas physique et géographique, fondé sur des observations astronomiques, des mesures trigonométriques et des nivellemens barométriques*, vol. 1, 66–70. Paris: F. Schœll, [1808–]1811.

"On the Volcanoes of Jorullo." *A Journal of Natural Philosophy, Chemistry, and the Arts* 26, no. 117 (June 1810): [81]–86.

Original publication: "Description du volcan de Jorullo, tirée de l'Essai politique sur le Royaume du Mexique, formant la troisième partie des Voyages d'Alexandre de Humboldt et Aimé Bompland. Troisième livraison." *Bibliothèque britannique* 14:41:4 (August 1809): 339–55.

Corresponding book publication: Alexander von Humboldt. *Essai politique sur le royaume de la Nouvelle-Espagne. Avec un atlas physique et géographique, fondé sur des observations astronomiques, des mesures trigonométriques et des nivellemens barométriques,* vol. 1, 248–55. Paris: F. Schœll, [1808–]1811.

"Account of the Character and present Condition of the different Classes of Inhabitants in Mexico, or New Spain." *The Scots Magazine, and Edinburgh Literary Miscellany* 72 (1810): 916–20; 73 (1811): 17–22.

Original publication: "Voyage de MM. Humboldt et Bonpland." *Gazette nationale ou le moniteur universel*, 47, February 16, 1809, p. 183–84; 48, February 17, 1809, p. 186–88; 49, February 18, 1809, p. 191–92.

Corresponding book publication: Alexander von Humboldt. *Essai politique sur le royaume de la Nouvelle-Espagne. Avec un atlas physique et géographique, fondé sur des observations astronomiques, des mesures trigonométriques et des nivellemens barométriques*, vol. 1, lxvii–lxxv, 3–10, 12–28, 39–41, 76–113, 114–31, 165–67, 169–71, 177–83, 187–95, 200–10, 282–88, 316–50, and 354–56. Paris: F. Schœll, [1808–]1811.

"Researches Concerning the Institutions and Monuments of the Ancient Inhabitants of America; with descriptions and views of some of the most striking scenes in the Cordilleras." *Supplementary Number to the Thirty-Eighth Volume of the Monthly Magazine* 38, no. 264 (30 January 1815): 608–21.

Original publication: "Pittoreske Ansichten in den Cordilleren." *Morgenblatt für gebildete Stände* 4, no. 5 (5 January 1810): 17–18; 4, no. 6 (6 January 1810): 23; 4, no. 28 (1 February 1810): 109–10; 4, no. 29 (2 February 1810): 115–16; 4, no. 86 (10 April 1810): 341–42; 4, no. 87 (11 April 1810): 346–47; 4, no. 211 (3 September 1810): 841–42; 4, no. 212 (4 September 1810): 847–48.

Corresponding book publication: Alexander von Humboldt. *Vues des Cordillères et monumens des peuples indigènes de l'Amérique,*

p. i–xvi, 2, 4–6, 9–41, 43–51, 56–86, 90–92, 96–99, 102–7, 118–12, 127–30, 136–39, 148, 160–62, 188–94, 200–5, 206–27, 238–39, 246–47, 270–73, 279–83, 297, 318–20. Paris: F. Schœll, 1810.

Alexander von Humboldt. *Relation historique du Voyage aux régions équinoxiales du Nouveau Continent*, vol. 1, 145, 207–9. Paris: F. Schœll, 1814[–1817]; N. Maze, 1819[–1821]; J. Smith et Gide Fils, 1825[–1831].

Alexander von Humboldt. *Vues des Cordillères et monumens des peuples indigènes de l'Amérique*, vol. 1, 87–89, 109–16, 116–117, 118–120, 122–123, 235–42, 251–52, 254–55, 265–66, 269–71, 383–84; vol. 2, 21–24, 123, 125–26, 175–78. Paris: Maze, 1816.

Alexander von Humboldt. *Essai politique sur le royaume de la Nouvelle-Espagne. Avec un atlas physique et géographique, fondé sur des observations astronomiques, des mesures trigonométriques et des nivellemens barométriques*, vol. 1, 92–94. Paris: F. Schœll, [1808–]1811.

"Yellow Fever." *Boston Intelligencer & Evening Gazette*, 6, May 13, 1820, p. [1].

Original publication: "Fragment d'un ouvrage ayant pour titre: Essais politiques sur le Mexique." *Journal général de médecine, de chirurgie et de pharmacie; ou Recueil Périodique de la Société de Médecine de Paris* 40 (January 1811): 210–25, 338–51, 413–45.

Corresponding book publication: Alexander von Humboldt. *Essai politique sur le royaume de la Nouvelle-Espagne. Avec un atlas physique et géographique, fondé sur des observations astronomiques, des mesures trigonométriques et des nivellemens barométriques*, vol. 2, 750–88. Paris: F. Schœll, [1808–]1811.

[Letter to the editor of the *Moniteur*]. *The Times*, 8432, October 28, 1811, p. 3.

Original publication: [Lettre à l'éditeur du Moniteur]. *Le moniteur universel.* 284 October 11, 1811, p. 1086.

"Extract from a report made to the class of phisico-mathematical sciences, on the 9th of Sept. 1811, by a committee, composed of Messrs. Hallé, Humboldt and Percy, on a memorial of Mr. Legallois, M. D. concerning the principle of action of the heart, and the seat of that principle." *The Baltimore Medical and Philosophical Lycæum* 1, no. 4 (October–December 1811): 342–53.

Original publication: "Rapport Fait à la Classe des Sciences Physiques et Mathématiques de l'Institut impérial de France, sur des Expériences relatives aux Mouvemens du Cœur; Par M. Percy."

Journal de physique, de chimie et d'historie naturelle 75 (November 1812): 329–58.

"Description of the Ascent of the Peak of Teneriffe." *The Scots Magazine, and Edinburgh Literary Miscellany* 77 (1815): 268–72.
Corresponding book publication: Alexander von Humboldt. *Relation historique du Voyage aux régions équinoxiales du Nouveau Continent*, vol. 1, 99–174. Paris: F. Schœll, 1814[–1817]; N. Maze, 1819[–1821]; J. Smith et Gide Fils, 1825[–1831].

"Personal Narrative of the Travels to the Equinoctial Regions of the New Continent, During the Years 1799–1804." *Supplementary Number to the Thirty-Eighth Volume of the Monthly Magazine* 38, no. 264 (30 January 1815): 587–608.
Corresponding book publication: Alexander von Humboldt. *Relation historique du Voyage aux régions équinoxiales du Nouveau Continent*, vol. 1, 63, 291, 306–18. Paris: F. Schœll, 1814[–1817]; N. Maze, 1819[–1821]; J. Smith et Gide Fils, 1825[–1831].

"Population of the Canaries." *Annals of Philosophy; or, Magazine of Chemistry, Mineralogy, Mechanics, Natural History, Agriculture, and the Arts* 6 (July–December 1815): 396.
Corresponding book publication: Alexander von Humbold. *Relation historique du Voyage aux régions équinoxiales du Nouveau Continent*, vol. 1, 195. Paris: F. Schœll, 1814[–1817]; N. Maze, 1819 [–1821]; J. Smith et Gide Fils, 1825[–1831].

"Temperature of the Atlantic." *Annals of Philosophy; or, Magazine of Chemistry, Mineralogy, Mechanics, Natural History, Agriculture, and the Arts* 6 (July–December 1815): 396.
Corresponding book publication: Alexander von Humboldt. *Relation historique du Voyage aux régions équinoxiales du Nouveau Continent*, vol. 1, 230. Paris: F. Schœll, 1814[–1817]; N. Maze, 1819 [–1821]; J. Smith et Gide Fils, 1825[–1831].

"Account of the Spermaceti Fishery on the Western Shore of America." *The Weekly Entertainer* 56 (20 May 1816): 401–2.
Corresponding book publication: Alexander von Humboldt. *Essai politique sur le royaume de la Nouvelle-Espagne. Avec un atlas physique et géographique, fondé sur des observations astronomiques, des mesures trigonométriques et des nivellemens barométriques*, vol. 2, 467–73. Paris: F. Schœll, [1808–]1811.

"Eye-Stones." *The London Medical and Physical Journal* 36, no. 209 (July 1816): 82–83.
Corresponding book publication: Alexander von Humboldt. *Relation historique du Voyage aux régions équinoxiales du Nouveau Conti*

nent, vol. 1, 346–47. Paris: F. Schœll, 1814[–1817]; N. Maze, 1819[–1821]; J. Smith et Gide Fils, 1825[–1831].

"Scenes at Cumana." *Farmers' Repository* 9, no. 432 (17 July 1816): [1].
Corresponding book publication: Alexander von Humboldt. *Relation historique du Voyage aux régions équinoxiales du Nouveau Continent,* vol. 1, 295, 297, 300, 303–4. Paris: F. Schœll, 1814[–1817]; N. Maze, 1819[–1821]; J. Smith et Gide Fils, 1825[–1831].

"On the Laws observed in the Distribution of vegetable Forms." *The Philosophical Magazine and Journal* 47, no. 218 (June 1816): 446–53.
Original publication: "Sur les Lois que l'on observe dans la distribution des formes végétales." *Annales de chimie et de physique* 1 (March 1816): 225–39.

"The Southern Constellations." *The Weekly Recorder* 3:6/110 (4 September 1816): 46.
Corresponding book publication: Alexander von Humboldt. *Relation historique du Voyage aux régions équinoxiales du Nouveau Continent*, vol. 1, 201–9. Paris: F. Schœll, 1814[–1817]; N. Maze, 1819 [–1821]; J. Smith et Gide Fils, 1825[–1831].

"To the Editor of the *Morning Chronicle*." *The Morning Chronicle*, 14592, February 8, 1816 [n.p.]

"On Isothermal Lines, and the Distribution of Heat over the Globe." *The Edinburgh Philosophical Journal* 3, no. 5 (July 1820): [1]–20; 3, no. 6 (October 1820): 256–74; 4, no. 7 (January 1821): 23–37; 4, no. 8 (April 1821): 262–81; 5, no. 9 (July 1821): 28–39.
Original publication: "Des lignes isothermes et de la distribution de la chaleur sur le globe." *Mémoires de physique et de chimie, de la société d'Arcueil* 3 (1817): 462–602; Errata, 516–18.

"Characteristic sketches of american vegetables." *The New Monthly Magazine and Universal Register* 7, no. 41 (1 June 1817): 397–98; 8, no. 43 (1 August 1817): 20–21; 8, no. 44 (1 September 1817): 109–10.
Original publication: "Pflanzenbilder. Die Orchiden-Familie." *Morgenblatt für gebildete Stände* 33 (7 February 1817): 129–31.
Corresponding book publication: Alexander von Humboldt. *De distributione geographica plantarum secundum coeli temperiem et altitudinem montium,* 216–40. Paris: Librairie grecque-latine-allemande, 1817.

"Cavern of Guacharo." *New-York Daily Advertiser*, 2, no. 398, July 21, 1818, p. [1].

Corresponding book publication: Alexander von Humboldt. *Relation historique du Voyage aux régions équinoxiales du Nouveau Continent*, vol. 1, 413–31. Paris: F. Schœll, 1814[–1817]; N. Maze, 1819[–1821]; J. Smith et Gide Fils, 1825[–1831].

"Curious Circumstance." *The Oxford University and City Herald, and Midland County Chronicle*, 13, no. 636, 1 August 1, 1818 [n.p.].
Corresponding book publication: Alexander von Humboldt. *Relation historique du Voyage aux régions équinoxiales du Nouveau Continent*, vol. 1, 370, 372–475, 531–32, 545, 590–94, 611. Paris: F. Schœll, 1814[–1817]; N. Maze, 1819[–1821]; J. Smith et Gide Fils, 1825[–1831].

[From De Humboldt's Travels, part 2.] *The Bath Chronicle* 61, no. 2926 (26 March 1818) [n.p.].
Corresponding book publication: Alexander von Humboldt. *Relation historique du Voyage aux régions équinoxiales du Nouveau Continent*, vol. 1, 373–74. Paris: F. Schœll, 1814[–1817]; N. Maze, 1819[–1821]; J. Smith et Gide Fils, 1825[–1831].

"On the Geographical Distribution of Ferns." *The Journal of Science and the Arts* 4, no. 7 (3 January 1818): [57–66].
Corresponding book publication: *Nova genera et species plantarum*, vol. 1, 33–37. Paris: Librairie grecque-latine-allemande, 1815.
De distributione geographica plantarum secundum coeli temperiem et altitudinem montium, 169–86. Paris: Librairie grecque-latine-allemande, 1817.

"Personal Narrative of Travels to the Equinoctial Regions of the New Continent, During the Years 1799–1804." *Monthly Magazine, or, British Register* 44, no. 307 (January 1818): 590–600.
Corresponding book publication: Alexander von Humboldt. *Relation historique du Voyage aux régions équinoxiales du Nouveau Continent*, vol. 1, 512–94. Paris: F. Schœll, 1814[–1817]; N. Maze, 1819[–1821]; J. Smith et Gide Fils, 1825[–1831].

"Remarks on the natural Family of the Grasses." *The Journal of Science and the Arts* 5, no. 9 (5 January 1818): 44–52.

"On the Milk extracted from the Cow Tree (l'Arbre de la Vache), and on Vegetable Milk in general." *The London Medical Repository, Monthly Journal, and Review* 9, no. 53 (1 May 1818): 424–29.
Original publication: "Sur le Lait de l'arbre de la Vache et le Lait des végétaux en général." *Annales de chimie et de physique* 7 (1818): 182–91.
Corresponding book publication: Alexander von Humboldt. *Relation historique du Voyage aux régions équinoxiales du Nouveau Conti*

nent, vol. 2, 106–14. Paris: F. Schœll, 1814[–1817]; N. Maze, 1819[–1821]; J. Smith et Gide Fils, 1825[–1831].

"Baron Humboldt's Last Volume. Extract from Baron Humboldt's Personal Narrative of Travels to the Equinoctial Regions of the New Continent. Vol. 4." *The Belles-Lettres Repository, and Monthly Magazine* 1, no. 2 (1 June 1819): 138–42.
Corresponding book publication: Alexander von Humboldt. *Relation historique du Voyage aux régions équinoxiales du Nouveau Continent*, vol. 2, 231–36. Paris: F. Schœll, 1814[–1817]; N. Maze, 1819[–1821]; J. Smith et Gide Fils, 1825[–1831].

"Baron Humboldt's Last Volume. Extract from Baron Humboldt's Personal Narrative of Travels to the Equinoctial Regions of the New Continent. Vol. 4." *The Belles-Lettres Repository, and Monthly Magazine* 1, no. 3 (1 July 1819): 209–12.
Corresponding book publication: Alexander von Humboldt. *Relation historique du Voyage aux régions équinoxiales du Nouveau Continent*, vol. 2, 29–35, 43–44, 47–48, 108–9. Paris: F. Schœll, 1814[–1817]; N. Maze, 1819[–1821]; J. Smith et Gide Fils, 1825[–1831].

"Baron Humboldt's Last Volume. Personal Narrative of Travels to the Equinoctial Regions of the New Continent. Vol. 4. London, 1819." *The Belles-Lettres Repository, and Monthly Magazine* 1, no. 2 (15 May 1819): 65–74.
Corresponding book publication: Alexander von Humboldt. *Relation historique du Voyage aux régions équinoxiales du Nouveau Continent*, vol. 2, 1–28, 173–90, 214. Paris: F. Schœll, 1814[–1817]; N. Maze, 1819[–1821]; J. Smith et Gide Fils, 1825[–1831].

"Baron Humboldt's Last Volume. Personal Narrative of Travels to the Equinoctial Regions of the New Continent. Vol. 4. London, 1819." *The Belles-Lettres Repository, and Monthly Magazine* 1, no. 4 (1 August 1819): 266–71.
Corresponding book publication: Alexander von Humboldt. *Relation historique du Voyage aux régions équinoxiales du Nouveau Continent*, vol. 2, 50–59. Paris: F. Schœll, 1814[–1817]; N. Maze, 1819[–1821]; J. Smith et Gide Fils, 1825[–1831].

"Famine." *Long-Island Star*, 10, no. 512, March 3, 1819 [n.p.].
Corresponding book publication: Alexander von Humboldt. *Essai politique sur le royaume de la Nouvelle-Espagne. Avec un atlas physique et géographique, fondé sur des observations astronom-*

iques, des mesures trigonométriques et des nivellemens barométriques, vol. 1, 72. Paris: F. Schœll, [1808].

"Scenery on the Rio Apure." *The Annual Register, or a View of the History, Politics, and Literature, for the Year 1819*, 1819, 553–56.
Corresponding book publication: Alexander von Humboldt. *Relation historique du Voyage aux régions équinoxiales du Nouveau Continent*, vol. 2, 211–30. Paris: F. Schœll, 1814[–1817]; N. Maze, 1819[–1821]; J. Smith et Gide Fils, 1825[–1831].

[Crocodiles of South America]. *The Cambridge Chronicle and Journal, and Huntingdonshire Gazette*, 2970, September 24, 1819 [4].
Corresponding book publication: Alexander von Humboldt. *Relation historique du Voyage aux régions équinoxiales du Nouveau Continent*, vol. 2, 191–214. Paris: N. Maze, 1819[–1821].

"On the Temperature of the different Mines in America." *The Edinburgh Philosophical Journal* 3, no. 6 (October 1820): 286–88.
Original publication: "Observations de M. de Humboldt sur les Températures de diverses mines d'Amérique." *Annales de chimie et de physique* 13 (1820): 207–12.

"On the Nocturnal Increase in the Intensity of Sound." *The Repertory of Arts, Manufactures, and Agriculture* 38, no. 228 (May 1821): 375–81.
Original publication: "Sur l'Accroissement nocturne de l'intensité du son. (Mémoire lu à l'Académie des Sciences le 13 mars 1820)." *Annales de chimie et de physique* 13 (1820): 162–73.
Corresponding book publication: Alexander von Humboldt. "Ueber die nächtliche Verstärkung des Schalles. Abhandlung, gelesen in der Akademie der Wissenschaften zu Paris am 13 März 1820, von Alexander v. Humboldt." In *Kleinere Schriften. Geognostische und physikalische Erinnerungen*, vol. 1, 371–97. Stuttgart and Tübingen: Cotta, 1853.

"Fish flour." *The Philosophical Magazine and Journal* 58, no. 281 (September 1821): 236.
Corresponding book publication: Alexander von Humboldt. *Relation historique du Voyage aux régions équinoxiales du Nouveau Continent*, vol. 2, 563. Paris: N. Maze, 1819[–1821].

"Humboldt's Personal Narrative." *The Literary Gazette; and Journal of Belles Lettres, Arts, Sciences, &c.* 234 (14 July 1821): 441–42.
Corresponding book publication: Alexander von Humboldt. *Relation*

historique du Voyage aux régions équinoxiales du Nouveau Continent, vol. 2, 436–38. Paris: N. Maze, 1819[–1821].

"Humboldt's Personal Narrative. Dirt Eaters." *The Westmorland Gazette, and Kendal Advertiser* 4, no. 177 (6 October 1821) [n.p.].
Corresponding book publication: Alexander von Humboldt. *Relation historique du Voyage aux régions équinoxiales du Nouveau Continent*, vol. 2, 608–23. Paris: N. Maze 1819[–1821].

"Humboldt's Travels." *The New Monthly Magazine and Literary Journal* 2 (1821): 314–18.
Corresponding book publication: Alexander von Humboldt. *Relation historique du Voyage aux régions équinoxiales du Nouveau Continent*, vol. 2, 336, 493–94, 504–5, 556–57, 608. Paris: N. Maze, 1819[–1821].

"Indian Rubber." *The Manchester Guardian* 10 (7 July 1821) [n.p.].
Corresponding book publication: Alexander von Humboldt. *Relation historique du Voyage aux régions équinoxiales du Nouveau Continent*, vol. 2, 408–9. Paris: N. Maze, 1819[–1821].

"Memoir of Don Jose Celestino Mutis." *The Medical Times. A Journal of English and Foreign Medicine and Medical Affairs* 8, no. 197 (1 July 1843): 219–21.
Original publication: "Mutis (Don Josef-Celestino)." In *Biographie universelle, ancienne et moderne, ou histoire, par ordre alphabétique, de la vie publique et privée de tous les hommes qui se sont fait remarquer par leurs écrits, leurs actions, leurs talents, leurs vertus ou leurs crimes*, vol. 30, 499–506. Paris: L. G. Michaud, 1821.

"New Inquiries into the Laws which are observed in the Dirstribution of Vegetable Forms." *The Edinburgh Philosophical Journal* 6, no. 12 (April 1822): 273–89; 7, no.13 (July 1822): 47–55.
Original publication: "Nouvelles Recherches sur les lois que l'on observe dans la distribution des formes végétales." *Annales de chimie et de physique* 16 (1821): 267–96.

"Personal Narrative of Travels to the Equinoctial Regions of the New Continent, during the years 1799–1804. By Alexander de Humboldt, and Aimé Bonpland, &c. &c. London, 1821, 8vo. 2 Vols. pp. 864." *The Literary Gazette; and Journal of Belles Lettres, Arts, Sciences, &c.* 229 (9 June 1821): 353–55; 230 (16 June 1821): 374–76.
Corresponding book publication: Alexander von Humboldt. *Relation historique du Voyage aux régions équinoxiales du Nouveau Continent*, vol. 2, 304–412. Paris: N. Maze, 1819[–1821.

"Religious Opinions of American Indians." *Evening Mail*, 8001, July 20–23, 1821 [n.p.].
Corresponding book publication: Alexander von Humboldt. *Relation historique du Voyage aux régions équinoxiales du Nouveau Continent,* vol. 2, 429–30. Paris: N. Maze, 1819[–1821].

"Shirt Trees." *The Philosophical Magazine and Journal* 58, no. 283 (November 1821): 395.
Corresponding book publication: Alexander von Humboldt. *Relation historique du Voyage aux régions équinoxiales du Nouveau Continent*, vol. 2, 562–63. Paris: N. Maze, 1819.

"Stories of Crocodiles." *The Literary Gazette; and Journal of Belles Lettres, Arts, Sciences, &c.* 248 (20 October 1821): 662–63.
Corresponding book publication: Alexander von Humboldt. *Relation historique du Voyage aux régions équinoxiales du Nouveau Continent*, vol. 2, 640–42. Paris: F. N. Maze, 1819[–1821].

"War Poison of the Indians." *The Philosophical Magazine and Journal* 58, no. 281 (September 1821): 231–34.
Corresponding book publication: Alexander von Humboldt. *Relation historique du Voyage aux régions équinoxiales du Nouveau Continent*, vol. 2, 547–50, 551, 553. Paris: F. N. Maze, 1819[–1821].

[The famous Ahauete]. *Daily National Intelligencer*, 9, no. 2688, August 24, 1821 [n.p.].
Corresponding book publication: Alexander von Humboldt. *Essai politique sur le royaume de la Nouvelle-Espagne. Avec un atlas physique et géographique, fondé sur des observations astronomiques, des mesures trigonométriques et des nivellemens barométriques*, vol. 1, 241. Paris: F. Schœll, [1808–]1811.

[The Guahibi mother]. *The Bath Chronicle*, 64, no. 3095, June 28, 1821 [n.p.].
Corresponding book publication: Alexander von Humboldt. *Relation historique du Voyage aux régions équinoxiales du Nouveau Continent,* vol. 2, 410–12. Paris: N. Maze, 1819[–1821].

"View of the various Productions of the different Countries of the Earth." *The London Medical Repository, Monthly Journal, and Review* 17, no. 97 (January–June 1822): 81–82.
Original publication: "A. v. Humboldts Ansicht über die Verschiedenheit der Erzeugnisse nach den Erdstrichen." *Notizen aus dem Gebiete der Natur- und Heilkunde* 2, no. 32/10 (April 1822): col. 145–47.
Corresponding book publication: Alexander von Humboldt. *Relation*

historique du Voyage aux régions équinoxiales du Nouveau Continent, vol. 2, 383–85. Paris: N. Maze, 1819[–1821].

“Description of the Muschelkalk and Quadersandstein.” In Henry Thomas De la Beche, *A Selection of Geological Memoirs Contained in the Annales des Mines, written by Brongniart, Humboldt, von Buch, and Others*, [313]–18. London: William Phillips, 1836.
Original publication: “Indépendance des formations.” In Georges Cuvier et al., *Dictionnaire des sciences naturelles*, vol. 23, 56–385. Strassburg: F. G. Levrault/Le Normant, 1822.
Corresponding book publication: Alexandre de Humboldt. *Essai géognostique sur le gisement des roches dans les deux hémisphères*, 1–379. Paris: Levrault, 1823.

“Visit of the ruins of Ataruipe.” In *The Cento: A Selection of Approved Pieces, from Living Authors*, 80–83. London: John Richardson/C. and J. White, 1822.
Corresponding book publication: Alexander von Humboldt. *Relation historique du Voyage aux régions équinoxiales du Nouveau Continent*, vol. 2, 596–600. Paris: N. Maze, 1819[–1821].

“Extent and population of America.” *The Morning Chronicle*, 17003, October 17, 1823 [n.p.].

“On Petrifactions, or Fossil Organic Remains.” *The Edinburgh Philosophical Journal* 9, no. 17 (July 1823): 20–35.
Corresponding book publication: Alexander von Humboldt. *Essai géognostique sur le gisement des roches dans les deux hémisphères*, 35–54. Paris: Levrault, 1823.

“On the Constitution and Mode of Action of Volcanoes, in different Parts of the Earth.” *Annals of Philosophy* 6, no. 2 (August 1823): 121–35.
Original publication: “Über den Bau und die Wirkungsart der Vulkane in verschiedenen Erdstrichen.” *Notizen aus dem Gebiete der Natur- und Heilkunde* 4, no. 4/70 (March 1823): col. 49–54.
Corresponding book publication: Alexander von Humboldt. *Ansichten der Natur*, Zweite, verbesserte und vermehrte Ausgabe, vol. 2, 126–86. Stuttgart and Tübingen: Cotta, 1826.

“Analysis of the Water of the Rio Vinagre, in the Andes of Popayan, by M. Mariano de Rivero; with geognostic and Physical Illustrations of some Phenomena which are exhibited by Sulphur, Sulphuretted Hydrogen, and Water, in Volcanoes.” *The Boston Journal of Philosophy and the Arts* 2, no. 5 (1825): 460–66; 2, no. 6 (1825): [513]–21.

Original publication: "Analyse de l'eau du Rio Vinagre, dans les Andes de Popayan, par M. Mariano de Rivero, avec des éclaircissemens géognostiques et physiques sur quelques phénomènes que présentent le soufre, l'hydrogène sulfuré et l'eau dans les volcans." *Annales de Chimie et de Physique* 27 (1824): 113–36.

"Scenery of the Upper Oroonoko." In John M'Diarmid, *The Scrap Book; a Collection of Amusing and Striking Pieces, in Prose and Verse, with Occasional Remarks and Contributions*, vol. 2, 52–54. Edinburgh: Oliver & Boyd, 1824.
Corresponding book publication: Alexander von Humboldt. *Relation historique du Voyage aux régions équinoxiales du Nouveau Continent*, vol. 2, 361–63. Paris: N. Maze, 1819[–1821].

"South America—missions in the interior." *The New Times*, 8101, July 16, 1824 [n.p.].
Corresponding book publication: Alexander von Humboldt. *Relation historique du Voyage aux régions équinoxiales du Nouveau Continent*, vol. 1, 371–462. Paris: F. Schœll, 1814[–1817].

The Periodicals

Alexander von Humboldt's writings appeared in more than 1200 periodicals during his lifetime. This list contains the more than 120 different newspapers and journals in which the texts of this edition were originally published in English. The titles are given as they appeared when Humboldt's texts were published. Editors are given if they were of major importance for their time.

The Adams Sentinel, edited by Robert G. Harper, Gettysburg, Pennsylvania, 1826–67.

Albany Evening Journal, Albany, New York, 1830–1925.

The American Journal of Science and Arts, edited by Benjamin Sillimann, New Haven, Connecticut, 1818.

The Analyst. A Quarterly Journal of Science, Literature, Natural History, and the Fine Arts, edited by Edward Mammatt, William Holl, and Neville Wood, London, 1834–40.

Annals of Botany, edited by John Sims and Carl König, London, 1805–6.

Annals of Philosophy; or, Magazine of Chemistry, Mineralogy, Mechanics, Natural History, Agriculture, and the Arts, edited by Thomas Thomson and Richard Philips, London, 1813–26.

The Annual Register, or a View of the History, Politics, and Literature, London, 1758–.

The Asiatic Journal and Monthly Register for British and Foreign India, China, and Australasia, London, 1830–43.

The Athenæum. Journal of English and Foreign Literature, Science, and the Fine Arts, London, 1828–1921.

The Atlas. A General Newspaper and Journal of Literature, London, 1826–62.

The Baltimore Medical and Philosophical Lycæum, Baltimore, Maryland, 1811.

Bangor Daily Whig and Courier, Bangor, Maine, 1834–1900.

The Bath Chronicle, Bath, England, 1760–.

The Belles-Lettres Repository, and Monthly Magazine, New York, 1819–20.

The Birmingham Journal, Birmingham, England, 1825–69.

Boston Daily Advertiser, edited by Nathan Hale and Nathan Hale Jr., Boston, 1813–1929.

Boston Daily Courier, Boston, 1851–64.

Boston Intelligencer & Evening Gazette, edited by William W. Clapp, Boston, 1818–21.

Boston Investigator, edited *by* Abner Kneeland, Francis Wright Darusmont, and Horace Seaver, Boston, 1831–1904.

The Boston Journal of Philosophy and the Arts, edited by John White Webster, John Ware, and Daniel Treadwell, Boston, 1823–26.

Boys' and Girls' Magazine and Fireside Companion, edited by Mark Forrester and Francis Forrester, Boston, 1848–52.

Brighton Gazette and Lewes Observer: Or, Chichester, Arundel, Horsham, Worthing, Eastbourne, Hastings, St. Leonard's, and Cumbridge Wells Fashionable Chronicle, Sussex and General Advertiser, Brighton, England, 1821–1985.

The Bristol Mercury. Western Counties, Monmouthshire, and South Wales Advertiser, Bristol, England, 1716–1877.

Buffalo Commercial Advertiser, Buffalo, New York, 1841–90.

Buxton Herald, and Gazette of Fashion, Buxton, England, 1842–97.

The Cambridge Chronicle and Journal, and Huntingdonshire Gazette, Cambridge, England, 1762–1934.

The Carlisle Patriot, edited by Robert Perring, Carlisle, England, 1815–1910.

Chambers's Edinburgh Journal, edited by William Chambers and Robert Chambers, Edinburgh, 1832–53.

Cheltenham Chronicle, and Gloucestershire Advertiser, Cheltenham, England, 1809–49.

The Cork Examiner, Cork, Ireland, 1841–.

The Country Gentleman. A Journal for the Farm, the Garden, and Fireside, Albany, New York, 1852–1955.

The Courier, London, 1804–42.

The Crayon, edited by William Stillman and John Duran, New York, 1855–61.

The Critic, edited by James Lowe, London, 1843–63.

The *Daily Crescent,* New Orleans, Louisiana, 1848–51.

The Daily Exchange, edited by Frank Key Howard, Baltimore, Maryland, 1858–61.

Daily National Intelligencer, Washington, DC, 1813–69.

The Daily News, London, 1846–1912.

The Durham Advertiser, Durham, England, 1830–54.

Durham Chronicle, Sunderland Times, and Darlington and Stockton Gazette, Durham, England, 1820–80.

The Economist, Weekly Commercial Times, Bankers' Gazette, and Railway Monitor, London, 1843–.

The Edinburgh Christian Magazine, Edinburgh, 1850–59.

The Edinburgh Journal of Natural and Geographical Science, edited by Henry H. Cheek and William Ainsworth, Edinburgh, 1830–31.

The Edinburgh Journal of Science, edited by David Brewster and William Blackwood, Edinburgh, 1824–32.

The Edinburgh Medical and Surgical Journal, Edinburgh, 1805–55.

The Edinburgh New Philosophical Journal, edited by Robert Jameson, William Jardine, Henry Darwin Rogers, and Laurence Jameson, Edinburgh, 1826–54.

The Edinburgh Philosophical Journal, edited by David Brewster and Robert Jameson, Edinburgh, 1819–26.

The Essex Standard, and General Advertiser for the Eastern Counties, Colchester, England, 1831–92.

Evening Mail, London, 1789–1868.

The Evening Post, New York, 1832–1919.

The Evening Star, Washington, DC, 1852–1981.

The Examiner, London, 1808–86.

The Examiner, edited by John C. Vaughan, F. Cosby, John H. Heywood, and Noble Butler, Louisville, Kentucky, 1847–49.

Farmers' Repository, Charleston, Virginia, 1808–27.

The Fife Herald. Kinross, Strathearn, and Clackmanan Advertiser, Cupar, Scotland, 1824–81.

The Friend. A Religious and Literary Journal, edited by Robert Smith, Philadelphia, 1827–1955.

The Gardener's Magazine, edited by John Claudius Loudon, London, 1826–43.

The Globe and Traveller, London, 1822–1921.

Greenock Advertiser, and Clyde Commercial Journal, Greenock, Scotland, 1802–84.

Huddersfield and Holmfirth Examiner, Huddersfield, England, 1851–60.

The Huddersfield Chronicle and West Yorkshire Advertiser, Huddersfield, England, 1850–1916.

The Hull Advertiser, and Exchange Gazette, Hull, England, 1794–1849.

The Jewish Herald, and Record of Christian Effort for the Spiritual Good of God's Ancient People, London, 1846–94.

A Journal of Natural Philosophy, Chemistry, and the Arts, edited by William Nicholson, London, 1797–1813.

The Journal of Science and the Arts, New York, 1816–19.

The Journal of the Royal Geographical Society of London, London, 1831–80.

The Journal of the Royal Institution of Great Britain, London, 1830–31.

Kentish Mercury, Gravesend Journal, and Greenwich Gazette, London, 1838–1964.

Launceston Examiner; Commercial and Agricultural Advertiser, Launceston, Australia, 1842–99.

The Literary Gazette; and Journal of Belles Lettres, Arts, Sciences, &c., edited by William Jerdan, London, 1817–63.

The Literary Magazine and American Register, edited by Charles Brockden Brown, Philadelphia, 1803–7.

Littell's Living Age, edited by Eliakim Littell and Robert Littell, Boston, 1844–96.

The London and Edinburgh Philosophical Magazine and Journal of Science, edited by David Brewster, Richard Taylor, Richard Phillips, London, 1832–40.

The London and Paris Observer; or, Chronicle of Literature, Science, and the Fine Arts, Paris, 1825–45.

The London, Edinburgh, and Dublin Philosophical Magazine and Journal of Science, edited by David Brewster, Richard Taylor, Richard Phillips and Robert Kane, London, 1840–1944.

The London Medical and Physical Journal, London, 1815–33.

The London Medical Repository, Monthly Journal, and Review, London, 1815–23.

The London Saturday Journal, edited by James Grant and Francis Ross, London, 1839–42.

Long-Island Star, edited by Thomas Kirk, New York, 1809–23.

Louisville Daily Courier, Louisville, Kentucky, 1851–68.

The Manchester Courier, and Lancashire General Advertiser, Manchester, 1825–1916.

The Manchester Guardian, Manchester, 1821–.

The Medical Repository, New York, 1797–1824.

The Medical Times. A Journal of English and Foreign Medicine and Medical Affairs, edited by Sidney Smith, London, 1839–50.

The Memphis Daily Appeal, Memphis, Tennessee, 1847–86.

Monthly Magazine, or, British Register, London, 1796–1843.

Monthly Notices of the Royal Astronomical Society, London, 1831–.

The Monthly Packet of Evening Readings for Younger Members of the English Church, edited by Charlotte Yonge, London, 1851–99.

Morning Advertiser, London, 1794–.

The Morning Chronicle, London, 1769–1862.

The Morning Post, London, 1803–1937.

The Musical World, a Weekly Record of Musical Science, Literature and Intelligence, London, 1836–90.

Nashville Daily Patriot, Nashville, Tennessee, 1855–57.

The Nation, Dublin, 1842–1900.

The New Monthly Magazine and Literary Journal, London, 1821–36.

The New Monthly Magazine and Universal Register, London, 1814–20.

The New Times, London, 1818–28.

New-York Daily Advertiser, New York, 1817–36.

New-York Daily Tribune, New York, 1842–66.

The New York Times, New York, 1851–.

The North American Review, Boston, 1815–.

The North Star, edited by Frederick Douglass, Rochester, New York, 1847–51.

The Oxford University and City Herald, and Midland County Chronicle, Oxford, England, 1806–21.

The Penny Magazine of the Society for the Diffusion of Useful Knowledge, edited by Charles Knight, London, New York, 1832–45.

The Philosophical Magazine, edited by Alexander Tilloch, London, 1798–1813.

The Philosophical Magazine or Annals of Chemistry, Mathematics, Astronomy, Natural History, and General Science, edited by Richard Taylor, London, 1827–32.

The Philosophical Magazine and Journal, edited by Alexander Tilloch and Richard Taylor, London, 1814–26.

The Practical Mechanic and Engineer's Magazine, Glasgow, Scotland, 1842–47.

Public Ledger, Philadelphia, 1836–1925.

The Quarterly Journal of Science, Literature, and Art, edited by William Thomas Brande, John Millington, and Michael Faraday, London, 1816–30.

Reading Mercury Oxford Gazette, Newbury Herald, and Berks County Paper, Reading, England, 1839–1960.

The Repertory of Arts, Manufactures, and Agriculture, London, 1802–25.

Reynolds's Miscellany of Romance, General Literature, Science, and Art, edited by George William MacArthur Reynolds, London, 1847–69.

The Scots Magazine, and Edinburgh Literary Miscellany, edited by John Leyden, Edinburgh, 1804–17.

Semi-Weekly Standard, Raleigh, North Carolina, 1853–58.
The Sheffield & Rotherham Independent, Sheffield, England, 1839–1901.
St. Thomæ Tidende, St. Thomas, Virgin Islands, USA, 1815–73.
The Sydney Morning Herald, Sydney, Australia, 1831–.
The Times, London, 1788–.
The Waterford Evening News. And General Weekly Advertiser, Waterford, Ireland, 1849–51.
The Weekly Entertainer, edited by Robert Goadby, Sherborne, England, 1783–1819.
Weekly National Intelligencer, Washington, DC, 1841–69.
The Weekly Recorder, Chillicothe, Ohio, 1814–21.
The Westmorland Gazette, and Kendal Advertiser, Kendal, England, 1818–.
The Youth's Instructer and Guardian, London, 1817–55.
The Zoological Magazine, or Journal of Natural History, edited by Richard Owen, London, 1833.

The Works of Alexander von Humboldt

Alexander von Humboldt's writings are not only very extensive but also extremely heterogeneous. They were created over eight decades, from the 1780s to the 1850s. They appeared internationally, especially in Europe and the Americas. Humboldt wrote his texts in German, French, and Latin, and they have been translated into numerous other languages. They touch on various fields of knowledge.

Four groups of works can be systematically distinguished. They have been made accessible in different ways over the past decades: two groups of manuscripts (travel journals and letters) and two groups of printed publications (books and nonbook publications). Humboldt's legacy also includes pictures (drawings, graphics) and material collections (e.g., objects and items from his travels).

I. DIARIES

Alexander von Humboldt kept journals on his research journeys (1799–1804 through the Americas, through Central Asia in 1829). His notes served as material for his subsequent publications. Editions of the American journals were published in four volumes by Akademie Verlag (1982–2000). The originals are kept in the Staatsbibliothek zu Berlin.

II. LETTERS

Alexander von Humboldt was a great communicator. He wrote and received thousands of letters. His correspondence with a number of individuals was successively published by the Alexander von Humboldt Research Centre of the Berlin-Brandenburg Academy of Sciences and Humanities. Correspondents include his brother Wilhelm; friends Aimé Bonpland, François Arago, and Varnhagen von Ense; the publisher Cotta; the artist Christian Daniel Rauch; the scientists Charles Darwin and Carl Friedrich Gauß; and the president of the United States, Thomas Jefferson.

III. NONBOOK PUBLICATIONS

Alexander von Humboldt published about 750 essays and articles in magazines and newspapers or as contributions to the books of other authors or editors. Together with their reprints, adaptations, and translations, these pieces were published approximately 3,600 times during his lifetime. Of these publications, approximately 250 texts were published in English, together with various versions for a total of over 1,530 prints. The English translations thus cover one third of Humboldt's writings. Humboldt's nonbook publications were first collected and edited at the University of Bern in Switzerland as his *Sämtliche Schriften* in ten volumes (published by dtv in 2019, online in 2021).

IV. BOOK PUBLICATIONS

Depending on one's way of counting, Alexander von Humboldt wrote 23 to 27 books in 49 to 52 volumes. These book publications can be divided into three groups: the early works (before the journey to the Americas), the works on America (especially the 29-volume *Voyage*), and the later works (on Asia especially). (The numerous reprints and new editions as well as the later excerpts and adaptations are not taken into account in the following overview of Humboldt's lifetime publications.)

Early Works

- *Mineralogische Beobachtungen über einige Basalte am Rhein. Mit vorangeschickten, zerstreuten Bemerkungen über den Basalt der Ältern und neuern Schriftsteller*, Braunschweig: Schulbuchhandlung, 1790.
- *Florae Fribergensis specimen plantas cryptogamicas praesertim subterraneas exhibens. Edidit Fredericus Alexander ab Humboldt. Accedunt aphorismi ex doctrina physiologiae chemicae plantarum. Cum tabulis aeneis*, Berlin: Heinrich August Rottmann, 1793.
- *Versuche über die gereizte Muskel- und Nervenfaser nebst Vermuthungen über den chemischen Process des Lebens in der Thier- und Pflanzenwelt*, 2 vols. Posen: Decker und Compagnie/Berlin: Heinrich August Rottmann, 1797.
- *Ueber die unterirdischen Gasarten und die Mittel ihren Nachtheil zu vermindern. Ein Beytrag zur Physik der praktischen Bergbaukunde*, Braunschweig: Friedrich Vieweg, 1799.
- *Versuche über die chemische Zerlegung des Luftkreises und über einige andere Gegenstände der Naturlehre. Mit zwei Kupfern*, Braunschweig: Friedrich Vieweg, 1799.

Works on the Americas

Voyage

Volumes 1–3: Travel account

- *Relation historique du Voyage aux régions équinoxiales du Nouveau Continent*, 3 vols. Paris: F. Schœll, 1814[–1817]; N. Maze, 1819[–1821]; J. Smith/Gide fils, 1825[–1831]. (English translations, 1814–31: *Personal Narrative of Travels to the Equinoctial Regions of the New Continent*, and 1852–53: *Personal Narrative of Travels to the Equinoctial Regions of America.*)

Volume 4: Anthropology

- *Vues des Cordillères et monumens des peuples indigènes de l'Amérique*. Paris: F. Schœll, 1810[–1813]. (English translation, 1814: *Researches, Concerning the Institutions & Monuments of the Ancient Inhabitants of America, With Descriptions and Views of Some of the Most Striking Scenes in the Cordilleras!.*)

Volume 5: Atlas

- *Atlas géographique et physique des régions équinoxiales du Nouveau Continent, fondé sur des observations astronomiques, des mesures trigonométriques et des nivellemens barométriques*. Paris: F. Schœll, 1814[–1838].

Volume 6: History of Geography

- *Examen critique de l'histoire de la géographie du Noveau Continent, et des progrès de l'astronomie nautique aux quinzième et seizième siècles*. Paris: F. Schœll, 1834[–1838].

Volumes 7–8: Zoology

- *Recueil d'observations de zoologie et d'anatomie comparée, faites dans l'Océan Atlantique, dans l'intérieur du Nouveau Continent et dans la Mer du Sud pendant les années 1799, 1800, 1801, 1802 et 1803,* 2 vols. Paris: F. Schœll/G.EL Dufour, 1811 [1812]; J. Smith/Gide, [1813–]1833.

Volumes 9–11: Mexico

- *Essai politique sur le royaume de la Nouvelle-Espagne. Avec un atlas physique et géographique, fondé sur des observations astronomiques, des mesures trigonométriques et des nivellemens barométriques*, 2 vols. Paris: F. Schœll, [1808–]1811. (English translation, 1811: *Political Essay on the Kingdom of New Spain.*)

- *Atlas géographique et physique du royaume de la Nouvelle-Espagne, fondé sur des observations astronomiques, des mesures trigonométriques et des nivellemens barométriques.* Paris: F. Schœll, [1808–]1811.

Volumes 12–13: Geography/Astronomy

- *Recueil d'observations astronomiques, d'opérations trigonométriques, et de mesures barométriques, faites pendant le cours d'un voyage aux régions équinoxiales du Nouveau Continent, depuis 1799 jusqu'en 1803, par Alexandre de Humboldt; Rédigées et calculées, d'après les Tables les plus exactes, par Jabbo Oltmanns. Ouvrage auquel on a joint des recherches historiques sur la position de plusieurs points importans pour les navigateurs et pour les géographes*, 2 vols. Paris: F. Schœll, 1810 [1808–1811], [1809–] 1810.

Volume 14: Plant Geography

- *Essai sur la géographie des plantes, accompagné d'un tableau physique des régions équinoxiales, Fondé sur des mesures exécutées, depuis le dixième degré de latitude boréale jusqu'au dixième degré de latitude australe, pendant les années 1799, 1800, 1801, 1802 et 1803. Avec une planche.* Paris: F. Schœll/Tübingen: J. G. Cotta, 1807.

Volumes 15–29: Botany

- *Plantes équinoxiales, recueillies au Mexique, dans l'île de Cuba, dans les provinces de Caracas, de Cumana et de Barcelone; aux Andes de la Nouvelle-Grenade, de Quito et du Pérou, et sur les bords du Rio-Negro, de l'Orénoque et de la rivière des Amazones*, 2 vols. Paris: F. Schœll/Tübingen: J. G. Cotta, [1805–]1808; Paris: F. Schœll, 1809 [1808–17].
- *Monographie des Melastomacées, comprenant toutes les Plantes de cet ordre recueillies jusqu'à ce jour, et notamment au Mexique, dans l'île de Cuba, dans les provinces de Caracas, de Cumana et de Barcelone, aux Andes de la Nouvelle-Grenade, de Quito et du Pérou, et sur les bords du Rio-Negro, de l'Orénoque et de la rivière des Amazones,* 2 vols. Volume 1: *Melastomes*, volume 2: *Rhexies.* Paris: Librairie grecque-latine-allemande, [1806–]1816; Gide fils, [1806–]1823.
- *Nova genera et species plantarum quas in peregrinatione orbis novi collegerunt, descripserunt, partim adumbraverunt Amat.*

Bonpland et Alex. de Humboldt. Ex schedis autographis Amati Bonplandi in ordinem digessit Carol. Sigismund. Kunth. Accedunt tabulæ æri incisæ, et Alexandri de Humboldt notationes ad geographiam plantarum spectantes, 7 vols. Paris: Librairie grecque-latine-allemande, 1815 [1816], 1817[–1818], 1818[–1820]; N. Maze, 1820, 1821[–1823]; Gide fils, 1823[–1824], 1825 [1824–1826] (title changed in volume 7: [...] *in peregrinatione ad plagam aequinoctialem orbis novis* [...]).

— As a part of it: *De distributione geographica plantarum secundum coeli temperiem et altitudinem montium, prolegomena*. Paris: Librairie grecque-latine-allemande, 1817.

- *Mimoses et autres plantes légumineuses du Nouveau Continent, recueillies par MM. de Humboldt et Bonpland, décrites et publiées par Charles-Sigismond Kunth. Avec figures coloriées*. Paris: Librairie grecque-latine-allemande, 1819[–1824].
- *Révision des Graminées publiées dans les Nova genera et species plantarum de Humboldt et Bonpland; précedée d'un travail général sur la famille des Graminées; par Charles-Sigismond Kunth. Dédié au Roi*, 3 vols. Volume 1: *Ouvrage accompagné de cent planches coloriées d'après les dessins de Madame Eulalie Delile*. Paris: Gide fils, 1829[–1830]; volume 2: *Supplément accompagné de cent planches coloriées d'après les dessins de Madame Eulalie Delile*. Paris: Gide fils, 1829 [1830–32]; volume 3: *Troisième et dernière partie, accompagnée de vingt planches coloriées d'après les dessins de Madame Eulalie Delile*. Paris: Gide fils, 1829 [1834].

Outside of the *Voyage* volumes:

- *Essai politique sur l'île de Cuba. Avec une carte et un supplément qui renferme des considérations sur la population, la richesse territoriale et le commerce de l'archipel des Antilles et de Colombia*, 2 vols. Paris: Gide fils, 1826. (English translation, 1856: *The Island of Cuba*.)
- *Ansichten der Natur mit wissenschaftlichen Erläuterungen*. Tübingen: J. G. Cotta, 1808.
- *Ansichten der Natur mit wissenschaftlichen Erläuterungen*, second improved and enhanced edition, 2 vols. Stuttgart and Tübingen: J. G. Cotta, 1826.
- *Ansichten der Natur mit wissenschaftlichen Erläuterungen*, third improved and enhanced edition, 2 vols. Stuttgart and Tübingen: J. G. Cotta, 1849. (English translations, 1849: *Aspects of Nature,*

In Different Lands and Different Climates; With Scientific Elucidations; 1850: *Views of Nature: Or Contemplations on the Sublime Phenomena of Creation; With Scientific Illustrations.*)

Later Works

- *Essai géognostique sur le gisement des roches dans les deux hémisphères*. Paris: F. G. Levrault, 1823. (English translation, 1823: *A Geognostical Essay on the Superposition of Rocks in Both Hemispheres.*)
- *Fragmens de géologie et de climatologie asiatiques*, 2 vols. Paris: Gide/A. Pihan Delaforest/Delaunay, 1831.
- *Asie centrale. Recherches sur les chaînes de montagnes et la climatologie comparée*, 3 vols. Paris: Gide fils, 1843.
- *Kleinere Schriften. Geognostische und physikalische Erinnerungen*. Stuttgart and Tübingen: J. G. Cotta, 1853.
    - — Supplement: *Umrisse von Vulkanen aus den Cordilleren von Quito und Mexico. Ein Beitrag zur Physiognomik der Natur*. Stuttgart and Tübingen: J. G. Cotta, 1853.
- *Kosmos. Entwurf einer physischen Weltbeschreibung*, 5 vols. Stuttgart and Tübingen: J. G. Cotta, 1845–62. (English translations, 1846–58: *Cosmos. Sketch of a Physical Description of the Universe*; 1849–58: *Cosmos: A Sketch of a Physical Description of the Universe.*)

Chronology—The Life and Times of Alexander von Humboldt

BEFORE THE EXPEDITION TO THE AMERICAS

1769	Born in Berlin on 14 September
1777–87	Educated by private instructors
1779	Death of his father, Alexander Georg von Humboldt
1787–92	Studies • University of Frankfurt an der Oder • University of Göttingen • Hamburg Academy of Commerce • Freiberg Mining Academy
1789	Study excursion with Steven Jan van Geuns in Germany • Results in the first book, *Mineralogische Beobachtungen über einige Basalte am Rhein*
1790	Travels with Georg Forster to Holland (April/May), England (May/June: London, Windsor, Reading, Bath, Bristol, Worcester, Droitwich, Gloucester, Birmingham, Cheadle, Buxton, Castleton, Middleton, Chatsworth, Matlock, Cromford, Derby, Coventry, Warwick, Stratford, Shipston, Chapel, Woodstock, Blenheim, Oxford, Dover), and revolutionary Paris (July).
1792–96	Civil service in the Prussian Mining Department
1792–93	Inspects mines in Bavaria, Franconia, Austria, Moravia, Silesia.
1794	Makes acquaintance of Johann Wolfgang von Goethe in Jena.
1795	Travels to Northern Italy and Switzerland.
1796	Death of his mother, Marie Elisabeth von Humboldt, née Colomb
1797	The first English translations of Humboldt's articles are published.
1798	Makes acquaintance of Aimé Bonpland in Paris. Journey to the Orient fails.

THE AMERICAN JOURNEY AND ITS AFTERMATH

1799 Audience with the King of Spain: Receives passport for America

1799–1804 Expedition

- Departure from La Coruña on 5 June 1799
- Tenerife
- Venezuela
- Cuba
- Colombia
- Ecuador
- Peru
- Mexico
- Cuba
- United States (20 May to 30 June): Philadelphia, Baltimore, Washington; has several meetings with President Thomas Jefferson and his cabinet
- On July 20, 1804, Humboldt becomes a member of the American Philosophical Society in Philadelphia.
- Returns to France on August 3, 1804.

1804 Makes acquaintance of Simón Bolívar in Paris.

1805 Travels to Italy, ascends Mount Vesuvius.

1807–27 Resides in Paris, even during the Napoleonic wars and the wars of liberation between Prussia and France.

1811 The English translation of *Essai politique sur le royaume de la Nouvelle-Espagne* appears under the title *Political Essay on the Kingdom of New Spain.*

1814 In June, stays in London with his brother Wilhelm von Humboldt and King Friedrich Wilhelm III of Prussia.

The English translation of *Vues des Cordillères* appears under the title *Researches, Concerning the Institutions & Monuments of the Ancient Inhabitants of America, With Descriptions and Views of Some of the Most Striking Scenes in the Cordilleras!.*

1814–31 The account of the American journey is published in three volumes as *Relation historique du Voyage aux régions équinoxiales du Nouveau Continent.* English

	translations appear in 1814–31 under the title *Personal Narrative of Travels to the Equinoctial Regions of the New Continent* and in 1852–53 under the title *Personal Narrative of Travels to the Equinoctial Regions of America.*
1817	With Arago, Humboldt visits his brother Wilhelm in London from October to December.
1818	Stays in London in September and October.
1822	Plans for an emigration to Mexico.
1823	The English translation of *Essai géognostique sur le gisement des roches dans les deux hémisphères* appears under the title *A Geognostical Essay on the Superposition of Rocks in Both Hemispheres.*
1827	Stays in London during April and May, dines with British Prime Minister George Canning. Relocates to Berlin.
1827–28	Lectures and speaks on the *Kosmos* project.

JOURNEY TO CENTRAL ASIA AND LATER WORKS

1829	Expedition to Central Asia: Königsberg, Riga, St. Petersburg, Moscow, Nizhny-Novgorod, Kazan, Yekaterinburg, Tobolsk, Barnaul, Orenburg, and Astrakhan, among other places
1830–48	Diplomatic missions in Paris
1835	Death of his brother, Wilhelm von Humboldt
1842	In January, Humboldt accompanies King Friedrich Wilhelm IV of Prussia to the christening of the future King Edward VII in London. He meets Charles Darwin on January 29.
1845–62	*Kosmos* is published in five volumes. English translations appear from 1846–58 under the title *Cosmos: Sketch of a Physical Description of the Universe* and in 1849–58 under the title *Cosmos: A Sketch of a Physical Description of the Universe.*
1848	Participates in the funeral procession for the killed revolutionaries in Berlin in March 1848.
1849	The first English translation of *Ansichten der Natur* appears under the title *Aspects of Nature, In*

	Different Lands and Different Climates; With Scientific Elucidations. The second follows a year later under the title *Views of Nature: Or Contemplations on the Sublime Phenomena of Creation; With Scientific Illustrations*.
1856	The English (American) translation of *Essai politique sur l'île de Cuba* appears under the title *The Island of Cuba*. Humboldt publicly protests the omission of the chapter on slavery.
1859	Dies in Berlin on May 6.

Index of Names